STATE-SELECTED AND STATE-TO-STATE ION–MOLECULE REACTION DYNAMICS
Part 1. Experiment

ADVANCES IN CHEMICAL PHYSICS

VOLUME LXXXII

EDITORIAL BOARD

STATE-SELECTED AND STATE-TO-STATE ION–MOLECULE REACTION DYNAMICS
Part 1. Experiment

Edited by

CHEUK-YIU NG
Ames Laboratory
U.S. Department of Energy and
Department of Chemistry
Iowa State University
Ames, Iowa

MICHAEL BAER
Department of Physics and
Applied Mathematics
Soreq Nuclear Research Center
Yavne, Israel

ADVANCES IN CHEMICAL PHYSICS
VOLUME LXXXII

Series Editors

ILYA PRIGOGINE
University of Brussels
Brussels, Belgium
and
University of Texas
Austin, Texas

STUART A. RICE
Department of Chemistry
and
The James Franck Institute
University of Chicago
Chicago, Illinois

AN INTERSCIENCE® PUBLICATION
JOHN WILEY & SONS, INC.
NEW YORK • CHICHESTER • BRISBANE • TORONTO • SINGAPORE

In recognition of the importance of preserving what has been written, it is a policy of John Wiley & Sons, Inc. to have books of enduring value published in the United States printed on acid-free paper, and we exert our best efforts to that end.

An Interscience® Publication

Library of Congress Catalog Number: 58-9935

ISBN 0-471-53258-4

Printed in the United States of America

10 9 8 7 6 5 4 3 2 1

CONTRIBUTORS TO VOLUME LXXXII
Part 1

SCOTT L. ANDERSON, Department of Chemistry, State University of New York at Stony Brook, Stony Brook, NY

JEAN-CLAUDE BRENOT, LCAM, Université Paris-Sud, Orsay, France

MARIE DURUP-FERGUSON, LPCR, Université Paris-Sud, Orsay, France

JEAN H. FUTRELL, Department of Chemistry & Biochemistry, University of Delaware, Newark, Delaware

DIETER GERLICH, Fakultat für Physik, Universität Freiburg, Freiburg, Germany

INOSUKE KOYANO, Department of Material Science, Himeji Institute of Technology, Himeji, Japan

CHEUK-YIU NG, Ames Laboratory, U.S. Department of Energy and Department of Chemistry, Iowa State University, Ames, Iowa

GEREON NIEDNER-SCHATTEBURG,* Max-Planck-Institut für Stromnungsforschung, Gottingen, F.R. Germany

KENICHIRO TANAKA, National Laboratory for High Energy Physics, Tsukuba, Japan

J. PETER TOENNIES, Max-Planck-Institut für Stromnungsforschung, Göttingen, F.R. Germany

JAMES C. WEISSHAAR, Department of Chemistry, University of Wisconsin-Madison, Madison, Wisconsin

*Present address: Institut für Physikalische und Theoretische Chemie, Technische Universität, München, Garching, F.R. Germany

INTRODUCTION

Few of us can any longer keep up with the flood of scientific literature, even in specialized subfields. Any attempt to do more and be broadly educated with respect to a large domain of science has the appearance of tilting at windmills. Yet the synthesis of ideas drawn from different subjects into new, powerful, general concepts is as valuable as ever, and the desire to remain educated persists in all scientists. This series, *Advances in Chemical Physics*, is devoted to helping the reader obtain general information about a wide variety of topics in chemical physics, a field which we interpret very broadly. Our intent is to have experts present comprehensive analyses of subjects of interest and to encourage the expression of individual points of view. We hope that this approach to the presentation of an overview of a subject will both stimulate new research and serve as a personalized learing text for beginners in a field.

ILYA PRIGOGINE
STUART A. RICE

PREFACE

Studies of the reaction dynamics involving gaseous ions are important for the fundamental understanding of ionic processes relevant to organic, inorganic, atmospheric, interstellar, plasma, and material chemistry.

As experimental systems to provide accurate cross-section data for first-principle theoretical developments in molecular dynamics, ion–molecule reaction studies offer many attractive features, the most noticeable being the relative ease of varying the collisional energy. For neutral–neutral studies, the seeded supersonic molecular beam techniques limit the highest collisional energy to about 10 eV and measurements of absolute reaction cross sections can only be made with great difficulty. Absolute total cross sections for ion–molecule reactions can be measured routinely from below thermal energy to well beyond 10 eV. The intensities of reactants used in neutral–neutral studies are usually much higher than those for reactant ions used in the study of ion–molecule reactions. The higher intensities of products resulting from a neutral–neutral reaction make possible the detailed product state detection by optical methods such as laser-induced fluorescence. The fact that product ions of an ion–molecule process can be detected with 100% efficiency compensates in part for the disadvantage of lower reactant and product ion intensities. Using mass spectrometric methods, both reactant and product ions can be unambiguously identified. For specific ion–molecule reaction systems, laser-induced fluorescence, laser photofragmentation, and differential reactivity methods have successfully been applied to measure rotational, vibrational, and electronic state populations of nascent product ions.

The demonstration by Chupka and coworkers in 1968 that photoionization techniques can be used for the preparation of reactant ions is a milestone in the field of state-selected ion–molecule reaction dynamics. Many recent technical developments are responsible for the rapid increase in activity in experimental studies of ion–molecule reactions. Much of the vitality in the field of state-selected ion chemistry is owed to the development of sophisticated photoelectron–photoion coincidence techniques. The introduction of the radio-frequency (rf) octopole ion guide has played an important role in absolute cross-section measurements. The maturity of vacuum ultraviolet laser and multiphoton ionization techniques and the establishment worldwide of many intense synchrotron radiation sources ensure a bright future for studies of state-to-state ion–molecule reaction dynamics. As a consequence of these advances, this field is in a period of broadening its current focus on reactions of atomic and small molecular ions to include more complex ion–molecule processes. Therefore, a thorough review of the recent experimental and theoretical accomplishments is timely. The contributors to this two-volume review are among the most active groups engaging in experimental and theoretical studies of state-selected and state-to-state ion–molecule reaction dynamics.

This experimental volume consists of eight chapters. The first chapter is by Gerlich, one of the original developers of the rf octopole ion-guide technique, and gives a comprehensive review on the theory and application of inhomogeneous rf fields for the study of the dynamics of low-energy ion–molecule processes. In the past decade, the use of an rf octopole ion-guide reaction gas cell has been accepted as the standard in the measurement of total absolute cross sections for ion–molecule collisions. The integration of rf ion optics with other techniques, such as merged beam, single- or multiphoton ionization, photofragmentation, chemiluminescence, chemiionization, coincidences, and so on, are also discussed.

In the second chapter Anderson reviews the application of multiphoton ionization (MPI) for the preparation of reactant ion states. For molecules with ideal spectroscopic properties, MPI can be a powerful state-selection method, giving control over ion electronic, vibrational, rotational, and doublet states. It also should be possible to use MPI to study angular momentum alignment effects on reactions. The essential principles of MPI and a prescription of molecular properties for good MPI state-selection are discussed. Molecules for which MPI state selection has been demonstrated are listed. The third chapter, by Weisshaar, reviews the application of MPI schemes for state-specific cross-section measurements involving transition-metal cations. These experiments have contributed to some understanding of how 3*d*-series transition-metal cations break C–H and C–C bonds of hydrocarbons. Electron spin is found to be a key quantum number in determining the outcome of transition-metal cation reactions. The high density of states of transition-metal ions demands the use of high-resolution laser-photoionization methods for reactant state selection.

Koyano describes the development of the threshold photoelectron secondary-ion-coincidence (TESICO) method for state-selected studies of a variety of ion–molecule reactions in the fourth chapter. The TESICO method, an extension of the well-established photoelectron photoion coincidence technique for state- or energy-selected studies of unimolecular-ion dissociation processes, is most versatile and promises to continue its significant role in studies of state-selected ion–molecule reaction dynamics. The fifth chapter also deals with coincidence method. Bernot and Durup-Ferguson present the conceptual and practical aspects of a multicoincidence technique. The multicoincidence scheme involves time-correlated measurements in a crossed-beam configuration, together with position-sensitive detectors and digital data handling. Experimental results are discussed for the anion–molecule system $X^- + H_2$, obtained using the multicoincidence methods. The multicoincidence scheme holds great promise for measurements of absolute state-to-state differential cross sections.

Using the simple photoionization method for reactant-ion preparation and the differential reactivity technique for product state detection, Ng and coworkers have obtained detailed state-to-state cross sections for several model ion–molecule reaction systems. For the $[Ar + N_2]^+$ and $[Ar + H_2]^+$ systems, charge transfer cross sections for the forward and backward reactions are determined. Ng summarizes the results of these experiments in the sixth chapter and compares them to theoretical predictions and to results of other laboratories.

The seventh chapter, by Futrell, is a review of several recent crossed-beam studies of charge transfer systems involving atomic and diatomic molecular ions. These experiments illustrate the importance of differential cross section measurements in the elucidation of charge transfer dynamics. New results on collision-induced dissociation of polyatomic ions show quite unexpected efficient coupling of translational and electronic energies in low-energy collisions. The review of Niedner-Schatteburg and Toennies in Chapter 8 gives a survery of 15 years of high-resolution crossed-beam scattering of protons with atoms, diatoms, and polyatomic molecules. This unique set of experiments represents the state-of-the-art application of time-of-flight techniques. State-to-state differential cross sections determined in these experiments have stimulated many theoretical investigations.

The studies of state-to-state reaction dynamics, which have been focused mostly on simple reaction systems, have often been considered to be of little relevance to "real" chemical problems involving more complex reactions. Judging from the progress in the field of state-selected and state-to-state ion chemistry, we may confidently conclude that much "real" chemical insights of ion–neutral interactions will be gained in the next cycle of developments.

CHEUK-YIU NG
MICHAEL BAER

Ames, Iowa
October 1991

CONTENTS

STATE-SELECTED AND STATE-TO-STATE ION–MOLECULE REACTION DYNAMICS
Part 1. Experiment

ADVANCES IN CHEMICAL PHYSICS

VOLUME LXXXII

INHOMOGENEOUS RF FIELDS: A VERSATILE TOOL FOR THE STUDY OF PROCESSES WITH SLOW IONS

DIETER GERLICH

Fakultät für Physik, Universität Freiburg, Freiburg, Germany

CONTENTS

State-Selected and State-to-State Ion–Molecule Reaction Dynamics, Part 1: Experiment, Edited by Cheuk-Yiu Ng and Michael Baer. Advances in Chemical Physics Series, Vol. LXXXII.
ISBN 0-471-53258-4

I. INTRODUCTION

Over the last four decades, the study of reactions between ions and neutral molecules has grown into a rather large field of research in chemistry and physics. Interest in the experimental investigation of the detailed dynamics of collision processes has resulted in the construction of a variety of novel devices, and considerable emphasis has been placed on the design and development of specific instrument components for ion generation (e.g., temperature-controlled ion sources, photoionization methods, and production of metal and cluster ions), interaction regions (scattering cells, molecular flows, and molecular beams), and product detection (mass, energy, and state analyzers). These devices have been combined in many different instruments such as flowing afterglow and drift tubes, ion–cyclotron resonance cells, ion traps, as well as ion-beam arrangements (e.g., beam–cell, crossed beam, and merged beam). Each of these instruments has proven useful for solving specific problems, but none is ideally suited for answering all questions of interest. For example, studies at low energies have been dominated by swarm and trapping techniques yielding predominantly thermal reaction-rate coefficients, while beam methods, which provide very detailed information on reaction dynamics, have been restricted to higher energies, typically above 1 eV. Different aspects of the field of gaseous ion chemistry have been reviewed often. A list of the most important reference up to 1979 is found in Franklin (1979), while more recent advances have been described in several extensive collections of reviews (Ausloos, 1979; Ausloos, and Lias, 1987; Bowers, 1979, 1984; Futrell, 1986).

In a recent review (Farrar, 1988) on *the major techniques for the study of ion–molecule reactions*, several authors have summarized advantages and shortcomings of different *bulk* and *beam* methods. Comparison of the variety of available techniques leads to the conclusion that there is still a large technology gap between the idealized experiment and the currently available methods. This is partly because electrostatic potentials, used in most of the apparatuses for handling charged particles, have the disadvantage that the stationary points are always saddle points, that is, there are no real minima, a direct consequence from Laplace's equation in a charge-free domain. Usually the situation becomes worse if space charges and field distortions are involved. Some of the resulting problems and other inherent difficulties encountered by most of the standard electrostatic techniques can be overcome by the use of *effective potentials* created by fast oscillatory electric or electromagnetic fields, since these fields allow one to create two- and three-dimensional potential minima. The principle of guiding or trapping charged particles by time-dependent forces has a longer history than its application in gas-phase ion chemistry and mass spectrometry. The most prominent examples of the use of *radio frequency* (*rf*) fields are the quadrupole mass filter and the Paul trap; however, the 1950s and 1960s have also seen many other related activities such as plasma confinement in rf fields, interaction of electrons with microwaves or light, and other applications, as will be briefly mentioned in Section II.

The first apparatus used to determine integral cross sections became operational 20 years ago (Gerlich, 1971; Teloy and Gerlich, 1974). This early instrument demonstrated several outstanding capabilities including a wide energy range, extreme sensitivity, and high accuracy for the determination of integral cross sections. These advantages (and also some weaknesses) have been discussed by Gentry (1979), who presented a semiquantitative comparison of different ion-beam instruments. In the past decade the *guided ion-beam method* has gained some popularity, but its incorporation into other laboratories has been rather slow. The number of research groups routinely using guided ion beams to determine integral cross sections is still limited to a few. This may, in part, be due to a lack of detailed theoretical information as well as missing technical and practical hints. Therefore, one aim of this chapter is to present a comprehensive overview of the technique and to provide some insight into the theory.

A second aim of this chapter is to demonstrate by several examples that the application of inhomogeneous rf fields is not limited only to precise integral cross-section measurements. Simultaneous recording of longitudinal and transverse velocities of guided product ions allows one to derive differential cross sections and product kinetic energy distributions. The combination of a guided ion beam with a coaxial supersonic beam has opened up the collision energy range down to 1 meV. Other examples, which illustrate the general use of rf fields in ion chemistry and physics, include several applications of the time-of-flight method, employment of lasers for ion preparation and/or product detection, use of coincidence techniques, and various combinations. Finally, we will show that rf storage of ions opens up a wide field of applications. Trapping in a large field-free region allows one to obtain very low ion temperatures and long storage times and to measure extremely small rate coefficients. On the other hand, trapping in quadrupole fields generally causes ion heating; this geometry is, however, better suited for mass analysis and therefore more frequently used in chemical applications (March and Hughes, 1989).

Section II provides information on the interaction between charges and fast oscillatory fields. After a historical overview of the development of related theories (mainly from the 1950s and 1960s), principles of ion motion in rf fields are explained within the adiabatic approximation. Computer simulations are used to illustrate the influence of potential distortions, to explain features of ion motion, and to derive realistic energy distributions.

Section III shows that effective potentials are generally applicable for trapping, storing, and confining ions in space, as well as selecting them according to energy and mass, focusing ion beams, and other purposes. Several instrument devices will be described including rf ion sources, energy and mass filters, and long ion guides for time-of-flight analysis or for studying ion–laser beam interactions. Technical and practical hints, as well as some design considerations, are also given. In addition, a few important performance tests of several rf devices are presented.

Several complete instruments in which a number of different rf components are assembled for special applications are described in Section IV. Here, we include a short list of applications from research groups besides our own. Some comments on kinematical averaging are included that are necessary for the characterization of the apparatuses and the interpretation of the experimental results.

Finally, Section V presents a series of measurements performed primarily in our laboratory over the past few years using different rf electrode arrangements. Included are some examples from previously unpublished work. The goal in selecting these examples is to demonstrate the technical variety of applications of inhomogeneous rf fields, rather than discuss in detail the dynamics of the chosen collision systems.

The conclusion will show that there remain many detailed improvements in the design of existing devices, and that through combinations with other established techniques the versatility of the apparatuses can be further advanced. A few speculations concerning new rf devices and applications to new problems are included.

II. MOTION OF CHARGED PARTICLES IN FAST OSCILLATORY FIELDS

The motion of a system under the influence of a force varying in time and space is a rather fundamental subject, well studied in theoretical and experimental physics. In general, most differential equations describing such systems are nonlinear and usually cannot be solved exactly. For special purposes, for example, for the characterization of solutions of mechanical and technical problems, analytical approximation procedures have been developed (Arnold, 1988, Hayashi, 1985; Lichtenberg and Lieberman, 1983), such as the perturbation, iteration, and averaging methods, the method of harmonic balance, or the guiding center approach (Littlejohn, 1979). It is beyond the scope of this chapter to discuss the modern mathematical developments of the analysis of nonlinear systems, and since we are more concerned with experimental applications, the principal aim of this section is to impart some basic understanding of the motion of a particle under the influence of a fast oscillatory force.

In the following we restrict the examples to the special case of the interaction between charges and time-dependent electric or electromagnetic fields, although a discussion of work performed in other areas would also contribute to the understanding of the mathematics relevant to our ion physics applications. The first part of this section briefly reviews a few related physical situations that are treated in the literature. Examples include the interaction of a free electron with a radiation field, applications in plasma and accelerator physics, mass spectrometry, and ion chemistry. Here, we report only the basic ideas and theoretical aspects, while pertinent experimental investigations will be mentioned in Section III. Following this overview we will introduce the adiabatic approximation and derive the effective potential. Particular attention is directed toward the question of stability of the ion motion and several practical and mathematical criteria for stability are discussed. Computer simulations are used both for illustrative purposes and for finding the limitations of approximate solutions.

A second purpose of this section, besides searching for solutions of the equation of motion, is to give support for the determination of electric fields from given boundary conditions or the search for appropriate electrode geometries. Although all related problems are well documented in electrodynamics textbooks, a few formulas are compiled in Section II C for calculating potentials for different geometrical arrangements of electrodes. A separate section is devoted to the special class of multipole fields. The last part of Section II is concerned with the analysis of the kinetic energy distribution of ions in different oscillatory fields. Some related questions can be answered analytically within the adiabatic approximation, while in the presence of collisions, analysis of kinetic energy distributions requires numerical simulations. These simulations will be used to derive realistic energy distributions of ions stored in the presence of a buffer gas.

A. Remarks on the Development of the Theory

One of the earliest treatments of the interaction of electrons with electromagnetic waves was Thomson's (1903) determination of the X-ray scattering cross section. His calculation is based on the classical non-relativistic motion of an electron in the field of a plane wave which is simply described by the equation of motion

$$m\ddot{\mathbf{r}} = q\mathbf{E}_0 \cos(\Omega t). \tag{1}$$

Here q is the charge of the electron, m is the mass, $\mathbf{E}_0$ is the peak electric field vector, and Ω is the angular frequency.

At high light intensities, electron–photon scattering is dominated by stimulated photon interactions known as the ponderomotive effect. This was first recognized by Kapitza and Dirac (1933), who suggested that an optical standing wave could scatter electrons. It was the availability of intense optical radiation, following the development of the laser in the 1960s, which prompted many new theoretical and experimental studies. The problem has been reexamined in detail in both relativistic and nonrelativistic regimes using classical or quantum-mechanical descriptions, or quantum field theory (Eberly, 1969).

It has been discussed by several authors [see, for example, Motz and Watson (1967)] that in special situations the properties of the field make it possible to obtain analytical solutions of the equation of motion without any approximation. An instructive example is the classical, nonrelativistic motion of an electron in a standing plane electromagnetic wave with circular polarization. In this case, it is an exact result (Motz and Watson, 1967) that the time-averaged energy of the classical wiggling motion of the electron, $q^2E_0^2/4m\Omega^2$, in response to the field $\mathbf{E}_0$, acts as a potential V^*. The resulting average force, $-\nabla V^*$, tends to confine the particle close to the nodes of the wave, where the motion is a harmonic oscillation with a frequency $\omega = qE_0/mc$.

A slightly different example is a standing wave with plane polarization. Here the trajectories close to the nodes obey the Mathieu equation. For the more general case, where $\mathbf{E}_0$ is some arbitrary but smoothly varying function of $\mathbf{r}$, for example, a laser beam with a realistic spatial profile, an approximate solution, called the guiding center theory, has been derived (Motz and Watson, 1967). The main result is that the smooth guiding center motion can be derived from a *quasipotential*, which is identical to the previously mentioned potential V^*. An early derivation of this quasipotential, which is also called the *ponderomotive potential*, the *pseudopotential*, or the *effective potential*, was given by Kapitza (1951) within the framework of classical mechanics. A comprehensive description of his treatment can be found in the textbook by Landau and Lifshitz (1960).

Other interesting contributions to the subject of this chapter were inspired by the goal to contain a nuclear fusion reaction by using rf confinement to isolate the plasma from the material walls of an apparatus. This was first proposed in the 1950s and discussed by several authors (Linhard, 1960; Weibel, 1959), and a comprehensive review has been given by Motz and Watson (1967). An important result from these studies is that one can create *two-* and *three-dimensional potential wells* with oscillatory fields, provided that certain stability conditions (see following discussion) are fulfilled. For the confinement of both electrons and ions in a neutral plasma, it has been recognized that the effective potential is proportional to q^2, that is, V^* is independent of the sign of the charge. It has also been proposed to make use of superimposed fields at different frequencies for simultaneous compression of electrons and ions as well as to cause plasma heating (Gapanov and Miller, 1958).

Orbits of electrons in the field of a cylindrical wave guide have been studied by Weibel (1959) and Weibel and Clark (1961) in the microwave region. As in the previous examples, the fields were treated as standing electromagnetic fields, and it was shown that particles can be bound to the nodal points, lines, or surfaces of these fields. The problem was again treated in a first-order approximation, taking just the time-averaged influence of the oscillating field. As an important contribution to the development of the theory, several conditions were studied under which the orbits remain bound for all times, and the criteria were tested by numerical integration of the equation of motion. The resulting stability condition [Eq. (3) in Weibel and Clark (1961)] is very similar to that which we are using (Gerlich, 1986; Teloy and Gerlich, 1974) (see Section II B 3).

More theoretical work related to this chapter was inspired by techniques and theories developed for high-energy particle accelerators. It was demonstrated in 1952 (Courant, 1952) that a series of quadrupole fields alternating in space can confine fast beams of protons. This strong-focusing alternating-gradient principle stimulated the invention and development of a new dynamic mass spectrometer by Paul and coworkers (Paul and Raether, 1955; Paul and Steinwedel, 1953; Paul et al. 1958), the well-known quadrupole mass filter. In this instrument, the spatial periodicity is replaced by quasistationary fields, which alternate in time, leading, under suitable conditions, to a focusing force for low-energy ions.

The principle of operation of quadrupoles, including the three-dimensional Paul trap (Fissher, 1959; Wuerker et al., 1959a), is well understood. Many details of the theory have been discussed in the literature, which, in turn, has been reviewed in several comprehensive monographs (Dawson, 1976; Dawson and Whetten, 1971; March and Hughes, 1989). It must, however, be emphasized that oscillating quadrupole fields occupy a very special place among all other field geometries. They are unique because the equation of motion describing the ion trajectory can be reduced to a set of *decoupled, one-dimensional* differential equations of the Mathieu type. Making use of the properties of their solutions, it is easy to see that the stability of a trajectory depends only on two dimensionless parameters and not on the initial conditions of the ion.

Some theoretical aspects that explore possible applications of higher-order multipole fields, for example, hexapoles, octopoles, or other field geometries, have been discussed by several authors. One of the earliest related proposals is due to Gaponov and Miller (Gaponov and Miller, 1958; Miller, 1958a, 1958b), who also investigated the nonrelativistic motion of charged particles in oscillatory electromagnetic fields. Using the high-frequency approximation, they derived the previously-mentioned effective potential and came to the conclusion that, under suitable conditions, such fields can be used to reflect charged particles from potential barriers or to squeeze and confine charges in bounded regions of space. For creating appropriate potential wells they proposed the use of two- and three-dimensional quasielectrostatic multipole fields.

Related ideas and suggestions were developed independently by Teloy (Bahr et al., 1969, 1970; Teloy and Gerlich, 1974) who also initiated the construction of various rf devices (see Sections III and IV) and gave many impulses to the work presented in this chapter. He has made contributions to the further development of the theory by generalizing the adiabatic approximation. In his treatment (Kohls, 1974; Teloy, 1980) the solutions of the equation of motion are expanded simultaneously in a power series in terms of $1/\Omega$ (the rf frequency) and a Fourier series in time. This theory allows the determination of higher-order correction terms for both the ion trajectories and the effective potential.

Dawson and Whetten (1971) have briefly prospected possible applications of higher-order multipole fields, trying to extend or generalize the quadrupole mass filter principle. They pointed out that, unlike the special case of the Mathieu equation, the equations describing the motion become *nonlinear coupled* differential equations, and that the stability of the ion motion depends strongly on the initial conditions (position, velocity, and rf phase). Fundamentals of ion motion in rf multipole fields have been treated by Friedman et al. (1982), as a "tutorial for mass spectrometrists and ion physicists." A similar textbook level presentation of the equation of motion and the solution of the Laplace equation has been published by Szabo (1986). In a series of papers, Hägg and Szabo (1986a, 1986b, 1986c) have tried to characterize the solutions of these differential equations based exclusively on the analysis of numerical simulations. They concluded that an overall classification of higher-order multipole fields within their proposed generalization of the quadrupole stability diagram is not possible. Unfortunately, these authors did not consider the adiabatic approximation, although those of their conclusions that have some general validity can be explained within this approximation. More details will be discussed in Section III D 3.

There are other theoretical aspects of rf containment of charged particles that have been treated in the literature. One example is the combination of an rf field with a magnetostatic field (Motz and Watson, 1967). A second is the inclusion of space charge, which can cause significant perturbations. This problem has been dealt with using many different theoretical approaches. Other perturbations through collisions or impairment by geometrical distortions have also been discussed. These problems have been treated in detail for the Paul ion trap, and a summary is given in the recent comprehensive review by March and Hughes (1989). Some of the developed analytical methods and conclusions make use of the linearity of the equation of motion and are therefore only valid for quadrupoles, while others could be extended to general rf devices.

Finally, and especially worthwhile mentioning, is a recent treatment of the motion of a particle in a high-frequency time-dependent potential by Cook et al. (1985). They have shown that the high-frequency approximation of the exact Schrödinger equation is identical with a Schrödinger equation that is based on the classically derived effective potential. These solutions are important for the quantum-mechanical description of very slow ions and ion crystals, prepared by laser cooling in rf fields (Diedrich et al., 1989; Neuhauser et al., 1978; Wineland and Dehmelt, 1975).

B. The Adiabatic Approximation

1. *The Equation of Motion*

Consider a particle of charge q and mass m moving in an external electromagnetic field $\mathbf{E}(\mathbf{r}, t)$ and $\mathbf{B}(\mathbf{r}, t)$. In the nonrelativistic regime, the classical equation of motion is (Gapanov and Miller, 1958; Motz and Watson, 1967)

$$m\ddot{\mathbf{r}} = q\mathbf{E}(\mathbf{r}, t) + q\dot{\mathbf{r}} \times \mathbf{B}(\mathbf{r}, t). \tag{2}$$

It has been mentioned in Section II A that this equation can be solved analytically only for special field geometries (Motz and Watson, 1967; Eberly, 1969). Here, we simplify this equation and derive an approximate but general solution. We begin by making use of the fact that most of the applications described in this chapter deal with weak electric fields and heavy particles with low kinetic energies (in contrast to electrons in intense laser fields). This means that the velocity $\dot{\mathbf{r}}$ is very small in comparison to the velocity of light, and, as a result, the weak force exerted by the magnetic field can be disregarded. Quasistationary magnetic fields are not discussed, since they are excluded so far in our experiments.

In addition, we assume that the quasistationary electric field $\mathbf{E}(\mathbf{r}, t)$ is composed of a static field, $\mathbf{E}_s(\mathbf{r})$, and a time-dependent part, $\mathbf{E}_0(\mathbf{r})\cos(\Omega t + \delta)$, where $\mathbf{E}_0(\mathbf{r})$ is the field amplitude, $\Omega = 2\pi f$ is a fixed angular frequency, and δ is a phase. The motion is then described by the differential equation

$$m\ddot{\mathbf{r}} = q\mathbf{E}_0(\mathbf{r})\cos(\Omega t + \delta) + q\mathbf{E}_s(\mathbf{r}). \tag{3}$$

The following derivation is easier to survey if, for the moment, we set the static field $\mathbf{E}_s(\mathbf{r})$ equal to zero and eliminate the phase δ by a shift in the time origin.

For illustrative purposes, we first solve this differential equation for the simple case of a charge q in the rf field of a parallel plate capacitor, as depicted in Fig. 1*a*. In this case, the field $\mathbf{E}_0$ is *homogeneous*, that is, independent of $\mathbf{r}$, and Eq. (3) can be solved directly. The special solution

$$\mathbf{r}(t) = \mathbf{r}(0) - \mathbf{a}\cos(\Omega t) \tag{4}$$

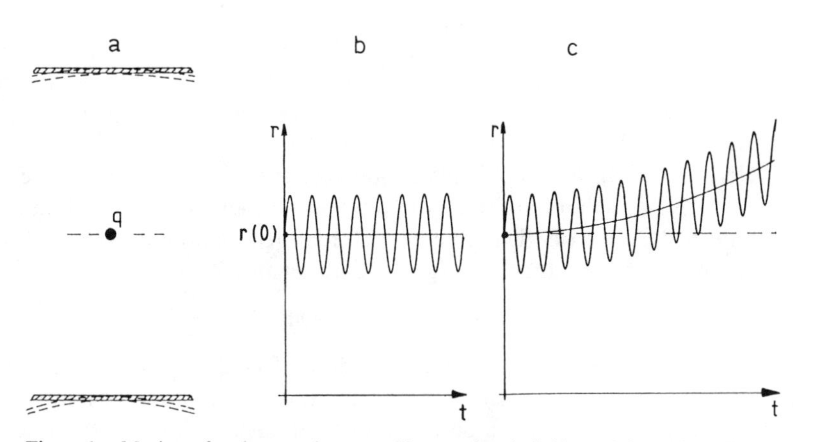

Figure 1. Motion of a charge q in an oscillatory electric field. (*a*) Schematic indication of the electrodes. For the parallel plate capacitor, the field is homogeneous; bending of the electrodes (dashed lines) results in an inhomogeneous field. (*b*) In the homogeneous field, the particle oscillates as a function of time with constant amplitude and without changing its mean position $\mathbf{r}(0)$. (*c*) The inhomogeneity of the field causes as additional slow drift motion toward the weaker field region.

can be verified immediately by inserting it into Eq. (3), and one obtains for the amplitude vector **a** the relation

$$\mathbf{a} = q\mathbf{E}_0/m\Omega^2. \tag{5}$$

This solution is plotted in Fig. 1*b* as a function of time. On average, the particle remains stationary at $\mathbf{r}(0)$, but it oscillates with an amplitude that is proportional to the field amplitude and inversely proportional to mass and to the square of the frequency.

The situation changes if the field becomes *inhomogeneous*. This is indicated in Fig. 1*a* by the dashed lines, in which the parallel plates are bent or replaced by a large-radius cylindrical capacitor. In general, the differential equation, Eq. (3), can no longer be solved analytically and the often complicated orbits of the particle can only be determined by numerical integration. However, in the present example, the field is only weakly inhomogeneous and the resulting trajectory, plotted in Fig. 1*c*, deviates only slightly from the homogeneous case. The amplitude is almost the same; however, the charge now experiences a varying field strength during its oscillatory motion, resulting in several modifications, the most important being a seeming force that causes the slow drift toward the outer electrode.

The goal of the following section is to make use of these observations to derive an approximate analytical solution of Eq. (3), and then to determine some general properties of the motion of a charge in an oscillatory field.

Related ideas and theories presented in the literature have already been reviewed in Section II A. For the purpose of the present chapter, it is sufficient to summarize the first-order results of the *adiabatic approximation* following essentially the derivations given by Dehmelt (1967), Eberly (1969), and Teloy and Gerlich (1974). A more fundamental discussion of the importance of adiabatic invariants and approximate integration methods can be found in the modern literature on the mathematical treatment of nonlinear dynamical systems (Arnold, 1988; Lichtenberg and Lieberman, 1983).

2. *The Effective Potential*

To describe the motion in an inhomogeneous rf field, we assume that the field varies smoothly as a function of the coordinate **r**, and that the frequency is high enough to keep the amplitude **a** in Eq. (5) small. These assumptions will be elaborated in Section II B 3. It can be expected that along the slow drift motion, as visualized in Fig. 1*c*, the amplitude $\mathbf{a}(t)$ and phase of the oscillatory motion vary slowly with time. We therefore seek a solution of Eq. (3) by superimposing a smooth drift term $\mathbf{R}_0(t)$ and a rapidly oscillating motion $\mathbf{R}_1(t)$:

$$\mathbf{r}(t) = \mathbf{R}_0(t) + \mathbf{R}_1(t), \tag{6}$$

with

$$\mathbf{R}_1(t) = -\mathbf{a}(t)\cos\Omega t. \tag{7}$$

We presume that slow spatial variation of $\mathbf{E}_0$ allows us to keep only the first two terms in the expansion

$$\mathbf{E}_0(\mathbf{R}_0 - \mathbf{a}\cos\Omega t) = \mathbf{E}_0(\mathbf{R}_0) - (\mathbf{a}\cdot\nabla)\mathbf{E}_0(\mathbf{R}_0)\cos\Omega t + \cdots, \tag{8}$$

and furthermore, that the slow time variation of **a** and $\mathbf{R}_0$ implies $\dot{\mathbf{a}} \ll \Omega\mathbf{a}$ and $\ddot{\mathbf{R}}_0 \ll \Omega\dot{\mathbf{R}}_0$. Substituting Eq. (6) and Eq. (8) into Eq. (3), and neglecting the small time derivatives, leads to an equation of the form

$$m\ddot{\mathbf{R}}_0 + m\Omega^2\mathbf{a}(t)\cos\Omega t = q\mathbf{E}_0(\mathbf{R}_0)\cos\Omega t - q[\mathbf{a}(t)\cdot\nabla]\mathbf{E}_0(\mathbf{R}_0)\cos^2\Omega t. \tag{9}$$

Assuming that the amplitude **a** changes in time only due to the motion along $\mathbf{R}_0$, we can replace $\mathbf{a}(t)$ with $\mathbf{a}(\mathbf{R}_0)$ using Eq. (5). In this way the amplitude factors of the two $\cos\Omega t$ terms in Eq. (9) become identical and cancel. To simplify the remaining equation we use the vector analysis relation

$$(\mathbf{E}_0\cdot\nabla)\mathbf{E}_0 = \tfrac{1}{2}\nabla\mathbf{E}_0^2 - \mathbf{E}_0 \times (\nabla \times \mathbf{E}_0) = \tfrac{1}{2}\nabla E_0^2, \tag{10}$$

because the rotation of $\mathbf{E}_0$ vanishes ($\nabla \times \mathbf{E}_0 = 0$) for quasistatic field conditions. Finally, replacing $\cos^2 \Omega t$ by its time averaged value $\frac{1}{2}$ leads to a differential equation for the nonoscillating motion $\mathbf{R}_0(t)$:

$$m\ddot{\mathbf{R}}_0 = -\frac{q^2}{4m\Omega^2}\nabla E_0^2. \tag{11}$$

This equation of motion shows the time-averaged effect of the oscillatory field. The charged particle experiences a force caused by the inhomogeneity of the field, the so-called *field gradient force*. This force is proportional to the square of the charge q, that is, independent of the sign, is mass dependent, and is inversely proportional to the square of the frequency Ω. The direction and the strength is determined by the gradient of E_0^2. Charged particles are always pushed into regions of weaker fields, as we have already seen in Fig. 1.

Now, if we account anew for the electrostatic force $q\mathbf{E}_s(\mathbf{r})$ in Eq. (3) and write the electrostatic field $\mathbf{E}_s$ as the negative gradient of the electrostatic potential $\mathbf{\Phi}_s$, that is,

$$\mathbf{E}_s = -\nabla\Phi_s, \tag{12}$$

the total average force acting on the ion can be derived from a time-independent *mechanical potential* V^*

$$V^*(\mathbf{R}_0) = q^2 E_0^2/4m\Omega^2 + q\Phi_s. \tag{13}$$

With this definition the equation of motion describing the smooth trajectory becomes simply

$$m\ddot{\mathbf{R}}_0 = -\nabla V^*(\mathbf{R}_0). \tag{14}$$

We shall subsequently refer to V^* as the *effective potential* (Eberly, 1969; Landau and Lifschitz, 1960). Other authors have designated V^* as the high-frequency potential (Gapanov and Miller, 1958; Miller, 1958a, 1958b), the quasipotential (Motz and Watson, 1967), or the pseudopotential (Dehmelt, 1967).

For most of our applications, this approximation allows a satisfying characterization of the properties of arbitrary field geometries based on the effective potential given in Eq. (13). Also, the important features of the resulting motion can be predicted easily. According to Eq. (6), $\mathbf{r}(t)$ is composed of only two terms, the smooth motion $\mathbf{R}_0(t)$ (also called the guiding center or secular motion) and the fast oscillating term $\mathbf{R}_1(t)$ (wiggling motion).

In order to determine the trajectory $\mathbf{R}_0(t)$ from Eq. (14), it is indeed necessary to solve a differential equation. However, this equation is much easier to discuss and to solve than the original one given in Eq. (3), since it does not contain time explicitly. The slowly varying amplitude of the superimposed sinusoidal oscillation is directly determined from Eq. (7) and Eq. (5) by inserting $\mathbf{E}_0(\mathbf{R}_0)$,

$$\mathbf{R}_1(t) = -q\mathbf{E}_0(\mathbf{R}_0)/m\Omega^2 \cos \Omega t. \tag{15}$$

Additional microoscillations and higher-frequency components, which can be seen in a Fourier analysis of the numerically exact trajectories, are corrections in higher order and are therefore neglected here. The resulting smooth motion is independent of the starting phase [see, for example, Fig. 4 of Teloy and Gerlich (1974)], provided that the trajectory starts in a sufficiently weak field region. Otherwise, the initial phase may influence the kinetic energy, as will be discussed in further detail in Sections II D and III C.

It has also been shown [see, for example, Motz and Watson (1967)] that the effective potential can be derived directly from Eq. (2) under inclusion of the magnetic part of an electromagnetic field, leading essentially to the same result given in Eq. (13). However, comparison of the velocity amplitude, $q\mathbf{E}_0/m\Omega$, obtained from Eq. (15), with the speed of light, shows quantitatively that the influence of the magnetic field and relativistic corrections can usually be disregarded, except for electrons in very strong fields.

The first integral of Eq. (14),

$$\tfrac{1}{2}m\dot{\mathbf{R}}_0^2 + q^2E_0^2/4m\Omega^2 + q\Phi_s = E_m, \tag{16}$$

demonstrates that the total energy E_m is a constant of the ion motion, or more precisely, an adiabatic constant of the motion. This equation expresses that kinetic energy of the smooth motion can be transferred into effective potential energy or electrostatic potential energy. It is easy to prove, using the time derivative of $\mathbf{R}_1(t)$ given in Eq. (15), that the second term of Eq. (16) is identical to the time-averaged kinetic energy of the oscillatory motion

$$\langle \tfrac{1}{2}m\dot{\mathbf{R}}_1^2 \rangle = q^2E_0^2/4m\Omega^2. \tag{17}$$

This proves that the so-called effective potential energy V^* is actually stored as kinetic energy in the fast oscillatory motion. Therefore, motion through an inhomogeneous field leads to a permanent exchange between three different forms of energy, the kinetic energies, $\tfrac{1}{2}m\dot{\mathbf{R}}_0^2$ and $\tfrac{1}{2}m\dot{\mathbf{R}}_1^2$, and the electrostatic potential energy $q\Phi_s$. For some calculations it is useful to express the momentary kinetic energy of the oscillatory motion using the effective potential V^*

$$\tfrac{1}{2}m\dot{\mathbf{R}}_1^2 = 2V^* \sin^2 \Omega t. \tag{18}$$

More details on the different forms of energy will be discussed in Section II E 1.

Some of the results obtained thus far are illustrated in Fig. 2, which depicts the motion of a charge in a cylindrical capacitor. The electrodes are sketched in the upper panel, where the applied voltage is $U_0 - V_0 \cos \Omega t$. The effective potential, defined in Eq. (13), can be calculated from the well-known electrostatic potential [see Eq. (33)]. Three results of $V^*(r)$ are plotted in the middle panel, each obtained for a different value of the dc voltage U_0, and all normalized to $V^*(r_0) = 0$. These potentials have minima, since the field gradient force always pushes the charge toward larger radii, independent of the sign of the charge, but it is counterbalanced by the electrostatic force, provided the sign of the dc voltage has been properly chosen in accordance with the sign of the charge. If an ion starts at the marked position (+), for simplicity with zero initial velocity and with phase $\delta = 0$ (e.g., created by photoionization), the trajectory depends on the applied dc voltage U_0, as can be seen in the lowest panel. The three resulting trajectories are easy to understand within the effective potential approximation. Different dc voltages correspond to different potential energies at the starting point (+) and therefore to different total energies E_m. These energies are indicated by the three dashed lines in the middle panel, and they determine the amplitudes of the pertinent secular motion. The amplitude of the wiggling motion is unaffected, since it depends only on the rf field, which is the same for all three cases.

A second example that illustrates the results of this section is the ion motion in an rf quadrupole. It is well known from earlier theoretical treatments (Dehmelt, 1967; Fischer, 1959) as well from an early experiment (Wuerker et al., 1959a) that, under certain conditions, the effective potential approximation can be used to clarify some of the features and working principles of quadrupoles. Although the motion of ions is fully described by the theory of the Mathieu equation and the well-known (a_2, q_2) stability diagram (Dawson and Whetten, 1971; March and Hughes, 1989), significant simplifications can be obtained for $q_2 \leqslant 0.4$ (to avoid confusion with the charge q, and to extend to higher-order multipoles, we use q_n to signify a $2n$-pole). Figure 3*a* depicts several ion trajectories for $q_2 = 0.1$. If a_2 is chosen such that $a_2 = \frac{1}{2}q_2^2$, that is, just on the border of the stability diagram, the smooth trajectory is a straight-lined steady motion. As will be demonstrated in Section II D, the effective potential V^* is equal to $\frac{1}{2}q_2^2 - a_2$, and therefore zero in this case. It can be seen from the growth of the amplitude of the wiggling motion that the electrostatic potential energy is completely converted into oscillatory motion. Increasing a_2 above $\frac{1}{2}q_2^2$ results in a net acceleration of the secular motion ($\frac{1}{2}m\dot{\mathbf{R}}_0^2$ increases), leading to the curved trajectories. Nonetheless, in this case, the total energy $\mathbf{E}_m$, as defined in

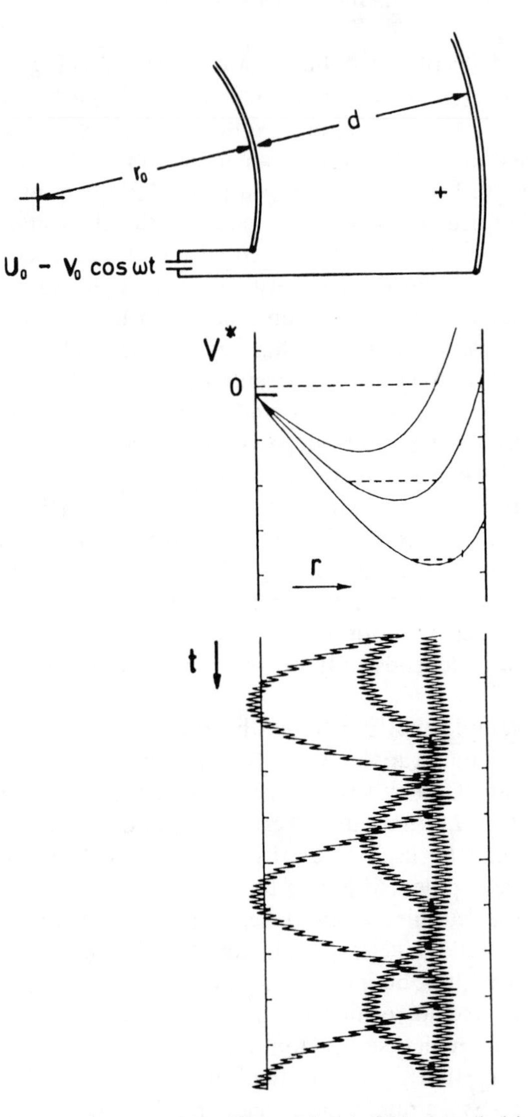

Figure 2. Motion of a charge in the oscillatory electric field of a cylindrical capacitor with different superimposed dc voltages. The upper panel sketches the geometry and the (+) marks the position where the ion is started with zero initial energy. The middle panel shows effective potentials V^* for three different dc voltages U_0, each leading to a different trajectory, plotted in the lower panel as a function of time. The dashed lines in the potential curves mark the respective total energies.

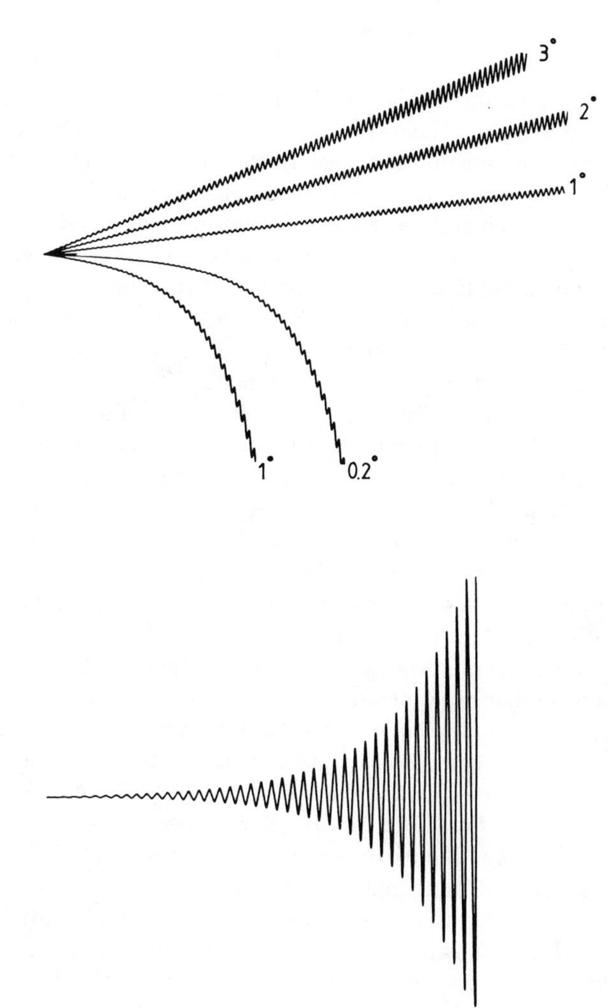

Figure 3. Illustration of unstable trajectories with different starting angles in a quadrupole for several (a_2, q_2) values (see text). In the upper panel the adiabatic approximation is valid ($q_2 = 0.1$); however, the effective potential is too weak to confine the ions. In the lower panel, the operating point $(a_2, q_2) = (0.2, 0.739)$ is outside of the stability triangle leading to an instability due to resonant energy build up.

Eq. (16), is an adiabatic constant of the motion and the loss of the ion is only due to the acceleration in the electrostatic field. Bound trajectories in a quadrupole will be discussed in Section III B (see Fig. 25). The situation in Fig. 3*b* is completely different, as will be seen subsequently.

3. *Adiabaticity*

The derivation presented in the preceding section leads to the important result that the average total energy is an adiabatic constant of the ion motion. This result is based on assumptions that allow one to separate the trajectories into fast and slowly varying terms. The range of validity of these assumptions will be further discussed and specified in this section. In general, a mechanical system, as described by Eq. (3), is not conservative, and it is conceivable that under certain conditions the oscillating force can continuously augment the energy of the particle and cause nonadiabatic behavior or instability.

Such a situation is illustrated in the lower part of Fig. 3, which depicts an unstable ion trajectory in a quadrupole mass filter. Here, the point of operation is $(a_2, q_2) = (0.2, 0.739)$, and therefore outside of the previously cited condition $q_2 < 0.4$. The effective potential approximation is no longer applicable. Indeed, this point is already past the right border of the lowest range of the stability diagram (for details see Section III B and Fig. 24). The result is a continuous augmentation of the kinetic energy, as indicated in the figure by the increase in the amplitude of the secular motion.

This example illustrates the well-known fact that for an ideal quadrupole (infinite length, etc.), the mathematical definition of stability can become equivalent to practical prescriptions of stability, which must be based on ion transmission or similar experimental requirements. Unfortunately, the simple (a_2, q_2) stability criterion *cannot be extended* or generalized to field geometries where the differential equations describing the motion are nonlinear and coupled. Related attempts (Dawson and Whetten, 1971; Friedman et al., 1982; Szabo, 1986) (see also Section II D3) to characterize the complete family of multipole fields were therefore unsuccessful. Also, according to our knowledge, no general mathematical criterion has been applied so far to rigorously characterize stable and unstable solutions from Eq. (3). Therefore, we have divided the following discussion into two parts. We search first for conditions for *adiabaticity*, and, in the following section, we discuss *stability*, briefly from a mathematical point of view but as motivated primarily from practical applications.

The range of validity of the effective potential approximation has been discussed in several publications, as indicated in Section II A. Typically, these approximations require a high frequency and other changes in time to be slow, that is, behave adiabatically. To quantify comparisons of *slow* and *fast*, such as $\dot{\mathbf{a}} \ll \Omega\, \mathbf{a}$ and $\ddot{\mathbf{R}}_0 \ll \Omega\, \dot{\mathbf{R}}_0$, it is common to introduce a dimensionless ratio that compares two characteristic features. For quasiperiodic motions, such as shown in Fig. 2, the frequency of the secular motion ω must be slow relative to the rf frequency, that is, ω/Ω must be small. Another possibility is to compare the maximum velocity acquired during one cycle with the unperturbed velocity. More complicated constructions have also been mentioned, for example $L|\nabla E|/E$, where L was designated as a characteristic length (Gapanov and Miller, 1958).

Our definition of such a characteristic parameter is motivated from Eq. (8). There, in deriving the adiabatic approximation, we assumed a smooth spatial variation of the electric field to justify keeping only the first two terms in the Taylor expansion of $\mathbf{E}_0$. We therefore demand that over the full distance of the oscillation, that is, over $2\mathbf{a}$, the change of the field be much smaller than the field itself

$$|2(\mathbf{a}\nabla)\mathbf{E}_0| < |\mathbf{E}_0|. \tag{19}$$

This basic requirement has already been imposed by Weibel and Clark (1961) for predicting safe confinement of charged particles in effective potential wells. Using this relation, we define (Teloy and Gerlich, 1974) our characteristic parameter η as

$$\eta = |2(\mathbf{a}\nabla)\mathbf{E}_0|/|\mathbf{E}_0|. \tag{20}$$

Inserting $\mathbf{a}$ from Eq. (5) and making use of the vector analytical identities given in Eq. (10), we obtain the definition of the *adiabaticity parameter* η in its final form

$$\eta = 2q|\nabla E_0|/m\Omega^2. \tag{21}$$

For a given mass, frequency, and charge, the parameter η depends only on the inhomogeneity of the field $\mathbf{E}_0$. For $|\nabla E_0| = 0$, we obtain the homogeneous field treated in Section II B, and the exact solution given in Eq. (4). The condition $\nabla E_0 = \text{const}$ leads to quadrupole fields, and, as will be seen in Section II D, in this case η is identical with the conventional stability parameter q_2. In general, $|\nabla E_0|$ is a function of the space coordinate $\mathbf{r}$, and as such the scalar field $\eta(\mathbf{r})$ must be known, at least along each individual ion trajectory, in order to check whether the adiabaticity parameter remains sufficiently small.

Before pursuing this idea further, it is necessary to find a generally applicable procedure to specify the maximum tolerable limit of the adiabaticity parameter, designated as η_m. Although the mathematical treatment of nonlinear mechanics is a rapidly growing field (Arnold, 1988; Lichtenberg and Lieberman, 1983), we have not yet found a general analytical means of determining η_m within the framework of the adiabatic theory. Based on the previous derivation of η and Eq. (19), we can only conclude that η_m must be smaller than 1. Therefore, we describe briefly in the following a few heuristic criteria and numerical methods for testing adiabaticity, and provide a few illustrative examples that support the already reported empirical result

(Gerlich, 1986; Teloy and Gerlich, 1974), that $\eta_m = 0.3$ guarantees adiabaticity for most practical applications. This limit has proven to be *sufficient* according to all of our numerical simulations and experimental observations. It is also in agreement with the restriction $\eta = q_2 < 0.4$, generally accepted for describing quadrupole fields using the effective potential (Dehmelt, 1967; Fischer, 1959; Wuerker et al., 1959a and b).

For lack of a rigorous mathematical characterization of the solutions from Eq. (3), we have determined η_m, the limiting value for safe application of the first-order adiabatic approximation, by computer simulation, that is, by numerically integrating Eq. (3). Of course it is not surprising that the fine details of limits determined this way are conditional on the specific criterion used to test the quality of the adiabatic approximation, and that the results depend as well on the field geometry and even on the initial conditions of individual trajectories. One of the simplest operational criteria to characterize adiabaticity in an arbitrary field geometry is to calculate the adiabaticity parameter $\eta(r)$ along a set of representative trajectories. This set of data is reduced by defining for each trajectory a series of numbers, η_{m_i}, using

$$\eta_{m_i} = \max_{t \in T_i} [\eta(\mathbf{r}(t))]. \tag{22}$$

The η_{m_i} values must be determined during consecutive time intervals T_i, which should be large enough that several local maxima of $\eta(r)$ are included. Since we are usually interested in stable confinement of ions in two- or three-dimensional effective potential wells, these maxima occur at the turning points of the trajectories. We have performed such calculations for different multipole fields and the ring electrode arrangement, and have found an empirical rule, that all values of η_{m_i} agree with each other within less than 1% if they are below 0.3. Therefore, such trajectories can be terminated if the first few values of η_{m_i} are below 0.3, leading to a significant saving of computer time. If the sequence of η_{m_i} lies between 0.3 and 0.5, the values fluctuate, typically by several percent. Above 0.5, a fast increase of η_{m_i} is often observed, which can be used as a typical indication of nonadiabaticity. Based on such observations, we can define η_m and formulate our rule for *safe operation* within the *validity of the adiabatic approximation* as

$$\eta_m = \max [\eta_{m_i}] = 0.3. \tag{23}$$

Another criterion for testing the quality of the adiabatic approximation is to check whether the mean kinetic energy of the ion is conserved within a given limit. This criterion is the most important for our experimental applications. However, as we have already seen in the last section, the momentary kinetic energy fluctuates, and it is not straightforward to define a numerical procedure for determing the adiabatic total energy E_m, as defined in Eq. (16). Related details will be discussed in Section II E. The problem is significantly simplified if one studies special trajectories, that is, those that pass periodically through a field-free region, since in this region their kinetic energy is well defined. Such a case is the oscillation of an ion along a symmetry axis of an octopole. We have performed one-dimentional trajectory calculations along such a line [$\varphi = 0$, that is, crossing a pole; see Eq. (39)]. The ions were started in the middle along the symmetry axis, the phase δ [Eq. (3)] was held fixed at zero, and the only remaining free initial condition was the starting velocity. Variation of the phase leads to insignificant changes in the results. Each trajectory was followed over 5000 rf periods, and each time it crossed the center its velocity was interrogated. The maximum of the relative changes, $\Delta v/v$, is plotted in Fig. 4 as a function of the nominal

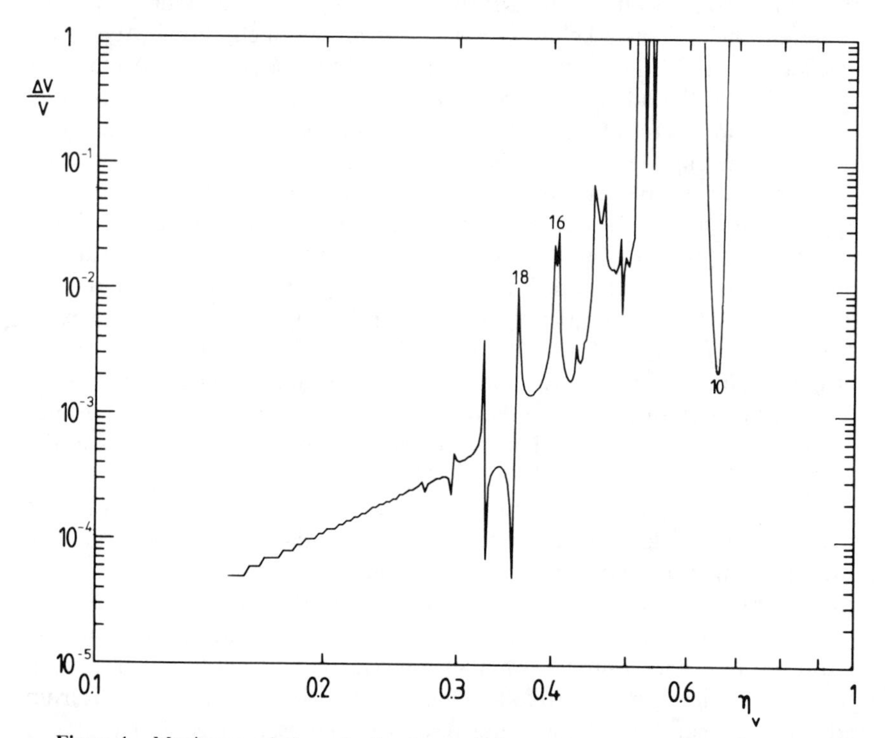

Figure 4. Maximum relative velocity change $\Delta v/v$ as a function of the nominal stability parameter η_v determined over 5000 rf periods. The difinition of η_v is given in Eq. (24). Details of the underlying one-dimensional motion along a symmetry line of an octopole are described in the text. The numbers on the peaks refer to the ratio of the rf frequency to the frequency of the secular motion. The stable region around $\eta_v = 0.65$ is in the vicinity of the special periodic orbit with $\Omega/\omega = 10$.

v-dependent adiabaticity parameter η_v in the turning point given by

$$\eta_v = 8^{1/2}(n-1)v/\Omega r_0. \tag{24}$$

This can be derived from Eq. (52) and Eq. (54), see Section II D, by setting $E_m = \frac{1}{2}mv^2$, $L = 0$, $\hat{r}_m = 1$, and eliminating V_0.

Although only a special class of trajectories has been selected in this example, the results demonstrate some interesting features. Below $\eta_v = 0.3$, $\Delta v/v$ is smaller than 10^{-3} and varies as a smooth function of η_v. Between 0.3 and 0.5, the change in ion velocity remains limited; however, the fluctuations of $\Delta v/v$ become larger. Above $\eta_v = 0.5$ there is a dramatic increase in the velocity, except for a few regions of stability. The connection of these regions with periodic orbits will be discussed subsequently. The most important conclusion to be drawn from Fig. 4 is the almost perfect conservation of energy below $\eta_v = 0.3$, that is, the fluctuations $(\Delta v/v)^2$ are smaller than 10^{-6}. This again supports our requirement of $\eta_m = 0.3$ for safe application of the adiabatic approximation. We wish to anticipate for our discussion in Section II E that the rf impairment of the ion energy is completely negligible in comparison to perturbations caused by collisions.

A very restrictive criterion for testing the quality of the adiabatic approximation can be formulated by measuring the deviation of the first-order approximation, $\mathbf{R}_0(t) + \mathbf{R}_1(t)$, from the exact solution, $\mathbf{r}(t)$, during one reflection from an rf wall. Typically, excellent agreement can be obtained for $\eta < 0.1$. A related comparison, analyzing the approximate and the exact energy, will be discussed in Section II E (Fig. 19). Other tests of the validity of the adiabatic approximation are based, for example, on the analysis of the Fourier components of the motion $\mathbf{r}(t)$, or on the conservation of orbital angular momentum for rotationally symmetric effective potentials, as in the case of multipole fields (see Section II D).

To summarize the discussion on adiabaticity, it is important to repeat that the scalar field $\eta(\mathbf{r})$, as defined by Eq. (21), charaterizes the rf field at each point in space and is independent of individual trajectories. Taking our empirical η restriction $\eta_m < 0.3$, one can mark out those regions in which the influence of the rf field can be described using the adiabatic approximation. However, in order to predict whether ions starting in such a region remain there, it is either necessary to follow their trajectories or to make profit from adiabatic constants of motion that restrict their phase space. For example, the accessible range of a trajectory is limited by the effective potential and the total energy E_m, which can be determined from the initial conditions. This will be further discussed in the next section and, for multipole fields, in Section II D.

4. *Stability*

In the preceding sections, we have applied the expression *stability* in several contexts. For example, the stability of the solutions of Mathieu's equation is characterized by the (a_2, q_2) stability chart, we have called the fast increase of η_{m_i} in Eq. (23) or of $\Delta v/v$ in Fig. 4 unstable behavior, and we have used stability in association with experimental requirements for safe ion transimission or long-time ion trapping. In this section we will make some general remarks about the *mathematical* meaning of stability and discuss briefly an illustrative example. Since, however, the mathematical theorems are not easy to apply to our concrete physical system, we simplify the problem by treating it only within the adiabatic approximation. As a result, we propose a simple definition of stability in a *practical* sense that can be directly applied to many experimental situations.

In general, it is difficult to define exactly what is meant by stability in a nonlinear and nonautonomous dynamical system. Typical mathematical definitions of stability (Arnold, 1988; Hayashi, 1985) either require that a trajectory remains forever in a bounded domain of the phase space (Lagrange stability) or are concerned with the question of what happens if the system is disturbed slightly near a stable stationary or oscillatory equilibrium (Liapunov stability). It has been discussed in Lichtenberg and Lieberman (1983) that the motion of a charged particle in a time-dependent electromagnetic field cannot be easily analyzed using standard methods of nonlinear mechanics. The main difficulty in this so-called guiding-center problem is to split the system into an unperturbed part and the perturbation. Several methods for obtaining such a separation using adequate coordinate transformations or multiple-scale averaging methods have been discussed in the literature (Kruskal, 1962; Lichtenberg and Lieberman, 1983; Littlejohn, 1979).

It is beyond the scope of this chapter to go into further details concerning the fundamental theory; however, we want to make a few illustrative remarks related to Fig. 4. On the one hand, we cannot exclude the possibility that there are unstable regions below $\eta_v < 0.3$, since the numerically determined limitation of $\Delta v/v$ is not a strict mathematical proof. More detailed numerical studies would have to search more carefully for regions of instability, which are expected (Lichtenberg and Lieberman, 1983) to become exponentially small with decreasing η_v [to order $\exp(-\text{const}/\eta_v)$]. On the other hand, it is evident from Fig. 4 that there are also regions of stability located significantly above our limit $\eta_v = 0.3$.

The stable region around $\eta_v = 0.65$ can be used to illustrate the definition of stability in the Liapunov sense. For this purpose one has to find a periodic orbit and to prove that this orbit is a stable oscillatory equilibrium. The

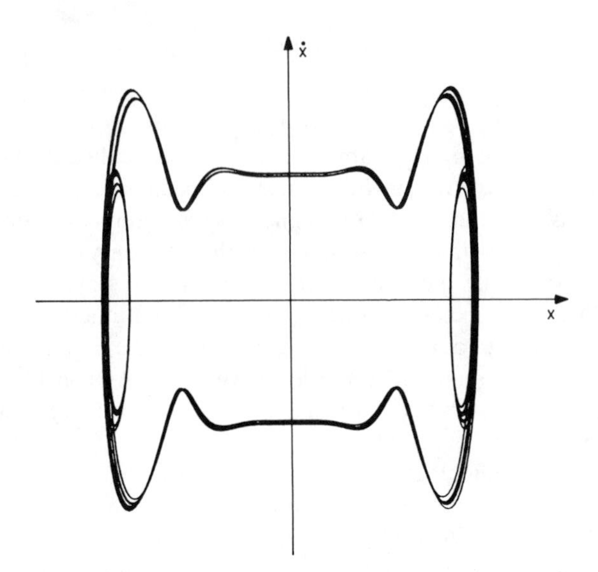

Figure 5. Phase space $(x, \dot{x})$ representation of a stable one-dimensional trajectory started with $\eta_v = 0.65$ in the vicinity of the periodic orbit $\Omega/\omega = 10$ ($\eta_v = 0.6454...$).

required periodic orbit, where the ratio of the rf frequency and the frequency of the secular motion is an integer, $\Omega/\omega = 10$, is obtained for $\eta_v = 0.64546355...$, as determined by a Fourth-order Runge-Kutta integration of Eq. (3). Instead of going into the details of stability investigations, we simply show in Fig. 5 a single trajectory that has been started in the neighborhood of this periodic orbit. The phase-space representation shows many cycles, and the fact that the slight change of initial conditions results in a quasiperiodic trajectory can be taken as a sign for stability.

The discussed examples are very special cases of one-dimensional trajectories in an octopole. This dynamical system, which is equivalent to an autonomous system of two degrees of freedom due to its explicit time dependence, could be studied in further detail using standard methods of nonlinear mechanics. However, a full characterization of ion confinement or transmission functions of rf devices requires treatment of the motion in two or three spatial dimensions. From the symmetry of a multipole arrangement, it can be concluded that there are many more periodic orbits, and one can expect that there are also many islands of stability that are localized in the unsafe η range. The question whether they are of some practical importance will be raised in Section II D 3.

Since we have no rigorous mathematical means to fully characterize stable solutions of Eq. (3), we restrict ourselves in the rest of this section to derive a simple condition for *stability* in a *practical* sense. For this purpose we make use of the effective potential and apply the experimentally necessary condition that the excursion of the ion does not exceed the internal dimensions of the electrodes. Adiabatic conservation of energy ensures that the accessible spatial range of an ion with a total energy E_m can be determined from Eq. (16) by setting $\dot{\mathbf{R}}_0 = 0$. To avoid ion loss by collision with the electrodes, the effective potential V^* close to the surfaces must be larger than E_m. Safe operating conditions must provide sufficient space for the small oscillatory motions $\mathbf{R}_1(t)$, Eq. (6). Therefore, an experimental condition for stability based on Eq. (16) can be formulated as

$$q^2 E_0(\mathbf{r}_m)^2/4m\Omega^2 + q\Phi_s(\mathbf{r}_m) > E_m, \tag{25}$$

where the vector $\mathbf{r}_m$ designates the closest allowed approach to the electrode surfaces. Note also that $\eta(\mathbf{r}_m) < 0.3$ is required for adiabatic behavior. An illustrative example is the periodic, large-amplitude oscillation in Fig. 2. Here the motion of the ion is unstable, since it exceeds the limitations imposed by the center cylinder. The total energy, marked by the dashed line, is too large to satisfy Eq. (25).

In order to predict whether certain initial conditions lead to stable trajectories as defined by Eq. (25), it is necessary to determine the energy E_m. If an electrode arrangement has a region where the field is zero, or sufficiently weak, E_m is simply equal to the kinetic energy, $\frac{1}{2}m\dot{\mathbf{r}}^2 = \frac{1}{2}m\dot{\mathbf{R}}_0^2$, of the ion in that region. If the ion starts somewhere in the rf field, the determination of the total energy E_m has to account for the local field strength $\mathbf{E}_0$ and the start phase δ. The condition given in Eq. (25) also allows one to determine experimentally the kinetic energy of stored ions. A detailed discussion of such critical operation modes for multipoles follows in Section III C 3.

We want to repeat that our practical criterion for stability, that is, adiabatic conservation of energy and stable confinement in a given geometrical region, is primarily motivated from experimental applications. The derived stability condition is a sufficient condition assuming the validity of the adiabatic approximation. The actual range of stable solutions of the dynamical system is larger and extends into regions $\eta > 0.3$.

Finally, from an experimental point of view, it seems to be necessary to draw attention to instabilities caused by sources other than the ideal rf field. For example, space-charge effects due to high ion densities can lead to plasma oscillations. An interesting study of chaos and order of Coulomb-interacting laser-cooled ions in a Paul trap has been reported recently (Blümel et al., 1989). Instabilities can also be induced through a superimposed low-frequency field or resonant modulation of the rf amplitude, as is well known from different applications of quadrupoles (Busch and Paul, 1961a, 1961b; March and

Hughes, 1989). Time-dependent effective potentials, for example, fast or slow variation of the amplitude, have been proposed for plasma acceleration (Miller, 1959) or can also be used for adiabatic cooling.

C. Special Field Geometries

1. *Laplace's Equation*

This section provides some explicit formulas for calculating the effective potential V^* and the adiabaticity parameter η for different boundary conditions. In addition, several formulas for practical applications are derived. In general, one has to determine the electromagnetic field from Maxwell's equations using appropriate boundary conditions. A thorough discussion of this subject is presented in many textbooks (Jackson, 1962; Magid, 1972; Simonyi, 1963; Sommerfeld, 1964). Related references can be found in the paper by Friedman et al. (1982), where several fundamental aspects of rf multipole fields are discussed. One typical example, which includes both the electric and the magnetic field components, has been given by Weibel (1959) who calculated the effective potential for a TE_{01} field in a circular waveguide. A more complete analysis of electromagnetic fields set up by arbitrary modes of excitation of square and circular section waveguides has been reported by Miller (1958a, 1958b). In both publications, it was shown that effective potential wells can be created in one, two, or three dimensions simply by choosing a suitable frequency and mode type, or by a linear combination of quasipotentials.

Since in most of our applications the charged particles are slow ions, the frequency of the oscillating field can usually be chosen low enough ($<30\,\mathrm{MHz}$) that the wavelength ($>10\,\mathrm{m}$) is much greater than typical geometrical dimensions ($<1\,\mathrm{m}$). As such, not only the static part of the field, $\mathbf{E}_s(\mathbf{r})$, but also the time-varying part of the field, $\mathbf{E}_{rf} = \mathbf{E}_0(\mathbf{r})\cos\Omega t$, can be derived from the simpler quasistatic solution. Furthermore, the electric field can be determined completely independently from the magnetic field. As has already been justified, the interaction of the magnetic field with the slow charged particles can be disregarded. In this *quasielectrostatic* limit, both fields $\mathbf{E}_s$ and $\mathbf{E}_{rf}$ are derived from the potentials Φ_s and Φ_{rf}, respectively. Because of the superposition principle, it is sufficient to search for a potential $\Phi = \Phi_s + \Phi_{rf}$, and the problem is therefore reduced to the solution of Laplace's equation under the inclusion of the appropriate boundary conditions, and possibly the space charge $\rho(\mathbf{r})$

$$\Delta\Phi(\mathbf{r}) = -4\pi\rho(\mathbf{r}). \tag{26}$$

For most of our applications, the problem can be treated in the limit of low ion densities. In the space-charge-free limit, Laplace's equation becomes simply

$$\Delta\Phi(\mathbf{r}) = 0. \tag{27}$$

Boundary conditions are imposed by the geometrical structure of a manifold of electrodes, which are assumed to be equipotential surfaces, and by the static and rf voltages applied to each of them.

There are several methods to find functions $\Phi(\mathbf{r})$ including the method of images, complex-variable theory, and infinite series expansion. This problem is well treated in the previously mentioned textbooks and in Friedman et al. (1982). For convenient future reference, some particular Laplacian solutions in different coordinate systems are listed here. Also, some hints for obtaining numerical solutions of Eq. (27) are given.

For certain symmetries it is convenient to write Laplace's equation in terms of spherical polar coordinates (r, ϑ, φ). Typical solutions are superpositions of functions such as

$$\Phi(r, \vartheta, \varphi) = (Ar^n + Br^{-n-1})P_n^m(\cos\vartheta)(C\sin m\varphi + D\cos m\varphi), \tag{28}$$

where A, B, C, and D are constants, P_n^m are the associated Legendre functions, and n and m are positive integers.

The problem of infinitely long cylindrical conductors, where the boundary conditions impose no z dependence, can be described in plane polar coordinates (r, φ). In this case, one obtains general solutions of Laplace's equation of the following type for $n > 0$:

$$\Phi(\mathbf{r}, \varphi) = (Ar^n + Br^{-n})(C\sin n\varphi + D\cos n\varphi), \tag{29}$$

and for $n = 0$:

$$\Phi(\mathbf{r}, \varphi) = (A + B\ln r)(C + D\varphi). \tag{30}$$

If the situation is adequately described in cylindrical coordinates (r, φ, z) and has azimuthal symmetry, general solutions containing the modified Bessel functions of the first kind, I_n and K_n, are obtained in the form

$$\Phi(r, z) = (A\cos kz + B\sin kz)(CI_n(r) + DK_n(r)). \tag{31}$$

2. *Special Solutions*

To achieve a simple mathematical solution represented by one of the exact potentials above, precisely machined and arranged electrodes are needed. Under such ideal conditions, the geometrical electrode surfaces follow the equipotential surfaces given by the selected solution, and the potential has a known and constant value at every surface, as given by the applied voltage. A very simple example of such a boundary value problem is the field of two line charges q and $-q$ and a cylindrical shield S, arranged according to Fig. 6. If in this special geometry the boundary condition $\Phi(\hat{r} = \alpha, \varphi) = \Phi_0$ is fulfilled, the potential at an arbitrary point in space is determined by

$$\Phi(\hat{r}, \varphi) = \Phi_0/\ln(\alpha^2) \ln \frac{\hat{r}^2 + 1 - 2\hat{r}\cos\varphi}{\hat{r}^2 + \alpha^4 - 2\alpha^2\hat{r}\cos\varphi}, \tag{32}$$

where α is a constant >1, r_0 is a characteristic length, and $\hat{r} = r/r_0$ is a reduced variable. This problem and the quadrupole potential of four line charges have already been discussed in detail in Sommerfeld (1964). The lower panel of Fig. 6 shows an arrangement of eight line charges that we have used to describe a wire octopole with shield S (Müller, 1983).

Another textbook example is the cylinder capacitor, shown schematically in Fig. 2. Two concentric conductions with potentials Φ_0 and 0 impose the boundary conditions $\Phi(r_0) = \Phi_0$ and $\Phi(r_0 + d) = 0$, and one obtains from

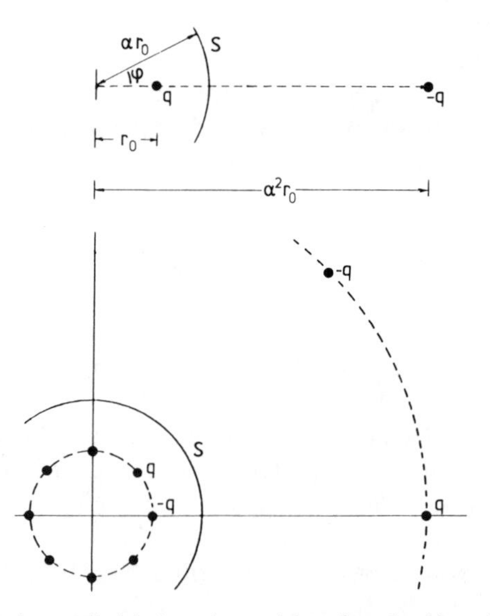

Figure 6. To obtain a cylindrical equipotential surface S with radius αr_0, the two line charges q and $-q$ must be situated at r_0 and $\alpha^2 r_0$, respectively. The potential is given in Eq. (32). The lower panel illustrates that superposition of eight cylindrical dipoles with line charges alternating in sign allows one to calculate the potential of an octopole made of eight thin wires surrounded by the cylindrical shield S.

Eq. (30) the r-dependent potential

$$\Phi(r) = \Phi_0 \ln(r/(r_0 + d))/\ln(r_0/(r_0 + d)). \tag{33}$$

A somewhat more complicated arrangement is a stack of ring electrodes, depicted in Fig. 7. In this case, it is appropriate to measure all distances in the unit z_0 and the reduced variables are $\hat{r} = r/z_0$, $\hat{r}_0 = r_0/z_0$, and $\hat{z} = z/z_0$. With the boundary conditions $\Phi(\hat{r}_0, \hat{z}) = \Phi_0 \cos \hat{z}$, which are shown as the parabolic shaped lines in Fig. 7, the following potential is obtained from Eq. (31).

$$\Phi(r, z) = \Phi_0 I_0(\hat{r})/I_0(\hat{r}_0) \cos \hat{z}. \tag{34}$$

Such an arrangement of similarly shaped electrodes is used as an ion trap, as discussed in Section III E.

Another simple potential that describes an rf grid and that can be assembled from a series of parallel rods has been used by Teloy (Kohls, 1974; Teloy, 1980) for many tests of his adiabatic theory

$$\Phi(\hat{x}, \hat{y}) = \Phi_0 \exp(-\hat{x}) \cos \hat{y}. \tag{35}$$

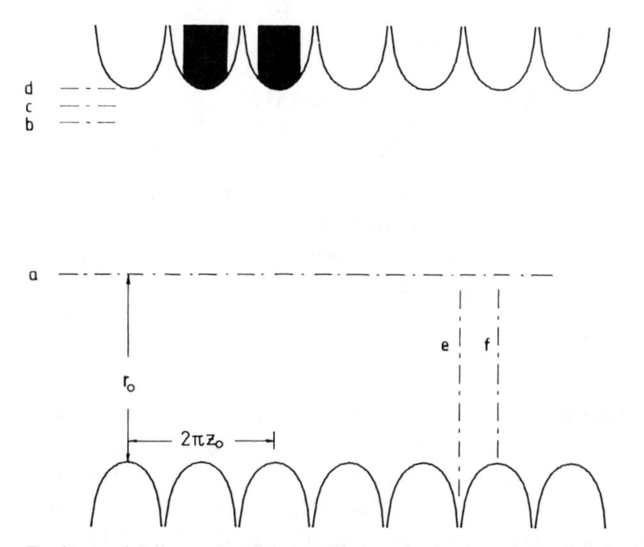

Figure 7. Equipotential lines of a rotationally symmetric ring electrode trap, defined in Eq. (34). The actual shape of the electrodes in marked in black (see Section III E). The letters a–f indicate locations along which Fig. 8 shows cuts through the effective potential.

The advantage of this potential is that it decreases exponentially with increasing $\hat{x}$, and, as such, one can start trajectories in an almost field-free region and study the influence of single reflections. Finally, Laplace's solutions of the very important class of multipole fields has been treated by Friedman et al. (1982) and Szabo (1986) and shall be discussed separately in Section II D.

Potentials for more complicated boundary value problems can be expressed as a infinite sum over a complete set of adequate general solutions making use of the superposition principle. A typical example is the calculation of the electrostatic potential and the resulting effective potential of a Paul trap with cylindrical rather than hyperbolical electrodes (Lagadec et al., 1988).

For those situations where simple analytical methods fail, many numerical methods have been developed for obtaining potential functions under given boundary conditions. Usually the problem is reduced to the solution of linear algebraic equations. One general method, based on N discrete surface-charge distributions, has been used recently to optimize a simple electrode structure for the approximation of a quadrupole trap (Beaty, 1987). Another method has been proposed and examined theoretically (Matsuda and Matsuo, 1977), whereby the parameters of a 12-wire arrangement were varied to obtain an optimal representation of a hyperbolic quadrupole field.

In order to calculate the potential of some realistic arrangement of electrodes with a circular profile and to account for other special laboratory situations like geometrical distortions and external field penetration, we have developed a computer program (Scherbarth, 1984) based on the superposition of $N = 96$ pairs of line charges. Some results of these calculations will be presented in Section II D.

3. *Effective Potentials*

To determine the effective potential we must first know the potential $\Phi = \Phi_s + \Phi_{rf}$, which itself depends on the geometry and the applied potential, Φ_0, as shown explicitly in several examples presented in Section II C 2. For the time and amplitude dependence of Φ_0, we generally use the form

$$\Phi_0 = U_0 - V_0 \cos \Omega t, \tag{36}$$

which means that a static voltage U_0 and a sinusoidal time-dependent voltage with amplitude V_0 are applied to the electrodes. The minus sign is in accordance with the convention used for the description of quadrupoles (Dawson, 1976). Concerning the amplitudes, there is some confusion in the literature, which has recently been discussed (March and Hughes, 1989). For our boundary conditions, used, for example, in the derivation of Eq. (34), Eq. (35), and Eq. (39), the potentials applied to two adjacent electrodes are

$+\Phi_0 = U_0 - V_0 \cos \Omega t$ and $-\Phi_0 = -U_0 + V_0 \cos \Omega t$. This leads to an electrode to electrode potential difference of $2U_0 - 2V_0 \cos \Omega t$. With these definitions and conventions we are consistent with March and Hughes (1989).

If the rf part, Φ_{rf}, of the potential is known, it is straightforward to determine $\mathbf{E}_0$, E_0^2, and $|\nabla E_0|$. With these expressions one obtains the effective potential V^* from Eq. (13) and the stability parameter η from Eq. (21). An illustrative example is the effective potential created by the ring electrode arrangement depicted in Fig. 7. The amplitude of the oscillatory field, $\mathbf{E}_0$, is calculated from the gradient of the potential given in Eq. (34), where Φ_0 is replaced by V_0:

$$\mathbf{E}_0 = (E_r, E_z) = \frac{V_0}{z_0}(-\mathbf{I}_1(\hat{r})/\mathbf{I}_0(\hat{r}_0)\cos\hat{z},\ \mathbf{I}_0(\hat{r})/\mathbf{I}_0(\hat{r}_0)\sin\hat{z}). \tag{37}$$

A detailed description of the the modified Bessel functions, and the relation $\mathbf{I}_0'(\hat{r}) = \mathbf{I}_1(\hat{r})$ used in the derivation of $\mathbf{E}_0$, can be found in Abramowitz and Stegun (1964). Inserting $|\mathbf{E}_0|^2$ into Eq. (13) leads immediately to the effective potential

$$V^* = \frac{q^2 V_0^2}{4m\Omega^2 z_0^2}[\mathbf{I}_1^2(\hat{r})\cos^2\hat{z} + \mathbf{I}_0^2(\hat{r})\sin^2\hat{z}]/\mathbf{I}_0^2(\hat{r}_0). \tag{38}$$

Several cuts through this potential, the positions of which are specified as *a–f* in Fig. 7 are plotted in Fig. 8. The left panel shows two radial cuts, *e* and *f*, the right panel the four axial cuts, *a–d*, at different distances from the centerline. For comparison, the effective potentials of a quadrupole and an octopole are also included in Fig. 8. Multipole devices and their analytical representations of V^* and η will be discussed in Section II D.

D. Two-Dimensional Multipoles

1. *The Ideal Multipole*

Among the most important electrode systems for the purpose of this chapter are the two-dimensional multipoles. The most prominent member, the quadrupole, is probably one of the best described and most thoroughly characterized devices. A general treatment of the entire class of multipoles, however, is rarely found in literature. Some theoretical background and various proposals for the use of multipoles to focus beams and confine ions were given in Gapanov and Miller (1958) and Miller (1958a, 1958b). The first experimental realization of an octopole was described by Gerlich (1971), and a general description of its function within the effective potential approximation was published in Teloy and Gerlich (1974). Additional details

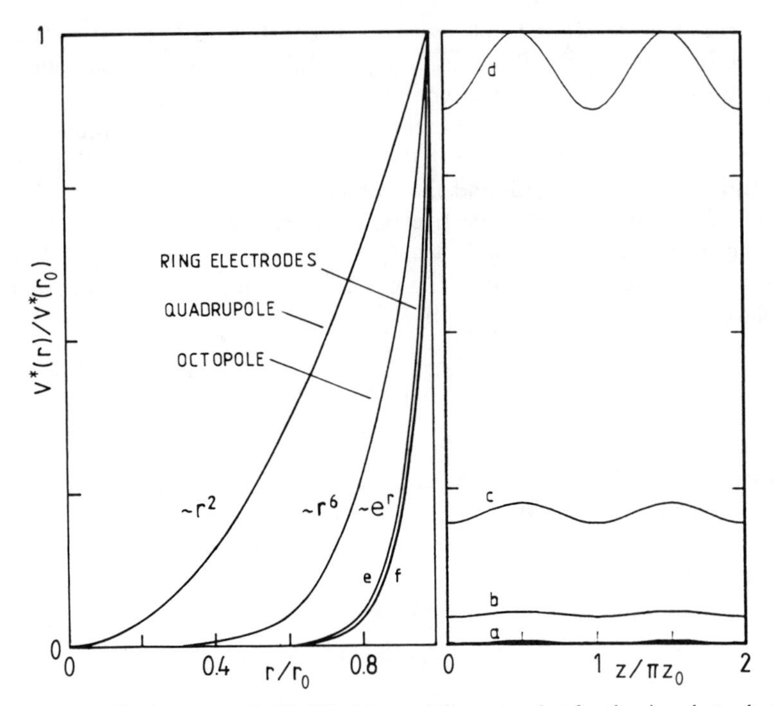

Figure 8. Effective potentials V^*. The left panel illustrates that for the ring electrode trap V^* increases proportional to e^r, in comparison to the r^2 and r^6 dependence of the quadrupole and the octopole. The letters *a–f* refer to locations marked in Fig. 7. In the axial direction (right panel) the effective potential of the ring electrode trap has a weak corrugation (*a–c*), which only becomes pronounced very close to the electrodes (*d*).

on multipole devices were reported more recently (Gerlich, 1986). There are a few additional references (Friedman et al., 1982; Szabo, 1986) that compile some fundamental facts and basic information about multiple potential functions and present the equation of motion of ions moving therein.

The potential of a multipole field can be obtained from Eq. (29) by setting $\Phi(r = r_0, \varphi) = \Phi_0 \cos n\varphi$ and $\Phi(r = 0, \varphi) = 0$, and by using the reduced variable $\hat{r} = r/r_0$:

$$\Phi(r, \varphi) = \Phi_0 \hat{r}^n \cos n\varphi. \tag{39}$$

The n dependence of these potentials, the hyperbolic equipotential lines, and the shape of the $2n$ electrodes have been discussed and depicted in several figures in Friedman et al. (1982) and Szabo (1986). The influence of deviations from the exact structure will be discussed in Section II D 4.

The polar components of the electric field, $\mathbf{E} = (E_r, E_\varphi)$, are obtained as the gradient of Φ;

$$\mathbf{E} = -\left(\frac{\partial}{\partial r}\Phi, \frac{1}{r}\frac{\partial}{\partial \varphi}\Phi\right) = \frac{\Phi_0}{r_0} n \hat{r}^{n-1}(-\cos n\varphi, \sin n\varphi). \tag{40}$$

The differential equation, Eq. (3), describing the motion of an ion in a two-dimensional multipole field, has been discussed in detail in Friedman et al. (1982) and Szabo (1986). Using the cartesian representation of the electric field,

$$(E_x, E_y) = \frac{\Phi_0}{r_0} n\hat{r}^{n-1}(-\cos(n-1)\varphi, \sin(n-1)\varphi), \tag{41}$$

and replacing Φ_0 by the sum of the static and rf potential from Eq. (36), we obtain the equation of motion in cartesian coordinates $\hat{\mathbf{r}} = (x/r_0, y/r_0)$

$$\ddot{\hat{\mathbf{r}}} + F(t)\hat{r}^{n-1}(-\cos(n-1)\varphi, \sin(n-1)\varphi) = 0. \tag{42}$$

For abbreviation and further reference we introduce the explicit time-dependent function

$$F(t) = n\frac{qU_0}{mr_0^2} - n\frac{qV_0}{mr^0_2}\cos\Omega t. \tag{43}$$

It is evident that for $n > 2$ this differential equation is nonlinear and, in contrast to the quadrupole, additional complications arise because the x and y motions are no longer decoupled. As such, it is advantageous to obtain a general characterization of the solutions within the adiabatic approximation.

To derive the effective potential V^* and the adiabaticity parameter η for a multipole field, we must determine $|E_0|$ and ∇E_0 according to the general definitions given in Eq. (13) and Eq. (21), respectively. The absolute value of the amplitude of the oscillatory field,

$$|E_0| = \frac{V_0}{r_0} n \hat{r}^{n-1}, \tag{44}$$

follows directly from Eq. (40), whereby Φ_0 is replaced by V_0 according to Eq. (36). Inserting Eq. (44) and the static multipole potential, $qU_0\hat{r}^n \cos n\varphi$, into Eq. (13) leads to the representation of the effective potential:

$$V^* = \frac{n^2}{4}\frac{q^2}{m\Omega^2}\frac{V_0^2}{r_0^2}\hat{r}^{2n-2} + qU_0\hat{r}^n\cos n\varphi. \tag{45}$$

With this potential we can rewrite Eq. (16) such that

$$E_m = \tfrac{1}{2} m\dot{\mathbf{R}}_0^2 + V^*(\mathbf{R}_0). \tag{46}$$

The adiabaticity parameter, defined in Eq. (21), can be easily derived with Eq. (44), since the vector

$$\nabla E_0 = \left(\frac{\partial}{\partial r}, \frac{1}{r}\frac{\partial}{\partial \varphi}\right)|\mathbf{E}_0|$$

has only a radial component, and one obtains

$$\eta = 2n(n-1)\frac{qV_0}{m\Omega^2 r_0^2}\hat{r}^{n-2}. \tag{47}$$

Finally, the reduced amplitude, $\hat{\mathbf{a}} = \mathbf{a}/r_0$, can be determined from Eq. (5) and Eq. (41).

$$\hat{\mathbf{a}} = \frac{qV_0}{m\Omega^2 r_0^2} n\hat{r}^{n-1}(-\cos(n-1)\varphi, \sin(n-1)\varphi). \tag{48}$$

Some of the results can be explained with the help of Fig. 9, which depicts a stable motion ($\eta_m = 0.3$) of an ion in an octopole. This trajectory corroborates the physical picture of a smooth motion of the guiding center under the influence of the quasipotential and a superimposed wiggling motion, which is seen especially during the reflections. The circle marks the radius where $\tfrac{1}{2}m\dot{\mathbf{R}}_0^2 = 0$, and the positions of the turning points near the circle indicate that the energy is conserved. Note the hole in the center, a region avoided by the trajectory, that is due to conservation of angular momentum of the ion motion. Since the absolute value of the electric field, $|\mathbf{E}_0|$, is independent of the azimuthal angle, φ, the rf part of the effective potential becomes rotationally symmetric. This has the important consequence that, within our high-frequency approximation, the interaction between the ion and the rf field results in a pure central force. In the absence of a superimposed static field U_0, there is a second constant of the motion, the angular momentum $\mathbf{L}$, and the total kinetic energy is composed of the radial term and a centrifugal term:

$$\tfrac{1}{2}m\dot{\mathbf{R}}_0^2 = \tfrac{1}{2}m\dot{R}_0^2 + \mathbf{L}^2/2mR_0^2. \tag{49}$$

It must be emphasized, however, that both the total energy and the total

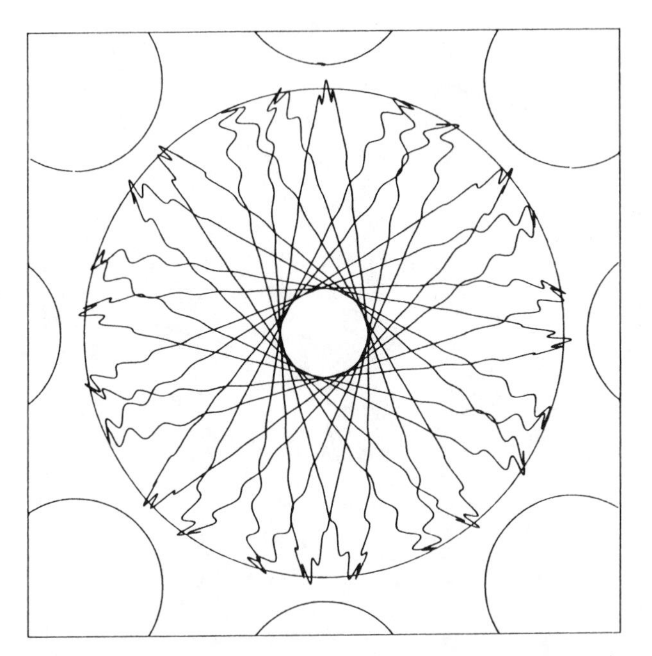

Figure 9. Typical trajectory of an ion in an octopole determined by direct numerical integration of the exact equation of motion, Eq. (42), for $\eta_m = 0.3$. The circle marks the turning radius, as determined from the effective potential approximation.

angular momentum are oscillating and that they are conserved only on average.

A trajectory which becomes unstable ($\eta_m \sim 0.55$) is plotted in Fig. 10 for an ion in a 32-pole. Here, the angular momentum is no longer conserved. The trajectory shown changes its closest approach to the center, and after a few more reflections, the ion will be lost. This figure further illustrate one of the most important features of higher-order multipoles, that is, the wide field-free region bordering on very steep potential walls. Most of the time the ion moves along straight lines and the interaction with the rf wall is very short ranged.

To evaluate numerically V^* and η in the case of multipoles, it has proven useful (Gerlich, 1986) to introduce an additional parameter, the characteristic energy ε, which combines the important parameters mass, frequency, and radius r_0:

$$\varepsilon = \frac{1}{2n^2} m\Omega^2 r_0^2. \tag{50}$$

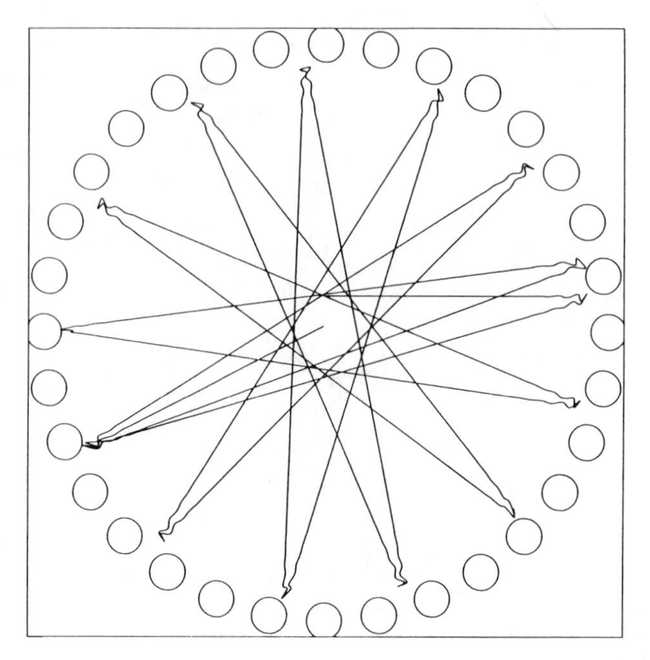

Figure 10. Trajectory in a 32-pole ($\eta_m \sim 0.55$). The ion is moving most of the time in a wide field-free region. After a few more reflections, the depicted trajectory will be lost. In contrast to Fig. 9, the closest approach of the trajectory to the center changes, indicating that here the orbital angular momentum is no longer conserved. This is a hint that the condition for the adiabatic approximation is no longer fulfilled.

This term allows one to rewrite Eq. (45) and Eq. (47) in the following simpler forms

$$V^* = \frac{1}{8} \frac{(qV_0)^2}{\varepsilon} \hat{r}^{2n-2} + qU_0 \hat{r}^n \cos n\varphi \tag{51}$$

and

$$\eta = \frac{n-1}{n} \frac{qV_0}{\varepsilon} \hat{r}^{n-2}. \tag{52}$$

The energy ε is not simply an abbreviation; it also has a physical interpretation. It corresponds to the kinetic energy of an ion cycling in phase with the rf on a radius r_0. It is obvious that for adiabatic behavior the actual ion energy must be much lower than ε.

2. *Safe Operating Conditions*

As was mentioned in our discussion on adiabaticity and stability, characterization of necessary or sufficient operating conditions of rf devices depends on the desired application. In this section we summarize a few analytical formulas that permit one to calculate conditions for operating multipoles in such a manner that the adiabatic approximation remains valid everywhere within given spatial boundaries, that is, those defined by the electrode surfaces. For a complete characterization, only the multipole parameters (n, r_0, Ω, V_0), the mass m, and the energy E_m are needed. Adiabatic conservation of energy ensures that safe transmission or storage of ions does not depend on the individual initial conditions, but only on their transverse energy.

The situation and the η restriction discussed in Section II B are illustrated in Fig. 11, which shows several trajectories, all starting in the center of an

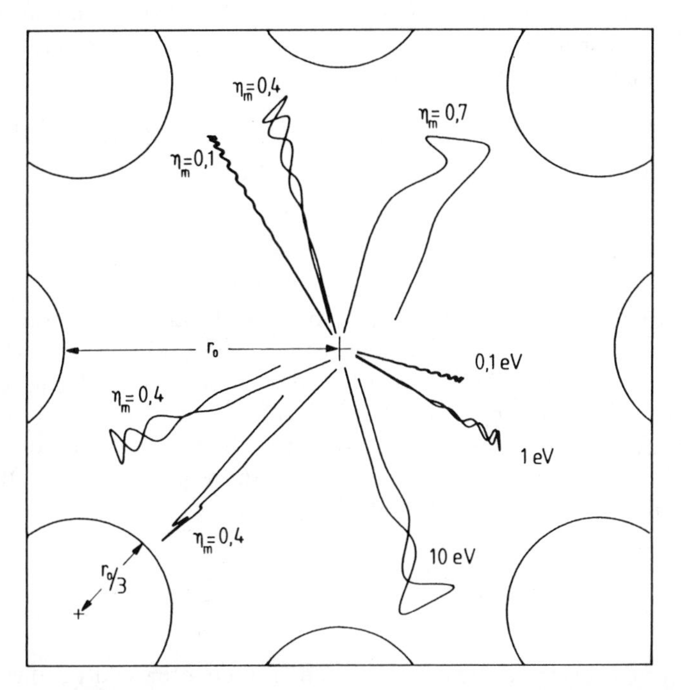

Figure 11. Characteristic ion trajectories in an octopole illustrating different operating conditions. All ion trajectories start in the center of the octopole. The examples with $\eta_m = 0.1$, 0.4, and 0.7 illustrate that with increasing η the amplitude of the wiggling motion increases and that the number of oscillations during one reflection from the rf wall decreases. For the three energies (0.1, 1, and 10 eV) the values ε and V_0 are constant. In the remaining trajectories ($\eta_m = 0.4$) only the starting angle has been changed. In one case, the small oscillations $\mathbf{R}_1(t)$ are parallel to the secular motion $\mathbf{R}_0(t)$; in the other, they are orthogonal.

octopole (i.e., $\mathbf{L} = 0$), but under different initial conditions. For $\eta_m = 0.1$, the energy and angular momentum are conserved. The ion with $\eta_m = 0.7$ is reflected once, but the energy and the angular momentum have already increased significantly. The three trajectories with $\eta_m = 0.4$ are all stable; however, one already sees small perturbations of the motion from nonradial forces. These results lend further support to the limitation $n_m = 0.3$ given in Section II B 3. The three remaining trajectories have been calculated for fixed values of ε and V_0, but with different starting energies $E_m = 0.1$ eV, 1 eV, and 10 eV. The turning radius of the 10-eV trajectory is so large that, with a different starting angle, this ion would hit the electrode. For the formulation of safe operating conditions, such trajectories must be excluded.

Since η increases proportionally to $\hat{r}^{n-2}$, according to Eq. (52), we can use the maximum turning radius, $\hat{r}_m$, for determining in good approximation the maximum of the adiabaticity parameter, $\eta(\hat{r}_m)$. To guarantee that the trajectories remain bound, $\hat{r}_m$ must remain smaller than 1, and since there must be some space for the wiggling motion, we normally require (Gerlich, 1986) for safe operating conditions that

$$\hat{r}_m < 0.8$$

and

$$\eta(\hat{r}_m = 0.8) < 0.3. \tag{53}$$

The correlation between the maximum allowed transverse energy E_m and $\hat{r}_m$ can be derived from adiabatic conservation of energy, which is expressed for multipoles in Eq. (46). For this purpose we apply Eq. (49), where the kinetic energy of the secular motion, $\frac{1}{2}m\dot{\mathbf{R}}_0^2$, is separated into its radial and centrifugal components, and make use of the fact that the radial component, $\frac{1}{2}m\dot{R}_0^2$, vanishes in the turning points. In the absence of a static field, the maximum energy E_m can be calculated from

$$E_m = \frac{1}{8}\frac{(qV_0)^2}{\varepsilon}\hat{r}_m^{2n-2} + L^2/2mr_m^2. \tag{54}$$

The centrifugal energy decreases with r^2 and is therefore usually rather rather small at the turning radius. In addition, the ions usually start very close to the centerline, resulting in a small angular momentum L. This allows us to neglect this energy term, especially since we wish to derive safe operating conditions.

From Eq. (52) and Eq. (54) the minimal characteristic energy ε and the smallest necessary amplitude V_0 for guiding an ion with a given maximal energy can be expressed by the three quantities $\hat{r}_m$, E_m, and η_m:

$$qV_0 = 8 \frac{n-1}{n} \frac{E_m}{\eta_m \hat{r}_m^n} \tag{55}$$

and

$$\varepsilon = 8 \frac{(n-1)^2}{(n\eta_m \hat{r}_m)^2} E_m. \tag{56}$$

Equation (55) shows the interesting result that the minimal amplitude expressed in terms of $\hat{r}_m$, η_m, and E_m is independent of m and Ω. Using, for example, $n = 4$, $\hat{r}_m = 0.8$, $\eta_m = 0.3$, and $E_m = 1$ eV, it is found that an amplitude above 48.8 V is needed to store a singly charged particle ($q = 1e$) in accordance with the formulated requirements. This is independent of whether it is an electron, a proton, or a heavy cluster ion. For a given mass m and radius r_0, the minimal frequency follows from Eq. (56) and Eq. (50). The specified numbers lead to $\varepsilon = 78.1$ eV. Solving Eq. (50) for Ω with $r_0 = 0.3$ cm and $m = 1\,u$, results in $f = \Omega/2\pi = 26$ MHz. For such numerical evaluations it is useful to use the units cm, MHz, and V, the mass unit u, the charge unit e, and the energy unit eV. In these units, Eq. (50) becomes

$$\varepsilon = 1.036 \frac{1}{2n^2} m\Omega^2 r_0^2 \quad \text{(units eV, } u\text{, cm, MHz).} \tag{57}$$

To guide an electron in a quadrupole ($n = 2$, $r_0 = 0.5$ cm) with a transverse energy $E_m = 0.1$ eV, an rf voltage of $V_0 = 2$ V is needed at a frequency $f = 70$ MHz. One must, however, pay attention here that the wavelength may become comparable to the dimensions of the device. An application at the other extreme of the frequency scale, the storage of charged microparticles in a quadrupole at $f = 53$ Hz, has been discussed by Wuerker et al. (1959a).

Safe conditions for simultaneous confinement of two masses $m_1 < m_2$ with a common maximum transverse energy E_m can also be calculated from Eq. (55) and Eq. (56) by replacing η_m with

$$\eta'_m = (m_1/m_2)^{n/(2n-2)} \eta_m, \tag{58}$$

and using in Eq. (57) the larger mass m_2. The optimum choice of the number of poles depends on the desired mass range, for example, for a relative range of a factor of 100, an octopole is best suited. Here we have used the obvious criterion of the lowest frequency and amplitude, and assumed the radius r_0 to be fixed.

As a typical numerical example for an octopole, where $m_1 = 1\,u$, $m_2 = 100\,u$, $n = 4$, $r_0 = 0.3$ cm, and $E_m = 0.1$ eV, the minimal values are $V_0 = 105$V and

$f = 17.8\,\mathrm{MHz}$. For a wider mass range, for example, a factor of 10,000, a decapole is preferred. Such a range might be needed for simultaneous storage of electrons and ions. In a hypothetical example, where $m_1 = 1\,u$, $m_2 = 10{,}000\,u$, $n = 5$, $r_0 = 0.3\,\mathrm{cm}$, and $E_m = 1\,\mathrm{eV}$, minimal values of $V_0 = 20\,\mathrm{kV}$ and $f = 110\,\mathrm{MHz}$ are needed. The derivation of conditions for the simultaneous

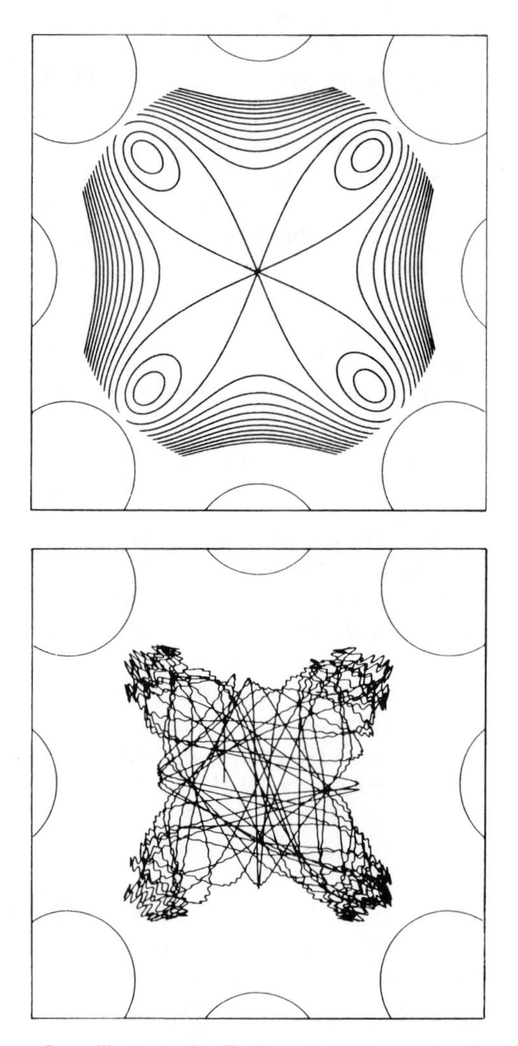

Figure 12. Upper figure illustrates the effective potential in an octopole with a superimposed static field, Eq. (45). The trajectory in the lower panel shows that the ion spends most of its time in the four potential minima located in the vicinity of the negatively biased rods.

storage of ions with different masses and with different energies is straightforward.

In nearly all applications described thus far, multipoles with $n > 2$ have been operated in the rf-only mode, that is, with no static potential U_0. The effective potential of a multipole with a superimposed dc potential [see Eq. (51)] is illustrated in the upper panel of Fig. 12. It is evident that the effective potential has lost its rotational symmetry. Since the static portion of the effective potential decreases proportional to $-\hat{r}^n$ in the direction of the negatively biased poles, and the rf portion increases as $\hat{r}^{2n-2}$, one obtains potential minima close to the negative rods for multipoles with $n > 2$. This is illustrated by the trajectory in the lower panel of Fig. 12, which shows that the ions avoid the positively biased poles. To ensure that ions started near the center of a multipole with $U_0 > 0$ do not hit the rods, it is necessary that $V^*(\hat{r} = 1) > 0$, and one obtains directly from Eq. (51) the relation

$$qU_0 < \frac{1}{8}\frac{(qV_0)^2}{\varepsilon}. \tag{59}$$

Since ε is proportional to the mass, Eq. (59) also shows that any multipole can be operated as a low-pass filter, transmitting only masses below a certain limit. Critical operating conditions will be discussed further in Section III C 4.

For determining safe operating conditions in the presence of a static field, the correlation between the maximum allowed transverse energy E_m and radius $\hat{r}_m$ can be established analogously to Eq. (54). It is also straightforward to derive corresponding formulas for stable confinement of ions that start in the vicinity of one of the potential minima located near the negatively biased rods (see Fig. 12).

3. *Is There an (a_n, q_n) Stability Diagram?*

It has already been mentioned in Section II A that for the special case of a quadrupole field, the region of stable trajectories is completely characterized by the well-known (a, q) stability diagram (see Fig. 24). In an attempt to generalize these parameters for multipoles, we define the parameters a_n and q_n,

$$q_n = 2n(n-1)\frac{qV_0}{m\Omega^2 r_0^2} = \frac{n-1}{n}\frac{qV_0}{\varepsilon} \tag{60}$$

and

$$a_n = 4n(n-1)\frac{qU_0}{m\Omega^2 r_0^2} = 2\frac{n-1}{n}\frac{qU_0}{\varepsilon}, \tag{61}$$

which leads to a simplification of Eq. (59):

$$a_n < \frac{n}{4(n-1)} q_n^2. \tag{62}$$

For a quadrupole this expression, $a_2 < \frac{1}{2}q_2^2$, is identical to the first term in the expansion of the lower (a_2, q_2) boundary limit of stability (Dehmelt, 1967; Wuerker et al., 1959a). Motivated by this similarity, one could be attempted to start from Eq. (62) and construct an (a_n, q_n) stability diagram for multipoles. However, one has to be aware of several restrictions that may render this approach impossible. First, Eq. (62) is only valid in the range of adiabaticity, and the question arises whether there is any meaningful generalization to extend the border lines to $q_n > 0.3$. Second, comparison of Eq. (60) with Eq. (47) reveals that q_n is only identical to the adiabaticity parameter η for $n = 2$, while to fully characterize the multipole one needs the $\hat{r}$ dependence of η, which can be expressed as

$$\eta = q_n \hat{r}^{n-2}. \tag{63}$$

Finally, it is important to remember that the underlying differential equations, Eq. (42), are nonlinear and coupled for $n > 2$.

An attempt to search for an (a_n, q_n) stability diagram, based on a similar analogy and numerical tests, was reported by Szabo (1986). His "analytical theory of multipole devices" began by comparing the canonical form of the Mathieu equation with the general equation of motion, Eq. (42), substituting the time t by the dimensionless parameter $\xi = \Omega t/2$. With this substitution, the function $F(t)$, given in Eq. (43), could be rewritten in the form $a_{n_s} - 2q_{n_s} \cos 2\xi$, and from this expression, Szabo defined the parameters (a_{n_s}, q_{n_s}). The index s was introduced here because for $n > 2$ there is a slight difference between the definition given by Szabo and our difinition given in Eq. (60) and Eq. (61). Taking our convention for U_0 and V_0 (see Section II C 3), the relation between (a_n, q_n) and (a_{n_s}, q_{n_s}) is given by

$$q_{n_s} = \frac{n^2}{4(n-1)} q_n \quad \text{and} \quad a_{n_s} = \frac{n^2}{4(n-1)} a_n. \tag{64}$$

This comparison shows that Szabo's "derivation" of the (a_{n_s}, q_{n_s}) parameters is not unequivocal because for $n = 2$, also, our definition is consistent with the canonical form of the Mathieu equation if we use the slightly different substitution, $\xi = \Omega t/2(n-1)^{1/2}$. Yet another possibility would be to introduce the time scale $\xi = \Omega tn/4(n-1)$, resulting in

$$q'_n = q_n \quad \text{and} \quad a'_n = \frac{4(n-1)}{n^2} a_n, \tag{65}$$

and in a somewhat simpler form of Eq. (62),

$$a'_n < \frac{1}{n} q'^2_n. \tag{66}$$

For all three substitutions of the time t by ξ, Eq. (42) is transformed to the same canonical form when $n = 2$, thus indicating an ambiguity in the definition of the characteristic parameters.

In a subsequent series of papers, Hägg and Szabo (1986a, 1986b, 1986c) examined the transmission properties of the hexapole and octopole in terms of their generalized parameters (a_{n_s}, q_{n_s}), numerically solving Eq. (42) for selected initial conditions where the ions were started suddenly, off-axis, at full local field. The large computational effort resulted in sets of diffuse (a_{n_s}, q_{n_s}) diagrams. Conclusions of general validity set forth by these authors are (i) a simple classification within such diagrams is not possible especially since they depend on the initial conditions and (ii) multipoles with $n > 2$ are suited for transporting, guiding, and collimating ion beams but probably not for mass analysis.

Unfortunately, there are other conclusions presented by these authors that are not due to properties of the multipoles themselves, but rather are a consequence of the selection of initial conditions used in their computations. One example is the reported correlation between the region of total stability and the size of the entrance hole. A second is the quoted dependence of the stability range on the number of poles. It is generally not true that an octopole or hexapole is better suited than a quadrupole for confining a beam close to the axis. In fact, for a specified transverse energy, the preceding statement must be completely reversed. There are other inferences reported that are simply a consequence of the initial conditions chosen. The sudden start of the ions at full local field leads to a phase-dependent additional transverse energy that increases with r and decreases with n. Such initial conditions have to be questioned, even from a practical point of view. Experimentally, it is advisable that the incoming ions see a slow, that is, adiabatic, rise of the rf amplitude. This will be discussed further in Section IV B 1.

Finally, there is a more conceptional problem related to Hägg and Szabo's attempt to extend the stability diagram. More serious than the ambiguity in the definition of (a_n, q_n) is their failure to observe the nonlinear $\hat{r}^{n-1}$

dependence in Eq. (42), which, in our adiabatic treatment, is reflected in the $\hat{r}$-dependent stability parameter, $\eta = q_n \hat{r}^{n-2}$. The fact that the region of stability in a multipole ($n > 2$) is better characterized with $\eta(r)$, rather than q_n, can be nicely illustrated by results presented in Hägg and Szabo (1986c). Figure 4 in this paper shows an ion-beam cross section with a maximum radius $\hat{r} < 0.2$. The operating conditions correspond to $\eta = 5.33\hat{r}^2$, leading to $\eta(\hat{r} < 0.2) < 0.21$, that is, one obtains safe and stable conditions. The conditions of their Fig. 6 result in $\eta = 37.5\hat{r}^2$. Here, many ions have access to regions $\hat{r} \geqslant 0.14$ where η is equal to 0.7 or higher. Therefore, one can expect further loss of ions before the residual cloud remains close enough to the center axis, where one always finds a finite region of stability, since $\eta(\hat{r} = 0) = 0$ for $n > 2$.

These two examples illustrate that our concept of a local stability parameter $\eta(r)$ is superior. The extremely large range of stability reported in Hägg and Szabo (1986b) is simply a consequence of defining the stability parameter only at the electrode surface, $\hat{r} = 1$. Even more important, our definition of η is based on a completely general analysis of the structure of solutions from the general equation of motion, and not on simple analogies. It can be concluded that the only adequate description of multipoles presently available is that based on the η field. Unfortunately, this method is only valid in the adiabatic limit, and it is not straightforward to extend the characterization of stability in multipoles toward higher η values. It is, however, mandatory that any generalization should merge into the conditions derived in this section, such as that given in Eq. (62).

One possible way to map out regions of stability at higher η values has been mentioned in Section II B 4. A systematic search for stable oscillatory equilibria (in the Liapunov sense) could be started from regions where the adiabatic approximation holds and where one can make use of the adiabatic constants of the motion, E_m and $\mathbf{L}$. To illustrate this idea, we have determined the period T of a full oscillation of the secular motion by integration of Eq. (46). If the motion goes through the center, that is, $\mathbf{L} = 0$, the frequency $\omega = 2\pi/T$ can be expressed by η

$$\omega/\Omega = C_n \eta(r_m), \tag{67}$$

where the constant C_n is obtained from an elementary integral, which must be evaluated numerically for $n > 2$. For the quadrupole, where $C_2 = 1/8^{1/2}$ and $\eta = q_2$, one obtains the well-known ratio between the harmonic secular motion with frequency ω and the rf frequency Ω. For $n > 2$, the potential V^* is no longer harmonic and the period T depends on the amplitude r_m, which enters into Eq. (67) via $\eta(r_m)$. Using this relation, initial conditions for stable periodic trajectories with prescribed Ω/ω can be narrowed down. For example, one predicts for an octopole with $C_4 = 0.152$ and $\Omega/\omega = 10$ an initial velocity corresponding to $\eta_v = 0.66$. This is in good agreement with the numerically determined value of $\eta_v = 0.645\cdots$ (see Section II B 4). Similar comparisons can be made to locate the positions of other integer Ω/ω values in Fig. 4 and to predict initial conditions for periodic trajectories with angular momenta $\mathbf{L} > 0$.

This example illustrates that there are mathematical methods (Arnold, 1988; Kruskal, 1962; Lichtenberg and Lieberman, 1983), which may be used for a more general description or classification of ion motion in multipoles or other rf devices. It is unlikely that such an approach will result in a simple (a_n, q_n) diagram and in sharp transitions from stability to instability, which could be easily used for mass or phase-space selection. However, it may lead to new theoretical insight, and to the development of new rf devices, or even to new applications of existing devices in an η range, which in this chapter has been deemed *unsafe*. At present we recommend for most applications operating under *safe* working conditions, that is, to restrict the ion trajectories to regions where $\eta(r) < 0.3$.

4. *Potentials of Realistic Multipoles*

Another question often discussed is the influence of deviations on the ideal electric field. There are several sources of deviations. These include the surfaces of the field defining conductors, which practically never coincide with the mathematically defined boundary conditions. Also, the relative alignment of the electrodes can be perturbed and, often, the metal surfaces are not exact equipotential surfaces. At high ion densities ($> 10^5\,\mathrm{cm}^{-3}$), space-charge effects can greatly contribute to perturbations, and, in this case the potential has to be determined explicitly from Eq. (26). A costly approach to closely approximating ideal fields is to use precisely machined metal pieces. Typically, however, mechanically simple and relatively low cost arrangements of electrodes are used, for example, rods, wires, metal plates, and so on, since many undesired imperfections can be minimized by adequate selection of geometrical and electrical parameters.

The ideal equipotential lines of multipoles were given in Eq. (39). In practice, $2n$-pole devices are usually composed of $2n$ circular rods, equally spaced on a inscribed circle with radius r_0. As a first approximation, the diameter d of the rod can be determined from the radius of curvature of the hyperbolic potential and one obtains

$$r_0 = (n - 1)d/2. \tag{68}$$

Use of nonideal electrode structures leads to perturbations in the field that can be expanded in a series of higher-order multipole terms, as given in

Eq. (29). This problem has often been discussed (Busch and Paul, 1961b) in the case of quadrupoles. By minimizing the lowest-order term in the multipole expansion, Dayton et al. (1954) have shown that for a quadrupole with round rods, the rod diameter should be slightly larger than that given by Eq. (68), that is, $d/2 = 1.148r_0$. Denison (1971) has shown that if one accounts for the cylindrical tube which commonly surrounds the rods, a diameter $d/2 = 1.1468r_0$ should be used. To accurately determine the effective potential or adiabaticity parameter, Eq. (45) and Eq. (47), respectively, the higher-order multipole terms should be taken into account. However, it is not necessary to minimize these terms if ion guidance or ion trapping rather than sharp mass selection is intended, since small geometrical deficiencies do not influence the general properties of these devices. This will be demonstrated using numerical examples.

Rather than using precisely positioned rods, very good approximations of given electric fields can be achieved by an arrangement of wires surrounded by a metal cylinder. This has been briefly described in Section II C 2 and has been used, for example, to approximate quadrupole fields (Matsuda and Matsuo, 1977). We have also used this method as a numerical means of studying the influence of field imperfections on the effective potential. Realistic octopole structures have been represented by 96 pairs of line charges, as illustrated in Fig. 13. The boundary conditions imposed by each rod are fulfilled at 12 points, indicated by the solid squares, while image charges are used to simulate the position and potential of the shielding cylinder (dashed line). This method reduces the boundary value problem to a simple algebraic problem, and thereby allows one to study many details of the electrostatic and the effective potential, for example, the influence of the pole shape, misalignment, electrical distortions, and also field penetration from surrounding ring electrodes.

Figure 14 shows the effective potential of a realistic octopole whose dimensions are given as defined in Eq. (68). The inner equipotential lines have almost perfect rotational symmetry and their positions are in excellent agreement with the values calculated from the analytical expression given by Eq. (45). There are some slight perturbations close to the electrodes, especially between the rods where the effective potential is somewhat weakened. This effect is even more pronounced in the lower part of Fig. 14, where the shape of four rods has been modified significantly, for example, for passing a crossed molecular beam. Note that in both cases ions with a transverse energy of less than 0.5 eV see almost circular equipotential lines. It is a general fact that the actual field converges to that of an ideal multipole as one moves farther away from the rods. One example illustrating the influence of a mechanical misalignment is depicted in Fig. 15. The significant shift in the position of one pole on the right perturbs the total symmetry

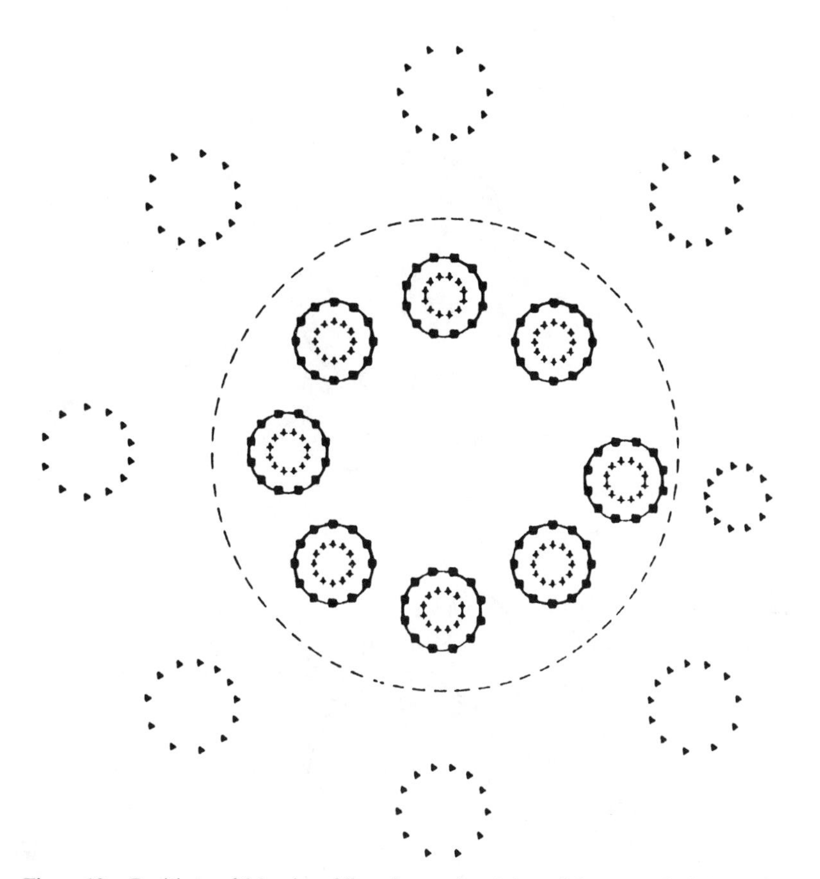

Figure 13. Positions of 96 pairs of line charges (+, ▶) used for numerical determination of realistic octopole potentials. The rods (circles) are approximated by imposing the desired boundary values in the positions (■). The dashed line represents the shielding cylinder. This example shows the arrangement of an octopole with a slightly distorted geometry; the resulting potential is plotted in Fig. 15.

and influences the inner potential lines. Nonetheless, this octopole can still be used to guide ions safely with transverse energies up to 4 eV. All three examples clearly demonstrate that the use of multipoles as ion guides does not require very accurate electrode shapes, and that the alignment is also rather uncritical.

More serious problems and significant distortions are usually caused by changes in the work function, contact potential imperfections, or contamination of the rods. The build up of charge on insulating layers of the electrode surface is less critical for an ion guide than for a mass filter,

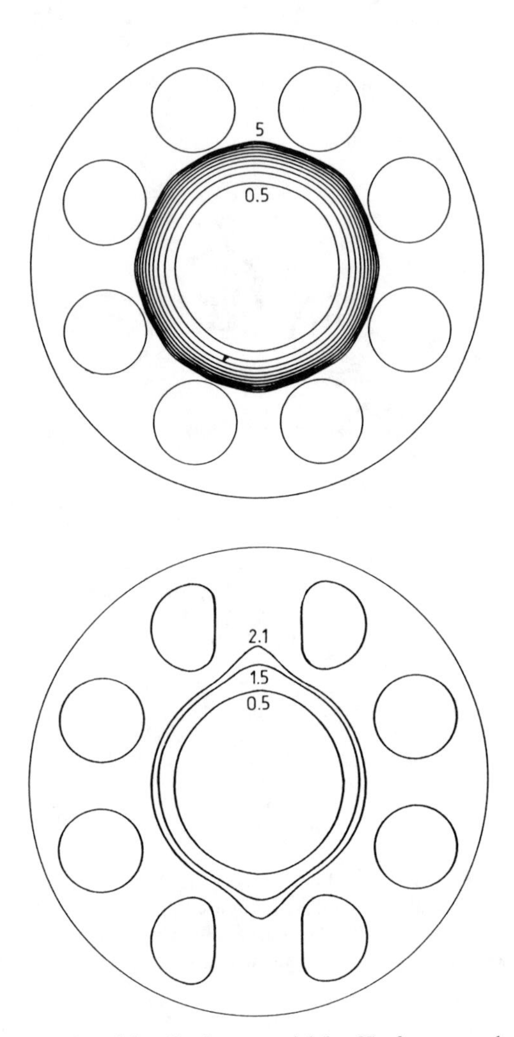

Figure 14. Contour plot of the effective potential (in eV) of an octopole with circular rods, diameter given in Eq. (68). The result illustrated is very similar to the potential obtained with ideal hyperbolic electrodes; the graphic resolution is not sufficient to demonstrate the differences. Only the equipotential lines close to the rods show slight deviations from the rotational symmetry. The lower panel shows the influence of a significant deformation of the electrodes (e.g., geometry for a crossed-beam arrangement).

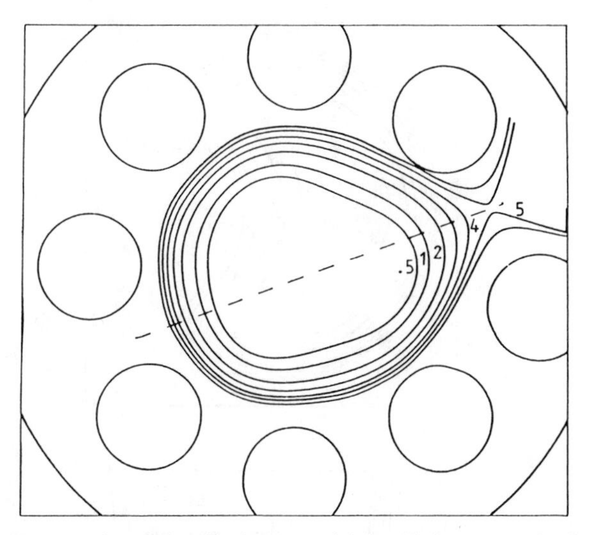

Figure 15. Contour plot of the effective potential (in eV) from a mechanically distorted octopole (see Fig. 13), illustrating that geometric precision is not required for using multipoles as ion guides.

since in the former all ions are usually transmitted, while in a quadrupole the surfaces are continuously under ion bombardment. The effect of an additional dc voltage applied to a single rod in an octopole is illustrated in Fig. 16. The effective potential becomes half-moon shaped, and the distortion pushes the ions toward the other half of the ion guide. A cut through the effective potential along the dashed line is shown in the lower part of Fig. 16 for several rf amplitudes. This figure illustrates that the minimum energy path is a function of V_0. This feature can be used to test a device for a surface distortion localized on a single rod. However, according to our experience, it is more reasonable to assume a patchiness of distortions distributed statistically both in the r and φ directions. Since the influence of the distortions is least significant in the center, operating the beam guide at high rf amplitudes so that the ions are compressed close to the center line, is usually superior. Further discussion of surface problems follows in Section III C 2.

A final example of a numerically determined potential distribution is given in Fig. 17 and shows the influence of a cylinder surrounding an octopole. In order to display the very small effect of the field penetration, the rf and the dc potentials have been set to zero. The equipotential lines are given as a fraction of the externally applied voltage, U_{ext}. It can be seen for the geometry shown that the center of the octopole is raised by 2.5 mV per 1 V applied to the cylinder. Using short cylinders, local potential barriers can be created.

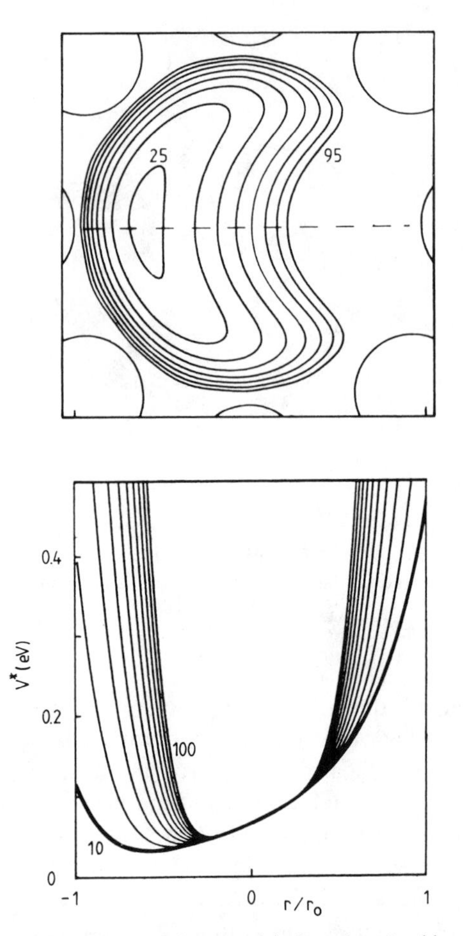

Figure 16. Influence of an electrostatic potential distortion. In this example one rod on the right is charged to +0.5 V. The upper panel shows a contour plot of the effective potential (in meV, 10-meV steps) for a low rf amplitude of 10 V. The lower panel is a cut through the potential along the dashed line and illustrates the influence of the rf amplitude as it is increased in 10-V steps.

Such ring electrodes have gained practical importance both for localizing potential distortions and for forming and trapping extremely slow ion beams. Here again we refer to Section III C.

In summary, it is not necessary that the contour of the cross section of the electrode have the exact shape given by Eq. (39) in order to achieve the best possible performance. The required accuracy of a multipole field depends strongly on the desired application. When using rf electrodes for guiding and

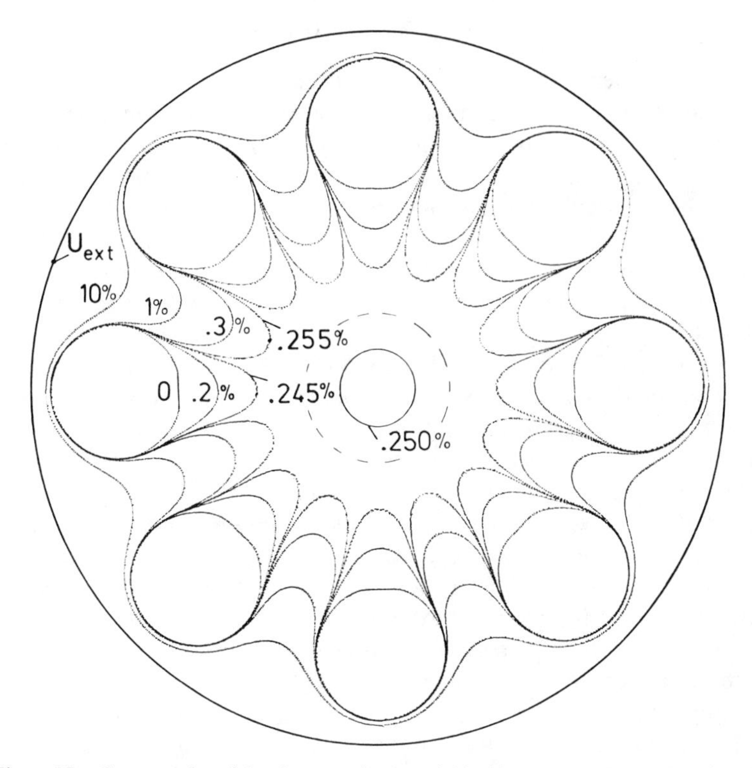

Figure 17. Contour plot of the electrostatic potential inside an octopole caused by a voltage U_{ext} applied to the surrounding cylinder. The potential of the eight rods is set to zero. The height of the equipotential lines is given relative to the external voltage. For the geometry shown, the potential on the centerline is shifted by 0.25%.

storing ions within the safe adiabatic limit, the approximation of the field geometries by real electrodes is not critical, in contrast to the quadrupole mass filter where sharp mass selection is intended. As a general rule, it is advisable to avoid sharp edges or small bending radii since they cause steep field gradients, and therefore large local $\eta(r)$ values, according to Eq. (21). Of course the situation becomes different if one finds, for a well-defined potential, some application where critical operation at the border between stability and instability is demanded. There, the requirements for the experimental realization of the ideal boundary conditions become the same as those for mass selective quadrupole fields, and may possibly be even more critical. Finally, one should be aware of the fact that problems due to potential distortions of the metal surfaces are often more critical than deviations from mechanically ideal electrodes.

E. Energy Distributions

To aid in the interpretation of experiments employing stored or guided ions, it is often necessary to understand the distortion of the ion kinetic energy distribution caused by the confining rf field. One extreme is the process of rf heating, which couples kinetic energy from the micromotion (i.e., from the rf field) into the secular motion. Such heating processes have been considered for acceleration of ions to high velocities (Motz and Watson, 1967) or for mass-selective ejection (Busch and Paul, 1961a), and are predominantly based on resonant approaches. In contrast, most of the applications in this chapter deal with slow ions and make use of adiabatic conservation of energy by working at high frequencies, as has been discussed in Section II B.

Several models and mathematical methods have been developed to examine kinetic energy distributions of ions stored in quadrupole traps, especially under the inclusion of collisions and space-charge effects. The results have been summarized in several reviews (March and Hughes, 1989; Todd et al., 1980; Vedel, 1991). Energy distributions in quadrupole fields are significantly inferior to those in higher-order multipole fields or other suitable electrode arrangements. This can be understood qualitatively based on several of the preceding figures. For example, Fig. 2 shows stable confinement; however, it is nearly a worst case example with respect to the kinetic energy distribution because the wiggling motion is always rather intense. Even in the minimum of the effective potential, the ion interacts strongly with both the dc and rf fields. Damping of this wiggling motion by inelastic collisions would lead to a drift into regions of weaker rf field, and finally, to a loss at the outer cylinder. All multipoles feature a weak-field region close to the center, which becomes broader with increasing number of poles, as can be seen from comparison of Fig. 9 and Fig. 10. Similar qualitative conclusions can be drawn from the steepness of the effective potential. For example, Fig. 8 shows that the ring electrode trap is superior to an octopole.

In the following section, we first describe the time dependence of the momentary kinetic energy of an ion during its reflection from an rf wall and derive time-averaged energy distributions. The influence of collisions with neutral molecules will then be treated for a few special cases using numerical simulations.

1. *Instantaneous and Time-Averaged Energy*

We have shown in Section II B that on average the system behaves like a conservative one if the condition for the adiabatic approximation, $\eta < 0.3$, is fulfilled. This means that the total energy, E_m, as defined by Eq. (16) or Eq. (46), is conserved within narrow limits. The example given in Fig. 4 nicely illustrated that the relative energy changes are typically below 10^{-6}

if η remains smaller than 0.3. It was discussed, however, in Section II B that the effective potential is in reality the average of the kinetic energy of the fast oscillatory motion. This means that there is a continuous exchange between the two different forms of translational energy, the wiggling motion and the secular motion.

For deriving an expression for the total instantaneous kinetic energy within the adiabatic theory, we assume for simplicity a constant static potential, that is, $\Phi_s = \text{const}$. The exact motion, $\mathbf{r}(t)$, is represented in first-order approximation by the superposition of the secular motion, $\mathbf{R}_0(t)$, and the wiggling motion, $\mathbf{R}_1(t)$, as discussed in detail in connection with Eq. (6). The relation between the effective potential energy, $V^*(\mathbf{R}_0)$, and the kinetic energy, $\frac{1}{2}m\dot{\mathbf{R}}_1^2$, has been emphasized by the representation chosen in Eq. (18) and is depicted in Fig. 18. This figure refers to an ion that has been started in the center of an octopole with an initial kinetic energy E_m. The smooth line indicates the effective potential energy, which reaches its maximum $V^* = E_m$ in the turning point of the secular motion, while the oscillatory line shows the kinetic energy $\frac{1}{2}m\dot{\mathbf{R}}_1^2$ along the same trajectory $\mathbf{R}_0(t)$. This kinetic energy fluctuates between 0 and $2V^*$ during the reflections from the rf wall. The absolute maximum can reach $2E_m$, depending slightly on the rf phase δ.

The *total* instantaneous kinetic energy has to be calculated from the sum of the two velocity vectors $\dot{\mathbf{R}}_0$ and $\dot{\mathbf{R}}_1$;

$$E_{\text{kin}}(t) = \tfrac{1}{2}m[\dot{\mathbf{R}}_0(t) + \dot{\mathbf{R}}_1(t)]^2. \tag{69}$$

Assuming the extreme case, that the smooth and the oscillatory motions are parallel, and expressing $\dot{\mathbf{R}}_0^2$ and $\dot{\mathbf{R}}_1^2$ by the effective potential using Eq. (16) and Eq. (18), respectively, one obtains for the momentary total kinetic energy the simple relation

$$E_{\text{kin}}(t) = [(E_m - V^*)^{1/2} + (2V^*)^{1/2} \sin \Omega t]^2. \tag{70}$$

It is obvious that the total instantaneous kinetic energy is identical to E_m in the field-free region, that is, for $V^* = 0$, and that it becomes equal to $2E_m \sin \Omega t$ for $V^* = E_m$. It is also easy to show that the absolute maximum of the total kinetic energy, $3E_m$, is not reached at the turning point of the secular motion, but rather at that position in the field where two-thirds of the total energy is converted into oscillatory motion and one-third remains in smooth motion. During one reflection, E_{kin} oscillates between 0 and $3E_m$, as can be seen from the dashed line in Fig. 19.

Note that Fig. 19 also allows one to gain an impression of the high quality of the first-order adiabatic approximation. The heavy line shows the exact kinetic energy, $\frac{1}{2}m\dot{\mathbf{r}}^2$, obtained by direct numerical integration of the equation

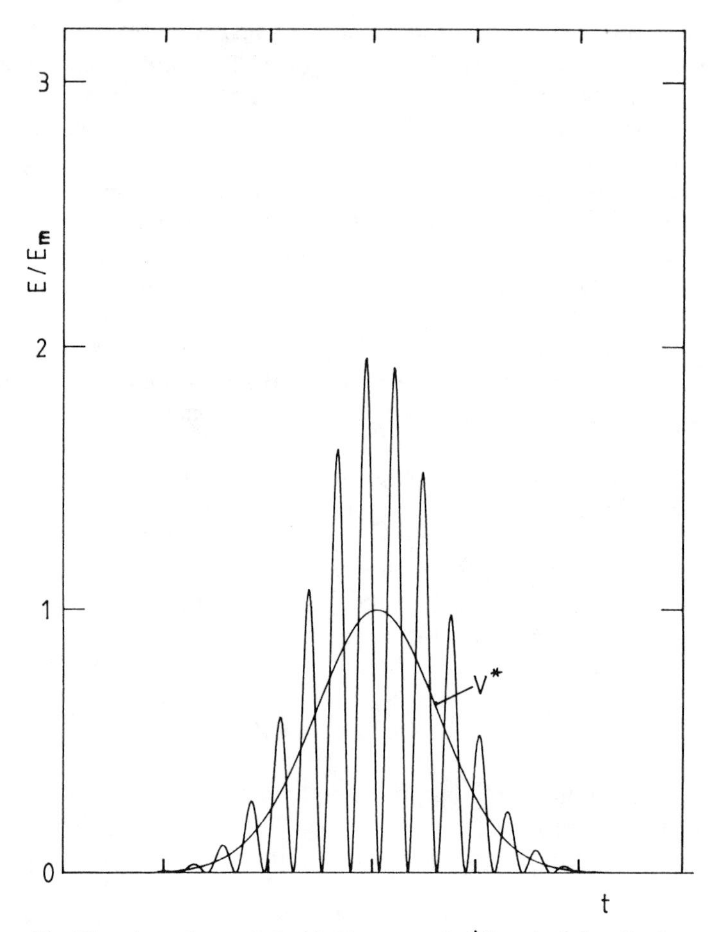

Figure 18. Time dependence of the kinetic energy ($\frac{1}{2}m\dot{\mathbf{R}}_1^2$) and of the effective potential energy V^* during one reflection from an rf wall. The oscillatory line represents the kinetic energy of the wiggling motion, which varies between 0 and $2E_m$. The smooth line shows the increase of the effective potential energy along $\mathbf{R}_0(t)$.

of motion. The dotted line, which has been calculated for $\eta_m = 0.3$, clearly shows that higher-order microoscillations contribute only with small corrections to the results derived within the first-order adiabatic approximation. The two curves become practically indistinguishable for $\eta_m < 0.2$.

The opposite extreme of relative orientation of $\dot{\mathbf{R}}_0$ and $\dot{\mathbf{R}}_1$ in Eq. (69) is obtained when the oscillatory motion is orthogonal to the smooth motion. This occurs, for example, in a multipole for a motion along those lines where

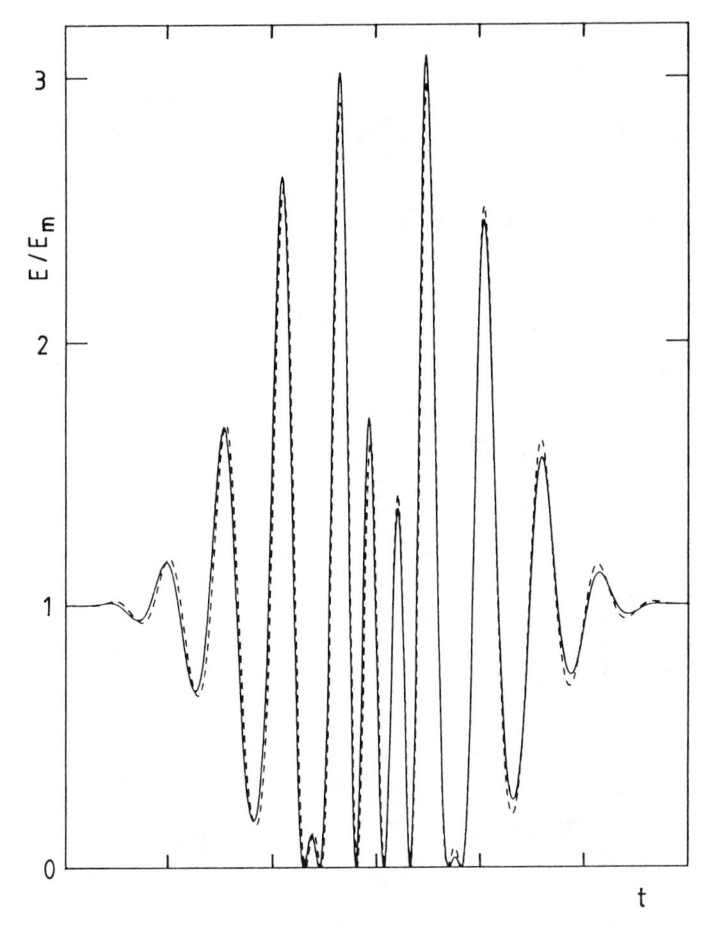

Figure 19. Total instantaneous kinetic energy of an ion during one reflection from an rf wall. The dashed line has been calculated within the adiabatic approximation, $\frac{1}{2}m(\dot{\mathbf{R}}_0 + \dot{\mathbf{R}}_1)^2$, while the solid line is the numerically exact solution, $\frac{1}{2}m\dot{\mathbf{r}}^2$.

$\cos n\varphi$ vanishes [see Fig. 11 or Eq. (39)], since in this case the two vectors $\mathbf{E}_0$ and ∇E_0 are orthogonal. In this situation, the total kinetic energy is just the sum of the two kinetic energies and oscillates only between 0 and $2E_m$. It should be emphasized that Eq. (70) is valid for all field geometries. Note, however, that the energy E_m only accounts for those degrees of freedom that are affected by the rf field, while energy contributions from other degrees of freedom, for example, the axial motion in an octopole, remain unperturbed.

For most applications, it is more important to know the time-averaged distribution of the momentary kinetic energy. Such results have been computed by following ion trajectories with different initial conditions and by continuously recording their energies (usually in 96 steps per rf period). A typical result, accumulated over 3×10^5 rf periods, is shown in Fig. 20 for an octopole using $\eta_m = 0.25$. The kinetic energy is measured relative to the initial energy E_m. The most probable value, which in this case is arbitrarily normalized to 1, is slightly below $E/E_m = 1$. Instances in which the energy reaches the predicted maximum of $3E_m$ are very rare, and in Fig. 20 are not

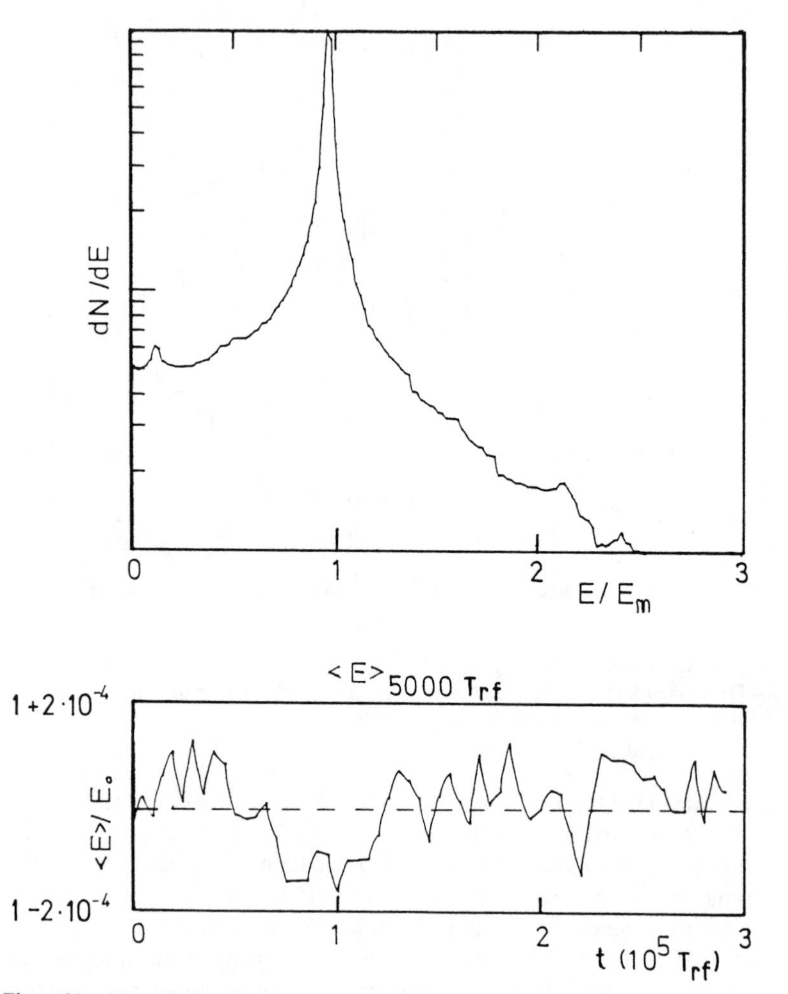

Figure 20. Numerically determined kinetic energy distribution of an ion moving in an octopole accumulated over 3×10^5 rf periods T_{rf} ($\eta_m = 0.25$). The lower panel shows the fluctuations of the mean energy averaged each time over 5000 rf periods.

on scale. The time dependence of mean values, $\langle E \rangle$, averaged in each case for 5000 rf periods, is plotted in the lower part of Fig. 20. The fluctuations are on the order of 10^{-4} and are due more to the statistics of the interrogation intervals rather than to a real fluctuation of the total energy. This result agrees qualitatively with that shown earlier in Fig. 4. Note, however, that Fig. 20 shows the energy averaged along the whole trajectory, while in Fig. 4 the energy is probed only at a single point in the field-free center.

As already indicated by the trajectories shown in Fig. 9 and Fig. 10, the storing field perturbs the ion energy to a smaller extent as the number of poles is increased. A quantitative comparison between numerically determined energy distributions for a 16-pole and a 32-pole is shown in Fig. 21. The extremely narrow widths of the distributions indicate that the ions spend most their time in field-free regions. The energy distributions have also been studied as a function of η_m. It was found that there are no significant changes as long as η remains smaller than 0.3.

In summary, the modulation of the kinetic energy by the rf field can often be neglected, especially in ion-beam guides where only the transverse part of the energy is affected and where usually either no or single-collision conditions prevail. The situation changes completely if the stored ions undergo many collisions with a buffer gas. This leads, on the one hand, to a thermalization of the ion energy if the energy exchange takes place in a field-free region. If, on the other hand, one of the collisions occurs in the rf field, the adiabatic energy exchange between the ion and the rf field breaks down. The consequences of collisions are discussed in the following section.

2. *Influence of Collisions*

As a means of determining the velocity and spatial distributions of stored ions under the inclusion of collisions, we have carried out a series of numerical simulations (Kaefer, 1989; Scherbarth, 1984). Since our primary goal was to obtain realistic energy distributions for H^+ and C^+ ions stored in a cold hydrogen buffer gas under actual experimental conditions, this work did not systematically study different parameter dependencies. Calculations have been performed for linear $2n$-pole traps ($n = 2, 4$, and 8) and for the ring electrode trap (see Section III E).

The computer program consists of our standard code for the numerically exact description of the ion motion in a particular time-dependent field and a subroutine for simulating collisions. The trajectory calculations take advantage of the fact that the potential of a cylindrical multipole field, Eq. (39), or a ring electrode trap, Eq. (34), depends only on two coordinates. During collisions, however, the relative motion is treated in three dimensions. The trapping fields have been assumed to be infinitely long, that is, effects from electrostatic gates or barriers used to close a real trap are ignored.

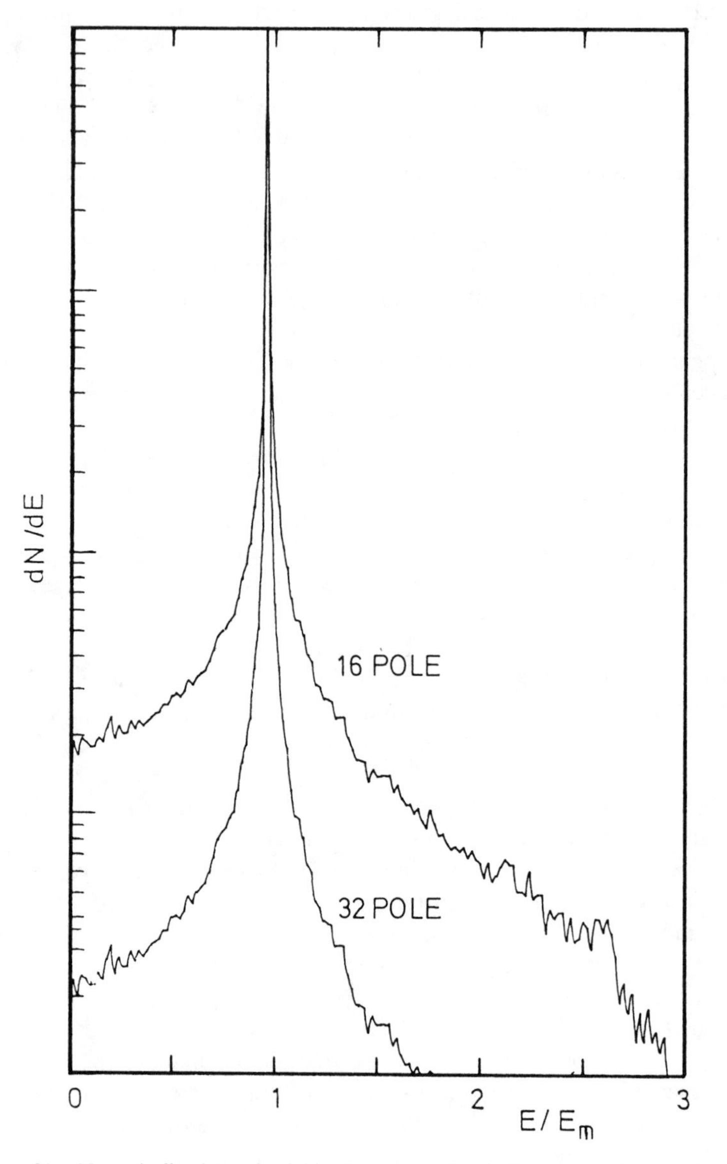

Figure 21. Numerically determined kinetic energy distributions of an ion moving in a 16-pole or a 32-pole averaged over 3×10^5 rf periods ($\eta_m = 0.25$).

The temperature of the hydrogen target gas was either 80 K or 300 K. The random selection of the target velocity, simulated by a three-dimensional Maxwell–Boltzmann distribution, and the thermal population of the individual rotational H_2 states were accounted for by properly weighted random selection. Usually we chose normal-H_2, although in some calculations pure para-H_2 was used. The time between collisions, on average 200 rf periods, was also chosen randomly, as was the rf phase. The collision rate was about 10 times higher than in typical experiments; however, it was kept low enough to avoid correlation between subsequent collisions.

The $H^+ + H_2$ collision process has been modeled in great detail using experimentally verified state-to-state differential cross sections, derived from a dynamically biased statistical theory (Gerlich 1977, 1982; Gerlich et al., 1980) and accounting for ortho–para transitions (Gerlich, 1990). $C^+ + H_2$ collisions were described in a similar way, since in this system, also, formation of a long-lived complex determines the collision mechanism at the low energies of our study (Gerlich et al., 1987). The laboratory motion of both collision partners was described independently and in all three dimensions. Details of the necessary forward–backward transformations between laboratory and center-of-mass frames are described in Gerlich (1989a).

To avoid additional expenditure with the initial conditions of the ion, we make use of the fact that these are essentially randomized after a few collisions. Therefore, it was sufficient to simulate the energy distribution by following a single ion over a long period of time and over many collisions. For low ion densities, the results were identical with an average over an ensemble of stored ions. The program was tested by switching off the rf field, leading to a perfect thermal energy distribution of the ions.

A few typical results are shown in Fig. 22 and Fig. 23. Each of the distributions is the result of 10^4 collisions. The parameters in mV correspond to potential distortions and are explained subsequently. The distributions obtained are not purely thermal; however, they can be composed in good approximation by a weighted sum of two Maxwell–Boltzman distributions, resulting in two characteristic temperatures T_1 and T_2, and related weighting factors.

The aim of the presentation in Fig. 22 is to emphasize the high-energy tail and the influence of the multipole order. It can be seen that the distribution of ions stored in an octopole with H_2 at 300 K has an extremely hot tail. The 16-pole also exhibits a high-energy tail, which is comparable to a 700 K Maxwell–Boltzmann distribution, shown for reference in the third panel of Fig. 22. Indeed, the fraction of very fast ions is rather small (10^{-6} and below); however, they can obstruct the observation of rare processes like radiative association if these products compete with products from endothermic

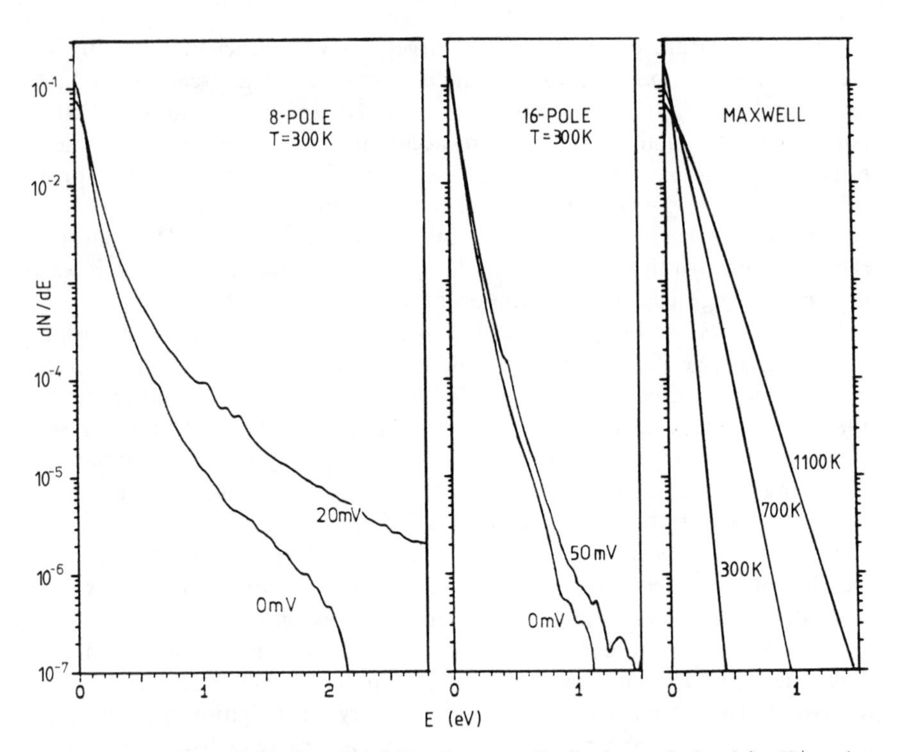

Figure 22. Numerically determined kinetic energy distributions calculated for H^+ under inclusion of collisions with room-temperature normal-H_2. The details of the model are described in the text. The two left panels show results for an 8-pole and a 16-pole. For comparison, the right panel depicts thermal distributions at three temperatures. The scales have been chosen to stress the high-energy tail of the distributions. For an octopole, a weak potential distortions of only 20 mV has a dramatic effect.

channels. One example is the formation of H_2^+ in fast $H^+ + H_2$ collisions via charge transfer.

A presentation that allows one to view simultaneously the low- and the high-energy regimes is given in Fig. 23 (although the logarithmic energy scale can lead to a somewhat erroneous visual weighting of the area normalized distributions). In these two examples, H^+ and C^+ ions are stored in a ring electrode trap ($r_0 = 0.5$ cm) with 80 K normal-H_2. An analysis of the H^+ distribution, plotted in the upper panel, indicates that 75% of the H^+ ions can be described by a 80 K thermal distribution. An additional 25%, however, resides in a 200 K hot tail. This heating is partly caused by too low of an rf frequency (27 MHz) and by the unfavorable mass ratio (light ion in a heavier buffer gas). Both effects are significantly reduced if one goes to C^+, as shown

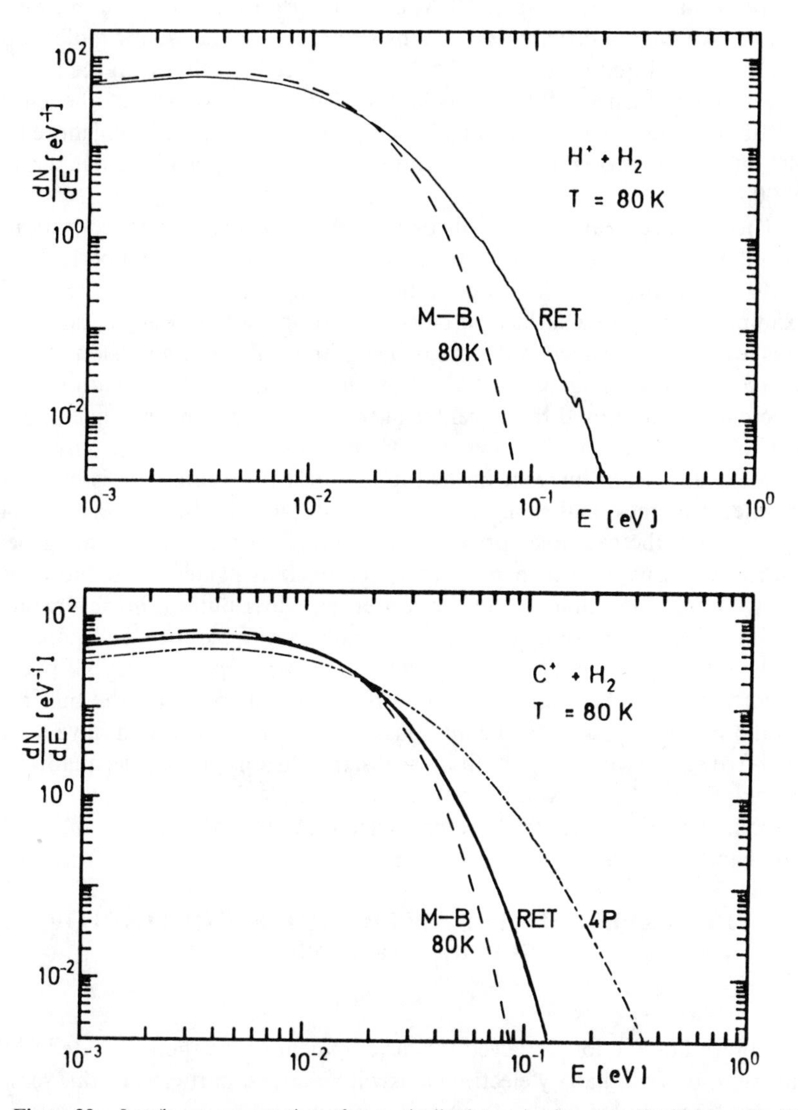

Figure 23. Log/log representation of numerically determined energy distributions for H^+ (upper panel) and C^+ (lower panel) stored in a ring electrode trap (RET) with 80 K H_2 buffer gas. The comparison with the 80 K thermal Maxwell–Boltzmann (M–B) and the quadrupole trap (4P) distributions is discussed in the text.

in the lower panel. To account for the small fraction of fast ions in this distribution, we only need a 160 K Maxwell distribution with a relative weight of 10%. An energy distribution of ions stored in a linear quadrupole trap is also compared quantitatively in this figure. Since in such a trap the ions are under the influence of the rf field for a much longer time, the energy distribution becomes significantly broader. In the example shown, more than 50% of the ions stored in the quadrupole are characterized by a temperature of 300 K.

There are several possible distortions that can increase the duration of the interaction of the ions with the rf field and cause additional perturbation of the energy distributions. Superimposed static field gradients, as shown for example in Fig. 12, or surface potential distortions, seen in Fig. 16, can attract ions toward the electrodes. A similar effect can result from repulsion by space charge. For modeling such an effect, we have performed calculations using a potential wall, raised in the center of the trap, with a mean range of $0.5r_0$. The left panel in Fig. 22 shows that in an octopole a barrier of only 20 mV causes a significant increase in the number of fast protons. If n is increased, the effect becomes smaller, as seen in the middle panel for the case of a 16-pole.

Although the examples presented are rather special cases, some general conclusions can be drawn from this section. It is evident that the rf field always has some influence on the ion energy distribution; however, under collision-free conditions, the effect is usually negligible. For collisional ion cooling, the rf frequency, the mass ratio, and the trap geometry all play an important role. The frequency should be as high as possible, the buffer gas should be light relative to the ion mass, and the rf trap should have a wide field-free region with steep confining walls. If collisions play a role, quadrupole ion traps should be avoided. However, for laser cooling of trapped ions, Paul traps are superior, since the harmonic effective potential simplifies the optical side-band spectrum.

III. EXPERIMENTAL APPLICATIONS AND TESTS OF SEVERAL RF DEVICES

A. Introduction

The preceding section surveyed the development of the theory related to the interaction of oscillatory electric fields with charged particles. In this section we describe detailed applications and experimental tests of several rf devices including the focusing quadrupole, octopoles, rf ion sources, and the ring electrode trap. As an introduction, we briefly review a few applications in addition to those just mentioned, some of which have a more historical significance. Included are experiments where laser fields or microwave fields interact with electrons or low-frequency fields with microparticles. The range of rf frequencies, $\Omega/2\pi$, of the selected examples extends from 10^2 to 10^{14} Hz, while the charge to mass ratio, q/m, ranges from 1.76×10^8 to 10^5C/g for electrons and heavy ions, to less than 10^{-6}C/g for microparticles. Characteristic lengths L of the field geometries range from cm to μm, and are either determined by a typical distance of the electrode structures, or given by the wavelength, as in the case of microwaves or optical waves. The interdependence of Ω, q/m, and L is given by Eq. (21), and a crude estimate of the relation between q/m and Ω can be obtained by using $\eta = 0.3$ and approximating the gradient of the field, ∇E_0, simply by E_0/L.

As in Section II A, we begin by examining the interaction between electrons and electromagnetic fields. One of the first experimental attempts to observe scattering of electrons by standing waves was reported in 1965 (Bartell et al., 1965), and later, with the availability of very intense laser beams, it became possible to provide safe experimental evidence (Bucksbaum et al., 1987) that an inhomogeneous laser field can create an effective potential wall, causing elastic scattering of free electrons. Many correlated problems occur in experiments on the ionization of atoms in strong radiation fields since the ejected photoelectron is created in a region of high laser field. Provided that the spatial and also the temporal variations of the field are slow, that is, change adiabatically, the electron–laser interaction can be explained on the basis of the conservative quasipotential derived in Section II. Nonconservative behavior has recently been demonstrated with ultrashort laser pulses (Bloomfield, 1990; Bucksbaum et al., 1988).

An interesting and illustrative experiment has been performed by Weibel and Clark (1961) who verified experimentally that a beam of electrons can be guided in stable orbits along the axis of a circular wave guide. This axis coincides with one nodal line of the microwave field and is therefore the locus of a two-dimensional effective potential minimum. The electron-beam guide consisted of a 20-cm-long section of a circular, properly terminated wave guide, operated with a frequency of $\Omega/2\pi = 9.29$ GHz. The cavity was fed with 250 kW of rf power in pulses of 2 μs duration. The well depth of the effective potential, achieved with a peak electrical field of 52 kV/cm, was as deep as 400 eV for electrons, and the frequency of the secular motion was reported as $0.033\,\Omega$ (corresponding to $\eta < 0.1$). Their measurement of the guided electron current was in good agreement with their estimate based on the adiabatic theory (Weibel, 1959). Other experiments using microwave resonance cavities have been surveyed by Motz and Watson (1967).

We now wish to shift our focus to rf confining devices that use quadrupole fields. The three-dimensional quadrupole, the Paul trap, has become very popular as a universal ion storage device. Early spectroscopical applications were pioneered by Dehmelt (1969) who employed a variety of different ingenious detection schemes. The last decade has been dominated by sensitive probing of a few ions or even a single ion via laser-induced fluorescence (Neuhauser et al., 1978). Applications to chemical problems have greatly expanded in the 1980s with the advent of the technique of mass-selective ejection and detection (March and Hughes, 1989). A reminiscence of the development of the quadrupole trap has been presented recently by Paul (1989) in his Nobel prize lecture. Among the first fundamental papers (Fischer, 1959) on the quadrupole ion trap is the experiment by Wuerker et al. (1959a), who suspended charged iron and aluminum particles ($q/m = 5 \times 10^{-6}$ C/g) in a low-frequency (100 Hz) rf field and was able to verify directly, by photographic observation, many details of the kinematics of the particle motion. Other experiments with microparticles and charged droplets (Owe Berg and Gaukler, 1959) have been summarized in Dawson (1976).

The question of the kinetic energy distribution of stored ions is a fundamental problem for spectroscopical as well as for collisional studies, and different attempts to determine or to reduce the ion motion have been made. Effective cooling of the ion motion can be obtained by interaction with a buffer gas (viscous "drag"), as was nicely demonstrated by the crystallization of macroscopic particles moving through air (Wuerker et al., 1959a). A necessary condition is that the neutrals are sufficiently lighter than the ions. Other interesting approaches include adiabatic cooling by slowly lowering the effective potential well (Dehmelt, 1969), or dissipation of the ion energy into an external resonant tank circuit, as has been successfully applied to cool stored protons (Church, 1969). Significant progress in obtaining ultracold ions has been made by laser sideband cooling of trapped ions (Neuhauser et al., 1980), leading to extremely low temperatures, for example, 50 μK (Diedrich et al., 1989), and also to interesting new physical phenomena such as the arrangement of the stored ions into crystals (Blümel et al., 1989; Wineland et al., 1987).

It is impossible in the context of this introduction to refer to the variety of operational modes, applications, and experimental tests that have been performed with the linear quadrupole. Note that in most cases the quadrupole is operated in the narrow band pass mode at q_2 values which are outside the range of validity of the adiabatic approximation, and, therefore, this mode is not considered in this chapter. There are, however, interesting applications and operational modes below $q_2 = 0.4$, which will be mentioned in Section III B.

Besides quadrupole fields, various other types of inhomogeneous rf fields have been used to create effective potential minima. Included are several quadrupole-like electrode structures that use spherical (Owe Berg and Gaukler, 1959) or cylindrically shaped electrodes (Lagadec et al., 1988). By bending the electrodes of a quadrupole into a circular or "racetrack" shape (Church, 1969), a closed structure has been obtained. In such a storage ring, ion-confinement times of several minutes have been achieved. A curiosity is a six-electrode rf trap that was driven with a three-phase ac generator (Wuerker et al., 1959b).

The use of more complex electrode arrangements for ion confinement was discussed in outlines by Gaponov and Miller (1958) and Miller (1958a, 1958b). As one of the simplest realizations, these authors proposed quasielectrostatic higher-order multipole fields. According to our knowledge, however, no such device was constructed and tested experimentally until the end of the 1960s. At this time, it was recognized by E. Teloy at the University of Freiburg that one can utilize multipole and other electrode geometries to store ions in wide field-free regions and to study low-energy ion-collision processes. The first experimental tests were performed with a ring electrode structure (Bahr et al., 1969; Wernes, 1968 see Section III E). In the subsequent year, several modifications of the electrode structure improved the performance of the trap as an ion source, the first octopole was constructed and tested, and these rf elements were assembled together resulting in the first version of the guided-ion-beam apparatus (Gerlich, 1971; Henchman, 1972). Further improvements and developments of the guided-ion-beam apparatus will be discussed in Section IV A.

B. Quadrupole

In Section II we occasionally referred to the quadrupole for illustrative purposes, since it is utilized in so many areas of science and is undoubtedly the best known rf device. Its features as a mass filter or an ion guide are well documented in many papers, review articles, and books (Dawson, 1976; March and Hughes, 1989). Nonetheless, there are several not generally recognized modes of operation and application that make use of dynamic focusing, time and spatial selectivities, long-time trapping and multiple traverses, and rf amplitude modulation or superimposed auxiliary rf fields. Some of these features are discussed in the following.

In general, the functioning principle of the quadrupole mass filter is fully described by the Mathieu equation, the solutions of which can be classified with the help of the (a_2, q_2) stability diagram. Of practical importance is only the lowest zone of stability, $q_2 < 1$. The triangularly shaped border lines, plotted in Fig. 24, enclose the range of (a_2, q_2) parameters where the ion motion is simultaneously stable in the x and y direction. Usually, this range is further restricted, either to region (1), which is defined by the condition $q_2 < 0.3$, or to region (2), which is close to the tip of the triangle, $(a_2, q_2) = (0.237..., 0.706...)$. Most commonly, the quadrupole is operated in this tip where one obtains high mass resolution, but at the expense of perturbation of the kinetic energy of the transmitted ions. This mode is thus well suited for mass analysis, but not for the preparation of a monoenergetic

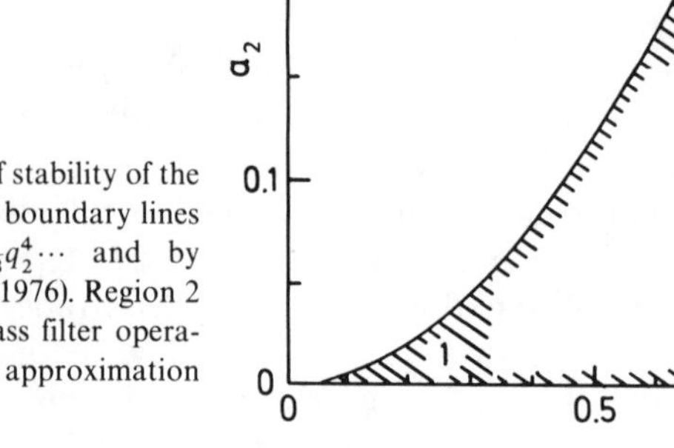

Figure 24. Lowest zone of stability of the Mathieu (a_2, q_2) diagram. The boundary lines are given by $a_2 = \frac{1}{2}q_2^2 - \frac{7}{128}q_2^4 \cdots$ and by $a_2 = 1 - q_2 - \frac{1}{8}q_2^2 \cdots$ (Dawson, 1976). Region 2 is used for high-resolution mass filter operation. In region 1 the adiabatic approximation is valid.

ion beam. Restriction of q_2 to region (1) has the advantage that the device operates within the range of validity of the adiabatic approximation. This ensures conservation of energy, as is discussed in detail in Section II E. In addition, it simplifies the understanding of specific transmission features of quadrupole fields since one can make use of the effective potential approximation.

1. *Low-Mass Band Pass*

One of the obvious applications of operating a quadrupole at low q_2 values is to use it as a low-mass band pass. Since the guiding force derived from the effective potential decreases with increasing mass, the device can be operated such that only ions with masses below a certain limit are transmitted. The condition,

$$m < qV_0^2/(\Omega^2 r_0^2 U_0), \tag{71}$$

can be derived directly from Eq. (59). The cut-off toward higher masses can be very sharp and efficient, as illustrated in the following example in which we have applied this method for the preparation of an extremely pure beam of protons (Scherbarth, 1984). In this case, a 20-cm-long quadrupole with $r_0 = 0.78$ cm was used. At an rf amplitude $V_0 = 100$ V, a dc difference $U_0 = 3$ V, and a frequency $\Omega/2\pi = 10$ MHz, one obtains for $m = 1$ u from Eq. (57), Eq. (60), and Eq. (61), $\varepsilon = 311$ eV, $q_2 = 0.1608$, and $a_2 = 0.0096$, respectively. The (a_2, q_2) parameters fulfill the condition given in Eq. (62); however, for $m = 2$ u, the point (0.0804, 0.0048) lies outside the stability triangle, and the H_2^+ ions are thus accelerated toward the rods and lost, as illustrated by the curved trajectories in the upper panel of Fig. 3. In an actual experiment performed with a pulsed beam (Scherbarth, 1984), a peak count rate of 10^9

H^+/s with an impurity of only 4 H_2^+/s was obtained. Accounting for the H^+/H_2^+ ratio in the storage ion source, this corresponds to a suppression of H_2^+ by more than nine orders of magnitude. It is important to note that for high resolution, a low axial energy of typically 50–100 meV is required.

2. *Focusing Properties*

Another important feature of quadrupole fields is their ability to create a phase-space-conserving image, for example, they are able to focus all trajectories starting from one point onto another point, a direct consequence of the harmonic electric (and also effective) potential. Conditions for obtaining periodic solutions of the Mathieu equation are well known (Meixner and Schäfke, 1954). It was suggested by von Zahn (1963), in connection with the monopole, to utilize focusing conditions to improve the resolution of the mass analyzer. Additional aspects of focusing and results from trajectory calculations were reported by Lever (1966) and reviewed by Dawson (1976), although in none of these references have experimental advantages been presented. A more recent discussion on spatial focusing in an rf-only quadrupole was given by Miller and Denton (1986). In their experiment, pronounced transmission minima and maxima were observed as a function of the rf amplitude, most probably caused by a combination of the imaging features of the ion guide with spatial discrimination of the subsequent detector. These authors provided several suggestions to minimize this unwanted effect. We, on the other hand, make use of these properties to achieve energy and mass filtering for ion creation and beam preparation.

Within the adiabatic approximation, it is straightforward to derive conditions for operating a quadrupole in the focusing mode. It can be seen from Eq. (51) that for $n = 2$, the effective potential V^* is harmonic. Therefore, one can separate Eq. (14) into two harmonic oscillator differential equations, describing independently the x and the y components of the smooth motion, $\mathbf{R}_0(t)$. Both oscillate with different secular frequencies,

$$\omega_{x,y} = \tfrac{1}{2}\beta_{x,y}\Omega, \tag{72}$$

with

$$\beta_{x,y} = (q_2^2/2 \pm a_2)^{1/2}. \tag{73}$$

In the absence of a superimposed dc difference, the two frequencies are identical and the value of β changes as a function of q_2 simply as

$$\beta = 2^{-1/2} q_2. \tag{74}$$

With an added dc potential that is, $a_2 > 0$, the oscillation in the x direction becomes faster than that in the y direction. Note that for $q_2 < 0.3$, Eq. (73) defines in very good approximation the well-known iso-β lines, and that Eq. (74) is in good agreement with the function $\beta(q_2, a_2 = 0)$, as reported by Miller and Denton (1986).

As discussed thus far, all transmission conditions depend only on the operating point in the (a_2, q_2) plane. However, if additional boundary conditions are imposed, for example, by using mechanical apertures, deflection plates, or time-of-flight selection, transmission through the quadrupole requires spatial and/or time focusing. An early version of such a mass and velocity filter was described in Teloy and Gerlich (1974). This filter was operated in the rf-only mode and employed two separate stages and three off-axis apertures. A mechanically simpler version of a focusing quadrupole with much higher transmission was described in Gerlich (1986). This device, assembled in several lengths and with different rod sizes, is used successfully in our apparatuses (see, for example, Fig. 45, Fig. 46, and Fig. 58). Ions are injected through a central entrance hole, the rf field focuses them onto a point at the exit of the quadrupole, and this point is then imaged onto an aperture using an electrostatic lens system, which is also used to pulse and correct the direction of the ion beam. The functioning principle of the focusing quadrupole is illustrated in Fig. 25, which shows the x and y projection of calculated trajectories for an (a_2, q_2) combination in which ω_x is ten times larger than ω_y. To obtain an erect or inverted image of the central entrance aperture onto the exit hole, both numbers of half cycles, N_x and N_y, must be integers. Therefore, the flight time t must obey the two equations,

$$t = N_x \pi / \omega_x$$

and

$$t = N_y \pi / \omega_y. \tag{75}$$

Both conditions are fulfilled in the illustration on the left-hand side of Fig. 25, since both foci coincide with the exit aperture. The figure on the right shows the effect of a slight increase in the kinetic energy. Here, the focus of the trajectories is shifted so that the ions are no longer transmitted. In contrast to the suggestion of this figure, which was calculated assuming ideal conditions, the experimental resolution is primarily determined by the fast x motion, rather than the y motion. This will be discussed below.

An experimentally determined transmission function for C^+ and CH^+ ions is shown in Fig. 26. In this example, the parameters $\Omega/2\pi = 12.25\,\mathrm{MHz}$, $r_0 = 0.42\,\mathrm{cm}$, and $U_0 = 0.247\,\mathrm{V}$ were held fixed and only the amplitude V_0 was varied. Based on these values, and on an axial energy of 0.25 eV and

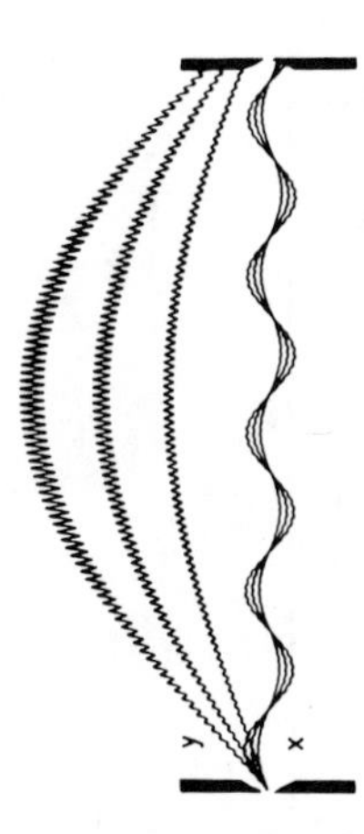

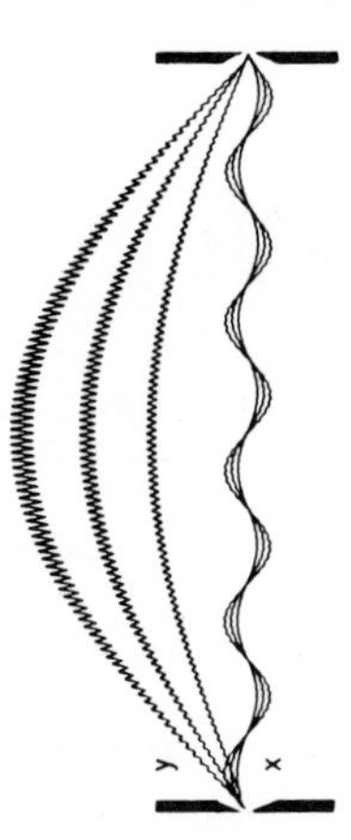

Figure 25. Illustration of the focusing properties of a quadrupole. Both panels depict trajectories, projected onto the x and y plane, respectively, for three different starting angles and for $N_x = 10$ and $N_y = 1$ half cycles. In the left panel, the common starting point is focused exactly into the exit hold. A slight increase of the axial energy leads to a loss of the ions, as illustrated in the right panel.

a quadrupole length of 25 cm, the numbers of half waves of the secular motions, N_x and N_y, have been calculated. The resulting values are shown in the two scales in the upper part of the figure. Although the dc difference is very small ($a_2 = 0.000152$), the two frequencies, ω_x and ω_y, differ significantly since the rf amplitude is varied in the vicinity of the border line, $a_2 = \frac{1}{2}q_2^2$. The observed C^+ intensity maxima can be assigned to $(N_x, N_y) = (27, 4)$ and $(29, 11)$. In the depicted example, a_2 and the injection conditions were optimized for the preparation of a C^+ beam in the $(27, 4)$ mode. The discrimination of the higher-mass CH^+ ion is simply achieved by operating so close to the border of the stability triangle that its y motion is unstable. If, on the other hand, one wishes to favor CH^+ ions and to suppress C^+, one could make use, for example, of the transmission gap between $N_x = 28$ and 29. High mass resolution requires additional energy

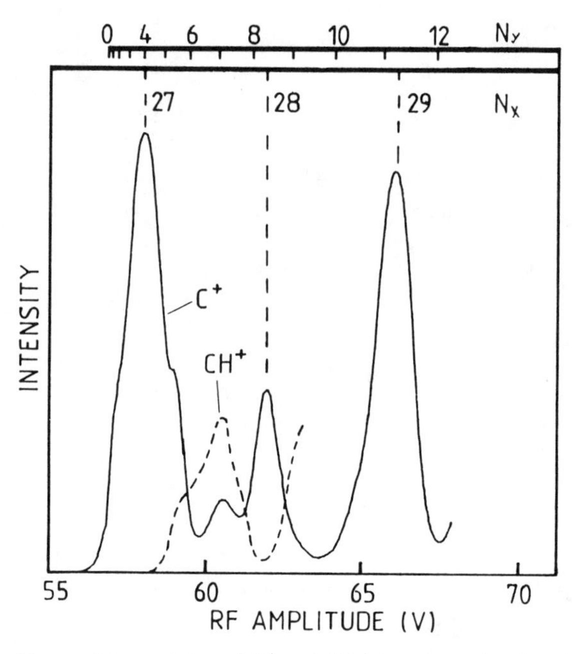

Figure 26. Measured transmission of C^+ and CH^+ ions through a focusing quadrupole as a function of the rf amplitude. The numbers on top refer to the number of half cycles of C^+ trajectories (compare Fig. 25). The C^+ intensity maxima are obtained for $(N_x, N_y) = (27, 4)$ and $(29, 11)$. Owing to its larger mass, the onset and the structure of the CH^+ intensity is shifted to higher amplitudes.

selection, and, therefore, we usually operate the device in a pulsed mode making use of time-of-flight selection. For preparing an ion beam with a very narrow energy spread, the axial energy usually must be below 100 meV.

It is important to note that the resolution is predominantly determined by the high-frequency component ω_x, while, as can be seen, for example, from the $N_x = 27$ peak in Fig. 26, contributions from $N_y = 2$–5 are only indicated as fine structure. This is primarily due to the much larger acceptance angle in the x plane, since, here, both the dc and the rf fields confine the ions close to the axis. In contrast, the guiding field in the y direction is very weak, and, as such, even small imperfections in the work function of the metal surfaces can cause significant perturbations. These are especially ciritical in regions where the ions come very close to the rods. This always leads to some asymmetries, and, therefore, operation of a focusing quadrupole at low energies and low guiding fields requires careful tuning of the injection and extraction conditions, as well as slow adjustment of the dc and rf fields to obtain stable surface conditions.

3. *Photoionization Source*

Single- and multiphoton ionization is gaining rapid acceptance as a technique for preparing state-selected ions. Combination of this method with the focusing properties of a quadrupole has lead to the development of an efficient ion source, shown in Fig. 27. Here, the focus of a pulsed laser is located in the interior of a quadrupole. Two cylindrical electrodes surrounding the quadrupole in the ionization region and one repeller allow one to shift the dc potential in the vicinity of the laser focus relative to the dc potential of the rest of the quadrupole and to adjust the kinetic energy in this region to typically 100 meV. Making use of the focusing properties of the guiding field,

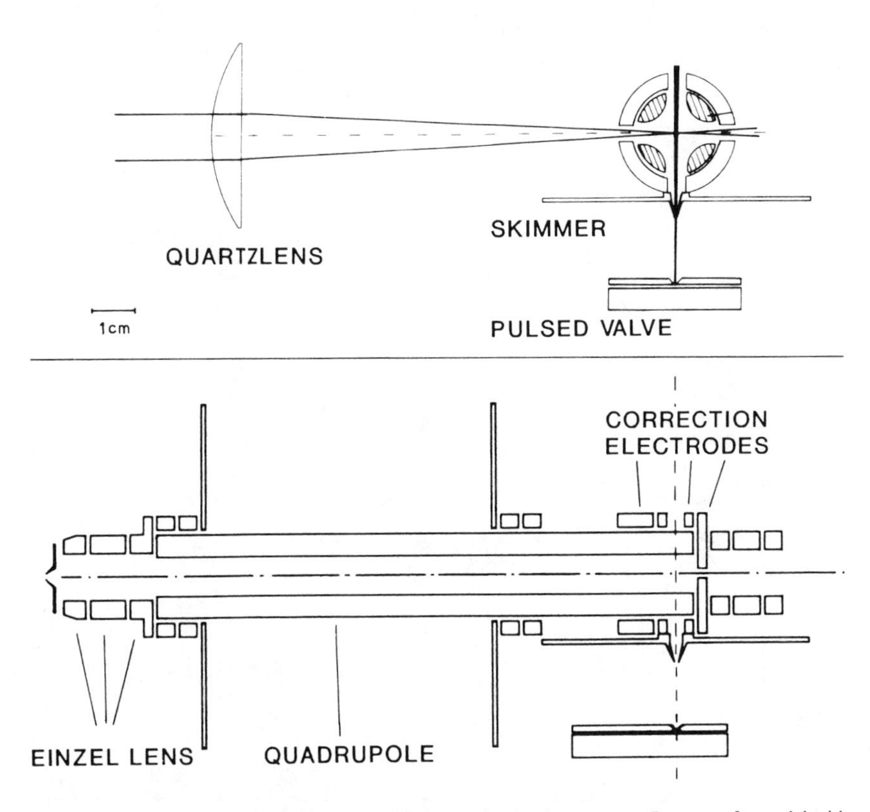

Figure 27. Quadrupole used as a multiphoton ionization source. Ions are formed inside the quadrupole in the intersection of the focused laser beam and the skimmed pulsed molecular beam. The correction electrodes allow one to create an axial dc potential gradient in the focal region, leading to a weak acceleration of the ions in the axial direction. Using the focusing properties of the quadrupole, the well-defined ionization volume can be mass selectively imaged to the exit hole.

the well-defined ionization volume can be imaged onto the exit hole at the other end of the quadrupole. This leads to a high collection efficiency without the need for strong extraction fields. In addition, the mass selectivity of the device can help suppress fragment ions. This is especially important for the ionization of molecules, where dissociative ionization or fragmentation often compete with the state-selective preparation of the molecular ion. The high time resolution of a laser (or of the light pulses from a synchrotron) also allows for time-of-flight selection. Further improvements can be imagined by synchronizing the ionization event with the rf field, leading to a well-defined starting phase.

In the ion source depicted in Fig. 27, a pulsed supersonic nozzle beam is used to create cold neutral precursors. The neutral beam, the quadrupole axis, and the laser are all orthogonal to each other, reducing the risk of collision-induced relaxation of the created ions. A skimmer plate separates

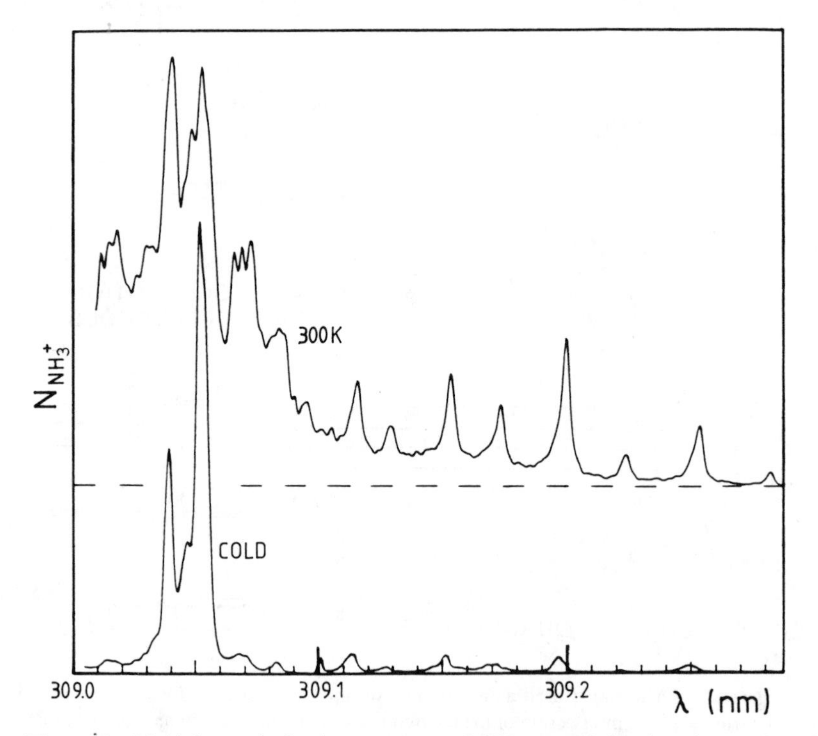

Figure 28. Multiphoton ionization spectrum of NH_3 recorded using the quadrupole photoionization source (Fig. 27). The lower curve shows NH_3^+ formed in the pulsed supersonic beam. The upper curve was recorded with such a long delay between the gas pulse and the laser firing that only the remaining 300 K effusive background gas was ionized.

the supersonic beam from the effusive background gas and eliminates the need for an additional differential pumping stage. Figure 28 shows a multiphoton ionization spectrum of NH_3 obtained with this source. The bottom spectrum was obtained by optimum synchronization of the laser and pulsed beam, while in the upper curve, the laser was delayed sufficiently long so that only the remaining NH_3 background gas was ionized.

4. *Resonant Excitation by Auxiliary Fields*

Another important modification of operating quadrupoles is to superimpose an auxiliary dipole or quadrupole field alternating at a lower frequency $\Omega' < \Omega$ with an amplitude V'_0. The mathematical problem of solving the resulting inhomogeneous Mathieu differential equation under the inclusion of such an additional term has been treated in Kotowski (1943). The principle has also been described in one of the first publications on quadrupoles (Paul et al., 1958). One early application of this technique was to separate different isotopes by resonant ejection of specific masses (Busch and Paul, 1961a). Recently, this method has found renewed interest (Watson et al., 1989) and has been coined as a notch rejection filter (Miller and Denton, 1990).

The procedure of resonant excitation is easy to understand if the quadrupole is operated within the limit of the adiabatic approximation. As described previously, the secular motion corresponds to a two-dimensional harmonic oscillator with frequencies ω_x and ω_y [see Eq. (72)]. If one superimposes on the quadrupole field a weak homogeneous field in the x direction oscillating at the frequency $\Omega' = \omega_x$, the corresponding motion becomes resonantly excited. This leads to a *linear increase* of the amplitude of the x component of the secular motion, and, finally, to the loss of all ions with the critical mass $m(\omega_x)$ determined by Eq. (72) and Eq. (73). This method of mass-selective ejection is a standard technique used in ICR cells, Penning traps, and also Paul traps (March and Hughes, 1989). On the other hand, superimposing a quadrupole distortion field leads to parametric excitation when $\Omega' = 2\omega_x$ and results in an *exponential increase* of the amplitude of the secular motion (Busch and Paul, 1961a). Parametric heating can also be achieved by modulating the rf amplitude with a frequency Ω' such that one of the two secular frequencies, $2\omega_x$ or $2\omega_y$, coincides with $\Omega - \Omega'$. Note that resonant excitation is restricted to harmonic potentials, that is, to quadrupoles. In all other potentials, the period of the ion secular oscillation depends on the amplitude, as has been derived for multipole fields in Eq. (67). As a result, accumulation of energy from the resonating field causes the ion motion to drift out of phase, and, for ejection, a continuous adjustment of the excitation frequency would be required.

Combining one of the previously described methods, for example, high mass cut-off and/or selective focusing, with resonant ejection usually leads to a very pure preparation of an ion beam. It is normally advisable to operate under conditions such that the desired mass m is just at the stability limit given by Eq. (71). This suppresses higher mass ions $m' > m$ very efficiently. In addition, the weak guiding field has the advantage that resonant excitation of an unwanted lower mass, for example, $m - 1$, perturbs only slightly the mass m since the two frequencies, $\omega_y(m)$ and $\omega_y(m - 1)$, are well separated, as can be derived from Eq. (73). Recall that it is important to operate at sufficiently low q_2 values to guarantee adiabatic conservation of energy.

C. Octopole Beam Guide

After quadrupoles, the octopole beam guide is the second most used rf device, and, although its use is not nearly as widespread as that of the quadrupole, it has seen expanded application in a number of laboratories. It is now applied routinely in many experiments where one wishes to guide very slow ions and where 4π collection efficiency of product ions is required. An overview of the different appratuses that employ octopoles will be given in Section IV A. In this section we begin by making a few general remarks concerning the characterization of transmission features of an octopole. We then discuss applications of ring electrodes to localize potential distortions and describe a procedure to calibrate the axial energy of a slow guided ion beam. The last part of this section describes an experimental test for obtaining a correlation between the transverse energy of ions and the minimum rf amplitude needed to confine them.

1. *Transmission Properties*

Recently, an experimental (Tosi et al. (1989b) and a theoretical (Hägg and Szabo, 1986c) attempt to characterize the transmission properties of an octopole ion guide have been reported. The theoretical approach has already been discussed in Section II D 3. Unfortunately it was based on trajectory calculations with experimentally unrepresentative initial conditions. Tosi et al. (1989b) have experimentally investigated the transport of different ions through an octopole as a function of the rf amplitude. They gave several interpretations of their measured transmission functions, and, in addition, they attempted to compare their findings to those limits we have recommended in Gerlich (1986) and in Section II D 2 for safe operation of an octopole as an ion beam guide. Unfortunately, these authors overlooked the fact that these proposed operating conditions are *sufficient* conditions, and their observation that transmission also occurs outside of these limits is therefore not surprising.

The principal difficulty in determining conditions that are simultaneously *sufficient* and *necessary* for ion transmission have already been discussed in Section II D 3 in context with adiabaticity and stability. Besides the theoretical uncertainties in defining border lines between stability and instability, there is the added experimental difficulty that the observed transmission is not only a feature of the octopole, but also a complicated function of the initial conditions and the acceptance of the detection system (see Section IV B). For a fundamental study of the transmission function of a multipole, it is mandatory to account for such perturbations.

It seems to be necessary to repeat here in this context that we do not yet know of any practical reason to use an octopole outside the range of validity of the adiabatic approximation, and therefore, we do not present any transmission functions extending outside this range. Operating below $\eta_m = 0.3$ ensures that the transmission depends only on the transverse energy, as defined in Eq. (54). From this equation we have derived the conditions for safe operation of the ion guide, that is, for 100% transmission. Equation (54) also allows one to derive critical operating conditions for the maximum confined transverse energy. This and a corresponding experimental test will be discussed in Section III C 4.

The strength of the effective potential, which determines the transmission features, depends on the shape and the position of the rf electrodes, and, in addition, on the potential of the metal surfaces that surround the octopole. In order to use directly the formula for V^*, as given by Eq. (51), it is advisable that the ratio of the rod diameter d and the radius r_0 be close to that given in Eq. (68). In some experiments (Ervin and Armentrout, 1985), smaller r_0/d ratios have been used. This weakens the effective potential between the rods (see Fig. 14) and, for a quantitative estimate of ion transmission, V^* must be calculated from the correct boundary conditions using, for example, the method described in Section II C 2. As an extreme case, we have performed an experimental test with an octopole constructed from eight thin wires with $d = 0.03$ cm arranged on a circle with $r_0 = 0.4$ cm (Müller, 1983). The effective potential was calculated as described in Section II C using the boundary conditions shown in Fig. 6. It is evident that with such an open structure the field penetration from surrounding electrodes becomes significantly larger than that depicted in Fig. 17. The transmitted ion intensity, measured as a function of amplitude and external field, was in good agreement with the calculated predictions (Müller, 1983).

2. *Potential Distortions, Ring Electrodes*

One of the most poorly characterized problems affecting the transmission properties of octopoles and other rf devices is that arising from nonuniformities of the surface potentials of the electrodes. It is well known that patchiness of the work function of metal surfaces can be caused by dust particles, insulating layers, chemisorption, or other sources. Experimentalists working with slow charged particles in electron-energy analyzers, electrostatic lens systems, ion traps, or other electrode arrangements have developed a variety of methods for treating the metal surfaces to reduce such impairments. In the following, we describe a procedure (Gerlich, 1984, 1986) that allows one to obtain detailed information about potential distortions along the axis of an octopole, and then make a few remarks on the treatment of the surfaces of the octopole rods.

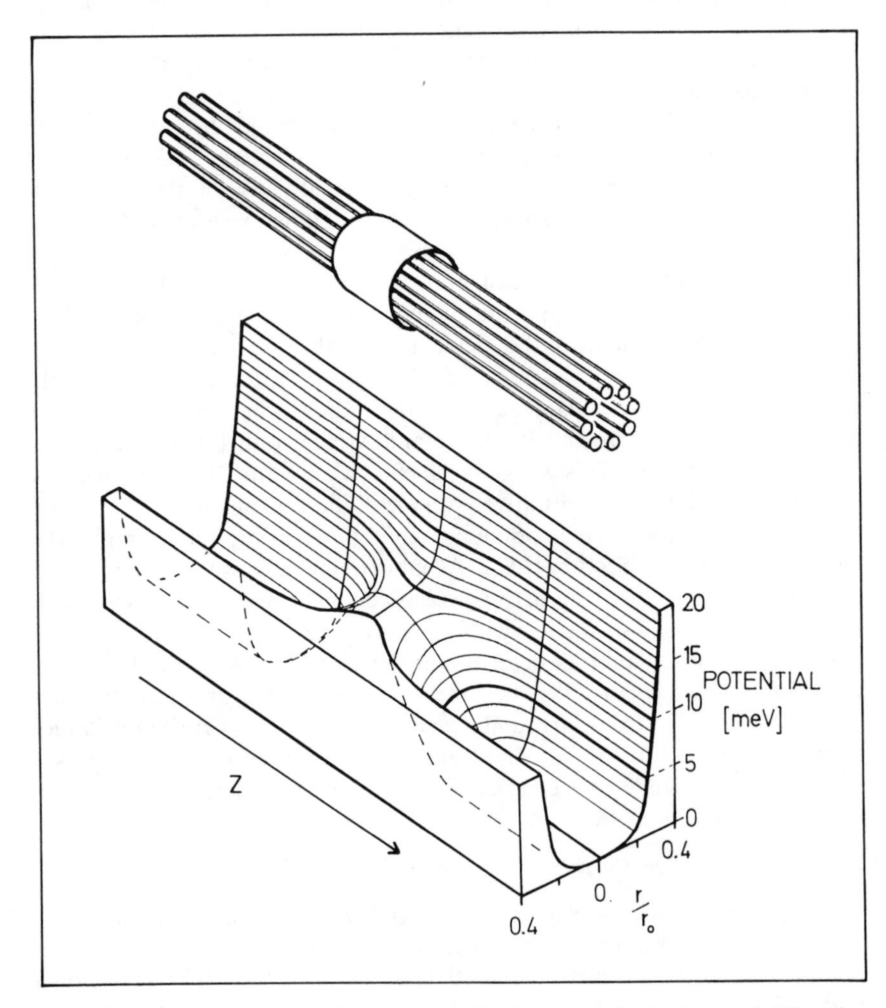

Figure 29. Perspective view of the sum of the effective potential and a dc potential distortion caused by a cylindrical ring electrode in an octopole. The penetrating field creates a local barrier of a few millivolts per volt applied to the ring.

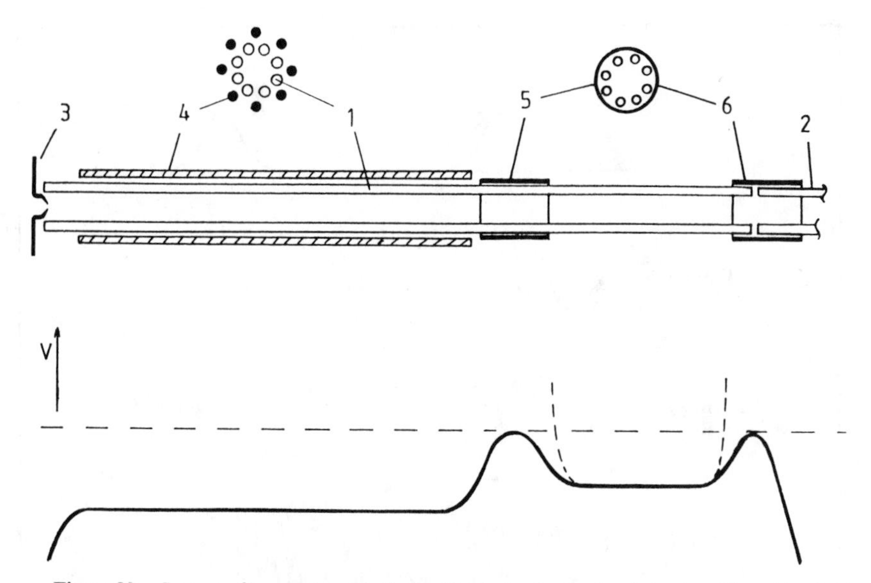

Figure 30. System of two octopoles (1, 2) with electrodes for ion injection (3) and for field correction (4–6). Correction electrode (4) consists of eight rods, staggered as indicated, (5) and (6) are cylinders. The lower panel shows a schematic representation of the dc potential along the axis of the system. In this example, the electrode (4) was used to lower the potential slightly, while (5) and (6) were used to create barriers.

The test procedure makes use of two or more cylindrical ring electrodes that surround the octopole; see Fig. 29 and Fig. 30. Choosing the dimensions as depicted in Fig. 17, a ring voltage of 1 V shifts the octopole potential in the vicinity of the axis typically by 2–3 mV. Superimposing this weak dc field with a rather strong rf guiding field leads to an effective potential with a variable local barrier, as plotted schematically in Fig. 29.

Figure 30 shows a part of the octopole system (1, 2) from our universal guided-ion-beam apparatus (Gerlich, 1986; Gerlich et al., 1987; Scherbarth and Gerlich, 1989), which is described further in Section IV B. Important for the present discussion is the region between the ring electrodes (5) and (6). Here, the electrodes determine the entrance and exit of the scattering cell. The lower part of this figure is a schematic representation of the dc potential along the axis of the ion guide. The electrode (4) is used to slightly lower the potential in the first part of the octopole, while the ring electrodes (5, 6) create two potential barriers. If an ion is injected into this octopole with a kinetic energy below the energy marked by the horizontal dashed line in Fig. 30, it will be reflected at (5) and lost. If the ion has a higher energy, it is transmitted

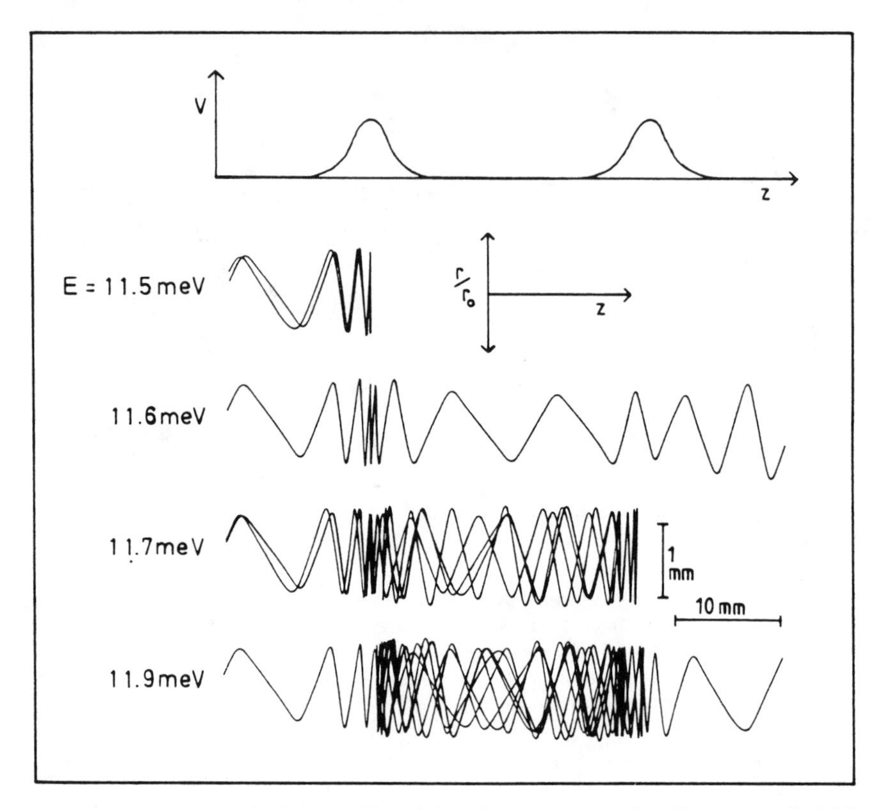

Figure 31. Model calculations of ion trajectories in an octopole with two equally high potential barriers. Several sequences of reflections at the barriers are obtained depending on the kinetic energy and the angle of the incoming ion.

directly. However, under certain conditions, there is also a small energy window where the ion can be trapped for a few reflections between two equally high potential barriers. This is illustrated in Fig. 31 based on a simple model calculation (Schweizer, 1988). In this numerical example, ions with an energy between 11.5 and 12 meV can be trapped, depending on the initial conditions. Necessary for this behavior is a weak coupling between the axial and the transverse motion.

Experimentally observed reflection signals of He^+ ions are plotted in Fig. 32. In this case, an intense, few microsecond short pulse of slow ions was injected into the octopole. From the approximately 50-meV-wide energy distribution, the slow part is reflected at electrode (5) and lost, while the fast part is transmitted directly (not shown in figure). The well-resolved individual peaks originate from those ions whose energies are close to the barrier height

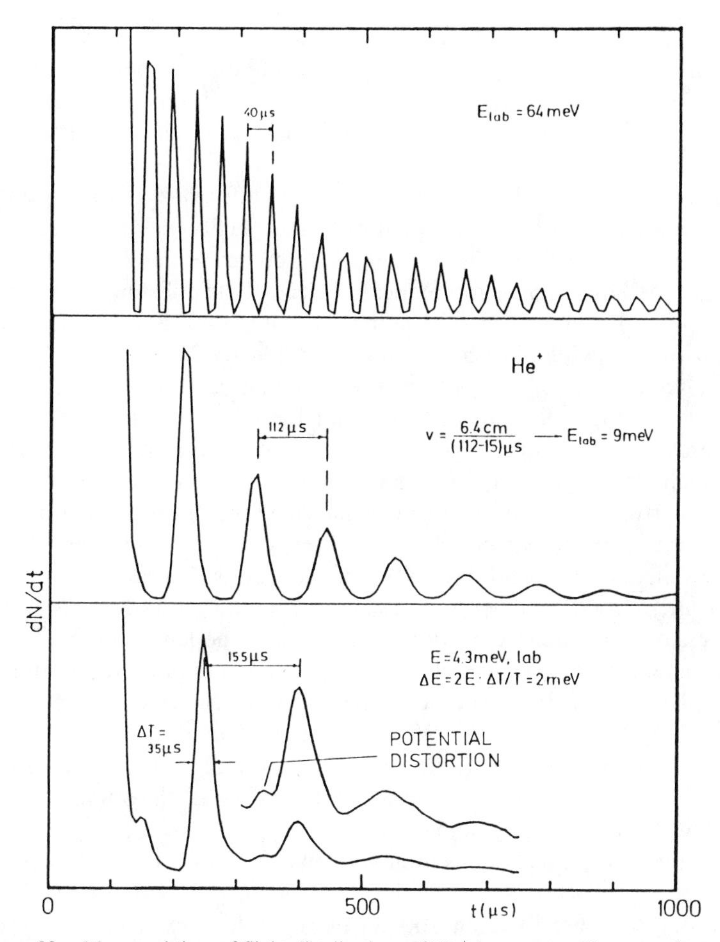

Figure 32. Measured time-of-flight distributions of He^+ ions trapped between the two ring electrodes (5) and (6) shown in Fig. 30. The external voltages correspond to barriers of 64, 9, and 4.3 meV. For a precise calculation of the energy, it is important to know that the potential at ring (5) is raised after passage of the ions, while ring (6) remains semitransparent. The effective distance between the rings is 3.2 cm and the turn around time in ring (6) is 15 μs. In the lowest panel, the additional structure is indicative of other potential distortions.

imposed by ring electrodes (5, 6) and are therefore trapped for a certain number of oscillations. The upper panel depicts a time-of-flight distribution recorded with a barrier corresponding to 64 meV. Here one can see contributions from ions that have been reflected back and forth more than 20 times. Most probably they escape because of a coupling of the axial energy to other sources of energy. Lowering both ring electrode potentials reduces the barriers and leads to the capture of slower ions, as shown in the lower two panels. The lowest panel shows an oscillating ion beam with a mean energy of only 4.3 meV. Here, the marked additional structure in the time-of-flight spectrum is due to an additional potential barrier caused by surface distortions.

Such a low-potential distortion is not straightforward to obtain owing to the previously mentioned surface problems. Usually, before an octopole is assembled, the surfaces (here stainless steel) are carefully treated using standard methods such as mechanical or electrolytic polishing and ultrasonic cleaning in different solvents. Nonetheless, distortions between the two rings (5) and (6) following such a cleaning are typically 30–50 meV. In order to further reduce these barriers, we have developed a method of cleaning the octopole—using a fine metal brush and blowing the region continuously with pressurized nitrogen—without removing it from the apparatus. After each such cleaning, the apparatus is pumped down and the positon of the highest perturbation is determined as described previously. According to our interpretation (Scherbarth, 1984; Schweizer, 1988), our treatment probably creates different random distributions of distortions that cancel with a certain probability close to the octopole axis. For obtaining very low distortions, it is necessary that the critical range, in our case the length of the scattering cell, be as short as possible and that the cleaning, pumping, and testing cycles be very fast in order to repeat them often. With the described octopole system, cycles of 1–2 h are possible. We have found that it is always possible to reduce the potential distortions below 10 meV after many iterations (24 in the depicted example). It is very important to note that the obtained potential distribution remains very stable, even over a period of months.

The described procedure employing two potential barriers is of course more than just a test for potential distortions. It is a unique method for selecting ions (typically a few 100 per pulse) with an extremely small energy spread of 1–2 meV, and for forming a pulsed, trapped ion beam with a well-defined kinetic energy that can be easily varied between 5 and 100 meV. For a precise determination of the kinetic energy, the semitransparent barriers (heavy line in the lower part of Fig. 30) have the disadvantage that the turning points are somewhat undefined within the region of the ring electrodes and that here the ions move very slowly. Therefore, the potential barrier (5) is usually raised as the desired ions are reflected from (6), while (6) is raised when the ion bunch is reflected from (5). This results in very steep and well-defined potential walls, as indicated by the dashed lines in Fig. 30. Also, under these conditions, the time spread of the enclosed ion bunch is very small. This has been tested by opening the trap after several round trips. Finally, it should be noted that the conservation of the time structure over a period of milliseconds, as can be seen for example from the upper panel in Fig. 32, indicates either that the coupling between the axial and the transverse motion is very weak or that the transverse energy is extremely small.

3. *Calibration of the Axial Energy*

In studying collision processes, the kinetic energy distribution of the ions in the interaction region must be known. The previously described method has the advantage that both energy selection and calibration are performed in the octopole itself, resulting in an overall accuracy of a few meV. To operate at high ion intensities, it is necessary to inject an externally prepared ion beam. In general, the energy distribution of the ions in the octopole depends on several factors such as ion source conditions, resolution of energy selectors, impairment by varying fields (pulsed electrodes, rf fields, low-frequency noise, etc.), and by potential inhomogeneities in the scattering cell. In some applications of guided ion beams, it may be sufficient to determine the beam energy by using the octopole as a retarding field analyzer; however, for studies at laboratory energies below 0.5 eV, we always employ a time-of-flight analysis.

A very precise calibration procedure consists of several steps. Intially, following the method described in Section III C 2, the potential inhomogeneities in the scattering cell are tested and reduced to the desired accuracy. Then, the potential of the external electrode (4) is adjusted such that the mean velocity is the same in both parts of the octopole (1), as can be tested by reflecting ions alternatively between (6) and (5), or (6) and (3). Next, one has to prepare externally a pulsed ion beam with a narrow energy distribution and a sufficiently small time width. Finally, one has to perform the following calibration routine, the attainable accuracy of which depends significantly on the quality of this beam.

At energies above a few tenths of an eV, it is usually sufficient to assume a linear dependence of the nominal laboratory energy E_1 on the voltage U applied to the field axis of the octopole

$$E_1 = -q(U + \Delta U). \tag{76}$$

It is advisable to measure U relative to the actual ion source potential, since then the correction ΔU only has to account for possible shifts due to space charge or differences in the work function. If only the voltage U of the octopole (1) is changed, while all other potentials are kept constant, the total flight time t of the ions through the apparatus can be separated into a voltage-dependent part t_u and constant part t_c

$$t = t_u + t_c. \tag{77}$$

The two constant parameters, ΔU and t_c, are determined experimentally by recording ion time-of-flight distributions, dN/dt, at different octopole voltages and by characterizing them with the mean flight time

$$\langle t \rangle = \int t(dN/dt)dt \bigg/ \int (dN/dt)dt. \tag{78}$$

Several potentials between 2 and 10 V are used to calculate an average value of t_c. A second set of measurements is performed between 0.2 and 2 V for obtaining ΔU. The attainable accuracy of ΔU is typically better than 10 mV.

At laboratory energies below a few tenths of an eV, where the mean energy of the ion cloud and its half width become comparable, it is no longer sufficient to characterize the beam simply by its mean flight time. In this case, each measured time-of-flight distribution dN/dt is transformed numerically into an energy distribution dN/dE_1 and the mean energy is determined from

$$\langle E_1 \rangle = \int E_1(dN/dE_1)dE_1 \bigg/ \int (dN/dE_1)dE_1. \tag{79}$$

Such obtained energy distributions are plotted in Fig. 33. In this example, the nominal energy E_1, as determined from Eq. (76), was in good agreement with $\langle E_1 \rangle$, since the energy half width of the ion beam was rather small (25 meV). Comparison of the distributions, however, reveals some changes in their shape with different energies.

To characterize the overall accuracy of the calibration routine, we usually record the mean energy $\langle E_1 \rangle$ and the total transmitted intensity as a function of the nominal energy E_1 between 10 mV and 1 V, as depicted in Fig. 34. The upper panel shows that under favorable conditions E_1 and $\langle E_1 \rangle$ are identical down to 50 meV. Further lowering of the octopole voltage leads to deviations due to cutting off of slow ions. In this example, the slowest ion beam with $\langle E_1 \rangle = 25$ meV was obtained at $E_1 = 10$ meV with a remaining intensity of 30%.

The lower panel of Fig. 34 illustrates that potential barriers in the octopole can lead to a significant shift of the retarding field curve. Here, the ring electrode (5) is used to create a 100 mV distortion. This leads to a corresponding shift of the onset of the intensity, although the energy of the ions is unchanged in the octopole with exception of the region at the localized distortion. The mean energy $\langle E_1 \rangle$ determined by time of flight is also slightly perturbed, since the total flight time in the octopole (1) is slightly increased by the barrier.

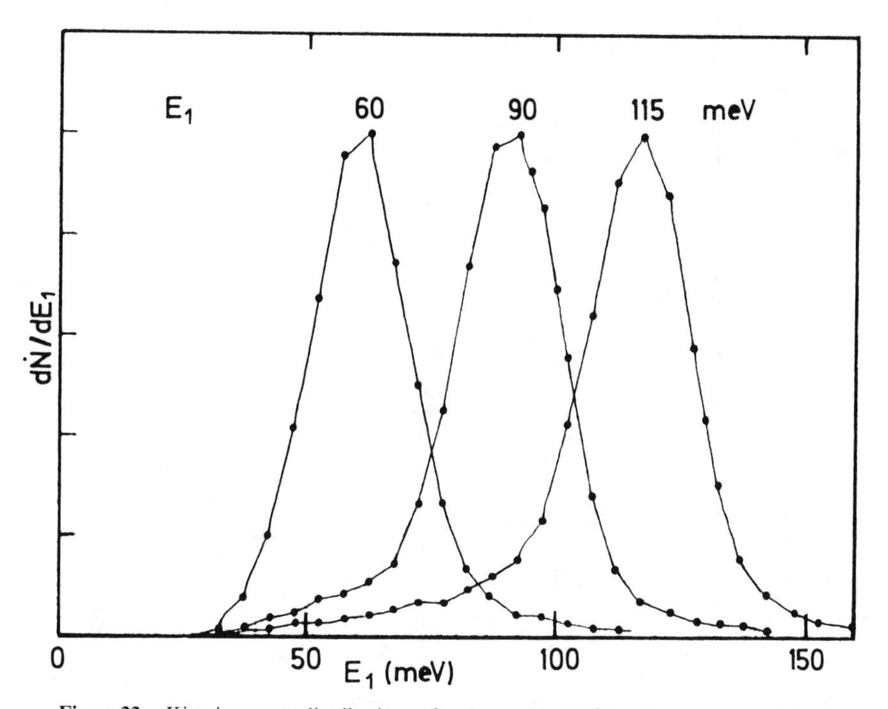

Figure 33. Kinetic energy distributions of a slow guided D_2^+ ion beam measured by time of flight at three octopole voltages U. The nominal energies $E_1 = -q(U + \Delta U)$ are in excellent agreement with the derived mean energies $\langle E_1 \rangle$. The energy half width is 24 meV.

This example shows that for a precise time-of-flight calibration of the ion energy the overall potential homogeneity is important. There are additional sources of error that must be considered. Since the flight time corresponds only to the axial component of the ion velocity, it is important to avoid injection conditions that lead to a wide angular spread, that is, it is not recommended to use a strong deceleration field between the injection electrode (3) and the octopole (1). Other problems can arise from time focusing effects caused by ion acceleration in the pulsed electrodes, which usually results in a nonlinear $\langle E_1 \rangle$-E_1 dependence.

4. *Maximum Confined Transverse Energy*

Since in most applications one is interested in operating at 100% collection and transmission efficiency, we have postulated in Section II D 2 safe conditions which ensure that the ions will not collide with the rods ($\hat{r}_m < 0.8$). However, under ideal conditions, the ions can be allowed to come much closer to the electrodes, and the maximum tolerable kinetic energy can be

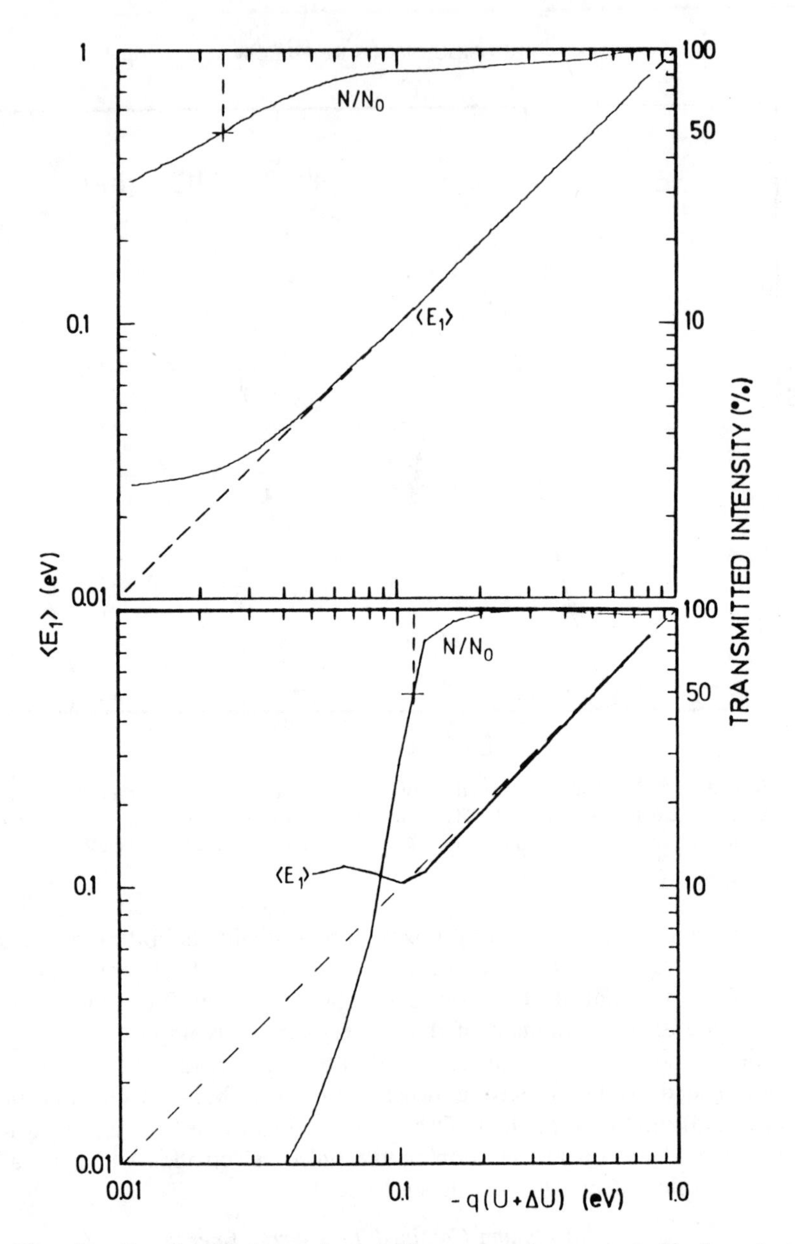

Figure 34. Experimental characterization of ion transport in an octopole. The figure depicts the dependence of the relative transmitted ion intensity N/N_0 (scale on right) and the mean kinetic energy $\langle E_1 \rangle$ on the nominal energy $E_1 = -q(U + \Delta U)$. The upper panel shows good agreement between $\langle E_1 \rangle$ and E_1 over a wide range, and that laboratory energies below 30 meV are accessible. The curves in the lower panel were recorded with a local potential distortion of 100 meV. Comparison with the upper curve reveals a corresponding shift of the retarding potential curve. The marked 50% value (+) is displaced from 24 meV to 115 meV.

significantly larger than given by Eq. (54) for $\hat{r}_m = 0.8$, since the effective potential increases steeply toward the rods. In determining critical operating conditions of an ion trap or an ion guide, it is necessary to find a relation between the initial conditions of ion trajectories and their closest approach to the electrodes. Within the adiabatic approximation, this correlation can be derived in general from Eq. (16). For multipoles the situation is even simpler, as discussed in detail in Section II D.

In first-order adiabatic approximation, the ion trajectory is given by $\mathbf{r} = \mathbf{R}_0 + \mathbf{R}_1$; see Eq. (6). The region where the risk of hitting an electrode is largest is that along the lines given by $\cos n\varphi = 1$. Here, the gap between a given equipotential line and the rod surface has its smallest value. In addition, the oscillatory motion, $\mathbf{R}_1$, in this region is directed toward the surface (see Fig. 9). In order to avoid colliding with the rod, the closest approach of the ion must fulfill the condition $R_0 + R_1 \leqslant R_0 + a < r_0$. Therefore, the secular motion R_0 must be restricted by $R_0/r_0 < \hat{r}_c$, where the critical radius is defined in reduced units by

$$\hat{r}_c = 1 - \hat{a}_c. \tag{80}$$

The critical amplitude $\hat{a}_c$ depends itself on $\hat{r}_c$, as given by Eq. (48):

$$\hat{a}_c = \frac{1}{2n} \frac{qV_0}{\epsilon} \hat{r}_c^{n-1}. \tag{81}$$

From these two equations, $\hat{r}_c$ can be calculated; for example, for an octopole, one obtains a third-order equation. Using the thus obtained critical radius, we define the critical effective potential

$$V_c^*(\hat{r}_c) = \frac{1}{8} \frac{(qV_0)^2}{\epsilon} \hat{r}_c^{2n-2}. \tag{82}$$

An exact, numerically determined amplitude dependence of this potential is shown in Fig. 35. For most practical applications it is easier to calculate the critical radius from the amplitude at $\hat{r} = 1$ and to use the simple approximation

$$\hat{r}_c \simeq 1 - c \frac{1}{2n} \frac{qV_0}{\epsilon}, \tag{83}$$

where c is a constant. Choosing $c = 0.81$ leads to excellent agreement with the exact result, as can be seen in Fig. 35. This figure also shows for

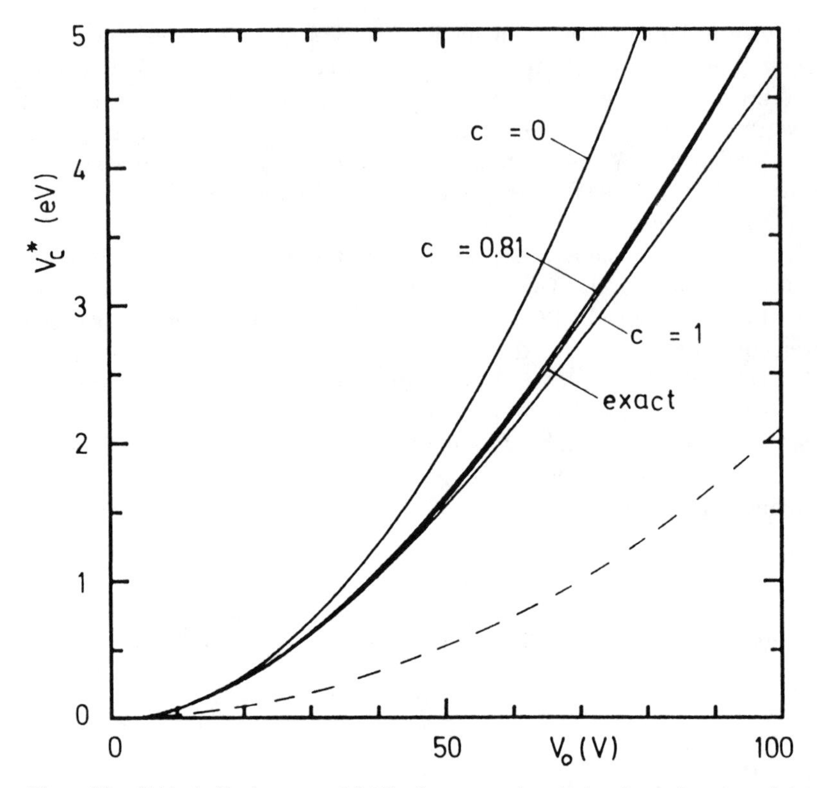

Figure 35. Critical effective potential V_c^* of an octopole calculated as a function of the rf amplitude V_0. The parameters used are $m = 20\text{u}$, $\Omega/2\pi = 8.3\,\text{MHz}$, and $r_0 = 0.3\,\text{cm}$. The curve $c = 0$ corresponds to the effective potential at $\hat{r} = 1$. For deriving safe guiding conditions, one has to account for the oscillatory motion. The different approximations are explained in the text. The dashed line corresponds to the safe operating condition, where $\hat{r}_m = 0.8$.

comparison $V_c^*(\hat{r}_c)$ for $c = 0$ and 1. The first case neglects the oscillatory motion, while $c = 1$ leads to a slight overestimation of the amplitude. At very low values of the adiabaticity parameter η (low amplitudes or high frequencies), the difference between the various approximations becomes very small. Recall that for high values of V_c^*, the operating conditions must always be maintained such that $\eta(\hat{r}_c) < 0.3$.

With the critical radius and the critical effective potential, specific limits for guiding ions can be derived from Eq. (45) and Eq. (46). For $U_0 = 0$, one obtains

$$E_m = \frac{1}{8}\frac{(qV_0)^2}{\epsilon}\hat{r}_c^{2n-2} + \frac{L^2}{2mr_c^2}. \tag{84}$$

In this case, the transmission depends only on the two adiabatic constants of motion, the total energy E_m and the orbital angular momentum L, that is, one obtains an (E_m, L) "stability diagram." As was discussed in Section II D 2, the centrifugal energy can be neglected in those cases where the ions start close to the centerline.

Setting $U_0 > 0$, and assuming for simplicity that the ion kinetic energy is zero at $\hat{r} = 0$, that is, $E_m = 0$, the ions will be accelerated in the direction of the negatively biased rods (see Fig. 12). Ion confinement requires at $\hat{r}_c$ the energy balance

$$qU_0\hat{r}_c^n = \frac{1}{8}\frac{(qV_0)^2}{\epsilon}\hat{r}_c^{2n-2}. \tag{85}$$

Comparing Eq. (84) for $L = 0$ with Eq. (85) reveals that the initial kinetic energy E_m and the kinetic energy gained from the electrostatic potential difference, $qU_0\hat{r}_c^n$, are equivalent. In both cases, the ion is lost as soon as this energy surmounts the critical effective potential $V_c^*(\hat{r}_c)$.

An experimental test has been performed to verify the validity of the derived relations. In this test ions were injected into an octopole close to the centerline. The axial energy was about 100 meV, corresponding to a negligible transverse energy with our injection conditions. The transmitted ion intensity was measured as a function of the rf amplitude V_0 at several different dc voltages U_0. Some typical results are shown in Fig. 36. As expected, the onset is shifted to higher amplitudes V_0 as U_0 is increased. A quantitative comparison is made in Fig. 37 where one can see directly that the experimentally determined correlation between V_c^* and U_0 is in good overall agreement with results (dashed line) from Eq. (85). The deviation at low guiding fields is most probably caused by potential distortions, which have been discussed in Section III C 2. Under favorable conditions they may indeed average out along the centerline of the octopole; however, it is reasonable to assume that on the surfaces there are still local distortions as high as 100 mV.

These uncertainties may also be responsible for the transverse energy resolution, which is on the order of 100 meV, as seen from the differentiated curve in the inset of Fig. 36. Because of this large uncertainty, we use the experimentally determined relation between U_0 and the rf amplitude for calibration purpose, rather than the theoretical dependence. For special ion trajectories, for example, those that are reflected from the rf wall between the rods, the transverse energy limit can be somewhat larger than given by Eq. (84). For transverse energy analysis it is therefore advisable to use low axial velocities, low orbital angular momenta L, and a long octopole in order to obtain many reflections from the guiding field.

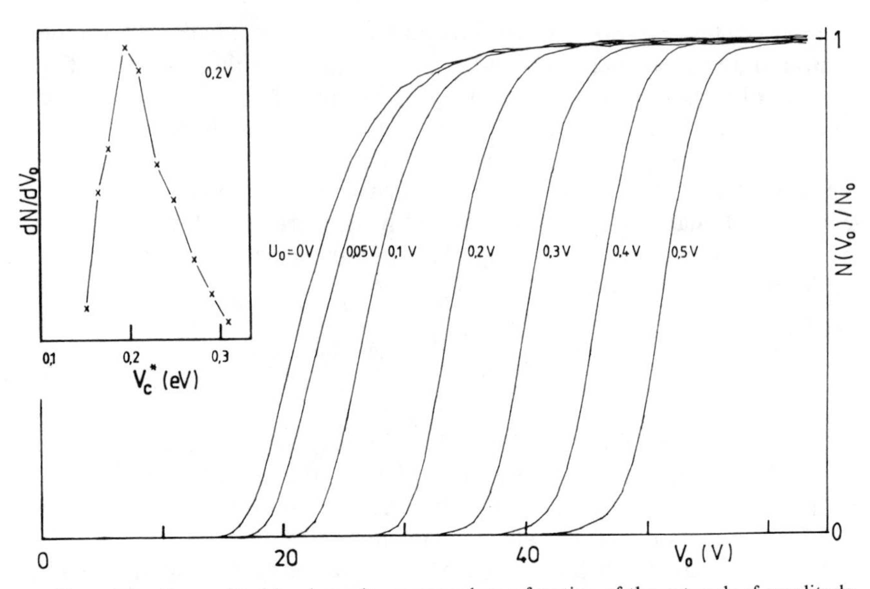

Figure 36. Transmitted ion intensity measured as a function of the octopole rf amplitude V_0 for several fixed dc voltages U_0. (The dc difference between neighboring rods is $2U_0$.) Parameters are $m = 40\,\text{u}$, $\Omega/2\pi = 12\,\text{MHz}$, and $r_0 = 0.3\,\text{cm}$. The inset shows the differentiated 0.2 V curve as a function of V_c^*, indicating a transverse energy resolution of about 100 meV.

D. Traps as Ion Sources

The combination of rf ion trapping with ion formation has lead to the development of different storage ion sources. The advantages of such sources include high collection efficiency without the use of energy-perturbing extraction fields, accumulation of ions for obtaining intense short pulses, thermalization by inelastic collisions, and chemiionization by secondary reactions.

Quadrupole ion traps have been used in some experiments as ion sources (Dawson, 1976), and they are now widely applied for analytical purposes (March and Hughes, 1989); however, they are not suited for preparing a well-defined ion beam owing to difficulties in ion extraction resulting from geometrical restrictions. Many of these inherent problems can be overcome by making use of the fact that the effective potential can be tailored according to specific experimental needs, as recognized by Teloy, who reported the first use of an rf ring electrode trap as an electron-impact ion source (Bahr, 1969; Bahr et al., 1969). Several improvements of the shape of the storage volume have lead to better suited arrangements. Especially important was the separation of the ionization region from the ion source exit by using a

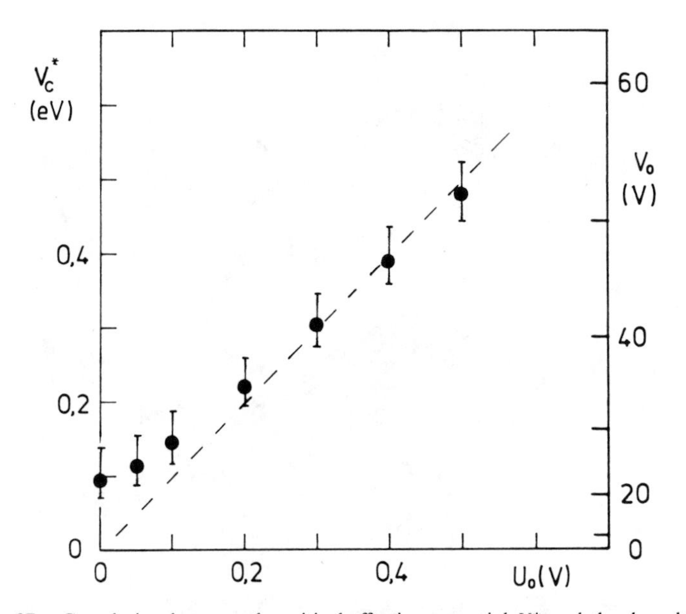

Figure 37. Correlation between the critical effective potential V_c^* and the dc voltage U_0. The rf amplitude, shown at the right-hand side, is converted into V_c^* using Eq. (82) and Eq. (83). The dashed line corresponds to the theoretical dependence given by Eq. (85). The dots are the 50% values of the transmission curves from Fig. 36, and the error bars indicate the 20–80% transmission range.

U-shaped storage volume (Gerlich, 1971). This rf storage ion source has been described thoroughly in Teloy and Gerlich (1974). A very similar version has been characterized in Sen and Mitchell (1986).

A slightly modified construction (Gerlich, 1977) is depicted in Fig. 38. The uppermost part shows the filament holder, the repeller, and the plate for electron extraction, which can also be used to pulse the electron beam. The U-shaped storage volume is defined by a stack of plates separated by ruby balls. The plates are alternately connected to the two phases of an rf generator. The top and bottom of this storage volume is limited by an adequate dc bias voltage applied to the two L-shaped endplates, each containing a small slot for the electron beam. In the exit region, the ions are guided in a field which changes smoothly to that of an octopole. Presently, we are using a storage ion source with a different electrode shape, as depicted in Fig. 46. This geometry has the advantage of being transparent in the axial direction for laser applications. In addition, the ion source has two equivalent filaments and ionization regions, and two symmetric ion exits that are useful for diagnostic purposes.

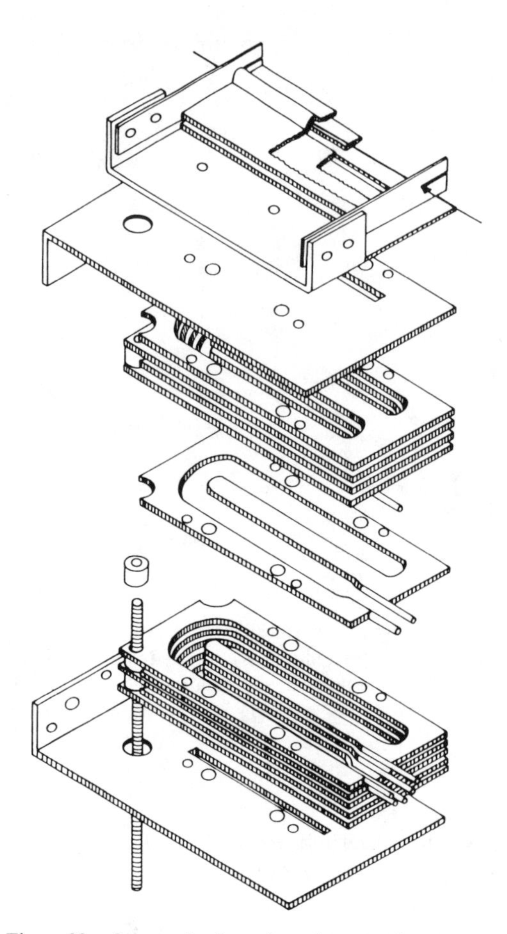

Figure 38. Schematic view of an rf storage ion source.

We limit our discussion here to a few examples that illustrate the advantageous features of the rf storage ion source. For the preparation of an intense, slow, and nearly monochromatic pulsed beam of protons (Gerlich, 1977; Teloy, 1978), we make use of the accumulating and mass-discriminating ability of the effective potential. Since in this case the trap is operated at the space-charge limit, it is important to avoid storage of undesired H_2^+ and H_3^+ ions by applying a weak dc difference and by properly adjusting the rf amplitude and H_2 pressure. On the other hand, the trap can be operated using a high pressure of H_2, such that more than 98% of the outgoing ions are H_3^+, formed by reactive $H_2^+ + H_2$ collisions. Collisional quenching of vibrationally excited ions has been demonstrated in several cases, for example, production of partially cooled H_2^+ ions was achieved by operating with a H_2/Ne mixture (Sen and Mitchell, 1986). Suppression of metastable atomic ions has also been demonstrated for N^+ (Frobin et al., 1977; Gerlich, 1984). Chemical quenching in the source can be used very efficiently, for example, to discriminate C_2H^+ from $C_2H_2^+$ formed by electron-impact ionization, as seen in the mass spectrum in Fig. 39. Using a mixture of 5% C_2H_2 and 95% H_2, the undesired species could be completely eliminated, since C_2H^+ reacts much faster with H_2 than $C_2H_2^+$ (Ikezoe et al., 1986). The resulting higher masses are suppressed by the high mass cut-off feature of the quadrupole following the ion source (see Fig. 46). In some cases, thermodynamical equilibrium cannot be reached in the trap within an acceptable storage time. One example is the population of the two $Ar^+(^2P_J)$ fine structure states (Scherbarth and Gerlich, 1989). Another is the quenching of metastable Ar^{2+}. Admixture of $Ar^{2+}(^1D)$ has been tested by single charge transfer in Ar^{2+} + He collisions. The Ar^+ product velocity distribution in Fig. 40 shows changes in the metastable population with storage conditions; however, the slow

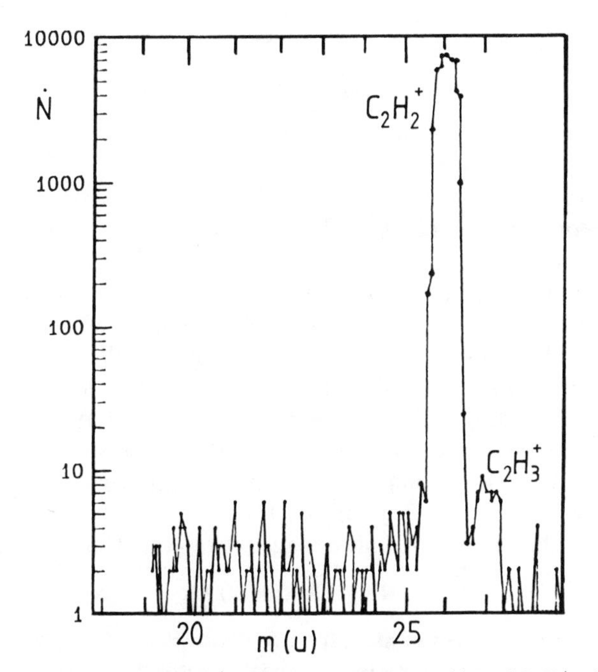

Figure 39. Preparation of a $C_2H_2^+$ beam by chemical quenching of C_2H^+ with H_2 in the storage ion source and by suppressing higher masses using the band-pass properties of the subsequent quadrupole.

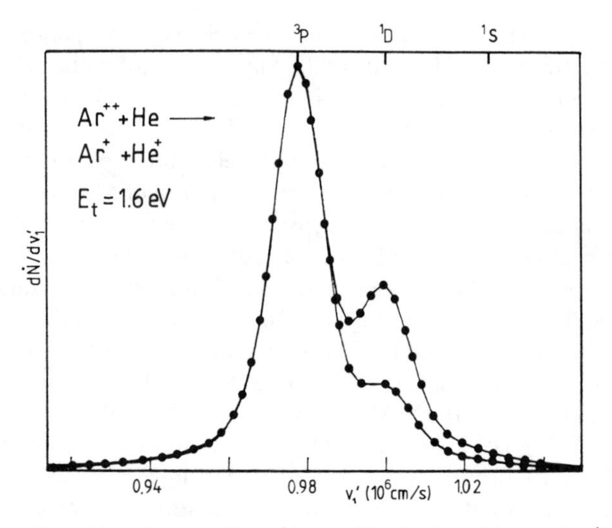

Figure 40. Quenching of metastable Ar^{2+} ions. The fraction of ions in the ^1D or ^1S state can be reduced by increasing the storage time in the rf source. Using the differential scattering apparatus, shown in Fig. 52, the composition of the initial Ar^{2+} beam has been analyzed by the charge transfer reaction $Ar^{2+} + He \rightarrow Ar^{+} + He^{+}$.

relaxation rate and the experimentally required duty cycle impeded complete relaxation.

In addition to the storage ion source and the previously mentioned combination of a quadrupole with photoionization, there are other possible combinations of guiding fields with ion formation. A photoionization source, combining a molecular beam of neutral precursors, monochromatized VUV light, and an oval 12-pole guiding field was described in Anderson et al. (1981). The oblong ionization volume, adapted to the exit slit of the monochromator, was converted smoothly into an octopole geometry using properly shaped electrodes. Another version of a stack-of-plates trap with a complicated labyrinth structure has been used by Anderson and coworkers (Hanley et al., 1987) for cooling metal-cluster ions in a He buffer gas. This group has also developed a method to produce boron-cluster ions by laser ablation in the interior of a specially designed rf ion trap (Hanley et al., 1988).

E. Ring Electrode Trap

The rf ring electrode ion trap has been proven over the past few years to be a very useful tool for obtaining low-temperature thermal rate coefficients of ion-molecule collisions owing to its unique sensitivity and efficient trapping capability (Gerlich and Kaefer, 1987, 1989; Gerlich et al., 1990). As was mentioned in the preceding section, the ring electrode device was pioneered by Teloy and Bahr (Bahr et al., 1969), who utilized the trap to obtain information on thermal rate coefficients from temporal changes of ion concentrations (Bahr, 1969). Progress in our applications (Kaefer, 1989; Paul, 1990) is due to mass-selected ion injection and developments that allow one to vary the trap temperature over a wide range. The axial symmetry of the trap provides easy access at both ends for ion injection and extraction. Moreover, the geometry is also well suited for studying ion–laser interactions and for product detection by optical methods. Integration of the trap in a complete instrument will be described in Section IV E.

Two different electrode arrangements are shown in Fig. 41. The dominant construction problem was the need to operate the trap at cryogenic

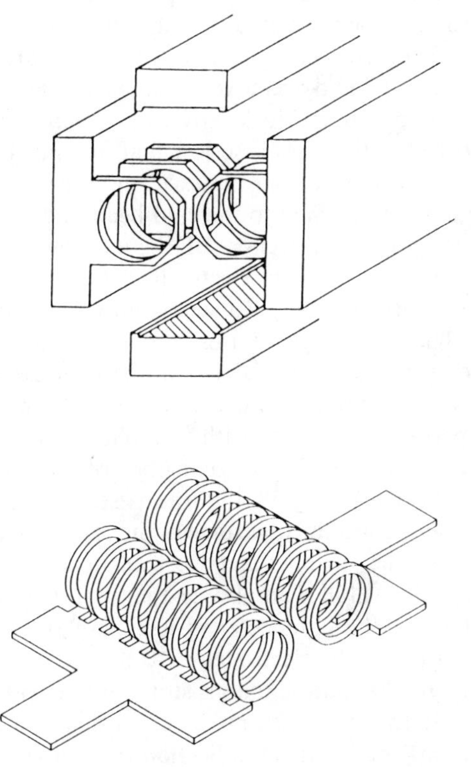

Figure 41. Illustration of two ring electrode traps. In the trap shown above the solid ring holders are cooled directly with liquid nitrogen. The lower trap has been designed for mounting on the head of a liquid helium cryocooler. The perspective views show the two groups of electrodes slightly removed from each other. The function of the two graphite-coated ceramic covers (hatched area in the upper figure) is explained in the text.

temperatures. An early version (Gerlich and Kaefer, 1987) not shown in Fig. 41, could only operate down to 170 K due to heating by absorbed rf power. To overcome this problem, a second version, upper structure in Fig. 41, was constructed (Gerlich and Kaefer, 1988, 1989), whereby the entire device is cooled directly by running liquid nitrogen through the massive ring holders using electrically insulated feedthroughs. The liquid nitrogen flows from one half of the trap to the other using stainless-steel tubing. The tubing is bent into the shape of a coil such that the inductance L of this coil and the capacitance c of the trap are used directly as a Lc-resonance circuit (Kaefer, 1989), which can be easily excited by inductive rf coupling. Typical frequencies and amplitudes are 25–35 MHz and up to 100 V, respectively.

1. *The 10–350 K Trap*

A third version (Paul, 1990) has been designed for mounting directly onto the cold head of a closed-cycle liquid helium refrigeration system (Leybold Heraeus, RD 210).The perspective view in the lower part of Fig. 41 shows the two identical halves slightly removed from each other. Each half can be machined from a solid copper block; however, experience has shown that the rf power absorbed in the rings is so low that one can simply hard solder or weld the rings onto the bottom plates using either copper or stainless steel. The two plates are attached to the cold head and a 0.1-mm-thin sheet of a special insulating material (Denka, BFG 20) is placed between the plate and the cold head to obtain good electrical insulation and sufficient heat conductivity, as has also been applied successfully for a similar purpose by Hiraoka (1987). The advantage of this material is its flexibility at low temperatures; however, the material is not best suited for our vacuum conditions and its heat conductivity decreases toward low temperatures. We are presently performing tests with sapphire insulation, the heat conductivity of which is even higher than that of copper at temperatures below 80 K. As indicated in Fig. 58, the trap is surrounded by a 80 K heat shield, which is connected to the first stage of the cryocooler. All connections to the trap, including the gas inlet and the wires to the external rf coil, are also precooled to 80 K. At full rf amplitude, the total required cooling power at the second stage of the cryocooler is less than 0.2 W and the lowest achievable temperature of the electrodes is 10 K.

The kinetic energy distribution of ions stored in the wide field-free region of the ring electrode trap has been discussed in detail in Section II E. The effective potential has been derived in Section II C 3 and is shown in Fig. 8. The relation between the actual shape of the electrodes and the mathematical boundary conditions used to describe this trap are compared in Fig. 7. Typical dimensions of the trap are $r_0 = 0.5$ cm, while the thickness of the rings and their separation is 0.1 cm, resulting in $2\pi z_0 = 0.4$ cm. The shape of the ion cloud has an almost cylindrical form with a volume of about 1 cm^3. The maximum observed ion density is 10^6 cm^{-3}, which is close to the space-charge limit. Most experiments are performed with typically 10^3–10^4 ions per cm^3. At these low ion densities space-charge effects and ion-ion collisions are negligible.

The problem of surface potential distortions, discussed in Section III. C 2, also perturbs the functioning of the ring electrode trap. In this case, the ions are thermalized and trapped; however, trapping can occur in regions of local potential minima, and, as a result, the ions are not distributed throughout the trap. This can lead to an augmentation of rf heating, as discussed in Section II E. In addition, this localized trapping can significantly increase the time needed for extraction. To aid in ion extraction, as well as correct the position of the trapped ion cloud, we make use of field penetration from external electrodes. As electrodes we use the graphite-coated ceramic plates, indicated in Fig. 41 by the hatched area, which are also used to confine the buffer gas. Applying a voltage of about 100 V across the ends of the graphite coating creates a constant axial potential gradient toward the exit. This voltage is pulsed on during the extraction period and shortens the time needed to evacuate the ions from the trap. With an additional voltage applied to a contact in the center of the correction electrode, a V-shaped field can be generated, which leads to a concentration of the ions close to the middle of the trap. This well can be made sufficiently deep to confine the ions in the axial direction.

The maximum achievable storage time of ions in the trap is usually limited by the extent of reactive-loss processes. Storage loss is very unlikely, since the confining potential wall is usually several eV and the chance that an ion can accumulate sufficiently high energy is rather low. Examples that illustrate the change of the ion composition in the trap as a function of storage time are given in Fig. 42. The upper panel shows a fast conversion of primary H^+ ions into D^+ products (Gerlich and Kaefer, 1987), the sum of these ions indicates no loss on the millisecond time scale. Ternary association of CD_3^+ with $2D_2$ (Gerlich and Kaefer, 1989) is plotted in the middle panel. Here, the time constants are in the range of seconds. The sum of the two ion intensities declines with a mean life time of 12 s, most probably due to formation of nonrecorded products. The lowest panel shows storage of He^+ ions over several minutes at 10 K. A dominant loss channel is due to reaction with background N_2 (density 5×10^5 cm^{-3}).

2. *Collision Temperature*

A general difficulty with all ion traps is the determination of the actual collision "temperature." The precise ion energy distribution is a complex function of parameters such as the temperature of the electrodes, the walls,

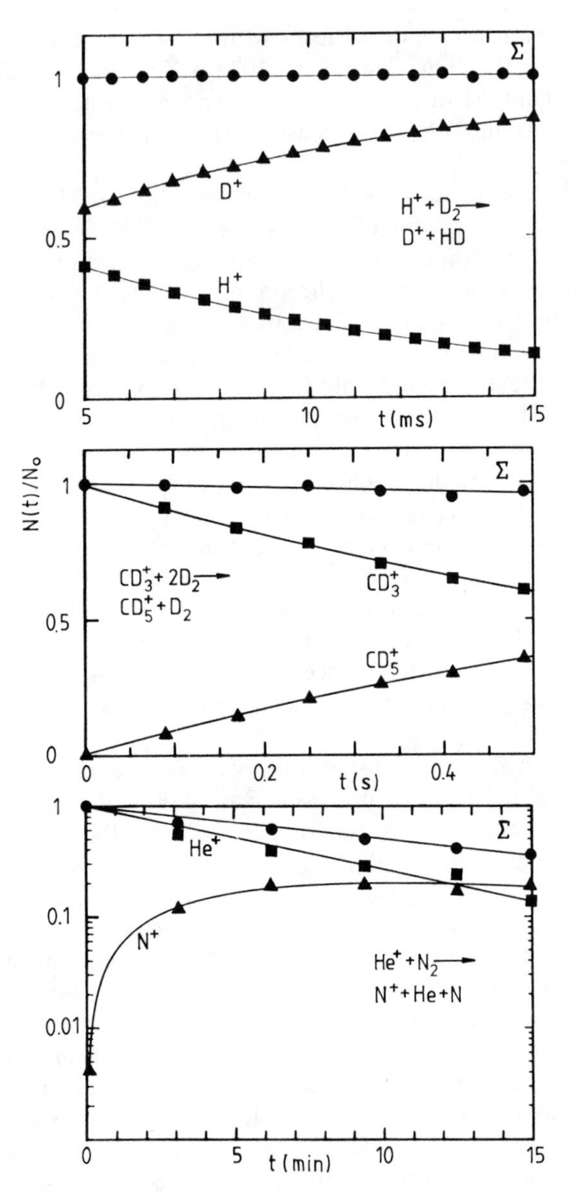

Figure 42. Change of ion composition in the trap as a function of storage time. The upper panel depicts a typical fast reaction; the time scale is in the millisecond range for target densities between 10^{12} and $10^{13}\,cm^{-3}$. The middle panel shows a ternary association reaction; time constants are in the range of seconds. The lowest figure shows storage of He^+ ions over several minutes. The loss of ions in this case is predominantly due to reactions with traces of the background gas, for example, N_2.

and the buffer gas. The rf influence on the ion energy has been discussed in Section II E. This can be investigated experimentally, based either on known temperature dependencies of rate coefficients or by determination of a thermodynamical equilibrium. One example is the slightly endothermic reaction $N^+ + H_2 \rightarrow NH^+ + H$, for which the endothermicity ΔH^0 has been established as 17 ± 2 meV from CRESU data (Marquette et al., 1988) and phase-space evaluation (Gerlich, 1989b). These results are compared to our ring electrode trap results (Gerlich et al., 1990) in Fig. 43. If we explain the slight deviation on the basis of a higher temperature in our experiment, we obtain a difference of 10 K between our nominal temperature and the actual temperature. However, the discrepancy may also be explained by incomplete thermalization of the $N^+(^3P_J)$ fine structure states. Our results are in perfect agreement with the statistical theory if we change the endothermicity ΔH^0 from 17 to 15 meV. Another test is the three-body association reaction $He^+ + 2He \rightarrow He_2^+ + He$ for which we have obtained a ternary rate coefficient $k_3 = 6 \times 10^{-31}$ cm^3/s at a nominal trap temperature of 10 K. This value compares well with results obtained from a liquid-helium-cooled drift tube (Böhringer et al., 1983). Accounting for possible errors in the determination of the He density and the absolute rate coefficient, the resulting uncertainty of the temperature is again 10 K.

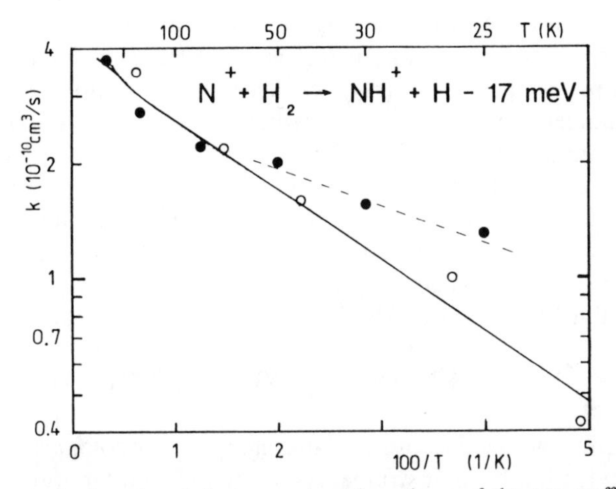

Figure 43. Arrhenius plot of the temperature dependence of the rate coefficient for the endothermic ion–molecule reaction $N^+ + H_2 \rightarrow NH^+ + H$. The phase-space calculation (Gerlich, 1989b) (solid line) has been adjusted to the CRESU results (Marguette et al., 1988) (○) by using $\Delta H^0 = 17$ meV. Possible explanations for the slight deviations from our trap measurements (●) are given in the text.

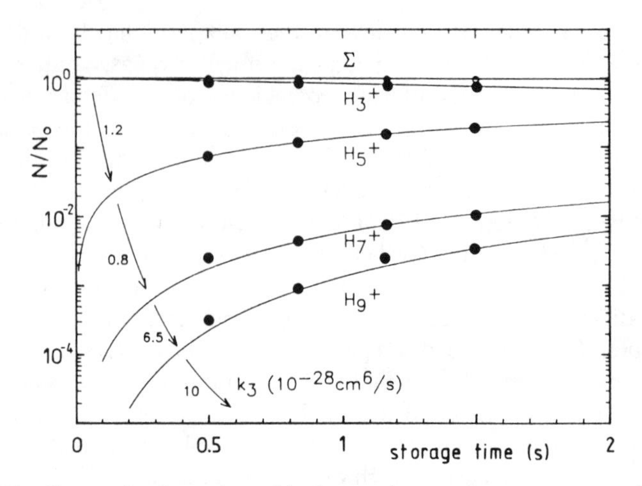

Figure 44. Consecutive formation of hydrogen clusters at a temperature of 25 K. H_3^+ primary ions were injected into the ring electrode trap containing normal $-H_2$ at a rather high density of $3.7 \times 10^{13}\,cm^{-3}$. The lines are solutions from the corresponding rate equation system. The numbers on the arrows are the resulting ternary rate coefficients in $10^{-28}\,cm^6/s$.

Determining the collision temperature based on thermodynamical equilibrium has the advantage that it is independent of the neutral density. At temperatures above 80 K, we have stored H^+ ions in a known H_2/D_2 mixture and measured the resulting H^+/D^+ equilibrium, the temperature dependence of which is well known (Gerlich, 1982; Henchman et al., 1981). At low temperatures, the $H_3^+ \cdot (H_2)_n$ cluster-size distribution can also be used as a thermometer over a wide range, based on recently published cluster binding energies (Hiraoka, 1987). Figure 44 shows the consecutive formation of hydrogen clusters following injection of H_3^+ into our trap containing H_2 at a nominal temperature of 25 K. From these data, the rate coefficient for cluster growth can be derived; however, the storage time was too short to reach thermodynamical equilibrium at this low temperature. The equilibrium can be established more quickly if one injects larger hydrogen cluster ions.

IV. DESCRIPTION OF SEVERAL INSTRUMENTS

In this section we begin with a short, mainly chronological, overview of various guided-ion-beam instruments used in different laboratories in which one or more of the rf devices described in Sections II and III are combined for stydyng processes in gas-phase ion physics and chemistry. Mostly these instruments are used to investigate low-energy ion–molecule collisions, but we will also refer briefly to a few spectroscopic applications. In Sections

IV B–IV E we describe in detail four apparatuses used in our laboratory, that is, an universal guided ion beam, a differential scattering, a merged beam, and a ring electrode ion trap apparatus. Some fundamentals of kinematic averaging and data analysis are compiled to compare the various experimental methods. Examples that illustrate special features of the instruments are also given in this section, while a more complete overview of specific applications follows in Section V.

A. Overview: Instruments Using rf Devices

One of the early uses of an rf field as an ion-beam guide was reported by von Zahn (Tatarczyk and von Zahn, 1965; von Zahn and Tatarczyk, 1964) for studying metastable decay of excited polyatomic ions in the time window between 0.1 and 1 ms. In this experiment, a mass-selected ion beam was injected into a 314-cm-long quadrupole operated in the rf-only mode ($a_2 = 0$). The field parameters were chosen such that a wide range of masses were transmitted, that is, to guide the parent ions and to simultaneously confine their fragments. The ions were analyzed in a subsequent quadrupole mass spectrometer. Decay rates on the order of $10^4\ s^{-1}$ were reported for several hydrocarbon ions.

Related experiments are now performed using so-called triple quadrupole mass spectrometers. These instruments were developed initially (McGilvery and Morrison, 1978; Vestal and Futrell, 1974) to study laser photodissociation processes in the center rf-only quadrupole, and their use has expanded to include among others the study of collision-induced dissociation of polyatomic ions. Such instruments and more complicate ones, for example, penta quadrupoles, are now available and are used routinely for analytical purposes, see, for example, Dolnikowski et al. (1988) and references therein. For obtaining accurate cross sections for collision processes, however, such arrangements are often insufficient. Unsuitable injection conditions lead to poorly defined collision energies, and inefficient product collection can result from unsafe ($\eta > 0.3$, see Section II D 2) operating conditions of the rf-only quadrupole or from a mismatch of the product phase space to the mass analyzer or to the ion detector.

The development of the first guided-ion-beam apparatus began in the late 1960s in the group of Ch. Schlier by E. Teloy (Bahr, 1969; Bahr et al., 1969, 1970; Werner, 1968), with the combination of a ring electrode trap ion source, described in Section III, and a magnetic mass spectrometer followed by the construction of the storage ion source and the first octopole (Gerlich, 1971). Early results obtained using this instrument have been mentioned in Henchman (1972), while a complete description of the first guided-ion-beam apparatus and several integral cross sections were published in Teloy and Gerlich (1974). Further results are given in Frobin et al. (1977) and Ochs and

Teloy (1974). An improved and more universal version of this apparatus will be described in Section IV B. A differential scattering apparatus consisting of several rf devices, including a 1-m-long octopole for product time-of-flight analysis, has also been constructed by the same group (Gerlich, 1977; Teloy, 1978). Details of this apparatus will be given in Section IV C.

The incorporation of rf octopoles and other special rf devices in different laboratories has proceeded rather slowly. At the end of the 1970s the guided-ion-beam technique was combined with the photoionization method in the group of Y.T. Lee (Anderson et al., 1981). This instrument consisted of a molecular-beam photoionization source, a He discharge lamp (Hopfield continuum), and vacuum ultraviolet monochromator, coupled with an octopole beam guide, scattering cell, and quadrupole mass spectrometer detector. With this apparatus effects of both reagent translational and vibrational energy have been studied for several reaction systems (Houle et al., 1982; Turner et al., 1984).

During the 1980s the guided-ion-beam technique became more popular and several groups successfully contributed to its further development. A discussion of the fundamental functioning of an octopole ion guide and some performance tests were published by Okuno (Okuno, 1986; Okuno and Kaneko, 1983) who studied one- and two-electron capture of Ar^{2+} and Kr^{2+} in their own gases.

Anderson and coworkers have extended the use of the guided-ion-beam technique into different fields. In their group they developed an apparatus for measuring integral cross sections of cluster ion reactions (Hanley and Anderson, 1985). A unique feature of their instrument is an rf storage trap with labyrinthine geometry, which is used to collisionally cool sputtered cluster ions (Hanley et al., 1987). In another instrument, state-selective preparation of ions via multiphoton ionization has been combined with a tandem guided-ion-beam spectrometer (Orlando et al., 1989). The influence of different vibrational modes on the reaction dynamics of several collision systems are discussed in more detail in Anderson's chapter of this book.

Also, the group of Armentrout started in the early 1980s (Ervin and Armentrout, 1985; Ervin et al., 1983), to utilize guided-ion-beam instruments for investigating various reactions, among others, those between first- and second-row atomic cations and isotopic hydrogen molecules. Some of their results have been reviewed in Farrar (1988). A recent experimental modification has been made to cool the octopole and scattering cell to near liquid nitrogen temperatures (Sunderlin and Armentrout 1990). In the past year, this group has also completed the construction of a new guided-ion-beam mass spectrometer, the design and capabilities of which have been described in Loh et al. (1989). This mew instrument has been developed to study collision-induced dissociation dynamics of cold metal-claster ions.

Ng and coworkers have made extensive use of photoionization coupled in various combinations with octopole beam guides and quadrupole mass filters (Flesch et al., 1990; Liao et al., 1990; Ng, 1988), including a crossed ion–neutral-beam photoionization instrument (Liao et al., 1984, 1985) and a triple-quadrupole double-octopole photoionization apparatus (Liao et al., 1986). A unique demonstration of the high sensitivity of the guided-beam technique is the differential reactivity method developed in this group (Liao et al., 1985). With this method it is possible to determine the internal energy of product ions by reacting them in the second octopole with a probing gas. The power of this method for obtaining state-to-state information, as well as further details of these instruments, are documented in Ng's chapter of this volume.

For several years the group of D. Zare (Conaway et al., 1987; Morrison et al., 1985) has applied resonance enhanced multiphoton ionization to the production of vibrationally state-selected ion beams (see Anderson's chapter, this volume). Using a tandem quadrupole arrangement with a static gas collision cell, ion molecule reactions have been studied. In order to improve the product collection efficiency and the accessible energy range, they have recently built a quadrupole-octopole-quadrupole system (Posey et al., 1991).

In order to complete, hopefully without omission, the list of guided-ion-beam instruments, we also wish to mention the investigations of Tosi et al. (1989b) who have recently completed the construction of such an instrument and have used it in a crossed beam arrangement to study $Ar^+ + H_2$ reactions (Tosi et al., 1989a). A very recent combination of photoionization with octopole guiding fields became operational in Orsay (Guyon et al., 1989). Special features of this apparatus include ion preparation using monochromatized synchrotron radiation in combination with the threshold photoelectron– photoion coincidence method, and also the ability to determine integral cross sections and to measure product velocity distributions using a short scattering cell in a system of two octopoles similar to that described in Section IV B. Finally, there are some spectroscopical applications which employ rf guiding fields. A molecular ion spectrometer consisting of a long rf octopole for guiding or storing ions has been developed in the group of Lee. An application of this device was to study the spin-forbidden radiative decay of $O_2^+(a^4\Pi_u)$ (Bustamente et al., 1987). Also, using tunable IR lasers, fragmentation of several cluster ions has been studied including $H_3^+(H_2)_n$ and $H_3O^+(H_2O)_n$ (Okumura et al., 1988; Yeh et al., 1989).

B. The Universal Guided-Ion-Beam Apparatus

The apparatus described here is, on the whole, comparable to the first guided-ion-beam apparatus (Teloy and Gerlich, 1974); however, there are many improvements and extensions that require elaboration. A fundamental capability of both instruments is the ability to precisely determine integral cross sections with high sensitivity. In the improved guided-ion-beam apparatus, the range of laboratory energies of guided ions has been extended to below 10 meV (Gerlich, 1984). Expansions of the capabilities of this apparatus include the ability to measure photons from chemiluminescent reactions and to detect coincidences between photons and mass selected product ions (Gerlich and Kaefer, 1985). Product angular distributions can also be obtained by a procedure that combines time-of-flight analysis with variation of the guiding field (Disch et al., 1985). A rather detailed description of this procedure and several applications have been reported in Gerlich (1986) and will be further discussed in Section IV B 3. The modular design of the apparatus allows for simple replacement of the scattering cell with a crossed pulsed supersonic beam, thereby improving the kinetic energy resolution. This led to the discovery of an oscillatory structure in the integral cross section for the reaction $C^+ + H_2 \rightarrow CH^+ + H$ (Gerlich et al., 1987). The geometry of the ion source and the off-axis detection system allow for ease of integration of various laser systems to state selectively prepare reactants and/or to analyze reaction products (Gerlich and Scherbarth, 1987). A detailed study of the $Ar^+ + O_2$ charge transfer, including the determination of rovibrational product state distributions by laser-induced predissociation, has been published recently (Scherbarth and Gerlich, 1989) and demonstrates the versatility of this instrument.

1. *Description of the Apparatus*

Figure 45 shows schematically the most important elements of a simple guided-ion-beam apparatus. An rf storage ion source is used (see Section III D) to trap and internally thermalize ions generated by electron bombardment. The trap is usually operated in a pulsed mode by opening the ion gate for a few μs leading to a rather high momentary peak current. For energy and mass filtering the ions must pass very slowly, typically 0.1 eV axial energy, through the focusing quadrupole, as described in Section III B 2. Additional time-of-flight selection is achieved by pulsing the beam a second time at the center element of the lens system following the quadrupole. In order to avoid interference with the rising and falling edges of the pulse, this lens element is divided into equal halves that are connected to two properly adjusted counter propagating pulses. Usually the lens system focuses the ions into the injection electrode; however, in some cases, to reduce the angular divergence of the injected ion beam, all lens elements are operated at the same voltage. Typical octopole dimensions are $d = 0.2$ cm and $r_0 = 0.3$ cm, fulfilling the relation given by Eq. (68). The scattering cell usually consists of a 3–15-cm-long cylinder containing the target gas. The vacuum conductance of this cell is reduced by two small tubes that closely surround the octopole. As described in Section III C 2 and illustrated in Fig. 30, the

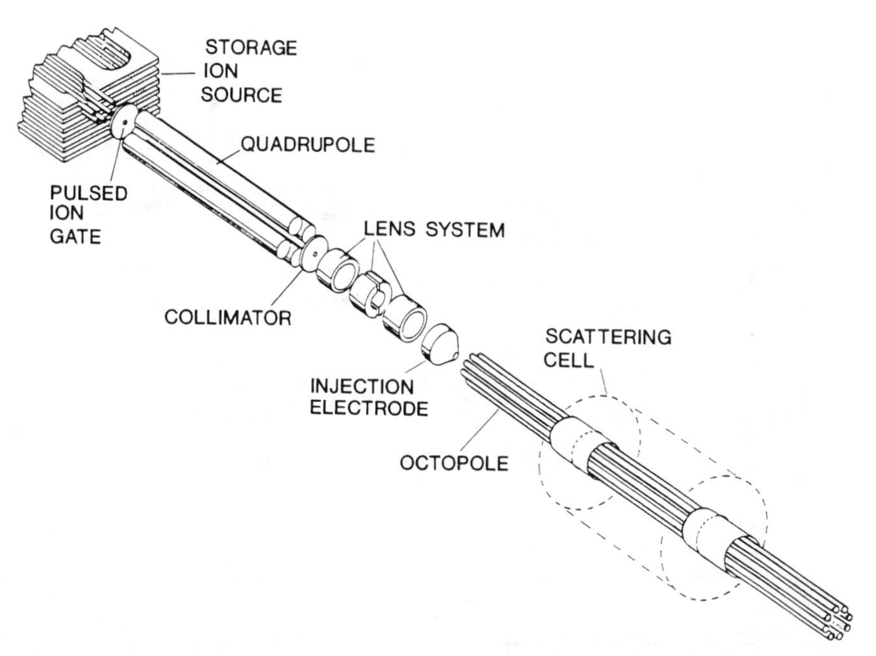

Figure 45. Perspective view of the major elements of a typical guided-ion-beam apparatus. Ions are created in a storage ion source, are mass and energy selected in a quadrupole and are injected into an octopole via a funnellike electrode. The scattering cell usually consists of a gas-containing cylinder surrounding the octopole. The ion beam is pulsed by opening the gate at the exit of the ion source and by pulsing the two halves of the middle element of the lens system.

field penetration of these cylindrical electrodes is used for diagnostic purposes, ion trapping, and so on.

A detailed illustration of our universal guided-ion-beam apparatus (Gerlich 1986; Scherbarth and Gerlich, 1989) is depicted in Fig. 46. It consists of the just described rf source for ion creation and preparation. Alternatively, this source can be easily exchanged with the multiphoton ionization source described in Section III B 3. The primary ions are injected into a tandem octopole arrangement with a short scattering cell. Primary and product ions are mass analyzed by a magnetic mass spectrometer and detected using a scintillation detector. The two octopoles are coupled mechanically such that the rf field does not change in the joint, yet the ions can be accelerated or decelerated by using different dc potentials.

Calibration of the kinetic energy of the ion beam in the octopole has been described in detail in Section III C 3. For some applications it may be sufficient to use the octopole as a retarding field analyzer; however, it is normally advisable to calibrate the difference between the onset of the ion

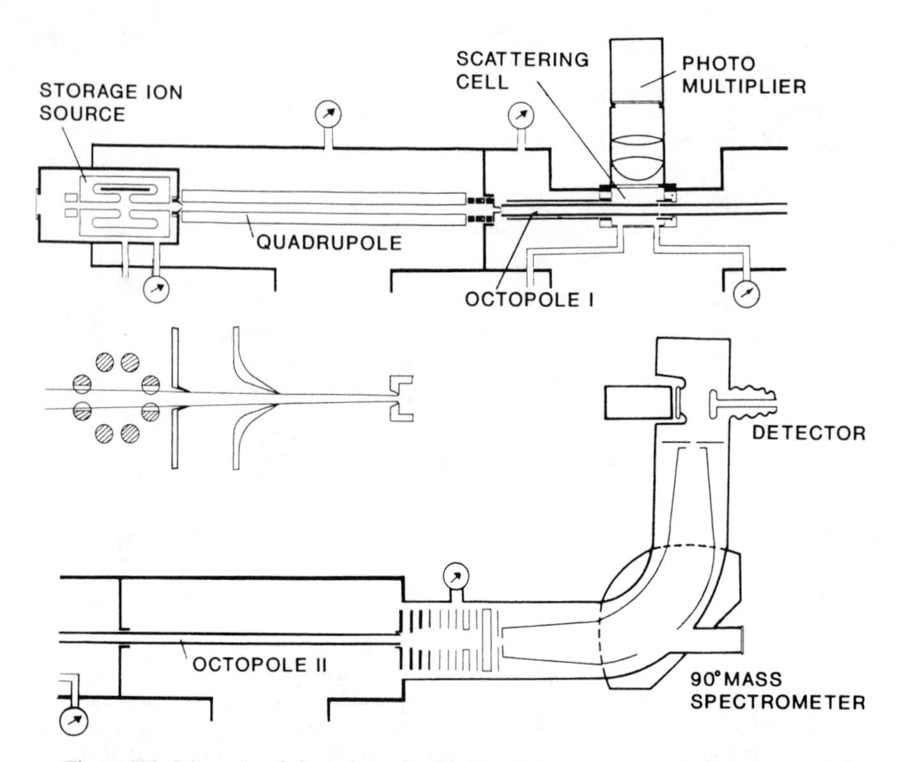

Figure 46. Schematic of the universal guided-ion-beam apparatus (the lower part of the figure is a continuation of the upper). The bakable vacuum system is pumped separately by three turbomolecular pumps. The crossed-beam arrangement, depicted between the two parts of the figure, can be used in place of the scattering cell/photomultiplier arrangement. Most electrodes are shown approximately to scale. The length of the quadrupole is 26 cm, octopoles I and II are 14 cm and 46 cm long, respectively. The 90° magnetic mass spectrometer has a 15 cm radius.

intensity and the energy zero point by time of flight. In order to avoid loss in accuracy, all measurements are made in the pulsed mode under the same conditions the time-of-flight calibration procedure was performed.

Problems related with the injection of ions into rf multipoles have been briefly mentioned in Sections II and III; however, a few additional comments and some practical hints are needed. Going from a pure dc field into an ideal multipole rf field, the ions traverse a region in which they are under the influence of a combined dc and rf field, which has both axial and transverse components. This always leads to perturbations of the energy and angular distributions of the injected ions. In order to minimize these effects, several precautions should be taken. An rf-field-free region along the multipole axis can be attained if the mechanical geometry is accurate and the amplitudes of the two rf voltages are symmetric and exactly opposite in phase. Note that it is also necessary to block the rf from the dc electrodes using capacitors. Under such conditions, an ion beam that is injected very close to the axis with a small angular divergence remains unperturbed. In cases where for intensity reasons, or for passage of a laser beam, one has to use entrance holes with dimensions up to $r/r_0 < 0.5$, it is important to avoid rapid entrance of the ions into the rf field, since the impairment of their kinetic energy is significantly increased if one works under nonadiabatic conditions in the transition region. A slow entrance can be achieved by a gradual spatial variation of the field using adequate electrodes or electrical-field-shielding devices, as known from quadrupoles. Alternatively, it is also possible to pass the critical region very slowly, that is, during several rf periods. In general, we operate in the transition region with ion energies of a few tenths of an eV. The injection field is defined by a conelike electrode, as can be seen in Fig. 30 or Fig. 45. This electrode is centered in a precisely machined ceramic part, which also holds the multipole rods in their exact position.

Neutral reactants are introduced either diffusively into a scattering cell or as a pulsed supersonic beam; see Fig. 46. The gas cell is rather short to reduce potential distortions, as discussed in Section III C 2, and for product time-of-flight analysis. Determination of the neutral gas density n_2 and the effective scattering length L have been described previously (Gerlich et al., 1987). The amplitude and the frequency of the rf field are always chosen such that all primary and product ions are confined in the octopole and guided toward the mass spectrometer detector. In the laboratory system, backward-scattered ions can be reflected forward by operating the ion beam and the injection electrode in a pulsed mode. Under such conditions, the octopole can guarantee 4π collection efficiency. Nonetheless, there are other possible sources of errors in determining absolute cross sections, for example, those caused by local potential distortions and the limited phase-space acceptance of the detection system.

Local potential barriers in the scattering cell can lead to trapping of slow product ions and their loss, for example, by secondary reactions. More critical, especially at low laboratory energies, is the trapping of scattered primary ions in the scattering cell. This often leads to a significant increase in the formation of products, especially if the cross section is large at low collision energies. These problems can be tested and partially avoided by applying adequate potentials to the two ring electrodes (5) and (6), depicted in Fig. 30. An external potential barrier at the entrance of the reaction zone reflects slow ions and reduces the extent of trapping. A barrier at the exit can increase the effect for diagnostic purposes. An example of a reaction that is very sensitive to potential barriers is the $N^+ + CO$ charge transfer, as discussed in Gerlich (1984 and 1986). Similar effects have been observed for the $Ar^+ + O_2$ charge transfer.

Difficulties in matching the phase space of the ions emerging from the octopole to the phase space accepted by the mass analyzer and detector have already been discussed in detail in Teloy and Gerlich (1974). Most of the newer guided-ion-beam instruments mentioned in Section IV A utilize quadrupole filters for mass analysis. However, according to our experience, it is not easy to avoid discrimination effects if the reaction products have large transverse velocities or if the rod diameter of the quadrupole filter is too small. In our universal guided-ion-beam apparatus, the ions are gradually accelerated via several lens elements with large apertures to 2–3 keV, then focused by two independent electrostatic einzel lenses to the entrance slit of a large magnetic mass spectrometer. The Daly-type detector is preferred over open ion multipliers owing to its large area and very uniform detection efficiency.

2. *Kinematic Averaging*

So far, the majority of the guided-ion-beam experiments has been performed by passing the octopole through a scattering cell. The influence of the random thermal motion of the target gas on the distribution of the collision energy has been treated often (Berkling et al., 1962; Chantry, 1971; Schlier, 1988; Teloy and Gerlich, 1974) and the impairment of the product energy distributions has recently been discussed in Gerlich (1989a). In a few experiments (Gerlich et al., 1987; Sunderlin and Armentrout, 1990), the target temperature has been reduced by operating with a cooled scattering cell. Significant improvement, however, can only be achieved by replacing the scattering cell with a supersonic beam. Two successful approaches in which a neutral beam was crossed with a guided ion beam at a right angle have been described in Gerlich et al. (1987) and recently in Tosi et al. (1989a). The kinematic conditions can be further improved if the crossed beams are replaced by merged beams, as will be described in Section IV D. To better compare the different arrangements, it is necessary to summarize a few facts concerning kinematic averaging.

In an ideal experiment, with two well-collimated monochromatic beams of velocities $\mathbf{v}_1$ and $\mathbf{v}_2$, the number of molecules reacting per unit time in the scattering volume $d\tau$ is given by (Gerlich, 1989a; Levine and Bernstein, 1974)

$$d\dot{N} = g\sigma(g)n_1 n_2 d\tau. \tag{86}$$

Here, $g = |\mathbf{g}| = |\mathbf{v}_1 - \mathbf{v}_2|$ is the relative velocity, n_1 and n_2 are the projectile and target densities, respectively, and $\sigma(g)$ is the intrinsic integral cross section. The corresponding intrinsic rate coefficient is denoted as $k = g\sigma(g)$. In contrast to the ideal case, the kinematic conditions of a realistic experiment are less well defined, and the actually obtained product rate $d\dot{N}$ is the average over the velocities of the reactants $\mathbf{v}_1$ and $\mathbf{v}_2$. The number of molecules reacting per unit time in the scattering volume $d\tau$ is given by the six-dimensional integral

$$d\dot{N} = \int_{\mathbf{v}_1} d\mathbf{v}_1 \int_{\mathbf{v}_2} d\mathbf{v}_2 g\sigma(g) n_1(\mathbf{r}, \mathbf{v}_1) n_2(\mathbf{r}, \mathbf{v}_2) d\tau. \tag{87}$$

Assuming that the velocity distributions are independent of the spatial coordinate $\mathbf{r}$, the density functions $n_i(\mathbf{r}, \mathbf{v}_i)$ can be factorized using normalized probability functions $f_i(\mathbf{v}_i)$ $(i = 1, 2)$

$$n_i(\mathbf{r}, \mathbf{v}_i) = n_i(\mathbf{r}) f_i(\mathbf{v}_i). \tag{88}$$

With $f_1(\mathbf{v}_1)$ and $f_2(\mathbf{v}_2)$, we define $f(g)$, the distribution of the relative velocity

$$f(g)dg = \int_{\mathbf{v}_1}^{*} d\mathbf{v}_1 \int_{\mathbf{v}_2} d\mathbf{v}_2 f_1(\mathbf{v}_1) f_2(\mathbf{v}_2). \tag{89}$$

$f(g)dg$ denotes the probability that the relative velocity lies in the interval $[g, g + dg]$. The asterisk indicates symbolically that the integration must be limited to that subspace of $(\mathbf{v}_1, \mathbf{v}_2)$ where $|\mathbf{v}_1 - \mathbf{v}_2|$ falls into this interval. With the mean relative velocity $\langle g \rangle$,

$$\langle g \rangle = \int_0^\infty g f(g) dg, \tag{90}$$

we define the *effective cross section* with a somewhat more general distribution function $f(g)$ than usual as

$$\sigma_{\text{eff}}(\langle g \rangle) = \int_0^\infty \frac{g}{\langle g \rangle} \sigma(g) f(g)\, dg. \tag{91}$$

With this definition of σ_{eff}, we obtain a result that is very similar to Eq. (86)

$$d\dot{N} = \langle g \rangle \sigma_{\text{eff}} n_1(\mathbf{r}) n_2(\mathbf{r}) d\tau. \tag{92}$$

Similarly,

$$k_{\text{eff}} = \int_0^\infty g\sigma(g) f(g)\, dg \tag{93}$$

corresponds to the effective rate coefficient.

For the description of the special case of a monoenergetic ion beam passing through a scattering cell with target gas at a temperature T_2, $f_1(\mathbf{v}_1)$ and $f_2(\mathbf{v}_2)$ in Eq. (88) are replaced by a δ-function and a Maxwellian $f_M(\mathbf{v}_2; m_2, T_2)$, respectively. The resulting integral can be reduced analytically, leading to the well-known generalized Maxwell–Boltzmann distribution (Chantry, 1971; Teloy and Gerlich, 1974)

$$f(g) = f^*(g; v_1, T_2) = (m_2/2\pi k T_2)^{1/2} \frac{g}{v_1}\left[\exp\left(-\frac{m_2}{2kT_2}(g - v_1)^2\right) - \exp\left(-\frac{m_2}{2kT_2}(g + v_1)^2\right)\right]. \tag{94}$$

At low ion velocities v_1, this function approaches a normal Maxwellian $f_M(g; \mu_2, T_c)$ with a reduced temperature $T_c = m_1/(m_1 + m_2)T_2$, as illustrated in Fig. 47. In most experiments, this limit is only reached for mass ratios

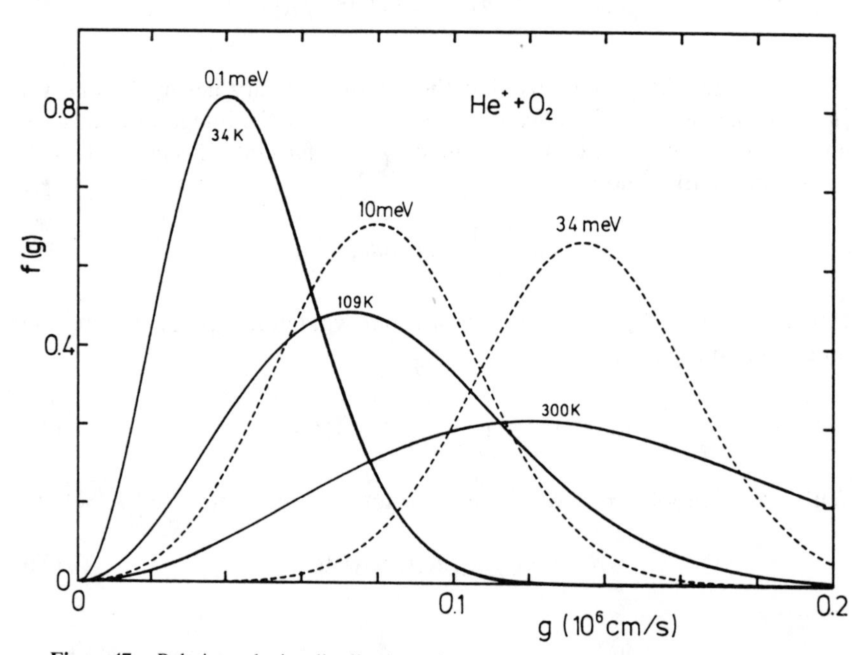

Figure 47. Relative velocity distributions $f(g)$ calculated using Eq. (89). The dashed lines are generalized Maxwell–Boltzmann distributions $f^*(g; v_1, T_2)$ [Eq. (94)] describing the relative motion between monoenergetic He^+ ions (0.1, 10, and 34 meV$_{lab}$) and $T_2 = 300$ K thermal O_2 neutrals. For comparison, the solid lines show thermal distributions $f_M(g; \mu, T)$ [Eq. (112)] having the same mean velocities $\langle g \rangle$.

$m_1 \gg m_2$ (Ervin and Armentrout, 1987; Teloy and Gerlich, 1974), and the lowest attainable collision temperature is therefore $T_c \sim T_2$. The trapped-ion-beam technique, described in Section III C 2, allows one to store nearly monoenergetic ions at such low energies that the velocities of the ions and the neutrals become comparable, even for mass ratios like He^+ and O_2. For this system, we have performed experiments with He^+ ions trapped at energies as low as 9 meV and obtained collision temperatures T_c that are significantly lower than T_2. The resulting effective cross sections are presented in Fig. 61.

3. *Low-Resolution Differential Cross Sections*

The direct combination of the short scattering cell surrounding the end of the first octopole with the second much longer octopole (see Fig. 30 and Fig. 46) allows one to extract information on product velocities with very high sensitivity and in an energy range that is inaccessible in standard crossed-beam experiments. Moreover, by varying the guiding field potential, information on the transverse velocity component can be obtained, and, thus, product angular distributions can be determined. The principle of this method is based on the fact that, under ideal conditions, the projection of the velocity of an ion onto the octopole axis is conserved. Therefore, this velocity component can be measured directly by time-of-flight analysis. The correlation between the transverse energy and the required rf amplitude was discussed in Section III C 4. Indeed, the angular resolution is limited, but several examples have shown that it is sufficient to gain important additional insight into the mechanisms of low-energy ion–molecule reactions.

The scattering process is illustrated schematically in the upper part of Fig. 48. A beam of primary ions with an angular divergence of less than 10° is moving in the axial direction with a velocity v_1. The products formed in reactive collisions with the target gas are scattered through certain angles and move with velocities $\mathbf{v}'_1$, which can be decomposed into parallel and transverse components (v'_{1p}, v'_{1t}). Since the primary ions remain close to the octopole axis, we ignore for simplicity the angular momentum L in Eq. (49). Owing to the axial symmetry of the guiding field, this method integrates over the azimuthal angle φ. This has the advantage of intensity gain and avoids the need for mathematical corrections for the azimuthal acceptance of the detector. Direct information on the forward–backward symmetry of all products is obtained by operating the octopole with a sufficiently high amplitude that all ions are guided. Using a very low guiding field, only the forward–backward cone of products is detected. In general, we measure an entire set of time-of-flight distributions for different rf amplitudes, spaced according to the desired resolution.

The overall resolution is limited by the time spread resulting predominantly from the finite length of the scattering cell and field imperfections of

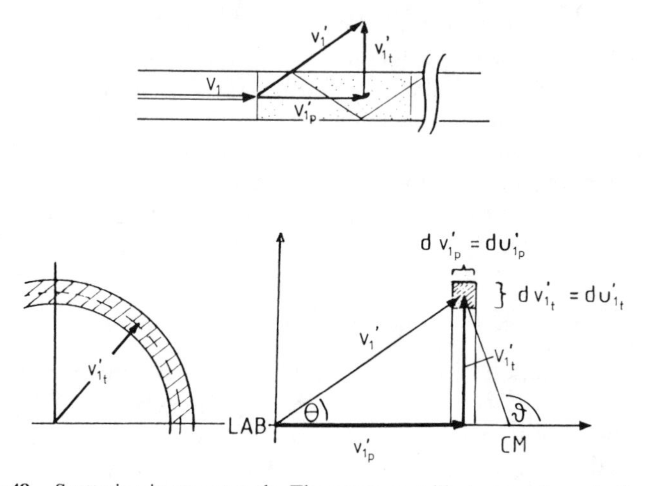

Figure 48. Scattering in an octopole. The upper part illustrates schematically the parallel and the transverse components of the product velocity vector $\mathbf{v}'_1 = (v'_{1p}, v'_{1t})$. The lower part shows the cylindrical ring volume element in the velocity space, needed to define the doubly differential cross section $d^2\sigma/du'_{1p}du'_{1t}$. In the figure on the right, θ and ϑ are the laboratory and center-of-mass scattering angles, respectively.

the octopole. These are partly due to mechanical inaccuracies and the connection between octopoles I and II, but the major perturbations are caused by surface potential distortions. The transverse energy calibration procedure described in Section III C 4 corrects for some of the uncertainty, but presently the resolution is limited to about 100 meV. The resolution in the axial direction depends strongly on the ion–target mass ratio and the energetics of the reaction. The thermal motion of the target gas can also have a significant influence on the product ion kinetic energy distribution. Several related details have been discussed recently in Gerlich (1989a). In order to simplify the following discussion, we assume the target to be at rest.

Since the experimental method differs from the standard differential scattering experiment, a few formulas must be compiled to explain the presentation of our results. The differences are primarily due to the fact that the previously described method is based on a two-dimensional differential detector $(dv'_{1p}dv'_{1t})$, shown schematically in the lower part of Fig. 48, while in a standard scattering experiment, a differential detector that selects in three dimensions is used. For example, in a spherical polar coordinate representation, the detector acceptance is usually given by $(dv'_1 d\Omega) = (dv'_1 \sin\Theta\, d\Theta d\varphi)$. In principle, however, both methods provide the same information as long as the reactants are randomly oriented, since the scattering is cylindrically symmetric with respect to the initial relative velocity.

One therefore usually speaks of a doubly differential cross section and writes $d^2\sigma/d\Omega dv'_1$ instead of $d^3\sigma/d\Omega dv'_1$. We shall use the last expression to emphasize the differences between the two methods.

The number of products detected per unit time within the cylindrical ring depicted in Fig. 48, that is, $\mathbf{v}'_1$ must lie between (v'_{1p}, v'_{1t}) and $(v'_{1p} + dv'_{1p}, v'_{1t} + dv'_{1t})$, is given by (Gerlich, 1989a)

$$\frac{d^2\dot{N}}{dv'_{1p}dv'_{1t}}\Delta v'_{1p}\Delta v'_{1t} = n_1 n_2 d\tau\, g\frac{d^2\sigma}{dv'_{1p}dv'_{1t}}\Delta v'_{1p}\Delta v'_{1t}$$

$$= n_1 n_2 d\tau\, g\frac{d^2\sigma}{du'_{1p}du'_{1t}}\mathfrak{J}\Delta v'_{1p}\Delta v'_{1t}. \tag{95}$$

These equations define the laboratory and center-of-mass differential cross sections, respectively. It can be easily seen from Fig. 48 that for the simple geometry of our detector the appropriate Jacobian is given by

$$\mathfrak{J} = \frac{du'_{1p}du'_{1t}}{dv'_{1p}dv'_{1t}} = 1. \tag{96}$$

As previously mentioned, in a standard differential scattering experiment a differential detector with a laboratory angular acceptance $\Delta\Omega = \sin\Theta\Delta\Theta\Delta\varphi$, positioned at the laboratory angles Θ and φ, is used together with a device for velocity filtering, selecting only products in the interval $(v'_1, v'_1 + \Delta v'_1)$. A typical example of such an experiment for measuring triply differential cross sections will be described in Section IV C. The correlation between the number of products detected per unit time and the differential cross section is then given by (Levine and Bernstein, 1987)

$$\frac{d^3\dot{N}}{d\Omega\, dv'_1}\Delta\Omega\Delta v'_1 = n_1 n_2 d\tau\, g\frac{d^3\sigma(\vartheta, u'_1)}{d\omega du'_1}\mathfrak{J}\Delta\Omega\Delta v'_1. \tag{97}$$

For a chemical reaction with many product states the Jacobian is given by

$$\mathfrak{J} = \frac{d\omega}{d\Omega}\frac{du'_1}{dv'_1} = \frac{v'^2_1}{u'^2_1}. \tag{98}$$

A comparison of the two different methods used to obtain differential cross sections is presented in Fig. 49. In both cases, the single electron charge transfer in $Ar^{2+} + N_2$ and $Ar^{2+} + CO$ collisions, respectively, leads to a narrow Ar^+ intensity ridge surrounding the center-of-mass origin. The upper

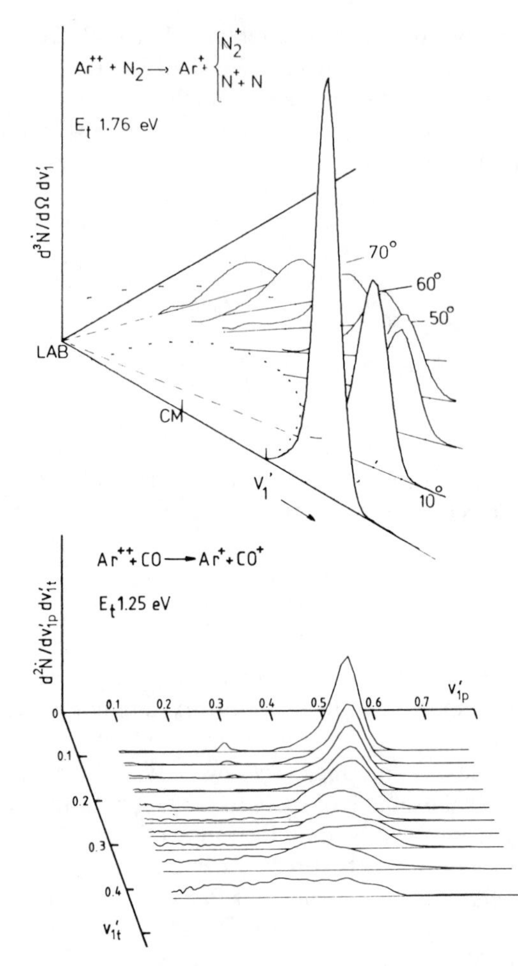

Figure 49. Illustration of product velocity distributions determined by two different experimental methods. The upper panel shows velocity distributions $d^3\dot{N}/d\Omega dv'_1$ [Eq. (97)] for Ar^+ products recorded with a traditional rotating detector at laboratory scattering angles $\theta = 0°–70°$. The velocity distributions $d^2\dot{N}/dv'_{1p}dv'_{1t}$ [Eq. (95)] shown in the lower panel were measured using the guided-ion-beam apparatus.

part of the figure shows results from our differential apparatus with a movable detector (Section IV C). Several Ar^+ time-of-flight distributions have been measured at laboratory scattering angles between $\Theta = 0°$ and $70°$. The results in the lower plot were obtained using our guided-ion-beam apparatus by varying the effective potential. Plotted are the differences of adjacent time-of-flight distributions recorded for a set of properly chosen rf amplitudes.

In order to derive a relation between the two methods, we must divide the measured intensity $(d^3\dot{N}/d\Omega dv'_1)\Delta\Omega\Delta v'_1$ by the velocity space volume element $v_1'^2\Delta v'_1\Delta\Omega$. This leads to the *density of the product flux in the velocity space*

$$D(v'_1, \Theta, \varphi) = \frac{d^3\dot{N}}{v_1'^2 d\Omega dv'_1}. \tag{99}$$

It is common to present $D(v'_1, \Theta, \varphi)$ as a contour map, called the *cartesian map*, which should not be confused with the *flux–velocity contour map* (Levine and Bernstein, 1987). Integration over the azimuthal angle results in a density function $D(v'_1, \Theta)$

$$D(v'_1, \Theta) = \int_0^{2\pi} D(v'_1, \Theta, \varphi) v'_1 \sin\Theta \, d\varphi. \tag{100}$$

It is easy to show that this two-dimensional product flux density is identical with

$$D(v'_{1p}, v'_{1t}) = \frac{d^2\dot{N}}{dv'_{1p} dv'_{1t}}. \tag{101}$$

Note that $D(v'_{1p}, v'_{1t})$ is measured directly in the guided-beam experiment without moving any mechanical parts and without any mathematical manipulation of the data. Figure 50 shows the distributions, depicted in the lower part of Fig. 49, as a contour plot of $D(v'_{1p}, v'_{1t})$, emphasizing that the mean value of the translational exoergicity is independent of the scattering angle. This graphical presentation has the further advantage of showing isotropic scattering, that is, $d\sigma/d\vartheta = \text{const}$, as circles concentric relative to v_c. In the depicted example, the forward-peaked distribution actually is due to the reaction dynamics. The more common presentation of such data in the form of a three-dimensional product flux density $D(v'_1, \Theta, \varphi)$, has the disadvantage of superimposing at $\Theta = 0°$, a $1/\sin\Theta$ singularity for an ideal detector.

Less detailed distributions can be readily derived from this distribution by integrating over a corresponding subspace. For example, integration over the center-of-mass scattering angle within the adequate velocity intervals leads to translational exoergicity distributions. Angular distributions are derived by integration over v'_1. Finally, integration over v'_{1p} and v'_{1t} leads to integral cross sections that are, of course, also measured directly. This method

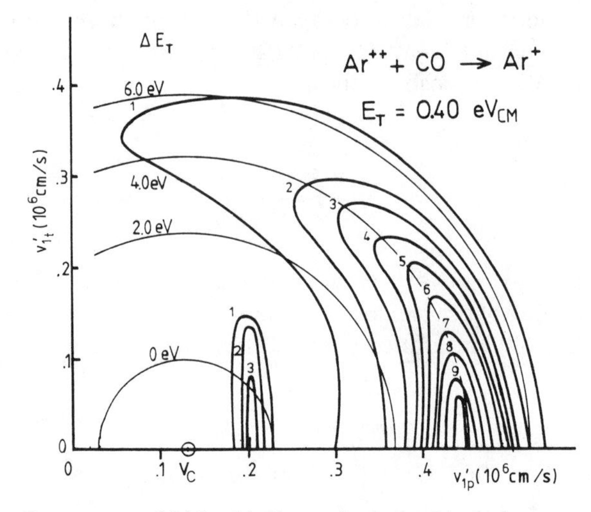

Figure 50. Contour map of $D(v'_{1p}, v'_{1t})$. The results depicted in the lower part of Fig. 49 are plotted here as Ar^+ equiflux contour lines. Shown is the two-dimensional flux as defined by Eq. (101). The circles mark loci of constant translational exoergicity ΔE_T, the most probable value being $\Delta E_T \sim 4$ eV. The small peak close to v_c is due to a small number of Ar^+ ions coming directly from the ion source.

is therefore uniquely capable of determining precise absolute values for differential cross sections.

4. *Combinations with Optical Methods*

The universal guided-ion-beam apparatus depicted in Fig. 46 can also be used in combination with different optical methods, as mentioned briefly in Section IV A. For example, photons created in the scattering cell, either by chemiluminescence or by laser-induced fluorescence, can be detected by a photomulitiplier via a quartz window. Two additional windows at the ion source and the mass spectrometer allow one to pass a laser beam coaxially through the entire apparatus. The narrowest beam restriction is usually dictated by the octopole entrance electrode, which has a typical diameter of 3 mm. For some applications, for example, for the multiphoton ionization source described in Section III B 3, the laser beam can also be directed transverse to the apparatus axis.

The instrument has been used to study chemiluminescent ion–molecule reactions at low collision energies. With an appropriate imaging system and optical filters, different wavelength ranges could be selected. Absolute integral cross sections for photon emission were determined by calibrating the optical system (Kaefer, 1984) based on known integral cross sections. The geometrical restrictions due to the octopole rods resulted in a light transmission of only a few percent. Nonetheless, the sensitivity was high enough to detect the photons in coincidence with mass-selected product ions. This simplified the observation of weak luminescent product channels in the presence of strong emitters and was also used to identify spectroscopically unknown emitters. Operating with a continuous ion beam and using the photon as start signal for a multichannel scaler, time-of-flight distributions of chemiluminescent products can be recorded. In this way, one can identify cascading transitions, since information is obtained simultaneously on the ion kinetic energy and the photon energy.

In addition to single-photon ionization and resonance-enhanced multiphoton ionization, there are several other schemes that can be used to prepare primary ions in selected states. We have demonstrated that the guided-ion-beam apparatus can be used to study reactions with electronically excited ions provided their lifetimes are at least in the μs range. In this case, the laser excites the ground-state primary ions during their passage through the scattering cell prior to reaction with the neutral collision partner. It is also conceivable to modify in a controlled manner the internal state population of the reactant ions created in the electron bombardment source by methods such as optical pumping, photofragmentation, or infrared excitation.

Laser methods can also be used for state-selective detection of product ions. We have explored the possibilities of laser-induced fluorescence (Scherbarth, 1988) using thermal CO^+ ions produced in the storage source. Unfortunately, with the present setup, the signal-to-noise ratio is only satisfactory if we store about 10^5–10^6 ions between the octopole ring electrodes, and this method is therefore not sensitive enough to determine nascent product distributions. We have, however, developed a very efficient state-detection scheme based on laser-induced predissociation of product ions. In contrast to the fluorescence photons, the photofragments can be formed throughout the entire length of the octopole and are confined and detected with nearly 100% efficiency. This method was used successfully for the study of O_2^+ charge-transfer products formed in the *a*-state (Scherbarth and Gerlich, 1989). Detecting the absorption of a laser photon via fragmentation has the additional advantage that one can operate in the regime of saturation. Extrapolation to total fragmentation results in a signal that is independent of optical transition probabilities. As an illustration, Fig. 51 shows a small portion of an O_2^+ fragment spectrum for three different laser pulse energies.

Predissociation of a molecular ion with a single photon is not a generally applicable method. Therefore, we are presently exploring the possibility of fragmenting product ions in subsequent steps with two photons. State

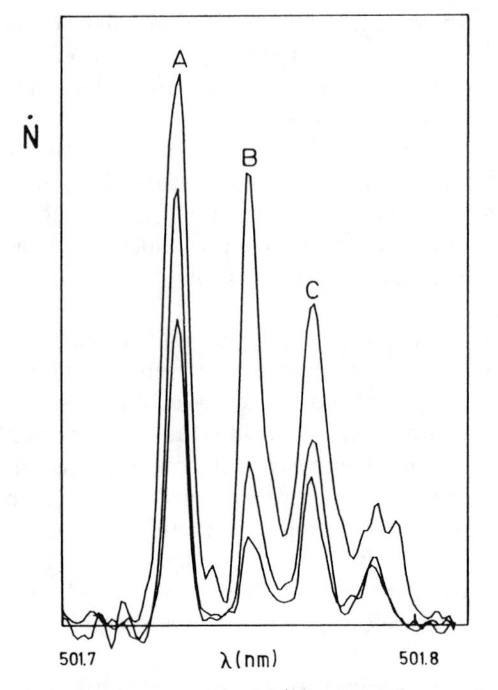

Figure 51. Laser fragmentation spectrum of O_2^+ for three laser energies. The spectra were recorded at pulse energies of 0.1 mJ, 0.2 mJ, and 1.7 mJ. Note that line *A* is strongly saturated. The energy necessary to reach 63.2% fragmentation is different for each of the depicted transitions, 0.13 mJ for *A*, 0.52 mJ for *B*, and 0.21 mJ for *C*.

selectivity is obtained in the first bound–bound transition and can be followed by any transition to a dissociative state.

C. Differential Scattering Apparatus

Over the past several years, significant progress has been made in experimental studies of high-resolution angular and velocity distributions of scattered product ions. Owing to continued improvements in energy analyzers and other instrument details, a kinetic energy resolution on the order of 10 meV is now possible. The chapter by G. Niedner-Schatteburg and P. Toennies on doubly differential proton scattering gives some insight into the present state of such high-resolution instruments. For chemical systems with closely packed rovibrational states, single-state resolution can only be achieved by optical methods. Nonetheless, angular and medium resolution velocity distributions are needed and provide important insight into the reaction dynamics, as can be seen from J. Futrell's chapter on crossed-molecular-beam studies of ion–molecule reactions.

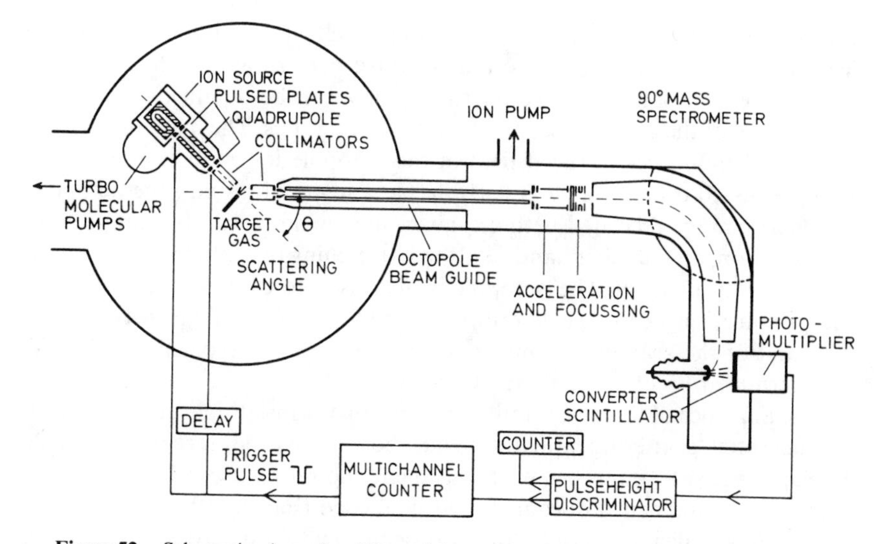

Figure 52. Schematic view of a differential scattering apparatus consisting of a rotating storage ion source and quadrupole, a scattering region, and a 1-m-long octopole for time-of-flight analysis of reaction products (Gerlich, 1977; Teloy, 1978).

In this section we present a differential scattering apparatus that employs several of the rf devices described in Section III. This apparatus, depicted in Fig. 52, has been characterized in detail in Gerlich (1977), and a short description can also be found in (Gerlich and Wirth, 1986; Teloy, 1978). Preparation of the primary ions is performed similarly to that described in Section IV B 1, using a storage ion source in combination with a focusing quadrupole and time-of-flight selection. With two 1-mm-wide collimation slits, the beam divergence is limited to 1°. Operating all elements at a low voltage avoids deceleration and facilitates the formation of an ion beam with a low laboratory energy. A typical lower limit is 0.2 eV, below which the angular profile of the beam becomes significantly broadened. The laboratory scattering angle Θ is varied by rotating the entire source. The accessible range is between $-20°$ and $+70°$ when a small scattering cell is used. In the case of a target gas beam, scattering angles up to $+100°$ can be reached. A second two-slit collimator restricts the angular acceptance of the detector to 3°. Surface potential distortions are most critical in the region of the two collimators and the interaction volume. These distortions are reduced by coating all electrodes with graphite and by baking this region with infrared radiation.

The energy of the scattered products is determined by recording their flight times through a 1-m-long octopole with a fast multichannel scaler. The use of an octopole guiding field results in a velocity analyzer with unique features. Since the ions are allowed to move very slowly in this region, a kinetic energy resolution of better than 5 meV can be obtained, even if the time spread of the primary beam amounts to several μs. The main distortions are due to influences of the rf field in the octopole injection region, as has been discussed in Section IV B 1. By accelerating the ions into the octopole, the energy range recorded within a given time window can be compressed. Deceleration, on the other hand, leads to a stretching of the time scale without a loss in ion intensity. The octopole is also used to calibrate the energy of the primary ion beam, as described in Section III C 3. Shifts between the potential of the scattering center and the octopole are typically ± 50 meV, as determined by other methods (Gerlich, 1977).

The influence of target motion on the product translational energy in such a differential scattering apparatus has been discussed recently (Gerlich, 1989a). Ideally, a beam–beam arrangement should be used. To obtain fast product ions the neutral beam should be seeded (Farrar, 1988). In order to achieve low collision energies, the two beams should be crossed at a small angle. In most cases, either for intensity reasons or experimental limitations, it is necessary to use a scattering cell or an effusive beam. In general, the thermal motion of the neutrals deteriorates the product energy resolution, with the magnitude of the broadening depending on the reaction system.

Using the described apparatus, we have obtained very detailed differential cross sections for proton deuteron exchange reactions in $H^+ + D_2$ collisions (Gerlich, 1977). By cooling the scattering cell to liquid nitrogen temperature and by taking advantage of the favorable kinematic conditions in this reaction, we have achieved an overall energy resolution sufficient to record partially resolved rovibrational distributions. Figure 53 gives an impression of the attainable energy resolution. It demonstrates also the long-time stability of the apparatus, which is very important for the study of reactive processes due to the long integration times required. A brief survey of the $H^+ + D_2$ results has been presented in Teloy (1978) and another example will be given in Section V B 3. The apparatus has also been used to study ortho–para transitions in $H^+ + H_2$ collisions, with hydrogen in different $j = 0$ and $j = 1$ mixtures at 80 K (Gerlich and Bohli, 1981). An analysis of these results based on a dynamically biased statistical model has been reported recently (Gerlich, 1990). Lower-resolution velocity distributions and flux–velocity contour maps have also been measured for reactions of N^+ with O_2 and CO (Gerlich and Wirth, 1986; Wirth, 1984).

D. Merged-Beam Apparatus

The first beam method capable of reaching meV collision energies was the traditional merged-beam apparatus pioneered by Trujillo et al. (1966) and further developed by Gentry et al. (1975). This method achieved very low

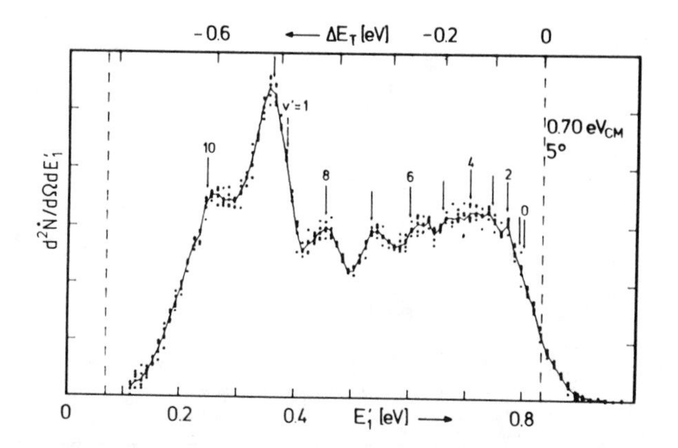

Figure 53. Laboratory energy distributions of D^+ ions produced in reactive $H^+ + D_2 \rightarrow D^+ + HD$ collisions (collision energy $0.70\,eV_{cm}$, scattering angle 5°_{lab}). The partially resolved structure of the translational spectrum reflects excitation of rotational states of $HD(v' = 0, j')$ products, with the rise below $E'_1 = 0.4\,eV$ due to contributions from $v' = 1$. The upper scale indicates the translational exoergicity ΔE_T. To demonstrate the overall stability of the instrument, five distributions are plotted, each accumulated over a period of 1 h. The line connects the average values of these measurements.

relative energies by merging two fast beams with parallel velocity vectors having the same magnitude. Operating at keV laboratory energies, kinematic compression allowed meV collision energies to be reached, even with beam energy spreads of several eV. Owing to these high energies, it was necessary to create the fast neutral target beam by charge exchange neutralization of an ion beam. Unfortunately, this led to a number of technical difficulties such as low reactant densities, a small and uncertain interaction volume, and a short interaction time. Other problems were due to the critical dependence of the energy resolution on the angular spread and the unknown internal excitation of the reactants, especially for molecular targets. All these drawbacks have impeded widespread use of this method. In this section, we describe a recently developed merged-beam instrument (Gerlich, 1989c; Kalmbach, 1989), where a slow ion beam is guided along the axis of a skimmed supersonic neutral beam. Despite the low laboratory energies involved, the kinematic compression resulting from the merged-beam geometry leads to surprising results. For advantageous systems, such as heavy ions and light targets, collision energies below 1 meV with corresponding resolution are obtainable.

1. *Description of the Apparatus*

A schematic diagram of the slow merged-beam apparatus is depicted in Fig. 54. For creation of the supersonic beam we use a piezoelectric driven pulsed valve (Gerlich et al., 1991). The fundamental principle is similar to

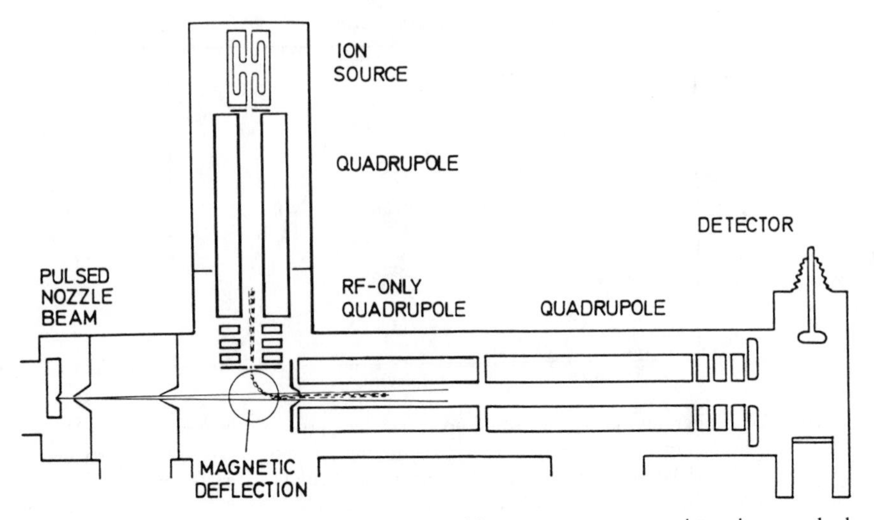

Figure 54. Schematic diagram of the merged-beam apparatus, superimposing a pulsed supersonic beam and a slow guided-ion beam. The ions are prepared in a storage source and merged into the neutral beam by deflection with a weak magnetic field. The interaction takes place in the weak guiding field of an rf-only quadrupole. Primary and product ions are detected via a quadrupole mass spectrometer.

that described by Cross and Valentini (1982); however, our valve can create pulses shorter than 20 μs and operate at repetition rates above 2 kHz. Two differential pumping stages are used to maintain a sufficiently low target-gas background pressure in the interaction chamber, while the density of the beam itself is on the order of $10^{11}\,cm^{-3}$. Owing to the limited pumping speed, the beam intensity has to be rather weak and the valve is typically operated with a pulse width of 40 μs at 100 Hz. The use of a nozzle beam leads to cold molecules with a well-defined internal energy, for example, it is possible to prepare a beam of pure ground-state hydrogen molecules. The well-collimated neutral beam has a typical angular divergence of 2°.

The established combination of a storage ion source and focusing quadrupole is used to create an ion beam that is pulsed synchronously with the neutral beam. The mass- and energy-selected ions are deviated 90° in a weak magnetic field of less than 1 kG and then injected at the desired energy into a weak rf guiding field. There, they run along with the neutral beam and can react at low relative energies. At present, a quadrupole ion guide is preferred, since the harmonic effective potential confines the ion beam so close to the axis that full overlap with the neutral beam is guaranteed. Calibration of the ion axial energy, E_1, and determination of the actual energy distribution is performed using the method described in Section III C 3. To measure the mean velocity $\langle v_2 \rangle$ and the spread of the neutral beam (FWHM, Δv_2), an additional differentially pumped ionization detector is mounted subsequent to the ion detector at a distance of about 1 m from the nozzle. In this arrangement we can operate routinely with ion laboratory energies down to 50 meV, and zero nominal relative velocity can be obtained for many mass ratios. In some cases, the neutrals must be seeded to operate at higher laboratory velocities.

Both primary and product ions are guided to the quadrupole mass filter and detected with a standard scintillation detector. In this experiment, the directed center-of-mass motion helps to reduce discriminating effects. Formation of products outside the quadrupole interaction region leads to an undesirable background signal. Contributions occurring in the short overlapping region in the magnetic deflection field and the injection electrode can be reduced by running the beams in these regions with a relative velocity at which the rate coefficient has a local minimum. Contributions from reactions in the quadrupole mass filter are usually small since it operates in the mass discriminating mode and at high axial energies between 5 and 20 eV; however, it causes some problems for endothermic reactions. Using very short neutral pulses (20 μs), and by properly adjusting the time overlap, these undesired reactions can be further reduced.

The reaction $H_3^+ + D_2 \rightarrow D_2H^+ + H_2$ has been studied by means of the traditional merged-beam technique (Dougless et al., 1982) and therefore offers an opportunity to compare the two experimental methods. The energy dependence of the integral cross section for this exothermic proton transfer is depicted in Fig. 55. Both sets of results show in good agreement that the reaction occurs with a cross section close to the Langevin limit. It is interesting to make a few quantitative comparisons of the two methods. In order to obtain a relative energy of 10 meV, we use a supersonic D_2 beam with a velocity $v_2 = 1.8$ km/s and an H_3^+ ion energy $E_1 = 0.13$ eV, in contrast to two fast beams with energies of 4427 eV and 4409 eV. The advantage of our experimental arrangement with respect to the density of the neutral target and the interaction time is clearly evident. Typical densities in our experiment are 3×10^{11} cm^{-3} compared to 10^6 cm^{-3}, and the interaction time is 35 μs rather than 0.27 μs. With these typical values and a reaction rate coefficient of $k = 10^{-9}$ cm^3/s, reaction probabilities of 10^{-2} and 10^{-10} are obtained for our slow merged-beam and the traditional merged-beam apparatus, respectively. To achieve a relative energy resolution of $\delta E_T/E_T = 0.3$ at $E_T = 10$ meV, the maximum tolerable angular divergences in the two instruments are 28° and 0.06°, respectively.

The minor deviation in the slopes of the cross sections in Fig. 55 may

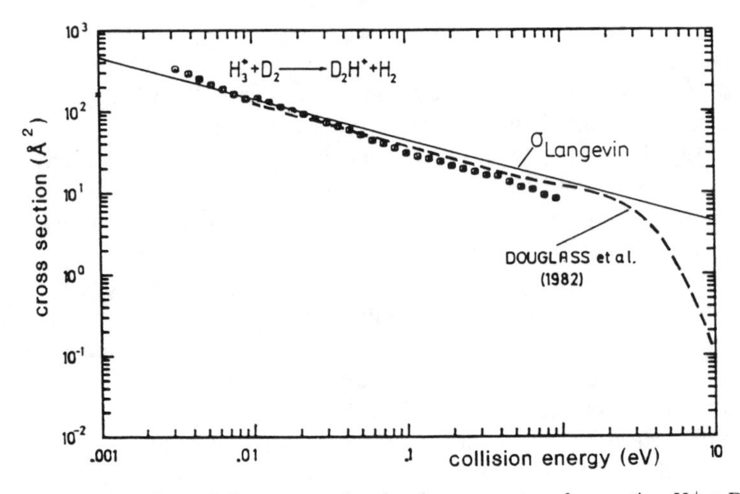

Figure 55. Comparison of the cross section for the proton transfer reaction $H_3^+ + D_2 \rightarrow D_2H^+ + H_2$ obtained with a conventional (dashed line) (Dougless et al., 1982) and our new slow merged beam apparatus (●).

be due to differences in the internal excitation of the D_2 target. In our experiment we use internally cold neutrals, while in the other, it is most probable that vibrationally excited D_2 is present. A related problem is the internal excitation of the ionic reactants. This becomes a real problem for the isotopically equivalent reaction $D_3^+ + H_2 \rightarrow H_2D^+ + D_2$, which is 29 meV endothermic. Figure 56 shows cross sections for this reaction using two different operational modes of our storage ion source. It can be seen that thermalization of the D_3^+ ions to the temperature of the ion source (~ 350 K) reduces the cross section. Nonetheless, the internal excitation of the reactants is larger than the endoergicity of the reaction. In order to observe the threshold onset, indicated schematically by the dashed line in Fig. 56, sufficiently cold ions have to be prepared.

To precisely determine the absolute integral cross section, the spatial overlap of the two merged beams and their local densities must be known. Here, the situation is simplified, since the ion beam is squeezed into the neutral beam and the neutral density n_2 is a simple function of z, the distance from the nozzle. Replacing in Eq. (92) the ion density n_1 by the ion flux $\dot{N}_1 = n_1 v_1 \Delta a$, where Δa is the cross section area of the ion beam, and substituting the volume element $d\tau$ by $\Delta a dz$, leads to the z-dependent differential product ion flux

$$d\dot{N} = \dot{N}_1(z)\frac{\langle g \rangle}{v_1}\sigma_{\text{eff}} n_2(z) dz. \tag{102}$$

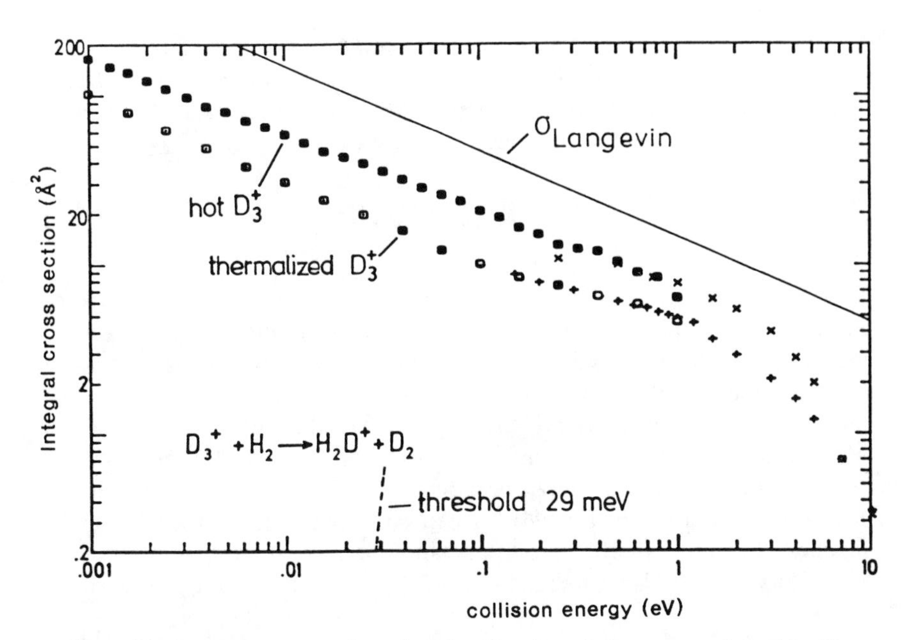

Figure 56. Integral cross sections for the 29 meV endothermic reaction $D_3^+ + H_2 \rightarrow H_2D^+ + D_2$ with hot D_3^+ (~ 2 eV internal energy) and ~ 350 K thermalized D_3^+. The merged beam results (●, ⊙) are, in the overlapping energy range, in good agreement with earlier guided ion-beam experiments (Piepke, 1980), recorded under similar storage ion source conditions (+, ×). The cross sections are significantly lower than the Langevin value.

For weak attenuation of the primary beam, we can use in good approximation $\dot{N}_1(z) \sim \dot{N}_1$, and the effective cross section can be simply calculated from

$$\sigma_{\text{eff}} = \frac{\dot{N}}{\dot{N}_1} \frac{1}{\langle n_2 \rangle L} \frac{v_1}{\langle g \rangle}, \tag{103}$$

where $\langle n_2 \rangle$ is the average density determined by integration of $n_2(z)$ from the beginning of the interaction region, z_0, to the end, $z_0 + L$

$$\langle n_2 \rangle = L^{-1} \int_0^{z_0 + L} n_2(z) dz. \tag{104}$$

The value $\langle n_2 \rangle L$ has been determined experimentally using reactions with known integral cross sections such as $Ar^+ + H_2$, D_2, and O_2.

2. *Kinematic Considerations*

The kinetic energy resolution of our merged-beam experiment can be characterized by three fixed parameters, the energy spread of the ion beam

ΔE_1, the velocity spread of the supersonic beam Δv_2, and the relative angular divergence between the two beams $\Delta\Lambda$. In general, we can assume rotational symmetry, and thus, the collision energy of the two reactants, colliding with velocities v_1 and v_2 under an angle Λ, can be expressed by

$$E_T = \tfrac{1}{2}\mu(\mathbf{v}_1 - \mathbf{v}_2)^2 = \tfrac{1}{2}\mu(v_1^2 + v_2^2 - 2v_1 v_2 \cos\Lambda). \tag{105}$$

As an illustrative example of the kinematic shifts caused by this vector addition, we compare in Fig. 57 guided-ion-beam results measured using a scattering cell and a crossed beam (Gerlich et al., 1987), with results obtained using the merged-beam arrangement. Shown is the threshold onset of the endothermic reaction $C^+ + H_2 \rightarrow CH^+ + H$ plotted as a function of the ion laboratory energy E_1. In the case of the scattering cell, the thermal random target motion causes an energy broadening of 300 meV, as can be derived from the distribution in Eq. (94). Crossing the two beams at a right angle significantly improves the resolution, and for $\Lambda = 90°$, the third term in Eq. (105) vanishes. Using merged beams, the rise of the product intensity is

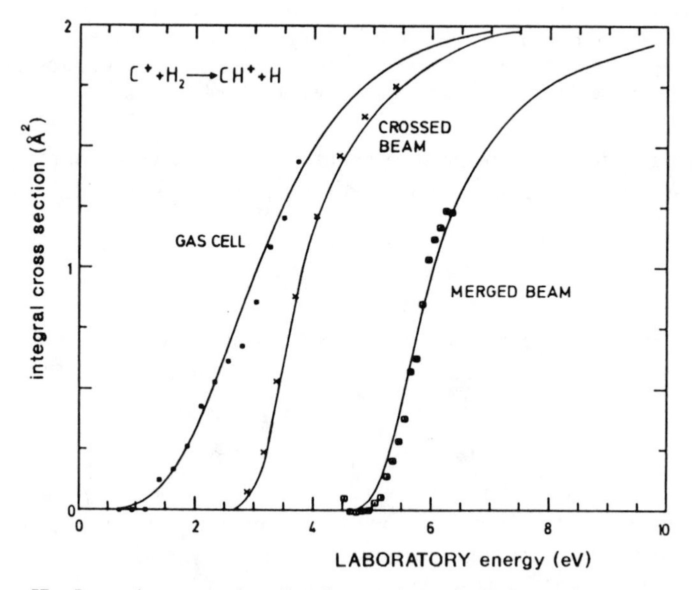

Figure 57. Integral cross sections for the reaction $C^+ + H_2 \rightarrow CH^+ + H$, measured in a beam gas cell, a crossed beam, and a merged-beam arrangement. In order to demonstrate the kinematic shift and energy broadening, the different results are plotted as a function of the C^+ laboratory energy.

again steeper and significantly shifted, that is, the laboratory energy of the C^+ ion has to be increased above $5\,eV_{lab}$ in order to reach the threshold. This is due to the fact that in the merged-beam arrangement the nominal value of Λ is $0°$ and the collision energy is reduced by $\mu v_1 v_2$.

To obtain better quantitative insight into the kinematic relations, we can calculate the broadening of the relative energy from Eq. (105) using error propagation:

$$\delta E_T = \left[\left(\frac{\partial E_T}{\partial v_1} \Delta v_1 \right)^2 + \left(\frac{\partial E_T}{\partial v_2} \Delta v_2 \right)^2 + \left(\frac{\partial E_T}{\partial \Lambda} \Delta \Lambda \right)^2 + \left(\frac{1}{2} \frac{\partial^2 E_T}{\partial \Lambda^2} \Delta \Lambda^2 \right)^2 \right]^{1/2}. \tag{106}$$

In the crossed-beam arrangement, a large contribution to δE_T is the angular divergence of the guided ion beam, entering in Eq. (106) for $\Lambda = 90°$ via the first-order term $(\partial E_T/\partial \Lambda)\Delta\Lambda$. In the depicted $C^+ + H_2$ example, one degree deviation from $90°$ causes an energy shift of $5\,meV$ at $E_T = 0.4\,eV$ (Gerlich et al., 1987). Merging the two beams, the same resolution is obtained with a much larger angular divergence of $\pm 8°$. For this geometry of parallel velocities, that is, for $\Lambda = 0°$, $\Delta\Lambda$ contributes only in second-order and we obtain from Eq. (106)

$$\delta E_T = \left[\left(\frac{\mu g}{m_1 v_1} \Delta E_1 \right)^2 + (\mu g \Delta v_2)^2 + \left(\frac{\mu}{2} v_1 v_2 \Delta \Lambda^2 \right)^2 \right]^{1/2}. \tag{107}$$

This formula can be used to get a quick estimate of the energy resolution based on the individual contributions and to determine the lowest relative energy attainable with the merged-beam arrangement. As an example, we assume for the system $N^+ + H_2$ an ion energy half width $\Delta E_1 = 50\,meV$ with an angular spread $\Delta\Lambda = 8°$, and a well-collimated supersonic hydrogen beam characterized by $v_2/\Delta v_2 = 20$ and $\langle v_2 \rangle = 2.7\,km/s$. A nominal relative energy of $E_T = 1\,meV$ is obtained for $E_1 = 0.57\,eV$, and at that energy, Eq. (107) predicts $\delta E_T = 1.4\,meV$ with equal contributions of $0.8\,meV$ for the three terms.

At low relative energies such as these it is more accurate to determine numerically the distribution of the relative velocity $f(g)$ from Eq. (89). The distributions $f_1(\mathbf{v}_1)$ and $f_2(\mathbf{v}_2)$ are determined experimentally, or can be approximated by Gaussians. Owing to the rotational symmetry, the six-dimensional integral over $(\mathbf{v}_1, \mathbf{v}_2)$ can be replaced by an integration over (v_1, v_2, Λ) as derived in detail in Gerlich (1989a). This leads to rather asymmetric velocity distributions at low collision energies. A few typical results are depicted in the lower part of Fig. 63.

E. Temperature-Variable Ion Trap Apparatus

The liquid-helium-cooled ring electrode trap, described in Section III E, is integrated into a complete instrument (Gerlich and Kaefer, 1987, 1989; Gerlich et al., 1990), as depicted in Fig. 58. The ion source and the reaction region are differentially pumped with two turbomolecular pumps, resulting in a residual gas pressure of typically 10^{-9} mbar. At low temperatures, cryopumping leads to a significant improvement of the vacuum in the region of the ion trap. Typical applications using this apparatus include the determination of very slow binary (10^{-17} cm^3/s) and ternary rate coefficients as a function of temperature, the nominal value of which can be varied between 10 K and 350 K. In addition, we are especially interested in radiative association of small collision complexes, both in its spontaneous and laser induced form. For this purpose, as well as other laser applications, the system is transparent in the axial direction.

1. *Description of the Apparatus*

As in the previously described instruments, ions are formed by electron bombardment in an rf storage ion source, prethermalized to the source temperature, mass selected in the quadrupole, and then injected at low energies into the trap. Filling of the trap is achieved by properly setting or pulsing the voltages at the input and exit-gate electrodes, and by retarding and capturing the ions via inelastic collisions. Preparation of the primary ion kinetic energy is not very critical in this experiment; however, high mass discrimination is required to eliminate traces of product ions in the primary

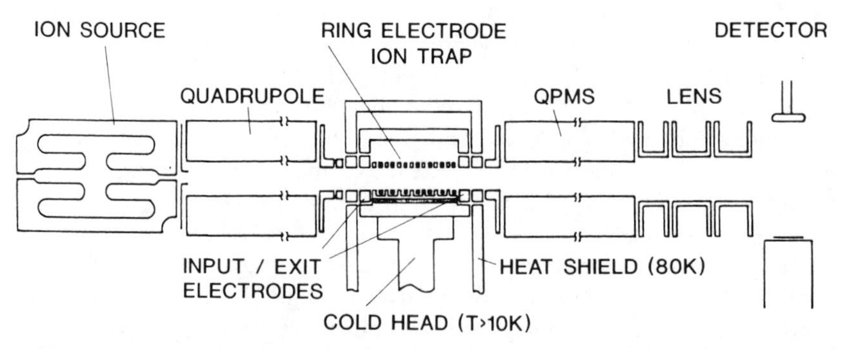

Figure 58. Schematic diagram of the ring electrode trap apparatus. Primary ions are created in a storage ion source, mass selected in a quadrupole, and injected into the trap by pulsing the input electrode. After a certain storage time (μs–min), the exit-gate electrode is opened, and the ions are mass analyzed with the quadrupole filter and detected with a scintillation detector. The temperature of the trap, which is mounted onto a cold head, can be varied between 10 K and 350 K.

ion beam. Filling times are between μs and ms and lead to typical ion densities ranging from 10^3 to 10^5 ions/cm^3. In the trap, the ions undergo many collisions with the target gas or an added buffer gas, usually He, thereby assimilating their translational and internal energy to the temperature of the trap environment. In certain cases, for example, for an endothermic reaction, an undesirable number of product ions is often formed during the filling and thermalization period. Provided these products are heavier than the primary ions, one can make use of the mass dependence of the effective potential and purify the contents of the trap. This method has been used very efficiently to reduce the H_3^+ background in the study of the radiative association reaction $H^+ + H_2 \rightarrow H_3^+ + h\nu$ (Kaefer, 1989; Gerlich et al., 1990).

After a selected storage time (μs–min), both the primary and product ions are ejected from the trap, mass analyzed, and detected with a scintillation detector. In the present apparatus, the acceptance and transmission of the detection system is less critical than in beam instruments, since the majority of the product ions are thermalized before they are extracted. This reduces the risk of discrimination in detecting the primary or product ions, which, in beam experiments, often occupy very different phase spaces. Nonetheless, it is necessary that the detection efficiency be independent of the storage time and that one always takes the same representative probe of the trap contents. For example, the change of the phase space of the injected ion cloud during thermalization can lead to a slight increase in the number of ions that reach the detector.

Opening the exit electrode usually results in an intense initial pulse of ions, and one has to be aware of the dead time of the counting system (200 MHz). The initial pulse is followed by a tail of slowly exiting ions that are delayed probably due to trapping by potential distortions. The emptying time can be significantly shortened by distributed acceleration of the ions using field penetration electrodes, as described in Section III E. In the case of endothermic reactions such as $C^+ + H_2 \rightarrow CH^+ + H$, one must take care that acceleration of the ions does not lead to additional formation of products. Even after a long opening time, it is conceivable that some ions remain in the trap. Therefore, it is advisable to completely empty the trap by switching off the rf amplitude for a short time prior to starting a new cycle.

The process of filling, storing, and extracting is repeated iteratively, with the delay time between injection and extraction being switched between a set of selected values. This delay defines the interaction time, which can be varied between μs and minutes. At the shortest reaction times the trap functions as a scattering cell, and the ions pass directly through as in the usual guided-ion-beam apparatus. The longest storage times are usually limited by primary ion loss due to reaction with the background gas or by loss due to operating at too low of an rf frequency and/or amplitude. Examples demonstrating the range of different time constants are shown in Fig. 42. Since in most cases reactions with slow rates are studied, the repetition rate is typically 1–100 Hz; in some applications it is dictated by a laser (10–30 Hz).

Extracting the ion cloud for detection has the advantage of very high sensitivity and a large dynamic range; however, with such a detection technique one loses the thermally prepared ensemble of primary ions, which, in the case of slow reactions, is almost unattenuated. A better detection scheme would be to observe product ion formation while conserving the primary ions. In principle this could be achieved by monitoring the trap contents via laser-induced fluorescence, or the products by laser-induced fragmentation, as described in Section IV B 4. Eestimates show, however, that the overall sensitivity of these techniques is inferior to our destructive detection method. Another approach would be to select ions based on the energy dependence of the trapping field. For favorable exothermic reactions, the gained kinetic energy could allow the products to escape the trap over a properly tuned potential barrier at the exit electrode, without losing the stored primary ions. An example illustrating this principle will be given in Section V A. Other possibilities, such as the use of a mass selective rf barrier at the outlet, are also conceivable.

2. *Determination of Rate Coefficients*

Reaction rate coefficients are determined by fitting the parameters of an appropriate rate equation system to the measured temporal change of the ion composition. In a simple case, as shown for $H^+ + D_2$ in Fig. 42, a two-channel model is sufficient. Using Eq. (92) and integrating over the storage volume leads directly to the differential equation for the attenuation of the primary ions

$$\frac{dN_1}{dt} = -k_{\text{eff}} n_2 N_1(t), \tag{108}$$

that is, $N_1(t)$ decreases exponentially with the time constant

$$\tau = (k_{\text{eff}} n_2)^{-1}, \tag{109}$$

while the number of product ions, $N_2(t)$, augments complementary. For a low conversion efficiency, the increase of products can be approximated by a linear dependence

$$N_2(t) = N_1 \frac{t}{\tau}. \tag{110}$$

A slight decrease in the total number of ions, $N_1(t) + N_2(t)$, as shown in the middle panel of Fig. 42, can be accounted for by introducing a global time constant for ion loss. In many situations, this loss is predominantly due to additional, unrecorded product channels, for example, in the lowest panel of Fig. 42, to the formation of N_2^+ and other background gas ions. Other extensions of the rate equation system are needed if the product ions undergo further reactions, as, for example, in the reactions of H_3^+ depicted in Fig. 44. For product ions formed in very low abundance, such as from radiative association, it is necessary to determine their rate of loss in the trap in a separate experiment. For this purpose, the product species is directly prepared in the ion source in a sufficiently large amount, and its residence time in the trap is determined under otherwise identical storage conditions. In most cases of interest, very stable products are formed and reconversion to reactants by collision-induced dissociation can be neglected. For weakly bound systems, for example, ionic clusters, the rate equations have to account for forward and backward reactions.

To determine k_{eff} from Eq. (109) or Eq. (110), the density n_2, which can be varied between $10^6\,cm^{-3}$ and $10^{15}\,cm^{-3}$, must be known. At low densities, n_2 is determined by injecting primary ions, which react with the target gas at a known, fast, and temperature independent rate coefficient. One example is the reaction of He^+ with N_2 (Fig. 42). To determine the density of H_2 and D_2, we use the reaction with Ar^+. At higher densities, the target pressure is determined with a Leybold VISCOVAC VM 210, connected via a short tube to the interior of the trap. The gauge itself operates at room temperature and is calibrated with a quoted accuracy of 5%. The density n_2 is calculated from the indicated pressure p (mbar) and the trap temperature T(K) using the relation

$$n_2 = 4.2 \times 10^{17} p T^{-1/2}\,cm^{-3}. \tag{111}$$

The influence of thermotranspiration (Miller, 1963) can be neglected. At very low temperatures, the attainable density is limited by the vapor pressure of the target gas. For this and other reasons, we use He has a buffer gas. In experiments employing H_2, the temperature is maintained above 25 K to prevent condensation. One of the problems of the presently used system is the large entrance and exit apertures, which result in a rather steep density gradient of the target gas from the center toward the ends of the trap. To determine absolute rate coefficients, the ion cloud is concentrated in the middle of the ion trap where the pressure is actually measured. This is achieved by using the correction electrodes described in Section III E. The overall attainable accuracy of the absolute values is better than 20%.

Finally, to calculate the intrinsic reaction cross section $\sigma(g)$ from the measured temperature dependence of the rate coefficient $k_{eff}(T)$ using Eq. (93), the distribution of the relative velocity $f(g)$ must be known. Ideally, both ions and neutrals are in thermal equilibrium in the trap at a common temperature T. Hence, $f(g)$, as defined by Eq. (89), is a Maxwell–Boltzmann distribution

$$f(g) = f_M(g; \mu, T) = (4/\pi^{1/2})(\mu/2kT)^{3/2} g^2 \exp\left(-\frac{\mu}{2kT} g^2\right). \qquad (112)$$

If the motion of the ions and the neutrals is characterized by different Maxwellians at temperatures T_1 and T_2, Eq. (112) still holds for a collision temperature given by

$$T = \mu(T_1/m_1 + T_2/m_2). \qquad (113)$$

The problem of determining the actual temperature in the trap has already been discussed in Section III E, while the influence of the rf field on the ion energy has been treated in Section II E 2.

3. *Association Rate Coefficients*

The high sensitivity and efficiency of the ion trap apparatus enables one to study association reactions at rather low densities. Under typical operating conditions of swarm experiments, the buffer-gas density is so high that ternary association reactions prevail. At the other extreme of very low densities, for example, conditions of interstellar space, stabilization of the collision complex can only occur via spontaneous emission of a photon. Denoting the ternary rate coefficient as k_3 and the radiative association rate coefficient as k_r, the apparent second-order rate coefficient is given by

$$k^* = k_r + n_2 k_3. \qquad (114)$$

This relation shows that to determine both k_r and k_3 it is advisable to vary n_2 in a range where the contributions of both processes have the same order of magnitude.

An example that shows the influence of temperature and target density on k^* is depicted in Fig. 59. The ternary rate coefficient k_3 for the association reaction $H_3^+ + 2H_2 \rightarrow H_5^+ + H_2$ increases by a factor of 2.5 as the nominal temperature is lowered from 80 K to 25 K. The absolute value $k_3(80\,K) = 2.8 \times 10^{-29}\,cm^6/s$ is in very good agreement with an earlier obtained value of $2.5 \times 10^{-29}\,cm^6/s$ using our liquid nitrogen cooled trap (Kaefer, 1989). In this example, extrapolation to zero density, such that $k^* = k_r$, indicates that the radiative association rate coefficient is significantly smaller than

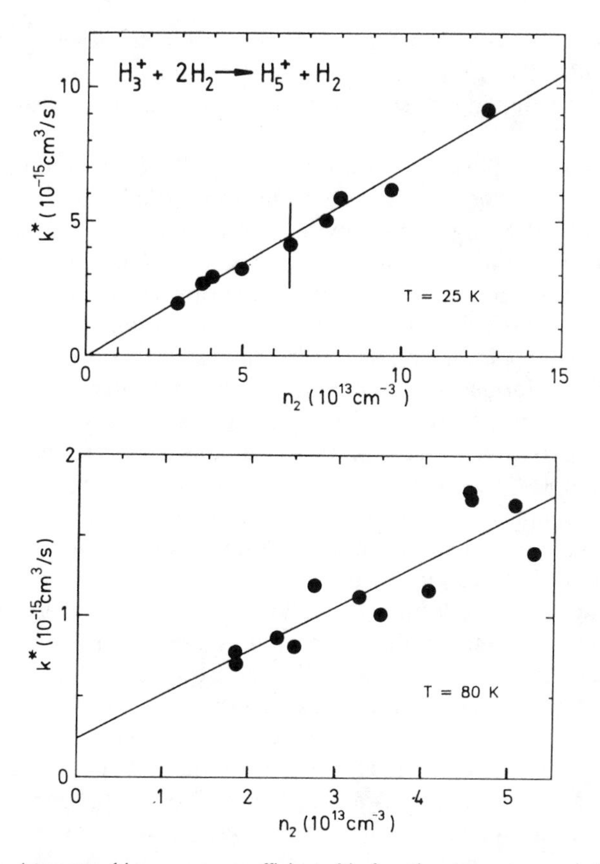

Figure 59. Apparent binary rate coefficient k^* for the ternary association reaction $H_3^+ + 2H_2 \rightarrow H_5^+ + H_2$, plotted as a function of the H_2 density n_2 (Paul, 1990). Increasing the nominal temperature of the ring electrode trap from 25 K to 80 K leads to a decrease of the ternary rate coefficient from 7×10^{-29} cm^6/s to 2.8×10^{-29} cm^6/s.

10^{-15} cm^3/s. For a more precise determination, measurements at densities below 10^{13} cm^{-3} are required. More results and other applications of this apparatus, for example, the measurement of radiative lifetimes, are described in Sections V A and V D.

V. STUDIES OF ION PROCESSES IN RF FIELDS: A SAMPLING

In Section III many of the technical details of special rf devices were described and various modes of operation were characterized and illustrated using selected examples. These rf devices have been combined in various instruments, as summarized in Section IV, and are used to study ion processes in fields such as mass spectrometry, spectroscopy, and thermochemistry, but, predominantly, they are used in the field of reaction dynamics. In this section we present a sampling of studies of ion processes performed using these instruments.

We begin by presenting absolute integral cross sections and rate coefficients for simple charge-transfer reactions, the energy dependence of which shows interesting changes in the collision energy range between 1 meV and 1 eV, or in the temperature range between 10 K and 350 K. Differential product velocity and angular distributions are shown for representative systems in Section V B and illustrate that in favorable cases, near state-to-state information can be obtained. A sampling of the variety of experiments that can be performed by combining optical and laser methods with our instruments is given in Section V C. The final section presents a few special results from radiative association processes and related reactions. Although several examples illustrating various aspects of the versatility of guided-ion-beam apparatuses could be taken from other groups mentioned in Section IV A, we restrict our presentation to data obtained using the instruments described in Sections IV B–IV E.

One criterion for selecting the following reaction systems is that they illustrate several features of the rf apparatuses. For example, for the prototype ion–molecule reaction of H^+ with H_2 and isotopic variants, integral cross sections, differential state-to-state cross sections, and reaction-rate coefficients for ternary and radiative association are presented. The atom–atom system, $Ar^{2+} + He$, is an ideal test case for the method used to determine differential cross sections in an octopole. A detailed analysis of O_2^+ product states formed in He^+ and Ar^+ collisions with O_2 is performed by measuring product velocities and using the method of laser predissociation. Chemiluminescence detection and coincidence between photons and mass and/or time-of-flight-selected ions is illustrated for CO^+ products from the $N^+ + CO$ reaction. The CO^+ ion is also used to demonstrate the combination of the guided-beam technique with laser-induced fluorescence and laser-induced charge transfer. It is beyond the scope of this chapter to discuss in detail the dynamics of these reactions or even to give a complete survey of the existing literature. Therefore, we restrict the theoretical interpretation of the results to a few remarks that are required for the overall understanding of the experimental procedures.

A. Integral Cross Sections and Thermal Rate Coefficients

1. *Charge Transfer to Rare Gas Ions*

The integral cross section for the atom–atom charge-transfer collision $Ar^{2+}(^3P) + He(^1S) \rightarrow Ar^+(^2P) + He^+(^2S)$ has been determined as a function

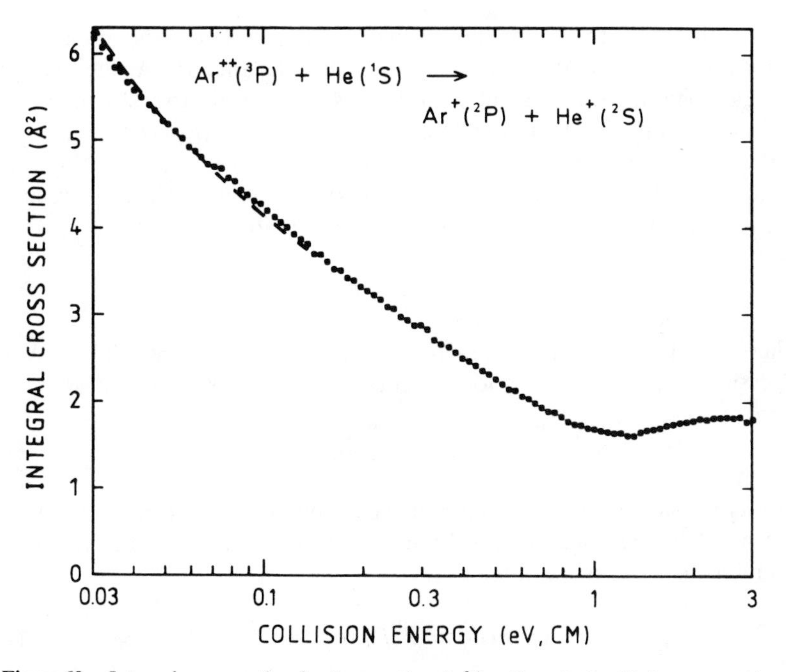

Figure 60. Integral cross section for the reaction $Ar^{2+} + He \rightarrow Ar^{+} + He^{+}$ measured in the universal guided-ion-beam apparatus. The dashed line corresponds to an effective cross section calculated using an analytical approximation for $\sigma(E_T)$ given in the text.

of collision energy using the guided-beam apparatus (Disch et al., 1985; Gerlich, 1986) and is depicted in Fig. 60. This system has been studied at low energies in the group of Z. Herman by crossed-beam scattering (Friedrich et al., 1984), and calculations of the differential cross section have been performed based on perturbation (Landau–Zener) techniques (Friedrich et al., 1986). The integral cross section, Fig. 60, has a shallow minimum at about 1 eV. The increase below 0.2 eV has been approximated by $\sigma(E_T) = 1.9\,E_T^{-0.345}$. This function and the almost identical corresponding effective cross section as defined by Eq. (91) are shown as a single dashed line. Our experimental results have stimulated a comparison with quantum-mechanical calculations performed by Braga et al. (1986). The ab initio determined cross sections are about a factor of 2 larger; however, these authors have demonstrated that a good agreement can be obtained by slightly changing the diabatic potential. This illustrates the need for precise absolute values of integral cross sections at low energies. The experimentally measured minimum at about 1 eV was not reproduced in the calculation, and it has been supposed that this could be due to the channel leading to metastable $Ar^{2+}(^1D)$.

For this and the following charge-transfer systems, the nonadiabatic interaction is localized at a large internuclear distance R_c. At high collision energies, the impact parameter method (straight trajectories) can be used, and the cross section is simply described by $\sigma(E_T) = P(E_T)\pi R_c^2$, where $P(E_T)$ is the total charge-transfer transition probability averaged over the impact parameter. At the low energies of our experiment, however, the two reactants attract each other, for example, by the polarization potential

$$V_p = -7.2\,\alpha/R^4, \tag{115}$$

where α is the polarizability in Å^3, and the units of V_p and R are eV and Å, respectively. Accounting for the curved trajectories, one obtains

$$\sigma(E_T) = P(E_T)[1 - V_p(R_c)/E_T]\pi R_c^2. \tag{116}$$

For $E_T = -V_p(R_c)$, the orbiting radius becomes identical to R_c. At even lower energies, the polarization (or Langevin) cross section σ_p determines the charge-transfer cross section

$$\sigma(E_T) = P(E_T)\,\sigma_p, \tag{117}$$

where σ_p is given by

$$\sigma_p = 16.86\,(\alpha/E_T)^{1/2}. \tag{118}$$

All of the charge-transfer cross sections presented in this section have been measured close to or in this transition-energy range. In the $Ar^{2+} + He$ reaction, the crossing radius $R_c = 4.7$ Å is readily determined from the known exoergicity of 3.14 eV, from the Coulomb repulsion of the products $V_{cb} = 14.4/R$, and from Eq. (115), with $\alpha(\mathrm{He}) = 0.22$ Å^3. Since $-V_p(R_c = 4.7\ \text{Å}) = 13$ meV is smaller than our lowest experimental collision energy (30 meV), the experimental cross section can be fitted by Eq. (116) (dashed line in Fig. 60). Comparison of $\sigma(E_T)$ with the measured values results in the probability function $P(E_T)$, which has a local minimum of only 3% at 1 eV and increases slowly to 6.4% at 30 meV.

Larger transition probabilities have been derived for the dissociative charge transfer reaction $He^+ + O_2 \rightarrow O + He$. In Fig. 61 our measured integral cross section is compared with results obtained by other groups at high collision energies. The data between 0.07 and 10 eV were measured using the octopole guided ion beam, while the values between 10 and 100 meV were determined by the trapped ion beam method. Here, a beam of translationally cold ions ($\Delta E_{\mathrm{lab}} = 2$ meV) is prepared in the octopole using

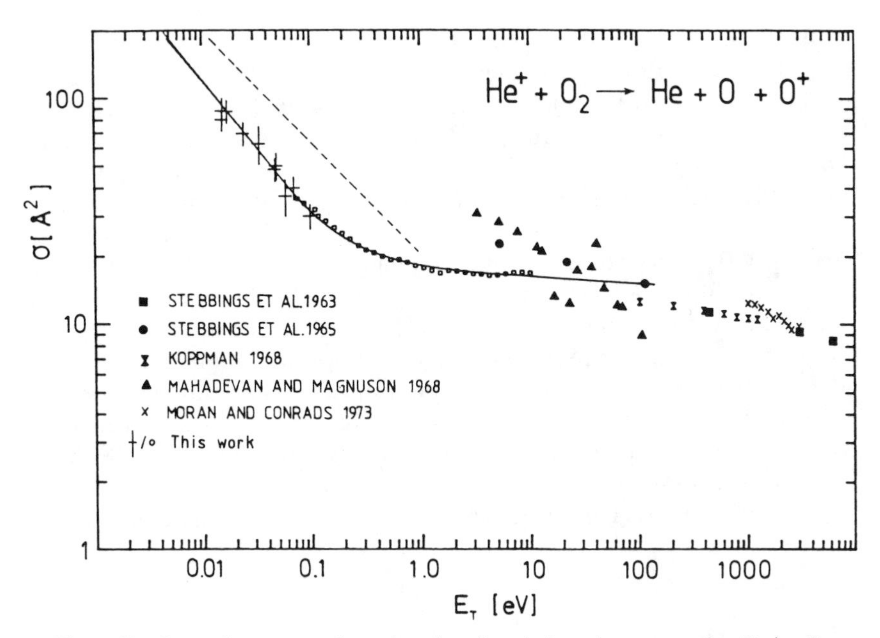

Figure 61. Integral cross sections for the dissociative charge transfer $He^+ + O_2 \rightarrow O^+ + O + He$. The guided-ion-beam results (○) have been extended to lower energies using the trapped ion beam method (+). The data at higher energies are from several other beam experiments, the references of which are compiled in Bischof and Linder (1986). The solid line is an effective cross section calculated using an analytical expression given in the text.

external potential barriers, as described in Section III C 2. This technique allows one to obtain effective rate coefficients at conditions corresponding to a translational temperature of about 100 K (see Fig. 47).

The measured integral cross sections have been evaluated based on the previously described model. A best fit to the measured effective cross section has been obtained for a crossing localized at $R_c = 3.5$ Å and with the transition probabilities

$$P(E_T > -V_p(R_c)) = 0.44 E_T^{-0.02}$$

and

$$P(E_T \leqslant -V_p(R_c)) = 0.358 E_T^{-0.1} \tag{119}$$

From Eq. (115) and the polarizability $\alpha(O_2) = 1.6$ Å^3, one obtains $-V_p(R_c = 3.5$ Å$) = 77$ meV. In contrast to the $Ar^{2+} + He$ system, here the change in the model cross section from Eq. (116) to Eq. (117) is within the energy range of our experiment, as can be clearly seen by the change in the slope in Fig. 61. For this atom–diatom system, the transition probability

$P(E_T)$ is almost energy independent at high energies (44%). Below 77 meV $P(E_T)$ increases slowly, and reaches a rather large value of 71% at 1 meV.

Thermal rate coefficients have also been measured for this system using the temperature-variable ring electrode trap. Rather than utilizing the standard technique for analysis, where all ions are extracted from the trap after a given interaction time (Section IV E 2), we have used here an alternative method. Since some of the created O^+ products are rather fast (see Section V B 2), one can adjust the trapping conditions such that these ions can escape from the trap, while the remaining He^+ ions are safely enclosed. Under these conditions the injected He^+ ions can be almost completely converted into reaction products. Figure 62 shows the O^+ intensity registered in a multichannel scaler after filling the trap with He^+ ions. This signal is directly proportional to the remaining number of He^+ ions, and therefore proportional to reactive decay rate kn_2 of the He^+ ion cloud. With this method, we have determined $k(350\text{ K}) = 0.92 \times 10^{-9}\text{ cm}^3/\text{s}$ and $k(80\text{ K}) = 1.0 \times 10^{-9}\text{ cm}^3/\text{s}$. These values agree, within the error limits, with the rate

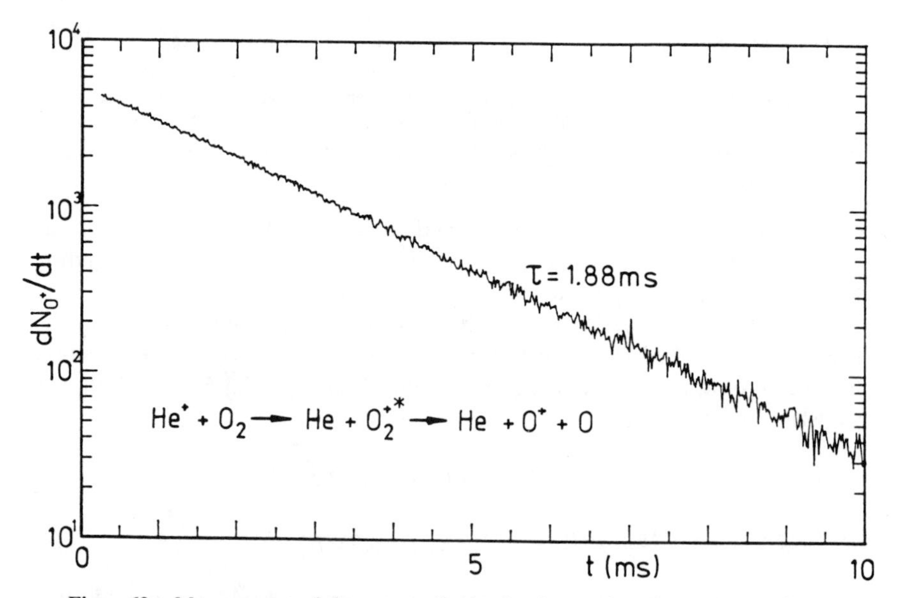

Figure 62. Measurement of the rate coefficient for the exothermic reaction $He^+ + O_2 \rightarrow O^+ + O + He$ by real-time registration of the O^+ products using a multichannel scaler. In this experiment, the He^+ ions are safely enclosed in the ring electrode trap, the only loss being due to reactions with O_2. A fraction of the fast O^+ products escapes over the properly adjusted dc barrier at the exit electrode. Their rate is directly proportional to the reactive decay rate kn_2 of the He^+ ion cloud.

coefficients $k_{eff} = 0.81 \times 10^{-9}\,cm^3/s$ and $0.92 \times 10^{-9}\,cm^3/s$ determined from Eq. (93) using the cross section σ defined by the parameters given in Eq. (119).

A third example, Fig. 63, shows the cross section for the $Ar^+ + O_2 \rightarrow O_2^+ + Ar$ charge transfer measured in our universal guided-ion-beam apparatus (Scherbarth and Gerlich, 1989) and our slow merged-beam instrument (Gerlich, 1989c; Kalmbach, 1989). Since the density of the O_2 beam was not determined directly, the relative cross sections measured in the merged-beam experiment were adjusted to the guided-ion-beam results at 0.1 eV. In the overlapping energy range above 40 meV the relative slopes are in excellent agreement. As in the case of the other two charge-transfer examples, the energy dependence deviates from the usual Langevin behavior

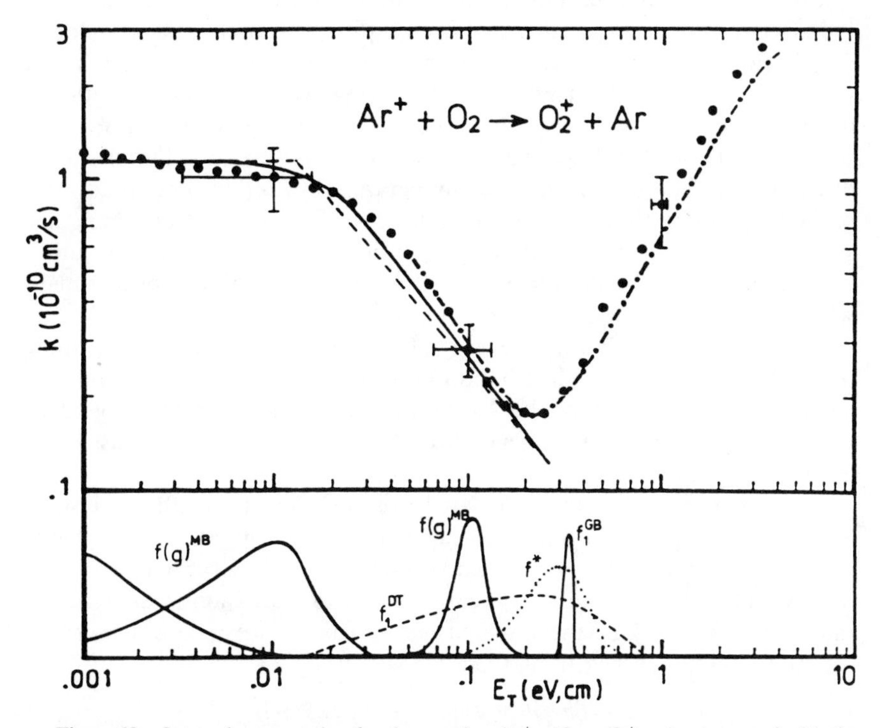

Figure 63. Integral cross section for the reaction $Ar^+ + O_2 \rightarrow O_2^+ + Ar$, measured with the merged beam (●) and the guided-ion-beam (●–●) methods (Scherbarth and Gerlich, 1989). The results are plotted as rate coefficients to emphasize the change in slope below 20 meV. The increase above 0.2 eV is due to the formation of $O_2^+(a\,^4\Pi_u)$. The low-energy behavior of the cross section has been approximated by the expression given in Eq. (120) (dashed line). Evaluation of Eq. (93) results in the effective rate coefficient (solid line). The lower part shows distributions of the relative velocity $f(g)^{MB}$ for the merged-beam conditions. For comparison the functions f_1^{GB}, f_1^{DT}, and f^* [see Eq. (94)] are given for the guided ion beam and DRIFT tube apparatuses.

and therefore gives some insight into the reaction mechanism. The increase above 0.2 eV is due to the formation of metastable $O_2^+(a^4\Pi_u)$ ions. This channel has been studied in detail using the method of laser fragmentation (Section IV B 4), a typical result of which is presented in Section V C 3.

The behavior of the cross section at low energies has been approximated by an expression predicted in Scherbarth and Gerlich (1989) on the basis of the guided-beam results and on thermal rate coefficients,

$$\sigma(E_T > 13\,\text{meV}) = 0.135 E_T^{-1.25}$$

and

$$\sigma(E_T \leqslant 13\,\text{meV}) = 3.5\, E_T^{-0.5}. \tag{120}$$

The dashed line in Fig. 63 shows this function, while the solid line represents the effective cross section, calculated from Eq. (91), where the relative velocity distribution, $f(g)$, was determined from the measured ion and nozzle beam velocities using Eq. (89). The almost perfect agreement of this fit with the merged-beam results nicely corroborates the predicted analytical cross section, Eq. (120). The lower part of Fig. 63 shows several collision energy distributions obtained by numerical integration of Eq. (89). The speed ratio of the O_2 beam in these experiments was $v_2/\Delta v_2 = 5$, at an assumed angular spread of $\pm 10°$. The lowest attainable collision energy and energy resolution was mainly limited by the spread of the ion beam, which in this experiment was only 100 meV. Nonetheless, a mean collision energy of 3 meV was obtained at an Ar^+ laboratory energy of 0.15 eV. For comparison, we show the estimated ion energy distribution of a drift tube at a nominal collision energy of 0.3 eV, as well as the Ar^+ ion energy distribution in the guided-ion-beam apparatus, and the generalized Maxwellian f^* for this mass ratio, as given by Eq. (94).

Thermal rate coefficients, calculated from $\sigma(E_T)$ in Eq. (120), are also in accord with experimental rate coefficients measured using the ring electrode trap, $k(80\,\text{K}) = 1 \times 10^{-10}\,\text{cm}^3/\text{s}$ and $k(400\,\text{K}) = 4.4 \times 10^{-11}\,\text{cm}^3/\text{s}$. More details on this charge-transfer system, including a discussion of the reaction mechanism that explains the observed energy dependence of the cross section, can be found in Scherbarth and Gerlich (1989).

2. *The Prototype System* $H^+ + H_2$

Reactive collisions between $H^+ + H_2$ and isotopic variants $H^+ + D_2$ have been studied quite extensively using several experimental techniques and different theoretical models. At energies below 1.8 eV, proton–proton or proton–deuteron exchange are the only open reactive channels. The reaction mechanism is determined by the deep potential well of the H_3^+ intermediate, and the formation of a strongly bound complex allows one to use statistical methods (Gerlich, 1982; Gerlich et al., 1980). Some restrictions of statistical behavior due to symmetry selection rules have been discussed recently (Gerlich, 1990). The low-energy behavior is also influenced by the differences in zero-point energies and the anisotropy of the long-range potential caused by the quadrupole moment of H_2.

Figure 64 shows the integral cross sections, plotted as rate coefficients, for the exothermic proton-deuteron exchange $D^+ + H_2 \rightarrow H^+ + HD$ at energies ranging from 1 meV to 1 eV. Above 30 meV, data from DRIFT measurements (Villinger et al., 1982) and guided-ion-beam results (Müller, 1983) are included. The different results agree with each other within the combined uncertainties. The heavy line has been calculated using a statistical model called the most dynamically biased (MDB) theory (Gerlich, 1982; Gerlich et al., 1980). The merged-beam apparatus has been used successfully

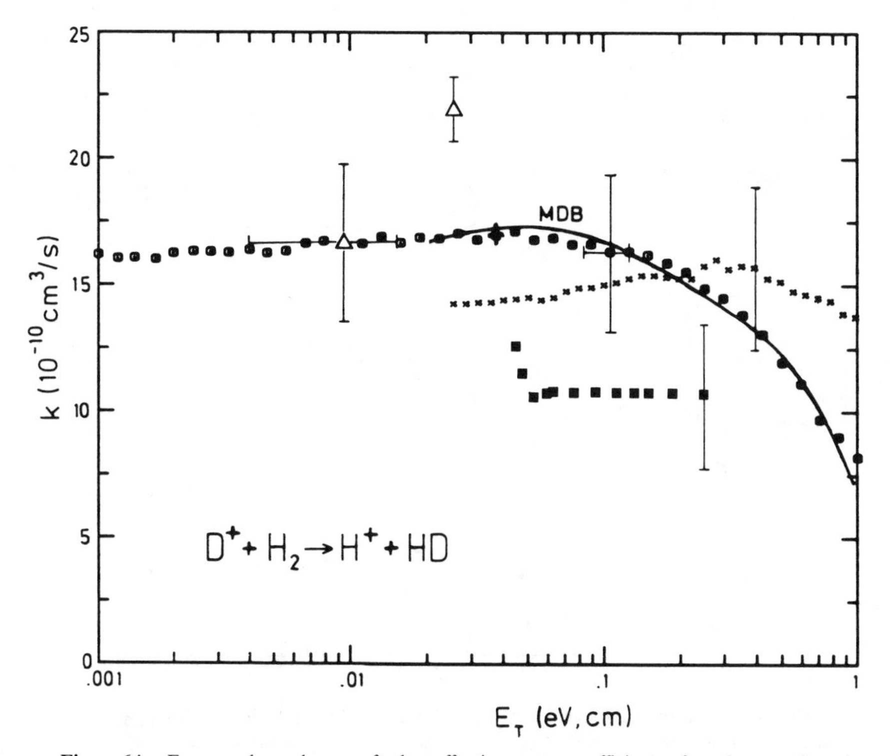

Figure 64. Energy dependence of the effective rate coefficients for the exothermic proton–deuteron exchange $D^+ + H_2 \rightarrow H^+ + HD$ measured with the merged beam (⊙), the guided ion beam (×) (Müller, 1983), and the DRIFT (■) methods (Villinger et al., 1982). The heavy line has been determined using the statistical MDB theory (Gerlich, 1982). The two triangles are thermal rate coefficients from a SIFT experiment (Henchman et al., 1981).

to extend the energy range down to a few meV. Here, the cross sections have been determined in absolute units by calibrating the density of the H_2 beam based on the well-known $Ar^+ + H_2$ reaction. The near perfect agreement with the theory is most probably fortuitous.

The decline of the cross section above 0.1 eV is understood within the MDB statistical theory. Below 0.1 eV, this reaction occurs with the typical Langevin cross section, resulting in a temperature independent rate coefficient. This can be seen from Fig. 65, which compares results from our

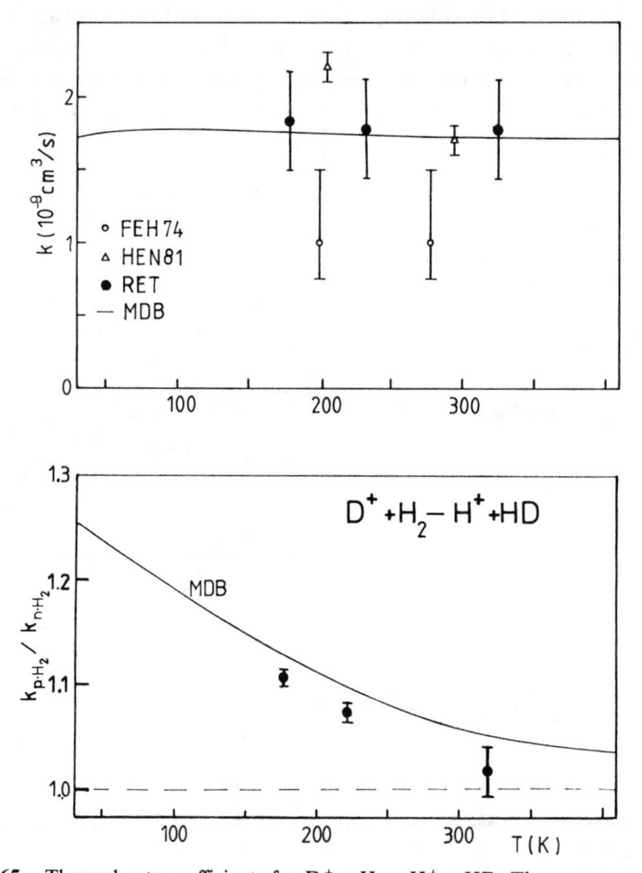

Figure 65. Thermal rate coefficients for $D^+ + H_2 \rightarrow H^+ + HD$. The upper panel shows a comparison of our ring electrode trap results (●) with those from a flowing afterglow (○) (Fehsenfeld et al., 1974) and a SIFT (Δ) (Henchman et al., 1981) apparatus. The solid line was determined using the statistical MDB theory (Gerlich, 1982). In all these experiments, normal-H_2 was used. The lower panel compares results obtained for 75% enriched para-H_2 and normal-H_2. The reaction rate with $H_2(j=0)$ increases with falling temperature due to the influence of the charge–quadrupole interaction, in reasonable accord with the statistical MDB theory.

first cooled ring electrode trap (Gaber, 1987) with those from a flowing afterglow (Fehsenfeld et al., 1974), and a SIFT (Henchman et al., 1981) apparatus. The reaction rate with nonrotating para-$H_2(j = 0)$ must increase with falling temperature due to the influence of the charge–quadrupole interaction. The lower part of Fig. 65 shows results obtained in the same trap with 75% enriched para-H_2 and normal-H_2 (Gaber, 1987). The observed increase is significant, although it is somewhat smaller than predicted from the statistical MDB theory.

The deuteron–proton exchange $H^+ + D_2 + D^+ + HD$ is 46 meV endothermic, which leads to a significant change in the low energy cross section, as seen from Fig. 66. Like the proton–deuteron exchange, this reaction has been studied extensively. The figure includes results obtained from two different guided ion beam apparatuses (Müller, 1983; Ochs and Teloy, 1974), DRIFT data (Villinger et al., 1982), and merged-beam results. Owing to the unfavorable mass ratio $m_1:m_2 = 1:4$, the lowest attainable energy of the merged-beam apparatus is 20 meV. This, together with incomplete rotational

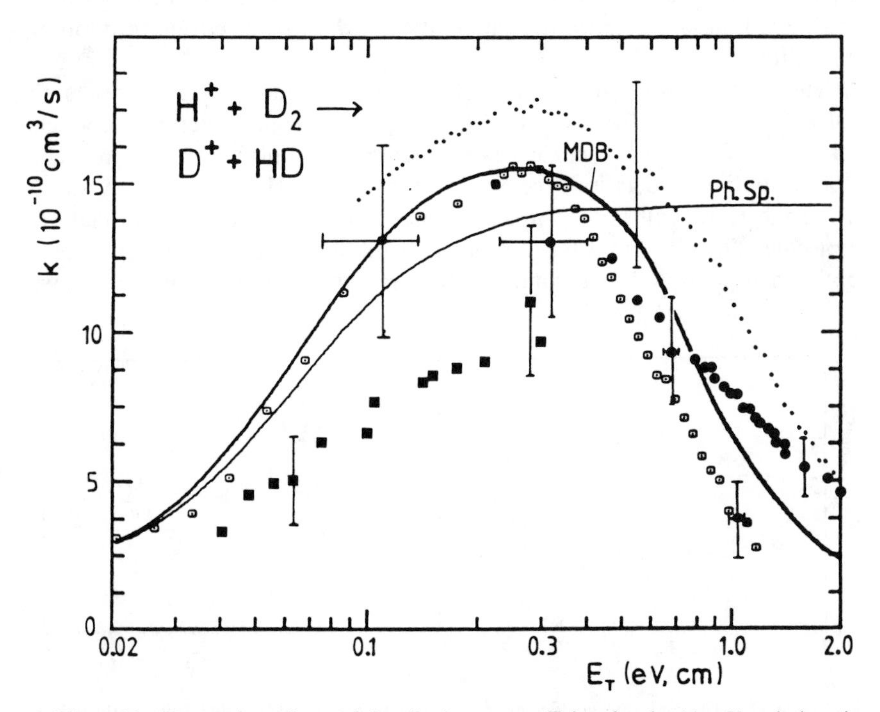

Figure 66. Energy dependence of the effective rate coefficient for the 46 meV endothermic proton–deuteron exchange $H^+ + D_2 \rightarrow D^+ + HD$ measured with different techniques [merged beam ⊙, guided ion beam ● (Ochs and Teloy, 1974), and • (Müller, 1983), DRIFT ■ (Villinger et al., 1982). The heavy line is the result from the statistical MBD theory (Gerlich, 1982), and the thinner line is the prediction based on conventional phase space theory.

relaxation of the D_2 target, leads to the broadening in the threshold region. At energies above 0.3 eV, the merged-beam cross sections may be somewhat too small owing to problems with the time overlap of the pulsed neutral and ion beam. The difference between our statistical MDB theory and the usual phase-space theory is small at low energies; however, there are significant deviations above 0.5 eV. More experimental results for the H_3^+ system will be presented in Sections V B 3 and V D 2.

3. *Small Rate Coefficients*

Most standard methods for determining thermal reaction rate coefficients are limited typically to values above $10^{-13}\,cm^3/s$. In order to obtain a sufficient product signal for a very slow reaction, the experiments are often performed at high neutral densities, which results in several perturbations, for example, by ternary processes. With ion trapping methods, much smaller rate coefficients can be determined through long interaction times, thereby avoiding the need for high neutral densities. This is briefly illustrated in this section with two slow binary reactions. Radiative association and other slow processes that characterize the capabilities of the ring electrode trap follow in Section V D.

The first example, depicted on the left-hand side of Fig. 67, refers to collisions between C^+ and H_2. For this system, the binary or ternary products formed are CH^+ from the 0.398 eV endothermic bimolecular reaction, CH_2^+ from radiative association (see Section V D 1), or CH_3^+ produced by stabilizing the CH_2^+ collision complex during its lifetime via an additional H_2 molecule. The CH^+ and CH_2^+ products cannot be observed directly, since both undergo very fast secondary reactions with the ambient H_2 to give

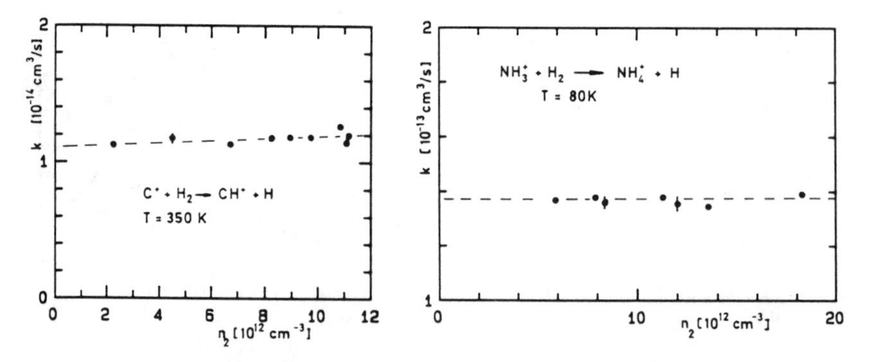

Figure 67. Thermal rate coefficients for two slow reactions as a function of the target density n_2. At the low density in the ring electrode trap, both reactions proceed predominantly via a bimolecular process. The slight increase in the rate coefficient for the $C^+ + H_2$ reaction indicates a minor contribution from ternary collisions.

CH_3^+. Owing to the large endothermicity, the binary reaction is very slow at room temperature. An estimate of the thermal rate coefficient can be obtained from the approximate formula $k(T) = 10^{-9} \exp(-4575\,K/T)\,cm^3/s$, which fits in good approximation the results of our phase-space calculations (Gerlich and Kaefer, 1987; Gerlich et al., 1987) between 100 K and 600 K.

At room temperature, the binary reaction is so slow that it competes with ternary association, even at the low densities of our experiment. In order to increase the contribution of the binary reaction, the ion trap was not cooled, and rf heating of the electrodes led to an estimated temperature of about 350 K. Under these conditions an almost density-independent rate coefficient $k = 1.1 \times 10^{-14}\,cm^3/s$ is observed, that is, binary processes prevail. This experimental value is larger than the theoretical rate $k(T)$, suggesting a somewhat higher trap temperature. The slight increase of the measured effective rate with the hydrogen density indicates a small contribution from ternary reactions with $k_3 < 10^{-28}\,cm^6/s$. More discussion of this reaction at lower temperature is given in Section V D 1.

The second example in Fig. 67 refers to the formation of the ammonium ion from the abstraction reaction $NH_3^+ + H_2 \rightarrow NH_4^+ + H$. Although this reaction is exothermic, it is very slow (Barlow and Dunn, 1987). At 300 K the rate coefficient is three orders of magnitude smaller than the Langevin rate coefficient. It has a minimum of $1.5 \times 10^{-13}\,cm^3/s$ at about 80 K (Böhringer, 1985), but it increases toward lower temperature. This has been explained by an energy barrier and a quantum-mechanical tunneling mechanism. Since the reaction must proceed via formation of an intermediate complex with a rather long lifetime, significant perturbations can occur due to the presence of buffer gas at high density. The influence of third-body-assisted binary collisions has been examined in a selected ion drift tube (Böhringer, 1985), however, at densities above $10^{16}\,cm^{-3}$. Our experiment has been performed at densities in the range of $10^{13}\,cm^{-3}$, further reducing the influence of ternary collisions by three orders of magnitude. The resulting rate coefficient $k(80\,K) = 1.4 \times 10^{-13}\,cm^3/s$ corroborates the previous results.

B. Differential and State-to-State Cross Sections

1. *Single-Electron Transfer in $Ar^{2+} + He$*

The integral cross section of the $Ar^{2+}(^3P) + He(^1S) \rightarrow Ar^+(^2P) + He^+(^2S)$ charge-transfer process was presented in Section V A and discussed on the basis of a localized crossing at 4.7 Å. This reaction is also well suited for characterizing the differential scattering guided-ion-beam apparatus and for testing the angular and energy resolutions of this new experimental method. A special feature of the one-electron charge transfer is that there is only one input and one exit channel, since the exothermicity of 3.14 eV is not sufficient to populate excited states at low kinetic energies. The Ar^{2+} and Ar^+ fine structure splitting, resulting in three reactant states and two product channels, are ignored, since they are not resolvable in the present experiment. For simplicity, we assume that metastable Ar^{2+} ions are quenched in the ion source (see Section III D). With these assumptions, the center-of-mass product velocity is well defined and the laboratory product velocity is easy to obtain, as illustrated in the Newton diagram in the upper part of Fig. 68.

To measure the differential cross section for this ideal collision system it is sufficient only to determine the projection of the product velocity, v'_{1p}, onto the axis defined by the primary velocity $\mathbf{v}_1$. In the guided-ion-beam experiment $\mathbf{v}_1$ coincides with the octopole axis and the distribution $d\dot{N}/dv'_{1p}$ is measured directly by time of flight. It is not necessary to determine the transverse component v'_{1t} of the product ions by varying the rf amplitude, as described in detail in Section IV B 3, since for this state-to-state process the scattering geometry is fully determined by v'_{1p} alone (see Newton diagram). A typical experimental Ar^+ product velocity distribution $d\dot{N}/dv'_{1p}$ is shown in the lowest panel of Fig. 68. In the laboratory frame, all products are forward scattered, while in the center-of-mass frame, the distribution is symmetric relative the nominal center-of-mass velocity v_c. A similar symmetric form is obtained at a collision energy of 0.2 eV, while at 0.5 eV and above, the symmetry is lost, indicating a change in the differential cross section (Disch et al., 1985).

The experimental product velocity distributions are broadened by the angular spread of the ion beam, the time resolution, and the finite length of the scattering cell. For this particular system, however, a detailed analysis shows that the overall resolution in our beam/cell arrangement is mostly affected by thermal broadening. The influence of thermal broadening and the dependence on the nominal scattering angle ϑ is illustrated in the middle panel of Fig. 68. The simulated Ar^+ product velocity distributions $d\dot{N}(\vartheta_0)/dv'_{1p}$ have been calculated using Eq. (95). For each distribution it was assumed that the differential cross section can be represented by a δ function, which peaks at different center-of-mass scattering angles $[d\sigma/d\vartheta = \sigma(E_T) \times \delta(\vartheta - \vartheta_0), \vartheta_0 = 0°, 20°, \text{etc.}]$. The energy dependence of $\sigma(E_T)$ has been given in Section V A 1. More details of the method can be found in Gerlich (1989a). These simulated results show that thermal broadening leads to rather narrow but closely spaced velocity distributions in the forward and backward directions. If sideways scattering occurs, the angular resolution is better, although the distributions are wider. This example also shows that, for this system, the angular resolution of our experimental method is on the order of $\pi/4$.

To fit the measured distributions we superimpose the simulated distributions $d\dot{N}(\vartheta_0)/dv'_{1p}$ with properly chosen weights. A typical result of

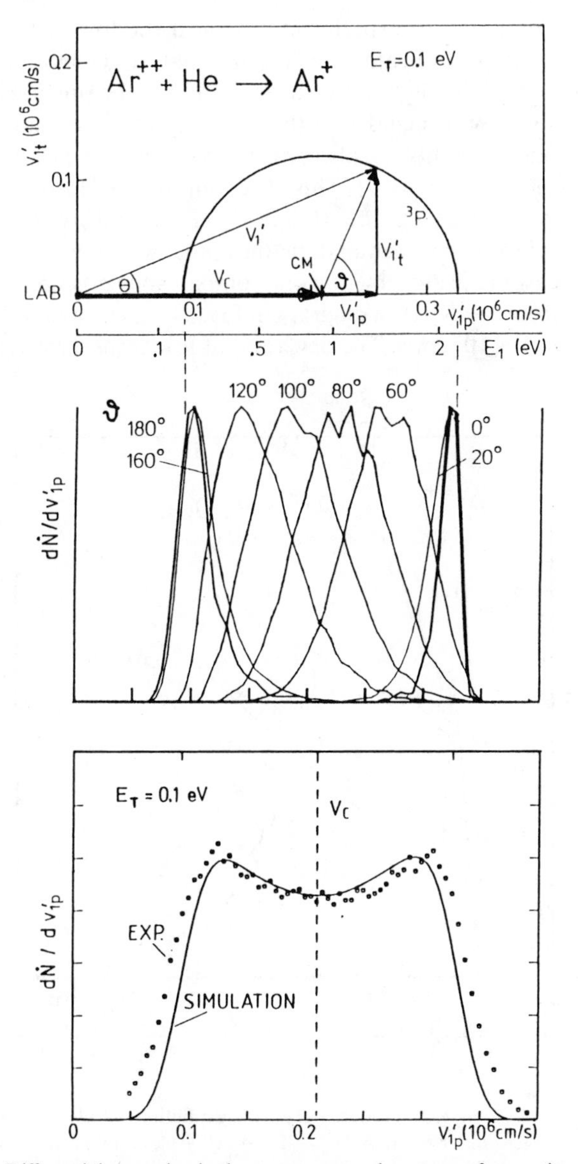

Figure 68. Differential scattering in the state-to-state charge-transfer reaction $Ar^{2+} + He \rightarrow Ar^{+} + He^{+}$ in the guided-ion-beam apparatus. The upper panel depicts a nominal Newton diagram scaled for a collision energy of 0.1 eV. CM marks the origin of the center-of-mass system; the He target is assumed to be at rest. The laboratory and center-of-mass scattering angles are θ and ϑ, respectively. The projection of the final velocity v'_1 onto the octopole axis, that is, the component v'_{1p}, is measured by time of flight. The product velocity distributions are broadened by the target motion, as illustrated in the middle panel. The lowest panel compares a measured velocity distribution with a simulation based on a weighted superposition of functions shown in the middle panel.

such a fit is compared to experimental data in the lowest panel of Fig. 68. The good agreement indicates that the thermal target motion causes most of the broadening. The slight differences are due to the previously mentioned effects, which were not included in this simulation. In an analogous manner these ϑ-dependent weights can also be used to simulate the transverse product velocity distribution $d\dot{N}/dv'_{1t}$. This distribution has also been measured independently by varying the rf guiding field and is given in Fig. 69. Comparison of the experimental data (dark line) with the simulation (dashed line) shows reasonable agreement if one compares the position of the maxima. As expected from the Newton diagram in Fig. 69 these maxima are positioned slightly above 1×10^5 cm/s. The deviation at low transverse velocities is due

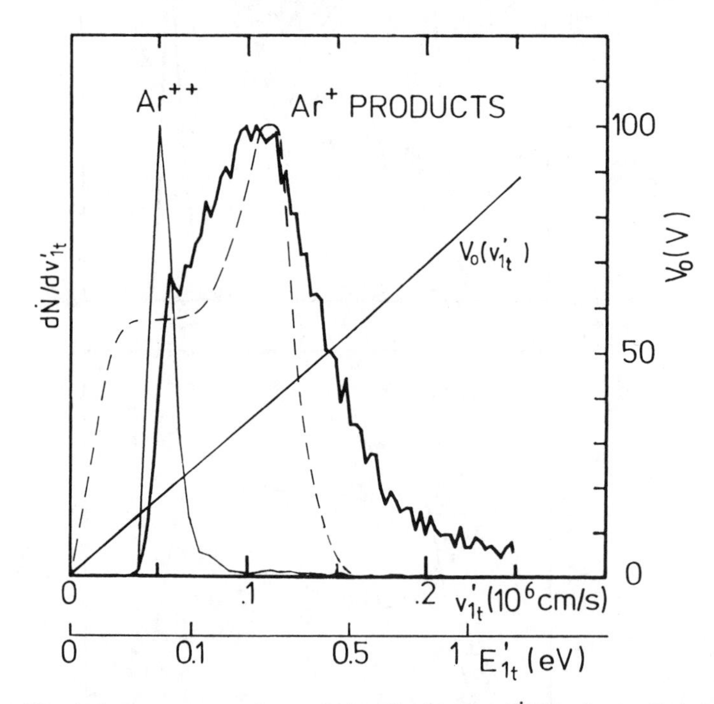

Figure 69. Ar^+ transverse product velocity distribution $d\dot{N}/dv'_{1t}$ (heavy line) for the state-to-state charge-transfer reaction $Ar^{2+} + He \rightarrow Ar^+ + He^+$, obtained by variation of the rf guiding field. The dashed line shows a simulation that only accounts for the thermal broadening. Deviations from the experimental Ar^+ curve are explained in the text. The maximum of the product distribution is situated slightly above $v'_{1t} = 1 \times 10^5$ cm/s in accordance with the Newton diagram in Fig. 68. The narrow Ar^{2+} distribution has been determined by differentiating the rf-dependent transmission function of the primary beam. The slightly curved line shows the relation between the maximum guided transverse velocity and the amplitude V_0, as obtained from Eq. (84).

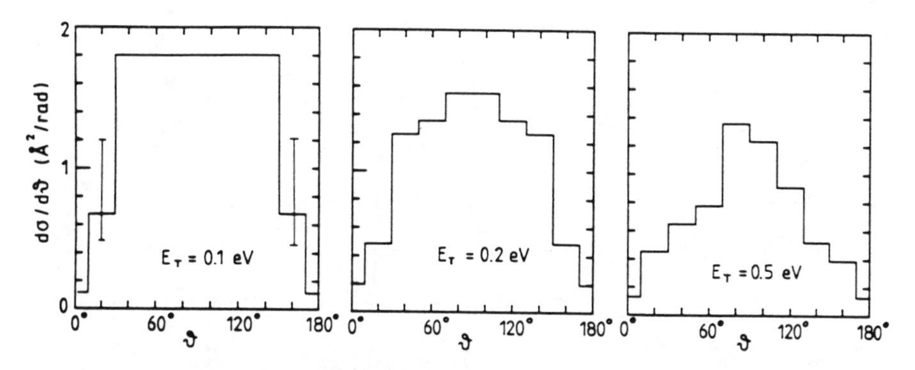

Figure 70. Differential cross sections $d\sigma/d\vartheta$ in absolute units (Å^2/rad) for the charge transfer reaction $Ar^{2+} + He \rightarrow Ar^+ + He^+$. At 0.1 eV isotropic scattering prevails, that is, $d\sigma/d\vartheta$ is nearly constant, while at 0.5 eV the Ar^+ products are mostly scattered sideways.

to potential distortions. It can also be seen from this figure that for guiding the primary Ar^{2+} beam, a minimum effective potential of about 50 meV is required. The additionl high-energy tail in the experimental distribution may be due to some unknown rf influence, but it can also be explained by the presence of metastable ions in our primary ion beam, which have been reported (Friedrich et al., 1984) to cause strongly sideways scattered products.

The most important results derived from the previously described fitting procedure are the ϑ-dependent weights, since they are directly proportional to the differential cross section. With these weighting factors and the measured integral cross section σ, one obtains the differential cross section $d\sigma/d\vartheta$ in absolute units. The angular dependence of $d\sigma/d\vartheta$ is depicted as a histogram in Fig. 70 at three collision energies. At 0.1 eV $d\sigma/d\vartheta$ is constant with exception of some uncertainty in the forward and backward directions, that is, isotropic scattering prevails. Increasing the collision energy to 0.2 eV and 0.5 eV leads to a preference for sideways scattering; however, the forward–backward symmetry is still maintained. At 1.6 eV, significant forward scattering prevails as can be seen immediately from the time-of-flight distribution in Fig. 71. These results can be compared with contour plots measured in a differential scattering experiment (Friedrich et al., 1984). Both the sideways scattering peak at $E_T = 0.53$ eV and the preference for forward scattering at $E_T = 1.6$ eV are in agreement with our observations. Some minor deviations may be due to an admixture of Ar^{2+} (^1D) ions which differ in the two experiments.

A self-consistency check of the differential guided-ion-beam method would be to detect the second ionic product, He^+. Since this light ion carries most of the reaction exothermicity (3.14 eV), one obtains a much larger

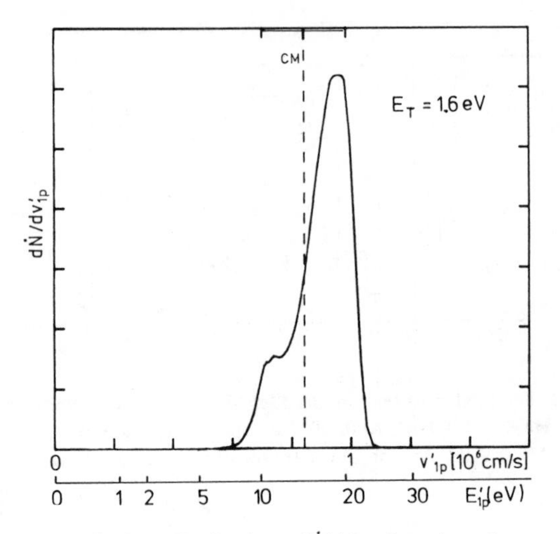

Figure 71. Product velocity distribution $d\dot{N}/dv'_{1p}$ for the charge-transfer reaction $Ar^{2+} + He \rightarrow Ar^{+} + He^{+}$ measured at a collision energy of 1.6 eV in the guided-ion-beam apparatus. The upper scale marks the position of the center-of-mass velocity and the range of the Newton circle. It is evident that at this energy forward scattering prevails.

Newton sphere and better angular resolution, but a large fraction of the products is scattered backward in the laboratory system. Owing to the high energy and low mass, one must operate the octopole at higher frequencies than used for guiding Ar^{2+} and Ar^{+}. The necessary operating conditions ($r_0 = 0.3$ cm, $E_m = 3$ eV, $\Omega/2\pi = 23$ MHz, $V_0 = 150$ V; see Section II D 2) are within the reach of our instrument, but the experiment has not yet been performed.

2. *Dissociative Electron Transfer in* $He^{+} + O_2$

The universal guided-ion-beam apparatus has also been used to study the kinetic energy release in the dissociative charge transfer $He^{+} + O_2 \rightarrow O^{+} + O + He$. It was previously established (Gentry, 1979) that more than 99% of the charge transfer leads to the dissociation of the highly excited O_2^{+}. In our experiment the amount of detected O_2^{+} was below 0.2%. Even after dissociation, an excess energy of 5.86 eV remains, which either results in fast fragments or can be lodged into exited products. In addition to the ground-state $O^{+}(^4S^0) + O(^3P)$ products, exited atoms $O(^1D)$ and $O(^1S^0)$, or excited ions $O^{+}(^2D^0)$ and $O^{+}(^2P^0)$, are accessible in different combinations.

Using a high-resolution crossed-beam apparatus, Bischof and Linder (1986) have studied the relative populations of the exited products, the accessible combinations of which have been denoted by I–VI. Integrating over the scattering angle, they have derived the branching ratios depicted in Fig. 74. If these branching ratios are extrapolated from the lowest collision energy of 0.5 eV to thermal energies, a result is obtained that is in conflict with conclusions from an ICR trap experiment (Mauclaire et al., 1979). More details are discussed in Bischot and Linder (1986).

In order to clarify this discrepancy, we have measured axial and transverse velocity distributions (Scherbarth, 1988) using our guided-ion-beam apparatus. A typical result is shown in Fig. 72. The details of the data presentation have been discussed in Section IV B 3. The lower part of the figure shows that the angular distribution is rotationally symmetric relative to the laboratory origin. This corresponds, within our limited resolution, to the predissociation of an almost unperturbed, but excited O_2^+ formed by a large distance electron jump. This is in accordance with the crossing radius $R_c = 3.5$ Å, determined from the integral cross section. Such rotational symmetry relative to $v_1' = 0$ is also observed at lower collision energies, since there the nominal center-of-mass velocity v_c is even closer to the laboratory origin. The dashed lines indicate the integration limits for deriving the intensity of the different product channels I–V. Since integration over our directly measured flux density $D(v_{1p}', v_{1t}')$ requires no weighting factors, it can be seen directly from Fig. 72 that formation of exited $O^+(^2P^0)$ are ground state $O(^3P)$ is the dominant channel (denoted as V).

In order to demonstrate the attainable energy resolution, Fig. 73 shows axial product velocity distributions $d\dot{N}/dv_{1p}'$ recorded at several collision energies. These were measured at a very low guiding field, corresponding to a transverse energy of about 80 meV. A common feature of all the curves is the dominant peak at low product velocities and a partially resolved structure at higher velocities. In the bottom curve, the dashed lines indicate the estimated contributions from the channels I–VI. It is obvious that our method cannot compete with the much higher angular and energy resolution of a crossed-beam experiment; however, it has the unique ability of operating down to thermal energies and provides a fast and reliable overview of the scattering dynamics.

Changes in the relative populations are also shown in Fig. 73. For example, channel II becomes more important at higher energies. For a more quantitative comparison, the angle-integrated branching ratios have been calculated for collision energies between 0.06 eV and 6.5 eV, with the results plotted as open circles in Fig. 74. In the overlapping energy range they are in good overall agreement with the crossed-beam data from Bischof and Linder (1986); however, they deviate significantly from the near-thermal results derived from the ICR trap experiment (Mauelaire et al., 1979). Our guided-ion-beam results indicate that there is no dramatic change in the reaction mechanism below 0.5 eV. Note that our lowest collision energy

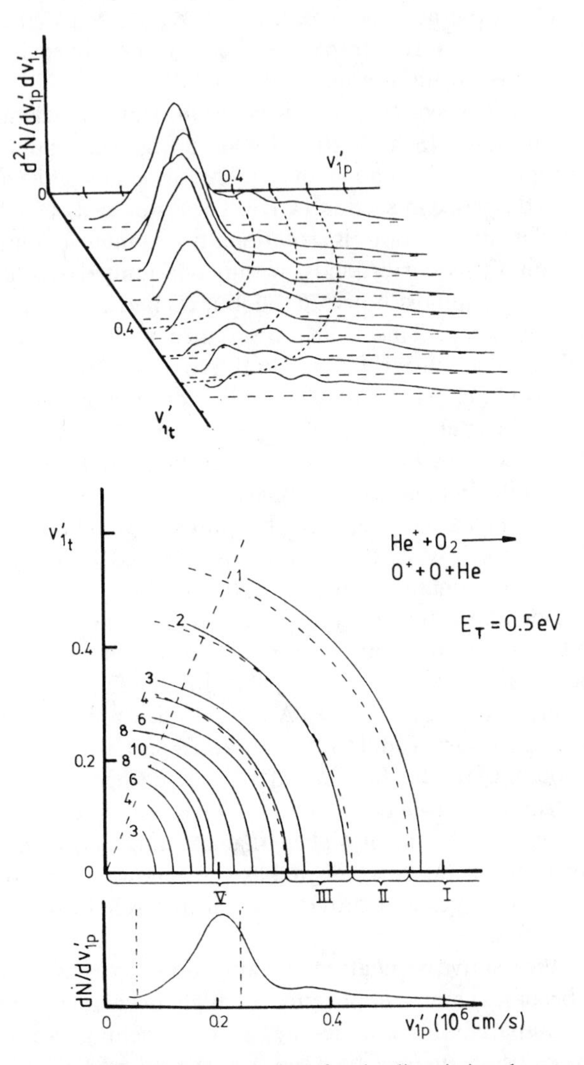

Figure 72. Doubly differential cross sections for the dissociative charge-transfer reaction $He^+ + O_2 \rightarrow O^+ + O + He$ determined in the guided ion beam apparatus. The O^+ product velocity distributions $d^2\dot{N}/dv'_{1p}\,dv'_{1t}$ are shown both perspectively and as a contour map of the two-dimensional product flux density, defined in Eq. (101). The dotted lines indicate the integration limits for deriving the intensity of the different product channels I–V; see text.

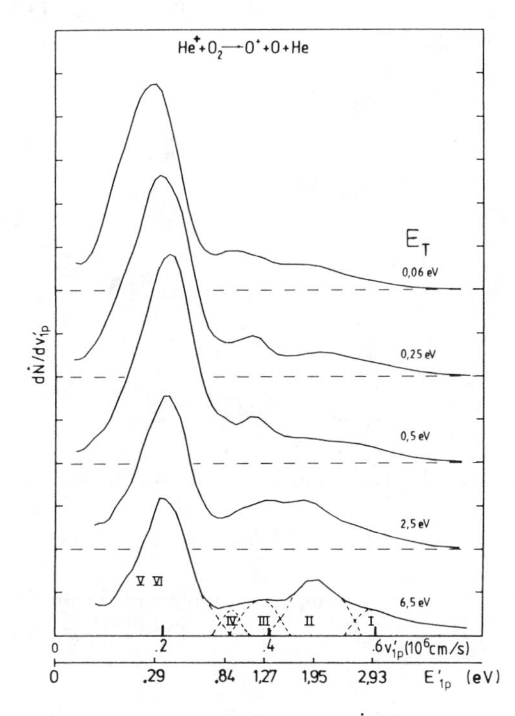

Figure 73. Axial product velocity distributions $d\dot{N}/dv'_{1p}$ measured for $He^+ + O_2 \rightarrow O^+ + O + He$ at a very low octopole guiding field. In the lowest curve, the dashed lines indicate the estimated contributions from channels I–V. Their relative populations changes slightly going from near thermal collision energies ($E_T = 0.06$ eV) to $E_T = 6.5$ eV; however, the slow peak is always dominant.

(60 meV) is already below the limit at 77 meV where the orbiting radius becomes larger than the crossing radius R_c, as shown in Section V A 1. For a detailed discussion of the reaction mechanism see Bischof and Linder (1986).

3. *Proton–Deuteron Exchange in* $H^+ + D_2$

It has been illustrated in Section V A 2 that integral cross sections and thermal rate coefficients for isotopic variants of the $H^+ + H_2$ reaction are in good agreement with the predictions of the MDB statistical theory (Gerlich, 1982; Gerlich et al., 1980). Here we want to demonstrate that the total energy available to the complex is distributed statistically among the open product channels. A survey of the results of this reaction, which has been studied in detail (Gerlich, 1977), has been reported in Teloy (1978).

High-resolution D^+ product velocity distributions for the reaction $H^+ + D_2 \rightarrow D^+ + HD$ have been measured by time of flight using the

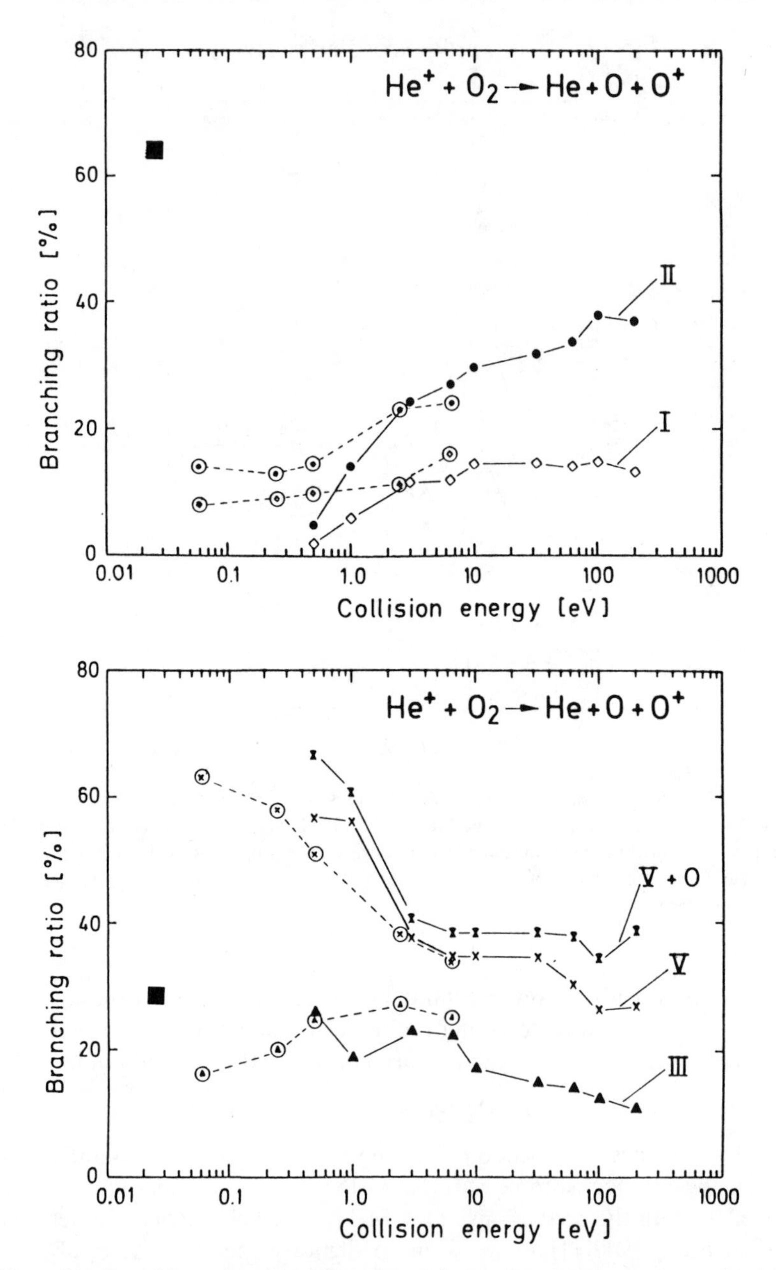

Figure 74. Branching ratios of the different $O + O^+$ channels (I–V) populated in dissociative charge-transfer collisions of He^+ with O_2. The data at high energies, connected with solid lines, were obtained with a differential scattering apparatus (Bischof and Linder, 1986). The two squares are near-thermal results, derived from an ICR trap experiment, combining channels (I + II) and (III + V) (Mauclaire et al., 1979). The guided-ion-beam results, connected with dashed lines, close the energy gap between thermal energies and those accessible in standard beam techniques.

differential scattering apparatus described in Section IV C. In this experiment, the target gas was cooled to liquid nitrogen temperature, the ion energy spread was 50 meV, and the angular resolution was 3°, resulting in an overall energy resolution of 80 meV. Product velocity distributions were determined by time of flight using a 1-m-long octopole as a flight tube. The recorded spectra were transformed numerically into laboratory energy distributions.

A selection of typical D^+ product distributions $d\dot{N}/dE'_1$ is plotted in Fig. 75, the lines representing a smooth interpolation of the data. As can be seen by comparing the position of the $\Delta E_T = 0$ mark, this reaction has a strong tendency to convert collision energy into internal excitation. The structure of the distributions is due to partially resolved rovibrational states of the HD product molecule, the nominal positions of which are marked by arrows. In the depicted examples, the collision energy is varied in the threshold range

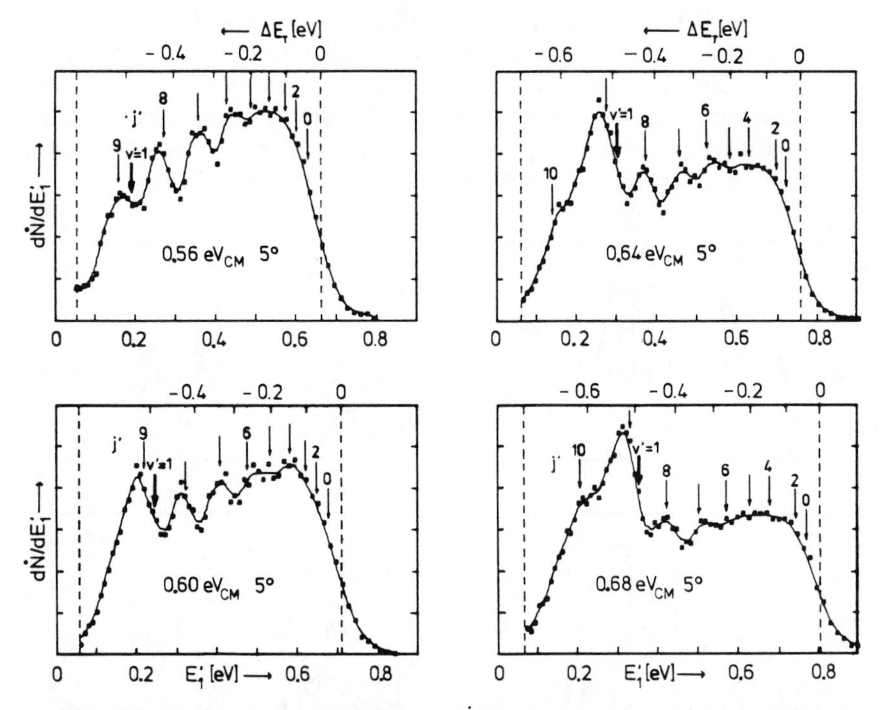

Figure 75. Product energy distributions $d\dot{N}/dE'_1$ from reactive $H^+ + D_2 \rightarrow D^+ + HD$ collisions measured with a differential scattering apparatus at a laboratory angle of 5°. The upper scale shows the translational exoergicity $\Delta E_T = E'_T - E_T$. The dashed lines mark the positions $\Delta E_T = 0$ and $\Delta E_T = -E_T$. The partially resolved structure corresponds to rotational excitation of HD products. The collision energy is varied in the range of the threshold for formation of HD($v' = 1$), resulting in the formation of additional slow products (left side of the $v' = 1$ arrow).

for formation of HD in the first vibrational state. Contributions from the $(v' = 1, j')$ states, superimposed on the $(v' = 0, j')$ states, result in a steplike structure in the kinetic energy distribution of the D^+ products, which are marked by the $v' = 1$ arrow. A further increase of the energy to 2 eV leads to additional steps due to population of the vibrational states $v' = 2$–4 and to a dense manifold of rovibrational states (Gerlich, 1977; Teloy, 1978).

To compare calculated rovibrational distributions $P(v', j')$ with the experimental data, the probabilities must be convoluted with $F_T(E'_1, \theta, \Delta E_T)$, the overall resolution function of the apparatus. This function combines the influence of the energy spread of the ion beam, the resolution of the detector, and the influence of the target motion. Details of this procedure have been mentioned in Section IV C and are discussed in Gerlich (1989a). Rovibrational distributions have been calculated with the MDB theory; however, it has been shown (Gerlich, 1977; Gerlich et al., 1980) that the results are very similar to the microcanonical equilibrium distribution $P_0(v', j')$. Therefore, we use here for simplicity these probabilities, shown as sticks in Fig. 76, to illustrate the influence of the resolution of the apparatus. The calculated resulting smooth distributions, plotted in Fig. 76, agree well with the measured curves depicted in Fig. 75. They also allow one to obtain a quantitative estimate of the contributions from the $v' = 1$ states.

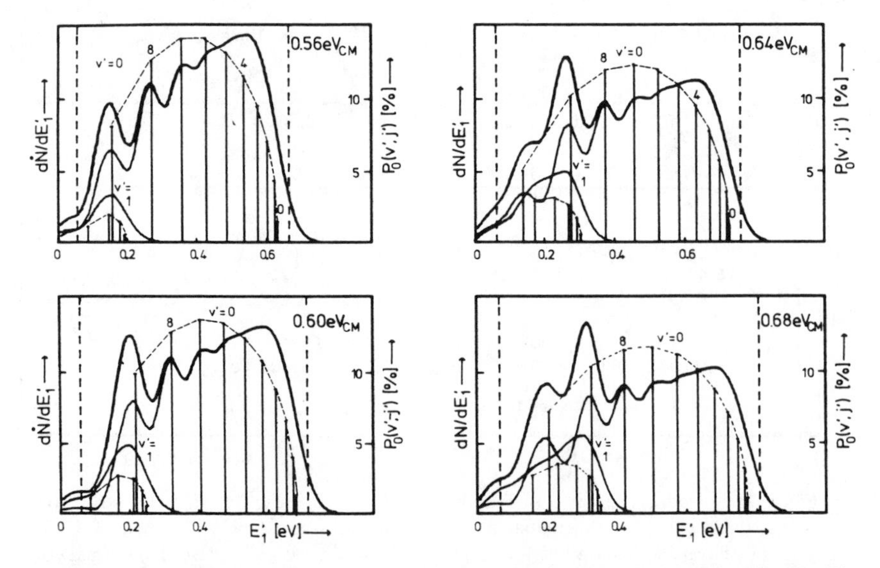

Figure 76. Simulation of the product energy distributions depicted in Fig. 75. The calculation is based on a statistical model that predicts the population probability $P_0(v, j)$ (stick diagram, scale at right). The smooth distributions were obtained by accounting for the overall resolution of the apparatus.

Corresponding high-resolution results have also been measured for $H^+ + H_2$ collisions, with emphasis on ortho–para transitions, and are reported in Gerlich (1990).

C. Applications of Optical Methods

Several possible combinations of the guided-ion-beam technique with optical methods have been mentioned in Section IV B 4. The following examples have been chosen to illustrate some of these capabilities. The $N^+ + CO$ reaction is used to demonstrate the sensitivity of the apparatus for detecting photons from a chemiluminescent reaction and to show two applications of the photon–ion coincidence technique. The use of intense laser pulses for multiphoton ionization is discussed in the chapter by Andersen; therefore, we present here only one brief example. Other laser applications described in this section include laser-induced charge transfer and laser-based analysis of products detecting fluorescence photons or fragment ions.

1. *Chemiluminescence*

The first application of a guided-ion-beam apparatus for the detection of photons from a chemiluminescent reaction was reported in Gerlich and Kaefer (1985). In this contribution it was shown that the observed B–X and D–X emission from N_2^+ products formed in $Ar^{2+} + N_2$ collisions (Neuschäter et al., 1979) are due to cascading transitions. This was proven directly by measuring photons, selected in wavelength using filters, in coincidence with product ions whose kinetic energies were determined by time of flight.

For the $N^+ + CO$ system, integral cross sections for several product channels have been reported based on a guided-ion-beam experiment (Frobin et al., 1977). The increase in the cross section toward higher energies, especially for CO^+, indicated contributions from exited products. By spectral analysis of the optical emission, Neuhauser et al. (1981) have shown that the dominant luminescent product is $CO^+(A\,^2\Pi)$. This reaction system has been studied in greater detail using luminescence and coincidence techniques in our universal guided-ion-beam apparatus (Kaefer, 1984).

Integral cross sections for luminescence in specific wavelength ranges, recorded using different filters, are shown in Fig. 77. The threshold onset at 2 eV coincides with the energy required for the formation of the exited $CO^+(A\,^2\Pi)$ state. At higher energies, the luminescence cross section increases to about 1 Å^2, corresponding to 30% of the total CO^+ cross section. Problems in calibrating the photon detection efficiency lead to an overall uncertainty of the absolute values of $\pm 50\%$. Accounting, in addition, for the differences in wavelength ranges, the agreement with previously published results (Neuschäfer et al., 1981) is satisfying.

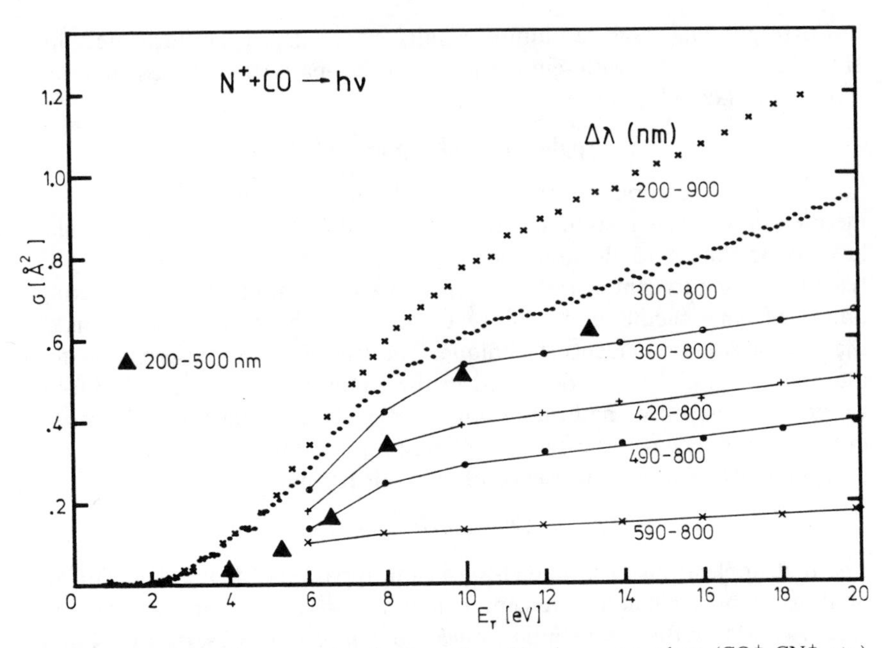

Figure 77. Integral cross sections for formation of luminescent products (CO^+, CN^+, etc.) in $N^+ + CO$ collisions recorded for selected wavelength ranges using the apparatus shown in Fig. 46. The triangles are from Neuschäfer et al. (1981).

A small contribution to the overall production of chemiluminescent photons is due to the formation of exited CN^{+*}. This specific products has been identified by detecting the photons in coincidence with mass-selected CN^+ ions. The integral cross section is plotted in Fig. 78 as a function of the total energy available for internal excitation of the products $[\Delta H^0(CN^+(a\,^1\Sigma)) = 2.97\,\text{eV}]$. Identification of the emitting state, based only on a threshold onset at about $E_T - \Delta H^0 = 3\,\text{eV}$, is not unequivocal since this threshold is significantly higher than the energy required to reach the $b\,^1\Pi$ state (1.03 eV), but below the $c\,^1\Sigma$ state (3.9 eV). A possible explanation is simultaneous excitation of both reaction products, for example, formation of the CN^+ ion in the $b\,^1\Pi$ state and the O atom in the ^1D state (1.97 eV).

Simultaneous excitation of both reactants has also been identified for the $CO^+(A\,^2\Pi) + N(^2D)$ product channel by using an even more complicated coincidence technique. In this case, photons, were detected in coincidence with ions that were selected according to their mass and flight time. The coincident velocity distributions of the luminescent products are compared in Fig. 79 with total velocity distributions measured for all CO^+ products. The transverse energy was limited to 80 meV to partially resolve some

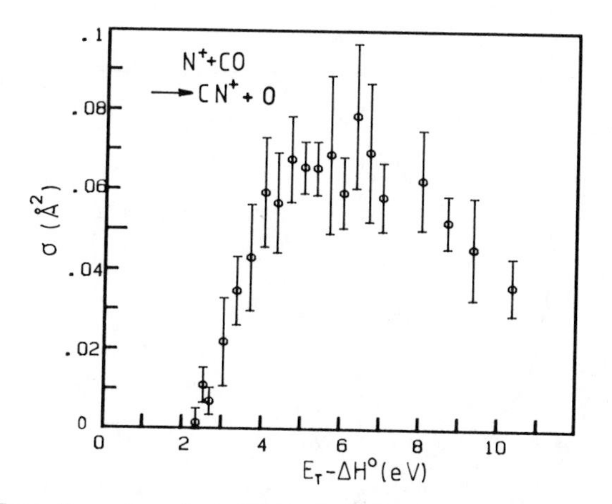

Figure 78. Formation of excited CN^{+*} ions in $N^+ + CO$ collisions. The specific chemiluminescent products are identified by detecting the photons in coincidence with the mass-selected CN^+ ions. The integral cross section is plotted as a function of the total available energy ($\Delta H^0 = 2.97$ eV).

structure, similar to Fig. 73. These axial product velocity distributions show a preference for formation of backward-scattered products. The areas of the luminescent and total intensity correspond to the partial integral cross sections. The double peak at laboratory energies below $E'_{1p} = 0.1$ eV is due to an experimental artifact; however, the structure at higher energies is real. Comparison of this structure with the ΔE_T scale reveals that at the collision energy of 2.66 eV, a significant fraction of products is formed with internal energies above 2 eV. At 7.3 eV (upper panel), a product group with a preference for converting more than 4 eV into internal energy can be identified. Comparison with the luminescent distributions reveals that at 2.66 eV collision energy, the dominant product channel is $CO^+(X\,^2\Sigma) + N(^2D) - 1.86$ eV, while at 7.3 eV, the luminescent distribution indicates excitation of both products, that is, formation of $CO^+(A\,^2\Pi) + N(^2D) - 4.39$ eV.

2. *Laser Preparation of Reactants*

Following the pioneering work of Chupka and Russell (1968), the endothermic reaction $H_2^+(v) + He \rightarrow HeH^+ + H$ has been extensively investigated using state-selective preparation by photoionization. Integral cross sections obtained with a guided-ion-beam apparatus were reported by Turner et al. (1984). We have used this reaction to test our multiphoton ionization source, described in Section III B 3. H_2^+ ions were created using 3 + 1 resonance-enhanced multiphoton ionization at 318.9 nm via the $B\,^1\Sigma_u^+(v' = 3, j')$ intermediate. At this wavelength the energy of the four photons does not

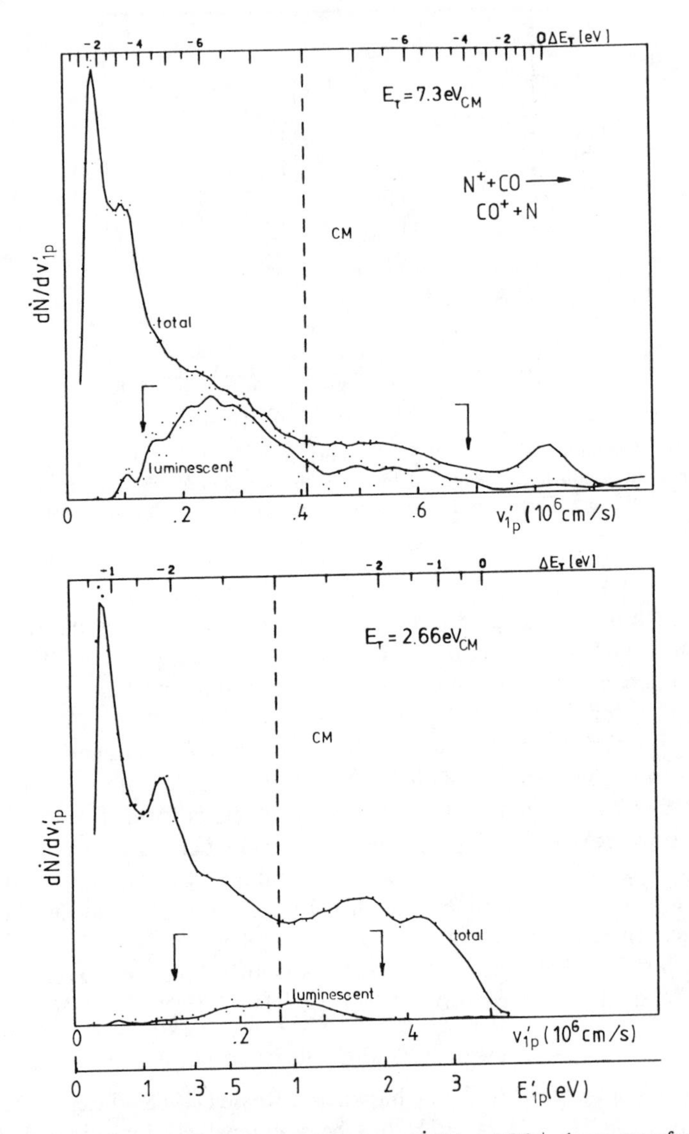

Figure 79. Axial product velocity distributions $d\dot{N}/dv'_{1p}$ of CO^+ charge-transfer products formed in $N^+ + CO$ collisions at two energies. In this experiment the transverse energy was limited to 80 meV using a low octopole guiding field. The dashed line marks the center-of-mass velocity. The upper scale refers to the translational exoergicity ΔE_T. Compared are the distributions of all CO^+ products (total) with those from luminescent CO^+ ions, distinguished by photon–ion coincidence.

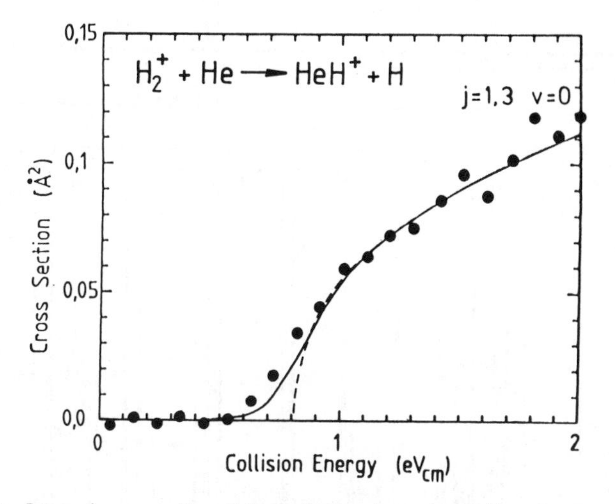

Figure 80. Integral cross section of the endothermic reation $H_2^+(v=0)+He \rightarrow HeH^+ + H$ in the threshold region. State-selective preparation of H_2^+ ions was achieved by multiphoton ionization using the quadrupole photoionization source in the guided-ion-beam apparatus. The data points are in reasonable agreement with the effective cross section (heavy line) calculated from an assumed threshold onset (dashed line) [Eq. (91)].

allow one to reach the first vibrational state of the H_2^+ ion. Thus the H_2^+ ions are all produced in their vibrational ground state. To test experimentally whether absorption of additional photons leads to higher excitation, the cross section for HeH^+ formation has been measured in the threshold region; see Fig. 80. The onset of the integral cross section below 0.8 eV is predominantly due to thermal broadening, as can be seen from a comparison of an assumed cross section (dashed line) and the convoluted result (solid line).

Another method for preparing ions in specific states is to excite them with a laser to a long-lived state. We have used this method to prepare CO^+ ions in the excited $CO^+(A)$ state in order to induce charge transfer with Ar, which for the ground-state $CO^+(X)$ ion is 1.66 eV endothermic. The optical laser excitation spectrum, shown in the upper part of Fig. 81, was recorded by detecting Ar^+ charge-transfer products from $CO^+ + Ar$ in the universal guided-ion-beam apparatus at a collision energy of 0.3 eV. The depicted wavelength range shows a rotational progression of $CO^+(X\,^2\Sigma^+(v=0) \rightarrow A\,^2\Pi(v=2))$ transitions. In this experiment CO^+ ground-state ions were prepared by electron bombardment, thermalized to the temperature of the trapping source, and injected into the octopole system. In the interaction region, the ion beam was crossed (see Fig. 46) with an intense, nonskimmed pulsed beam of Ar gas, creating a momentary and local target density above $10^{13}\,cm^{-1}$. Synchronously, the ions were excited with a pulsed laser (energy

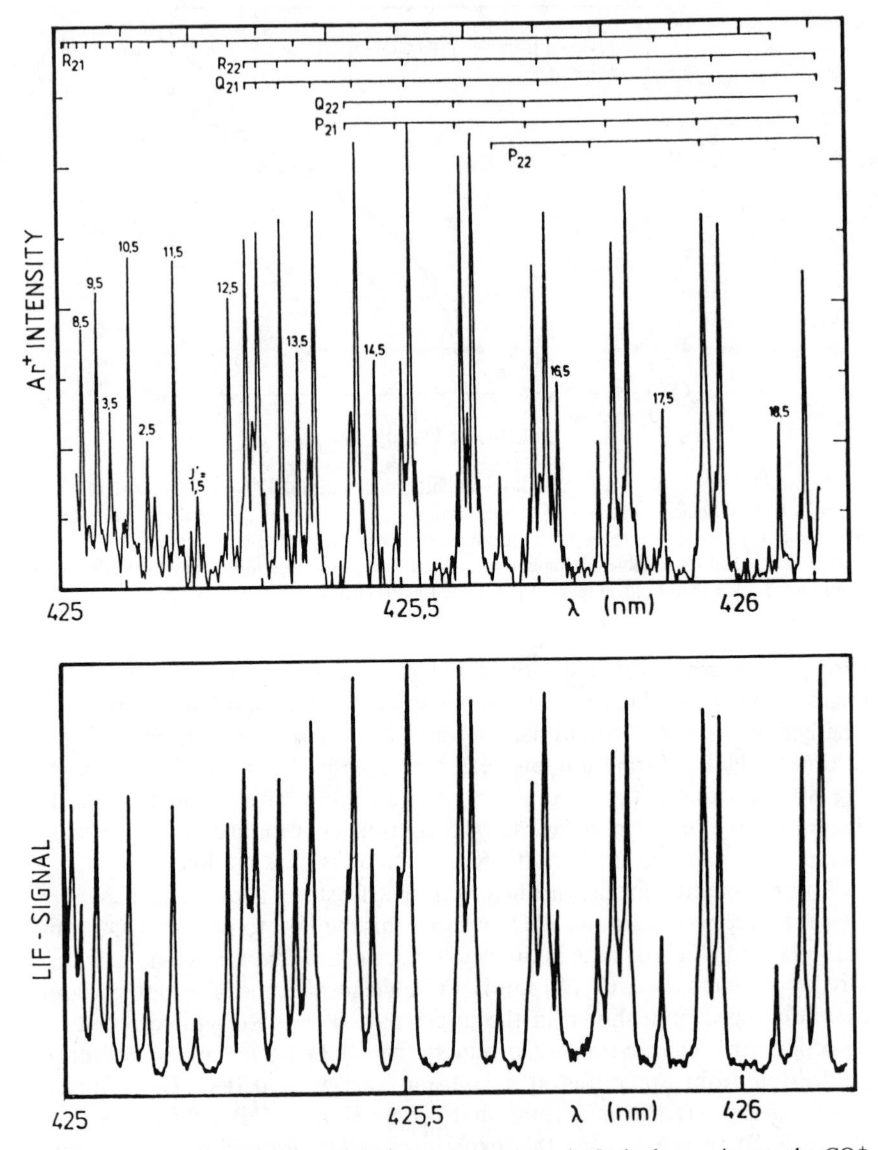

Figure 81. Laser excitation of CO^+ ions in an octopole. In both experiments the CO^+ ions were formed by electron bombardment in the rf storage ion source of the guided-ion-beam apparatus. The upper panel shows the wavelength dependence of Ar^+ charge-transfer products formed in low-energy collisions between laser excited $CO^+(A\ ^2\Pi)$ and Ar in the scattering cell. The lower panel depicts, in the same wavelength range, the laser-induced fluorescence signal from CO^+ ions trapped in the scattering cell between two potential barriers.

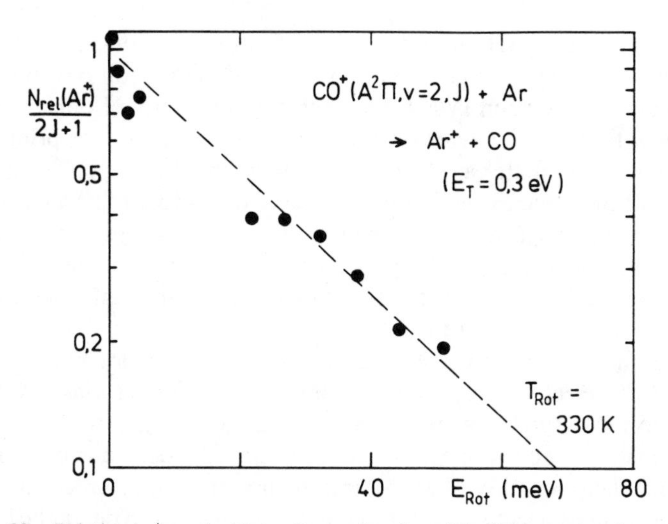

Figure 82. Relative Ar^+ product intensity as a function of the CO^+ rotational states, derived from the upper spectrum in Fig. 81 using $2J + 1$ weighting coefficients. The slope mirrors the 330 K thermal rotational distribution of the $CO^+(X)$ primary ions, that is, there is no significant rotational dependence of the charge transfer cross section at the collision energy of 0.3 eV.

4 mJ, beam diameter of 1 mm). Saturation of the transition and the rather long lifetime of the $CO^+(A)$ (3 μs) led to a significant fraction of excited ions. At a typical primary intensity of 5×10^4 CO^+ ions per laser shot, a product rate of 180 Ar^+ ions was detected.

The product signal is proportional to the thermal population of the rovibrational ground-state CO^+ ions and to the charge-transfer cross section. An evaluation of intensities measured for different rotational states is plotted in Fig. 82. Within the accuracy of our experiment, the relative Ar^+ product intensity can be fitted with a 330 K thermal rotational distribution of the $CO^+(X)$ primary ions. This is so close to the ion source temperature that one can exclude a significant rotational dependence of the charge-transfer cross section at our collision energy of 0.3 eV. This example illustrates that the combination of laser excitation with guided ion beams can be used to study reactions with state-prepared reactants. Of course, this method is also well suited for spectroscopical studies. A similar method, using charge-exchange detection in an ion flow tube, has been reported by Kuo et al. (1986).

3. *Laser Analysis of Products*

CO^+ product-state distributions in the reverse charge-transfer reaction, $Ar^+ + CO \rightarrow CO^+ + Ar$, were examined using laser-induced fluorescence by the group of Leone (Hamilton et al., 1985; Lin et al., 1985). We have tried to apply this method in combination with the guided-ion-beam apparatus. As explained in Section IV B 4, however, the sensitivity of our arrangement is too low to detect the reaction products. Only when we use primary CO^+ ions and accumulate a sufficient number in the region seen by the photomultiplier (see Fig. 46), can we observe the induced fluorescence signal. A corresponding spectrum, recorded in the same wavelength range as in the case of the laser-induced charge transfer, is depicted in the lower half of Fig. 81. In both spectra, the intensities of the individual lines reveal the rotational temperature of the CO^+ ions.

A significantly more sensitive laser-based technique for product analysis is based on selective photofragmentation of the product ions. As described in Section IV B 4, this method has several advantages including a large laser–ion interaction volume, efficient collection and detection of the fragment ions, and the possibility of saturating the fragmentation process.

We have used selective photofragmentation to study in detail the formation of metastable $O_2^+(a\,^4\Pi_u)$ formed in $Ar^+ + O_2$ collisions (Scherbarth and Gerlich, 1989). The integral cross section, depicted in Fig. 63, shows the threshold onset of this endothermic channel. State-selective analysis of the metastable $O_2^+(a\,^4\Pi_u)$ state has been performed by fragmentation via the $b\,^4\Sigma_g$ state. A small section of the experimental $\Delta v = 3$ photofragment spectrum of the $O_2^+(a\,^4\Pi_u)$ products is plotted in the lowest panel of Fig. 83. Using a set of molecular constants (Hansen et al., 1983), and assuming a microcanonical rovibrational population of the O_2^+ products, we have simulated the O_2^+ fragmentation spectra. Comparison of these results, shown for $v'' = 0$–3 in the upper panels of Fig. 83, with the measured spectrum reveals that most of the experimental lines are due to the $v'' = 0$ products. An analysis of all measured spectra (Scherbarth and Gerlich, 1989) corroborates this nonstatistical preference of the $v'' = 0$ state (more than 90% at $E_T = 0.55$ meV). In contrast to specific excitation of the vibrational ground state, a wide range of rotational states is excited. The population of these $(v'' = 0, j'')$ states is in accordance with a statistical distribution. This can be seen from Fig. 84 where two populations measured at collision energies of 0.53 eV and 1.05 eV are compared with the microcanonical equilibrium distribution. As described briefly in Section IV B 4, we are presently exploring the possibilities of extending the versatility of this method by using a two-photon fragmentation scheme.

D. Radiative Association and Fragmentation

Radiative association of positive ions with molecules is considered to be an important process in the synthesis of complex molecules in low-density interstellar clouds. One of the first applications of an ion trapping method

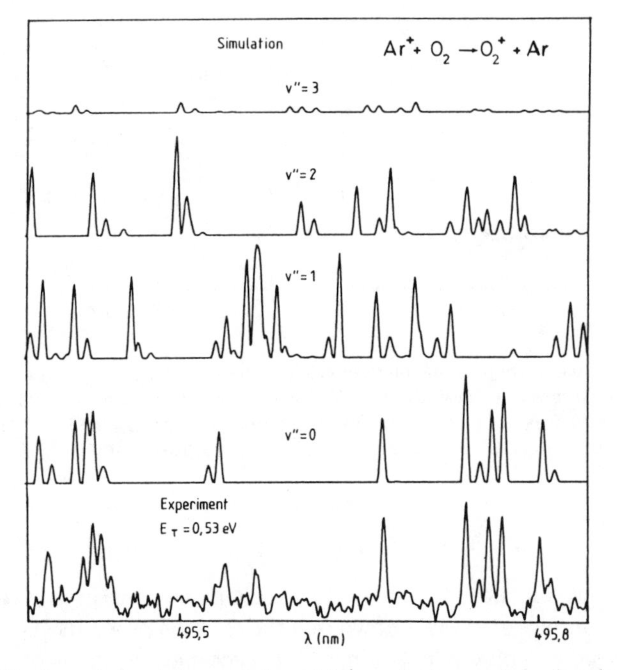

Figure 83. Laser-induced fragmentation of O_2^+ products recorded using the guided-ion-beam apparatus shown in Fig. 46. The lowest panel shows a small section of the experimental $\Delta v = 3$ photofragment spectrum of $O_2^+(a\,^4\Pi_u)$ products populated in the $Ar^+ + O_2$ charge-transfer reaction. The upper curves are simulated spectra calculated separately for the accessible product vibrational states $v'' = 0$–3. For the rovibrational states a statistical population was assumed.

at very low temperatures (11–20 K) for studying radiative association was reported by the group of Dunn (Barlow et al., 1986; Luine and Dunn, 1985). They determined a radiative association rate coefficient of $k_I = 1.1 \times 10^{-13}\,\mathrm{cm^3/s}$ for $CH_3^+ \cdot H_2$; however, for the astrophysically important $C^+ \cdot H_2$ radiative stabilization they could only measure an upper limit of $k_I = 1.5 \times 10^{-15}\,\mathrm{cm^3/s}$. The higher sensitivity of our temperature variable ring electrode trap has extended the range for determining radiative association rate coefficients from $10^{-15}\,\mathrm{cm^3/s}$ to $10^{-17}\,\mathrm{cm^3/s}$. Results for the stabilization of the $CH_3^+ \cdot H_2$ complex were reported recently (Gerlich and Kaefer, 1989). In this section, we present similar results for the prototype system $H^+ \cdot H_2$ and for $C^+ \cdot H_2$. Another very recent application of the rf ring electrode trap is the determination of the radiative decay rate of highly excited molecules using photofragmentation with a CO_2 laser. This method will be illustrated using H_3^+ and CH_5^+ as examples.

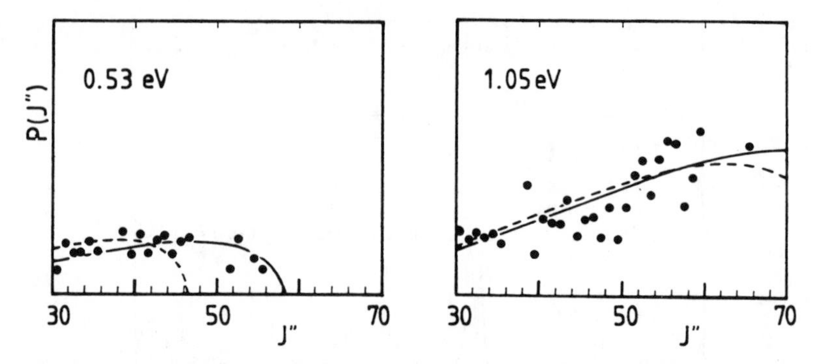

Figure 84. Relative populations of rotational levels for $O_2^+(a\,^4\Pi_u, v=0)$ product ions formed in $Ar^+ + O_2$ collisions at 0.53 eV and 1.05 eV. The depicted probabilities were extracted from photofragmentation spectra [see, for example, Fig. 83 and Scherbarth and Gerlich (1989)]. The lines are the microcanonical equilibrium distributions for the two $Ar^+(^2P_J)$ states.

1. *Association of $H^+ \cdot H_2$ and $C^+ \cdot H_2$*

The prototype reaction $H^+ + H_2$ has been mentioned in several of the preceding examples. It is well known that the reaction proceeds via formation of a long-lived complex. The lifetime of this complex, which is on the order of several 10^{-10} s, has been studied using classical trajectory calculations (Schlier and Vix, 1985). Such a highly excited H_3^+ molecule can emit infrared photons and corresponding calculated radiation spectra, based also on classical trajectories, have been reported (Berblinger and Schlier, 1988). The radiative and three-body stabilization of $H^+ \cdot H_2$ and $D^+ \cdot D_2$ has been studied in the ring electrode trap at 350 K and 80 K. A typical result is shown in Fig. 85 for formation of D_3^+. The procedure used to determine the apparent binary rate coefficient $k^* = k_r + n_2 k_3$ has been described in Section IV E. For $D^+ + D_2$, the slope of the line in Fig. 85 yields a three-body rate coefficient $k_3(80\,\mathrm{K}) = 7.6 \pm 1 \times 10^{-29}\,\mathrm{cm}^6/\mathrm{s}$ and the intersection at $n_2 = 0$ can be taken as an estimate for the radiative stabilization rate coefficient $k_r(80\,\mathrm{K}) = 1 \times 10^{-16}\,\mathrm{cm}^3/\mathrm{s}$. Similar results have been obtained for $H^+ + H_2$ [$k_3(80\,\mathrm{K}) = 5.4 \pm 1 \times 10^{-29}\,\mathrm{cm}^6/\mathrm{s}$ and $k_r(80\,\mathrm{K}) = 1.3 \times 10^{-16}\,\mathrm{cm}^3/\mathrm{s}$].

Based on assumptions that have been discussed in Gerlich and Kaefer (1989), one can obtain from the two values k_r and k_3 an estimate of the complex and the radiative lifetimes. An uncertain parameter in this estimation is the stabilization factor f_s, that is, the efficiency of deactivating the intermediate complex via collision with a third body. Assuming $f_s = 0.17$ for $H^+ \cdot H_2$ in a pure H_2 environment, one obtains a complex lifetime of 0.65×10^{-10} s and a radiative lifetime of 1.2×10^{-3} s. These values deviate from those determined from classical trajectory calculations that estimate a

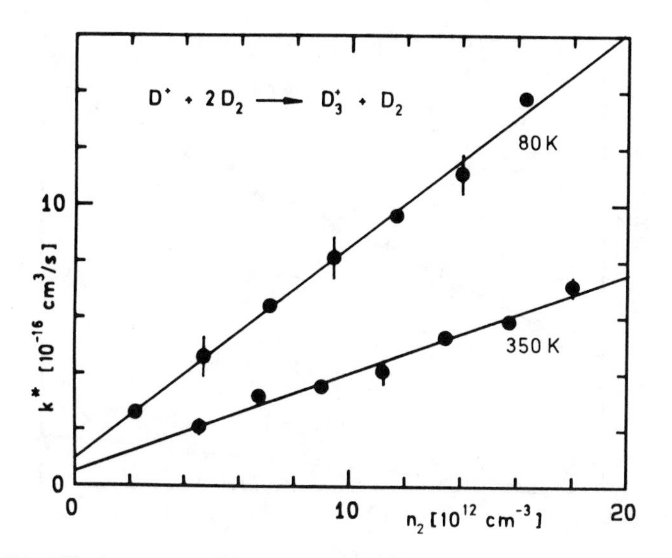

Figure 85. Effective rate coefficient $k^* = n_2 k_3 + k_r$ for D_3^+ formation via ternary and radiative association of $D^+ \cdot D_2$ complexes measured at 80 K and 350 K in the ring electrode trap. Extrapolation of the target density to zero results in an upper limit for the radiative association rate coefficient k_r.

complex lifetime of 6.8×10^{-10} s (Schlier and Vix, 1985) and a radiative lifetime of 0.14×10^{-3} s (Berblinger and Schlier, 1988). This comparison indicates that, even for this simple prototype system, more detailed experiments and theoretical studies are required.

An interesting extension of the $H^+ \cdot H_2$ system is to increase the degrees of freedom by adding additional hydrogen molecules. In Sections III and IV, this was used predominantly to illustrate the low-temperature capabilities of the ring electrode trap. Figure 44, for example, gives several ternary rate coefficients for the growth of hydrogen clusters up to H_g^+ at 25 K, while Fig. 59 demonstrates the increase of the ternary rate coefficients for $H_3^+ \cdot H_2$ stabilization as the temperature is lowered from 80 K to 25 K.

The radiative association of $C^+ \cdot H_2$ is the important first step in the interstellar gas-phase formation of hydrocarbons. This process has been studied in the liquid nitrogen cooled ion trap (Kaefer, 1989). The 80 K data are plotted in Fig. 86. The resulting rate coefficients are $k_3(80\,\text{K}) = 5.9 \times 10^{-29}\,\text{cm}^3/\text{s}$ and $k_r(80\,\text{K}) = 7 \times 10^{-16}\,\text{cm}^3/\text{s}$. Assuming a third-body stabilization efficiency $f_s = 1$, one obtains a radiative lifetime of $2 \times 10^{-5}\,\text{s}^{-1}$, which indicates that stabilization occurs via an optical transition. Very recent measurements with the liquid-helium-cooled trap at a nominal temperature of 25 K are also included in Fig. 86. The resulting radiative rate coefficient

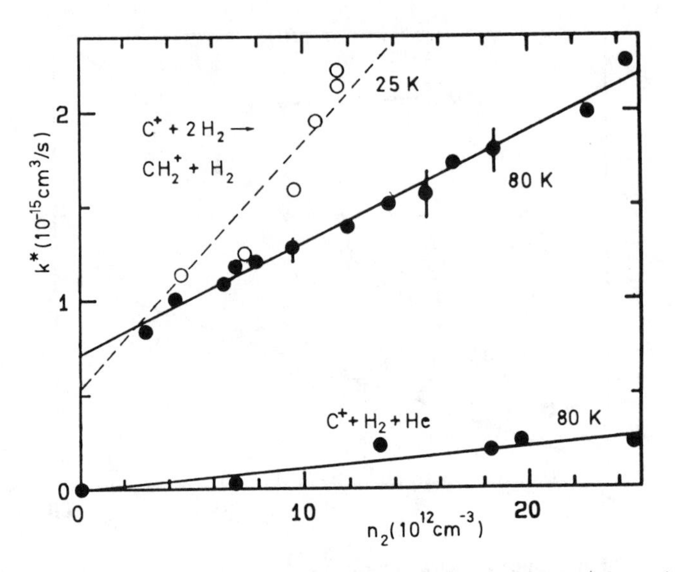

Figure 86. Effective rate coefficients $k^* = n_2k_3 + k_r$ for associative CH_2^+ formation in the liquid-helium- (25 K) and the liquid-nitrogen-cooled (80 K) ring electrode traps. The two upper curves were recorded using pure hydrogen and include $k_3(H_2)$ and k_r. The data suggest that the radiative association rate coefficients k_r(25 K) and k_r(80 K) are smaller than 8×10^{-16} cm³/s. The lowest curve shows the net influence of additional He acting as the third body.

has the same magnitude, $k_r(25\text{ K}) = 5.5 \times 10^{-16}$ cm³/s, but ternary association is two times faster [$k_3(25\text{ K}) = 13 \times 10^{-29}$ cm⁶/s]. The estimated error limits of the two radiative association rates are $\pm 50\%$. The main problem in determining these values arises from reactions with gas impurities, and we have observed variations for data sets obtained under different conditions. Additional complications are due to the fact that the CH_2^+ products, as well as possibly formed CH^+ ions, are not directly observable because they are converted into CH_3^+ by fast secondary reactions with H_2 as has been discussed in Section V A 3.

2. *Radiative Lifetimes of* $H^+ \cdot H_2$ *and* $CH_3^+ \cdot H_2$

As described previously, intermediate complex and radiative lifetimes can be inferred by measuring the apparent rate coefficient k^* as a function of n_2. In order to obtain more direct insight into the radiative decay of highly exited molecular complexes, we have probed the population of states in the vicinity (0.1 eV) of the dissociation limit via photofragmentation with a CO_2 laser. Owing to a low laser intensity of a few watts, multiphoton transitions can be ruled out. Therefore, only those ions whose internal energy deviates at most by $h\nu$ from the dissociation limit can be affected by a single CO_2 laser photon.

Highly exited H_3^+ and CH_5^+ ions have been prepared in the storage source, but with high-energy electrons and short storage times to avoid quenching collisions. Under such conditions, a sufficiently large number of excited ions could be created. These ions were injected into the trap, this time under ultrahigh vacuum conditions ($< 10^{-9}$ mbar). After a variable storage time of several ms, a typically 100-μs-short CO_2 laser pulse was sent through the trap leading to the formation of fragment ions. With increasing storage time, the number of ions that could be fragmented decreased, while the total number of trapped ions remained unchanged. The net signal is plotted in Fig. 87. It can be seen that the number of excited ions decreases exponentially. Since under our vacuum conditions collision rates are on the order of $1\ s^{-1}$, this decay must be due to radiative relaxation.

The radiative lifetime determined for H_3^+, 0.37×10^{-3} s, is larger than the calculated value of 0.14×10^{-3} s, but smaller than that deduced from the association measurement, 1.2×10^{-3} s. The deviation of our two experimental values is most probably due to differences in the angular momentum distributions of the two prepared H_3^+ ions, that is, the collision complex $H^+ \cdot H_2$ and the highly excited molecule $(H_3^+)^{**}$. The lifetime ratio,

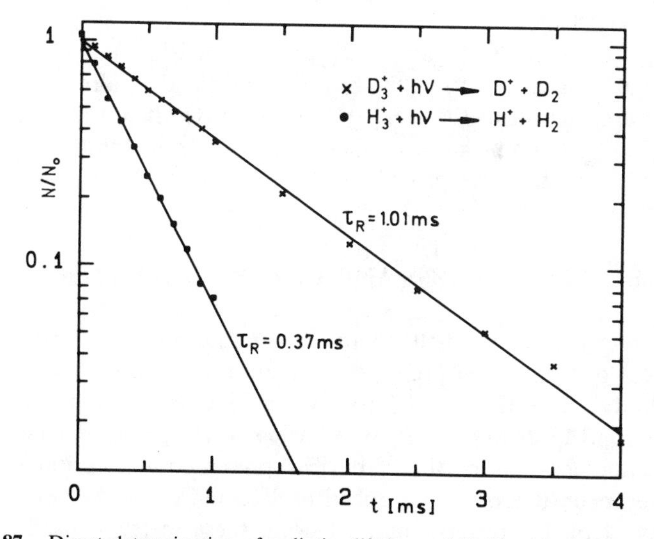

Figure 87. Direct determination of radiative lifetimes of highly exited H_3^+ and D_3^+ ions. Externally created ions were injected into the ring electrode trap, and their spontaneous radiative decay was probed by delayed CO_2 laser-induced fragmentation. Loss by processes other than radiative decay is excluded on the depicted time scale and at the low pressure ($<10^{-9}$ mbar).

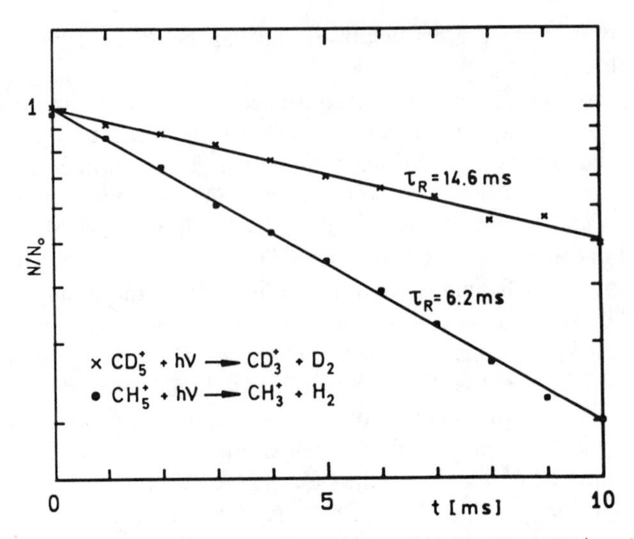

Figure 88. Determination of the radiative lifetimes of highly exited CH_5^+ and CD_5^+ ions using CO_2 laser fragmentation. The lifetimes are significantly longer than for H_3^+ and D_3^+, shown in Fig. 87.

$\tau(D_3^+)/\tau(H_3^+) = 2.4$ is in good agreement with the mass ratio, $(m(D_3^+)/m(H_3^+))^{3/2} = 2.8$.

Similar results for CH_5^+ and CD_5^+ are depicted in Fig. 88. The lifetimes of 6.2 ms and 14.6 ms support the recently published results on ternary and radiative association of the $CH_3^+ \cdot H_2$ complex (Gerlich and Kaefer, 1989). These examples show that combinations of laser methods with the ring electrode trap can also provide interesting spectroscopic and dynamic information.

VI. CONCLUSIONS AND FUTURE DEVELOPMENTS

This chapter has reviewed the theory of the motion of charged particles in fast oscillatory fields and presented several applications of electric rf fields for the study of processes involving slow ions. The major aim of the theoretical part, Section II, was to provide some basic knowledge for understanding the experimental features of the rf devices presented in Section III. Several complete apparatuses were described in Section IV, and Section V gave an overview of typical results obtained using these instruments. In this final section we make a few remarks to the state of the theory and their developmental possibilities, we discuss several improvements of the presented instruments, and add a few speculations.

Our theoretical treatment was dominated by the adiabatic approximation, since conservation of the mean kinetic energy of the ion is required for most experimental application and the adiabatic constants of the motion simplify the determination of conditions for safe confinement of ions in any rf field. Adiabatic theories are well established in theoretical physics and mathematics, the only blemish of the presented theory is the empirical determination of the border line between "safe" and "unsafe" regions, based on $\eta_m(\mathbf{r}) = 0.3$. Here, additional theoretical work is needed for a rigorous justification. More generally, it would be desirable to find a mathematical procedure to completely map out regions of stability and their adiabatic subregions for any rf field. This would also allow one to find out whether there is any suitable rf device, besides the quadrupole, in which one could take experimental advantage of sharp transitions from stability to instability, for example, for mass or phase-space selection. It is suspected that the available modern mathematical methods developed for characterizing nonlinear dynamical systems, provide the framework for a fundamental description of the motion of charged particles in inhomogeneous rf fields.

The presented variety of rf devices has illustrated that there are almost no geometrical restrictions to create two- or three-dimensional potential minima. The applicability of these devices can be further extended by superimposing additional temporal or spatial varying fields (auxiliary low-frequency field, pulsed electrodes, field penetration from correction electrodes, etc.). For example, a series of ring electrodes surrounding a multipole could be used to create an axial, near harmonic dc potential and a weak resonant excitation of the axial motion could result in a mass-selective ejection. Other extensions can be imagined by adding a magnetostatic field in which the cyclotron motion of specific ions may be resonantly excited. Beyond all these discussed quasielectrostatic fields also the nodal points or lines of standing or traveling electromagnetic waves could be used to store ions. A possible application is the making of a well-defined lattice of laser-cooled ions in a microwave guide, for example, for studying collective effects.

The versatility of instruments based on rf devices has been illustrated through a variety of applications, and it can be imagined that these developments will stimulate the invention of new apparatuses or other rf devices. However, many possibilities still remain for improving the existing instruments. Significant progress can be expected from combination of ion guiding or trapping with various optical methods. Especially, laser based methods such as multiphoton ionization, one- or two-color fragment spectroscopy, laser-induced fluroescence, laser cooling, infrared excitation of ions or neutrals, and so on, will lead to many new applications in spectroscopy, reaction dynamics, analytical chemistry, and other fields.

The capability of the new merged-beam arrangement for the study of ion reaction dynamics at very low energies can be further enhanced, for example, by adding a temperature variable ion trap for thermalization of the ionic reactants or by using supersonic beams with various expansion conditions. Other extensions of this instrument could make use of an atomic beam, of a beam of free radicals, or of a second slow ion beam. A more speculative extension of the merged-beam method is the superposition of an electron and an ion beam used to study dielectronic recombination, whereby the beam of slow electrons could be confined in a microwave guide.

Similar improvements can also be imagined for ion-trapping devices. Pulsed gas valves would allow one to utilize different gases for thermalization and reaction. The advantage of time-of-flight mass selection for simultaneous analysis of the entire trap contents is clearly evident. Progress can also be made by mass-selective extraction or mass-selective ejection of product ions without perturbing the thermalized ensemble of primary ions. Finally, laser cooling of stored molecular ions may even be feasible by using stimulated-emission pumping or infrared transitions, and in combination with cold trapped neutrals, could lead to μeV collision energies.

It can be expected that the described methods of trapping and guiding charged particles are not only limited to those applications mentioned in this chapter. Therefore, it is obvious that there is much space for inventiveness and that we can look forward to a much broader application of inhomogeneous rf fields.

Acknowledgements

This work involved the cooperation and assistance of many people. I am indebted especially to Professor Ch. Schlier, who rendered possible a large fraction of this research by providing the scientific environment in his group and by financial support, and who contributed by many stimulating and fruitful discussions and by his continuous interest. I am particularly grateful to Dr. E. Teloy who introduced me into the fundamentals of the theory and applications of inhomogeneous rf fields and with whom I frequently discussed various technical and scientific questions. A large fraction of the recent experiments has been performed by two former students, Dr G. Kaefer and Dr. S. Scherbarth, and in cooperation with several Diplom–students whom I want to thank for their efforts. It is also a pleasure to acknowledge the help and assistance of the technical staff from the Fakultät für Physik, Freiburg, and especially the skills of Dipl. Ing. U. Person. I am especially indebted to Dr. S. Horning for his careful reading of the manuscript and for many suggestions and very helpful hints. Various aspects of this research were supported by the Deutsche Forschungsgemeinschaft (Schwerpunktprogramm "Dynamik zustandsselektierter chemischer Primärprozesse").

References

M. Abramowitz and I. Stegun (1964), *Handbook of Mathematical Functions*, National Bureau of Standards, Washington, DC.

S. L. Anderson, F. A. Houle, D. Gerlich, and Y. T. Lee (1981), *J. Chem. Phys.* **75**, 2153.

V. I. Arnold (Ed.) (1988), *Dynamical Systems III*, Springer-Verlag, Berlin.

P. Ausloos (Ed.) (1979), *Kinetics of Ion–Molecule Reactions*, Plenum Press, New York.

P. Ausloos and S. G. Lias (Eds.) (1987), *Structure/Reactivity and Thermochemistry of Ions*, Reidel, Dordrecht.

R. Bahr (1969), Diplom Thesis, Univ. Freiburg.

R. Bahr, D. Gerlich, and E. Teloy (1970), *Verhandl. DPG (VI)* **5**, 131.

R. Bahr, D. Gerlich, and E. Teloy (1969), *Verhandl. DPG (VI)* **4**, 343.

L. S. Bartell, H. B. Thomson, and R. R. Roskos (1965), *Phys. Rev. Letters* **14**, 851.

S. E. Barlow, J. A. Luine, and G. H. Dunn (1986), *Int. J. Mass Spectrom. Ion Processes* **74**, 97.

S. E. Barlow and G. H. Dunn (1987), *Int. J. Mass Spectrom. Ion Processes* **80**, 227.

E. C. Beaty (1987), *J. Appl. Phys.* **61**, 2118.

K. Berkling, R. Helbing, K. Kramer, H. Pauly, Ch. Schlier, and P. Toschek (1962), *Z. Physik* **166**, 406.

M. Berblinger and Ch. Schlier (1988), *Mol. Phys.* **63**, 779.

W. Beyer (1976), Diplom Thesis, Univ. Freiburg.

G. Bischof and F. Linder (1986), *Z. Phys.* **D1**, 303.

L. A. Bloomfield (1990), *J. Opt. Soc. Am.* **B7**, 472.

R. Blümel, C. Kappler, W. Quint, and H. Walther (1989), *Phys. Rev. A* **40**, 808.

H. Böhringer (1985), *Chem. Phys. Lett.* **122**, 185.

H. Böhringer, W. Glebe, and F. Arnold (1983), *J. Phys. B, At. Mol. Phys.* **16**, 2619.

M. T. Bowers (Ed.) (1979), *Gas Phase Ion Chemistry*, Academic Press, New York, Vols. 1 and 2.

M. T. Bowers (Ed.) (1984), *Gas Phase Ion Chemistry*, Academic Press, New York, Vol. 3.

J. P. Braga, D. B. Knowles, and J. N. Murrell (1986), *Mol. Phys.* **57**, 665.

P. H. Bucksbaum, M. Bashkansky, and T. J. McIlrath (1987), *Phys. Rev. Lett.* **58**, 349.

P. H. Bucksbaum, D. W. Schumacher, and M. Bashkansky (1988), *Phys. Rev. Lett.* **10**, 1182.

F. v. Busch and W. Paul (1961a), *Z. Phys.* **164**, 581.

F. v. Busch and W. Paul (1961b), *Z. Phys.* **164**, 588.

S. W. Bustamente, M. Okumura, D. Gerlich, H. S. Kwok, L. R. Carlson, and Y. T. Lee (1987), *J. Chem. Phys.* **86**, 508.

P. J. Chantry (1971), *J. Chem. Phys.* **55**, 2746.

W. A. Chupka and M. E. Russell (1968), *J. Chem. Phys.* **49**, 5426.

D. A. Church (1969), *J. Appl. Phys.* **40**, 3127.

W. E. Conaway, T. Ebata, and R. N. Zare (1987), *J. Chem. Phys.* **87**, 3447.

R. J. Cook, D. G. Shankland, and A. L. Wells (1985), *Phys. Rev.* **A31**, 564.

E. D. Courant (1952), *Phys. Rev.* **88**, 1190.

J. B. Cross and J. J. Valentini (1982), *Rev. Sci. Instr.* **53**, 38.

P. H. Dawson (1976), *Quadrupole Mass Spectrometry*, Elsevier Scientific Publishing, Amsterdam.

P. H. Dawson and N. R. Whetten (1971), in *Dynamic Mass Spectrometry*, editted by D. Price, Heyden & Son, London, Vol. 2.

I. E. Dayton, F. C. Shoemake, and R. F. Moseley (1954), *Rev. Sci. Instr.* **25**, 485.

H. G. Dehmelt (1967), *Advan. At. Mol. Phys.* **3**, 53.

H. G. Dehmelt (1969), *Advan. At. Mol. Phys.* **5**, 109.

D. R. Dension (1971), *J. Vac. Sci. Technol.* **8**, 266.

F. Diedrich, J. C. Bergquist, W. M. Itano, and D. J. Wineland (1989), *Phys. Rev. Lett.* **62**, 403.

R. Disch, S. Scherbarth, and D. Gerlich, in *Conference on the dynamics of Molecular Collisions*, Snowbird, Utah, edited by D. G. Truhlar and P. J. Dagdigian.

G. G. Dolnikowski, M. J. Kristo, C. G. Enke, and J. T. Watson (1988), *Int. J. Mass Spectrom. Ion Processes* **82**, 1.

C. H. Dougless, G. Ringer, and W. R. Gentry (1982), *J. Chem. Phys.* **76**, 2423.

J. H. Eberly (1969), in *Progress in Optics*, edited by E. Wolf, North-Holland, Amsterdam, Vol. VII, p. 382.

K. M. Ervin and P. B. Armentrout (1985), *J. Chem. Phys.* **83**, 166.

K. M. Ervin and P. Armentrout (1987), *J. Chem. Phys.* **86**, 2659.

K. M. Ervin, S. K. Loh, N. Aristow, and P. B. Armentrout (1983), *J. Phys. Chem.* **87**, 3593.

J. M. Farrar (1988), in *Techniques for the Study of Ion–Molecule Reactions*, edited by J. M. Farrar and W. H. Saunders, Jr., Wiley, New York.

F. C. Fehsenfeld, D. L. Albritton, Y. A. Bush, P. G. Fournier, T. R. Govers, and J. Fournier (1974), *J. Chem. Phys.* **61**, 2150.

E. Fischer (1959), *Z. Phys.* **156**, 26.

G. D. Flesch, S. Nourbakhsh, and C. Y. Ng (1990), *J. Chem. Phys.* **92**, 3590.

J. C. Franklin (Ed.) (1979), *Ion and Molecule Reactions, Kinetics and Dynamics*, Dowden, Hutchinson, & Ross, Stroudsburg, PA.

B. Friedrich, S. Pick, L. Hladek, Z. Herman, E. E. Nikitin, A. I. Reznikov, and S. Ya. Umanskii (1986), *J. Chem. Phys.* **84**, 807.

B. Friedrich, W. Trafton, R. Rockwood, S. Howard, and J. H. Futrell (1984), *J. Chem. Phys.* **80**, 2537.

M. H. Friedman, A. L. Yergey, and J. E. Campana (1982), *J. Phys. E Sci. Instrum.* **15**, 53.

W. Frobin, Ch. Schlier, K. Strein, and E. Teloy (1977), *J. Chem. Phys.* **67**, 5505.

J. H. Futrell (Ed.) (1986), *Gaseous Ion Chemistry and Mass Spectrometry*, Wiley, New York.

H. Gaber (1987), Diplom Thesis, Univ. Freiburg.

A. V. Gaponov and M. A. Miller (1958), *Soviet Phys. JETP* **7**, 168.

W. R. Gentry (1979), in *Gas Phase Ion Chemistry*, edited by M. T. Bowers, Academic Press, New York, Vol. 2.

W. R. Gentry, D. J. McClure, and C. H. Douglas (1975), *Rev. Sci. Instr.* **46**, 367.

D. Gerlich (1971), Diplom Thesis, Univ. Freiburg.

D. Gerlich (1977), PhD Thesis, Univ. Freiburg.

D. Gerlich (1982), in *Symposium on Atomic and Surface Physics*, edited by W. Lindinger et al., Univ. Innsbruck, Salzburg, p. 304.

D. Gerlich (1984), in *Symposium on Atomic and Surface Physics*, edited by F. Howorka et al., Univ. Innsbruck, Salzburg, p. 116.

D. Gerlich (1986), in *Electronic and Atomic Collisions*, edited by D. C. Lorents et al., Elsevier, Amsterdam, p. 541.

D. Gerlich (1989a), *J. Chem. Phys.* **90**, 127.

D. Gerlich (1989b), *J. Chem. Phys.* **90**, 3574.

D. Gerlich (1989c), in *XII International Symposium on Molecular Beams*, edited by V. Aquilanti, Perugia, p. 37.

D. Gerlich (1990), *J. Chem. Phys.* **92**, 2377.

D. Gerlich and H. Bohli (1981), in *European Conference on Atomic Physics*, edited by J. Kowalski, G. zu Putlitz, and H. G. Weber, Heidelberg.

D. Gerlich, R. Disch, and S. Scherbarth (1987), *J. Chem. Phys.* **87**, 350.

D. Gerlich and G. Kaefer (1985), in *Second European Conference on Atomic and Molecular Physics*, edited by A. E. Vries et al., Amsterdam.

D. Gerlich and G. Kaefer (1987), in *5th Int. Swarm Seminar, Proceedings*, edited by N. G. Adams and D. Smith, Birmingham University, 133.

D. Gerlich and G. Kaefer (1988), in *Symposium on Atomic and Surface Physics*, edited by A. Penelle et al., La Plagne, France, 115.

D. Gerlich and G. Kaefer (1989), *Ap. J.* **347**, 849.

D. Gerlich, G. Kaefer, and W. Paul (1990), in *Symposium on Atomic and Surface Physics*, edited by T. D. Märk and F. Howorka, Obertraun.

D. Gerlich, G. Jerke, U. Mück, and U. Person (1991) (unpublished).

D. Gerlich, U. Nowotny, Ch. Schlier, and E. Teloy (1980), *Chem. Phys.* **47**, 245.

D. Gerlich and S. Scherbarth (1987), in *Electronic and Atomic Collisions*, edited by J. Geddes et al., Contributed papers, p. 809.

D. Gerlich and M. Wirth (1986), in *Symposium on Atomic and Surface Physics*, edited by F. Howorka et al., Univ. Innsbruck, Obertraun, p. 366.

P. M. Guyon, G. Bellec, O. Dutuit, D. Gerlich, E. A. Gislason, and J. B. Ozenne (1989), *Bull. Soc. Roy. des Sciences de Liege*, 58e annee 3–4, pp. 187–198.

C. Hägg and I. Szabo (1986a), *Int. J. Mass Spectrom. Ion Processes* **73**, 237.

C. Hägg and I. Szabo (1986b), *Int. J. Mass Spectrom. Ion Processes* **73**, 277.

C. Hägg and I. Szabo (1986c), *Int. J. Mass Spectrom. Ion Processes* **73**, 295.

C. E. Hamilton, V. M. Bierbaum and S. R. Leone (1985), *J. Chem. Phys.* **83**, 601.

J. C. Hansen, J. T. Moseley, and P. C. Cosby (1983), *J. Mol. Spectrosc.* **98**, 48.

L. Hanley and S. L. Anderson (1985), *Chem. Phys. Lett.* **122**, 410.

L. Hanley, S. A. Ruatta, and S. L. Anderson (1987), *J. Chem. Phys.* **87**, 260.

L. Hanley, J. L. Whitten, and S. L. Anderson (1988), *J. Phys. Chem.* **92**, 5803.

C. Hayashi (1985), *Nonlinear Oscillations in Physical Systems*, Princeton University Press, Princeton, NJ.

M. Henchman (1972), in *Ion Molecule Reactions*, edited by J. L. Franklin, Butterworth, London.

M. J. Henchman, N. G. Adams, and D. Smith (1981), *J. Chem. Phys.* **75**, 1201.

K. Hiraoka (1987), *J. Chem. Phys.* **87**, 4048.

F. A. Houle, S. L. Anderson, D. Gerlich, T. Turner, and Y. T. Lee (1982), *J. Chem. Phys.* **77**, 748.

Y. Ikezoe, S. Matsuoka, M. Takebe, and A. Viggiano (1986), *Gas Phase Ion–Molecule Reaction Rate Constants Through 1986*, published by Ion Reaction Research Group of The Mass Spectroscopy Society of Japan, Tokyo.

J. D. Jackson (1962), *Classical Electrodynamics*, Wiley, New York.

G. Kaefer (1984), Diplom Thesis, Univ. Freiburg.

G. Kaefer (1989), PhD Thesis, Univ. Freiburg.

H. Kalmbach (1989), Diplom Thesis, Univ. Freiburg.

P. L. Kapitza (1951), *Zn. Eksperim. i. Teor. Fiz.* **21**, 588.

P. L. Kapitza and P. A. M. Dirac (1933), *Proc. Cambridge Phil. Soc.* **29**, 297.

R. Kohls (1974), Zulassungsarbeit, Univ. Freiburg.

G. Kotowski (1943), *Z. angew. Math. Mech.* **23**, 213.

M. D. Kruskal (1962), *J. Math. Phys.* **3**, 806.

C. H. Kuo, I. W. Milkman, T. C. Steimle, and J. T. Moseley (1986), *J. Chem. Phys.* **85**, 4269.

H. Lagadec, C. Meis, and M. Jardino (1988), *Int. J. Mass Spectrom. Ion Processes* **85**, 287.

L. D. Landau and E. M. Lifschitz (1960), *Theoretical Physics*, Pergamon, Oxford, Vol. 1, p. 93.

R. F. Lever (1966), *IBM J. Res. Develop.* **10**, 26.

R. D. Levine and R. B. Bernstein (1974), *Molecular Reaction Dynamics*, Oxford University Press, Oxford.

R. D. Levine and R. B. Bernstein (1987), *Molecular Reaction Dynamics and Chemical Reactivity*, Oxford University Press, NY.

C. L. Liao, C. X. Liao, and C. Y. Ng (1984), *J. Chem. Phys.* **81**, 5672.

C. L. Liao, C. X. Liao, and C. Y. Ng (1985), *J. Chem. Phys.* **82**, 5489.

C. L. Liao, J. D. Shao, R. Xu, G. D. Flesch, Y. G. Li, and C. Y. Ng (1986), *J. Chem. Phys.* **85**, 3874.

C. L. Liao, R. Xu, G. D. Flesch, m. Baer, and C. Y. Ng (1990), *J. Chem. Phys.* **93**, 4818.

A. J. Lichtenberg and M. A. Lieberman (1983), *Regular and Stochastic Motion*, Springer, NY.

J. G. Linhard (1960), *Plasma Physics*, North Holland, Amsterdam.

G. H. Lin, J. Maier, and S. R. Leone (1985), *J. Chem. Phys.* **82**, 5527.

R. G. Littlejohn (1979), *J. Math. Phys.* **20**, 2445.

S. K. Loh, D. A. Hales, L. Lian, and P. B. Armentrout (1989), *J. Chem. Phys.* **90**, 5466.

J. Luine and G. Dunn (1985), *Ap. J.* **299**, L67.

L. M. Magid (1972), *Electromagnetic Fields, Energy and Waves*, Wiley, NY.

J. Marquette, C. Rebrion, and B. Rowe (1988), *J. Chem. Phys.* **89**, 2041.

R. E. March and R. J. Hughes (1989), *Quadrupole Storage Mass Spectrometry*, Wiley, NY.

H. Matsuda and T. Matsuo (1977), *Int. J. Mass Spectrom. Ion Processes* **24**, 107.

G. Mauclaire, R. Derai, S. Fenistein, and R. Marx (1979), *J. Chem. Phys.* **70**, 4017.

D. C. McGilvery and J. D. Morrison (1978), *Int. J. Mass Spectrom. Ion Processes* **28**, 81.

J. Meixner and F. W. Schäfke (1954), *Mathieusche Funktionen und Sphäroidfunktionen*, Springer, Berlin.

M. A. Miller (1958a), *Soviet. Phys. JETP* **8**, 206.

M. A. Miller (1958b), *Radiofizika* **1**, 110.

M. A. Miller (1959), *Soviet. Phys. JETP* **36**, 1358.

G. Miller (1963), *J. Phys. Chem.* **67**, 1359.

P. E. Miller and M. B. Denton (1986), *Int. J. Mass Spectrom. Ion Processes* **72**, 223.

P. E. Miller and M. B. Denton (1990), *Int. J. Mass Spectrom. Ion Processes* **96**, 17.

R. J. S. Morrison, W. E. Conaway, and R. N. Zare (1985), *Chem. Phys. Lett.* **113**, 435.

H. Motz and C. F. Watson (1967), *Advan. Electron. Electron. Phys.* **23**, 153.

D. Müller (1983), Diplom Thesis, Univ. Freiburg.

W. Neuhauser, M. Hohenstatt, P. Toschek, and H. Dehmelt (1978), *Appl. Phys.* **17**, 123.

W. Neuhauser, M. Hohenstatt, and P. E. Toschek (1980), *Phys. Rev.* **A22**, 1137.

D. Neuschäfer, Ch. Ottinger, and S. Zimmermann (1981), *Chem. Phys.* **55**, 313.

D. Neuschäfer, Ch. Ottinger, S. Zimmermann, W. Lindinger, F. Howorka, and H. Störi (1979), *Int. J. Mass Spectrom. Ion Processes* **31**, 345.

C. Y. Ng (1988), in *Techniques for the Study of Ion–Molecule Reactions*, edited by J. M. Farrar and W. H. Saunders, Wiley, New York.

K. Okuno (1986), *J. Phys. Soc. Japan* **55**, 1504.

K. Okuno and Y. Kaneko (1983), in *Electronic and Atomic Collisions*, edited by J. Eichler, I. V. Hertel, and N. Stolterfoth, Verlag, Berlin, p. 543.

G. Ochs and E. Teloy (1974), *J. Chem. Phys.* **61**, 4930

M. Okumura, L. Yeh, and Y. T. Lee (1988), *J. Chem. Phys.* **88**, 79.

T. M. Orlando, B. Yang, and S. L. Anderson (1989), *J. Chem. Phys.* **90**, 1577.

T. G. Owe Berg and T. A. Gaukler (1959), *Am. J. Phys.* **37**, 1013.

W. Paul (1989), Nobel Lecture, Stockholm.

W. Paul (1990), Diplom Thesis, Univ. Freiburg.

W. Paul and M. Raether (1955), *Z. Phys.* **140**, 262.

W. Paul, H. P. Reinhard, and U. von Zahn (1958), *Z. Phys.* **152**, 143.

W. Paul and H. Steinwedel (1953), Z. Naturforsch. **8a**, 448.

G. Piepke (1980), Diplom Thesis, Univ. Freiburg.

L. A. Posey, R. D. Guettler, N. J. Kirchner, B. A. Keller, and R. N. Zare (1991), *to be published.*

S. Scherbarth (1984), Diplom Thesis, Univ. Freiburg.

S. Scherbarth (1988), PhD Thesis, Univ. Freiburg.

S. Scherbarth and D. Gerlich (1989), *J. Chem. Phys.* **90**, 1610.

Ch. Schlier (1988), *Chem. Phys.* **126**, 73.

Ch. Schlier and U. Vix (1985), *Chem. Phys.* **95**, 401.

M. Schweizer (1988), Diplom Thesis, Univ. Freiburg.

A. Sen and J. B. A. Mitchell (1986), *Rev. Sci. Instrum.* **57**, 754.

K. Simonyi (1963), *Foundations of Electrical Engineering*, Pergamon Press.

A. Sommerfeld (1964), *Elektrodynamik*, Akademische Verlagsgesellschaft, Leipzig.

L. S. Sunderlin and P. B. Armentrout (1990), *Chem. Phys. Lett.* **167**, 188.

I. Szabo (1986), *Int. J. Mass Spectrom. Ion Processes* **73**, 197.

H. Tatarczyk and U. von Zahn (1965), *Z. Naturforsch.* **20**, 1708.

E. Teloy (1978), in *Electronic and Atomic Collisions*, edited by G. Watel, North Holland, Amsterdam, p. 591.

E. Teloy (1980), *Verhandl. DPG (VI)* **15**, 567.

E. Teloy and D. Gerlich (1974), *Chem. Phys.* **4**, 417.

J. F. J. Todd, R. M. Waldren, D. A. Freer, and R. F. Bonner (1980), *Int. J. Mass Spectrom. Ion Processes* **34**, 17.

P. Tosi, F. Baldo, F. Eccher, M. Filippi, and D. Bassi (1989a), *Chem. Phys. Lett.* **164**, 471.

P. Tosi, G. Fontana, S. Longano, and D. Bassi (1989b), *Int. J. Mass Spectrom. Ion Processes* **93**, 95.

S. M. Trujillo, R. H. Neynaber, and E. W. Rothe (1966), *Rev. Sci. Instr.* **37**, 1655.

T. Turner, O. Dutuit, and Y. T. Lee (1984), *J. Chem. Phys.* **81**, 3475.

F. Vedel (1991), *Int. J. Mass Spectrum. Ion Processes* **106**, 33.

M. L. Vestal and J. H. Futrell (1974), *Chem. Phys. Lett.* **28**, 55.

H. Villinger, J. M. Henchman, and W. Lindinger (1982), *J. Chem. Phys.* **76**, 1590.

J. T. Watson, D. Jaouen, H. Mestdagh and C. Rolando (1989), *Int. J. Mass Spectrom. Ion Processes* **93**, 225.

E. S. Weibel (1959), *Phys. Rev.* **114**, 18.

E. S. Weibel and G. L. Clark (1961), *J. Nucl. Energy, Part C, Plasma Physics* **2**, 112.

R. Werner (1968), Diplom Thesis, Univ. Freiburg.

D. J. Wineland, J. C. Bergquist, W. M. Itano, J. J. Bollinger, and C. H. Manney (1987), *Phys. Rev. Lett.* **59**, 2935.

D. J. Wineland and H. G. Dehmelt (1975), *Bull. Am. Phys. Soc.* **20**, 637.

M. Wirth (1984), Diplom Thesis, Univ. Freiburg.

R. F. Wuerker, H. M. Goldenberg, and R. V. Langmuir (1959a), *J. Appl. Phys.* **30**, 342.

R. F. Wuerker, H. M. Goldenberg, and R. V. Langmuir (1959b), *J. Appl. Phys.* **30**, 441.

L. I. Yeh, M. Okumura, J. D. Myers, J. M. Price, and Y. T. Lee (1989), *J. Chem. Phys.* **91**, 7319.

U. von Zahn (1963), *Rev. Sci. Instr.* **34**, 1.

U. von Zahn and H. Tatarczyk (1964), *Phys. Lett.* **12**, 190.

MULTIPHOTON IONIZATION STATE SELECTION: VIBRATIONAL-MODE AND ROTATIONAL-STATE CONTROL

SCOTT L. ANDERSON*

Department of Chemistry,
State University of New York at Stony Brook,
Stony Brook, NY

CONTENTS

*Alfred P. Sloan Foundation Research Fellow, Camille and Henry Dreyfus Foundation Teacher–Scholar.

State-Selected and State-to-State Ion–Molecule Reaction Dynamics, Part 1: Experiment, Edited by Cheuk-Yiu Ng and Michael Baer. Advances in Chemical Physics Series, Vol. LXXXII.
ISBN 0-471-53258-4

I. INTRODUCTION—UNIQUE FEATURES OF MULTIPHOTON ION SOURCES

A number of methods for preparing ions with controlled excitation in various internal states have been developed and some of the most powerful are reviewed in this volume. Single-photon ionization is a simple method that produces relatively high intensities of state-selected ions; however, it is only applicable to a few (but important) reactant ions. Use of photoionization with coincidence detection of energy-resolved photoelectrons allows purer state selection and extends considerably the range of molecular ions that can be state selected.

Multiphoton ionization (MPI) state selection is basically a variant of the simple photoionization method, but one with substantial advantages. The limitation in single-photon ionization is that the distribution of ionic states produced is basically determined by the photon energy and the Franck–Condon factors connecting the neutral and ionic potential surfaces. With multiphoton excitation through various intermediate vibronic states, it is possible to "tune" the ionization Franck–Condon factors to produce purer selection of a wider variety of ionic states. MPI is by no means the simplest state selection method to apply. Its use frequently requires considerable optical and photoelectron spectroscopic effort in working out suitable ionization pathways, and successful state selection is not possible for all ions. Nonetheless, MPI state selection has several unique capabilities that make the effort well worthwhile in appropriate cases.

One powerful feature of MPI state selection is the ability to produce small polyatomic ions with controlled excitation in several different vibrational modes. MPI provides clean selection of modes which are Franck–Condon active in ionization, and in addition, some non-Franck–Condon vibrations that cannot be selected by other methods. This is an important contribution because it allows study of the effects of different symmetry vibrations on collisional phenomena. Photoion–photoelectron coincidence methods can also produce ions with some non-Franck–Condon excitations (e.g., high vibrational levels of O_2) by using resonance autoionization. For more complicated molecules MPI has the advantage that the ionic internal state produced is controlled by optical rather than electron spectroscopy, and thus the resolution for closely spaced vibrations is better.

Another major capability of MPI is production of ions in narrow and variable rotational-state distributions. This allows direct measurement of the effects of rotational energy and angular momentum on reactive and non-reactive scattering and on unimolecular decay rates. Rotational effects are particularly important at low reaction temperatures (as in interstellar chemistry), where energies are low and collision times are long. For some molecules, rotational-state control can be combined with vibrational state selection, and MPI thus allows control over all nuclear degrees of freedom for the reactant ion.

A property of MPI that to my knowledge has not yet been exploited for ion–molecule scattering experiments is the fact that the ion rotational angular momentum can be highly aligned with respect to the polarization vector of the laser. Varying the angle of polarization with respect to the ion–molecule relative velocity direction would allow probing of alignment (and indirectly orientation) effects on collisions.

There are several other features of multiphoton ion production that can be useful in some experiments. It is possible to selectively ionize one component of a gas mixture, which can be useful in kinetics measurements. For example, in one technique to be discussed, MPI is used to produce cold reactant ions of one molecule entrained in a supersonic jet of the neutral reactant.

Another useful property is that ions are created in a relatively intense, short pulse with small translational energy spread. This can be useful in experiments that rely on time-of-flight techniques for ion mass selection or for product velocity measurements. The combination of state selection, high translational energy resolution, and good time structure make MPI an ideal method to produce ions for scattering dynamics experiments.

This chapter is focused on MPI state-selection methods and their applications. The first part discusses various MPI schemes and gives a list of successful state selection applications. The second part reviews a number of state selected ion chemistry studies to illustrate the type of information which MPI can provide. For more general information the reader is referred to recent reviews of MPI[1] and MPI photoelectron spectroscopy.[2]

II. MPI STATE-SELECTION METHODS

A. Single-Color Ionization

A schematic energy-level diagram for resonance-enhanced MPI (REMPI) of a typical small polyatomic molecule is shown in Fig. 1. In the most general

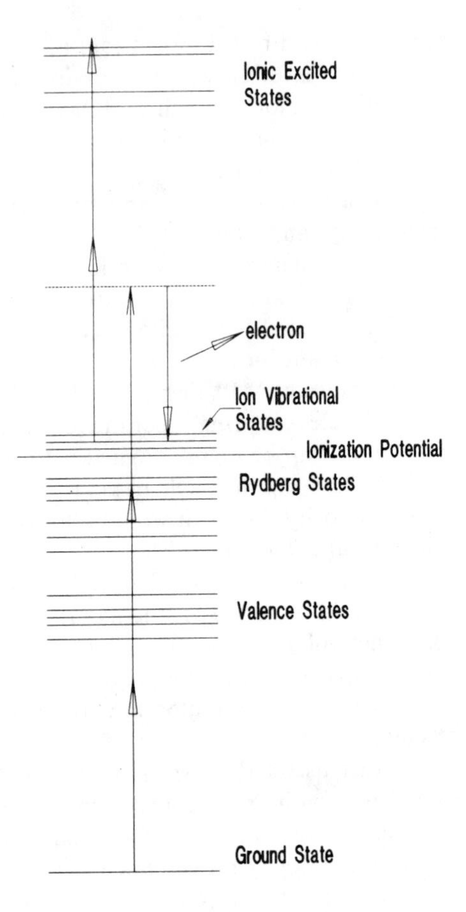

Figure 1. Schematic of MPI for a typical small polyatomic molecule. In the 2 + 1 MPI process shown, two photons are used to pump the molecule to a Rydberg intermediate state, and one more photon excites above the ionization limit.

form of REMPI, the neutral molecule absorbs one or more photons from a pulsed laser and is excited to an intermediate electronic state. The electronically excited molecule then absorbs additional photons from the laser field (or perhaps from a second laser) and is excited above its ionization potential. Ionization normally occurs as soon as the energy of the absorbed photons exceeds the ionization limit, producing ions in a distribution of internal energy states, along with photoelectrons that carry away the excess energy.

I will use the most common notation to describe various MPI schemes. A particular ionization process is specified by the number of photons needed to reach the intermediate level, plus the number needed to ionize the intermediate state. For example, the scheme shown in Fig. 1 is 2 + 1 photon ionization. The laser power needed for MPI depends largely on the number of photons required to reach the resonant intermediate level, and varies from on the order of a megawatt (for one photon to resonance) to gigawatts ($\geqslant$ three photons). These laser intensities are easily produced by common nanosecond-pulsed dye lasers, and by picosecond and femtosecond laser systems with amplification.

If MPI is to result in state selection of the nascent ions, only special excitation schemes are useful. Some of the requirements are included in Fig. 1. Ideally, the resonant intermediate level is a particular ro-vibrational level of a Rydberg electronic state. It is important that the intermediate state lies within a single-photon energy of the ionization limit. Since the excited molecule is in an intense laser field, absorption of a single photon results in ionization of the Rydberg molecule with high probability.

By definition, Rydberg states consist of a molecular ion core weakly interacting with an excited electron. This means that the vibrational wave functions of the Rydberg molecule and the resulting ion are nearly orthogonal, and thus the Franck–Condon factors for ionization of the Rydberg state are diagonal, that is, the ion is formed predominantly in the same vibrational level that was selected in the intermediate state. The ejected photoelectron carries away the excess energy imparted in the multiphoton excitation process, and measurement of the photoelectron spectrum allows determination of the actual ionic state distribution.

An example of this is given in Fig. 2, which shows photoelectron energy distributions resulting from MPI of C_2H_2. Three-photon excitation was used to pump the molecules to one of three different vibrational levels of the $G\,^1\Pi_u$ Rydberg intermediate state,[3] which then was ionized by one more photon from the same laser. Note that in each case, nearly all the photoelectrons are in a single peak, indicating that the ions are in a single vibrational state—the same state that was selected in the Rydberg intermediate. It is also clear that for the $v_2 = 1$ intermediate level (middle frame), the ion state selection is not perfect—a small peak corresponding to ions in the ground vibrational state is observed in addition to the main peak of ions with one quantum of v_2 excited.

Since it is the diagonal Franck–Condon factors for ionization of our Rydberg intermediate state that we rely on for state selection, it is important that the intermediate state be high enough in energy that it will ionize with absorption of one additonal photon. If two or more photons are required, then there may be accidental resonances with higher excited states as the molecule is ionized in a multiphoton transition. This of course changes the Franck–Condon factors that control ionization and will produce ions in a state distribution quite different from that expected.

For some atoms and molecules it is also possible to select cleanly the electronic state of the ions created by MPI. This works if the MPI spectrum of the neutral has bands due to Rydberg states in which the ion core has an

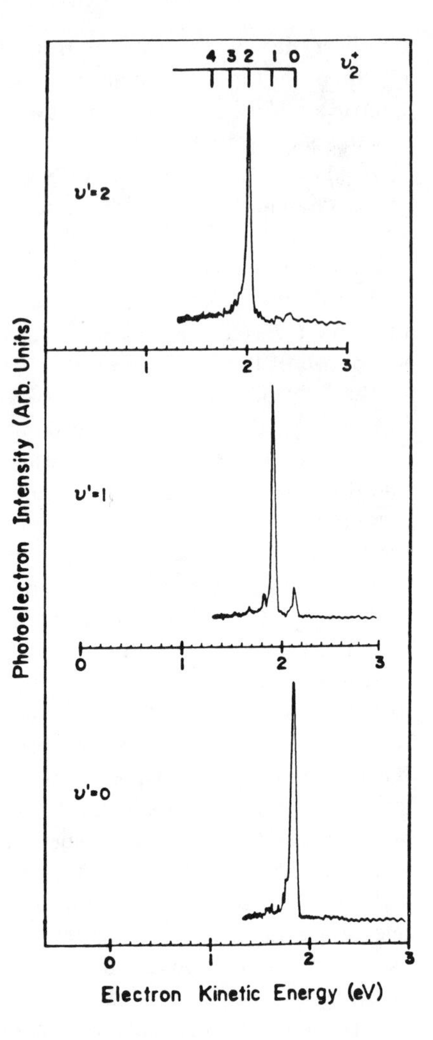

Figure 2. MPI photoelectron spectra for ionization of C_2H_2 through the 0_0^0, 2_0^1, and 2_0^2 bands of the $G\ ^1\Pi_u$ state. Note that the ion vibrational state is the same as that pumped in the intermediate.

excited electronic configuration. If the state is a "good" Rydberg state (i.e., weak interaction between the core and the Rydberg electron), ionization will leave the ion in the excited electronic configuration. For molecular ions this core preservation also leads to simultaneous vibrational state selection.

Finally, for molecules that have rotationally resolvable MPI spectra, it is possible to achieve control over the ionic rotational level. A single rotational level is excited in the transition to the resonant intermediate state. A propensity for ejection of photoelectrons in low J partial waves, together with applicable conservation laws, typically results in a narrow ion rotational distribution. In favorable cases MPI allows simultaneous selection of electronic, vibrational, rotational, and λ-doublet state. Some control over the molecular alignment should also be obtainable. No other ion state selection method can achieve this state resolution.

B. Limitations on MPI State selection

Alas, there are a number of problems that can make MPI state selection considerably more difficult than might be anticipated from the rosy description given previously. The basic problems include the following.

For many molecules, some intermediate levels, which from their spacings in excitation spectra appear to be good Rydberg states, are actually of mixed character. Interactions with both valence states and other Rydberg states (e.g., vibronic coupling of *ns* and *nd* Rydbergs or Σ and Δ components of an *np* Rydberg) are possible, and both cause problems for state selection. In mixed intermediate states, the vibrational wave functions are perturbed and ionization is unlikely to produce ions in a single vibrational level.

Not infrequently, mixing leads to rapid non radiative decay of the intermediate level. If the decay rate is comparable to the ionization rate (which depends on laser intensity), ionization results in broad (thus useless) ion vibrational distributions. In some states, mixing leads to such rapid non radiative decay that these bands are actually absent in the MPI spectrum except at very high laser intensities.

If the geometries of the neutral ground state, Rydberg states, and ionic ground state are all similar, there will be at least some vibrational modes that cannot be excited by MPI through Rydberg intermediate states. This is simply because if there is no geometry change along a vibrational coordinate, then the Franck–Condon factors for that vibrational mode strongly favor diagonal (i.e., $\Delta v = 0$) transitions in ionization. For example, it is not easy to excite bending vibrations in a molecule for which the relevant neutral and ionic states are all linear.

Several problems result from the high laser intensities needed to pump multiphoton transitions efficiently. If the laser field is intense enough, or if there are discrete states in the continuum that provide a stepping stone, there is some probability that ionization will not occur as soon as the molecule has absorbed enough photons to exceed the ionization potential. Ionization after absorption of photons in excess of the minimum needed is known as above threshold ionization (ATI) and can degrade the purity of MPI state selection by allowing access to higher ionic excited states. ATI can usually be avoided by reducing laser power, if necessary choosing an ionization route that requires fewer photons.

A problem that is not so easy to eliminate is shown in Fig. 1. The state-selected ions that are produced by MPI are irradiated by the high intensity laser that created them. If there are ionic excited states that can be reached by optically allowed one-photon transitions, the nascent ions have a high probability of being excited. Some ion photoexcitation may also occur even if two photons are required, especially if the MPI process requires intense lasers. The probability can be reduced to a large extent (but never completely eliminated) by using $1 + 1$ ionization schemes where both the excitation and ionization steps require only one photon, and thus low laser power. Regardless of whether the excited ions can decay optically back to the ground electronic state or not, the initial state selection will be degraded. Note that even a tiny fraction of highly excited ions in the reactant distribution of a chemistry experiment may cause significant errors if their reaction cross sections or rates are much higher than those of ions in the desired state. This will almost certainly be the case for most endoergic reactions at low collision energies.

There are several other potential problems that are less serious but nonetheless must be kept in mind when designing experiments. In rotational-selection experiments, strong magnetic fields in the ionization volume should be avoided since they can induce angular momentum state mixing.[4] For some (mostly small) molecules, there may be strong autoionizing states in the continuum at the energy level where the ionizing photon pumps the molecule. These can perturb the ionization step (and thus the ionic-state distribution), and may also provide a quasi stable state that can absorb additional photons, leading to ATI. Finally, a particular concern for ion-chemistry experiments is that space-charge effects from the high density of ions and electrons in the laser focus can produce translationally hot ions. To minimize this, it is necessary to avoid producing too many ions per laser pulse. The number that can be tolerated depends on how tightly the laser is focused and how sensitive the experiment is to hot ions. In our experiments, we typically work with about one to five thousand ions/shot.

For all these reasons it is by no means safe to assume that it will be straightforward (or even feasible) to state select a molecule for which the spectroscopic groundwork has not been done. It is possible, however, to formulate a few guidelines that may be helpful in choosing molecules and designing experiments using MPI state selection:

1. Because it is not possible to predict *a priori* when various spectroscopic and experimental problems may result in poor state selection, it is *absolutely essential* to measure the actual ion-state distributions using photoelectron- or laser-induced fluorescence spectroscopy.
2. The neutral precursor molecule should have at least some spectroscopically accessible Rydberg states that are not of mixed character. Note, however, that in some cases intermediate-state mixing

can actually be a significant advantage, since it mixes in vibrational modes that otherwise would not be accessible. Several examples are given below in discussions of specific molecules.

3. There should be a significant difference in geometry between the neutral ground state and the MPI intermediate states. Ideally the geometry should change along all of the vibrational coordinates. Vibrational coordinates that do not change are not Franck–Condon active in the multiphoton-excitation process, and therefore cannot be populated in the ion using unmixed intermediate states. In the worst case it may only be possible to produce ground pure vibrational state ions. This situation is not a total loss, since it may still be possible to rotationally state select the ions.
4. Any ionic electronic state that can be reached by one-photon transitions from the ionic ground state should be dissociative or predissociative. Even at low laser intensities it is difficult to avoid one-photon excitation of the state-selected ions, and, as discussed previously, this seriously degrades state selection. The problem is eliminated if the excited states simply dissociate. In this case any ionic photoexcitation will result in fragment ions that can easily be removed by a mass filter. Otherwise, it is essential to work under conditions in which photoexcitation of the nascent ions is negligible. This is probably only really possible for MPI by $1 + 1$ photon excitation at low laser intensities.
5. If experimentally feasible, it is best to work with ionization schemes that minimize the number of photons the molecule must absorb. This allows operation at low laser powers, which minimizes a number of problems, and also usually results in higher ion intensities. If high enough energy photons can be generated, $1 + 1$ ionization processes are generally superior.

C. Two-Color Ionization Schemes

Several problems with one-color MPI can be ameliorated with multicolor ionization. Consider, for example, the one-photon-resonant, two-photon-ionization scheme typically used for aromatic molecules, which is shown in Fig. 3*a*. Here the intermediate state used is the first excited singlet state (S_1), which is not really a Rydberg state, but has geometry intermediate between those of the neutral and ion ground states. For most aromatic molecules, the S_1 state lies more than half way to the ionization limit, and therefore a laser tuned to pump the molecule to a particular vibrational level of the S_1 state will then pump the molecule well above the energy of that vibrational level in the ion. This increases the likelihood that state selection purity will be degraded by ionization to unwanted vibrational levels (since the

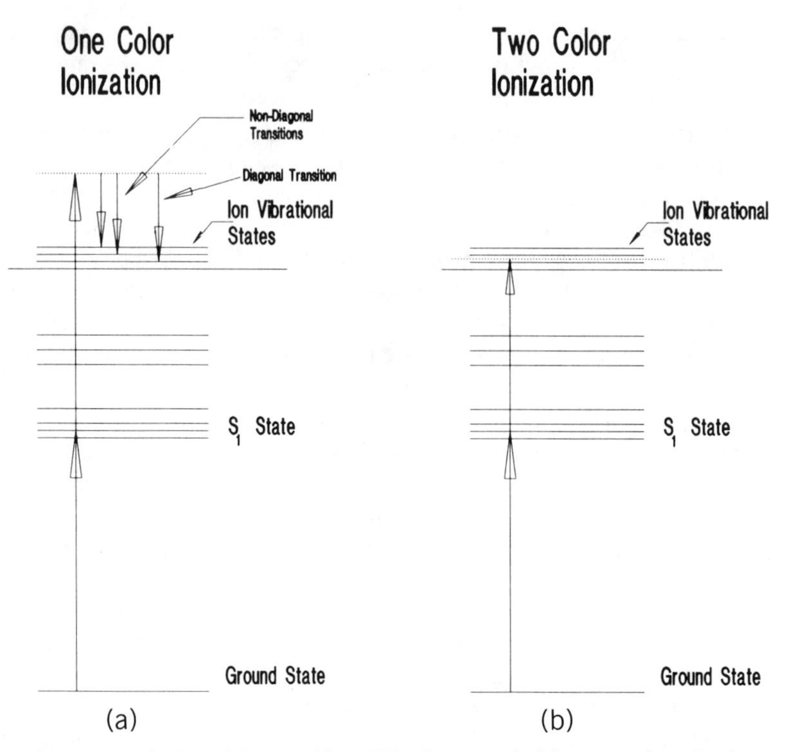

Figure 3. (a) Single and (b) two-color MPI schemes typical for aromatic molecules.

intermediate is not really a Rydberg state), by photoexcitation to ionic electronic states, and by autoionization.

Figure 3*b* shows a simple two-color modification that reduces these effects. The first laser (ω_1) is tuned to pump a transition to a particular vibrational level in the S_1 intermediate state, and has low enough intensity that the ionization rate of the excited state due to the first laser alone is small. A second, higher-intensity laser is used to ionize the vibronically excited molecules. Its frequency (ω_2) is tuned just above the threshold for production of the desired ionic vibrational state, and thus conservation of energy prevents contamination, at least from higher-energy vibrations. The energy of the second laser is too low to allow ionization by two ω_2 photon.

There obviously are limitations on this scheme. In order that the desired $\omega_1 + \omega_2$ process is dominant, it is necessary that the intensity of ω_2 be significantly greater than that of ω_1. If the second laser is too intense, however, it may induce ionization by $3\omega_2$ absorption, producing ions in unknown and perhaps highly excited states. This must be carefully avoided, since even a small fraction of highly excited ions can cause serious problems in ion chemistry experiments. In addition, at any reasonable intensity there will always be some one-color ionization by the first laser alone. This rules out use of this type of two-color scheme in cases where excitation to the resonant intermediate state requires two or more photons (the situation for most small molecules), since high intensities are required to give significant multiphoton excitation rates.

As a way around this last problem it is possible to replace the two- or three-photon excitation to the intermediate state with single-photon excitation by a vacuum-ultraviolet (VUV) laser, followed by ionization with a visible or UV laser. MPI in this manner (with presumed state selection) has recently been demonstrated for H_2[5] and N_2,[6] although in neither case was the actual ionic-state distribution measured spectroscopically. As VUV lasers become increasing easy to build, this approach should become more widespread for state selection. The advantage of using strictly one-photon transitions, requiring only low laser intensities, is significant. In addition to reducing problems with unwanted multiphoton excitations, there are the practical considerations that eliminating the need for tight laser focusing makes it much easier to overlap laser beams for two-color ionization schemes and may also increase ion signal by irradiating a larger volume of the precursor gas.

Two-color ionization schemes add greatly to the flexibility of MPI for ionic-state selection, but of course the advantages come at some cost. Obviously, two dye lasers are needed (although both may be pumped by the same pump laser for visible or near UV generation). VUV lasers require one or more powerful tunable laser sources together with gas cells and/or pulsed jets for mixing or harmonic generation, and in addition a standard pulsed dye laser is needed to produce the second color for ionization. From the ion–molecule chemistry perspective, two-color schemes increase the complexity of the ion source. The two lasers must be overlapped spatially, and temporal overlap is also necessary when the intermediate state has a short lifetime. This can be troublesome if one or both lasers need to be focused, especially if the ion source has a geometry that restricts access.

D. Molecular Ions That Have Been MPI State Selected to Date

This section is an attempt to give a reasonably complete list of systems for which MPI state selection has been demonstrated, and a couple of examples that illustrate problems. Please note that in nearly all cases the main emphasis of these studies has been on the spectroscopy and photoionization dynamics of the neutral excited states and not on state selection. For many of the molecules discussed, there probably are unexplored excitation routes that are better for state selection than the ones reported. For some molecules, many research groups have studied similar excitation pathways, and I give references to only one or two representative papers.

The list of molecules is rather short at this time, mostly because the MPI-photoelectron spectroscopic work needed to find good state selection routes is quite time consuming. As noted, most of the people working in the field are spectroscopists, and not very many dedicated efforts to find state selection routes have been mounted. My experience has been that it is really not possible to predict *a priori* whether a particular molecule or excitation scheme will be successful. For that reason I have not attempted to suggest a list of likely candidate molecules for MPI state selection. There is simply too little known about vibronic coupling and autoionizing levels in small polyatomic molecules.

1. *Rare Gases and Other Atoms*

For a complete listing of the extensive MPI-photoelectron spectroscopic work on atoms, see the recent review by Compton and Miller.[2] State selection of transition-metal cations by MPI and other methods is discussed by Weisshaar in another chapter in this volume. The ground states of the rare gas cations are $^2\Pi$, and thus are spin–orbit split into $j = 3/2$ and $1/2$ states. Since the ion cores of rare-gas Rydberg states are also spin–orbit split, it is possible to provide state-selected rare-gas cations by MPI through the appropriate intermediate states. This has been experimentally demonstrated in many labs for Xe,[7] Kr,[7] and Ar,[3] and is routinely used to calibrate photoelectron spectrometers.

Another example of a non metal atomic ion that can be state selected is iodine, where it is possible to achieve selection of the ^2P_j levels.[8]

2. *Hydrogen*

There have been numerous MPI-photoelectron spectroscopy studies of H_2 with both single- and multicolor excitation through several different intermediate states. Vibrational state selection with MPI is not impressive compared to single-photon methods; however, it is possible to achieve good rotational selection. Using single-color $2 + 1$ photon ionization through the mixed E, F state,[9] it is possible to produce 85% pure $v = 0$: however, higher v^+ levels are badly contaminated. Single-color $3 + 1$ ionization through the $B'\,^1\Sigma_u^+$,[10] $C\,^1\Pi_u$,[11] and $D\,^1\Pi_u$[10] states gives somewhat better results, with only small contamination for $v^+ = 0$ and 1.

Two photon excitation to the E,F state followed by excitation to autoionizing levels with a second laser was observed to give excellent rotational state selection[4] of the resulting H_2^+.

3. *Nitrogen*

N_2^+ would appear to be a good candidate for MPI state selection, however, mixing in the easily accessible Rydberg states makes life difficult; 3 + 1 photoionization through single vibrational levels of the *b*, *c*, and *c′* states, for example, results in ions with messy vibrational distributions in both the *X* and *A* electronic states.[12] 3 + 1 ionization through the o_3 state actually produces almost pure *A* state ions, and with good vibrational state selection.[13]

Several different two-color ionization schemes have been used to suppress the *A* state formation. Opitz et al.[14] were able to obtain ground vibrational and electronic state N_2^+ with rotational state selection (as determined by laser-induced fluorescence). Trickl et al.[15] used VUV 1 + 1 ionization to produce ions just above the thresholds for the $v^+ = 0$ and 1 states. In principle, this should give good vibrational state selection; however, it was not verified spectroscopically.

4. *Oxygen*

2 + 1 ionization through 4*s*-3*d* and 5*s*-4*d* Rydbergs results in production of ions primarily in the same vibrational level that is excited in the intermediate state.[16] For the best intermediates studied, selection purity is only about 70–90% for $v^+ = 0$–3; thus MPI (at least for this set of intermediate states) is not competitive with photoion–photoelectron coincidence methods.

5. *Nitric oxide*

NO has been the favorite small molecule for MPI studies, and a wide variety of single- and multiple-color schemes (with and without double resonance) have been explored.[2] For state selection, the simplest scheme is probably 1 + 1 single-color excitation through the *A* state, which gives resonable vibrational[17] and rotational[18] selection. Better vibrational selection was achieved by 2 + 1 MPI through the *E* and *C* states.[19] Two-color 1 + 1 ionization through the *A* state, with the second laser tuned to ionize just above threshold for the desired v^+ level, should give excellent selection of both vibration and rotation.

6. *Carbon Monoxide*

MPI photoelectron spectroscopy has been applied to 3 + 2 and 3 + 3 ionization CO through the *A* state.[20] This is not a Rydberg state, and thus vibrational selection is not expected. Coincidentally, however, the photon energy when pumping the *A* $v = 3$ band is just high enough to allow two-photon ionization to the ionic ground vibrational level, and nearly pure CO^+ $v = 0$ results. This transition has been used for production of vibrationally cold ions with variable rotational energy by Gerlich and Rox.[21]

2 + 1 ionization of CO through the B[22] and E[23] intermediate states, with laser-induced fluorescence detection of the CO^+ rotational distributions, has demonstrated good rotational selection, although in neither study was the vibrational selectivity addressed. Both states are Rydberg, and thus it is likely that simultaneous rotational and vibrational state selection could be achieved by MPI.

7. *Hydrogen Bromide*

Xie and Zare[24] recently reported a study of 2 + 1 ionization of HBr through several Rydberg states, with laser-induced fluorescence detection of the HBr^+ state distribution. They showed that it was possible to achieve simultaneous control over the vibrational, fine structure, rotational, and λ doublet state of the ions. Vibrational selectivity[25] is better than 95% for $v = 0$ and 1, and it appears likely that $v = 2$ and 3 can also be selected with good purity.

8. *Carbonyl Sulfide*

At least for the moment, this molecule is the best case for vibrational mode control by MPI state selection. We[26] reported a 2 + 1 MPI-photoelectron spectroscopy study of the several Rydberg states of OCS. Using various intermediates states it is possible to produce ions with variable excitation in ν_1 or ν_3 with high purity. The $\nu_2 = 2$ state can also be populated, allowing comparison of C–O stretch, C–S stretch, and bending vibrations. OCS^+ is a $^2\Pi$ molecule, and MPI allows population of various vibrational levels in both the upper and lower fine-structure states.

9. *Ammonia*

NH_3 was one of the first molecules for which MPI state selection was demonstrated.[17,27] By single-color MPI through the C' and B Rydberg intermediate states it is possible to produce NH_3^+ with up to 10 quanta of ν_2 (umbrella bend).[28] Selection purity is 70–80%, but there may be highly excited ions present when selecting some ion vibrational levels. This would appear to be a case where two-color ionization might improve state-selection purity considerably. Selection of other vibrational modes has not been reported, and will be difficult since only ν_2 is very Franck–Condon active in ionization.

10. *Acetylene*

$C_2H_2^+$ is probably typical of small polyatomic systems in that vibrational mode control is possible, but only for a subset of the molecule's vibrational modes. 3 + 1 photon ionization through several different ungerade Rydberg states allows production of $C_2H_2^+$ with pure (> 90%) excitation of 0, 1, or 2 quanta of the Franck–Condon active ν_2 (C–C stretch) vibration.[3] 2 + 1 ionization through gerade Rydbergs[29] also allows ν_2 selection, but in addition permits production of ions with two quanta of a bending vibration excited. This is possible because this vibration (probably the cis–bend) vibronically couples two Rydberg states and thereby becomes optically active. Actually the bending excited ions are produced together with 44% ground-state $C_2H_2^+$, but since pure ground-state ions can be studied independently, the chemical effects of the bend can be extracted by subtraction.

11. *Aromatics*

MPI photoelectron spectroscopic studies have been reported for a wide variety of aromatic molecules, particularly substituted benzenes,[2] and this class of molecules appears to be well suited for two-color MPI state selection. Aromatic molecules generally are easily ionized by 1 + 1 MPI using readily available UV radiation in the 250–300 nm range. Typically the intermediate state used is the first excited singlet (S_1), which is not really a Rydberg state, but still has reasonably diagonal ionization Franck–Condon factors. This usually makes it possible to select which vibrational mode the ions have excited. For example, if the 1_0^1 band in S_1 is used as the intermediate level, ions are typcially produced with excitation in a short progression in ν_1. If a different mode is pumped in the intermediate, that mode will also dominate the ion vibrational distribution. The degree of selectivity varies from molecule to molecule.

Single-color ionization works well for some vibrational levels of some aromatics. For example, Schlag and co-workers[30] have produced ions in narrow and variable vibrational distributions for use in photodissociation spectroscopy. Neusser[31] was able to produce benzene cations with control over the rotational distribution in several low-lying vibrational states. The problem with one-color MPI in this context is that for many aromatic molecules, the S_1 state is considerably more than halfway to the ionization limit, and thus the second photon carries the molecule far above the desired cation state. This tends to result in broad cation vibrational distributions. Take, for example, the case of phenol.[32] When ionized through a variety of intermediate vibronic levels, the phenol cations are formed in vibrational distributions dominated by progresions in the mode that was excited in the intermediate state. The distributions are over 0.5 eV wide, however, and thus useless for state selection purposes. The same problem occurs to different degrees in other benzenoid species, and becomes increasingly bad as higher energy vibrations are pumped.

Recently, Botter and co-workers[33] have shown that use of two-color MPI can greatly improve the state selection purity in this class of molecules. They used the scheme illustrated in Fig. 3*b* where the first laser pumps the molecule to a particular vibrational level of the intermediate state, and the second laser is tuned to have just enough photon energy to reach the threshold for producing that vibrational mode of the ion. As this second laser is tuned through the threshold for the $\Delta v = 0$ ionization, they observe sharp photoionization onsets, indicating production of pure mode-selected ions. These ions are then used for spectroscopy and chemistry studies. Judging from the results to date, this two-color ionization scheme appears to be the method of choice for state selection of benzenoid cations.

12. *Problem Molecules*

To be fair, let's look at a few examples of molecules for which MPI state selection has problems. Koenders et al.[34,35] studied Rydberg states of molecular chlorine by MPI photoelectron spectroscopy. In the two-photon excitation spectrum they observed a number of apparently unperturbed Rydberg states, with a vibrational progression up to $v = 15$. In the photoelectron spectra, instead of the expected $\Delta v = 0$ behavior, they observed broad vibrational distributions. This was attributed to electronic autoionization of core-excited Rydberg states that were accidently resonant with the ionization photon. This problem might well be solvable by two-color ionization, where the first laser excites the Cl_2 to a Rydberg level, and the second laser ionizes with just enough energy to reach the desired final state.

Another example is H_2S, which has been studied in detail by Miller et al.,[36] Achiba et al.,[37] and Steadman et al.,[38,39] Steadman et al. examined both $2 + 1$ and $3 + 1$ photon ionization through a variety of Rydberg levels by MPI photoelectron spectroscopy. For a few intermediate levels the photoelectron spectra were dominated by a single band, indicating reasonably diagonal ionization, and giving limited state selection of the ion. For most intermediate levels, however, the electron spectrum contained large numbers of vibrational bands, indicating that the intermediate levels were strongly perturbed. Considerable signal due to ionization of S atoms from dissociation of the intermediate states was also observed. This problem of vibronic mixing in the intermediate Rydberg states appears to be quite common in small polyatomic molecules. From a spectroscopic point of view this is very interesting; however, for easy state selection, uninteresting spectroscopy is to be preferred.

In summary, considerable progress has been made in developing multiphoton excitation routes for state selection of systems ranging in complexity from atoms to moderate size polyatomic molecules. In the process, a tremendous amount of spectroscopic information on the properties of intermediate electronic states and the dynamics of ionization has been obtained. At least for favorable cases, MPI gives the most precise control obtained to date over the internal state of ground-electronic-state ions. Not all molecules can be state selectively ionized with MPI; however, it is clear that many systems can added to the repertoire by spectroscopic studies dedicated to finding state selection routes.

III. MPI STATE SELECTED ION CHEMISTRY STUDIES

Because MPI is a new and somewhat complicated state selection technique, relatively few MPI ion chemistry studies have been published to date. A representative sample is discussed in the following sections. The ordering is related to the complications inherent in the chemistry involved, and does not follow the chronological development of the field.

A. Studies of Ultracold Chemistry

There has been considerable effort recently in extending the temperature/energy range of ion–molecule reaction studies to as near absolute zero as possible. This interest is due partly to the possibility of testing theoretical models for ion–molecule capture processes, and also because this energy range is important in interstellar chemistry. Since photoionization is a good way to produce ions without translational heating and since MPI allows control over internal state down to the rotational level, it is a logical way to study ultracold ion processes.

MPI of precursor molecules in the core of a supersonic free-jet expansion provides a simple way to produce cold ions in a cold and reasonably well–controlled environment. MPI can also selectively ionize one component of a mixed jet, which extends the range of this technique beyond only reactions of ions with their neutral precursors.

Smith and co-workers have done the most work with this method to date. In their arrangement the ions are produced early in the expansion then allowed to react as the expansion proceeds. Products are extracted at right angles after a variable time, and then mass analyzed. Under their conditions there are multiple collisions and the ions are assumed to be internally and translationally equilibrated with the neutral gas. Extraction of rate information is complicated by the fact that both the density and temperature of the gas change continuously during the reaction time. Smith and co-workers[40] have done extensive experimental and modeling work in developing the analysis needed to extract reaction rates and temperatures from these experiments. So far they have reported studies of the reaction of $C_2H_2^+$ with H_2,[41] three-body association of N_2^+ with N_2,[42] and rotational relaxation of N_2^+ in collision with N_2.[43]

Figure 4 shows the type of data obtained in Smith and co-workers' experiments on the reactions

$$C_2H_2^+ + H_2 \rightarrow C_2H_3^+ + H,$$

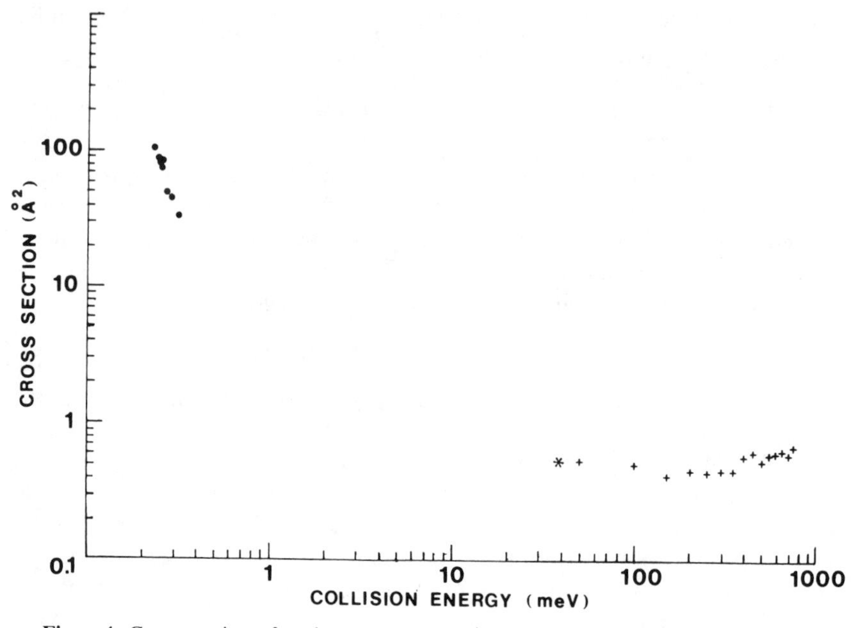

Figure 4. Cross sections for the reaction $C_2H_2^+ + H_2 \longrightarrow C_2H_3^+ + H$ as a function of collision energy. The meaning of the different points is given in the text.

in which the $C_2H_2^+$ reactant has been prepared in the ground vibrational state by MPI. In the figure[44] the data have been plotted as cross sections versus average collision energy. The single point at about 35 meV is taken from a rate constant measurements by Adams and Smith[45] using a room-temperature flow tube, and the other points at high energy are from the photoionization ion beam experiments of Turner and Lee[46] (scaled to fit the absolute value from the flow-tube results). The very-low-collision-energy points are from Hawley and Smith.[41] Note the tremendous increase in cross section at very low collision energies. This indicates that a change in reaction mechanism takes place at some point below 30 meV collision energy, and also requires a revision of the literature ΔH_f for $C_2H_3^+$, which previously suggested that this reaction would slightly endothermic.

Gerlich and Rox[21] recently reported a free-jet MPI study of the three-body association reaction

$$CO^+ + 2CO \rightarrow (CO)_2^+ + CO.$$

In their configuration the ions were also produced in the expansion, but were accelerated immediately by a small applied field. The gas density and acceleration rate were adjusted so that few of the ions could undergo multiple collisions before the ion velocity relative to the neutral molecules increased significantly. Since the association reaction turns off rapidly with increasing collision energy, this arrangement effectively allowed Gerlich and Rox to work under single-collision conditions (i.e., small probability of state-changing collisions before acceleration to energies high enough to stop association). This allowed them to work with state selected ions, and they studied the rotational and collision energy dependence of the association rate. It was found that the effect of rotation was the same as an equivalent amount of translational energy—both inhibit association.

Another example of rotational effects as measured by Gerlich, Jerke, and Schweizer[47] is shown in Fig. 5, which compares the effects of rotational and vibrational energy on the reaction

$$H_2^+(v,j) + H_2 \rightarrow H_3^+ + H$$

at a fixed collision energy below 10 meV. The experiment was carried out in the pulsed expansion apparatus described previously. H_2 was ionized through the B state, which allows preparation of ground-vibrational-state

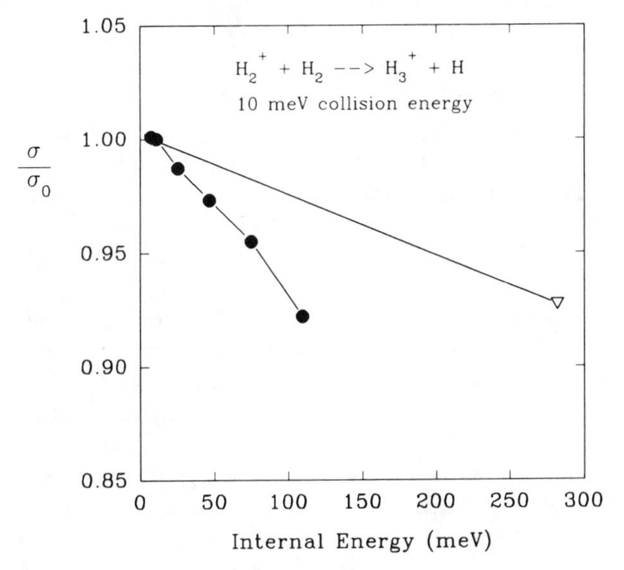

Figure 5. Rotational and vibrational effects on the reaction $H_2^+ + H_2 \longrightarrow H_3^+ + H$ at about 10 meV collision energy. The circles give the ratio of the cross sections for different ($v = 0$, J) states to that for the ($v = 0$, $J = 0$) ground state. The traingle gives the cross section ratio for H_2^+ ($v = 1$, $J = 0$, 1) relative to the ground state.

ions with variable rotational energy, and also known distributions of $v = 1$ and $v = 0$. Since no ternary collisions are required for this reaction, the ionization was done well downstream of the nozzle to ensure that reactions occurred under single-collision conditions. Note that both vibration and rotation inhibit the reaction, but that the effect of rotation is roughly three times larger. Gerlich and co-workers suggest that the rotational effect may involve formation of a long-lived H_2–H_2^+ complex at meV collision energies.

Recently Gerlich and co-workers have begun experiments on J-selected ion–molecule reactions in a *low-energy* merged-beam apparatus. In this experiment a low-energy ion beam is merged with a supersonic neutral beam and then directed into an octapole ion guide where the relative energy can be adjusted. This method has several advantages over the free-jet experiments. The ion–neutral relative energy can be varied independently of the both the neutral and ion temperatures, energy resolution is improved, there is no question of multiple collisions, and, in principle, it is possible to get some time-of-flight information about the scattering dynamics.

B. Beam Studies of Internal Energy Effects

The MPI state selected molecular-ion-beam studies reported to date have all been concerned with reactions of polyatomic ions and molecules. This is a natural combination because MPI provides the possibility to control ion vibrational energy and mode, and beam experiments allow control over collision energy, which is the other high-energy mode of nuclear motion. Rotational effects have not yet been addressed in beam experiments, although this situation should change soon.[25] The dynamics of polyatomic reactions are complex, and unlike the well-studied case of $A + BC$ reactions, almost no insight exists into the role of different modes of nuclear motion in driving polyatomic reactions. Several groups are using MPI state selection to attack this problem, and enough data are being generated that some qualitative insights are beginning to emerge.

Zare and co-workers pioneered the use of MPI in ion chemistry studies, and the first MPI state-selected ion reaction reported was the beam study by Morrison, Conaway, and Zare[48] of the reaction of $NH_3^+(\nu_2)$ with D_2. They used a single quadrupole mass spectrometer system and examined the effects of NH_3^+ umbrella vibration and collision energy on the relative intensities of D atom addition and isotope exchange/fragmentation products. D addition was almost independent of ν_2 excitation. At high energies the exchange/fragmentation channel increases dramatically, and is enhanced about a factor of 2 by ν_2 excitation up to seven quanta.

Conaway, Ebata, and Zare studied reactions of $NH_3^+(\nu_2)$ with methane[49] and ammonia.[50] For these studies the mass spectrometer was upgraded to a tandem quadrupole configuration to eliminate fragment ions from the beam. For methane the major channel at all collision energies is

$$NH_3^+ + CH_4 \rightarrow NH_4^+ + CH_3.$$

This reaction is nearly independent of collision energy, but is enhanced by about a factor of 2 by nine quanta of NH_3^+ umbrella motion. They proposed that the umbrella vibration facilitates the transition to the NH_4^+ product geometry. At low collision energies they also observed condensation products, and these channels are strongly suppressed by both vibration and collision energies.

For reaction of NH_3^+ with ammonia, isotopic labeling was used to distinguish between charge, proton, and H-atom transfer. Charge transfer is modestly enhanced by umbrella vibration, which they discuss in terms of a Franck–Condon factor/energy gap model.[51] The rationale is that since NH_3 undergoes a large change in geometry along the ν_2 (umbrella) coordinate when ionized (or neutralized), resonant charge transfer between ground-state reactants has poor vibrational overlap and thus low cross section. Vibrational excitation of the reactant ion improves the overlap, at least for one of the charge-transfer partners.

Proton transfer was observed to be inhibited by about a factor of 2 by umbrella motion, which is only in qualitative agreement with VUV photoion–photoelectron coincidence measurements,[52] which show larger vibrational effects. This discrepancy is probably at least partly due to the fact that as the vibrational level is increase, the state selection purity obtained by single-color MPI becomes increasingly poor. Conaway, Ebata, and Zare speculated that the inhibition is due to vibration, making it difficult for a NH_3–H^+–NH_2 complex to remain in the favorable linear NHN geometry.

In contrast, the H-atom transfer reaction is enhanced by a factor of 5 as the umbrella vibration increases from 0 to 10 quanta. The enhancement was attributed to the same mechanism that was put forward to explain a similar enhancement in reaction with methane: ν_2 vibration distorts the planer NH_3^+ ion toward a more favorable transition state for production of the tetrahedral NH_4^+ ion.

Ebata and Zare[19] reported a study of charge transfer in the $[NH_3 + NO]^+$ system, where they can state select both NH_3^+ and NO^+ by MPI. They found that charge transfer from NH_3^+ to NO (exoergic by 0.88 eV) is essentially unaffected by the umbrella vibration of NH_3^+, while charge transfer from NO^+ to NH_3 (endoergic by 0.88 eV) is strongly enhanced by NO^+ vibration, at least at low collision energy. They assumed a model in which charge transfer proceeds through an intermediate complex, and used phase-space theory to estimate the expected vibrational dependence. Qualitative agreement was obtained, and the authors concluded that complex formation is important at least in the lower part of their collision energy range 1.5–16 eV).

In these pioneering experiments, Zare and co-workers did not report reaction cross sections; instead, they reported the ratio of the signal for a given reactant ion vibrational state divided by the signal for ground-state reactant ions. In many cases, the magnitude and collision energy dependence of the actual cross sections would provide additional clues that can be vital to understanding the reaction mechanisms.

Zare and co-workers have incorporated an octapole ion guide into the most recent version of their instrument and have reexamined the reactions of NH_3^+ with ND_3 and D_2. In these new experiments, they have used the timing characteristics of MPI ion production together with TOF velocity analysis to measure the recoil energy distributions of the charge-, atom-, and proton-transfer channels in these systems. They are able to estimate the internal energy distributions of the products and have observed that these distributions change with the internal excitation of the NH_3^+ reactant ions.

In our group we have concentrated on reactions of ions for which it is possible to state select at least two different vibrational modes. These include OCS^+ and $C_2H_2^+$ for which we have worked out the excitation pathways, and also various benzenoid cations that can be selected by two-color methods. Reactions are studied in a tandem guided beam mass spectrometer,[53] one incarnation of which is shown in Fig. 6.

A differentially pumped, pulsed beam of the neutral precursor molecule is injected from the top and passes through an octapole ion guide, in the

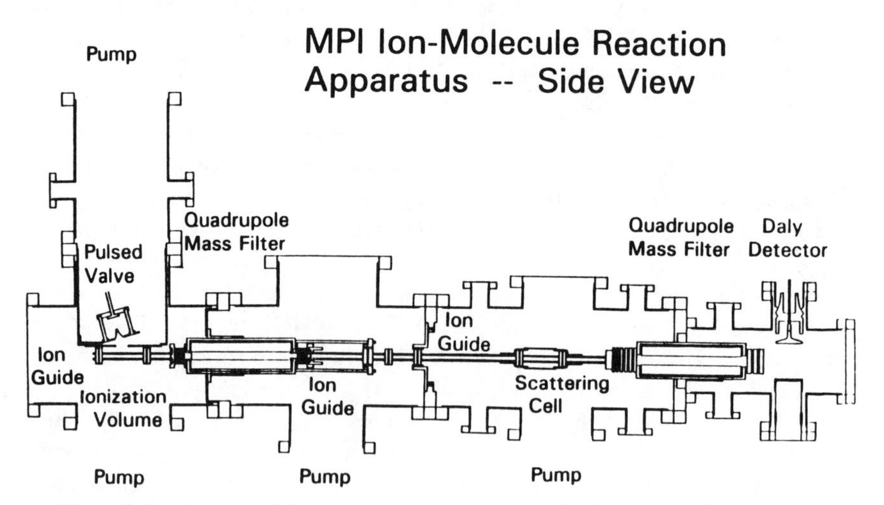

Figure 6. Tandem guided-beam mass spectrometer used for MPI state-selected ion–molecule reaction studies at stony Brook.

center of which is focused the ionizing laser. The guide collects the state-selected reactant ions and transports them to a quadrupole mass filter that removes any fragment ions from the beam, which is then injected into a second set of octapole guides. At this point, we set the desired collision energy, then guide the beam either through a scattering cell or a pulsed jet (not shown) containing the neutral reactant. Reactions occur inside the guide, which collects reactant and product ions and guides them to the second quadrupole mass filter, where they are mass analyzed and counted. This apparatus allows us to generate accurate absolute cross sections at collision energies nearly down to the thermal range ($\Delta E_{collision} \lesssim 0.04$ eV).

To date we have studied reactions of $C_2H_2^+$ with methane[53] and D_2, collision-induced dissociation of OCS^+, and reactions of both charge states of the $[C_2H_2 + OCS]^+$ system.[54] In the later system we can probe the effects on reactivity of six different nuclear degrees of freedom (C–C stretch and bend of $C_2H_2^+$, C–S stretch, C–O stretch, and bend for OCS^+, and collision energy) as well as the OCS^+ spin–orbit state. Here I will give a couple short examples of the effects observed.

For the reaction of $C_2H_2^+$ with methane, the product channels are

$$
\begin{aligned}
C_2H_2^+ + CH_4 \rightarrow \quad & C_3H_5^+ + H & \Delta H = -0.92\,\text{eV} \\
& C_3H_4^+ + H_2 & \Delta H = -1.33\,\text{eV} \\
& C_3H_3^+ + H + H_2 & \Delta H = +0.41\,\text{eV} \\
& C_2H_3^+ + CH_3 & \Delta H = +0.02\,\text{eV}
\end{aligned}
$$

Figure 7 shows the cross sections we measure for the $C_3H_5^+$ and $C_2H_3^+$ channels, plotted as a function of collision energy for four different $C_2H_2^+$ reactant vibrational states. ν_2 is the C–C stretch with energy of 1814 cm^{-1}, and two quanta of ν_{bend} (probably the cis–bend) is 1253 cm^{-1}. The cross sections for the $C_3H_4^+$ channel are not plotted because they are simply a factor of 3.1 lower than those for $C_3H_5^+$, but otherwise have identical collision and vibrational energy dependence. $C_3H_3^+$ is a minor channel that primarily results from decomposition of $C_3H_5^+$ and $C_3H_4^+$ at high collision energies.

Isotopic-labeling experiments show that the $C_3H_n^+$ channels all involve a common intermediate complex. This suggests that any mode-specific nuclear motion effects (collision or vibrational) must come from effects on the complex formation cross section, since presumably the complex rapidly loses memory of the initial mode of excitation. What we observe is that both collision and vibrational energy suppress the reaction; however, different modes are not equally effective. Close examination of the figure shows that the inhibition from adding one or two quanta of ν_2(C–C stretch) is similar

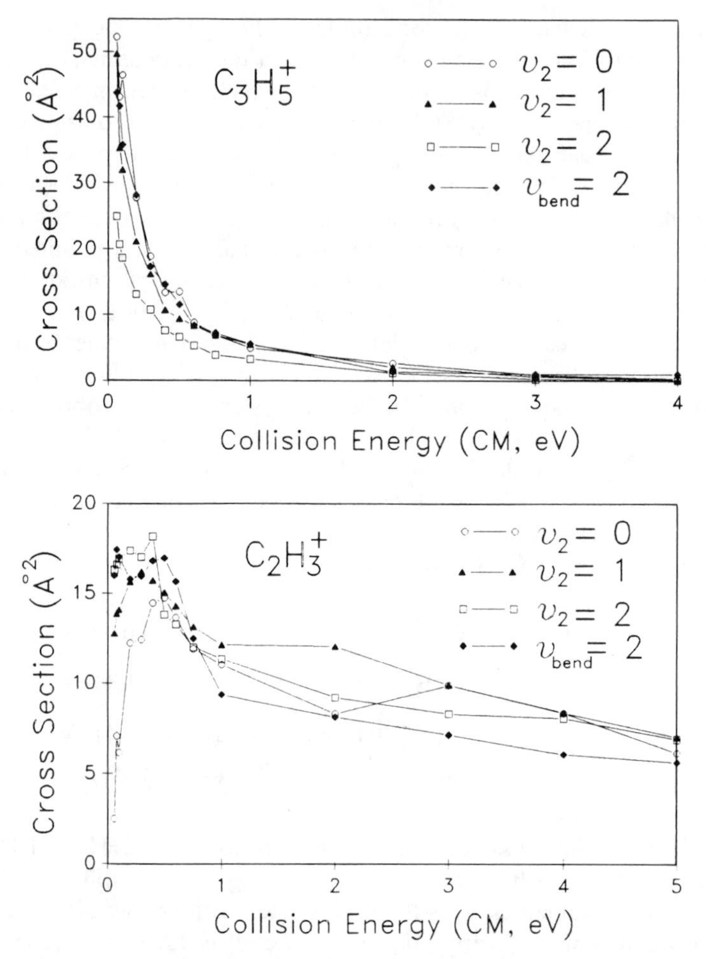

Figure 7. Cross sections for reaction of $C_2H_2^+$ with methane as a function of collision energy and reactant ion vibrational state. Cross sections are given for the $C_3H_5^+$ and $C_2H_3^+$ product channels. The cross sections for $C_3H_4^+$ formation are identical to those for $C_3H_5^+$, except for a factor of 3.1 reduction in scale.

to the reduction due to an equivalent increase in collision energy. On the other hand, 1253 cm^{-1} of bending excitation has essentially no effect—in other words, bending excited ions are more reactive than would be expected based on their energy content.

The $C_2H_3^+$ channel does not show much isotopic scrambling, and thus appears to be mostly due to direct H-atom transfer from methane to $C_2H_2^+$.

This reaction is slightly endoergic and the cross section for ground-state $C_2H_2^+$ has an almost resolved threshold. At low energies, H-atom transfer is strongly enhanced by both collision energy and vibration. The effects from adding one or two quanta of ν_2 excitation are about 20% and 30% larger that the enhancements that result from adding equivalent amounts of collision energy for ground-state $C_2H_2^+$. Bending excitation is much more effective. Note that two quanta of the bend (1253 cm^{-1}) have nearly the same effect as two quanta of the C–C stretch (3624 cm^{-1}) even though the energy is three times lower. At high energies, bending actually decreases reactivity slightly.

At least at low energies, bending excited acetylene cations appear to be generally more reactive than ions with equivalent amounts of collision energy, while C–C stretching has an effect similar to collision energy. Our rationalization for this mode selectivity is as follows. For any of the reaction channels it is necessary to make additional bonds to $C_2H_2^+$, and this requires rehybridization of one or both acetylenic carbon atoms. C–C stretch and collision energy are not particularly helpful in this electronic promotion; however, bending energy goes directly into lowering the energy gap between the *sp* and sp^2 states. Bending is thus exactly all type of energy that can promote reactivity.

This rehybridization argument is closely related to the transition-state explanation given by Zare and co-workers to explain why umbrella bending of NH_3^+ enhances NH_4^+ formation in reaction with methane and ammonia. Whether these "freshman chemistry" ideas have anything to do with reality remains to be seen as additional experimental and theoretical results accumulate.

Consider, for example, the reactions of NH_3^+ or $C_2H_2^+$ with hydrogen, which in both cases lead to H-atom addition products. The transition-state/rehybridization picture would suggest that both these reactions should be enhanced by bending vibration. In the case of NH_3^+, Morrison, Conaway, and Zare[48] observed no effect from umbrella bending. In a preliminary study of $C_2H_2^+$ reactions with hydrogen, we observed that channels involving C–H bond formation are actually inhibited by bending excitation. These results appear to conflict with the simple transition-state-rehybridization picture, however, it should be noted that these reactions with hydrogen are both very inefficient processes (i.e., $\sigma_{reaction}$ is less than 10% of $\sigma_{collision}$). It therefore seems quite likely that some factor that is not simply related to the ion-bending motion controls reactivity for these systems.

We have done considerable work on the $[C_2H_2 + OCS]^+$ system, measuring cross sections for all the product channels as a function of collision energy. Since we can state select both $C_2H_2^+$ and OCS^+, we have measured reactivity for a total of 12 different reactant internal energy states. As an example of the effects observed, I will summarize the results for just the charge-transfer channels.

The top half of Fig. 8 gives the cross sections for charge transfer (CT) from $C_2H_2^+$ to OCS, a process that is 0.226 eV exoergic. Results are given for

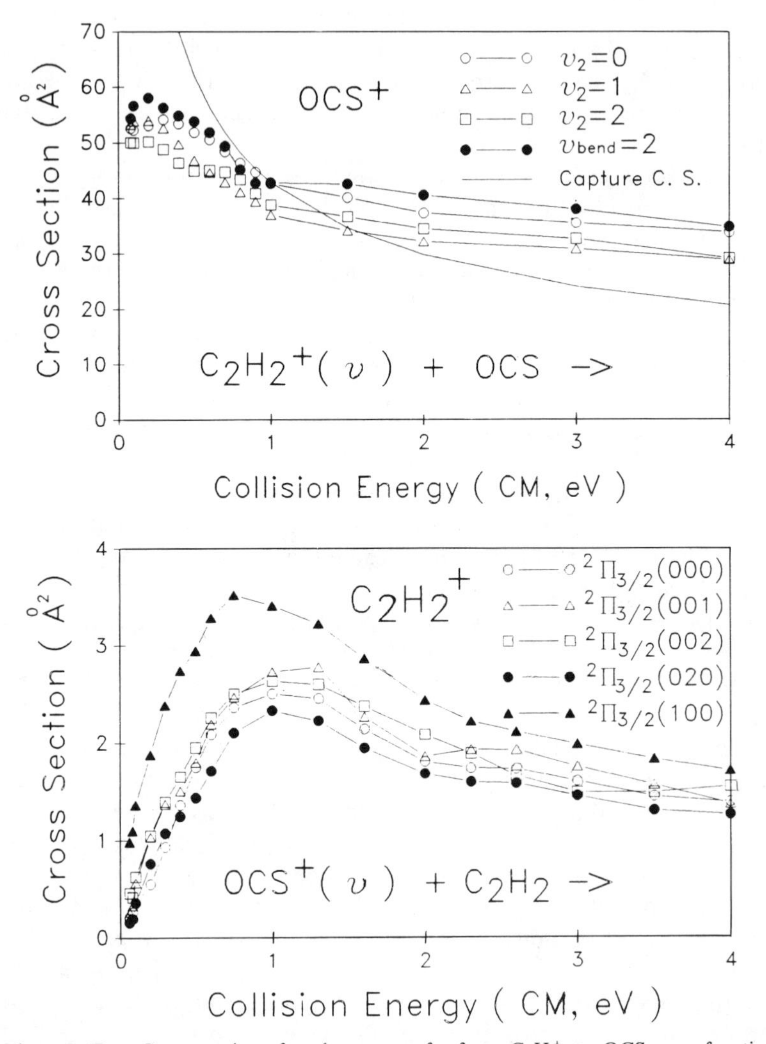

Figure 8. Top: Cross sections for charge transfer from $C_2H_2^+$ to OCS as a function of collision energy. Data for four reactant ion vibrational states are plotted, together with the classical ion–dipole capture cross section. Bottom: cross sections for charge transfer from OCS^+ to C_2H_2. Data are shown for ground electronic state OCS^+ in five different vibrational levels.

$C_2H_2^+$ reactant ions in their ground state, with one and two quanta of C–C stretch (ν_2), and with two quanta of bending excitation (ν_5). Note that the CT cross section is large and only weakly dependent on collision energy. In particular, the CT cross section is larger than the ion–dipole capture cross section (plotted as a solid line) for collision energies above about 1 eV. This behavior is typical of CT that occurs at least partly by long-range electron transfer, which in turn is only efficient if the initial and final states are near resonant.

In this high-energy range, the vibrational effects are clearly state specific. CT probability is reduced by ν_2 excitation, but one quantum has a larger effect than two. Bending excitation actually enhances CT. These effects are at least qualitatively consistent[54] with a crude energy-gap/Franck–Condon picture in which it is assumed that the probability of long-range CT is enhanced for a particular reactant vibrational state if there are near resonant final states with good vibrational overlap.

Note that as the capture cross section shoots up at low collision energies, the CT cross section increases and the vibrational effects change. Bending excited ions are still anomalously reactive; however, the effect of ν_2 excitation changes to a monotonic inhibition. These changes are evidence for a new CT mechanism growing in at low energies in addition to long-range electron transfer. The obvious explanation is that at low energies, attractive ion-dipole/induced-dipole forces become more effective. This increases the total CT cross section, since it increases the range of impact parameters that are brought to short enough intermolecular separation to allow electron transfer. In addition, many collisions are captured into intimate contact, where CT can occur by a completely different mechanism with different vibrational effects. One possibility might be CT mediated by complex formation.

Note that at our lowest collision energies the CT cross sections begin to decrease. This is due to competition with the reaction

$$C_2H_2^+ + OCS \rightarrow C_2H_2S^+ + CO,$$

which has a cross section that is large only at very low energies.[54]

The bottom frame of Fig. 8 give the CT cross sections for the reverse process:

$$OCS^+ + C_2H_2 \rightarrow C_2H_2^+ + OCS.$$

The cross sections are small and energy dependent as expected, since energy transfer is required to drive this endoergic reactions. Clearly the mechanism must involve intimate collisions, which may involve complex formation at low energies. The five cross sections plotted are all for OCS^+ in the ground fine structure state. In our notation (*jkl*) gives the number of quanta for ν_1(C–O stretch, 2041 cm^{-1}), ν_2(bend, two quanta $\approx$ 1000 cm^{-1}), and ν_3 (CS stretch, 700 cm^{-1}). We also measured cross sections for OCS^+ in the upper fine structure state (360 cm^{-1} splitting) but found essentially no effects outside our error limits.

At collision energies below 0.5 eV the vibrational effects are basically just a monotonic enhancement with increasing vibrational energy, as might be expected for an endoergic reaction. At higher collision energies mode specificity becomes apparent. Excitation of ν_2 continues to give a small enhancement; however, ν_1 gives a much larger effect, and bending excitation actually results in a small inhibition. Perhaps the real surprise here is that ν_1 excitation does not result in a much larger enhancement. This reaction is 0.226 eV endoergic, and one quantum of ν_1 is 0.253 eV, that is, CT for OCS^+ with ν_1 excited is actually slightly exoergic. One might think that this would allow long-range CT with large cross section to occur, but evidently this is not the case.

At this point only a handful of polyatomic, bimolecular ion–molecule reactions have been studied with state selection techniques. In most cases vibrational energy and mode effects have been observed, but as is apparent from the preceding discussion, the dynamical basis of the effects is mostly unknown. Results on a wider range of systems would clearly be useful in trying to understand these effects, but there is also a need for theoretical studies. For example, scattering studies on a series of model potential surfaces would give a better notion of the origin of different types vibrational effects that might be observed.

C. Jet Studies of Angular Distributions

Pollard, Lichtin, and Cohen[55] recently developed a novel, MPI-based technique for measuring ion–molecule angular distributions. They make use of MPI both to state select the reactant ions and to provide high intensity pulses of ions with good time structure. In their method a pulsed molecular beam of precursor molecules is multiphoton ionized. Immediately after ionization, a pulsed electric field gives an impulsive acceleration to the ions, leaving them with a controlled kinetic energy. This acceleration field is *rotatable*; thus, the ions can be directed at variable angles with respect to the direction of the neutral beam. Reactions occur as the ions pass through the beam, then product ions are collected by a fixed-angle mass spectrometer detector, and their velocities are determined by time of flight. Angular distributions are measured by varying the angle of acceleration for the reactants, rather than by rotating the detector or ion source as is more typically done. After appropriate data analysis, center-of-mass flux contour maps can be extracted. The nice feature of this arrangement is that the experimental setup is very simple. Only a single pulsed beam is needed, and it is unnecessary to rotate large parts of the vacuum system (i.e., the detector or ion source). The problem is that analysis is complicated by variations in reaction volume and detector collection efficiency with incident angle.

Their initial study was of several isotopic variants of the reaction:

$$H_2^+ + H_2 \rightarrow H_3^+ + H.$$

They observed that both atom and proton transfer are direct reactions, giving scattered product peaks near the spectator stripping limit. This technique appears promising, and Pollard and co-workers have been carefully developing the tools needed to fully analyze the results.

D. ICR Studies of Lifetimes and Vibrational Effects

Bowers and co-workers have recently combined the Fourier-transform ion-cyclotron-resonance (FTICR) method with MPI state selection to study radiative lifetimes and chemistry of excited ions. In their apparatus,[56] ions are created by MPI in the strong magnetic field of an ICR trapping cell. Application of a perpendicular electric field causes the ions to be trapped in circular orbits, with cyclotron frequencies that are dependent only on the ion mass. The mass distribution in the trap is monitored by measuring the image current the cycling ions induce in a pair of electrodes. Storage times can be as long as seconds, which lends itself to measurement of vibrational lifetimes that can be hundreds of milliseconds. The neutral precursor molecule is admitted as a pulsed jet, and the gas is quickly pumped away so that the state selected ions do not suffer unwanted collisions which would perturb the state distribution.

To measure vibrational lifetimes, ions are created in a particular vibrational state by MPI, then allowed to cycle without collisions for a variable time, during which some radiative decay can occur. At that point a monitor reactant is injected by a second pulsed valve. This reactant is chosen to have an ionization potential high enough that it will only undergo efficient CT with ions in the initial vibrational state. By measuring the decrease in CT as a function of delay time, the vibrational lifetime is directly obtained. To date, they have measured lifetimes for $v = 1$–5 of NO^+,[57,58] and for decay of metastable states of O_2^+ and NO^+.[59]

The technique can also be applied to measurements of vibrational effects on reaction rate constants. Wyttenbach et al.[58] have presented preliminary measurements of vibrational effects on rate constants for reaction of NO^+ with two dozen medium size organic molecules. Mostly they observe CT, with rates that depend strongly on whether the energy of NO^+ vibrational state is greater than the molecule's ionization potential. In reaction with CH_3CHO, hydride transfer is observed, with efficiency that is strongly suppressed by NO^+ vibrational excitation.

E. Unimolecular Dissociation Studies

Space limitations preclude doing justice to this category of MPI studies and I will only mention a small selection. Basically the objective here is to produce molecular ions with precisely controlled excitation energy in excess of the dissociation limit, then to measure the unimolecular decay rate. This type of experiment has long been done by photoion–photoelectron coincidence techniques[60]; however, there are several advantages of MPI for energy selection at least for some systems. Since bound–bound transitions are involved, it is possible to avoid transferring any of the neutral ground-state thermal vibrational distribution to the ion. Intensities are usually higher. Finally for molecules with resolvable rotational structure in the MPI spectrum it is possible to directly examine the influence of angular momentum on dissociation rate.

Single-color MPI was used in a study of chlorobenzene dissociation rates by Durant et al.[61] The neutral was ionized through a series of vibronic intermediates, producing ground-electronic-state ions in known (from photoelectron spectra) vibrational distributions. Absorption of an additional photon pumped the cations to a narrow and tunable band of excitation energies and the dissociation rate was monitored by the metastable peak-shape method.

Single-color-ionization experiments have also been reported by Schlag and co-workers[30] and by Neusser.[31] In the studies by Schlag's group, the main emphasis has been on ion spectroscopy monitored by dissociation, rather than kinetics. Neusser's work on $C_6H_6^+$ was able to demonstrate that for constant total energy, increasing the rotational angular momentum of the ions resulted in a small decrease in total dissociation rate, which was attributed to increased density of states in the excited ion.

A serious disadvantage of the single-color scheme is that as the excitation energy is increased, the ionic vibrational distributions produced become increasingly broad, thus the energy resolution is degraded. This limits both the molecules and range of vibrational excitation that can studied.

In multicolor schemes this problem is reduced by allowing ionization into a narrow vibrational distribution, followed by photoexcitation with an independently tunable laser. Experiments of this type have been done by Botter and co-workers, who have reported both dissociation spectroscopy work[33] and an improved study of the unimolecular decay kinetics of chlorobenzene cation.[62]

IV. FUTURE DIRECTIONS

A. Ideas for New MPI State-Selection Schemes

The single- and two-color MPI schemes previously discussed have been demonstrated to produce state selected ions, at least for some molecules. The two-color method in which the ionizing laser is set to provide just enough energy to produce the desired ion state is already being applied with great success for aromatic molecules. For smaller molecules, this technique should also be useful; however, this will frequently require a VUV pump laser to allow single photon access to high-energy Rydberg states. As laser technology evolves, these techniques should become routine. This section presents ideas for different approaches to MPI state selection that might significantly improve both the number of molecular ions that can be state selected and the range and purity of the excitations that can be produced. The schemes are really not technologically feasible with the lasers available at present, however, it is reasonable to assume that this problem will be solved with time.

1. *IR–UV Double Resonance*

A potentially powerful multi-color excitation scheme that has not been exploited for state selection purposes is infrared–UV double resonance. The ground-state molecule would be excited by infrared radiation to an excited vibrational level. The vibrationally excited molecules would then be multiphoton excited to a Rydberg intermediate state, then ionized by one addition photon. Since the transition to the Rydberg level is discrete, it should be possible to select only those molecules that have been excited vibrationally. This scheme has a number of significant advantages. The only limit on the range of vibrations that can be populated in the ion is that they must either be infrared active in the ground state or Franck–Condon active in excitation to the Rydberg intermediate. Since it is no longer necessary to use mixed intermediate states to populate non-Franck–Condon active vibrations, state selection purity should be better.

Ideally, this should be a three-color experiment: infrared to excite the ground-state vibration, UV and VUV to excite to the Rydberg intermediate state, and near UV or visible to ionize with just enough energy to populate the desired ion-vibrational level. There are several disadvantages of this scheme (aside from cost). It complicates the ion chemist's life by requiring spatial and temporal overlapping of two or three lasers in the ion source. At the moment, it is difficult to produce intense tunable IR in the energy range needed. In the long run the real price is the requirement for additional spectroscopic work to find appropriate MPI excitation routes for the vibrationally excited molecules. Of course, there are people who regard this sort of problem as an enjoyable challenge.

2. *MPI–ZEKE Coincidence*

The other MPI state selection scheme I would like to propose is fundamentally different. Perhaps the ultimate method for vibrational state selection of polyatomic ions will be MPI by high-repetition-rate lasers in coincidence with detection of zero-kinetic-energy electrons. This MPI–ZEKE method should in principle give pure selection of several levels of nearly all vibrational modes for most molecules. Depending on the molecule and the type of laser used, rotational selection should also be possible. This would really allow systematic studies of the effects of different modes of nuclear motion on reaction dynamics. The idea is very closely related to photoion–photoelectron-coincidence state selection methods that have been used successfully for years in conjunction with single-photon ionization (see the chapter by Koyano in this volume). In particular, the experiment described below is patterned on the CERISES instrument developed by Guyon, Dutuit, Gerlich, and co-workers using synchrotron radiation for ionization.[63]

The outline of the method is as follows: Two pulsed dye lasers are synchronized to provide tunable output pulses at a repetition frequency in the range from 0.2 to 1 Mhz. One laser is used to pump the molecule of interest to an intermediate vibronic state, and the second is tuned to pump the excited molecule just to the threshold for producing ions in the internal energy state desired. The laser intensities are kept low enough so that the total ionization probability is around one event per 10 laser shots. The zero-electron-kinetic-energy (ZEKE) spectroscopy method developed by Müller-Dethlefs and co-workers[64] is used to identify ionization events that produce zero energy electrons, which by conservation of energy correspond to ions produced in the desired vibrational state. These ions are then injected into a guided-ion-beam mass spectrometer system where state-selected chemistry can be studied.

ZEKE in the low-resolution form need for this experiment is very simple. During the laser pulse, the ionization region is field free. If an ionization event occurs, the photoelectron moves out of the focal volume with a velocity that is small only if the ion was created in the desired internal energy state. After a delay time of around 100 ns a uniform electric field is applied to the ionization region. This extracts the electron towards a detector, and accelerates the ion toward the ion chemistry part of the apparatus. ZEKE identifies low-energy electrons in two ways. As in traditional threshold electron detectors, the geometry used strongly favors detection of low-energy electrons, since those with energies above about 50 meV are only detected if they happen to have initial velocities at small angles to the detector axis. ZEKE also uses the time of arrival of the electron at the detector to idendify positively low-velocity electrons. If a zero energy electron is detected, then the ion is allowed to pass into the chemistry part of the instrument, otherwise a transverse field is applied to reject the ions. In this way reactions are only studied for ions in the desired internal energy state.

In coincidence schemes, the ion-state selection purity is controlled by the detection of threshold electrons, and indeed one might ask what advantages are gotten by using MPI rather than single-photon ionization in this case. The major advantage comes in increasing the range of vibrational levels that can be accessed. In single-photon methods the distribution of vibrational levels produced is basically determined by the Franck–Condon factors connecting the neutral ground state to the ionic potential surface. (Some additional ion vibrations can be populated by resonance with autoionization structure in the continuum.) MPI through different intermediate states allows some tuning of the ionization Frank–Condon factors, which increases the range of ionic vibrations that can selected even in non coincidence experiments. Furthermore, in a coincidence experiment where we can rely on the electrons to tell us when we have the right ion state, MPI can make use of excited valence states that can have very broad ionization Franck–Condon factor distributions. This should allow selection of otherwise inaccessible vibrational modes.

The ideal laser system for MPI–ZEKE coincidence state selection does not yet exist. To avoid high false coincidence levels, it is necessary to keep the ionization probability per laser shot at 10% or less. To obtain state selection ion intensities high enough for detailed ion–molecule reaction experiments, the ideal laser repetition frequency should be around 0.2–1 MHz. At present the best candidate would be a cavity dumped picosecond dye laser system. The problem is that some amplication of the dye laser pulses is needed to give sufficient intensity for MPI; however, current pulse amplifiers do not run at such high repetition frequencies. Copper-vapor laser-pumped dye amplifiers should give sufficient gain (currently only at around 5 KHz repetition rate), and it is reasonable to expect that the technology for generating intense rapid pulse trains will become available. ZEKE with picosecond lasers (30 Hz amplification) has demonstrated by Knee and co-workers,[65] and appears to work very nicely.

B. Conclusions

I hope that I have shown some of the benefits that MPI can bring to the study of molecular ion chemistry. At the moment, the list of molecules for which effective MPI state selection routes have been worked out is small. This situation should improve as MPI–photoelectron spectroscopy is extended to more polyatomic systems, and as high-power VUV and IR laser systems become more routinely available. Since the actual use of MPI in state selected chemistry experiments is quite new, it is reasonable to suppose that its impact on the ion dynamics field will grow considerably in the next few years.

Two particular areas where I believe that MPI state selection will be important are in high-resolution studies of state-to-state ion reaction dynamics and in studies of vibrational mode effects on polyatomic ion reactions. In the first case the experiments will take advantage of MPI to produce beams of ions that are really in a single quantum state. As the necessary spectroscopic work evolves, there should be a growing list of diatomic and small polyatomic systems in which vibrational, fine structure, rotational, Λ-doublet, and perhaps angular momentum alignment can be simultaneously selected using MPI. This will also allow production of beams with narrow translational energy spreads, and with temporal widths short enough to make product state detection possible by either time-of-flight or spectroscopic means.

For most polyatomic molecules the main contribution will be in allowing studies of the effects of different vibrational modes on reactions. Lack of detailed spectroscopic information will continue to be a problem here, but I expect that the list of state-selectable systems will continue to grow now that MPI–photoelectron spectroscopy has become a standard spectroscopic tool. For the near future, ion chemists wanting to state select new molecules will undoubtedly have to get involved in spectroscopic survey work.

Acknowledgments

I would like to thank the many people in the MPI field who made suggestions and communicated recent data prior to publication, particularly Mark Smith (Tucson) and Dieter Gerlich (Freiburg), who contributed unpublished figures for this review. Dieter Gerlich and Joe Knee (Wesleyan University) helped develop the MPI–ZEKE coincidence idea.

Much of the work for the review was done while I was on sabbatical leave at the Institute for Molecular Science in Okazaki, and at the Physics Department of the University of Freiburg, with generous support from the Japan Society for Promotion of Science and Deutsche Forschungsgemeinschaft (SFB276), respectively.

I also am grateful for fellowship support from the Alfred P. Sloan and Camille and Henry Dreyfus Foundations. Our own work on MPI state selected ion chemistry is supported by the U.S. National Science Foundation (CHE8903765).

References

1. For example, see
S. H. Lin, Y. F. Fujimura, H. J. Neusser, and E. W. Schlag, *Multiphoton Spectroscopy of Molecules*, Academic Press, New York, 1984; *Resonance Ionization Spectroscopy 1988*, T. Lucatorto and J. E. Parks, eds., British Institute of Physics Conference Series 94, 1988.
2. R. N. Compton and J. C. Miller, in *Laser Applications in Physical Chemistry*, D. K. Evans, ed., Marcel Dekker, New York, 1989, p. 221.
3. T. M. Orlando, S. L. Anderson, J. R. Appling, and M. G. White, *J. Chem. Phys.* **87**, 852 (1987).
4. S. T. Pratt, E. F. McCormack, J. L. Dehmer, and P. M. Dehmer, *J. Chem. Phys.* **92**, 1831 (1990).

5. A. H. Kung, T. Trickl, N. A. Gershenfeld, and Y. T. Lee, *Chem. Phys. Lett.* **144**, 427 (1988).
6. T. Trickl, E. F. Cromwell, Y. T. Lee, and A. H. Kung, *J. Chem. Phys.* **91**, 6006 (1989).
7. See, for example, K. Sato, Y. Achiba, and K. Kimura, *J. Chem. Phys.* **80**, 57 (1984).
8. S. T. Pratt, *Phys. Rev. A* **33**, 1718 (1986).
9. S. L. Anderson, G. D. Kubiak, and R. N. Zare, *Chem. Phys. Lett.* **105**, 22 (1984).
10. S. T. Pratt, P. M. Dehmer, and J. L. Dehmer, *J. Chem. Phys.* **86**, 1727 (1987).
11. S. T. Pratt, P. M. Dehmer, and J. L. Dehmer, *Chem. Phys. Lett.* **105**, 28 (1984).
12. S. T. Pratt. P. M. Dehmer, and J. L. Dehmer, *J. Chem. Phys.* **81**, 3444 (1984).
13. S. T. Pratt, P. M. Dehmer, and J. L. Dehmer, *J. Chem. Phys.* **80**, 1706 (1984).
14. S. Opitz, D. Proch, T. Trickl, and K. L. Kompa, *Chem. Phys.* **143**, 305 (1990).
15. T. Trickl, E. F. Cromwell, Y. T. Lee, and A. H. Kung, *J. Chem. Phys.* **91**, 6006 (1989).
16. H. Park, P. J. Miller, W. A. Chupka, and S. D. Colson, *J. Chem. Phys.* **89**, 6676 (1988).
17. Y. Achiba, K. Sato, K. Shobatake, and K. Kimura, *J. Chem. Phys.* **78**, 5474 (1983).
18. S. W. Allendorf, D. J. Leahy, D. C. Jacobs, and R. N. Zare, *J. Chem. Phys.* **91**, 2216 (1989).
19. T. Ebata and R. N. Zare, *Chem. Phys. Lett.* **130**, 467 (1986).
20. S. T. Pratt, E. D. Poliakoff, P. M. Dehmer, and J. L. Dehmer, *J. Chem. Phys.* **78**, 65 (1983).
21. D. Gerlich and T. Rox, *Z. Phys. D* **13**, 259 (1989).
22. L. F. DiMauro and T. A. Miller, *Chem. Phys. Lett.* **138**, 175 (1987).
23. A. Fujii, T. Ebata, and M. Ito, *Chem. Phys. Lett.* **161**, 93 (1989).
24. J. Xie and R. N. Zare, *Chem. Phys. Lett.* **159**, 399 (1989).
25. J. Xie, D. J. Leahy, and R. N. Zare (private communication).
26. B. Yang, M. H. Eslami, and S. L. Anderson, *J. Chem. Phys.* **89**, 5527 (1988).
27. J. H. Glownia, S. J. Riley, S. D. Colson, J. C. Miller, and R. N. Compton, *J. Chem. Phys.* **77**, 68 (1982).
28. W. E. Conaway, R. J. S. Morrison, and R. N. Zare, *Chem. Phys. Lett.* **113**, 429 (1985).
29. M. N. Ashfold, B. Tutcher, B. Yang, Z. Jin, and S. L. Anderson, *J. Chem. Phys.* **87**, 5105 (1987).
30. K. Walter, R. Weinkauf, U. Boesl, and E. W. Schlag, *Chem. Phys. Lett.* **155**, 8 (1989) K. Walter, U. Boesl, and E. W. Schlag, *Chem. Phys. Lett.* **162**, 26 (1989).
31. H. J. Neusser, *J. Phys. Chem.* **93**, 3897 (1989).
32. S. L. Anderson, L. Goodman, K. Krogh-Jespersen, A. G. Ozkabak, R. N. Zare, and C. Zheng, *J. Chem. Phys.* **82**, 5329 (1985).
33. J. Lemaire, I. Dimicoli, F. Piuzzi, and R. Botter, *Chem. Phys.* **115**, 119 (1987); J. Lemaire, I. Dimicoli, and R. Botter, *Chem. Phys.* **115**, 129 (1987); X. Ripoche, I. Dimicoli, J. le Calve, F. Piuzzi, and R. Botter, *Chem. Phys.* **124**, 305 (1988).
34. B. G. Koenders, D. M. Wieringa, K. E. Drabe, and C. A. De Lange, *Chem. Phys.* **118**, 113 (1987).
35. B. G. Koenders, S. M. Koeckhoven, G. J. Kuik, K. E. Drabe, and C. A. De Lange, *J. Chem. Phys.* **91**, 6042 (1989).
36. J. C. Miller, R. N. Compton, T. E. Carney, and T. Baer, *J. Chem. Phys.* **76**, 5648 (1982).
37. Y. Achiba, K. Sato, K. Shobatake, and K. Kimura, *J. Chem. Phys.* **77**, 2709 (1982).
38. J. Steadman, S. K. Cole, and T. Baer, *J. Chem. Phys.* **89**, 5498 (1988).
39. J. Steadman and T. Baer, *J. Chem. Phys.* **89**, 5507 (1988).
40. M. Hawley, T. L. Mazely, L. K. Randeniya, R. S. Smith, X. K. Zeng, and M. A. Smith, Int. J. Mass Spectrom. Ion Proc. **97**, 55 (1990).

41. M. Hawley and M. A. Smith, *J. Am. Chem. Soc.* **111**, 8293 (1989).
42. L. K. Randeniya, X. K. Zeng, R. S. Smith, and M. A. Smith, *J. Phys. Chem.* **93**, 8031 (1989).
43. T. L. Mazely and M. A. Smith *J. Phys. Chem.* **94** 6930 (1990).
44. Mark A. Smith (unpublished results).
45. N. G. Adams and D. Smith, *Chem. Phys. Lett.* **47**, 383 (1977).
46. T. Turner and Y. T. Lee, *J. Chem. Phys.* **81**, 5638 (1984).
47. D. Gerlich (private communication).
48. R. J. S. Morrison, W. E. Conaway, and R. N. Zare, *Chem. Phys. Lett.* **113**, 435 (1985).
49. W. E. Conaway, T. Ebata, and R. N. Zare *J. Chem. Phys.* **87**, 3447 (1987).
50. W. E. Conaway, T. Ebata, and R. N. Zare, *J. Chem. Phys.* **87**, 3453 (1987).
51. T. Ebata, W. E. Conaway, and R. N. Zare, *Int. J. Mass Spectrom. Ion Proc.* **80**, 51 (1987).
52. T. Baer and P. T. Murray, *J. Chem. Phys.* **75**, 4477 (1981).
53. T. M. Orlando, B. Yang, and S. L. Anderson *J. Chem. Phys.* **90**, 1577 (1989).
54. T. M. Orlando, B. Yang, Y. Chiu, and S. L. Anderson, *J. Chem. Phys.* **92**, 7356 (1990).
55. J. E. Pollard, D. A. Lichtin, and R. B. Cohen, *Chem. Phys. Lett.* **152**, 171 (1988).
56. C. G. Beggs, C. H. Kuo, T. Wyttenbach, P. R. Kemper, and M. T. Bowers (unpublished).
57. C. H. Kuo, T. Wyttenbach, C. G. Beggs, P. R. Kemper, M. T. Bowers, D. J. Leahy, and R. N. Zare, *Chem. Phys. Lett.* **163**, 291 (1989).
58. T. Wyttenbach, C. G. Beggs, C. H. Kuo, and M. T. Bowers, abstracts of the 38th ASMS conference on Mass Spectrometry and Allied Topics, Tucson, AZ, June 3–8, 1990.
59. C. H. Kuo, T. Wyttenbach, C. G. beggs, P. R. Kemper, and M. T. Bowers, *J. Chem. Phys.* **92**, 4849 (1990).
60. See for example, T. Baer, B. P. Tsai, D. Smith, and P. T. Murray, *J. Chem. Phys.* **64**, 2460 (1976); H. M. Rosenstock, R. Stockbauer, and A. C. Parr, *J. Chem. Phys.* **71**, 3708 (1979).
61. J. L. Durant, D. M. Rider, S. L. Anderson, F. D. Proch, and R. N. Zare, *J. Chem. Phys.* **80**, 1817 (1984).
62. X. Ripoche, I. Dimicoli, and R. Botter (unpublished).
63. P. M. Guyon, G. Bellec, O. Dutuit, D. Gerlich, E. A. Gislason, and J. B. Ozenne, *Bull. Soc. Roy. Sci. Liège,* **58** (3-4), 187 (1989).
64. K. Kuller-Dethlefs, M. Sander, and E. W. Schlag, *Z. Naturforsch. Teil A* **39**, 1089 (1984); *Chem. Phys. Lett.* **112**, 291 (1984).
65. J. Smith, C. Lakshminarayan, and J. L. Knee (unpublished).

CONTROL OF TRANSITION-METAL CATION REACTIVITY BY ELECTRONIC STATE SELECTION

JAMES C. WEISSHAAR

Department of Chemistry,
University of Wisconsin–Madison,
Madison, WI

CONTENTS

State-Selected and State-to-State Ion–Molecule Reaction Dynamics, Part 1: Experiment, Edited by Cheuk-Yiu Ng and Michael Baer. Advances in Chemical Physics Series, Vol. LXXXII.
ISBN 0-471-53258-4

I. INTRODUCTION

Given the goal of controlling the rate and product branching of bimolecular chemical reactions, a key question is how different kinds of reactant energy—kinetic, vibrational, and electronic—affect total reaction cross sections and product branching ratios. It seems that ion–neutral reactions lend themselves naturally to reactant state selection, one of the primary themes of this volume. In principle, reactant *electronic* states offer the most powerful means of controlling bimolecular chemistry, since the electrons govern the shape of the potential energy surfaces on which the nuclei move. Organic and inorganic photochemists routinely use electronic excitation to alter the course of *unimolecular* reactions in ever more subtle fashion. However, we have only begun to study effects of initial electronic state on *bimolecular* reactions,[1] partly because many excited states relax radiatively on time scales short compared to collision times. In addition, most excited-state reactions involve two or more potential energy surfaces, and such nonadiabatic reactions are inherently complex.[2]

The gas-phase transition-metal cations (M^+) are an appealing class of reactants for the study of electronic-state effects. Each M^+ atom has a rich set of low-lying metastable states that differ in electron configuration and spin multiplicity. More than 10 years ago, Ridge and co-workers[3] discovered the ability of gas-phase Fe^+ to break both C–H and C–C bonds of linear alkanes, eliminating H_2 or CH_4 in low-energy collisions. Solution-phase chemists have recently found reactions somewhat analogous to the C–H bond-breaking channel.[4] Despite intense interest in low-energy cracking of C–C bonds in alkanes, to our knowledge the gas-phase Fe^+, Co^+, and Ni^+ reactions remain unique. Meanwhile, qualitative M^+ chemistry in the gas phase has been extensively explored. More quantitative experiments have begun to unravel the mechanisms of a few reactions in greater detail.

In this chapter we review our recent efforts to learn how electronic structure (Fig. 1) controls chemical reactivity in open *d*-shell systems by measuring electronic state-specific reaction cross sections for low-energy collisions of 3*d*-series M^+ with small hydrocarbons. One-color, resonant two-photon ionization (R2PI, Fig. 2) of neutral M produces low-kinetic-energy M^+ beams in a variety of state distributions, some of them remarkably specific. Time-of-flight photoelectron spectroscopy (TOF–PES) measures the state distribution from each ionizing wavelength. A crossed-beam experiment then measures total reactions cross sections as a function of M^+ state distribution.

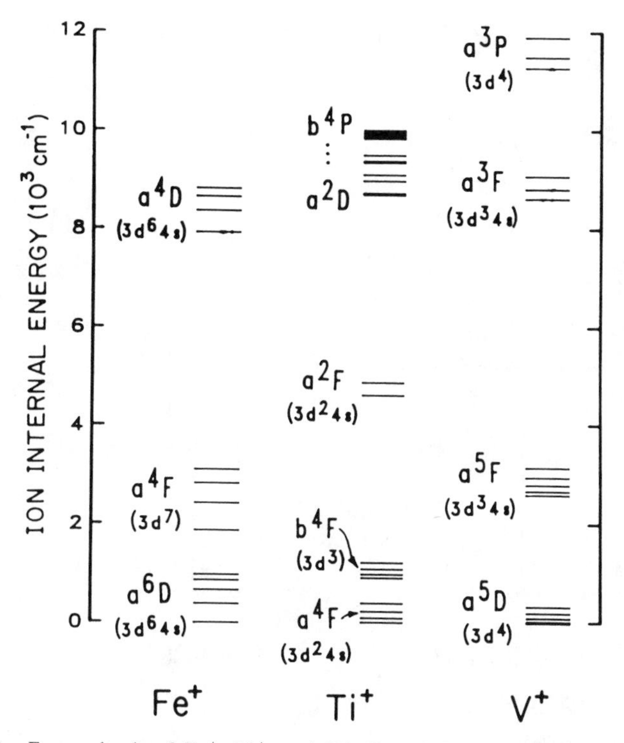

Figure 1. Energy levels of Fe^+, Ti^+, and V^+ from Ref. 6. Each horizontal line is a spin–orbit level.

Our research applies tunable dye-laser, molecular-beam, and mass-spectrometric techniques to much more complex chemical reactions than those traditionally studied by gas-phase chemical physicists. The work touches on the traditional areas of atomic physics, gas-phase molecular dynamics, and inorganic and organometallic chemistry. We hope that by studying electronic effects in a few atomic M^+ reactions under carefully controlled conditions we can both inspire and calibrate new *ab initio* calculations for transition metal species. Atomic M^+ chemistry is simple at least in the sense that the reactant electronic structure is well known (Fig. 1). A good understanding of how electronic structure controls M^+ chemical reactivity should also help inform studies of the chemistry of small metal clusters,[5] for which the electronic structure is more complex and seldom known.

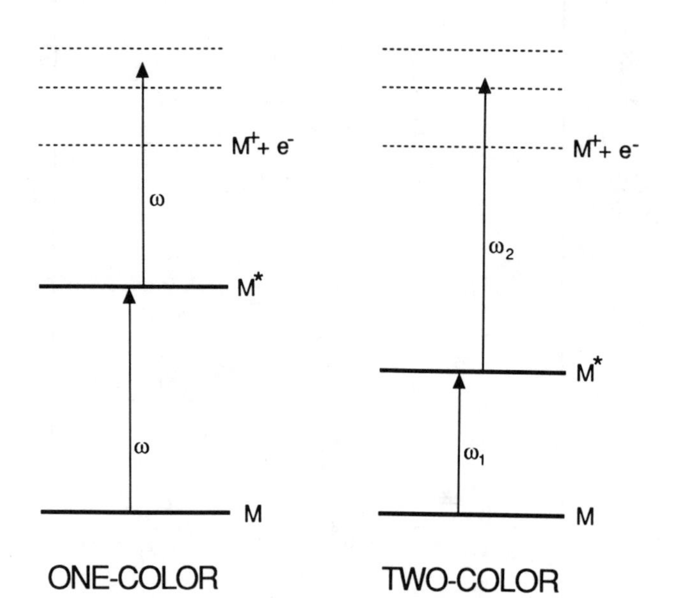

Figure 2. Schematic of one- and two-color resonant two-photon ionization schemes.

II. OVERVIEW OF TRANSITION-METAL-CATION CHEMISTRY WITH HYDROCARBONS

Figure 1 displays the low-energy states[6] of Ti^+, V^+, and Fe^+, the three cations we have studied. The energy levels are listed in Table I. Russell–Saunders coupling is a good approximation in the 3*d*-series, since spin–orbit splittings are measured in hundreds of cm^{-1}, while energy gaps between electronic terms are measured in thousands of cm^{-1}. Accordingly, each M^+ level can be labeled by an *electron configuration* and the cation quantum numbers $|L^+S^+J^+\rangle$, where L^+ (total orbital angular momentum) and S^+ (total electron spin) define *an electronic term*, and $J^+ = |L^+ + S^+|$ labels individual *spin–orbit levels* by their total angular momentum. We sometimes use "+" superscripts to denote cation quantum numbers when they improve clarity. *In principle, each M^+ level is potentially a different chemical reactant!* Since all of the low-energy states have even parity, radiative relaxation of excited states is very slow, probably measured in seconds. We have the opportunity to explore how M^+ electron configuration, electron

TABLE I
Energy Levels of Ti^+, V^+, and Fe^+ from Ref. 6

Atom	Electronic Term	Range of J^+	Energy (cm^{-1})
Ti^+	$3d^24s(^4F)$	3/2–9/2	0–393
	$3d^3(^4F)$	3/2–9/2	908–1216
	$3d^24s(^2F)$	5/2–7/2	4628–4898
V^+	$3d^4(^5D)$	0–4	0–339
	$3d^34s(^5F)$	1–5	2605–3163
	$3d^34s(^3F)$	2–4	8640–8841
	$3d^4(^3P)$	0–2	11,296–11,908
Fe^+	$3d^64s(^6D)$	9/2–1/2	0–977
	$3d^7(^4F)$	9/2–3/2	1873–3117
	$3d^64s(^4D)$	7/2–1/2	7955–8847
	$3d^7(^4P)$	5/2–1/2	13,474–13,905

spin, and spin–orbit level affect reactivity. We will see that the overall *pattern* of low-lying states affects the reactivity of each individual state, since potential energy surface intersections are commonplace and very important in M^+ collisions.

In the $3d$-series transition-metal cations (Sc^+ through Zn^+), the similarity of $3d$ and $4s$ orbital energies accounts for the electronic complexity. Figure 3 shows the low-lying term energies for the entire $3d$ series. The ground-state configuration varies between $3d^n$ and $3d^{n-1}4s$ across the series owing to the interplay of electron exchange effects and orbital energy changes. As atomic number increases, $3d$ is stabilized relative to $4s$; concomitantly, the spatial extent of both $3d$ and $4s$ decreases, but $3d$ shrinks more rapidly than $4s$.[7] For a given M^+, all of the electronic states below 3 eV have either $3d^n$ or $3d^{n-1}4s$ configuration. Both high-spin and low-spin electronic terms appear, with high-spin terms more stable than low-spin terms of the same configuration owing to electron exchange stabilization.

Armentrout, Squires, Ridge, Russell, Jarrold, Buckner and Freiser, MacMillan and Gross, Beauchamp, and Allison have recently written excellent reviews[8] of the gas-phase chemistry of transition-metal species, including atomic M^+, small metal clusters M_x^+, and ligated metal $M(AB)^+$ and anions $M(AB)^-$. We will not attempt even to outline the scope of current exploration of gas-phase transition-metal chemistry.

The groups of Armentrout and of Ridge have made important contributions to our understanding of electronic state-specific M^+ chemistry. Armentrout and co-workers[9–12] have used a hot-filament, surface ionization source to vary the gas-phase M^+ electronic-state distribution by changing the surface temperature of the filament. In an "ion-beam-plus-gas-cell"

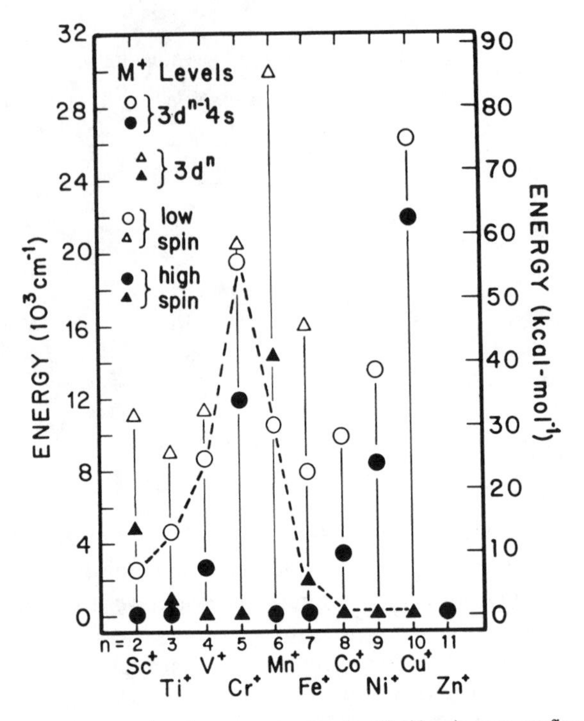

Figure 3. Transition-metal-cation term energies classified by electron configuration type and electron spin. Dashed line connects the lowest-energy term in each M^+ that can insert into a C–H bond in spin-allowed fashion.

configuration similar to that described by Anderson in this volume, they measure total reaction cross section versus kinetic energy for different filament temperatures. The ability to measure endothermic reaction cross sections by increasing kinetic energy and extract ligand bond energies to M^+ is an important strength of the technique. The assumption that the M^+ cations are emitted in a Boltzmann distribution of electronic states at the measured filament temperature permits extraction of excited-state cross sections from the change in effective cross section with filament temperature. Reactions for which cross sections are sensitive to filament temperature include many $M^+ + H_2$ examples;[9] $V^+ + CH_4$ and C_2H_6;[10] $Fe^+ + CH_4$, C_2H_6, and C_3H_8;[11] and many others.[12] The $M^+ + H_2$ studies are very important in showing that simple molecular orbital arguments can explain the pattern of M^+ reactivity across the 3*d*-series.

Ridge and co-workers[13] have used electron impact on Cr $(CO)_6$ and $Mn_2(CO)_{10}$ to produce metastable excited states of Cr^+ and Mn^+ trapped in an ion-cyclotron-resonance (ICR) mass spectrometer. They subsequently study the electronic quenching and chemical reactivity of the metastables. The availability of long trapping times, a unique advantage of the ICR experiment, permits measurement of a lower bound of 6 s on the radiative lifetime of the $3d^5\ 4s(^5S)$ state of Mn^+, which lies 1.2 eV above the $3d^54s(^7S)$ ground state. Radiative decay of ^5S by electric dipole transitions is both parity and spin forbidden.

We have developed resonant two-photon ionization (R2PI) in conjunction with time-of-flight photoelectron spectroscopy (TOF–PES) as a means of creating low-kinetic-energy beams of M^+ in well-characterized electronic-state distributions.[14] Figure 4 provides an overview of the experiment. Laser vaporization[15] of a metal target rod entrains metal vapor in a pulsed Ar expansion; subsequent collisions in the expanding gas produce a M-atom beam with low velocity dispersion. By tuning ω_1 to excite different M^* intermediate states, we can influence the M^+ distributions, as described in Section II. In favourable cases, R2PI produces highly non-Boltzmann state distributions in which a single-electron configuration, a particular spin multiplicity, and sometimes even a single spin–orbit level dominates. R2PI–PES determines the M^+ state distribution produced by a given ionization wavelength.

In a subsequent experiment (Section III), [16–18] we cross the state-specified M^+ beam with a neutral reactant beam of hydrocarbon gas in the field-free extraction region of a time-of-flight mass spectrometer (TOF–MS). After a reaction time delay of several μs, we apply pulsed extraction fields to the TOF–MS and collect a mass spectrum of product cations and unreacted M^+. The ratio S_p/S_r of product to reactant integrated peak intensities can be converted to a total reaction cross section averaged over the known M^+ electronic-state distribution created at a particular ionizing wavelength. By measuring cross sections averaged over widely varying reactant-state distributions, we can extract electronic state-specific cross sections.

Direct *measurement* of the reactant M^+ state distributions for each ionizing wavelength is a key advantage of the R2PI–PES technique. In addition, certain R2PI schemes produce highly non-Boltzmann, narrow distributions of M^+ excited states, often permitting good sensitivity in measurements of cross sections of highly excited states. Finally, the center-of-mass collision energy, currently determined by the fixed M^+ and hydrocarbon beam velocities, is well defined, typically 0.20 ± 0.04 eV. Creation of the ions in the reaction zone avoids many problems often involved in transporting low-energy ions without broadening their kinetic energy distribution. Coupling an R2PI source to a variable kinetic energy apparatus (as described in Anderson's chapter of this volume) equipped with a cold collision cell would provide independent control of reactant kinetic energy and electronic state.

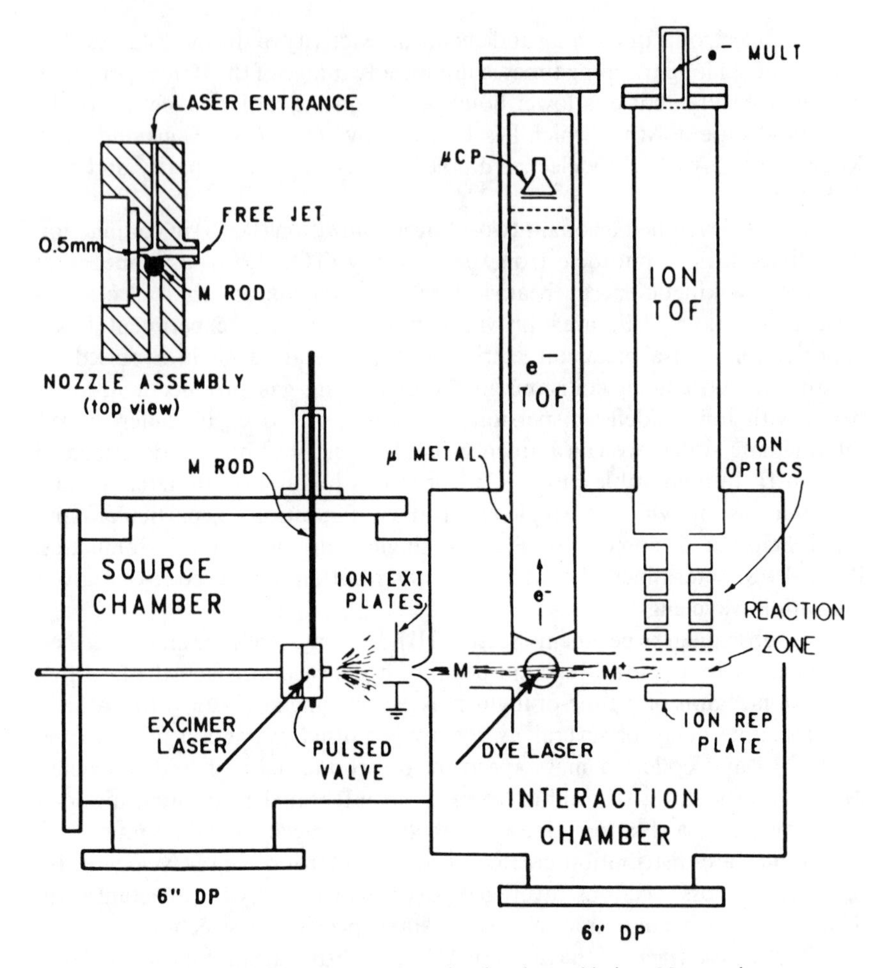

Figure 4. Overview of beam apparatus showing laser-ablation M-atom beam source, time-of-flight photoelectron spectrometer, and time-of-flight mass spectrometer. See Fig. 8 for details of the reaction zone.

III. PREPARATION OF ELECTRONIC STATE-SPECIFIC M^+ BEAMS BY RESONANT TWO-PHOTON IONIZATION

A. Photoionization Physics

The concept of M^+ state selection by resonant two-photon ionization is closely related to the use of molecular Rydberg states to create vibrationally selected molecular cations, as described elsewhere in this volume. Each M^+

electronic state characterized by an electron configuration and the quantum numbers $|J^+L^+S^+\rangle$ gives rise to multiple Rydberg series of the neutral atom M. To selectively create a particular cation term $|L^+S^+\rangle$, we tune the dye laser to a convenient member of a Rydberg series (M^*) whose cation core matches the desired M^+ term. We view the R2PI process (Fig. 2) as the sequential absorption of two photons (ω_1 and ω_2): $M \rightarrow M^* \rightarrow M^+ + e^-$. To the extent that the cation core quantum numbers are well defined in the Rydberg state and direct, one-electron transitions dominate the photoionization cross section, absorption of ω_2 may gently remove the Rydberg electron and selectively create the desired M^+ term. Viewed the other way around, measurement of M^+ electronic-state branching fractions by photoelectron spectroscopy sheds light on the electronic character of the intermediate state in an R2PI scheme. This simple picture ignores possible complications from configuration interaction in the intermediate state and from excitation of autoionizing states.[19]

The lowest three terms of M^+ are typically high-spin $3d^{n-1}4s$; high-spin $3d^n$; and low-spin $3d^{n-1}4s$ (Fig. 3). The energy ordering of $3d^{n-1}4s$ and $3d^n$ high-spin terms varies across the $3d$ series. Since $3d$ orbitals are much smaller than $4s$ and $4p$ orbitals, which are of comparable size, the $4s$–$4p$ electrostatic coupling is much stronger than either the $3d$–$4s$ or $3d$–$4p$ coupling. Thus, the $3d^7$, ^{4}F term of Fe^+ gives rise to neutral Rydberg states labeled $3d^7(^4F)4p(^2P)$; $^3G^o$ and $3d^7$ (^4F) $4p$ (^2P); $^5G^o$ (among many others) by Corliss and Sugar.[6] This notation implies a zeroth-order model with a $3d^7$ Fe^+ cation core with the electrons strongly coupled as ^{4}F due to $3d$–$3d$ exchange interactions. The remaining $4p$ Rydberg electron is more weakly coupled to the core to produce the electronic terms $^3G^o$ and $^5G^o$, with the superscript "o" designating odd parity. A shorthand notation would be $3d^7$ $(S_cL_c)4p; SL$, where S_c and L_c are the nominal Fe^+ core quantum numbers and $SL$are the overall term quantum numbers. The same zeroth-order core and $4p$ Rydberg electron also produce $^{3,5}F^o$ and $^{3,5}D^o$ terms. Another important kind of Fe excited state is $3d^6(^5D)$ $4s4p(^{1,3}P)$, which produces the terms $^7D^o$, $^7F^o$, $^5D^o$, $^5F^o$, and so on. Here we first couple the *dipositive* Fe^{2+} core electrons to form L_c and S_c; we then couple the 'outer shell' $4s$ and $4p$ electrons to form $L_o = 0$ and $S_o = 0$ or 1; finally, we couple L_c and L_o to form L and S_c and S_o to form S. Atomic physicists have labeled most of the observed M and M^+ states in this fashion by fitting observed energy levels to model Hamiltonians. When seeking to produce a particular M^+ term, we use these nominal core and Rydberg designations to guide our choice of M^* intermediate states in the R2PI scheme.

We use photoelectron spectroscopy to measure the actual outcome of a particular R2PI scheme. For two-photon ionization with θ the angle between laser linear polarization axis and the electron detection axis, the most general photoelectron angular distribution is given by[19]

$$ds_i/d\theta = (s_i/4\pi)[1 + \beta_{2,i}P_2(\cos\theta) + \beta_{4,i}P_4(\cos\theta)]. \quad (1)$$

Here s_i is the total (angle-integrated) photoionization cross section into resolved electronic feature i of a photoelectron spectrum, either a M^+ term or a particular spin–orbit level. P_2 and P_4 are the second and fourth Legendre polynominals, and $\beta_{2,i}$ and $\beta_{4,i}$ are electronic feature-specific anisotropy coefficients to be determined by experiment. We use s_i for photoelectron cross section to distinguish from $\sigma(i)$, which will denote the state-specific chemical reaction cross sections below.

Simply stated, each photon absorbed by an electric dipole transition adds additional anisotropy to the cation-plus-electron system. Neglecting nuclear hyperfine interactions, the anisotropy must be "distributed" between the electron and cation, that is, in the photoelectron angular distribution or in alignment of the total angular momentum vector J^+ of the cation. Since there is no angle for which $P_2(\cos\theta)$ and $P_4(\cos\theta)$ are simultaneously zero, there is no magic detection angle that provides a single photoelectron spectrum whose intensities are proportional to the proper angle-integrated cross sections s_i. We must therefore measure complete angular distributions at each ionization wavelength, fit them to Eq. (1), and extract electronic branching fractions from the fitted parameters.

B. Experimental Technique: Atomic Beam Source and Time-of-Flight Photoelectron Spectroscopy

Other groups [20,25] have studied resonant multiphoton ionization spectra of transition-metal atoms generated by focusing tunable dye lasers into bulb samples of stable transition-metal carbonyls such as $Fe(CO)_5$. Kimura has reviewed the photoelectron spectroscopy of atoms.[21] Berry and co-workers measured the first angular distributions for photoelectron spectra of atoms ionized by R2PI and made important contributions to the theory.[22] In our experiments, a beam of metal atoms with narrow velocity distribution and few excited states is formed by laser vaporization of a rotating metal rod in the throat of a pulsed nozzle expansion of Ar. Clustering of metal atoms is efficient in He expansions, but we observe almost exclusively atomic metal species in Ar expansions. Typical conditions are 1.1 atm of Ar behind a 0.5 mm orifice in a pulsed nozzle that opens for 150 μs. The vaporization laser is an excimer laser (10–15 mJ/pulse on the target, 308 or 248 nm, 10 Hz). A pair of charged plates upstream of the skimmer extracts any ions formed in the source. A 2-mm-diam conical skimmer separates the source chamber from the interaction chamber (Fig. 4) and forms a well-collimated, neutral atomic beam.

The interaction chamber includes a time-of-flight mass spectrometer (TOF–MS) with a two-field, Wiley–McLarin[23] extraction region and a mu-metal-shielded time-of-flight photoelectron spectrometer (TOF–PES, 52 cm long). The TOF–PES is modeled after the early designs of Reilly and co-workers[24] and of Zare, Anderson, and co-workers.[25]

We first measure resonant ionization spectra by intersecting the M beam with a gently focused dye laser in the TOF–MS extraction region with DC extraction fields applied. Typical dye laser conditions are 0.1 mJ/pulse of frequency doubled, near-UV light focused by a 25 cm cylindrical lens. The intersection of laser and atomic beams is a rectangular prism of dimensions 5 mm × 3 mm × 0.25 mm, with the long side parallel to the M-atom beam velocity and the short side parallel to the electron detection axis. Mass-selected M^+ current versus laser wavelength yields sharp spectra that allow location and assignment of the well-known neutral M resonances. Figure 5 shows an example from one-color, two-photon ionization of V.[14] The line intensities reveal that the V atoms are internally quite cold, with the state distribution dominated by the $^4F_{3/2}$ ground level.

Time-of-flight photoelectron spectra are recorded by positioning the dye laser in the electron flight tube and tuning to a particular atomic transition

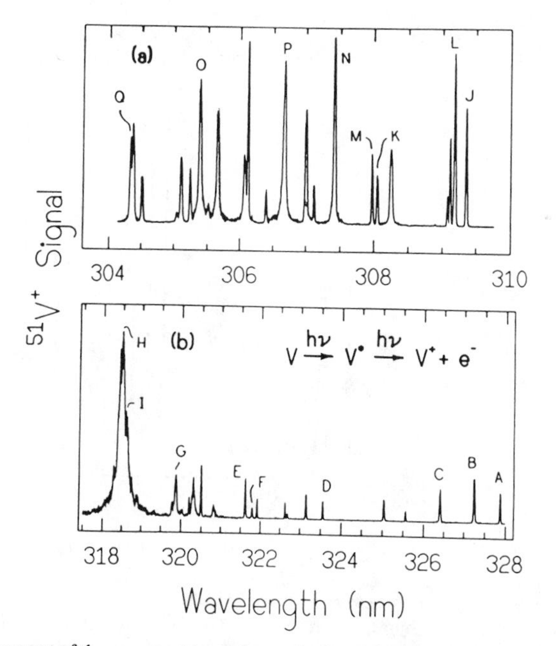

Figure 5. Segment of the resonant two-photon ionization spectrum of V atom. See Ref. 14 for assignments of transitions.

in the resonant ionization spectrum. We use start–stop, time-to-amplitude conversion analog electronics and a multichannel analyzer in pulse height mode to collect a histogram of electron arrival times at a microchannel plate detector 52 cm from the interaction region. Detection efficiency is $\sim 10^{-4}$ owing to the small detection solid angle. Adjustment of dye laser intensity limits the count rate to less than 0.3 electron per laser shot. We correct intensities for paralysis of the detection system, a minor effect.

An example of a TOF photoelectron spectrum for V is shown in Fig. 6. Owing to the time-of-flight detection, the electron energy resolution varies roughly as $\Delta E/E = 2\Delta t/t$, where E is electron kinetic energy and t is arrival time. The most useful energy range of the spectrometer is 0.2–2.0 eV. The FWHM energy resolution can be 30 meV = 240 cm^{-1} near 2 eV (where peak widths in time are often as narrow as 5 ns FWHM, the laser pulse duration) and 1–2 meV = 8–16 cm^{-1} near 200 meV (where residual fields in the apparatus often broaden peaks beyond the laser width).

Since the energy resolution is best for slow electrons, we often resolve individual spin–orbit levels in the highest M^+ term energetically accessible, but only individual terms at lower M^+ energy. In the example of Fig. 6, we easily resolve individual ^3F_J levels, partially resolve ^5F_J levels, and poorly resolve ^5D_J levels. In this case, we can measure J-resolved angular

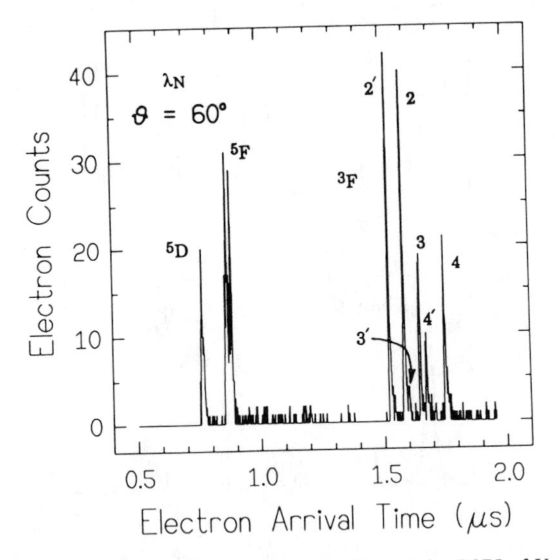

Figure 6. Time-of-flight photoelectron spectrum obtained by R2PI of V at 307.5 nm. V^+ term and spin–orbit level assignments as shown. The laser excites two different transitions originating from $J = 7/2$ and $J = 5/2$ of the V ground state; corresponding contributions are resolved (primed and unprimed features) within $3d^4 4s(^3F)$ of V^+.

distributions for ^3F_J levels and term-resolved angular distributions (averaged over undetermined J distributions) for the two quintets. The PES of Fig. 6 also reveals that the laser excites two different overlapped transitions originating from two different levels of the $3d^34s^2$, ^4F term.

The transmission of the spectrometer tube may vary with electron kinetic energy, since stray electric and magnetic fields may perturb the flight of the slowest electrons. When the spectrometer is performing well, slow electron peaks ($t \sim 1.5\ \mu s$) are 5–10 ns wide and arrive at times predicted by the simple TOF equation $E=(1/2m(L/t)^2$ to within about 5%. We estimate from classical trajectory modeling and experimental work using extraction fields that systematic errors in branching ratio due to the electron transmission function are $\leqslant 10\%$ over the useful energy range of 0.2–2.0 eV.[14]

The linearly polarized, frequency-doubled dye laser traverses a double Fresnel rhomb before entering the interaction chamber. We measure photoelectron angular distributions by manually rotating the rhomb and measuring a complete spectrum at each of 12 angles equally spaced in the range $\Theta = 0$–$180°$. The anisotropy coefficients β_{2i} and β_{4i} and a relativeintensity parameter $A_i = As_i$ (with A an unknown proportionality constant) are obtained from each angular distribution by linear least squares fitting to Eq. (1.1). Integration of Eq. (1.1) over the entire sphere yields the branching fractions for production of particular electronic feature i as $f_i = A_i/(\Sigma_i A_i)$.

C. Results: M^+ State Distributions

The one-color R2PI spectra (see V example in Fig. 5) generally exhibit sharp, readily assigned atomic lines. We create $\sim 10^4\ V^+$ per laser shot on a peak. Essentially all features can be assigned to atomic transitions from the V ground term, $3d^34s^2(^4F)$. There is no evidence in the mass spectrum or in the M^+ R2PI spectrum of van der Waals molecules such as VAr or of metal dimers or clusters V_x. It is quite feasible to ionize via spin-changing transitions and orbitally forbidden, two-electron transitions such as $3d^{n-1}4s^2 \rightarrow 3d^n4p$ with only several tenths of a mJ/pulse.

We have used TOF–PES to study perhaps 50 different one-color R2PI transitions through $3d^{n-1}4s4p$ and $3d^n4p$ intermediate states in Ti, V, and Fe at varying levels of detail. Since we were limited to one-color ionization and our goal was selectively to create low-lying terms of M^+, we chose intermediate states located slightly more than halfway to the ionization continuum. More limited experience with one-color, *three*-photon ionization schemes through $5s$ Rydberg states of Fe comes from the work of Kimura and co-workers[21] and Anderson, Zare, and co-workers.[25]

The substantial body of electronic branching fraction data can be summarized as follows. For ionization through $4p$ Rydberg states, when core-conserving, one-electron ionization steps are energetically allowed the *electron-configuration selectivity* is generally good, about 80–90%. That is, ionization of nominal $3d^n4p$ Rydberg states produces primarily $3d^n$ cations, while ionization of $3d^{n-1}4s4p$ states produces primarily $3d^{n-1}4s$ cations. This indicates that the configuration labels of the intermediate states are typically correct and that one-electron transitions dominate when available.

When both low- and high-spin $3d^{n-1}4s$ terms are energetically accessible by ionization through an intermediate of designation $3d^{n-1}(S_cL_c)4s4p(^1P)$; (SL), selection of $3d^{n-1}4s$ cations is good but the *electron-spin selectivity* is poor. This does not violate one-electron selection rules; since 4*s* and 4*p* electron spins are more strongly coupled to each other than to S_c, the spin of 4*s* relative to S_c in a singlet-coupled 4*s*4*p* outer shell is ill-defined. Intriguingly, we almost always observe more low-spin $3d^{n-1}4s$ cations than high-spin $3d^{n-1}4s$ cations. This is a nonstatistical result, since the degeneracy factor $(2S^+ + 1)$ favors high spin. Ionization through terms with triplet-coupled 4*s*4*p* outer shells *and* with overall high-spin configurations [e.g., $3d^6(^5D)4s4p(^3P)$; $^7D°$] should select high-spin $3d^{n-1}4s$ cations, since in these states the 4*s* spin must be parallel to S_c. This idea has not been tested experimentally as yet.

The limited information available for three-photon ionization via two-photon resonances[20,21] (mostly 5*s* Rydbergs) suggests that these schemes are less configuration selective than the 4*p* schemes, perhaps because the 5*s* orbital penetrates to the nucleus and mixes core configurations more extensively than 4*p*. Again the spin of $3d^{n-1}4s$ cation terms is poorly selected in the few examples studied.

Complete tables of branching fractions and anisotropy coefficients are presented in the literature.[14,16] These fractions and their uncertainties serve as input data for extraction of state-specific chemical reaction cross sections. Table II contains selected examples of electronic branching fractions for Ti^+, V^+, and Fe^+. Nominally core-conserving ionization steps in Ti and Fe behave similarly. We can create high-spin $3d^n$ beams with selectivity on the order of 90%. We can also create ~90% pure $3d^{n-1}4s$ beams, but the electron-spin selectivity is poor when both high- and low-spin terms are accessible. There is little J selectivity within $Ti^+(3d^24s, ^4F)$. In Fe^+, we have found several examples of remarkable state selectivity in which an R2PI transition produces >90% of a particular spin-orbit level J in the $3d^7, ^4F$ term. This suggests that even J_c is quite well defined in certain Fe* Rydberg states. Spin–orbit level splittings in Fe are typically about twice as large as those in Ti, which may explain why a 4*p* electron acting as a perturbation of the cation core mixes different J_c less effectively in Fe.

In V, there are relatively few intermediate states with energy slightly higher than half the ionization potential that also conserve nominal cation core on

TABLE II
Selected Electronic Branching Percentages from Angle-Integrated Photoelectron Specta[a]

Ti^+	λ(nm)	$3d^24s(^4F)$	$3d^3(^4F)$
	353.2	9	91
	351.3	15	85
	349.7	17	83

V^+	λ(nm)	$3d^4(^5D)$	$3d^34s(^5F)$	$3d^34s(^3F)$		
				$J=2$	$J=3$	$J=4$
	321.6	91	9	X	X	X
	321.8	20	80	X	X	X
	309.4	14	23	22	23	18
	308.0	9	6	14	20	51

Fe^+	λ(nm)	$3d^64s(^6D)$	$3d^7(^4F)$			
			$J=9/2$	$J=7/2$	$J=5/2$	$J=3/2$
	287.5	10	89	1	0	0
	282.7	3	6	91	0	0
	284.1	26	0	7	64	3
	282.9	10	2	9	77	2

[a]See Ref. 14 and 18 for detailed descriptions of R2PI schemes and error limits. The entry X indicates that the energy level was not energetically accessible. The Ti^+ percentages are only illustrative estimates based on photoelectron specta taken at three angles.

ionization. All of the intermediates investigated have nominal $3d^34s4p$ configurations, but many of these have the wrong coupling within the $3d^3$ dipositive core (e.g., ^2G, ^2P, or ^2H) to match the low-lying $3d^3(^4F)4s$; $^{3,5}F$ terms of V^+. When core-conserving, one-electron ionization steps are available, V behaves much like Ti and Fe. The $3d^34s(^5F)$ and $3d^34s(^3F)$ terms of V^+ dominate the electronic branching but spin selectivity is poor. Perhaps not surprisingly, transitions through intermediates whose $3d^3$ dipositive cores do not match the accessible V^+ states lead to unpredictable cation branching. In one example, a nominal $3d^3(^2G)4s4p(^3P)$; $^4F^o$ intermediate produces 91% $3d^4(^5D)\,V^+$!

Thus far we have assumed that only the character of the intermediate level determines the cation state branching when one-electron transitions to the continuum are available (direct ionization). This is certainly a useful first approximation, but of course autoionizing structure in the electron continuum can affect the cation branching as well.[19] In a one-color R2PI experiment, we are usually oblivious to autoionizing effects, since both the intermediate state and the total energy vary as the dye laser scans.

However, we sometimes excite the same intermediate level J' via two different $J' \leftarrow J''$ transitions. In these cases, we ionize through the same intermediate with different total energy. Differences in cation branching or in angular distributions then very likely indicate the influence of autoionizing states on the electron-ejection dynamics. In four such comparisons for V and two such comparisons for Fe, we find only one example of dramatically different electronic *term* branching. In general, small but statistically significant differences in term branching are observed. Larger effects on spin–orbit level branching within a cation term are common, however. The angular distributions also differ significantly from one $J' \leftarrow J''$ excitation scheme to the other. This suggests that we often excite autoionizing states, but the effects on the electronic branching are subtle compared with the influence of the intermediate-state character. Of course, two-color R2PI experiments with ω_1 fixed and ω_2 varied would provide a much more definitive means of studying autoionizing effects.

We have measured many term-resolved and J-resolved photoelectron angular distributions from R2PI of V and Fe[14]. Figure 7 illustrates the data. To our chagrin, the V data are equally well fit using both $\beta_{2,i}$ and $\beta_{4,i}$ as parameters or using only $\beta_{2,i}$ as a parameter with $\beta_{4,i}$ fixed at zero. None of the fitted $\beta_{4,i}$ is significantly different from zero. This means that accurate V^+ electronic branching fractions could have been obtained from single spectra at the magic angle for which $P_2(\cos\theta) = 0$. Either V^+ carries away substantial anisotropy as alignment of J^+ or the anisotropy of the intermediate-state J' is effectively washed out by hyperfine coupling[26] to the nuclear spin, $I(^{51}V) = 7/2$. In the angular distributions from Fe, our diligence is rewarded by several clear examples of nonzero β_4. Since ^{56}Fe has zero nuclear spin, it is tempting to infer that hyperfine coupling is indeed responsible for the negligible values of β_4 in V. Since this review focuses primarily on M^+ chemical reactivity, we refer the reader to the primary literature for detailed discussions of the angular distributions themselves.

IV. DETERMINATION OF STATE-SPECIFIC M^+ REACTION CROSS SECTIONS

A. Experimental Technique: Measurement of State-Averaged Reaction Cross Sections

1. *Overview*

Angle-integrated photoelectron spectroscopy (Section III) yields a set of electronic branching fractions $f_{x,i}$ for each resolved M^+ electronic feature

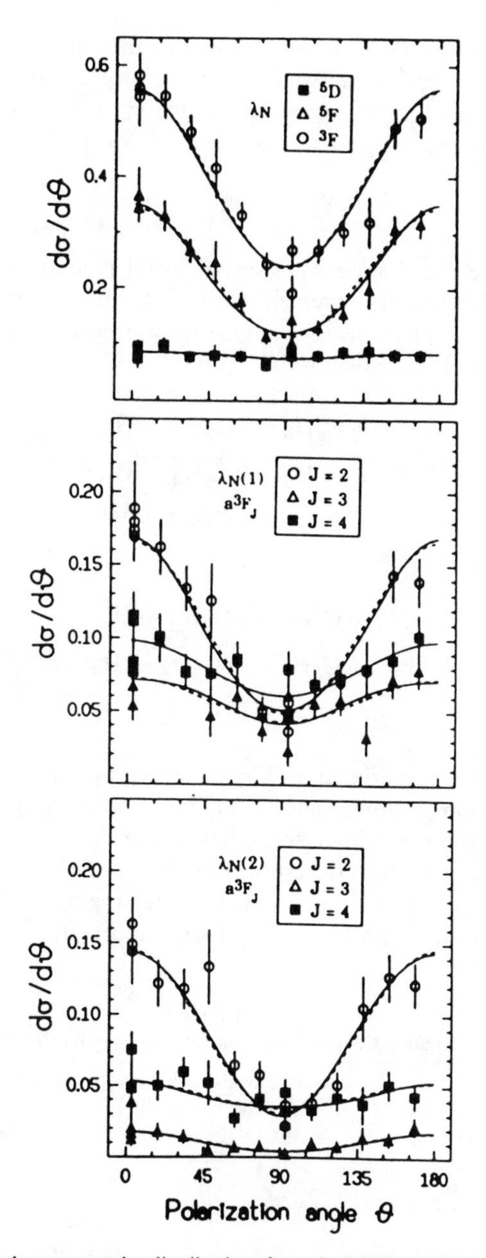

Figure 7. Photoelectron angular distributions from the R2PI transition of Fig. 6. Top panel shows term-resolved distributions summed over J^+; lower two panels show J^+-resolved distributions for the two transitions excited by the laser.

(either term or individual level) at each R2PI ionizing wavelength λ_x. In the second part of the experiment, we use a crossed ion-neutral beam technique with pulsed time-of-flight mass spectrometry (TOF–MS) to measure electronic *state-averaged* total reaction cross sections:

$$\langle\sigma\rangle_x = \Sigma_i f_{x,i}\sigma(i). \tag{2}$$

Here the $\sigma(i)$ are the desired *state-specific* total reaction cross sections. In this section we describe measurement of the $\langle\sigma\rangle_x$. In Section IV B we describe extraction of the $\sigma(i)$ by a least-squares-fitting procedure and the results.

We have studied these reactions:

$$V^+ + C_2H_6 \rightarrow VC_2H_4^+ + H_2; \tag{3}$$

$$V^+ + C_3H_8 \rightarrow VC_3H_6^+ + H_2; \tag{4a}$$

$$\rightarrow VC_3H_4^+ + 2H_2; \tag{4b}$$

$$\rightarrow VC_2H_4^+ + CH_4; \tag{4c}$$

$$V^+ + C_2H_4 \rightarrow VC_2H_2^+ + H_2; \tag{5}$$

$$Fe^+ + C_3H_8 \rightarrow FeC_3H_6^+ + H_2; \tag{6a}$$

$$\rightarrow FeC_2H_4^+ + H_2. \tag{6b}$$

Table III collects experimental kinetic energies for each reaction and estimated reaction exothermicities for the dominant product channels. Figure 1 shows the low-lying electronic states of V^+ and Fe^+.

Figure 8 shows details of the crossed-beam experiment.[16–18] Three pulsed beams collide in space and time in the field-free region of the TOF–MS: the skimmed atomic M beam formed by laser vaporization in the throat of a

TABLE III
Enthalpy Changes[a] and Kinetic Energy of Reactions

	Hydrocarbon Reactants and Ionic Products					
	$V^+ + C_2H_6$	$V^+ + C_3H_8$		$V^+ + C_2H_4$	$Fe^+ + C_3H_8$	
Parameter	$VC_2H_4^+$	$VC_3H_6^+$	$VC_2H_4^+$	$VC_2H_2^+$	$FeC_3H_6^+$	$FeC_2H_4^+$
ΔH^0(eV)	−0.7	−0.9	−1.3	−0.4	−0.6	−0.9
Collision Enery	0.21 ± 0.04	0.21 ± 0.04		0.18 ± 0.04	0.22 ± 0.04	

[a] ΔH^0 is the standard enthalpy change at 298 K as in Refs. 11 and 17. The bond energies of M^+ to alkenes are not well known.

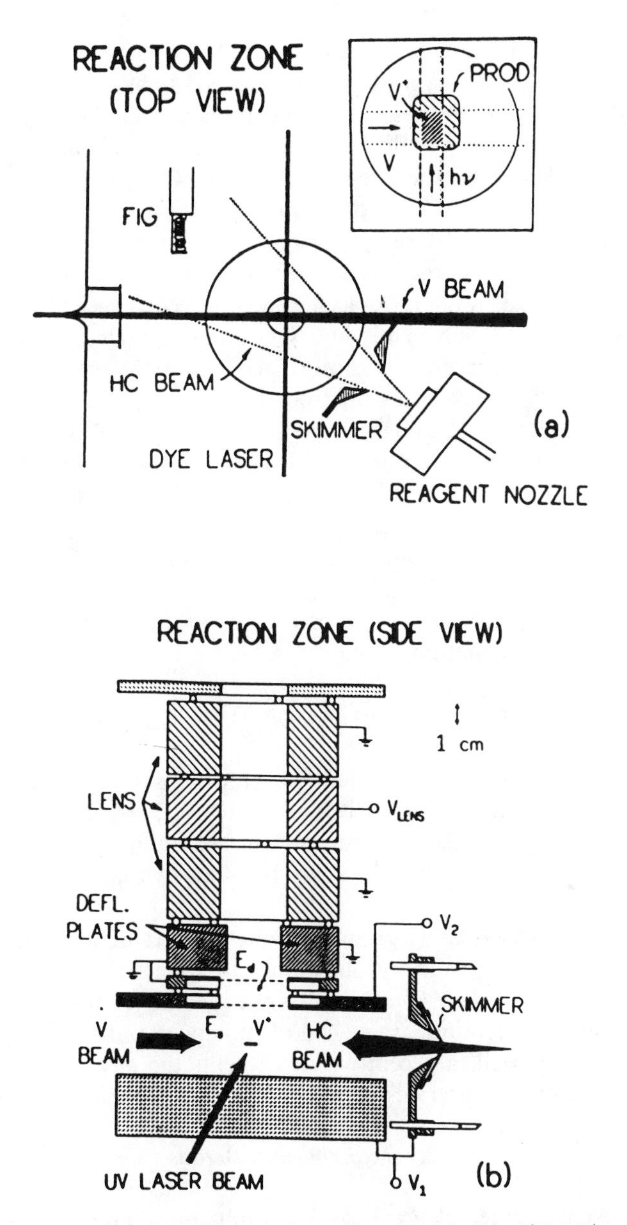

Figure 8. (*a*) Top view of reaction zone. Inset shows the size of the initial V^+ packet at $t = 0$ and of the VC_2H_4 products after $t_{rxn} = 6\,\mu s$ relative to the extraction plate aperture, assuming maximum possible kinetic energy release. (*b*) Side view showing pseudoskimmer, the three beams, and the extraction optics.

pulsed nozzle expansion of Ar; the tunable UV laser beam at λ_x, which creates a packet of M^+ cations at a well-defined time in a known electronic state distribution; and the neutral hydrocarbon reactant beam from a second pulsed nozzle, which is "pseudoskimmed" by a set of homebuilt knife edges.

The principle of the experiment is straightforward. At $t = 0$, the dye laser creates the M^+ packet. The initial velocity distribution of the M^+ beam is that of the neutral M beam, since the electron mass is so small and we carefully avoid space-charge effects. A small fraction of the M^+ cations collide with a neutral hydrocarbon molecule in field-free space; the hydrocarbon number density n_{hc} is sufficiently low to ensure the single-collision limit. The relative M^+ + hydrocarbon collision velocity v_{rel} in the center-of-mass frame is determined by the M^+ and hydrocarbon beam velocities and the 145° angle between them. During a reaction time t_{rxn} of 0–10 μs, M^+ + hydrocarbon collisions occur and may lead to product cations at a new mass. Fast rise-time, high-voltage pulses applied to the TOF–MS extraction region quench the reaction at time t_{rxn} and initiate collection of a mass spectrum comprising both the remaining reactant cations (intensity S_r) and all product cations (intensity S_p).

The state-averaged cross sections $\langle\sigma\rangle_x$ are related to experimental parameters by the equation

$$\langle\sigma\rangle_x = [-\ln(1 - S_p/S_r)]/(n_{hc} v_{rel} t_{rxn}). \tag{7}$$

We measure S_p and S_r as integrated intensities of well-resolved peaks in the TOF–MS. A fast ion gauge (FIG) measures n_{hc}. We calculate v_{rel} from measurements of the reactant beam velocities v_{M^+} and v_{hc}. The reaction time is accurately determined by the delay between dye laser and the edge of the ion-extraction pulse.

Absolute cross sections $\langle\sigma\rangle_x$ are accurate to an estimated $\pm 55\%$, limited by the accuracy of n_{hc} and by our ability to collect reactant and product cations on equal footing. *Relative* cross sections $\langle\sigma\rangle_x/\langle\sigma\rangle_y$ are determined to $\pm 10\%$, since uncertainties in n_{hc} cancel and uncertainties in S_p/S_r probably tend to cancel as well. A recent article presents the measurements in great detail.[17] In the next section we describe some of the important considerations.

2. *Experimental Details*

a. Pulsed, Skimmed Metal-Atom Beam. Results for V and Fe beams using a 308-nm vaporization laser are similar. FIG measurements versus distance from the nozzle for the free expansion in an open chamber show Ar pulses of 190μs FWHM; the FIG response time is 0.5 μs. Measurements of peak Ar number density versus distance yield $v_{Ar} = (5.7 \pm 0.8) \times 10^4$ cm/s, in accord with the standard continuum model of a free jet expansion. With skimmer in place, we use R2PI and TOF–MS to measure the distribution of V atom arrival times from the source to the center of the reaction zone 27 cm downstream, and also to a point 44 cm downstream. The V atom packet is $47 \pm 10\,\mu s$ FWHM, much narrower than the Ar pulse, indicating efficient trapping and cooling of the metal. From the arrival times and the absence of spreading of the packet from 27 to 44 cm, we estimate the V atom velocity as $v_V = (6.3 \pm 0.5) \times 10^4$ cm/s, some 12% larger than v_{Ar}, presumably due to deposition of energy by the ablation laser.

More important is the velocity distribution of the V^+ reactants formed by R2PI. The V^+ and V velocities will be the same unless space charge distorts the cation velocities. The TOF–MS technique with DC or delayed extraction provides a sensitive estimate of the maximum V^+ velocity component perpendicular to the expansion axis from the observed width in time of the $^{51}V^+$ peak. Under typical experimental conditions ($\sim 10^4$ V^+ per laser shot), the observed 20–40 ns FWHM peaks at arrival time of $12\,\mu s$ imply a *maximum* perpendicular velocity component of 5.5×10^3 cm/s. Therefore we take $v_{V^+} = (6.3 \pm 0.7) \times 10^4$ cm/s in the laboratory frame.

b. Velocity and Number Density of Neutral Reactant Beam. A second pulsed nozzle (Lasertechnics LPV, $200\,\mu s$ FWHM) delivers the neutral reactant as a "pseudoskimmed" beam. A homebuilt, rectangular skimmer separates the hydrocarbon pulsed valve from the reaction zone (Fig. 8), deflecting off-axis flux toward the chamber walls until long after extraction of reactant and product cations is complete. There is no true differential pumping of the neutral source. The geometry is chosen so that all straight-line paths from the nozzle through the skimmer pass cleanly through the ion-extraction region, missing both extraction plates. The horizontal dimension allows the nozzle to bathe the entire reaction zone in an essentially uniform density of neutral reactant.

Typical hydrocarbon stagnation pressures are 20–200 torr, which creates peak number densities in the reaction zone in the range $n_{hc} = 3 \times 10^{11}$–$3 \times 10^{12}\,cm^{-3}$. We have used the FIG to measure hydrocarbon gas pulses with and without the skimmer and with and without the ion-extraction plates. The homebuilt skimmer attenuates the on-axis peak of a gas pulse by a factor that depends on the expansion gas (0.73 for C_2H_6, 0.73 for C_2H_4, 0.70 for C_3H_8, 0.86 for Ar). The *shape* of a hydrocarbon arrival time distribution does not vary from 11–21 cm downstream, indicating that the skimmer deflects part of the gas pulse off-axis but does not substantially alter the on-axis beam velocity distribution. From plots of peak arrival time versus distance and from the absence of arrival time spread with distance, we obtain the velocities $(8.0 \pm 1.0) \times 10^4$ cm/s for C_2H_4, $(9.0 \pm 1.3) \times 10^4$ cm/s for C_2H_6, and $(7.6 \pm 1.0) \times 10^4$ cm/s for C_3H_8. We estimate that at least 90% of the flux through the reaction volume is direct flux whose velocity is known; less than 10% of the flux is due to slower molecules scattered by the skimmer and extraction plates. The observed proportionality between product ion signal and hydrocarbon stagnation pressure shows that the scattered flux has negligible effect on the reaction cross sections.

The $\pm 30\%$ accuracy of the hydrocarbon number density measurement is an important limitation on the accuracy of absolute cross sections. Individual measurements of n_{hc} are precise to $\pm 5\%$, so the number density contributes only modestly to uncertainties in relative cross sections.

c. Kinematics and Calculation of Collision Energy. Since the V^+ and C_2H_6 velocity vectors are nearly antiparallel and the masses are similar, the velocity of the center of mass is small. For example, for $V^+ + C_2H_6$ we calculate $v_{cm} = (2.3 \pm 0.6) \times 10^4$ cm/s. The most probable relative collision velocity is $v_{rel} = (14.6 \pm 1.4) \times 10^4$ cm/s, corresponding to a center-of-mass frame collision energy of $E = 0.21 \pm 0.04$ eV (Table III). The uncertainty in E includes the uncertainty estimates in v_{hc} and v_{V^+}. If an exothermic reaction cross section scaled as $E^{-1/2}$, as the Langevin classical capture cross section does,[27] the $\pm 20\%$ uncertainty in E would correspond to a range of $\pm 11\%$ in cross section.

Given the exothermicity of each reaction, we can calculate the maximum laboratory-frame velocity of the product cation. For example, the kinematics and exothermicity restrict the velocity of $VC_2H_4^+$ products from $V^+ + C_2H_6$ to $\leqslant 6 \times 10^4$ cm/s. These kinds of limits serve as useful input to a numerical model[28] of the product-ion-extraction and -collection efficiency, but they almost surely overestimate actual product ion velocities owing to the large number of degrees of freedom available for deposition of energy.

d. Pulsed Time-of-Flight Mass Spectrometer and Ion-Detection. The double-field Wiley–McLarin[23] extraction region (Fig. 8) provides excellent space focusing at the detector of ions born over the 0.25 mm height of the ionization region. Typical mass resolution is $m/\Delta m = 300$. The 2.9 cm gap between the lower two extraction plates permits the direct hydrocarbon beam to pass through the interaction region without striking a surface, but also dictates a large diameter of the lower two plates (12 cm) to produce piecewise constant electric fields. The 3 cm aperture of the ion optics allows us to extract and image reactant and product cations from an extended reaction zone (roughly 4 mm deep $\times$ 5 mm along the molecular beam axis) onto the detector. The large reaction zone is in turn dictated by the need to avoid space-charge forces on reactant cations and to permit sufficient reaction time.[28]

A variable reaction time of $t_{rxn} = 0.1–10\ \mu s$ follows the ionizing laser pulse. At t_{rxn}, we apply fast, high-voltage steps to the lower two extraction plates, creating piecewise constant electric fields that accelerate product and reactant cations up the 1 m flight tube to the ion detector. A thyratron circuit designed in the UW Chemistry Electronics Shop produces a + 3 kV step that rises to full amplitude in 50 ns and remains constant to 10% during the extraction of ions.

The ion detector (EMI EMT-9643/2B, nominal gain 10^6) is a 17 dynode CuBe Venetian blind design with 2 cm active diameter. We accelerate cations to 6 kV kinetic energy at the first dynode. The literature[29] suggests that mass discrimination over the typical range 50–100 amu is minimal ($\leqslant 25\%$) under these conditions. In recent work, we gate the detector off during the large reactant cation signal to improve baseline recovery prior to the much smaller product signals. Detector output current drops directly over the 50 Ω input terminator of a 100 Ms/s signal averaging, digital oscilloscope. The oscilloscope records TOF mass spectra, typically averaging 500 laser shots, and performs background subtraction and peak intergration. Figure 9 shows examples from the $V^+ + C_3H_8$ and C_2H_4 reactions.

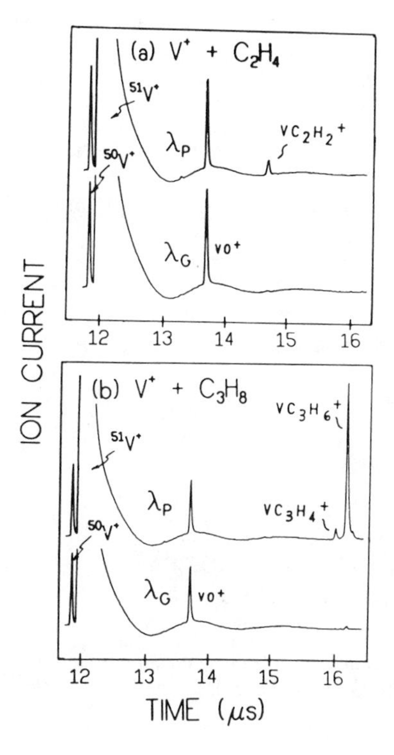

Figure 9. Raw time-of-flight mass spectra from the $V^+ + C_2H_4$ and C_3H_8 reactions. For each reaction mass spectra are shown for wo ionization wavelengths; λ_P creates 58% $3d^34s(^3F)$ reactants, while λ_G creates a mixture of $3d^4(^5D)$ and $3d^34s(^5F)$. The VO^+ peak is due to ionization of an impurity in the beam.

For a single $\langle\sigma\rangle_x$ measurement, we select the laser wavelength and set a hydrocarbon backing pressure and a reaction time. We then measure the hydrocarbon number density with the FIG; obtain a mass spectrum for M^+ reactant and all product channels by averaging 500 laser shots; collect a background spectrum with neutral reactant valve turned off; and cycle through these steps two or three times to check for stability. The oscilloscope subtracts background from signal and integrates the chosen peaks. We calculate $\langle\sigma\rangle_x$ from Eq. (7) by combining the measurements of S_p/S_r, n_{hc}, t_{rxn}, and v_{rel}. The reproducibility of $\langle\sigma\rangle_x$ within a run is typically $\pm 7\%$.

e. Performance Tests. We have carried out extensive tests of the apparatus to check the performance of the TOF–MS, the ion source, and the collection efficiency of reactant and product cations. A detailed description of these tests is given elsewhere.[17]

Perhaps most important is the behavior of S_p/S_r as a function of hydrocarbon reactant backing pressure, of the number of M^+ cations formed, and of the reaction time t_{rxn}. Examples of such test results are shown in Figs. 10 and 11. The ratio S_p/S_r is constant versus the number V^+ ions

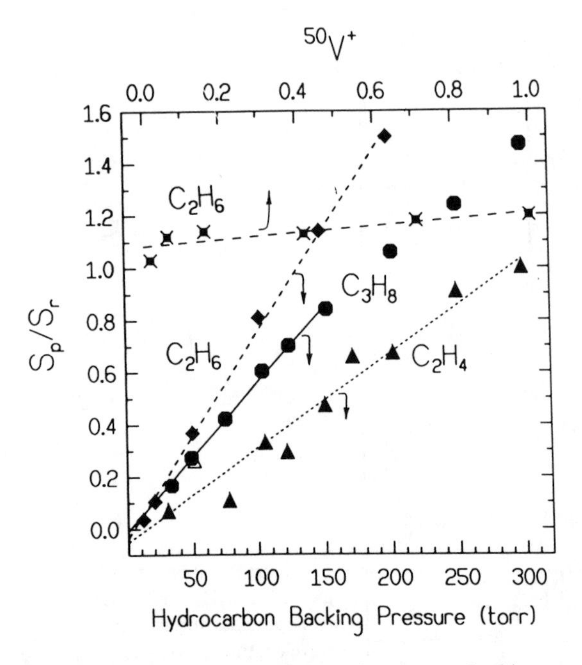

Figure 10. Plot of S_p/S_r [Eq. (7)] versus hydrocarbon reagent backing pressure behind the nozzle (lower scale) and versus the number of V^+ cations produced per laser shot (upper scale, arbitrary units).

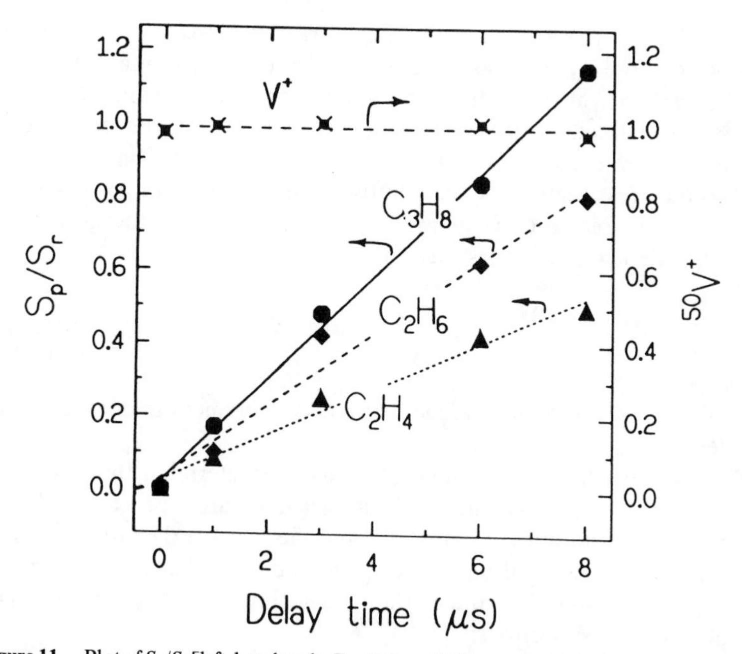

Figure 11. Plot of S_p/S_r [left-hand scale, Eq. (7)] and V^+ intensity (right-hand scale, arbitrary units) versus t_{rxn}, the delay time between ionizing laser and ion-extraction pulses.

produced over the range 350–40,000 ions/shots. This rules out space-charge effects on the kinetic energy distribution and thus the cross section, indicates linear detector response, and shows that the laser pulse energy does not significantly affect the ionization dynamics. The linearity of S_p/S_r vs n_{hc} confirms the single-collision limit and dismisses possible contributions from hydrocarbon dimers. The proper behavior of S_p/S_r vs Ar stagnation pressure and vs n_{hc} rules out collision-induced dissociation of product ions during extraction into the TOF drift tube. We estimate that less than 1% of the reactant or product ions suffer a high-energy collision during extraction and field-free flight. Constant S_r (V^+ signal) vs t_{rxn} from 0 to 8 μs, the time delay between laser and ion-extraction pulses, and linear S_p/S_r vs t_{rxn} over the same range provide strong evidence of good collection efficiency for both reactants and products.

B. Results: State-Specific Reaction Cross Sections

1. *Extraction of State-Specific Cross Sections from Data*

For each laser wavelength λ_x, Eq. (2) relates the measured electronic branching fractions $f_{x,i}$ and the measured *state-averaged* absolute cross sections $\langle\sigma\rangle_x$ to the desired *state-specific* cross sections $\sigma(i)$. In principle, we could extract the $\sigma(i)$ by measuring $\langle\sigma\rangle_x$ and a set of $f_{x,i}$ for many wavelengths and fitting the $\sigma(i)$ as overdetermined, adjustable parameters in a least-squares procedure. In practice,[17,18] we follow a more complicated procedure designed to obtain more accurate *relative*, state-specific cross sections $\sigma_{\text{rel}}(i)$ as well as absolute cross sections, since relative cross sections can be very useful in understanding reaction mechanisms. Consequently, we measure a set of *relative*, state-averaged cross sections:

$$\langle R\rangle_x \equiv \langle\sigma\rangle_x/\langle\sigma\rangle_{\text{ref}} = \sum_i f_{x,i}\sigma(i)/\langle\sigma\rangle_{\text{ref}} \equiv \sum_i f_{x,i}\sigma_{\text{rel}}(i). \tag{8}$$

Here $\langle\sigma\rangle_{\text{ref}}$ is the state-averaged cross section measured at a reference wavelength λ_{ref}.

We measure enough different $\langle R\rangle_x$ to overdetermine the $\sigma_{\text{rel}}(i)$, which are extracted by a weighted, linear least-squares procedure,[30] which propagates uncertainties in both $\langle R\rangle_x$ and the $f_{x,i}$. We then use the available absolute measurement of $\langle\sigma\rangle_x$, the $\sigma_{\text{rel}}(i)$, and the $f_{x,i}$ as inputs to a second least-squares fit that determines the best scaling factor C between relative and absolute cross sections:

$$\langle\sigma\rangle_x \equiv \sum_i f_{x,i}\sigma(i) = C\sum_i f_{x,i}\sigma_{\text{rel}}(i). \tag{9}$$

This procedure yields relative, state-specific cross sections $\sigma_{\text{rel}}(i)$ whose accuracy is limited by uncertainties in the $\langle R\rangle_x$ and the $f_{x,i}$, but not by uncertainties in the number density, mass discrimination of the detector, or deficiencies in ion collection (to the extent that these are similar at different wavelengths). The $\sigma_{\text{rel}}(i)$ are very useful tools for understanding electronic state effects in M^+ chemistry. The absolute cross sections $\sigma(i) = C\sigma_{\text{rel}}(i)$ are less accurate; the procedure funnels average uncertainties, which are typically $\pm 55\%$, in number density and collection efficiency into the error estimate for the scale factor C.

2. V^+ + *Hydrocarbon Results*

The resolved electronic features in R2PI photoelectron spectra of V are the $3d^4(^5D)$ ground term (J not resolved), the $3d^3 4s(^5F)$ first excited term (J not resolved), and the individual spin–orbit levels $J = 2, 3$, and 4 of the $3d^3 4s(^3F)$ second excited term. The index i in Eqs. (2), (8), and (9) then labels these five electronic features. Table IV displays state-specific cross sections for the $V^+ + C_2H_6, C_3H_8$, and C_2H_4 reactions at 0.2 eV kinetic energy. Both relative and absolute total reaction cross sections are reported. Uncertainty limits

TABLE IV

Absolute and Relative State-Specific Total Reaction Cross Sections[a] for V^+ + Hydrocarbons at 0.2 eV

	Hydrocarbon Reactant; Langevin Cross Section					
	C_2H_6; $\sigma_L = 77$ Å^2		C_3H_8; $\sigma_L = 99$ Å^2		C_2H_4; $\sigma_L = 82$ Å^2	
V^+ State	$\sigma(i)$ (Å^2)	$\sigma_{rel}(i)$	$\sigma(i)$ (Å^2)	$\sigma_{rel}(i)$	$\sigma(i)$ (Å^2)	$\sigma_{rel}(i)$
$3d^4(^5D)$						
$J = 0$–4	$\leqslant 0.11$	$\leqslant 0.0037$	0.44 ± 0.32	0.012 ± 0.006	$\leqslant 0.7$	$\leqslant 0.16$
$3d^34s(^5F)$						
$J = 1$–5	$\leqslant 0.11$	$\leqslant 0.0035$	0.18 ± 0.17	0.005 ± 0.004	$\leqslant 0.3$	$\leqslant 0.06$
$3d^34s(^3F)$						
$J = 2$	20 ± 12	1.02 ± 0.24	38 ± 25	1.04 ± 0.38	—	—
$J = 3$	22 ± 16	1.10 ± 0.51	43 ± 32	1.17 ± 0.54	—	—
$J = 4$	17 ± 10	0.87 ± 0.22	29 ± 21	0.79 ± 0.38	—	—
$\langle \sigma(^3F_J) \rangle$[b]	20 ± 11	$\equiv 1.00$	37 ± 19	$\equiv 1.00$	2.7 ± 1.6	$\equiv 1.00$

[a]State-specific absolute cross sections $\sigma(i)$ defined in Eq. (9); state-specific relative cross sections $\sigma_{rel}(i)$ defined in Eq. (8). Uncertainty limits are estimates of 95% confidence limits.
[b]Mean cross section of the three ^3F_J spin–orbit levels, which is defined as 1.00 for the relative cross section scale.

are nominal 95% confidence intervals estimated from the covariance matrix of the least-squares-fitting procedure.[17]

For all three hydrocarbons, the $3d^34s(^3F)$ triplet term of V^+ reacts much more efficiently than either of the lower-energy quintet terms $3d^4(^5D)$ or $3d^34s(^5F)$. The ratio of the term-specific triplet cross section to the larger of the quintet cross sections is $\geqslant 270$ for C_2H_6, 80 for C_3H_8, and $\geqslant 6$ for C_2H_4. The absolute cross sections of the V^+ $3d^34s(^3F)$ reactions with C_2H_6, C_3H_8, and C_2H_4 are 20 ± 11 Å^2, 37 ± 19 Å^2, and 2.7 ± 1.7 Å^2, respectively. For the lower-energy quintets, we can only set upper bounds on the cross sections with C_2H_6 and C_2H_4, while the C_3H_8 cross sections are small but discernible. A measure of the absolute efficiency of an ion–molecule reaction is the ratio σ/σ_L of the total reaction cross section to the Langevin cross section,[27] the classical mechanical cross section for ion-induced dipole "capture" collisions. The reaction efficiencies are 0.26, 0.41, and 0.03, for C_2H_6, C_3H_8, and C_2H_4, respectively. J-specific cross sections within $3d^34s(^3F)$ are equal within our uncertainty of $\pm 25\%$. The use of term-specific cross sections of $3d^4(^5D)$ and the $3d^34s(^5F)$, rather than individual J-specific cross sections, cannot affect the conclusions about $3d^34s(^3F)$ reactivity because the quintet reactivity is so small in comparison.

Elimination of a single H_2 molecule from the $[V(\text{hydrocarbon})]^+$ adduct is the dominant product channel for reaction of the $3d^34s(^3F)$ term with C_2H_6, C_2H_4, and C_3H_8. For the C_3H_8 reaction, we also observe about 6% elimination of $2H_2$ and about 1% elimination of CH_4.

3. $Fe^+ + C_3H_8$ Results

The Fe^+ level structure is quite different from V^+ (Fig. 1, Table I). The ground term is $3d^64s(^6D)$, the first excited term is $3d^7(^4F)$, and the second excited term is $3d^64s(^4D)$. We have found R2PI wavelengths that create 90–95% pure beams of individual spin–orbit levels $3d^7(^4F_J)$ (Table II), so it is possible to look closely for spin–orbit effects on total reaction cross sections. Other one-color R2PI schemes create mixtures of $3d^64s(^6D)$ and$3d^64s(^4D)$. The photoelectron spectra typically resolve spin–orbit levels within ^4D but not within ^6D. Unfortunately, we have not yet found R2PI schemes that create predominantly the ground term, so extraction of accurate $3d^64s(^6D)$ cross sections remains difficult.

The Fe^+ reactivity work is in progress as of this writing. We have obtained relative, state-specific cross sections $\sigma_{\text{rel}}(i)$ and product branching ratios for the $Fe^+ + C_3H_8$ reaction [Eqs. (6a) and (6b)] at 0.22 ± 0.04 eV.[18] Table V displays the relative cross sections extracted from a data set of five R2PI transitions that create only the lowest two Fe^+ terms, $3d^64s(^6D)$ and $3d^7(^4F)$. While we have not yet measured absolute reaction cross sections, the small product signals ($\leqslant 1$ ion per shot) indicate that the reaction efficiency is $\ll 1$ for all state distributions.

The most important result is a definite spin–orbit level dependence of the reaction cross section within $3d^7(^4F)$. We used two different models of Eq. (9) to fit the $\langle R \rangle_x$ and $f_{x,i}$ data. Model I includes four separate cross sections—a combined, term-specific $\sigma(3d^64s, {}^6D)$ and J-specific cross sections for $J = 9/2$, $7/2$, and $5/2$ within $3d^7(^4F)$, the only levels substantially populated. Model II includes only term-specific cross sections for $3d^64s(^6D)$ and $3d^7(^4F)$. The reduced chi-squared statistic $\chi_\nu^2 = \chi^2/\nu$ is 0.07 for the first model (indicating an excellent fit) and 6.4 for the second. Here ν is the number of degrees of freedom in the fit; $\nu = 1$ for Model I and 3 for Model II. If the errors in the input data are normally distributed, the probability is 0.78 that the first model is correct *and* that random errors in the data set will product $\chi_\nu^2 \geqslant 0.07$; the probability is $< 10^{-4}$ that the second model is correct *and* that $\chi_\nu^2 \geqslant 6.4$. Hence we reject Model II and conclude that Model I is sufficiently flexible to fit the data.

We can calculate the following ratios of J-specific cross sections: $\sigma(^4F_{7/2})/\sigma(^4F_{9/2}) = 1.12 \pm 0.06$ and $\sigma(^4F_{5/2})/\sigma(^4F_{9/2}) = 1.89 \pm 0.23$. The error limits are $\pm$ one standard deviation of the cross-section ratios with off-diagonal elements of the covariance matrix included in the error

TABLE V
State Specific Relative Reaction Cross Sections[a] $\sigma_{rel}(i)$ for $Fe^+ + C_3H_8 \rightarrow FeC_2H_4^+ + CH_4$ at 0.22 ± 0.04 eV

	Model I[b] (*J* Specific)			Model II[c] (Term Specific)	
	$3d^7$ ^4F_J				
$3d^64s$ ^6D	$J = 9/2$	$= 7/2$	$= 5/2$	$3d^64s$ ^6D	$3d^7$ ^4F
0.3 ± 0.7	0.92 $\pm .012$	1.03 ± 0.04	1.74 ± 0.22	2.48 ± 0.40	0.96 ± 0.04
	Model I: $\chi_v^2 = 0.07$[d]			Model II: $\chi_v^2 = 6.4$[d]	

[a]State-specific relative reaction cross sections from least-squares fit to Eq. (8). Error limits are plus or minus the square root of the digonal elements of the formal covariance matrix of the fits, analogous to plus or minus one standard deviation. (See Ref. 18 for the covariance matrix of the fit to Model I.)

[b]Model I uses four states in Eq. (8), the ground term ^6D and three spin–orbit levels in the first excited term: $^4F_{9/2}$, $^4F_{7/2}$, and $^4F_{5/2}$, The least-squares fit involves five equations and four parameters.

[c]Model II uses two states in Eq. (8), the electronic term ^6D and ^4F with no *J* specification. The least-squares fit involves five equations and two parameters.

[d]The chi-squared statistic χ_v^2 is χ^2 divided by v, the number of degrees of the freedom in the fit. Based on these values, we reject Model II as insufficiently flexible to fit the data.

propagation. Within the $3d^7(^4F)$ term, $J = 9/2$ and $7/2$ have equal cross sections, but $J = 5/2$ reacts almost twice as efficiently. We obtain the identical result if we freeze $\sigma(^6D)$ at any value between 0 and 1 and solve the resulting least-squares equations or if we delete any one of the five equations and solve directly for the four cross sections. The ability of three of the R2PI transitions to create remarkably pure beams of specific ^4F_J levels (89%, 91%, and 77% for $J = 9/2$, $7/2$, and $5/2$) is the key feature that enables us to cleanly extract *J*-specific cross sections.

Because no R2PI transition creates predominantly the ground term, the data provide only a rough estimate of the ratio of $3d^64s(^6D)$ to $3d^7(^4F_{9/2})$ cross sections as $\sigma(^6D)/\sigma(^4F_{9/2}) = 0.3 \pm 0.8$ ($\pm$ one standard deviation). The data support the inequality $\sigma(^6D) \leqslant \sigma(^4F_{9/2})$ and do not exclude the possibility that $\sigma(^6D) \ll \sigma(^4F_{9/2})$. New two-color experiments designed to create pure beams of $^6D_{9/2}$ Fe^+ should provide a definitive measurement of the ground-state reactivity.

Preliminary results[31] indicates that the second excited term, $3d^64s(^4D)$, reacts two or three times more efficiently than $3d^7(^4F)$ at 0.22 eV. There is evidence of spin–orbit level effects on cross sections of $3d^6 4s(^4D)$ reactants, but the case is not as clear-cut as for $3d^7(^4F)$. The qualitative picture emerging for $Fe^+ + C_3H_8$ at 0.22 eV kinetic energy is similar to that for $V^+ + C_2H_6$, C_3H_8, and C_2H_4. The low-spin Fe^+ terms are more reactive than the high-spin ground term.

We and others[32,33] observe two product channels, H_2 and CH_4 elimination [Eqs. (6a) and (6b)]. The branching fraction of CH_4 elimination products is $\sigma_x(FeC_2H_4^+ + CH_4)/\sigma_{tot} = 0.83 \pm 0.05$, independent of reactant state distribution within the precision of the data, a result that has mechanistic implications.

4. *Comparisons with Previous Results*

a. V^+ Reactions. The variable kinetic energy, ion-beam-plus-collision-cell experiments of Armentrout and co-workers and Beauchamp and co-workers[34] provide the most quantitative data for comparison with our own work. In particular, the Armentrout group[10] has used hot-filament sources of M^+ and observed variations in cross section with filament temperature. By assuming a Boltzmann distribution of gas-phase M^+ electronic states *at the measured temperature of the hot filament*, they extract excited-state cross sections. The best opportunities for measuring excited-state cross sections occur when a low-lying excited state is orders of magnitude more reactive than the ground state, since the beam contains predominantly ground-state M^+ at all filament temperatures. A different source uses electron-impact ionization of volatile ligated metal species and a drift cell whose purpose is to thermalize the M^+ electronic states to 300 K.

The best comparison of absolute reaction cross sections between experiments would use a reaction with large ground-state cross section so that the R2PI and hot-filament sources are most likely producing the same states. The $V^+(3d^3 4s, {}^5D) + C_3H_8$ reaction, while not ideal because of its small cross section, provides the best quantitative comparison at present. We obtain 0.4 ± 0.3 Å^2 at 0.22 ± 0.04 eV compared with Armentrout and Aristov's 1.0 ± 0.3 Å^2 at nominal kinetic energy of 0.2 eV.[10] The measurements agree to about a factor of 2; the error estimates just overlap. This is not unreasonable agreement given the very different experiments and the general difficulty of measuring absolute cross sections. Differences in kinetic energy distribution between experiments may complicate the comparison. The very large $V^+ + C_3H_8$ cross section of 13 Å^2 at 0.5 eV reported in early work[34] seems incorrect, perhaps due to a large fraction of triplet excited states in the beam.

It is difficult to check the Armentrout group's assumption of a Boltzmann distribution of V^+ reactants at the filament temperature. For $V^+ + C_2H_6$, Aristov and Armentrout[10] observe an increase in effective cross section with filament temperature that they attribute to highly reactive excited states. The specific states responsible were not identified nor was an excited-state cross section measured. We can synthesize Aristov and Armentrout's results at 0.2 eV and filament temperatures 1800, 1925, and 2200 K from our state-specific cross sections if we use 0.1 $Å^2$ for both quintet cross sections (the largest value consistent with our data) and make the further assumption that excited states above $3d^3 4s(^3F)$ react at the Langevin cross section. However, that model contradicts the earlier assertion[10] that the quintet cross sections are < 0.01 $Å^2$ below 0.2 eV. For $V^+ + C_2H_4$, Aristov and Armentrout observed no effect of filament temperature on cross section, which is sensible given the relatively small $3d^3 4s(^3F)$ cross section.

Combining Aristov and Armentrout's measurements of V^+ $3d^4(^5D) + C_2H_6, C_3H_8$, and C_2H_4 cross sections at $E = 1.3$ eV with our $3d^3 4s(^3F)$ cross sections at $E = 0.2$ eV allows us to compare the efficiency of electronic and kinetic energy in promoting H_2 elimination reactions at equal total reactant energies of 1.3 eV. For all three hydrocarbons, electronic excitation to the triplet state facilitates reaction much more effectively than the same amount of kinetic energy. Comparing reaction *efficiencies* (i.e., taking account of the factor of 2.5 decrease in σ_L from 0.2 to 1.3 eV kinetic energy), electronic energy is $\geqslant 130$ times, 200 times, and 20 times more effective than kinetic energy in driving the C_2H_6, C_3H_8, and C_2H_4 reactions, respectively.

b. $Fe^+ + C_3H_8$. Our observation of J-dependent cross sections[18] could affect the interpretation of previous work. Schultz, Elkind, and Armentrout (SEA)[11] have studied electronic and kinetic energy effects on the $Fe^+ + C_3H_8$ reaction. Use of two different Fe^+ sources (hot filament at 2300 K and 100 eV electron-impact source with cooling in a 300 K collision cell of Ar) provides reactant cations in two different electronic-state distributions. In each case, SEA assume a Boltzmann distribution of Fe^+ electronic states at the source temperature, either 2300 K or 300 K. A 2300 K distribution would comprise 70% $3d^6 4s(^6D)$ and 21% $3d^7(^4F)$ cations with broad J distributions peaked at $J = 9/2$ within each term. A 300 K distribution would be dominated by $3d^6 4s(^6D)$ with an 86% contribution from the single level $^6D_{9/2}$.

For the exothermic H_2 and CH_4 products of the $Fe^+ + C_3H_8$ reaction, SEA report state-averaged cross sections for the two Fe^+ beams versus kinetic energy over the range 0.04–4.0 eV. Assuming Boltzmann distributions and J-independent cross sections, they extract term-specific cross sections for ^6D and ^4F. Below 0.3 eV, the resulting ratio $\sigma(^4F)/\sigma(^6D)$ is 0.33 ± 0.80. The ratio increases with reactant kinetic energy. Above 0.5 eV, the relative reactivity reverses; at 1.0 eV, $\sigma(^4F)/\sigma(^6D) \simeq 4$.

Our new $Fe^+ + C_3H_8$ data at 0.22 eV (Table V) indicate $\sigma(^6D) \leqslant \sigma(^4F_{9/2})$ and do not exclude the possibility that $\sigma(^6D) \ll \sigma(^4F_{9/2})$. The most probable ratio $\sigma(^4F)/\sigma(^6D)$ at 0.2 eV kinetic energy is 0.3 from SEA; the most probable ratios from our data are $\sigma(^4F_{9/2})/\sigma(^6D) \simeq \sigma(^4F_{7/2})/\sigma(^6D) = 3$ and $\sigma(^4F_{5/2})/\sigma(^6D) = 6$. The two experiments[11,18] are not necessarily inconsistent, since neither determines both ^6D and ^4F cross sections accurately at present. Our observation of a spin–orbit effect within ^4F raises the possibility that interpretation of SEA's experiment is complicated by J-dependent cross sections. The two Fe^+ sources likely have very different J distributions within ^6D.

Definitive experiments may now be feasible. Creation of Fe^+ in specific ^6D_J levels will require two-color resonant two-photon ionization, since we have explored the useful one-color schemes. We are working on this experiment.

IV. REACTION MECHANISMS

A. Overview

One of the most striking features of M^+ + hydrocarbon chemistry is the diverse behavior of the different metals.[35] For example, restricting the discussion to low-energy (less than ~ 0.4 eV), ground-state $M^+ + C_3H_8$ reactions in single-collision conditions, Ti^+ and V^+ effect H_2 elimination[10,17,36]; Co^+ and Ni^+ and probably Sc^+ and Fe^+ effect both H_2 and CH_4 elimination[37]; and Cr^+, Mn^+ Cu^+, and Zn^+ are inert.[35] There are large variations in cross sections among the chemically active ground states. Apparently subtle changes in M^+ structure switch the chemistry on or off and change the product branching. The purpose of this section is to seek simplicity in this apparent complexity by fashioning a working model of how the myriad M^+–hydrocarbon potential surfaces control the outcome of low-energy collisions. Ultimately, we need to understand not only why some metals are inert and others reactive but also why some metals break only C—H bonds while others break both C—H and C—C bonds.

To build the model, we draw on a wide variety of experimental results, on qualitative molecular-orbital arguments, on a small but growing body of *ab initio* quantum-chemical calculations, and on the mature field of the gas-phase reaction dynamics of simpler systems. Offsetting the lack of rigorous theory is the increasing specificity of experimental measurements, which now significantly constrain the range of plausible models and sharpen the remaining questions.

For a small subset of M^+ + hydrocarbon reactions in single-collision conditions at 0.2 eV kinetic energy, our crossed-beam experiment has revealed dramatic effects of the initial M^+ electronic state on reaction cross section. For $V^+ + C_3H_8$, C_2H_6, and C_2H_4, we find the lowest-energy *triplet* term $3d^34s(^3F)$ far more reactive than the two low-lying *quintet* terms, $3d^4(^5D)$

and $3d^3 4s(^5F)$. The $V^+(^3F) + C_3H_8$, C_2H_6 reactions are highly efficient ($\sigma/\sigma_L = 0.41$ and 0.26, respectively), while the $V^+(^3F) + C_2H_4$ reaction is inefficient ($\sigma/\sigma_L = 0.03$). This is a surprising result since we would expect stronger long-range attraction between V^+ and alkene than between V^+ and alkane. Within $3d^3 4s(^3F)$, spin–orbit level effects on reaction cross section are less than $\sim 25\%$. The term-specific data for $Fe^+ + C_3H_8$ are less precise as yet, but it is clear that the excited $3d^6 4s(^4D)$ term is two to three times more reactive than the $3d^7(^4F)$ term and probable that $3d^7(^4F)$ is more reactive than the ground term, $3d^6 4s(^6D)$. Neither $Fe^+(^4F)$ nor $Fe^+(^4D)$ attacks C_3H_8 efficiently at 0.2 eV, and $Fe^+(^6D)$ may be highly inert. Within $3d^7(^4F)$, we observe a clear factor of 2 dependence of cross section on spin–orbit level.

The cleavage of two C—H bonds by M^+ and the subsequent elimination of H_2 is surely a multistep, highly indirect process. M^+ + alkane reaction rates and product branching ratios measured in the *multicollision* environment of a fast-flow reactor show efficient collisional stabilization of long-lived M^+–alkane complexes at 0.7 torr He.[35] A simple kinetics model indicates lifetimes in excess of 10 ns for internally hot $[VC_3H_8^+]^*$ and $[FeC_3H_8^+]^*$ complexes and in excess in 1 ns for $[VC_2H_6^+]^*$ complexes formed from ground-state M^+ at 300 K (mean collision energy 0.04 eV). The time scale of H_2 or CH_4 elimination reactions may be measured in tens or even hundreds of nanoseconds. Although this suggests bottlenecks or small barriers along low-energy paths to elimination products, additional kinetic energy does not increase reaction efficiency. Armentrout and co-workers[10–12] consistently observe exothermic cross sections that decrease more rapidly with E than the $E^{-1/2}$ dependence of the Langevin cross section.

There is increasing evidence that the "decision point" between reactive and non reactive M^+–alkane encounters comes early in the reaction path. As we will see, state-specific cross sections and the pattern of reactivity across the 3*d*-series M^+ ground states can be qualitatively understood by assuming that the rate-controlling step is insertion of M^+ into a C–H bond to form the intermediate $[H–M^+–R]^*$. For the more reactive M^+ cations, it is plausible based on M^+–H and M^+–CH_3 gas-phase bond energies that $[H–M^+–R]^*$ lies below $M^+ + RH$ in energy.[38]

We know much less about what factors control product branching between H_2 and CH_4 elimination. M^+ chemists have suggested[8] that H_2 elimination from C_2H_6 and larger alkanes occurs by initial M^+ insertion into a C–H bond, followed by migration of a β-H to the metal (through a four-center transition state) to form a $M^+(H)_2$ (alkene) intermediate, which subsequently eliminates H_2. In a similar vein, CH_4 elimination was often postulated to occur by intial M^+ insertion into a C–C bond, again followed by β-H migration and CH_4 elimination. Considering the steric difficulties, this author has always found initial C–C insertion implausible. Recent kinetic energy distributions, isotope effects, and phase-space calculations[39] support an alternative route to CH_4 elimination involving initial C–H insertion and β-CH_3 migration to the metal. The independence of the H_2/CH_4 branching for various initial Fe^+ states suggests a common precursor for both products. Either CH_4 mechanism can explain why only the H_2 channel is observed in C_2H_6 recations, since the H–M^+–C_2H_5 intermediate lacks a β-CH_3 group and the CH_3–M^+–CH_3 intermediate lacks a β-H. New experiments designed to probe the structures, energies, and lifetimes of such intermediates are needed.

We focus our mechanistic discussion on factors controlling initial M^+ insertion into a C–H bond of a linear alkane, since our results and many others can be understood in that context. Discussions of the overall mechanisms can be found elsewhere.[8,10–12,37]

B. Orbital Symmetry and Electron Spin Conservation in C–H Bond Insertion

1. *The Simpler Case of $M^+ + H_2$*

Following Rappe and Upton,[40] consider the simplest relevant example of M^+ insertion into a σ bond, the approach of Sc^+ toward H_2 in C_{2v} symmetry. Sc^+ has seven electronic terms below 2.5 eV (Fig. 3) that give rise to 65 different multiplet components. Mercifully, the ${}^1\Sigma_g^+$ ground state of H_2 is well isolated from any excited states. In addition, in this high-symmetry example, there are severe restrictions on which Sc^+ orbital occupancies correlate to an H–Sc^+–H intermediate with two σ bonds. Since ground-state ScH_2^+ doubly occupies both symmetric (a_1) and antisymmetric (b_2) bonding molecular orbitals, it correlates only to the $3d_{yz}{}^2$ component of the $3d^2({}^1D)$ excited term of Sc^+ (Fig. 12). Clearly the ground state of ScH_2^+ must be a singlet to form both bonds.

In donor–acceptor language,[41] two interactions must be optimal for barrierless insertion of Sc^+ into H_2. The H_2 ground state, which is $(a_1)^2$ in C_{2v} symmetry, must find an unoccupied 4*s* acceptor orbital on Sc^+; this necessitates a $3d^2$ configuration on the metal. (An empty $3d_{z^2}$ orbital has the proper a_1 symmetry but too small a spatial extent to overlap well with the H_2 σ orbital at long range.) In addition, the two electrons on Sc^+ must occupy d_{yz}, the only *d* orbital of the proper b_2 symmetry to donate into the antibonding σ^* orbital on H_2. This necessitates singlet coupling of the two metal electrons.

These qualitative molecular-orbital arguments indicate that in C_{2v} symmetry, only one of the six nondegenerate *diabatic* $Sc^+ + H_2$ potential energy surfaces arising from the $3d^2({}^1D)$ *excited-state* asymptote will fall

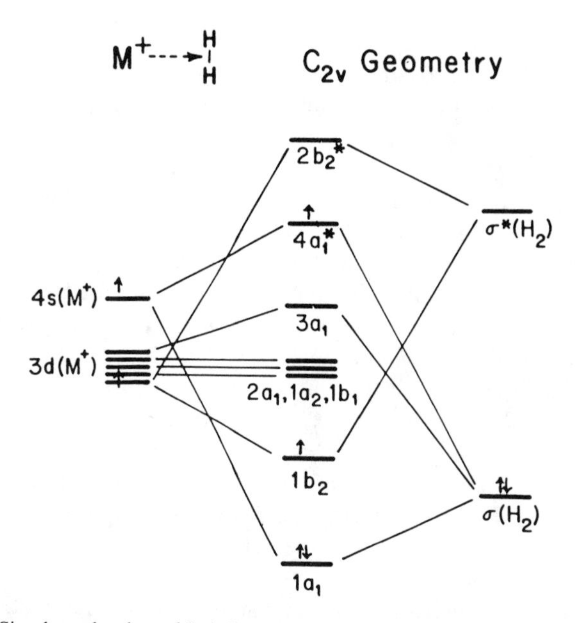

Figure 12. Simple molecular-orbital diagram for the C_{2v} approach of M^+ to H_2.

monotonically into an H–Sc^+–H potential well. We use *diabatic* in the sense of *conserving orbital symmetry and electron spin.* The remaining five potentials from $3d^2(^3D)$ will be repulsive, since they occupy d orbitals of the wrong symmetry to donate into σ^* of H_2. All of the diabatic potentials arising from the $3d4s(^3D)$ ground term, the $3d4s(^1D)$ first excited term at 0.31 eV, and the $3d^2(^3F)$ second excited term at 0.60 eV may have small attractive wells at long range owing to ion-induced dipole forces, but are highly repulsive at short range. The $3d4s(^{3,1}D)$ asymptotes lead to repulsive interactions between metal $4s$ and doubly occupied σ and H_2. The $3d^2(^3F)$ asymptote produces only triplet ScH_2^+ surfaces, which are repulsive because they cannot form two σ bonds.

In an important study, Rappe and Upton[40] carried out high-level *ab initio* calculations of the potential energy surfaces arising when the low-lying terms of Sc^+ approach H_2 in C_{2v} and in collinear geometries. The qualitative features expected from simple molecular-orbital arguments are calculated in quantitative detail.

One key lesson from $Sc^+ + H_2$ is that *the lowest-energy terms, high-spin $3d^{n-1}4s$, low-spin $3d^{n-1}4s$, and high-spin $3d^n$, produce only repulsive potentials along paths to bond insertion.* In addition, *only a small subset of the potentials arising from the low-spin, $3d^n$ excited term produce attractive potentials.* Referring to Fig. 3, we see that in the $3d$-series metal cations, the lowest-energy terms are generally ill suited to insert in H_2. A second lesson is that in order for Sc^+ in the $3d4s(^3D)(^3D)$ ground state to reach the singlet ScH_2^+ potential well, the collision must access the intersection between the repulsive, *triplet* reactant surface and the only attractive, *singlet* surface arising from $3d^2(^1D)$ excited-state asymptotes. We use the words "surface intersection" rather than "curve crossing" because of the multidimensionality of the problem,[2] an important feature later on. Effective mixing of triplet and singlet surfaces by the spin–orbit interaction is limited to a local region of configuration space near the intersection of the diabatic surfaces. *Surface intersections between diabatic surfaces of different electron spin are only weakly avoided and highly localized in configuration space.*

Rappe and Upton[40] use a local spherical-potential approximation to estimate that the spin–orbit matrix element between the singlet and triplet surfaces is $\leqslant 2\xi_{3d}$ and thus the splitting between adiabatic surfaces is $\leqslant 4\xi_{3d} = 0.035$ eV, where ξ_{3d} is the atomic spin–orbit radial integral for a Sc^+ $3d$ orbital.[42] In C_{2v} geometry, only one space–spin component of the five in the $^3D_2$ ground level of Sc^+ interacts with the attractive singlet surface in this approximation.

What happens in real $M^+ + H_2$ collisions of lower symmetry than C_{2v}? Previously degenerate surfaces will split, and new orbital interactions will arise in lower symmetry. However, as usual in quantum mechanics, we expect the rigorous arguments from high-symmetry limits to carry over qualitatively into lower symmetries. Matrix elements that are zero in high symmetry will be small in lower symmetry, and matrix elements that are large in high symmetry will remain substantial in lower symmetry. In an important series of experiments. Elkind and Armentrout[9] showed that endothermic $M^+ + H_2 \rightarrow MH^+ + H$ reaction cross sections indeed follow a simple set of rules derived from the high-symmetry limits of collinear or C_{2v} approach. Lessons gleaned from rigorous $Sc^+ + H_2$ calculations apparently transfer to other $3d$-series $M^+ + H_2$ reactions.

In summary, much of the complexity of M^+ + alkane collision dynamics is already present, and is relatively well understood, in the much simpler case of $Sc^+ + H_2$. The key concepts are the existence of many repulsive potentials arising from low-lying Sc^+ states and a relatively few attractive potentials from excited states. This results in *multidimensional* surface intersections that are weakly avoided when they involve two surfaces of different electron spin and much more strongly avoided when of the same spin. The important role played by electron spin is clear already in $Sc^+ + H_2$.

2. *Generalization to M^+ + Alkane: The Importance of Electron Spin Conservation*

We are now prepared to leap from $M^+ + H_2$, which is quite well understood,[9] to the more complex case of M^+ + alkane. To what extent can we apply orbital and spin conservation arguments from $M^+ + H_2$ to the *much*-lower-symmetry case of M^+ insertion into a C–H bond of an alkane? At present, we must try to infer the answer from experimental results. For M^+ + alkane, we lack the quantitative theoretical underpinning that Rappe and Upton[40] have provided for the interpretation of $M^+ + H_2$ experiments. No calculations are available even for the complete reaction path from $M^+ + CH_4$ to H–M^+–CH_3, to name the next simplest relevant case.[43]

To gain qualitative insight, we consider the approach of M^+ toward a C–H bond of C_2H_6 as occurring with a plane of symmetry, the H–M^+–C plane. This is not an approximation for certain approach angles of M^+ and certain conformations of the alkane. The relevant orbitals are now classified as symmetric or antisymmetric with respect to reflection in the plane. Both σ_{CH} and $\sigma_{CH}{}^*$ are symmetric, so there are fewer restrictions on which metal orbitals can be acceptors from σ_{CH} or donors to $\sigma_{CH}{}^*$. In particular, 4s, $3d_{z^2}$, $3d_{yz}$, and $3d_{x^2-y^2}$ are all symmetric orbitals that could function as either donor or acceptor. Thus, even in the highest-symmetry approach of M^+ to a C—H bond of an alkane, there are no orbital symmetry *rules* dictating that repulsive surfaces must arise from $3d^{n-1}4s$ states. Of course there will still be favorable and unfavorable orbital interactions, but quantitative estimation of these "orbital propensities" requires rigorous calculations. It remains an open and important question to what extent remnants of the $M^+ + H_2$ rules may carry over to M^+ + alkane.

On the other hand, the Pauli exclusion principle remains in full force no matter the symmetry, and electron spin is just as good a quantum number for M^+ collisions with H_2 or with alkanes. Consequently, we expect only repulsive diabatic surfaces from certain high-spin M^+ terms. For example, the two low-lying *quintet* terms of V^+, $3d^4(^5D)$ and $3d^34s(^5F)$ (Fig. 1), must produce only repulsive potentials on paths toward C–H bond insertion. A σ-bonded H–V^+–R intermediate has only two nonbonded electrons that could be coupled to form a triplet or a singlet; quintet states will lie high in energy and presumably be unbound. By the same argument, the lowest-energy excited *triplet* terms of V^+, $3d^34s(^3F)$ and $3d^4(^3P)$, *could* produce attractive potentials that fall monotonically into a triplet H–V^+–R well *if* orbital interactions are sufficiently favorable. Similarly, we presume an H–Fe^+–R intermediate with two σ bonds formed from two 4s–3d hybrid orbitals on Fe^+ and thus three unpaired electrons that couple to form low-lying states of *quartet* spin. It follows that the $3d^64s(^6D)$ ground term of Fe^+ produces only repulsive diabatic potentials along paths toward C–H bond insertion. Both excited quartet terms, $3d^7(^4F)$ and $3d^64s(^4D)$, *could* produce attractive potentials leading to low-energy states of H–Fe^+–R.

Regardless of the quantitative details of the orbital interactions, electron spin conservation on diabatic surfaces dictates that few M^+ ground states can insert in C–H or C–C bonds without first gaining access to an attractive surface of lower spin. Accordingly, Fig. 3 highlights the lowest-energy term of each M^+ that can insert in a C–H bond in a"spin-allowed" process. Only the Co^+, Ni^+, Cu^+ ground states have the correct spin, and the filled $3d$ subshell of Cu^+ $3d^{10}4s(^2S)$ prevents formation of two σ bonds. Fe^+, Sc^+, Ti^+, and perhaps V^+ have sufficiently low-lying excited terms that can conserve spin during bond insertion, while Cr^+ and Mn^+ do not.

Conservation of electron spin along paths to C–H insertion can qualitatively explains the observed pattern of ground state M^+ reactivity with alkanes. The chemically active M^+ cations either have a ground state that can conserve spin during insertion (Co^+, Ni^+) or have a low-lying excited state that can do so (Sc^+, Ti^+, and the much less reactive V^+ and Fe^+); see Fig. 3. The higher in energy the first low-spin asymptote, the larger will be the potential barrier on the *adiabatic* surface arising from the avoided intersection between high-spin and low-spin diabatic surfaces. When the spin-conserving state is an excited state, the spin multiplicity must change on the low-energy adiabatic surfaces connecting high-spin reactants to low-spin bond insertion intermediate. Apparently ground-state Sc^+, Ti^+, V^+, and Fe^+ reactants at low collision energy can sometimes find the avoided intersection and change spin. The other cations are inert because they have well-isolated, high-spin ground states (Cr^+, Mn^+) or because they cannot form two σ bonds (Cu^+).

3. $V^+ + C_2H_6$, C_3H_6, C_2H_4 Mechanisms

Now consider the electronic-state dependence of the $V^+ + C_3H_8$, C_2H_6, and C_2H_4 cross sections.[17] Based on the same arguments, we sketch diabatic potentials from the low-lying asymptotes as shown in Fig. 13. The quintet surfaces are repulsive, but a subset of the surfaces from the *triplet* $3d^34s(^3F)$ asymptotes are attractive along the assumed reaction path toward the triplet H–V^+–R bond insertion intermediate. The resulting weakly avoided surface intersections lead to "spin barriers" on low-lying *adiabatic* surfaces from quintet reactants. Correspondingly, we observe no definite reaction for quintet $V^+ + C_2H_6$ and very small cross section for quintet $V^+ + C_3H_8$.

Moreover, reactant kinetic energy in quintet V^+ does *not* promote H_2 elimination. Aristrov and Armentrout's experimental $V^+(^5D) + C_3H_8$ cross section[10] decreases roughly as $E^{-1.1}$, significantly more rapidly than the $E^{1/2}$ dependence of the Langevin cross section. In addition, we know from the multicollision kinetics studies[35] that even $V^+(^5D) + C_3H_8$ collisions at 300 K form long-lived complexes with lifetimes measured in nanoseconds. In direct atom abstraction reactions, we would expect kinetic energy to enhance passage over a barrier. However, in polyatomic systems with deep potential energy wells prior to the barrier, reaction probability is determined by a competition between barrier passage and return to reactants. We suggest[35]

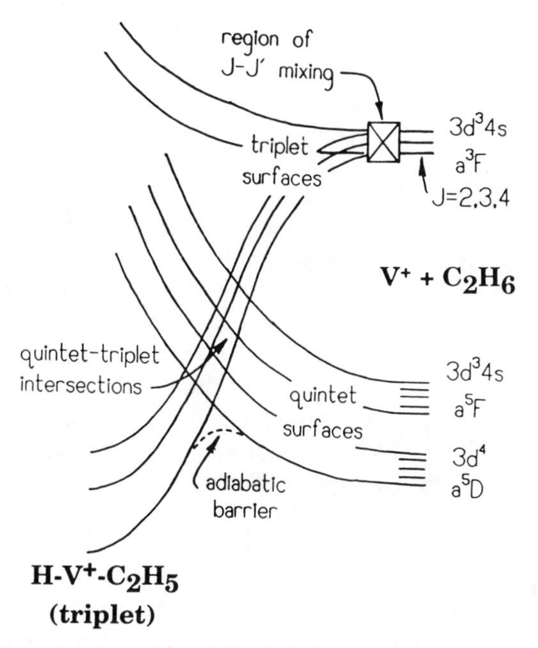

Figure 13. Schematic of the $V^+ + C_2H_6$ diabatic potential energy surfaces (solid lines) leading from quintet and triplet asymptotes toward an $H–V^+–C_2H_5^+$ bond insertion intermediate. Crossing of triplet and quintet surfaces cause small barriers on adiabatic surfaces d$3d^34s(^3F)$ are strongly mixed at long range to form electrostatic surfaces.

that the inefficiency of the quintet V^+ reactions may be due to the long time required for the reaction complex to find the seam where quintet and triplet surfaces intersect, rather than to a true energetic constraint. The effect of increased kinetic energy may be to increase the rate of redissociation of the complex back to reactants, which competes with the rate of passage over the barrier. Beauchamp, Bowers, Armentrout, and co-workers[39] have recently used this same idea in phase-space calculations of $Co^+ + C_3H_8$ cross sections.

When V^+ + alkane reactants approach on the triplet surfaces arising from $3d^34s(^3F)$ V^+, the reaction efficiency is large, 26% for $V^+ + C_2H_6$ and 41% for $V^+ + C_3H_8$. In the context of Fig. 13, this suggests that a large fraction of the initial flux passes *through* the weakly avoided quintet–triplet surface intersections to the triplet $H–V^+–R$ intermediate, *remaining on the diabatic (triplet) surfaces.*

In atom + atom collisions, the Landau–Zener model[44] of hopping between adiabatic surfaces predicts that the probability of remaining on the initial diabatic surface becomes large in the limit of large approach velocity to the crossing point and in the limit of small quintet–triplet interaction strength (weakly avoided crossing). However, we resist the temptation to label the observed behavior as Landau–Zener-like. Even the $[VC_2H_6^+]^*$ complex has 21 vibrational degrees of freedom and an enormous number of accessible electronic states as well. In contrast to the diatomic case, in a polyatomic system the attractive triplet well results not in rapid acceleration of reactants toward each other as the crossing point is approached, but rather in an especially long $[VRH^+]^*$ lifetime for triplet reactants.[35] Hence, we prefer a model in which triplet reactant pairs that evolve onto attractive triplet surfaces form very long-lived complexes that sample the region of the weakly avoided surface intersections many times (perhaps thousands of times) before dissociating back to reactants. Each encounter with a quintet–triplet seam provides a small probability for hopping[2] from the upper adiabatic surface to the lower one, from which evolution to $[H–V^+–R]^*$ and on to products is irreversible. The long complex lifetime, which is due to the deep well into which triplet reactants fall, allows time for the surface hop to occur. This model makes the definite prediction that changes in the kinetic energy of triplet V^+ reactants should have little effect on reaction efficiency, since the $[H–V^+–R]$ potential well looks very deep from the triplet asymptote.

Now recall the orbital restriction that led to repulsive C_{2v} interaction between $M^+(3d^{n-1}4s)$ and H_2. Does our experimental result that $3d^34s(^3F)$ V^+ reacts efficiently indicate that the reduced symmetry of V^+ + alkane has allowed diabatic surfaces from $3d^{n-1}4s$ configuration of the metal to become *attractive*? Not necessarily. A set of $3d^4(^3P)$ levels lie at 1.40–1.48 eV, just above the $3d^34s(^3F)$ levels at 1.07–1.13 eV (Fig. 3). An alternative picture[11] to Fig. 13 would draw only repulsive diabatic surfaces from $3d^34s(^3F)$ and some attractive diabatic surfaces from $3d^4(^3P)$. The resulting surface intersections will now be *strongly* avoided, since both sets of diabatic surfaces have triplet spin. In this alternative model, low-energy collisions originating on surfaces from $3d^34s(^3F)$ acquire $3d^4(P)$ character by remaining on the lower-energy *adiabatic* surface at the $^3F–^3P$ intersection and then retain that character during their subsequent *diabatic* passage through the weakly avoided quintet–triplet intersections.

Again, the $Sc^+ + H_2$ calculations of Rappe and Upton[40] provide some quantitative guidance. Diabatic surfaces of the same spin and overall symmetry from different Sc^+ asymptotes lead to avoided intersections with electronic gaps on the order of 0.3–0.5 eV, about 10 times larger than the estimated gaps from singlet–triplet intersections. If the ^3F *diabatic* surfaces are repulsive but the $^3F–^3P$ surface intersections are avoided by about 0.5 eV, it becomes plausible that *adiabatic* surfaces from $3d^34s(^3F)$ could be barrierless up to the triplet–quintet intersections (Fig. 13).

An experimental test of whether or not remnants of orbital symmetry rules make $3d^{n-1}4s$ low-spin term interact repulsively with alkanes may be feasible. We need to measure the reactivity of $3d^4(^3P)$ V^+. If the $3d^34s(^3F)$ and $3d^4(^3P)$ diabatic surfaces intersect and repel strongly, then V^+ in $3d^4(^3P)$ should be *unreactive*, which would be a striking result. Unfortunately, we have not yet devised a means of creating $3d^4(^3P)$ sufficiently selectively, but two-color R2PI may work.

The asymptotic quantum number J is destroyed as the alkane approaches M^+.[45] The absence of discernible spin–orbit level dependence of the $3d^34s(^3F)$ reaction cross sections with C_2H_6 and C_3H_8 suggests that scrambling of flux among the different adiabatic surfaces during the collision is essentially complete. That is, reactants lose all memory of their initial spin–orbit quantum number before deciding whether or not to react, and the 460 cm^{-1} total energy difference between $J = 4$ and $J = 2$ is inconsequential. The nanosecond lifetimes of $[V(alkane)^+]^*$ complexes provide the time necessary for loss of memory to occur.

Finally, it is intriguing that the $3d^34s(^3F)$ reaction with C_2H_4, while still more efficient than the quintet reactions, is much less efficient than the $3d^34s(^3F)$ reactions with the saturated alkanes C_2H_6 and C_3H_8. We can suggest that $V^+ + C_2H_4$ fall into a deep attractive well due to $p\pi \rightarrow d\pi$ donor–acceptor interactions. The large density of states in this well may divert reactants away from paths to C–H bond insertion, as suggested by others. Alternatively, the presence of the C=C double bond may destabilize some crucial intermediate along the reaction path to elimination products. It would be interesting to extend our work to larger alkenes to see if the cross sections converge with those of alkanes.

4. $Fe^+ + C_3H_8$ *Mechanism*

For $Fe^+ + C_3H_8$, we would draw qualitative diabatic surfaces completely analogous to the $V^+ + C_2H_6$ surfaces in Fig. 13.[46] The ground term $3d^64s(^6D)$ gives rise only to repulsive surfaces, the first excited term $3d^7(^4F)$ gives rise to some attractive surfaces, and the resulting intersections are weakly avoided since they involve a change of electron spin. The adiabatic surfaces from the $3d^64s(^6D)$ ground term will have small spin barriers to bond insertion. We mentally sketch the diabatic surfaces from the second excited $3d^64s(^4D)$ term as attractive, although once again they may borrow attractive character from the nearby $d^7(^4P)$ term during the early part of a collision.

The simplest qualitative interpretation[18] of our $Fe^+ + C_3H_8$ results to date is completely analogous to the interpretation of the more comprehensive $V^+ + C_2H_6$, C_3H_8 results. In particular, we infer the same qualitative behavior when the complex approaches the weakly avoided, spin-changing surface intersections, which is satisfying. The bond insertion intermediate

$H-Fe^+-C_3H_7$ has two σ bonds formed from $3d$–$4s$ hybrid orbitals. The remaining five metal electrons occupy four $3d$ orbitals; high-spin coupling leads to a *quartet* ground state of the intermediate. The $3d^7(^4F)$ and $3d^64s(^4D)$ terms are therefore reactive because they can conserve spin during bond insertion. We assert that low-energy collisions from quartet asymptotes sometimes find their way *through* the quartet–sextet surface intersections owing to the long complex lifetime and the weak quartet–sextet spin–orbit interaction. The $3d^64s(^4D)$ term may be two to three times more reactive than $3d^7(^4F)$ because it falls into a deeper potential well and forms complexes of longer lifetime, which assists diabatic passage through the quartet–sextet surface intersections. Of course, ^4D also has 0.7 eV additional internal energy compared with ^4F. The $3d^64s(^6D)$ term is less reactive than $3d^7(^4F)$, and perhaps highly unreactive, owing to the spin barrier between reactants and the C–H bond insertion intermediate on the lowest-energy adiabatic surfaces.

The experimental result of a factor of 2 dependence of the ^4F_J cross section on spin–orbit level is surprising. While much larger spin–orbit effects are well documented in direct atom abstraction reactions,[47] we might have expected the long lifetime of the $[FeC_3H_8^+]^*$ collision complex to permit complete loss of memory of initial J, as occurred in the V^+ reactions. The larger spin–orbit splittings in Fe^+ $3d^7(^4F)$ than in V^+ $3d^34s(^3F)$, 400–560 cm^{-1} compared with 200–250 cm^{-1}, may begin to explain why we observe these effects in Fe^+. In addition, the fact that the highest-energy spin–orbit level studied has the largest cross section suggests that the 400 cm^{-1} increase in total energy from $J = 7/2$ to $J = 5/2$ may be important in overcoming a small barrier along the reaction path. The overall inefficiency of the reaction for both $3d^64s(^6D)$ and $3d^7(^4F)$ agrees qualitatively with this suggestion. It would be interesting to see if spin–orbit effects persist in more efficient reactions such as $Fe^+ + C_4H_{10}$.

While appealing for its consistency with the V^+ interpretation, our picture of what happens to $3d^64s(^6D)$ and $3d^7(^4F)$ reactants at low kinetic energy conflicts with Schultz, Elkind, and Armentrout's (SEA's) interpretation of their own $Fe^+ + C_3H_8$ data[11] (reviewed in Section II B 4). By using two different ion sources *assumed* to produce Fe^+ at electronic temperatures of 2300 K and 300 K, SEA inferred that at *low* kinetic energy (below 0.3 eV) ^6D reacts more efficiently than ^4F. The ratio of ^4F to ^6D cross sections increases with increasing energy. At *high* kinetic energy (above 0.5 eV), SEA infer that ^4F reacts more efficiently than ^6D. While our results *suggest* that ^4F reacts three times more efficiently than ^6D at 0.22 eV, in apparent conflict with SEA's results, we emphasize that *neither* experiment has accurately measured both cross sections as yet.

SEA suggest the same qualitative character of the diabatic surfaces, repulsive from ^6D and attractive from ^4F, due to the same spin considerations described previously. If the SEA data are correct, we must draw very different conclusions about how a complicated reaction system behaves in the vicinity of a weakly avoided surface intersection. SEA quite naturally interpret their results as indicating that at low energy some ^6D reactants follow adiabatic surfaces, penetrate the spin barrier, insert in a C–H bond, and ultimately react. At low energy, all ^4F reactants follow adiabatic surfaces, acquire repulsive sextet character at close range, and fail to react. At high kinetic energy the reactants maintain their *diabatic* (spin-conserving) character, so the reactivity reverses and ^4F becomes more efficient than ^6D. *Qualitatively, this is the behavior of diatomic (atom + atom) collisions, as in the Landau–Zener*[44] *model*!

Better experiments will be necessary to choose definitely between these two competing pictures of the $Fe^+ + C_3H_8$ mechanism. We are currently exploring near-threshold, two-color R2PI schemes as a means of creating pure beams of Fe^+ $^6D_{9/2}$ in order to accurately compare ^6D and ^4F reactivity. In the meantime, we comment that our observation of a spin–orbit effect within Fe^+ ^4F raises the possibility that interpretation of SEA's experiment is complicated by J-dependent cross sections within ^6D as well. SEA's two different Fe^+ sources probably have very different J distributions within ^6D.

The invariance with initial Fe^+ electronic state of the $Fe^+ + C_3H_8$ product branching between H_2 and CH_4 elimination products [Eqs. (6a) and (6b)] is mechanistically significant. An $[FeC_3H_8^+]^*$ complex formed from $3d^64s(^4D)$ reactants has 0.7 eV more initial energy than a complex from $3d^7(^4F)$ reactants. Fe^+ in $3d^64s(^4D)$ reacts two to three times more efficiently than $3d^7(^4F)$, yet electronic excitation has no apparent effect on the product branching. This suggests that the decision between H_2 and CH_4 elimination channels occurs relatively late in the collision, after the decision point between reaction or nonreaction. Remember that the lifetime of these complexes is probably measured in nanoseconds. We presume that eventually the available internal energy is distributed roughly statistically in the complex. An additional 0.7 eV distributed among 30 vibrational degrees of freedom and a large number of electronic degrees of freedom may be quite insignificant after memory of the initial electron configuration is lost. In contrast, if the reactants branched between *initial* C–H and C–C bond insertion, we would expect a substantial effect of initial state and total energy on product branching. Other work[30] has provided evidence that an initial C–H insertion intermediate later branches between H_2 and CH_4 elimination channels in $Co^+ + C_3H_8$.

Finally, we comment that we have no great insight into why $Fe^+ + C_3H_8$ is much less efficient than $V^+(3d^34s, ^3F) + C_3H_8$, even from the $3d^7(^4F)$ and $3d^64s(^4D)$ excited states. *Ab initio* calculations on $M^+ + CH_4$ may be able to shed light on this issue in the near future.

VI. SUMMARY AND PROGNOSIS

By combining our state-specific cross-section measurements with a variety of important results from other groups, we are beginning to learn how $3d$-series transition-metal cations break C–H and C–C bonds of hydrocarbons. While much of our suggested mechanism is necessarily speculative, precise experimental results and chemical intuition set increasingly strict limits on the range of acceptable models. Clearly electron spin is a key quantum number in determining the outcome of M^+ collisions.

The high density of reactant electronic states and the very different chemical behavior of these states make avoided surface intersections commonplace in M^+ collisions. Only the Co^+ and Ni^+ ground states have the proper electron spin and configuration to insert directly in C–H or C–C bonds without barriers. More generally, it is the overall *pattern* of electronic states (Fig. 3) that determines the reactivity of a particular state. We believe we have substantial intuition about when surface intersections will occur; this is only a matter of qualitatively predicting the shapes of diabatic potential energy surfaces. It should be clear, however, that we have only begun to understand how to predict what happens when complex reaction systems approach an avoided surface intersection at low, medium, and high kinetic energy. It also remains unclear how M^+ electron configuration, $3d^{n-1}4s$ vs $3d^n$, affects energetics at large M^+–alkane distances.

We cannot overemphasize the need for high-quality *ab initio* calculations that include all the configurations necessary to follow surfaces from reactants to bond insertion intermediates and beyond. Calculations on $Sc^+ + C_2H_6$, the electronically and vibrationally simplest system that exhibits exothermic H_2 elimination in low-kinetic-energy collisions, remain a difficult prospect. Important quantitative insights could be gained from a simpler model system such as $Sc^+ + CH_4$, which presumably has many of the same steric features of $Sc^+ + C_2H_6$. We need to understand how the quantitative details change from $Sc^+ + H_2$ to $Sc^+ + C_2H_6$. These calculations are not out of the question, as witness recent work on bound molecules such as M^+–CH_3, $M^+(H_2O)_x$, and $M(CO)_x$.[43]

On the experimental front, the strengths and limitations of resonant two-photon ionization in selectively creating M^+ states are becoming clear. We have intentionally limited the total energy in our one-color R2PI schemes because the selectivity decreases when more than three excited terms are accessible. New possibilities (and complications) open up when a second tunable dye laser is added. We are currently working to prepare pure beams of Fe^+ in the lowest energy level, $^6D_{9/2}$, by two-color R2PI with ω_1 selectively exciting one initial level out of the neutral state distribution and ω_2 tuned just above the Fe^+ threshold. By varying ω_2, we can study the effects of autoionizing states on the cation branching, a possible source of nonselective ionization in the one-color scheme. A strategy worth investigating uses ω_1 to excite the lowest energy odd parity state with the desired core; the tunability of ω_2 provides great flexibility in choice of resonant state. This may improve selectivity by using intermediates in the sparser region of the neutral-atom spectrum, where zeroth-order descriptions should be more accurate because perturbations by nearby levels should be less important.

Coupling of a two-color R2PI source of M^+ to a variable kinetic energy machine can provide independent control of both electronic and kinetic energy. The kinetic energy spread of the R2PI source depends primarily on space-charge repulsion, which can be minimal compared with the broad distribution of kinetic energies from a hot-filament source. This is a modest extension of current technology, since only the molecular-beam source need be changed to carry out such experiments in the apparatus described by Scott Anderson in this volume.

The use of threshold electron–M^+ coincidence techniques to select the reactant state is more ambitious but much more general than R2PI. The chapters by Marie Durup-Ferguson and by Cheuk Ng describe the state of the art in coincidence detection for ion–molecule reactions. Recent advances using pulsed electron extraction[48] make it possible to detect threshold electrons with energy resolution on the order of $3\,cm^{-1}$. It follows that two-color R2PI or laser one-photon ionization can be combined with threshold detection of electrons to select essentially *any* $|J^+L^+S^+\rangle$ level of any transition-metal cation. Detection of the desired threshold electron signals the existence of the desired M^+ electronic state, whose reactivity is then measured.

The difficulty with laser-based coincidence schemes is the need to create less than one electron–cation pair per laser shot. The repetition rate of nanosecond dye lasers that are sufficiently intense to frequency double efficiently into the near-UV is currently about 1 kHz in the new excimer lasers, which seems adequate. Synchronously pumped Nd:YAG systems operating at several kilohertz may be a better idea. We hope to try this experiment with a 30 Hz Nd:YAG system soon. Demonstation of state selection by coincidence detection looks feasible at 30 Hz, but measurement of reaction cross sections seems formidable. Alternatively, electronic term selectivity could probably be achieved using lamps and monochromators for one-photon ionization. A quasicontinuous metal-atom source[49] is required for high-repetition-rate experiments.

State-selected reaction cross sections are only part of the picture. While the initial state profoundly affects reaction probabilities, in the $Fe^+ + C_3H_8$ example the initial state seems not to affect product branching at all. Studies of isotope effects, kinetic energy release distributions,[50] and collisional stabilization of long-lived adducts[35] will provide more information on the dynamics of the later stages of M^+ reactions in the near future.

Finally, we need better information about the energetics and the geometric and electronic structures of polyatomic reactants, products, and reaction intermediates. There has been substantial progress on the important question of the energetics of *two* ligands (e.g., H and C_2H_5) binding to M^+.[38] In general, even mono-ligated transition-metal species pose important structural questions and formidable experimental challenges. *Ab initio* calculations are now available for many simple transition-metal species[43]; there are often no spectroscopic data for comparison. In M^+ + hydrocarbon chemistry, we really do not even know that the structure of the $MC_2H_4^+$ products is an ethylene molecule bound to M^+ in T-shaped geometry. Collision-induced dissociation (CID)[51] is certainly a valuable structural tool that may reveal atomic connectivity in complex systems, but it is impossible to prove that CID occurs prior to "CIR," collision-induced rearrangement!

In this author's view, the absence of spectroscopic data even for many *stable* cation reactants and products, not to mention key reaction intermediates (which are often stable molecules with high internal energy), is the biggest problem in gas-phase ion chemistry. Several groups have begun to attack this difficult problem using photoelectron spectroscopy[52] or high-resolution absorption spectroscopy.[53] Pulsed detection of threshold photoelectrons[47] offers the possibility of $3\,\text{cm}^{-1}$ resolution vibrational spectra of a variety of cations in the foreseable future. Structural data for simple ligated species will calibrate the quality of *ab initio* calculations, which will increasingly be a primary tool for understanding the structure of unstable reaction intermediates in all of chemistry. We can all look forward to a new era of ion–molecule chemistry in which the structures of at least our stable chemicals will no longer be a topic for speculation.

Note added in proof. Recent progress in our lab and Prof. Armentrout's lab has improved the agreement of $Fe^+ + C_3H_8$ state-specific cross sections. We have succeeded in preparing pure beams of the Fe^+ ground level ($3d^64s$, $^6D_{9/2}$) by two-color R2PI with $\omega_1 + \omega_2$ tuned just above threshold. We have obtained several relative and absolute reaction cross sections at two kinetic energies, 0.22 and 1.0 eV. The higher kinetic energy was obtained by seeding the Fe beam in He. The resulting state-specific, relative cross section ratio $\sigma(^4F_{5/2})/\sigma(^6D_{9/2})$ is 3.6 ± 0.5 at 0.22 eV. The ratio $\sigma(^4F_{9/2})/\sigma(^6D_{9/2})$ 2.0 ± 0.3 at 0.22 eV and 3.5 ± 1.0 at 1.0 eV. Error estimates are $\pm$ one standad deviation. The absolute cross section $\sigma(^6D_{9/2})$ is quite small: $0.31 \pm 0.17\,\text{Å}^2$ at 0.22 eV and $0.11 \pm 0.06\,\text{Ǎ}^2$ at 1.0 eV ($\pm$ two standards deviations).

Meanwhile, Armentrout and co-workers have developed a much more intense flowing afterglow source of Fe^+ at 300 K and remeasured $\sigma(^6D)$ and

$\sigma(^4F)$ versus kinetic energy. The two labs now agree that ^6D reacts two to four times less efficiently than ^4F over the energy range 0.2–1 eV, as our original data suggested. The *ratios* of ^6D to ^4F cross sections measured by the two techniques are in good accord at both kinetic energies. Our view of multidimensional surface-hopping phenomena is unchanged from the description above.

There remains a large quantitative difference between *absolute* cross sections measured by the two techniques. For both Fe^+ terms and both collisions energies, Armentrout's cross sections are larger than ours by a factor of 5–10. This is well outside the estimated uncertainties in either measurement. We note that the two techniques agree within a factor of 2 for the ground state $V^+ + C_3H_8$ absolute cross section at 0.2 eV, but again our absolute cross section is smaller than Armentrout's.

One possible cause of the discrepancy is a substantial effect of C_3H_8 internal energy on reaction cross section. Our C_3H_8 reactants are internally cooled by the pulsed expansion that delivers them to the reaction zone, while Armentrout's reactants are at 300 K. We estimate the resulting difference in internal energy (vibration plus rotation) to be 2.7 kcal/mol. Perhaps this internal energy enhances the $Fe^+ + C_3H_8$ reaction cross section for 300 K reagents by allowing penetration of a potential barrier whose height is comparable to the initial energy, as suggested earlier for the $Co^+ + C_3H_8$ reaction.[39] The barrier may occur subsequent to the surface intersections we have described above. Vibrational energy may accelerate reaction efficiently because it is entirely available to the reaction coordinate. In contrast, a fraction of additional kinetic energy must be stored as centrifugal motion during the collision. By heating the pulsed nozzle, we hope to vary the C_3H_8 internal energy somewhat independently of the center-of-mass collision energy and test these ideas.

We thank Prof. Armentrout for supplying these new data and for many stimulating discussions over the years.

Acknowledgments

I want to thank my graduate students for the hard work that has gotten us this far. In particular, Dr. Lary Sanders had the tenacity and experimental talent to develop the R2PI-PES and reaction cross section techniques. Dr. Scott Hanton and Mr. Robert Noll have continued to improve these experiments. Dr. Andy Sappey designed the molecular beam apparatus, which borrows many ideas from Professor Fleming Crim, Dr. Carl Hayden, and Dr. Stephen Penn. Dr. Russ Tonkyn developed multi-collision kinetics measurements for M^+ + alkane reactions and provided important new ideas used to interpret the molecular beam results. We have all benefitted from the insights of Dr. David Ritter and Mr. Joel Harrington. Generous support of this work has come from the National Science Foundation Chemistry Division, from the ACS Petroleum Research Foundation, and from the University of Wisconsin Chemistry Department and Graduate School. We would be lost without the outstanding technical support of the Chemistry Department Machine Shop and Electronics Shop.

References

1. R. D. Levine and R. B. Bernstein, *Molecular Reaction Dynamics and Chemical Reactivity*, Oxford University, New York, 1987.
2. J. Tully, In *Dynamics of Molecular Collisions Part B*, W. H. Miller, ed., Plenum, New York, 1976.
3. J. Allison, R. B. Freas, and D. P. Ridge, *J. Am. Chem. Soc.* **101**, 1332 (1979); R. B. Freas and D. P. Ridge, *J. Am. Chem Soc.* **102**, 7129 (1980).
4. For a recent example of C–H bond activation in ethylene in solution phase, see M. H. Chisholm and M. Hampden-Smith, *J. Am Chem. Soc.* **109**, 5871 (1987).
5. See, for example, R. L. Whetton, D. M. Cox, D. J. Trevor, and A. Kaldor, *J. Phys. Chem.* **89**, 566; E. K. Parks, G. C. Nieman, L. G. Pobo, and S. J. Riley, *J. Chem. Phys.* **88**, 6260 (1988).
6. C. E. Moore, NBS Circ. No. 467 (U.S. Dept. of Commerce, Washington, D. C., 1949, 1952); C. Corliss and J. Sugar, *J. Phys. Chem. Ref. Data* **14**, (Supp. 2), 407 (1985).
7. C. W. Bauschlicher and S. P. Walch, *J Chem. Phys.* **76**, 4650 (1982).
8. P. B. Armentrout in *Gas Phase Inorganic Chemistry*, D. H. Russell, ed., Plenum, New york, 1989, pp. 1–42; S. W. Buckner and B. S. Freiser, *ibid.*, pp. 279–322; D. P. Ridge and W. K. Meckstroth, *ibid.*, pp. 93–113; R. R. Squires and K. R. Lane, *ibid.*, pp. 43–91; D. H. Russell, D. A. Fredeen, and R. E. Teckleburg, *ibid.*, pp. 115–135; M. F. Jarrold, *ibid.*, pp. 137–192; MacMillan and M. L. Gross, *ibid.*, pp. 369–401; P. B. Armentrout and J. L. Beauchamp, *Acc. Chem. Res.* **22**, 315 (1989); J. Allison in *Progress in Inorganic Chemistry*, S. J. Lippard, ed., Wiley-Interscience, New York, 1966.
9. J. L. Elkind and P. B. Armentrout, *J. Phys. Chem.* **91**, 2037 (1987); P. B. Armentrout, *Annu. Rev. Phys. Chem.* **41**, 313 (1990).
10. N. Aristov and P. B. Armentrout, *J. Am. Chem. Soc.* **108**, 1806 (1986); N. Aristov, Ph.D. Thesis, Univ. of California-Berkeley, Dept. of Chemistry (1988).
11. R. H. Schultz and P. B. Armentrout, *J. Phys. Chem.* **91**, 4433 (1987); R. H. Schultz, J. L. Elkind, and P. B. Armentrout, *J. Am. Chem. Soc.* **110**, 411 (1988).
12. E. R. Fisher, R. H. Schultz, and P. B. Armentrout, *J Phys. Chem.* **93**, 7382 (1989); P. B. Armentrout and R. Georgiadis, *Polyhedron* **7**, 1573 (1988); R. Georgiadis, E. R. Fisher, and P. B. Armentrout, *J. Am. Chem. Soc.* **111**, 4251 (1989): S. K. Loh, E. R. Fisher, L. Lian. R. H. Schultz, and P. B. Armentrout, *J. Phys. Chem.* **93**, 3159 (1989); R. Georgiadis and P. B Armentrout, *J. Phys. Chem.* **92**, 7067 (1988); **92**, 7060 (1988); L. S. Sunderlin and P. B. Armentrout, *J. Phys. Chem.* **92**, 1209 (1988); R. H. Schultz and P. B. Armentrout, *J. Phys. Chem.* **91**, 4433 (1987); J. L. Elkind and P. B. Armentrout, *J. Phys. Chem.* **90**, 5736 (1986).
13. W. D. Reents, F. Strobel, R. G. Freas, J. Wronka, and D. P. Ridge, *J. Phys. Chem.* **89**, 5666 (1985); F. Strobel and D. P. Ridge, *J. Phys. Chem.* **93**, 3635 (1989); F. Strobel and D. P. Ridge, *J. Am. Soc. Mass Spec.* (to be published).
14. L. Sanders, A. D. Sappey, and J. C. Weisshaar, *J. Chem. Phys.* **85**, 6952 (1986); L. Sanders, S. D. Hanton, and J. C. Weisshaar, *J. Chem. Phys.* **92**, 3485 (1990).
15. M. D. Morse, J. B. Hopkins, P. R. R. Langridge-Smith, and R. E. Smalley, *J. Chem. Phys.* **79**, 5316 (1983), and references therein.
16. L. Sanders, Ph.D. Thesis, Univ. of Wisconsin–Madison, Dept. of Chemistry (1988).
17. L. Sanders, S. D. Hanton, and J. C. Weisshaar, *J. Phys. Chem.* **91**, 5145 (1987); S. D. Hanton, L. Sanders, and J. C. Weisshaar, *J. Phys. Chem.* **93**, 1963 (1989); L. Sanders, S.D. Hanton, and J. C. Weisshaar, *J. Chem. Phys.* **92**, 3498 (1990).
18. S. D. Hanton, R. J. Noll, and J. C. Weisshaar, *J. Phys. Chem.* **94**, 5655 (1990).

19. G. Leuchs and H. Walther, in *Multiphoton Ionization of Atoms*, S. L. Chin and P. Lambropoulos, eds., Academic, New York, 1984, Chap. 5, and references therein.
20. A. Achiba, K. Sato, K. Shobatake, and K. Kimura, *J. Chem. Phys.* **78**, 5474 (1983).
21. K. Kimura, *Adv. Chem Phys.* **82**, 5329 (1985).
22. S. Edelstein, M. Lambropoulos, J. Duncanson, and R. S. Berry, *Phys. Rev. A.* **9**, 2459 (1974).
23. W. B. Wiley and I. H. McLarin, *Rev. Sci. Instrum.* **26**, 1150 (1955).
24. S. R. Long, J. T. Meek, and J. P. Reilly, *J. Chem. Phys.* **79**, 3206 (1983).
25. S. L. Anderson, L. Goodman, K. Krogh-Jesperson, A. G. Ozkabak, R. N. Zare, and C. Zheng, *J. Chem. Phys.* **82**, 5329 (1985).
26. O. C. Mullins, R. Chien, J. E. Hunter III, J. S. Keller, and R. S. Berry, *Phys, Rev. A* **31**, 321 (1985); O. C. Mullins, R. Chien, J. E. Hunter III, D. K. Jordon, and R. S. Berry, *Phys. Rev. A* **31**, 3059 (1985); J. E. Hunter III, J. S. Keller, and R. S. Berry, *Phys. Rev. A.* **33**, 3138 (1986).
27. G. Gioumousis and D. P. Stevenson, *J. Chem. Phys.* **29**, 294 (1958).
28. We have used the program SIMION to build two-dimensional model electrostatic fields that guided design of the extraction optics. See D. A. Dahl and J. E. Delmore, *The SIMION PC/AT Users Manual, Version 3.0*, EC&G Idaho, Idaho Falls, Idaho, 1987.
29. See, for example, K. H. Krebs, *Fortschritte der Physik* **16**, 419 (1968); B. L. Schram, A. J. H. Boerboom, W. Kleine, and J. Kistemaker, *Physica* **32**, 749 (1966).
30. We use a singular value decomposition algorithm to solve the least squares equations. See W. H. Press, B. P. Flannery, S. A. Teukilsky, and W. T. Vetterling, *Numerical Recipes*, Cambridge, New York, 1987.
31. S. D. Hanton, R. Noll, and J. C. Weisshaar (unpublished).
32. D. B. Jacobson and B. S. Freiser, *J. Am. Chem. Soc.* **105**, 5197 (1983); G. D. Byrd, R. C. Burnier, and B. S. Freiser, *J. Am. Chem. Soc.* **104**, 3565 (1982).
33. L. F. Halle, P. B. Armentrout, and J. L. Beauchamp, *Organometallics* **1**, 963 (1982); see also Refs. 11 and 35.
34. M. A. Tolbert and J. L. Beauchamp, *J. Am. Chem. Soc.* **108**, 7509 (1986).
35. R. Tonkyn and J. C. Weisshaar, *J. Phys. Chem.* **90**, 2305 (1986); R. Tonkyn, M. Ronan, and J. C. Weisshaar, *J. Phys. Chem.* **92**, 92 (1988).
36. L. S. Sunderlin and P. B. Armentrout, *Int. J. Mass Spectrom. Ion Proc.* (to be published).
37. G. D. Byrd, R. C. Burnier, and B. S. Freiser, *J. Am. Chem. Soc.* **104**, 3565 (1982); D. B. Jacobson and B. S. Freiser, *J. Am. Chem. Soc.* **105**, 5197 (1983); P. B. Armentrout and J. L. Beauchamp, *J. Am. Chem. Soc.* **103**, 784 (1981); L. F. Halle, P. B. Armentrout, and J. L. Beauchamp, *Organometallics* **1**, 963 (1982); R. Houriet, L. F. Halle, and J. L. Beauchamp, *Organometallics* **2**, 1818 (1983); M. L. Mandich, L. F. Halle, and J. L. Beauchamp, *J. Am. Chem. Soc.* **106**, 4403 (1984); T. C. Jackson, T. J. Carlin, and B. S. Freiser, *J. Am. Chem. Soc.* **108**, 1120 (1986); M. A. Tolbert and J. L. Beauchamp, *J. Am. Chem. Soc.* **106**, 8117 (1984).
38. P. B. Armentrout and R. Georgiadis, *Polyhedron* **7**, 1573 (1988).
39. P. A. M. van Koppen, J. Brodbelt-Lustig, M. T. Bowers, D. V. Dearden, J. L. Beauchamp, E. R. Fisher, and P. B. Armentrout, *J. Am. Chem. Soc.* (to be published).
40. A. K. Rappe and T. H. Upton, *J. Chem. Phys.* **85**, 4400 (1986).
41. J. Y. Saillard and R. Hoffmann, *J. Am. Chem. Soc.* **106**, 2006 (1984).
42. H. Lefebvre-Brion and R. W. Field, *Perturbations in the Spectra of Diatomic Molecules*, Academic, New York, 1986.
43. See, for example, E. A. Carter and W. A. Goddard III, *J. Phys. Chem.* **92**, 5679 (1988): J. B.

Schilling, W. A. Goddard III, and J. L. Beauchamp, *J. Phys. Chem.* **91**, 5616 (1987); K. L. Kunze and J. F. Harrison, *J. Phys. Chem.* **93**, 2983 (1989); C. W. Bauschlicher, S. R. Langhoff, H. Partridge, and L. A. Barnes, *J. Chem. Phys.* **91**, 2399 (1989).

44. L. D. Landau, *Phys. Z. Sowjetunion* **2**, 46 (1932); C. Zener, *Proc. R. Soc. A* **137**, 696 (1932); E. C. G. Stueckelberg, *Helv. Phys. Acta.* **5**, 369 (1932).
45. M. H. Alexander, in *Gas Phase Chemiluminescence and Chemi-ionization*, A. Fontijn, ed., Elsevier, Amsterdam, 1985.
46. See Ref. 11 for a schematic of the $Fe^+ + C_3H_8$ potential surfaces.
47. N. Furio, M. L. Campbell, and P. J. Dagdigian, *J. Chem. Phys.* **84**, 4332 (1986); P. J. Dagdigian and M. L. Campbell, *Chem. Rev.* **87**, 1 (1987); K. M. Ervin and P. B. Armentrout, *J. Chem. Phys.* **90**, 118 (1989); K. M. Ervin and P. B. Armentrout, *J. Chem. Phys.* **85**, 6380 (1986).
48. K.Müller-Dethlefs, M. Sander, and E. W. Schlag, *Z. Naturforsch.* **39a**, 1089 (1984); W. Habenicht, R. Baumann, K. Müller-Dethlefs, and E. W. Schlag, *Ber Bunseges Phys. Chem.* **92**, 414 (1988).
49. S. K. Loh, D. A. Hales, and P. B. Armentrout, *Chem. Phys. Lett.* **129**, 527 (1986).
50. M. Hanratty, J. L. Beauchamp, A. J. Illies, P. A. M. van Koppen, and M. T. Bowers, *J. Am. Chem. Soc.* **110**, 1 (1988).
51. D. A. Peake, M. L. Gross, and D. P. Ridge, *J. Am. Chem. Soc.* **106**, 4307 (1984); B. S. Larson and D. P. Ridge, *J. Am. Chem. Soc.* **106**, 1912 (1984).
52. A. D. Sappey, G. Eiden, J. E. Harrington, and J. C. Weisshaar, *J. Chem. Phys.* **90**, 1415 (1989); J. E. Harrington and J. C. Weisshaar, *J. Chem. Phys.* **93**, 845 (1990).
53. See, for example, M. W. Crofton, M. F. Jagod, B. D. Rehfuss, and T. Oka, *J. Chem. Phys.* **91**, 5139 (1989); C. S. Gudeman, M. H. Begemann, J. Pfaff, and R. J. Saykally, *Phys. Rev. Lett.* **50**, 727 (1983).

STATE SELECTED CHARGE TRANSFER AND CHEMICAL REACTIONS BY THE TESICO TECHNIQUE

INOSUKE KOYANO

Department of Material Science, Himeji Institute of Technology, Himeji, Japan

KENICHIRO TANAKA

National Laboratory for High Energy Physics, Tsukuba, Japan

CONTENTS

State-Selected and State-to-State Ion–Molecule Reaction Dynamics, Part 1: Experiment, Edited by Cheuk-Yiu Ng and Michael Baer. Advances in Chemical Physics Series, Vol. LXXXII.
ISBN 0-471-53258-4

I. INTRODUCTION

One of the main objectives of chemistry is to understand how chemical reactions occur. Because a chemical reaction is a reorganization process of chemical bonds in the presence of reactants, it would be essential for a full understanding of chemical reactions to "see" the reorganization event itself. In the absence of any technique that enables such an observation, however, the experimental study of chemical reactions or reaction dynamics has traditionally been conducted by "seeing" only reactants and products, that is, by seeing only "before" and "after" of a chemical reaction, regrading the "during" as occurring in a black box. A well developed extreme of this study is the so-called "state-to-state" chemistry, in which a single quantum state (translational, rotational, vibrational, and electronic) of a reactant system is selected and the resulting quantum state of the product system is analyzed[1,2]. Although various recent techniques, such as femtosecond spectroscopy[3] and negative-ion photodetachment spectroscopy[4], have opened a new age of chemical reaction dynamics in which all efforts are directed to "seeing" intermediate states of a chemical reaction directly, the "state-to-state chemistry" still remains a technique that yields most useful and much information for understanding the "during" of a chemical reaction.

Ion–molecule reactions have received increasing attention in recent years from the viewpoint of the state-to-state or state-selected reaction dynamics[5], since it has increasingly been recognized that, for various reasons, they are particularly suited for such studies. First, in ion–molecule reactions, collision energies can easily be changed continuously over a wide range, by simply applying a suitable electric field. Second, the reactant molecular ions can readily be prepared state selectively or with well-defined distribution of states, by utilizing the technique of photoionization, the information on the state distribution in the latter case being provided by photoelectrons. Third, the detection sensitivity of ions is much higher than that of neutrals, enabling one to conduct studies with very low densities of state-selected species.

About 10 years ago, we took advantage of these properties to initiate an extensive study of state-selected ion–molecule reactions utilizing a novel photoionization/coincidence technique named TESICO[6]. Our idea was that, if the product ions of a reaction were measured in coincidence with energy-selected photoelectrons ejected when the reactant ions were produced by photoionization at a fixed wavelength, then the signals obtained would be the direct information on the reaction of ions in a single internal state.

We always utilize threshold electrons as the energy-selected photoelectrons, thus the wavelength alone determining the internal state to be selected. The name TESICO (threshold electron–secondary-ion coincidence) emphasizes that the secondary ions (products of bimolecular reactions) still retain the memory of the birth of the reactant ions, making our technique possible.

This chapter is a brief review of recent results obtained with this technique. The rest of the sections are organized as follows. In Section II, a brief description of the technique and apparatus is given. In Sections III and IV, the relation between the charge transfer and chemical (rearrangement) reactions in the three-atom and four-atom systems, respectively, are discussed. Section V deals with some interesting aspects seen in more complex systems. In Section VI, we show another capability of the technique, that is, the capability of distinguishing two microscopic reaction mechanisms in the reaction of the type $MH^+ + MH \rightarrow MH_2^+ + M$. Possible future developments will be mentioned in the final concluding section.

This area has also been briefly reviewed in recent publications[5,7-9].

II. EXPERIMENTAL TECHNIQUES

A. General Consideration

As briefly mentioned in Section I, the essence of our technique is the use of unique properties of photoionization. As is well known from, for example, HeI photoelectron sectroscopy, the single-photon photoionization of atoms and molecules, in general, produces ions of all internal states that are energetically attainable with the photons used, with ejected photoelectrons carrying away the excess energy. Thus, although the ions produced by photoionization are a mixture of those in various internal states and although there is no way to spatially separate ions in a single internal state from others, the kinetic energy of each photoelectron can tell us in which internal state its counterpart ion was produced. Analysis of kinetic energies of photoelectrons is a well-established technique of photoelectron spectroscopy and usually has sufficient energy resolution to resolve vibrational states of most of the diatomic and triatomic ions, those of some favorable polyatomic ions, spin–orbit states of some atomic and molecular ions, and even the rotational states of some favorable molecular ions.

On the other hand, the signals of the photoelectron and photoion separated from a single molecule have a definite time correlation with each other on the microsecond time scale, that is, the electron and ion are detected with a fixed time interval between them, whereas the signal of a photoelectron originating from one molecule does not have any correlation with the signal of a photoion originating from another molecule, because photoionization events take place with random intervals on the microsecond scale. Thus, the ions in a particular internal state can be selected from the mixture "temporally," although their "spatial" separation is not possible, by utilizing this time correlation, that is, by taking coincidence measurements of ions with energy-selected photoelectrons at a fixed wavelength. This is the basis of the PEPICO (photoelectron–photoion coincidence) technique, which has been used widely in the studies of unimolecular decomposition of molecular ions[9,10] and also in some studies of bimolecular processes, such as charge transfer[11] and collision-induced dissociation[12].

If the product ions of a chemical reaction of the photoions still have some correlation with the photoelectrons, then a similar coincidence measurement of these secondary ions with the photoelectron would also provide the means to select "reactions" of the ions in a particular internal state. What we have shown in the studies reviewed here is that this is indeed the case and that the technique based on this principle works very well and has a wide applicability. In addition, we have found that this coincidence measurement, when used in a beam-chamber arrangement, can also provide information on the microscopic reaction mechanisms in some types of reactions.

B. Apparatus and Procedure

A schematic diagram of our apparatus for this coincidence measurement[13] is shown in Fig. 1. It consists of six major parts: a helium Hopfield continuum light source (*LS*), a 1 m Seya–Namioka vacuum monochromator (*M*), an ionization chamber (*I*), a reaction chamber (*R*), a threshold electron analyzer (*EA*), and a quadrupole mass spectrometer (*Q*), these being assembled together and coupled to a six-stage differential pumping system (*P*1–*P*6).

For selecting various internal states of interest successively, we adopt a threshold-electron method, in which threshold electrons are always detected for use as gating signals, while the wavelength of the incident photon is set each time at the threshold wavelength for each internal state to be selected. In the context of state selection, this is equivalent to the conventional photoelectron spectroscopic method, in which electrons of various kinetic energies are selected successively, while the wavelength of the ionizing photon is fixed at, for example, 58.4 nm. However, the threshold method has several advantages over the conventional method for the present purpose, as will be discussed in the next section. A hemispherical electrostatic electron energy analyzer was originally used[13] for the selection of threshold electrons, as drawn in Fig. 1, but this has since been replaced by a non-line-of-sight steradiancy analyzer[14] of our own design[15], a schematic and a detailed description of which are given in Fig. 2 and Ref. 15, respectively. This latter analyzer is found to give four to five times higher transmission for the threshold electrons than the previous analyzer, although the resolution is

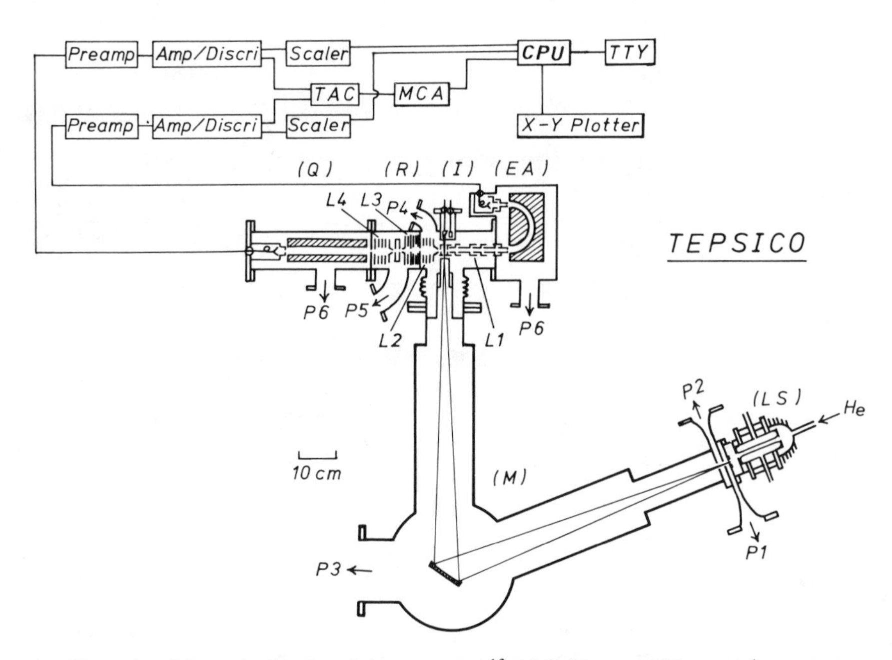

Figure 1. Schematic diagram of the apparatus.[13] *LS*: light source; *M*: monochromator; *I*; ionization chamber; *R*: reaction chamber; *Q*: quadrupole mass spectrometer; *EA*: electron energy analyzer; *L*1–*L*4: electron and ion lens systems; *P*1–*P*6: ports for differential pumping.

somewhat lower. Some reactions with very small cross sections could only be studied with this analyzer[15].

The primary ions A^+ and corresponding photoelectrons are produced in the ionization chamber *I* by photoionization of parent molecules *A* by the monochromatic vacuum ultraviolet radiation from the monochromator *M*. The ions and electrons are repelled out of the chamber in directions perpendicular to the incident photon beam and opposite to each other. Either the hemispherical electrostatic electron energy analyzer or the steradiancy analyzer *EA* separates threshold electrons from those having finite kinetic energies and allows them to pass to the channel multiplier. The ions, on the other hand, are extracted from the ionization chamber by a lens system *L*2, formed into a beam of desired velocity, and focused into the reaction chamber *R* by another lens system *L*3. Reactions take place here with neutral reactant *B*. Product ions C^+, as well as unreacted primary ions, are extracted from the reaction chamber in the same direction as the primary ion beam, are mass analyzed by the quadrupole mass spectrometer *Q*, and are detected by another channel multiplier. These ion signals are then counted in coincidence

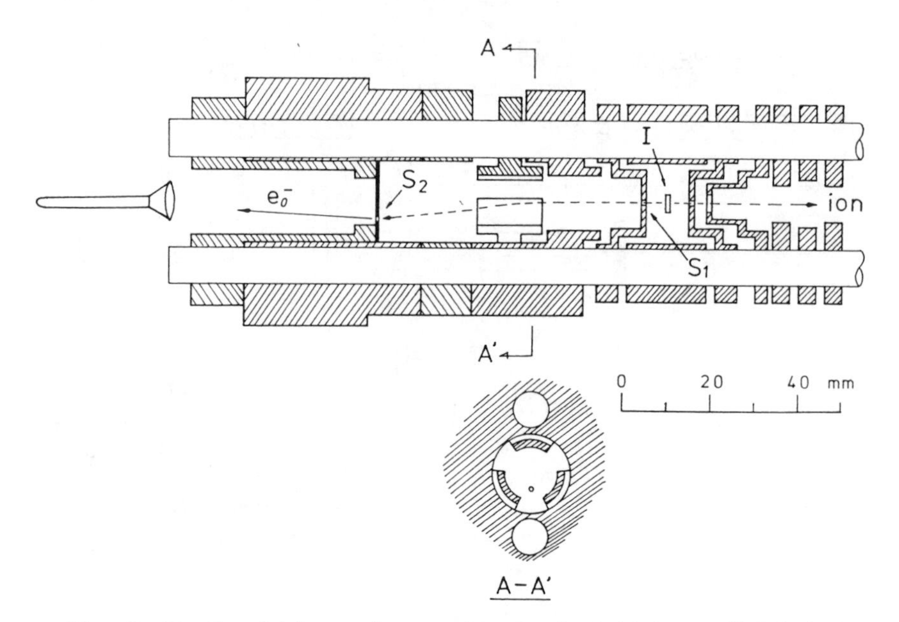

Figure 2. Non-line-of-sight steradiancy analyzer for threshold electrons.[15] I: ionization region; S_1 and S_2: entrance and exit hole, respectively, of the analyzer.

with the threshold electron signals using a standard technique involving a time-to-pulse height converter and a multichannel analyzer.

When the wavelength of the incident photon corresponds to the threshold energy of the production of the ith internal state of the A^+ ion [$A^+(i)$], the ions produced are $A^+(i)$, $A^+(i-1)$,, . . . , $A^+(0)$, among which only $A^+(i)$ gives threshold electrons. Thus, although all of the above ions react with B to produce C^+, we can observe the reactions of $A^+(i)$ selectively by measuring C^+ in coincidence with the threshold electrons.

C. Threshold-Electron Spectra (TES) and the Applicability of theTechnique

As mentioned in the preceding section, the threshold electron method as adopted here has several advantages over the conventional photoelectron spectroscopic method. First, the collecting efficiency for threshold electrons is much higher than that for energetic electrons and, in addition, does not suffer the effect of different angular distributions for different photoelectrons. Second, we do not have to calibrate for the transmission coefficient of the analyzer, which is a function of energy of electrons to be transmitted. Third, in the threshold photoionization, we can often produce molecular ions efficiently in states that are not produced in the fixed-wavelength photoionzation because of the unfavorable Franck–Condon factors.

As is well known, the distribution of internal states of ions in molecular photoionization are governed by the Franck–Condon factors between the ground vibrational state of the neutral molecule and the ionic states, unless the wavelength of the photon coincides with an autoionization state of molecule or the process is affected by some other phenomenon, such as shape resonance. When the wavelength of the ionizing photon coincides with an autoionization state, the autoionizing transitions from that state often populate unfavorable states effectively, because the autoionizing transition is governed by a factor other than the Franck–Condon factors for the direct ionization. Such autoionizing states occur very often near the ionization thresholds, even in diatomic molecules, and are conveniently utilized in the threshold-electron technique with variable-wavelength light source. Some examples will be given below.

In any case, the kinetic energies of photoelectrons at a fixed wavelength give exact information on the internal states of the corresponding ions, so that the threshold-electron spectrum (TES) is an important factor in considering the applicability of the technique. In principle, any internal state of an atom or a molecule resolved in the TES can be studied with our technique. In this sense, the technique is very versatile. Only other prerequisite for a state to be studied with this technique is that the lifetime of that state must be sufficiently long compared with the flight time of the primary ions between the ionization and reaction chambers. This statement is particularly important when we apply the technique to electronically excited states. We find that the electronically excited ions with a lifetime longer than about 10 μs can be studied using the preceding beam-chamber mode of operation. Besides, we can operate the apparatus in a single-chamber mode[13], in which both the source gas A and neutral reactant B are introduced into the ionization chamber, keeping the reaction chamber empty. In this case, the lower limit of the lifetime of the states that can be studied is found to be a few microseconds.

It should also be added that, even with the molecules whose TES does not show any resolved structure, we can study "energy-selected," rather than "state-selected," ion–molecule reactions most unambiguously with this technique. In the following, we present examples of several types of TES to illustrate the variety of the internal states that can be studied with this technique.

Figure 3 shows a TES of N_2, taken with energy resolution of the analyzer of about 14 meV (FWHM) and photon bandwidth of 0.5 Å (FWHM)[16]. The spectrum is typical of TES of diatomic molecules at moderate resolution, in which *vibrational* levels of the ground- and low-lying excited electronic states are generally well resolved but rotational levels are not. In addition, the spectrum illustrates one of the advantages of the threshold technique

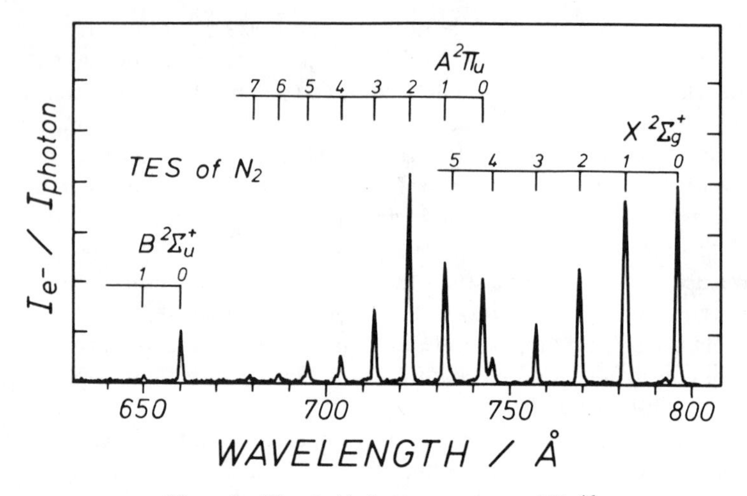

Figure 3. Threshold-electron spectrum of N_2.[16]

mentioned previously. If one looks at a HeI PES of N_2,[17] it is seen that the N_2^+ $X^2\Sigma_g^+$ ions produced at this wavelength are overwhelmingly dominated by $v=0$, with $v=1$ of less than 10% and essentially none of $v=2$, in agreement with the Franck–Condon factors. Thus, if one were to use HeI photoionization, only the reactions of $v=0$ and 1 could be studied as long as the $X^2\Sigma_g^+$ state are concerned. In contrast, considerable intensities are seen for the $v=1$–4 states in Fig. 3. These are evidently produced through autoionization, allowing us the studies of N_2^+ reactions up to $v=4$.[16]

In Fig. 3, the vibrational states of the first electronically excited states $A^2\Pi_u$ are also well resolved and populated with almost the same intensities as those of the ground states. These states have radiative lifetimes somewhat longer than 10 μs,[18] despite the fact that the transtion $A^2\Pi_u \rightarrow X^2\Sigma_g^+$ is optically allowed, and represent the case of *vibronic* states that can be studied with the present technique.

For polyatomic molecules, TES are usually broad and vibrational structures are not well resolved. However, there are several favorable cases in which the vibration of some mode shows a progression with sufficient resolution for the study of state selected reactions. Two examples are given in Figs. 4 and 5. Acetylene (Fig. 4) represents a case in which the progression of only one mode of ionic vibration (ν_2 in this case) appears in the TES, quite resembling the typical diatomic case.[19] Its peaks are fairly sharp and well resolved, indicating the sufficiently long lifetimes of these states.

Ethylene (Fig. 5), on the other hand, shows progressions of two modes (ν_2 and ν_4) of ionic vibration, although the peaks are rather broad.[20] The

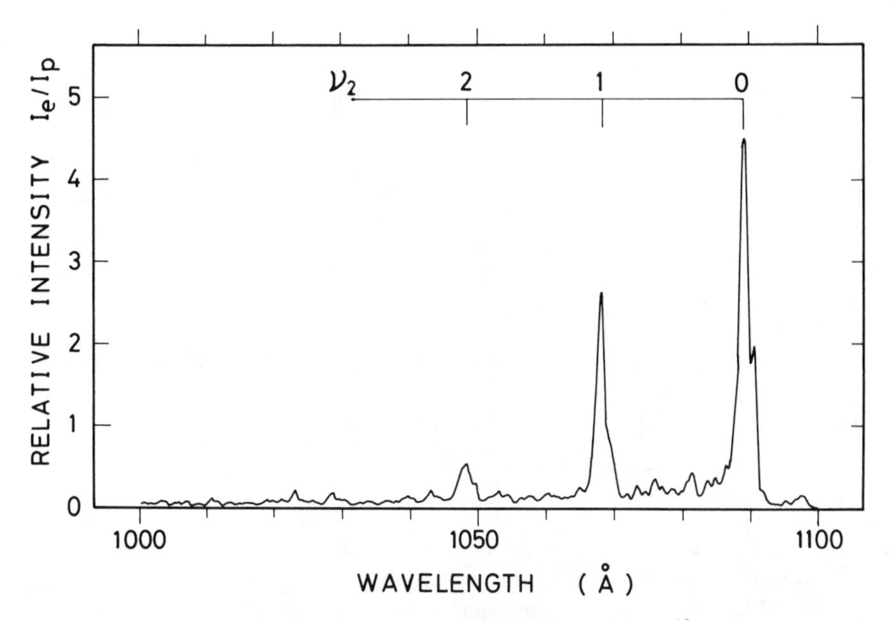

Figure 4. Threshold electron spectrum of acetylene.[19]

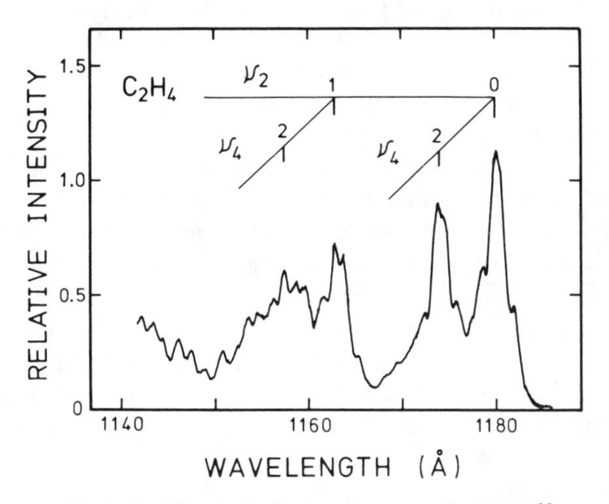

Figure 5. Threshold electron spectrum of ethylene.[20]

four major peaks are assigned to the (0000), (0002), (0100), and (0102) states of the $C_2H_4^+$ ion[21] (the four numbers in each parenthesis represent the quantum numbers of the v_1–v_4 modes, respectively), although there seems to be considerable overlapping of the (0010) and (0012) states on the third

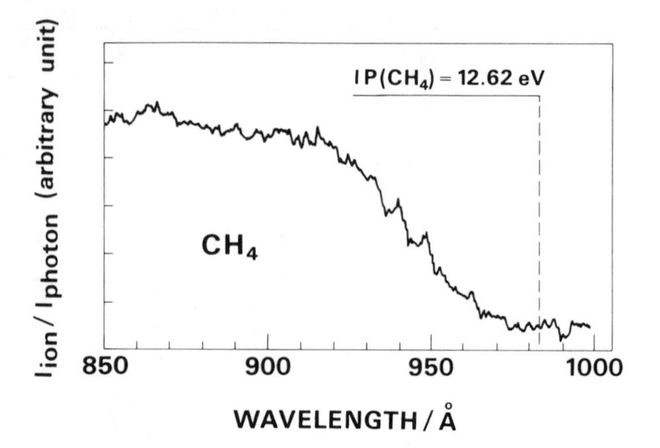

Figure 6. Threshold electron spectrum of methane.[22]

and fourth peaks, respectively.[21] Thus, the reaction system involving ethylene ion provides a unique opportunity to investigate roles of vibrational energies in two different modes, that is, *mode specificity* in bimolecular reactions.

Methane is an example of molecules whose TES does not show any resolved structure, as can be seen from Fig. 6,[22] implying that the internal energy deposited in the molecule is not localized in any particular mode. In such a case, state-selected reactions are not feasible, but instead the reactions of *internal-energy-selected* ions can be studied. Such reactions are also most clearly studied with this technique.

The *spin–orbit states* of some atomic and molecular ions constitute another important class of internal states that can be selected conveniently with the present technique. Reactions of rare-gas ions in each of the two spin–orbit states, $^2P_{3/2}$ and $^2P_{1/2}$, especially those of Ar^+, are examples of reactions that have been studied most extensively with state selection, including those by other techniques. In Fig. 7, we present a TES of Ne, which was taken using the same type of steradiancy analyzer as described previously, and synchrotron radiation,[23] to illustrate the capability of our technique in selecting the two spin–orbit states having as small a mutual separation as 0.09 eV.

III. CHARGE-TRANSFER AND CHEMICAL REACTIONS IN THREE-ATOM SYSTEMS: $(Ar + XY)^+$

Ion–molecule reactions in the $(A + B)^+$ systems in general are characterized by the involvement of at least two closely lying potential energy surfaces (PES), one correlating with the $A^+ + B$ state and the other with the $A + B^+$

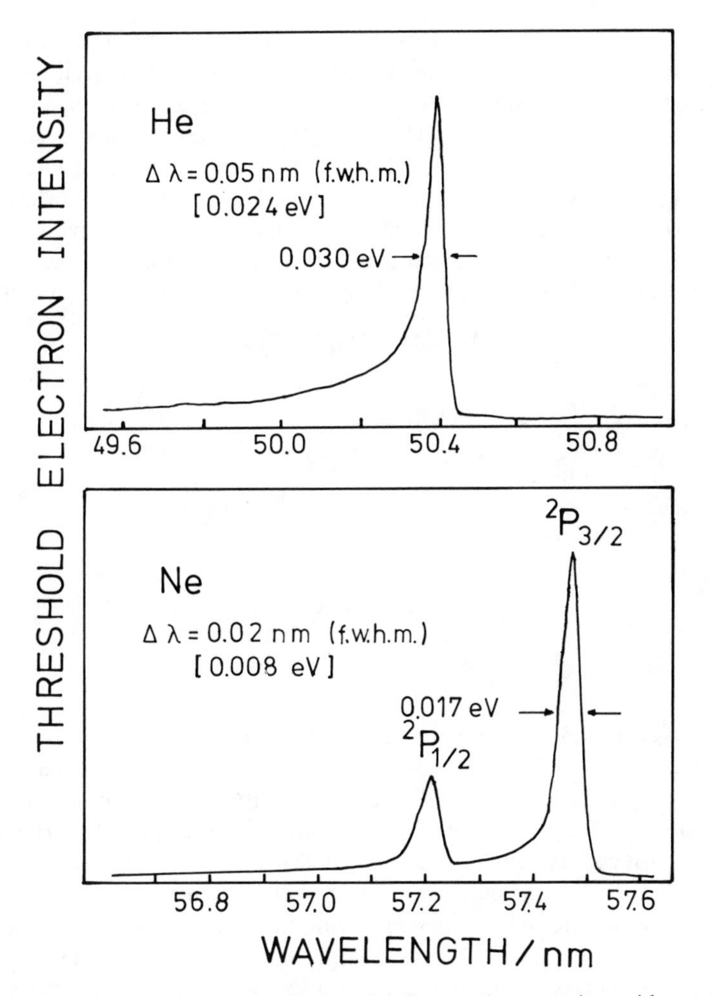

Figure 7. Threshold electron spectrum of helium and neon taken with synchrotron radiation.[23]

state at large intermolecular separation. As the ionization potentials of many molecules lie in a comparatively narrow energy range, the two PES generally begin to interact soon as the reagents start to approach each other, giving an avoided crossing early in the entrance channel. While undergoing nonadiabatic transitions (electron jumps) several times in this region, the reactants continue to approach each other until they reach the region of intimate interaction, where rearrangements of atoms (chemical reactions) take place. Thus, the charge-transfer and chemical reactions are nonseparable processes in ion–molecule systems in which the latter reaction is possible. The importance of studying the two processes together, as well as that of studying the $A^+ + B$ and $A + b^+$ reactions together, is stressed.

The $(\mathrm{Ar} + XY)^+$ system, where XY is a typical diatomic molecule, is most suited for such studies with state selection, because both the spin–orbit states of Ar^+ and the vibrational states of XY^+ are usually well resolved, as mentioned in the previous section, and, in addition, often lie very close to each other. Indeed, this system constitutes a category that has been studied most extensively. Some selected results are given in the following subsections.

A. $(\mathrm{Ar} + \mathrm{H}_2)^+$ System

The reactions

$$\mathrm{Ar}^+(^2P_{3/2}, {}^2P_{1/2}) + \mathrm{H}_2 \rightarrow \mathrm{ArH}^+ + \mathrm{H} \quad (1)$$

$$\rightarrow \mathrm{H}_2{}^+ + \mathrm{Ar} \quad (2)$$

and

$$\mathrm{H}_2{}^+(\mathrm{v}) + \mathrm{Ar} \rightarrow \mathrm{ArH}^+ + \mathrm{H} \quad (3)$$

$$\rightarrow \mathrm{Ar}^+ + \mathrm{H}_2 \quad (4)$$

are early examples that have been studied extensively with this and also other techniques. Absolute cross sections have been determined for the individual quantum states indicated, using the beam-chamber mode of operation.[24,25] Results obtained for reactions (1) and (2) in the 0.05–0.5 eV c.m. collision energy range[24] and for reactions (3) and (4) at a c.m. collision energy of 0.77 eV[25] are shown in Figs. 8 and 9, respectively.

Main features found in these measurements are as follows. The cross sections of reaction (1) are 1.5 times larger for the $J = 1/2$ state than for the $J = 3/2$ state, whereas those of reaction (2) are about 7 times larger for $J = 1/2$ state than for $J = 3/2$ state, regardless of collision energy in the range studied. The cross sections of reaction (4), which is endoergic for $v = 0$ and 1, show a resonance like enhancement at $v = 2$ at all collision energies studied (0.48–19.0 eV c.m.). On the other hand, the cross section of exoergic reaction (3) is found to be almost independent of the vibrational quantum number V or slightly increasing with increasing v at collision energies of 1.24 and 0.77 eV. However, at the still lower collision energy of 0.48 eV, it shows a dip at $v = 2$ in agreement with the peak in the cross section of reaction (4), indicating the competitive nature of the two processes at this collision energy.

Both the much larger $J = 1/2$ cross sections than the corresponding $J = 3/2$ ones in reaction (2) and the occurrence of the $v = 2$ enhancement in

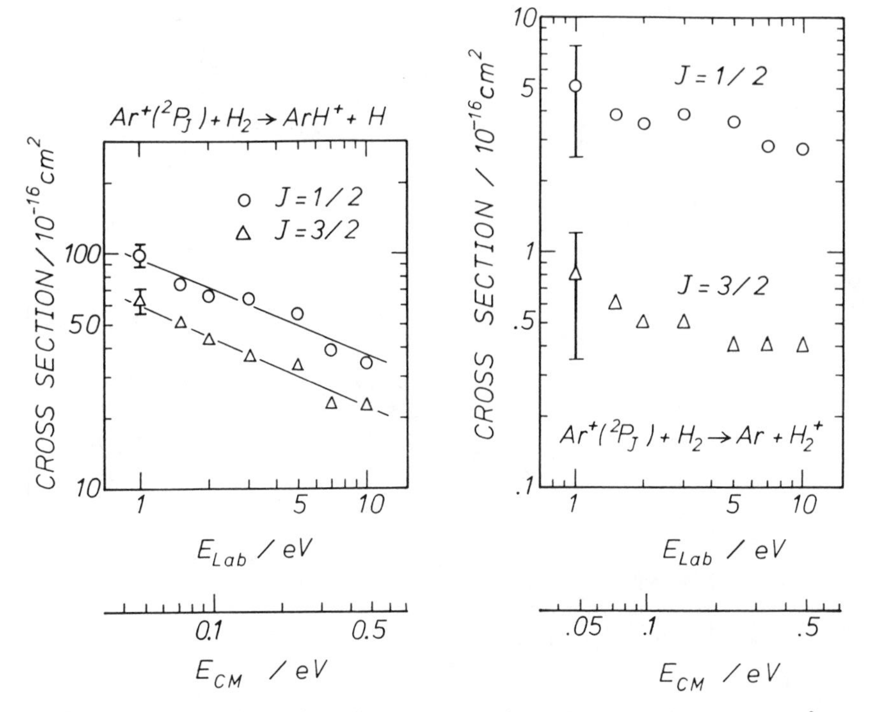

Figure 8. Cross sections of reactions (1) and (2) for each of the spin–orbit states $^2P_{3/2}$ and $^2P_{1/2}$ of Ar^+.[24]

the cross sections of reaction (4) are attributed to the close energy resonance between the $Ar^+(^2P_{1/2}) + H_2(X,v=0)$ and $H_2{}^+(X,v=2) + Ar(^1S_0)$ states (their energies at asymptotic limits differ only by 16 meV). In fact, the recent extensive state-to-state study of the $(Ar + H_2)^+$ system by Ng and co-workers[26,27] have unambiguously shown that the preferential product channels formed in reactions (2) and (4) are $H_2{}^+(X,v=2) + Ar$ and $Ar^+(^2P_{1/2}) + H_2$, respectively, at low collision energies, indicating that the dominant factor governing these processes is the energy resonance between the reactant and product states. However, this resonance effect becomes less important as the collision energy is increased. The same authors,[27] who covered much wider collision energy ranges, indeed report that, at $E_{c.m.}$ greater than 6 eV, reaction (2) is more favored with $J = 3/2$ than with $J = 1/2$. Correspondingly, at $E_{c.m.} \geqslant 5$ eV, the product intensity of $Ar^+(^2P_{3/2})$ is found to be greater than that of $Ar^+(^2P_{1/2})$ in reaction (4).[26]

This system has also been extensively studied theoretically.[28–31] Theory and experiment generally agree in qualitative features, although the agreement is not completely satisfactory quantitatively.

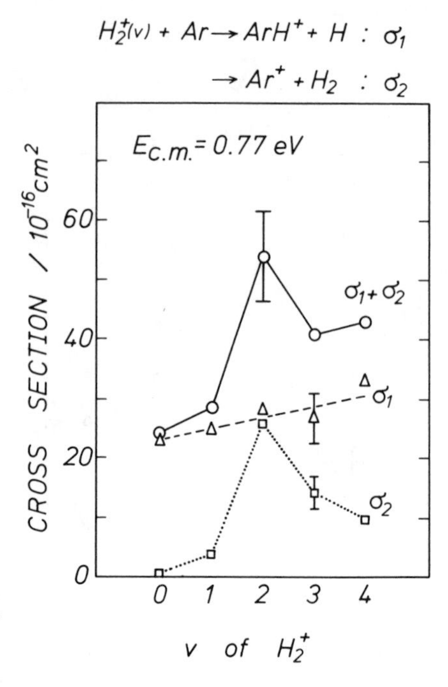

Figure 9. Cross sections of reactions (3) and (4) for the vibrational states $v = 0$–4 of $H_2{}^+$ at $E_{c.m.} = 0.77\,eV$.[25]

B. $(Ar + O_2)^+$ System

Reactions in the $(O_2 + Ar)^+$ system have been studied by selecting high vibrational states, $v = 19$ and 20, of the ground electronic state $O_2{}^+(X^2\Pi_g)$ and $v = 0$–7 of the metastable excited state $O_2{}^+(a^4\Pi_u)$, as well as the two spin–orbit states of Ar^+.[32] This is because the lowest ionization potentials of Ar fall in the energy range of the above $O_2{}^+$ ionic states. The experimental results for the reactions

$$O_2{}^+(X^2\Pi_g, v;\ a^4\Pi_u, v) + Ar \rightarrow Ar^+ + O_2 \tag{5}$$

are shown in Fig. 10. Here, only relative cross sections are obtained and the two curves corresponding to two different collision energies ($E_{c.m.} = 1.4$ and $5.8\,eV$) are normalized at $v = 5$ of the $a^4\Pi_u$ state.

Interesting features seen in Fig. 10 are as follows. The cross sections for reaction (5) are extremely small for the $v = 19$ and $v = 20$ states of $X^2\Pi_g$ despite the fact that reaction (5) is exoergic for $v = 19$ of $X^2\Pi_g$ and states lying above, whereas the cross sections become significant for $v = 0$ of $a^4\Pi_u$, which lies only $0.17\,eV$ above the $v = 20$ state of $X^2\Pi_g$. When the vibrational quantum number is increased in the $a^4\Pi_u$ state, a sharp rise in the cross section is observed at $v = 2$, followed by a gradual decrease except

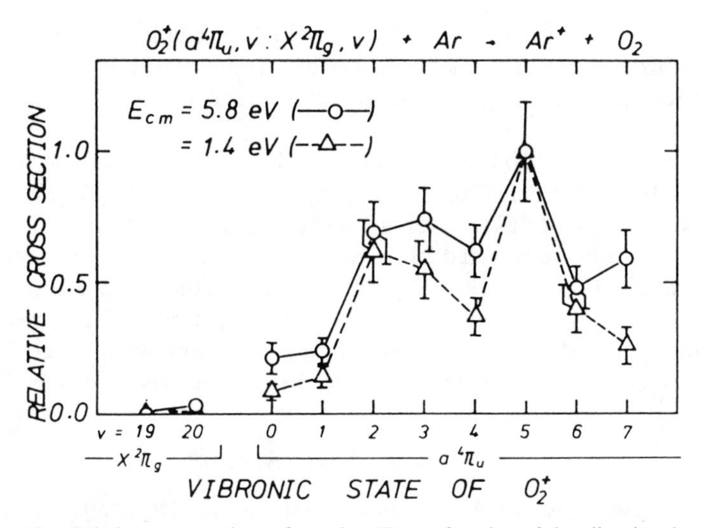

Figure 10. Relative cross sections of reaction (5) as a function of the vibrational quantum number v of $O_2{}^+(X^2\Pi_g)$ and $O_2{}^+(a^4\Pi_u)$ at $E_{c.m.} = 1.4$ and 5.8 eV.[32]

at $v = 5$, where the cross sections show a resonancelike enhancement. This enhancement may also be attributed to the close energy resonance between the $O_2{}^+(a^4\Pi_u, v = 5) + Ar(^1S_0)$ and $Ar^+(^2P_{3/2}) + O_2(X^3\Sigma_g{}^-, v = 5)$.

In contrast to the preceding results for reaction (5), no significant spin–orbit state dependence is observed in the cross section of the reaction

$$Ar^+(^2P_{3/2}, {}^2P_{1/2}) + O_2 \rightarrow O_2{}^+ + Ar. \qquad (6)$$

The experimental results for the ratios of the $J = 3/2$ and $J = 1/2$ cross sections are essentially unity at the two collision energies studied ($E_{c.m.} = 1.4$ and 5.8 eV).[32] These results suggest that the charge transfer in this system does not proceed by a simple electron-jump mechanism as in the $(Ar + H_2)^+$ system, but proceeds via a complex mechanism.

Reaction (6) has recently been studied by Ng and co-workers[33] using their triple-quadrupole–double-octopole photoionization (TQDO) apparatus in the c.m. collision energy range 0.044–133.3 eV. Their results show that the product $O_2{}^+$ ions of reaction (6) is mainly formed in the $a^4\Pi_u$ state in the $E_{c.m.}$ range of 0.4–3.4 eV, supporting the high efficiency of $O_2{}^+(a^4\Pi_u)$ in the reverse reaction (5). They also report that the vibrational distribution of the product $O_2{}^+(a^4\Pi_u)$ between the $O_2{}^+(a^4\Pi_u, v = 0)$ and $O_2{}^+(A^2\Pi_u, v = 0)$ thresholds is similar to that of $O_2{}^+(a^4\Pi_u)$ produced by photoionization of O_2. However, it is seen from Fig. 10 that the relative cross sections of these vibrational states in undergoing the reverse reaction (5) do not follow this distribution: there is a dip in the cross section curve at $v = 4$. This dip occurs at all collision energies studied and is also reproduced by a model calculation of the reaction cross section.[32] Evidently some dynamical effect is reflected here.

Reaction (6) has also been studied using the guided-ion-beam technique by Scherbarth and Gerlich,[34] covering the $E_{c.m.}$ range from 0.04 to 3 eV. All these studies indicate that this reaction in these energy ranges is really complicated, showing various unexpected phenomena, such as preferential excitation of high rotational states of the $O_2^+(a^4\Pi_u)$ products at the expense of vibrational excitation. It has also been shown that the reaction changes mechanism drastically according to the collision energy, for example, between 0.05 and 0.5 eV.[34] The long-range electron jump mechanism seems to prevail above 14 eV.

C. $(Ar + NO)^+$ System

In the reaction $NO^+ + Ar$, the internal states selected are the vibrational states of the electronically excited metastable states $NO^+(a^3\Sigma^+)$ and $NO^+(b^3\Pi)$ since, again, these states fall in the energy range of the first ionization potentials of Ar. Namely, we studied the reactions

$$NO^+(a^3\Sigma^+, v;\ b^3\Pi, v) + Ar \rightarrow Ar^+ + NO \text{ (Ref. 35)} \tag{7}$$

as well as the reaction

$$Ar^+(^2P_{3/2};\ ^2P_{1/2}) + NO \rightarrow NO^+ + Ar \text{ (Ref. 32).} \tag{8}$$

Experimental results for reaction (7) obtained at $E_{c.m.} = 1.4$ eV is shown in Fig. 11. Reaction (7) is slightly (by 0.09 eV) endoergic for $v = 0$ of the $a^3\Sigma^+$ state and exoergic for $v = 1$ and above.

Here, again, a pronounced resonancelike enhancement of the cross section is observed at $v = 2$ of the $a^3\Sigma^+$ state. Except for this point, the cross section shows a tendency to increase slightly but steadily with increasing v in the $a^3\Sigma^+$ state. It is notable that a cross section of substantial magnitude occurs for $v = 0$ in spite of the fact that reaction (7) is endoergic for this state. Undoubtedly, the reaction is caused by the collision energy of 1.4 eV. It is also noteworthy that the reaction cross section sharply (by a factor of 3) goes down in going from $v = 5$ of the $a^3\Sigma^+$ state to $v = 0$ of the $b^3\Pi$ state, despite the larger overall internal energy of the latter state. The magnitude of the cross section of the latter state is almost the same as (or even smaller than, in the case of 5.8 eV of collision energy[35]) that of the $v = 0$ of the $a^3\Sigma^+$ state, whose reaction with Ar is endoergic. This may indicate the importance of the vibrational energy and the inefficiency of the electronic energy in

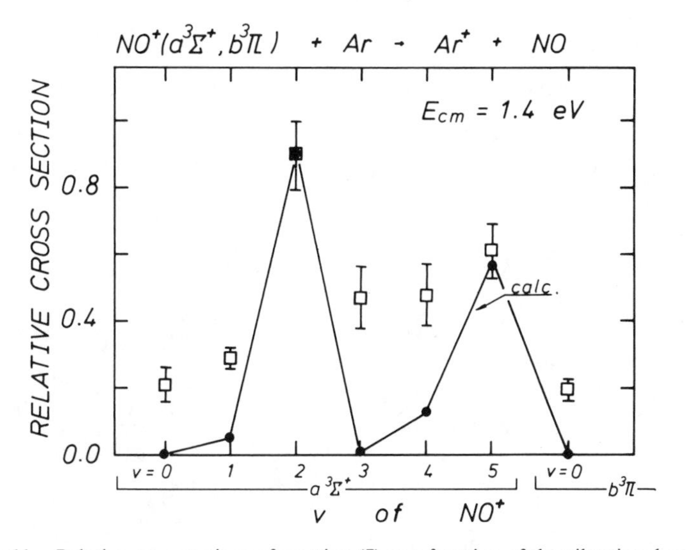

Figure 11. Relative cross sections of reaction (7) as a function of the vibrational quantum number v of $NO^+(a^3\Sigma^+)$ and $NO^+(b^3\Pi)$ at $E_{c.m.} = 1.4\,eV$.[35]

promoting the reaction. The latter inefficiency is undoubtedly related to the symmetry of the electronic states or the molecular orbitals involved in the reactions of these states.[35]

In contrast to reaction (7), which shows a pronounced enhancement of the cross section at a particular vibrational state, reaction (8) shows only very weak dependence of the cross section on the spin–orbit states of Ar^+.[32] The ratios of the $J = 3/2$ and $J = 1/2$ cross sections are essentially unity at all collision energies studied (0.2, 1.4, and 5.8 eV). These results, together with the results on reactions (4) and (6),[32] are shown in Fig. 12.

D. $(Ar + N_2)^+$ System

This system has been the subject of various types of experimental and theoretical studies. Concerning the state selection, we first studied the reactions

$$Ar^+(^2P_{3/2}, {}^2P_{1/2}) + N_2 \rightarrow N_2{}^+ + Ar \text{ (Ref. 36)} \tag{9}$$

and

$$N_2{}^+(X^2\Sigma_g{}^+, v) + Ar \rightarrow Ar^+ + N_2 \text{ (Ref. 16)} \tag{10}$$

by selecting the two spin–orbit states $J = 3/2$ and $J = 1/2$ in reaction (9) and $v = 0$–3 in reaction (10). Later, the studies of reaction (10) were extended to

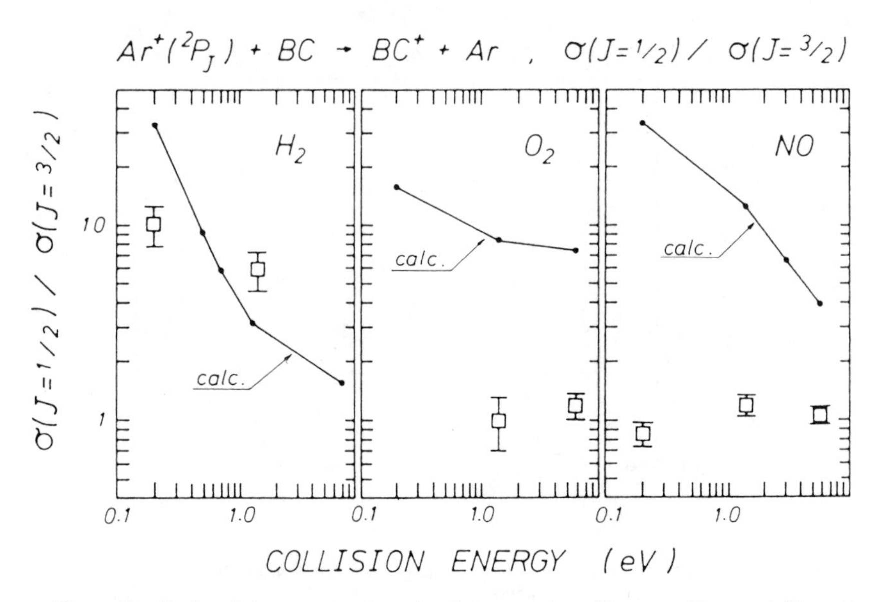

Figure 12. Ratio of the cross sections for the two spin–orbit states $^2P_{3/2}$ and $^2P_{1/2}$ of Ar^+ for reactions (2), (6), and (8), as a function of $E_{c.m.}$.[32]

the vibronic states $A^2\Pi_u$, $v = 0$–2 of the $N_2{}^+$ ion.[37] Relative cross sections of reaction (10) obtained at three collision energies are shown in Fig. 13.[16] It is clearly seen that the reaction, which is endoergic for $v = 0$, is considerably enhanced by adding one vibrational quantum of $N_2{}^+$ at all collision energies, whereas the enhancement by collision energy is much smaller. While one vibrational quantum enhances the cross section substantially, addition of further quanta (up to $v = 3$) still enhances the cross section. This additional enhancement (increment), which is much smaller than that gained in going from $v = 0$ to $v = 1$, is most conspicuous at the highest collision energy studied (11.8 eV c.m.).

State-selective studies of reaction (10) have since been performed by two groups of researchers, one with a direct technique quite similar to ours but using synchrotron radiation,[38] and the other with an indirect technique.[39] Both of these groups determined absolute cross sections and compared with the results of theoretical calculation performed using estimated potential energy surfaces of this system.[40,41] Their results generally agree with the theoretical predictions using the more accurate potential energy surfaces.[41] The latter theory indicates that the cross sections of both relations (9) and (10) depend strongly on the orientation of the reactant ions.

As to reaction (9), on the other hand, the cross section for $Ar^+(^2P_{3/2})$ was found always to be greater than that for $Ar^+(P_{1/2})$ at all collision energies

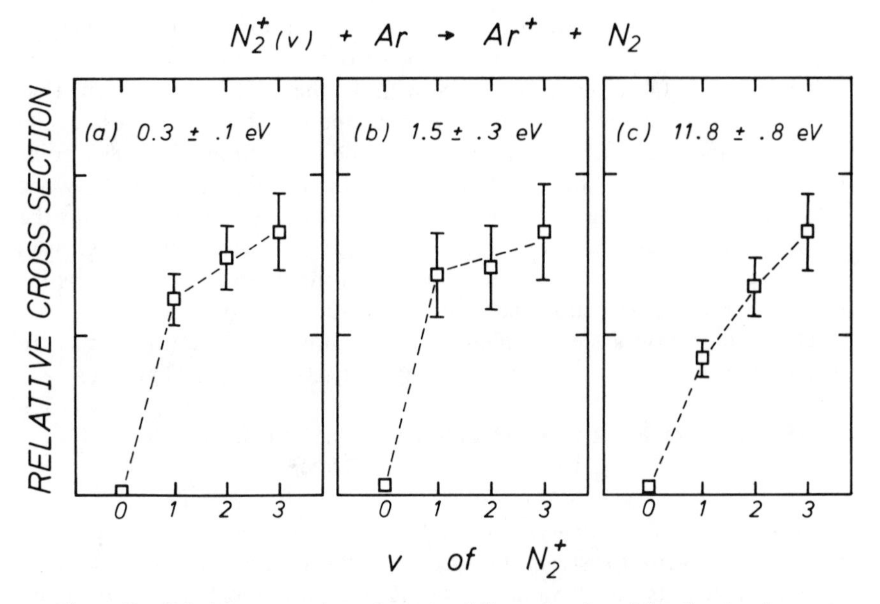

Figure 13. Relative cross sections of reaction (10) as a function of the vibrational quantum number v of N_2^+ $(X^2\Sigma_g^+)$ at three collision energies $E_{c.m.} = 0.3$, 1.5, and 11.8 eV.[16]

studied (0.2–5.8 eV), the ratio $\sigma(1/2)/\ \sigma(3/2)$ slightly changing (0.5 to 0.8) with the collision energy.[36] This result is consistent with the simple model calculation based on the two state impact parameter theory of Rapp and Francis,[42] combined with the Franck–Condon factors between the ground vibrational state of N_2 and the relevant vibrational states of the product N_2^+ ion. This model is believed to apply to the charge-transfer processes that proceed via an electron-jump mechanism at large interreagent distances. However, it is likely that the preceding agreement is rather accidental and reaction (9) involves some intimate interaction mechanism.[36]

Reaction (9) has also been studied by many groups of researchers using various techniques. The state selected study by Liao et al.[43] using an indirect selection technique has shown that the ratio $\sigma(1/2)\ \sigma(3/2)$ varies smoothly as a function of the c.m. collision energy over the wide range of 0.41–164.7 eV, having a broad minimum at around 10 eV. While the results definitely demonstrate the smaller cross sections for $Ar^+(^2P_{1/2})$ than for $Ar^+(^2P_{3/2})$, in agreement with our results, the ratios at $E_{c.m.} = 1.4$ and 5.8 eV are much smaller than our values. Similarly, their ratio at $E_{c.m.} = 8$ eV is smaller than the value of Guyon et al.,[44] which falls between the Liao et al. value and our value at $E_{c.m.} = 5.8$ eV.

The Liao et al. ratios between 10 and 25 eV are in good agreement with the theoretical values[40] but the values at higher and lower collision energies are in considerable disagreement with the theoretical values. While our large value at $E_{c.m.} = 1.4\,eV$ agrees very well with the theory, Liao et al. argue that the large theoretical value at this energy is due to the incorrectness of the estimated potential energy surfaces used in the calculation. They also point out that the purity of the reactant state is liable to be destroyed by the fine-structure changing collisions with the background Ar gas in the ionization chamber. Our larger values of the ratio, as well as that of Govers et al.,[44] could be due to this effect. The surprisingly large cross sections for this fine-structure transitions have also been shown theoretically.[45]

Other important systems which we have studied but are not given here include the $(Ar + CO)^+$,[36,46] $(H_2^+ + He)$,[47] and $(H_2^+ + Ne)$[48] systems.

IV. CHARGE-TRANSFER AND CHEMICAL REACTIONS IN FOUR-ATOM SYSTEMS

Four-atom ion–molecule systems $(AB + CD)^+$ are even more complicated than the three-atom systems discussed previously. For each state-selected pair of reactants, for example, $AB^+(0) + CD(0)$, there exist so many close-lying charge-transferred states $AB(v') + CD^+(v'')$ that interact with the reactant state in different manner, all of which must be taken into account. Several of these states often fall in close resonance with the reactant state at infinite reagent separation. Furthermore, the relative orientation of the two molecules, which is difficult to control experimentally, is more complicated but is more important a factor in the four-atom reactions than the atom–diatom orientation in the three-atom systems.

For these and other reasons, the dynamical studies of four-atom ion–molecule reactions have been meager until recently. As to state-selected or state-to-state studies, only reactions that have been investigated using an indirect[49-51] or direct[13,52,53] selection technique are those of H_2^+. We have applied our TESICO technique to a variety of four-atom ion–molecule systems $(AB + CD)^+$ to study both reactions $AB^+ + CD$ and $CD^+ + AB$ in each case, with the hope to obtain information on the potential energy surface and dynamics of these reactions.[54]

A. $(O_2 + H_2)^+$ System

In this system, both charge transfer and chemical reactions occur. Thus, the reactions studied are

$$O_2^+(X^2\Pi_g, v; a^4\Pi_u, v) + H_2 \rightarrow H_2^+ + O_2 \tag{11}$$

$$\rightarrow HO_2^+ + H \tag{12}$$

$$H_2^+(X^2\Sigma_g^+, v) + O_2 \rightarrow O_2^+ + H_2 \quad (13)$$

$$\rightarrow HO_2^+ + H. \quad (14)$$

The internal states selected for the ground electronic state of O_2^+ are again very high vibrational states $v = 19$ and 20 for the same reasons as in the O_2^+ + Ar reaction. Those for $O_2^+(a^4\Pi_u)$ and $H_2^+(X^2\Sigma_g^+)$ are $v = 0$–9 and $v = 0$–4, respectively. Experimental results for reactions (11) and (12)[54,55] are shown in Fig. 14, while those for reactions (13) and (14)[54] are given in Fig. 15. The center-of-mass collision energy $E_{c.m.}$ is 0.25 and 1.05 eV, respectively, for Figs. 14 and 15. The former energy is much smaller than the endoergicities of these two reactions.

Features seen in Figs. 14 and 15 are summarized as follows. Reaction (12), which is endoergic by 1.75 eV with the O_2^+ ion in the ground electronic state, proceeds at collision energies below the translational threshold if a sufficient vibrational energy is provided. The cross sections are not so large, however, despite the fact that the energy of the $v = 19$ state far (by 1.94 eV) exceeds the energetic threshold of the reaction. Comparison of the $v = 19$ and $v = 20$ cross sections indicates that further addition of 0.16 eV of vibrational energy (corresponding to the energy difference between $v = 19$ and $v = 20$) does not have a significant effect on the reaction cross section. In contrast, an order of magnitude increase in the cross section is observed on going from $v = 20$ of the $X^2\Pi_g$ state to $v = 0$ of the $a^4\Pi_u$ state, in spite of the fact that the additional energy supplied to the system by this change (0.17 eV) is almost the same as that added in going from $v = 19$ to $v = 20$ in the $X^2\Pi_g$ state.

When vibrational energy is added in the $a^4\Pi_u$ electronic state, the cross section shows an interesting variation as a function of the vibrational quantum number. While the cross-section for $v = 1$ is almost the same as that for $v = 0$, it increases sharply at $v = 2$ by a factor of 2 and then decreases less sharply, reaching at $v = 4$ almost the same value as that for $v = 1$. This relatively large cross section of the $a^4\Pi_u$ state persists up to $v = 7$, above which the cross section decreases to less than half this value. The relatively low values for $v = 8$ and 9, however, do not necessarily mean that these states actually have a smaller cross section than the $v = 4$–7 states, since the $v = 8$ and $v = 9$ states of $a^4\Pi_u$ are overlapped by the $v = 0$ and $v = 1$ states, respectively, of the second excited electronic state $A^2\Pi_u$.

The cross-section for reaction (11), which is endoergic by 3.38 eV for the ground state O_2^+, is essentially zero even for $v = 19$ and 20 of that state within the experimental error, but has a significant value for $v = 0$ of the $a^4\Pi_u$ state. Within the latter electronic state, the cross section again shows an interesting variation as a function of vibrational quantum number; the variation follows that of the cross section of reaction (12).

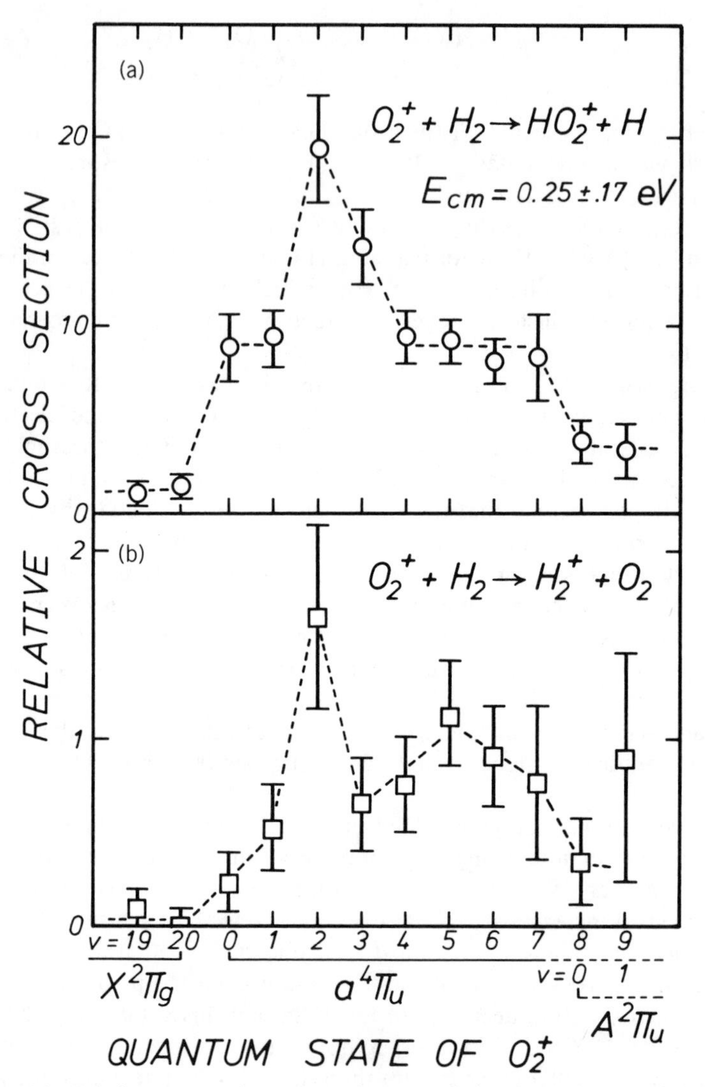

Figure 14. Relative cross sections of reactions (11) (*b*, lower panel) and (12) (*a*, upper panel) as a function of the vibrational quantum number v of $O_2^+(X^2\Pi_g)$ and $O_2^+(a^4\Pi_u)$ at $E_{c.m.} = 0.25$ eV.[54,55]

In sharp contrast to the preceding cases, the vibrational-state dependences of reactions (13) and (14) are monotonous and rather weak, as can be seen from Fig. 15. In particular, it should be noted that the charge-transfer and chemical reactions show vibrational-state dependences of opposite trends to each other, distinct from the relation between reactions (11) and (12).

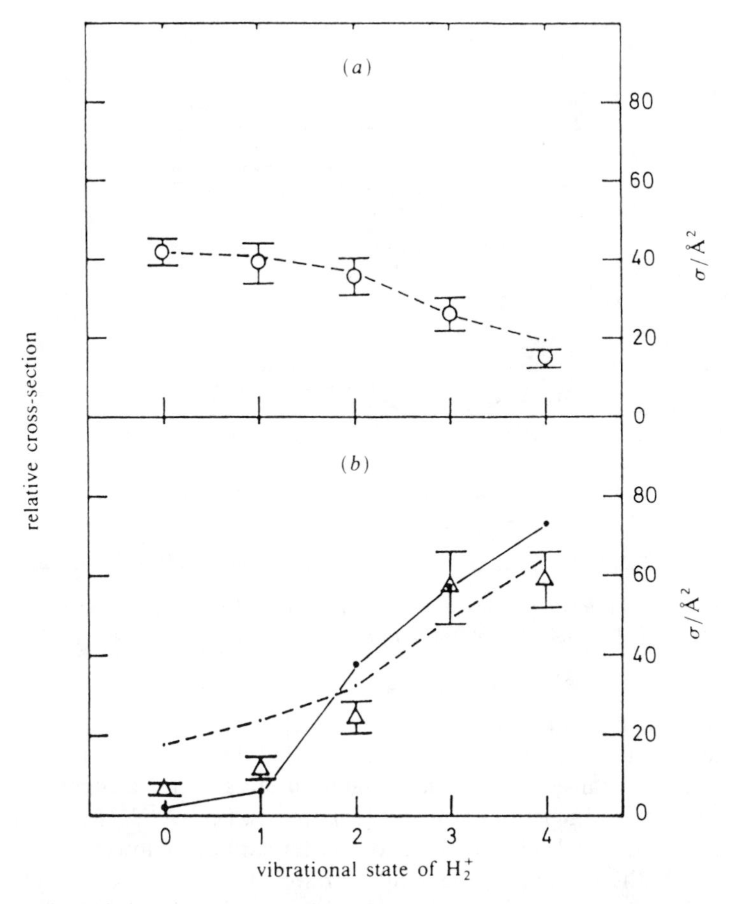

Figure 15. Relative cross sections of reactions (13) (*b*, lower panel) and (14) (*a*, upper panel) as a function of the vibrational quantum number v of $H_2{}^+(X^2\Sigma_g{}^+)$ at $E_{c.m.} = 1.05\,eV$.[54] Broken lines show the absolute cross section (scale on the right-hand ordinate) obtained by Anderson et al.,[50] while the dots connected by solid lines in the lower panel are values obtained by the model calculation. Our $v = 0$ cross section of reaction (14) is normalized to that of Anderson et al. at 10 eV.

The marked contrast of the vibrational-state dependence between the set of reactions (11) and (12) and that of reactions (13) and (14) is interpreted by considering reaction paths connecting known states of reactant and products. A schematic drawing for this[54] is given in Fig. 16. The reactant state $O_2{}^+({}^4\Pi_u) + H_2({}^1\Sigma_g{}^+)$ can only correlate with quartet surfaces, and thus reaction (11) with $O_2{}^+(a^4\Pi_u)$ must proceed via nonadiabatic transitions between two quartet surfaces. The similarity of the internal-state dependence between reactions (11) and (12) strongly suggests that the state of the

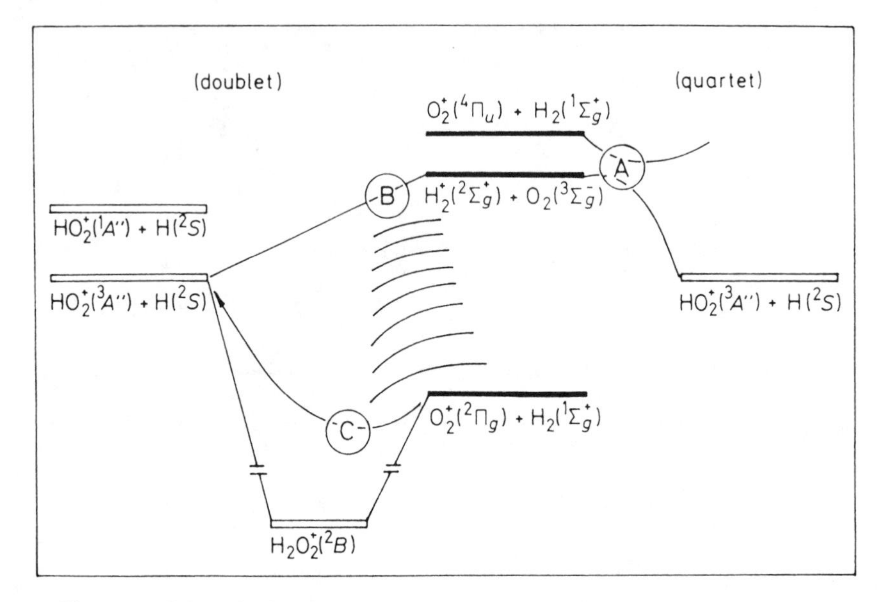

Figure 16. Schematic drawing showing energy levels of the reactant and product states of reactions (11)–(14) and possible reaction paths.[54]

rearrangement products $HO_2^+ + H$ correlate with the same adiabatic surface as the charge-transfer products, that is, that reaction (12) also involves nonadiabatic transitions between quatet surfaces. On the other hand, the reactant states of reactions (13) and (14), that is, $H_2^+(X^2\Sigma_g^+) + O_2(X^3\Sigma_g^-)$, can correlate with both doublet and quartet surfaces, allowing at least two reaction paths leading to the same (ground-state) products $HO_2^+(^3A') + H(^2S)$. This difference, combined with a consideration of the transition probabilities on both sufaces as a function of the vibrational quantum numbers, is found to explain all experimental results consistently.[54,55]

B. $(NO + H_2)^+$ System

In this system the reactions studied are

$$NO^+(a^3\Sigma^+, v;\quad b^3\Pi, v) + H_2 \rightarrow H_2^+ + NO \tag{15}$$

and

$$H_2^+(X^2\Sigma_g^+, v) + NO \rightarrow NO^+ + H_2 \tag{16}$$

with the selection of $v = 0$–5 for $NO^+(a)$, $v = 0$–5 and 1 for $NO^+(b)$, and $v = 0$–5 for $H_2^+(X)$.[54] No detectable rearrangement products, such as

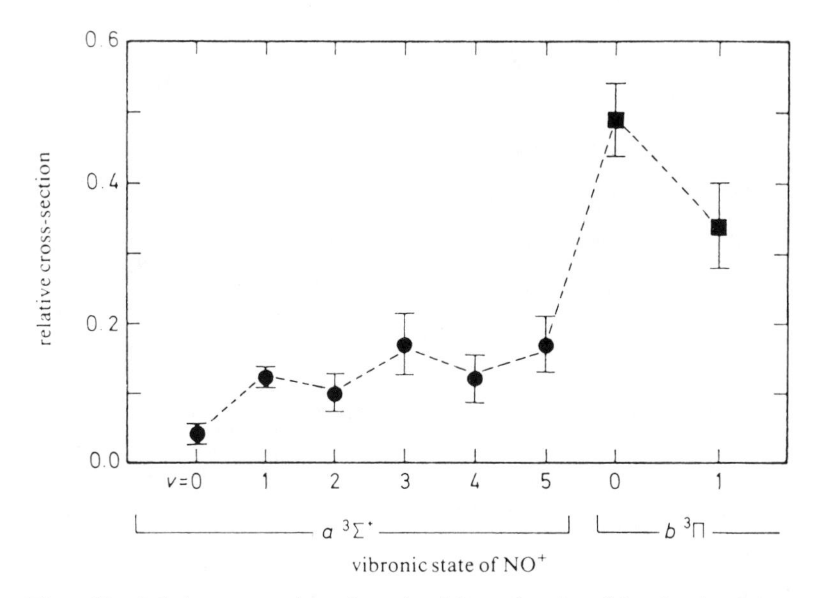

Figure 17. Relative cross sections of reaction (15) as a function of the vibrational quantum number v of $NO^+(a^3\Sigma^+)$ and $NO^+(b^3\Pi)$ at $E_{c.m.} = 0.8\,eV$.[54]

HNO^+, are observed in both combinations of reactants. Experimental results for these two reactions at $E_{c.m.} = 0.8\,eV$ are shown in figs. 17 and 18, respectively.

A salient feature of reaction (15) is the comparatively low reactivity of $NO^+(a^3\Sigma^+)$, even when they have high vibrational energies, and the high reactivity of $NO^+(b^3\Pi)$. This feature contrasts markedly with the results of reaction (7) described in the preceding section, in which the $a^3\Sigma^+$ state shows a higher reactivity than the $b^3\Pi$ state. Another feature is that the reaction with $NO^+(a^3\Sigma^+)$ does not depend greatly on the vibrational state, in spite of the fact that the model calculation[54] predicts cross sections that are heavily dependent on the vibrational states. This also contrasts with the results of reaction (7), which shows a resonancelike enhancement of the cross section at $v = 2$, in agreement with the results of the same model calculation.

The cross-section of reaction (16) also does not show any marked dependence on the vibrational quantum number of $H_2{}^+$. This feature is again contrary to the expectation from the same model calculation as above. These facts, concerning both reactions (15) and (16), may indicate that charge transfer in the $(NO + H_2)^+$ system takes place via an intimate collision, probably by forming an intermediate complex, and that the cross section is determined by the characteristics of the potential energy surfaces in this

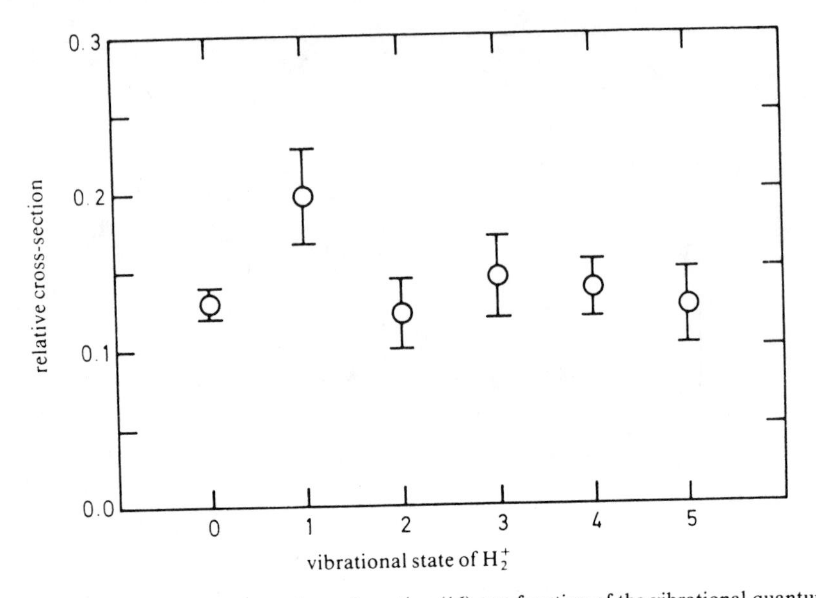

Figure 18. Relative cross sections of reaction (16) as a function of the vibrational quantum number v of $H_2^+(X^2\Sigma_g^+)$ at $E_{c.m.} = 0.8$ eV.[54]

intimate interaction region, rather than by the transition probability in the avoided crossing region at large interreagent distances.

No detailed information is available on the potential energy surfaces of this system, nor on the structure and electronic states of the expected complex molecule H_2NO^+. However, we may assume, by analogy with the isoelectronic molecule H_2CO,[56] that the lowest triplet state of H_2NO^+ is $3A''$. When reactants approach each other in the C_s symmetry, the $NO^+(b^3\Pi) + H_2(X^1\Sigma_g^+)$ pair forms $3A'$ and $3A''$ states, while the $NO^+(a^3\Sigma^+) + H_2(X^1\Sigma_g^+)$ pair forms only $3A'$ state. As the intermolecular distance is decreased, the $3A''$ component of the former reactant pair is considered to be stabilized toward the lowest triplet state of the complex, making an avoided crossing on its way with another $3A''$ state that correlates with the $H_2^+(X^2\Sigma_g^+) + NO(X^2\Pi)$ state. Reaction (15) with $NO^+(b^3\Pi)$ would proceed along the $3A''$ path through this avoided crossing. On the other hand, reaction (15) with $NO^+(a^3\Sigma^+)$ can proceed only along the $3A'$ path. The difference in the reactivity between the $a^3\Sigma^+$ and $b^3\Pi$ states may probably be ascribed to this difference in the potential energy surfaces involved.

C. $(N_2 + H_2, HD, D_2)^+$ Systems

In this system, both charge-transfer and rearrangement reactions (producing N_2H^+/N_2D^+) occur. Experimantal results[20,54] for the reactions

$$N_2^+(X^2\Sigma_g^+, v; \quad A^2\Pi_u, v) + D_2 \rightarrow N_2D^+ + D \tag{17}$$

$$\rightarrow D_2^+ + N_2 \tag{18}$$

at $E_{c.m.} = 1.3\,\text{eV}$ and for the reactions

$$D_2^+(X^2\Sigma_g^+, v) + N_2 \rightarrow N_2D^+ + D \tag{19}$$

$$\rightarrow N_2^+ + D_2 \tag{20}$$

at $E_{c.m.} = 2.5\,\text{eV}$ are shown in Figs. 19 and 20, respectively. Internal states selected for $N_2^+(X)$ and $N_2^+(A)$ are $v = 0$–3 in each case and those selected for $D_2^+(X)$ are $v = 0$–7. The $N_2^+(A^2\Pi_u)$ states have radiative lifetimes ranging 10–17 μs[18] and represent a case of nonmetastable electronically

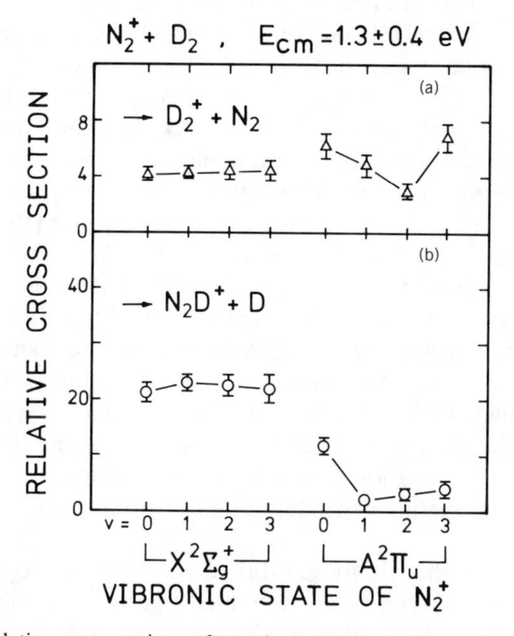

Figure 19. Relative cross sections of reactions (17) (*b*, lower panel) and (18) (*a*, upper panel) as a function of the vibrational quantum number v of $N_2^+(X^2\Sigma_g^+)$ and $N_2^+(A^2\Pi_u)$ at $E_{c.m.} = 1.3\text{eV}$.[20,54]

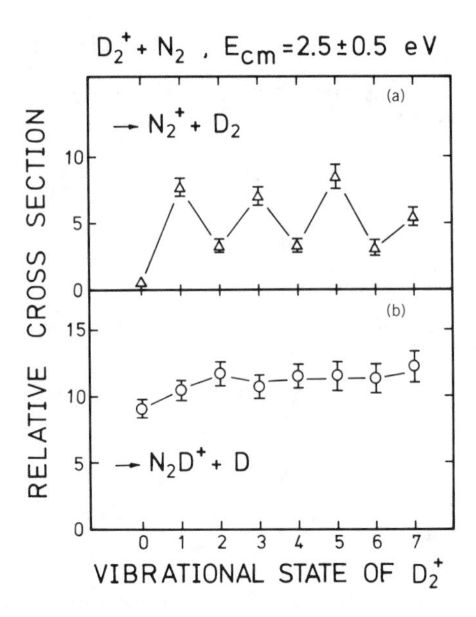

Figure 20. Relative cross sections of reactions (19) (*b*, lower panel) and (20) (*a*, upper panel) as a function of the vibrational quantum number v of D_2^+ $(X^2\Sigma_g^+)$ at $E_{c.m.} = 2.5\,eV$[20,54]

excited state that can be studied using the double-chamber mode of operation, as mentioned in Section II. In determining the relative cross sections of these states, corrections are made for their radiative decay during the flight time between the ionization and reaction chambers, using the lifetimes of the individual vibrational states. The Franck–Condon factors for the $A^2\Pi_u$–$X^2\Sigma_g^+$ bands[57] are used to determine what percentage of the decayed molecules end up in each vibrational state of $N_2^+(X^2\Sigma_g^+)$.

An outstanding feature of this system is that reactions (19) and (20) show quite different vibrational-state dependences, whereas reactions (17) and (18) with $N_2^+(X^2\Sigma_g^+)$ exhibit exactly the same dependence. It is remarkable that the cross section of reaction (20) oscillates as a function of vibrational quantum number, increasing at odd quantum numbers and decreasing at even quantum numbers. In contrast, the cross section of reaction (19) is almost independent of the vibrational quantum number. It is also to be noted that while cross sections for both reactions (17) and (18) are almost independent of the vibrational state in the ground electronic state $X^2\Sigma_g^+$, they exhibit an interesting dependence on the vibrational state in the $A^2\Pi_u$ electronic state.

The similarity in the vibrational-state dependence between reactions (17) and (18) suggests that chemical reaction (17) proceeds along a path whose critical part is the same as that of charge-transfer reaction (18), that is, it proceeds via a nonadiabatic transition to the potential energy surface that asymptotically correlates with $D_2^+ + N_2$. On the other hand, the distinct vibrational-state dependence of the cross sections for reactions (19) and (20) would indicate that chemical reaction (19) proceeds quite differently from charge-transfer reaction (20), that is, it proceeds without undergoing nonadiabatic transition to the surface correlating with $N_2^+ + D_2$. Thus, the reactions from both reactant pairs consistently point to the conclusion that the rearrangement product state $N_2D^+ + D$ correlates with the surface that correlates with the $D_2^+(X^2\Sigma_g^+) + N_2(X^1\Sigma_g^+)$ and not with the surface correlating with the $N_2^+(X^2\Sigma_g^+) + D_2(X^1\Sigma_g^+)$.

Then the preceding results may indicate that the probability of the nonadiabatic transitions between these two surfaces strongly depends on the vibrational motion of D_2^+ but not on that of N_2^+. However, the detailed mechanism for this is not clear at present. We have also studied the same set of reactions using HD and H_2 instead of D_2, to obtain further information on these points.[54,58] Results show a subtle change in the vibrational-state dependence of cross sections, especially in that of reaction (20), in going from D_2 through H_2. For example, the oscillating structure of reaction (20) seen in Fig. 20 disappears in the reactions of HD^+ and H_2^+ owing to the largely reduced $v = 3$ cross sections. Instead, the $v = 1$ cross sections are generally (at all collision energies) more enhanced in the latter reaction systems than in the D_2^+ system. However, the following basic features are the same throughout the three systems: strong dependence of reaction (20) and comparatively weak dependence of reaction (19) on the vibrational quantum number of D_2^+ (HD^+, H_2^+), and independence of reactions (17) and (18) on the vibrational quantum number of $N_2^+(X^2\Sigma_g^+)$. Moreover, the results on reaction (20) are found to be consistently reproduced by the same model calculation as that used several times previously, except some aspects of the D_2^+ reaction. This model considers only energy defects between the reactant state and all energetically accessible product states and the Franck–Condon factors, and accordingly applies only to the cases in which charge transfer takes place at large intermolecular separations. Thus, from the basic agreement between the experimental results and the model calculation, this system is considered to be the one in which electron jump takes place at large intermolecular separations and the characteristics of this nonadiabatic transition determine the gross features of the reactions.

D. $(O_2 + N_2)^+$ System

In this system, only charge-transfer reactions take place. The reactions studied are

$$O_2^+(X^2\Pi_g, v; \quad a^4\Pi_u, v) + N_2 \rightarrow N_2^+ + O_2 \tag{21}$$

and

$$N_2^+(X^2\Sigma_g^+, v; \quad A^2\Pi_u, v) + O_2 \rightarrow O_2^+ + N_2, \tag{22}$$

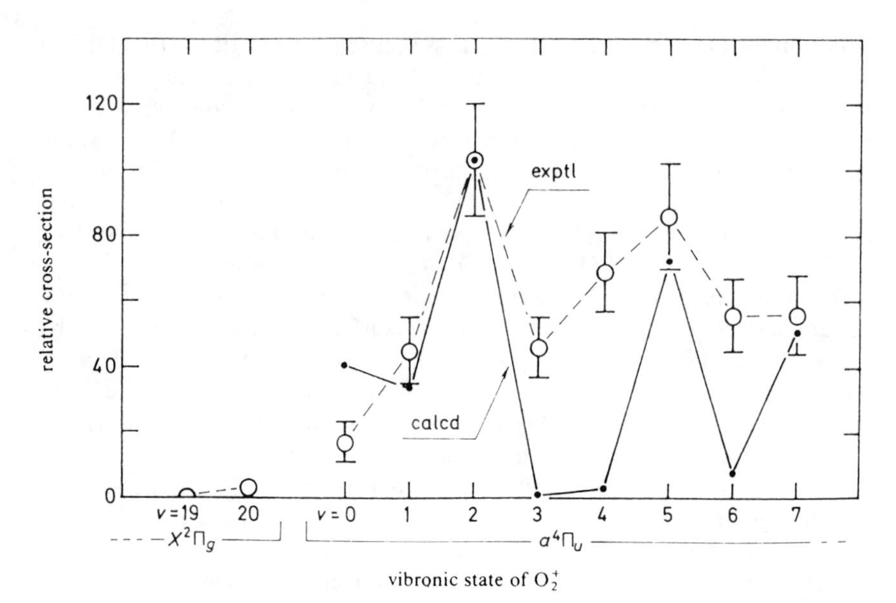

Figure 21. Relative cross sections of reaction (21) as a function of the vibrational quantum number v of $O_2{}^+(X^2\Pi_g)$ and $O_2{}^+(a^4\Pi_u)$ at $E_{c.m.} = 2.1$ eV.[54]

with the selected vibrational states $v = 19$ and 20 for $O_2{}^+(X)$, $v = 0$–7 for $O_2{}^+(a)$, and $v = 0$–3 for both $N_2{}^+(X)$ and $N_2{}^+(A)$. Experimental results[54] are summarized in Figs. 21 and 22 for reactions (21) and (22), respectively.

Figure 21 shows that the charge-transfer cross section of $O_2{}^+$ to N_2 is extremely small for $v = 19$ and 20 of the $X^2\Pi_g$ state, although the reaction is exoergic if the product $N_2{}^+$ ion is produced in the ground electronic state. The small cross section of these high vibrational states of $O_2{}^+$ is common to all reactions studied, reactions (5), (11), (12) and (21). Thus the main reason for this must be the very small Franck–Condon factors between these states and the expected product states of neutral O_2. The cross section is seen to be greately enhanced in the $a^4\Pi_u$ state and, in addition, vary drastically as a function of vibrational quantum number in that state. Examining the energy levels of possible product states, the particularly large enhancement of the $v = 2$ and $v = 5$ cross sections are understood as being due to the near resonance of these reactant states with the product states $N_2{}^+(X^2\Sigma_g{}^+, v = 0) + O_2(X^3\Sigma_g{}^-, v = 4)$ and $N_2{}^+(A^2\Pi_u, v = 0) + O_2(X^3\Sigma_g{}^-, v = 0)$, respectively.

The solid line in Fig. 21 shows the results of the same model calculation as in the preceding section (normalized at $v = 2$). It is seen that, while the calculation reproduces the main features of the experimental results described previously, there exists considerable discrepancy between the experimental

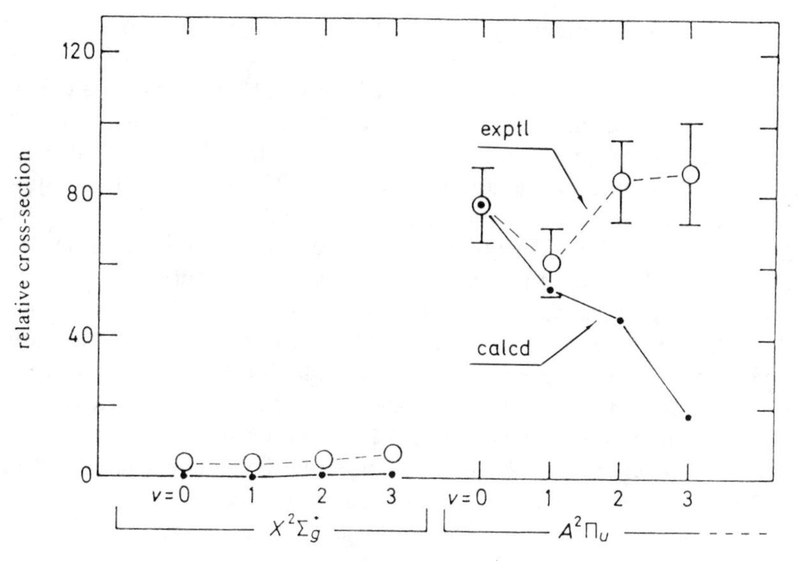

Figure 22. Relative cross sections of reactions (22) as a function of the vibrational quantum number v of $N_2^+(X^2\Sigma_g^+)$ and $N_2^+(A^2\Pi_u)$ at $E_{c.m.} = 2.1$ eV.[54]

and calculated curves in a quantitative sense. A similar situation was met previously[35] in reaction (7) with $NO^+(a^3\Sigma^+)$. This sort of "half" (not "moderate") agreement between experiment and the present model calculation is considered to mean the occurrence of two concurring microscopic reaction mechanisms, one giving the cross sections that are determined by the energy defect and the Franck–Condon factors, and the other giving the cross sections governed by (a) quite different factor(s). These mechanisms might be an electron jump at large intermolecular distances and an electron transfer in an intimate collision state, respectively. However, we would need further detailed experiments as well as the knowledge of the detailed potential energy surfaces of these systems.

Figure 22 shows that $N_2^+(X^2\Sigma_g^+)$ has very small but nontrivial charge-transfer cross sections against O_2, independent of the vibrational quantum number in the range $v = 0$–3. No noticeable change in the cross section is seen between the $v = 2$ and $v = 3$ states, where energetic threshold for the formation of $O_2^+(a^4\Pi_u)$ product occurs, indicating that the product of the $N_2^+(X^2\Sigma_g^+)$ reaction is not $O_2^+(a^4\Pi_u)$, in spite of the high reactivity of the latter state in reverse reaction (21). When we go to the $A^2\Pi_u$ state, on the other hand, cross section increases by more than an order of magnitude and shows an interesting variation as a function of the vibrational quantum number.

The dots connected by the solid line in Fig. 22 again show calculated values based on the same model as above (normalized at $v = 0$). As can be seen, the calculated results for all vibrational states of $X^2\Sigma_g^+$ (extremely small, actually falling on the abscissa) and those for the $v = 0$ and 1 states of $A^2\Pi_u$ follow the experimental values quite well, but those for the $v = 2$ and $v = 3$ states of $A^2\Pi_u$ deviate from the experimental values considerably. The reason for this discrepancy is not clear immediately, but could be related to the opening of a new product channel involving the formation of $O_2^+(A^2\Pi_u)$, since the energetic threshold for the formation of the latter product lies between the $v = 1$ and $v = 2$ states of $N_2^+(A^2\Pi_u)$. However, those product states are already included in the model calculation as energetically attainable states. Thus, if the opening of the new channel is indeed the reason, the discrepancy would imply that the mechanism of the newly opened channel is of the intimate collision type.

Another system which we have studied is the $CO^+ + D_2$ reaction.[59]

V. MORE COMPLEX SYSTEMS

All chemical reactions discussed so far involve transfer of only a light species, such as proton, hydrogen atom, or hydride ion, whether these are transferred in a direct or a complex mechanism. In this section, we present and discuss some results on more complex systems, in which more drastic rearrangements take place. Such reactions are usually (at thermal energies) believed to proceed via an intermediate complex. Specific effects of internal and collisional energies on such reactions are of great interest.

A. $O_2^+ + CH_4$ Reaction: A Case Involving Deep Rearrangement

This reaction system is of particular interest for several reasons. First, several experimental studies[60-64] revealed that its overall rate constant varies in an interesting manner as a function of temperature or collision energy, over a very wide range. Second, although there are many conceivable exoergic reaction channels for this system, only one reaction

$$O_2^+ + CH_4 \rightarrow CH_3O_2^+ + H \tag{23}$$

actually takes place at low energies and, when the temperature or collision energy is increased, two endoergic channels

$$O_2^+ + CH_4 \rightarrow CH_3^+ + HO_2 \tag{24}$$

$$\rightarrow CH_4^+ + O_2 \tag{25}$$

begin to take place.[15,63] No other reactions occur. There has been considerable controversy concerning the nature of the $CH_3O_2^+$ product ion of reaction (23), but now there is an agreement that it is methylene hydroperoxy cation H_2COOH^+.[64] This structure gives the least exoergicity for reaction (23) among all possible structures of $CH_3O_2^+$, and also makes reaction (23) least exoergic among all conceivable exoergic reaction channels. Third, it has been suggested in some drift or flow-drift experiments[62] that possible vibrational excitation of O_2^+ by collisions with buffer-gas molecules could have additionally enhanced the reaction rate constant of the $O_2^+ + CH_4$ reaction measured with some kinds of buffer gas. However, there have been no direct information on the reactivities of the vibrationally excited O_2^+ ions with CH_4.

Figure 23 shows the vibrational-state-selected ($v = 0$–3) relative reaction cross sections for each of the three reaction channels (23)–(25), determined by the TESICO technique at $E_{c.m.} = 0.27$ eV.[15,20] As can be seen, the results demonstrate, in a direct manner, that the vibrational excitation of the O_2^+ ions up to $v = 3$ indeed enhances the overall reactivity of these ions with methane. At the same time, however, the results indicate that this enhancement for $v = 1$ and 2 is primarily due to the enhancement of the exoergic channel forming $CH_3O_2^+$, and not to the opening of a new channel that becomes accesible for these vibrational states as was assumed in previous studies.[60-63] The enhancement at $v = 3$ is indeed seen to be due to the opening of the endoergic channels, especially charge-transfer channel (25). However, considering the relative populations of the $v = 1$–3 ions expected in the drift or flow-drift tubes, it would be obviously incorrect to assume that the enhancement of the reactivity by possible vibrational excitation in some drift experiments is due to the opening of the endoergic channel(s).

The formation of the methylene hydroperoxy cation $CH_3O_2^+$ from the $O_2^+ + CH_4$ reaction requires a drastic rearrangement of the constituent atoms. In order to interpret this rearrangement and other features of the reaction, especially the behavior of the rate constant at very low temperatures, a dynamical model has been proposed,[15,64] which considers a potential energy surface having two minima along the reaction coordinate. Thus, two forms of intermediate complex might be involved in this reaction, making its dynamical aspects complicated.

B. $C_2H_4^+ + C_2H_4$ Reaction: Mode Specificity in Bimolecular Reaction

As mentioned in Section II, ethylene ion provides a unique opportunity to investigate mode specificity in bimolecular reactions, since the TES of

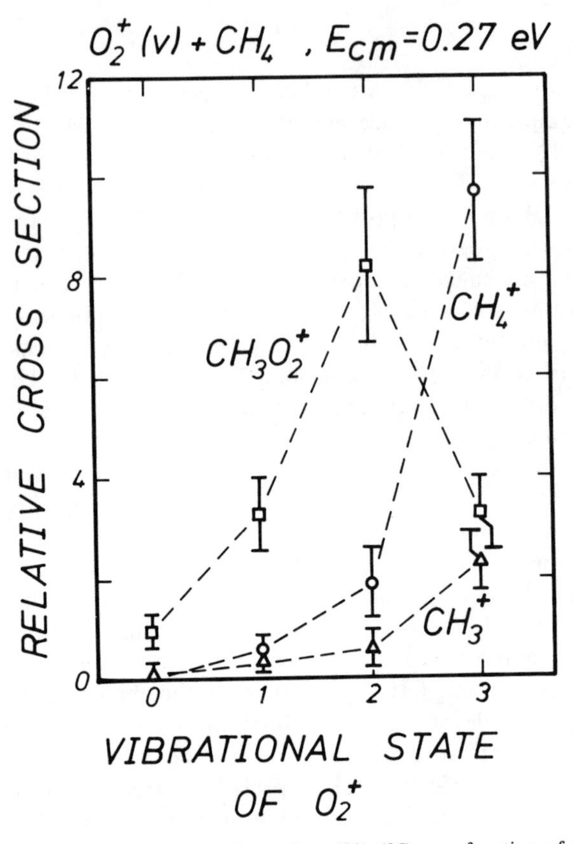

Figure 23. Relative cross sections of reactions (23)–(25) as a function of the vibrational quantum number v of $O_2^+(X^2\Pi_g)$ at $E_{c.m.} = 0.27$ eV.[15,20]

ethylene (Fig. 5) shows progressions of two different modes of vibration, namely, ν_2 (C = C stretching) and ν_4 (twisting). Figure 24 shows the results for the reaction

$$C_2H_4^+(\nu_2, \nu_4) + C_2H_4 \rightarrow C_3H_5^+ + CH_3 \tag{26}$$

in which the vibrational states selected are $(\nu_2, \nu_4) = (0,0)$, (0,2), (1,0), and (1,2).[20] This reaction could only be studied using the single mode of operation because of the very small cross sections. The three sets of data points, each connected by a dotted line, correspond to the average collision energies of 0.1, 0.2, and 0.8 eV, as indicated.

A salient feature of the result in Fig. 24 is the variation of the vibrational-state dependence of the cross section as a function of collision

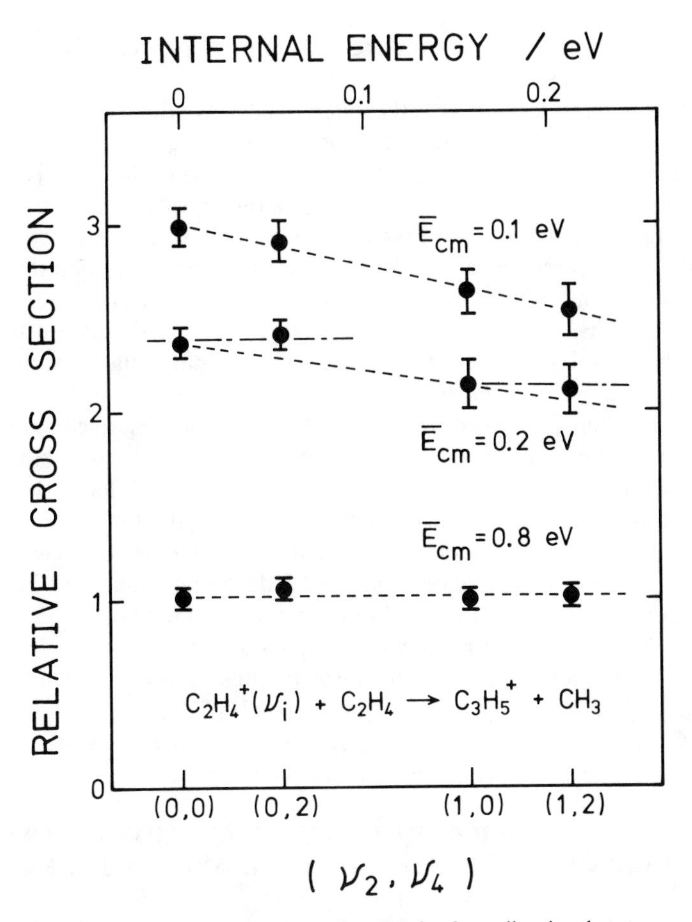

Figure 24. Relative cross sections of reaction (26) for four vibrational states concerning the ν_2 and ν_4 modes of $C_2H_4^+$.[20] The abscissa indicates the total internal energy (sum of the ν_2 and ν_4 vibrational energy) of the four states.

energy. At the lowest collision energy studied (0.1 eV), the cross section decreases linearly with increasing vibrational energy, regardless of whether these energies are carried in by the ν_2 or ν_4 mode. When we go to a somewhat higher collision energy of 0.2 eV, however, it is found that the excitation of the ν_4 vibration to $v = 2$ becomes to have no effect on the reaction cross section [compare between the (0,0) and (0,2) states as well as between the (1,0) and (1,2) states], while the excitation of the ν_2 vibration to $v = 1$ still has an inhibiting effect for the reaction [compare between the (0,0) and (1,0) states as well as between the (0,2) and (1,2) states]. At still higher collision energy of 0.8 eV, the cross section is completely independent of the vibrational energy of both modes.

Of particular interest here is the different effects of the ν_2 and ν_4 vibrational energies on the reaction cross section observed at $E_{c.m.} = 0.2$ eV. Although the deviation of the (0,2) and (1,2) data points from the dashed line [connecting the (0,0) and (1,0) data points] is very small, the repeated measurements with long data-collecting times always gave the same result and the error bar for the (0,2) data point never reached the dashed line. The somewhat longer error bar for the (1,2) data point is ascribed to the poorer resolution of this state in the TES. Thus, the observed mode specificity is believed to be real. Then the experimental results indicate that this particular situation occurs only in a narrow, intermediate collision energy range.

The reaction (26) is known[65] to proceed via a long-lived intermediate complex in the collision energy range covered in the present study. Thus, the observed mode specificity may be related to the different behavior of the ν_2 and ν_4 vibration in this complex. Probably, the time required for the energy migration or randomization in the complex, which determines the lifetime of the complex against backward decomposition, may be different for the energies initially localized in the ν_2 and ν_4 mode of the ionic moiety, thus giving the different probabilities for the forward reaction.

Other complex systems studied, except those given in the next section, include $C_2H_2^+(\nu_2) + D_2$,[19] $C_2D_2^+(\nu_2) + H_2$,[66] $CH_4^+/CD_4^+(E_{int}) + CH_4/CD_4$,[22] and $C_3H_4^+(E_{int}) + C_3H_4$,[67] C_3H_4 in the last system including the three isomers allene, propyne, and cyclopropene.

VI. ANOTHER CAPABILITY: SEPARATION OF TWO MICROSCOPIC REACTION MECHANISMS IN THE REACTION $MH^+ + MH \rightarrow MH_2^+ + M$

Ion–molecule reactions of the type $MH^+ + MH \rightarrow MH^+ + M$, for example,

$$H_2^+ + H_2 \rightarrow H_3^+ + H \tag{27}$$

and

$$CH_4^+ + CH_4 \rightarrow CH_5^+ + CH_3, \tag{28}$$

could in principle have two distinct microscopic reaction mechanisms, namely, proton transfer from ion to neutral and hydrogen atom transfer (or abstraction) from neutral to ion. These mechanisms can be either of direct type or that involving intermediate-complex formation. These two mechanisms are usually discriminated by using deuterated compounds either in ionic or neutral reactant. Such experiments, however, suffer possible kinetic isotope effect, which often prevents the correct determination of the important quantities, such as mechanism branching ratios, relevant to the unlabeled systems. The only quantitative information available on the mechanism (branching ratio) of the unlabeled system is that from the beam-scattering experiments of Herman *et al.*[68] on reaction (28). Qualitative or semiquantitative information on reaction (28) is also available from simple discrimination experiments[69] and pulsed tandem mass-spectrometer studies.[70]

Our TESICO technique is found to be able to discriminate these two microscopic reaction mechanisms proceeding in a direct-type process,[71–73] in addition to its capability in the state selection. This is possible by utilizing the time-of-flight (TOF) difference that exists between the product ions produced by these two mechanisms. Because our technique uses a beam-chamber arrangement, in which the primary ions MH^+ are injected into the reaction chamber with a velocity larger than the thermal velocity, the secondary ions $MH_2{}^+$ produced by the H atom abstraction from the neutral reactant MH, which is in thermal motion, have a greater initial forward velocity than that of the $MH_2{}^+$ ions produced by the proton transfer to the netural. This difference in the initial forward velocity makes difference in the subsequent TOF of the product ions, allowing our TOF coincidence technique to discriminate these two mechanisms. Since the TOF measurements are gated by the energy-selected photoelectrons, the most detailed information obtained are state- or energy-selected cross sections for the individual microscopic mechanisms.

The use of threshold electrons for defining the time of ion formation in a TOF experiment was pioneered by Baer and co-workers.[11] With a single component gas in the ionization/acceleration region of their PIPECO apparatus,[74] they recognized that the product ions of the symmetric-charge-transfer reactions of Xe^+, NO^+,[73] $O_2{}^+$,[75] and $NH_3{}^+$,[76] are distinguishable from the respective primary ions in the TOF spectra, owing to the much smaller (thermal) initial velocities of the product ions compared with those of the unreacted primary ions.

Figure 25 shows an example of the TOF coincidence spectra for the reaction

$$CH_3F^+ + CH_3F \rightarrow CH_4F^+ + CH_2F, \qquad (29)$$

taken with incident photons of 95.48 nm at 3.5 eV c.m. collision energy.[72] It is clearly seen that the spectrum of the mass-selected product ions CH_4F^+ consists of two TOF peaks. The peak position of the faster component almost coincides with that of the primary ions and thus the peak is assigned to the product ions produced by the H atom abstraction mechanism. This is consistent with the estimation of the TOF using the classical mechanics with

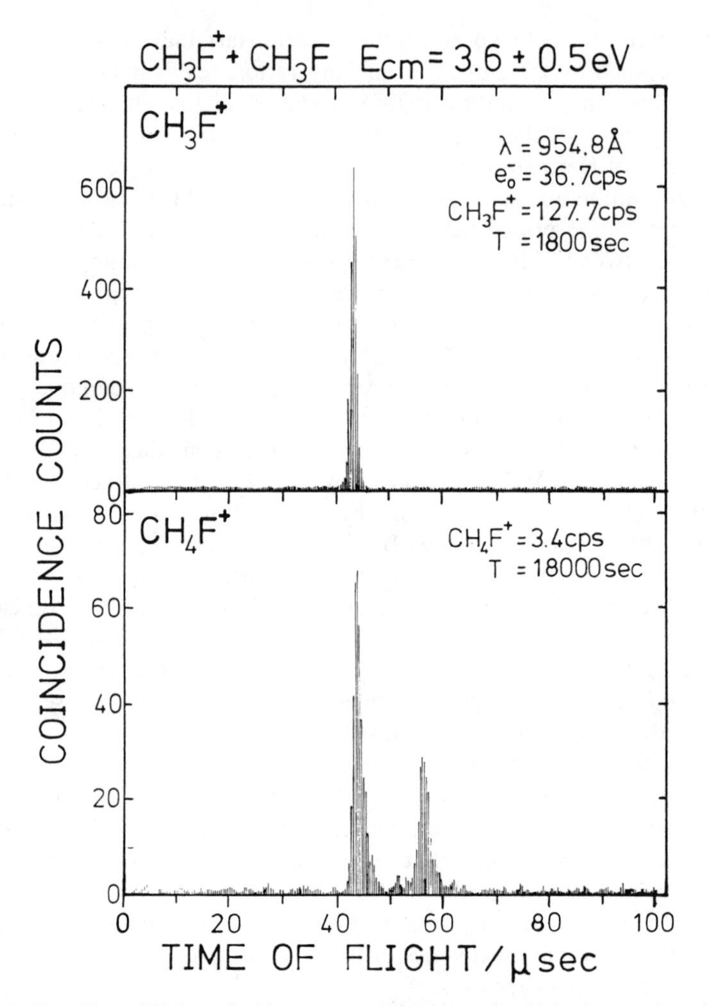

Figure 25. Time-of-flight coincidence spectra for the primary and secondary ions of reaction (29) at $E_{c.m.} = 3.6\,eV$.[72] The figure shows raw data without correcting for the secondary ions produced in the ionization chamber (see text).

approximate field gradients along the ion flight path.[72] The slower peak is then assigned to the product ions produced by the proton-transfer mechanism. The separation between the two peaks as a function of collision energy (primary ion velocity) is also consistent with the preceding estimation.[72]

Thus, the relative reaction cross sections for the individual reaction mechanisms are determined from the integrated intensities of these two peaks. (To do this, we need a correction for the contamination of the faster peak by the product ions produced in the ionization chamber. Although we use as low a pressure in the ionization chamber as is practically possible, the reactions occurring there are not completely negligible when product signals are accumulated over the long data collection time.) In this way, Suzuki[72] could determine the relative cross sections of the two microscopic mechanisms for reactions (28) and (29), as well as the reaction

$$CH_3Cl^+ + CH_3Cl \rightarrow CH_4Cl^+ + CH_2Cl,^{70,71} \quad (30)$$

as a function of the internal and collisional energies. The results are summarized in Fig. 26 in the form of the ratio of the proton-transfer cross section (σ_{PT}) to the hydrogen-abstraction cross section (σ_{HA}) as a function of the total available energy (sum of the internal and collisional energies).

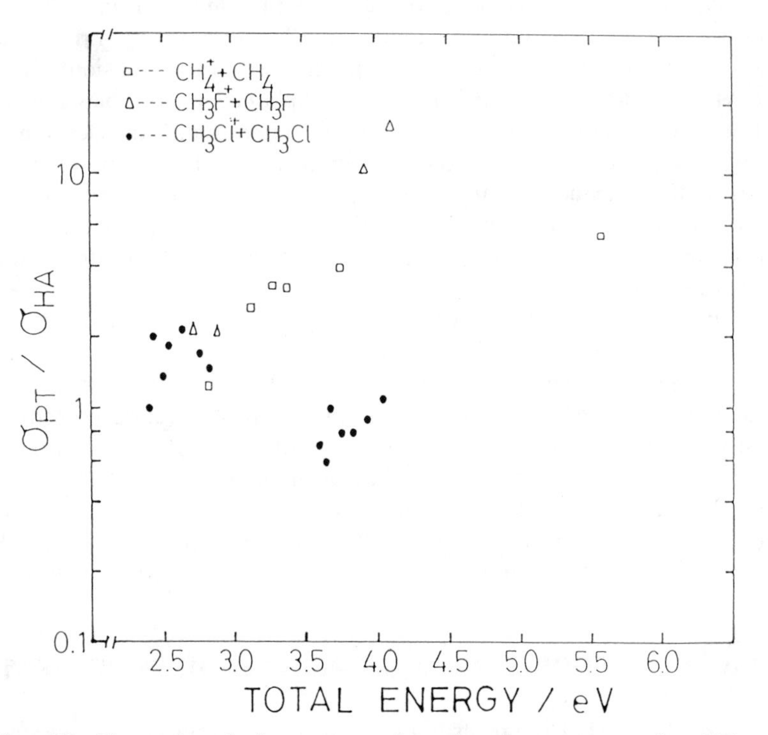

Figure 26. The ratio of the cross sections for the proton-transfer (PT) and hydrogen-abstraction (HA) mechanisms for reactions (28)–(30) as a function of the total energy (sum of the internal energy of the reactant ion and the center-of-mass collision energy).[72]

From Fig. 26, it is found that the reactions are divided into two types according to the behavior of this ratio: Type I in which the ratio is almost unity and independent of the total energy (CH_3Cl system) and type II in which the ratio is generally larger than unity and increases with increasing total energy (CH_4 and CH_3F systems). Besides these, the merging-beam[77] and crossed-beam[78] studies have revealed that reaction (27) also belongs to type I. We have also studied the reaction $D_2^+ + D_2 \rightarrow D_3^+ + D$, which seems to belong to type II[79]. These behavior of the ratios, especially the differences between the CH_3Cl and CH_3F systems and between the H_2 and D_2 systems, are of particular interest.

In order to interpret these experimental results and other facts concerning these reactions consistently, Suzuki[72] proposed a model for the reaction $MH^+ + MH \rightarrow MH_2^+ + M$ in general. In this model, the nuclear entity actually transferred is considered always to be proton, although both proton transfer and hydrogen abstraction are observed phenomenologically. Namely, the phenomenological hydrogen abstraction is considered to be the proton transfer that occurs following the electron jump at large intermolecular separation in the entrance channel of the collision. Then the two types of behavior of the σ_{PT}/σ_{HA} ratio is explained in terms of the efficiency of this electron jump at large intermolecular separation: If the electron jump is very efficient and the transition between the two surfaces takes place many times in the entrance channel, then $\sigma_{PT}/\sigma_{HA} = 1$ and the type I behavior results. If the electron jump is inefficient and the net probability of the transition is smaller than 1/2, the ratio σ_{PT}/σ_{HA} is larger than unity and the type II behavior results.

A similar model was also used to interpret the experimental results on the reactions $NH_3^+(\nu_2) + ND_3$ and $ND_3^+(\nu_2) + NH_3$, where the proton transfer and hydrogen atom abstraction, as well as the charge transfer, were distinguished utilizing the different isotopic composition for the products of these processes.[80] It has been found that the interesting variation of the ratio σ_{PT}/σ_{HA} as a function of the ν_2 vibrational quantum number (shown in Fig. 27) is consistently explained by considering the electron-jump probabilities at large intermolecular distances (approximated by the products of the Franck–Condon factors) between the reactant and each of all possible product surfaces.[80]

VII. CONCLUDING REMARKS—FUTURE DEVELOPMENTS

We have shown that the measurements of product ions (secondary ions) of bimolecular ion–molecule reactions in coincidence with the threshold electrons ejected when the reactant ions are produced by photoionization of the parent molecule provide a powerful technique for the study of

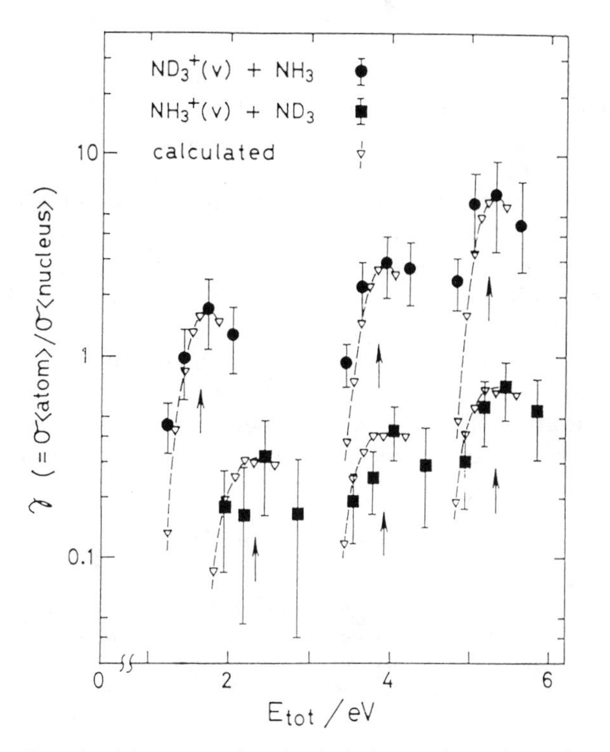

Figure 27. The ratio of the cross sections for the hydrogen-abstraction and proton-transfer mechanisms of the reactions $NH_3^+ + ND_3$ and $ND_3^+ + NH_3$ as a function of the total energy (sum of the internal energy of the reactant ion and the center-of-mass collision energy).[80] The three groups of data points for each reaction correspond to $E_{c.m.} = 1.5$, 3.1 and 4.5 eV in the case of the $NH_3^+ + ND_3$ reaction and to $E_{c.m.} = 0.9$, 3.1, and 4.5 eV in the case of the $ND_3^+ + NH_3$ reaction. The four data points in each group correspond to $v(\nu_2) = 4$, 6, 8, and 11 in the case of NH_3^+ and $v(\nu_2) = 4$, 6, 9, and 12 in the case of ND_3^+.

state-selected ion–molecule reaction dynamics. A variety of internal states that can be selected with this technique have been demonstrated. Obviously missing from our study is the selection of rotational states, the effect of which on reaction dynamics is least understood. In principle, the present technique is capable of selecting rotational states as well, but practically would require have much higher resolution of the threshold electron analyzer. However, it should also be noted that very few rotationally resolved single-photon PES or TES have been reported, even under the "high-resolution" conditions. The application of the present technique to rotational-state selection will largely be limited to a few favorable systems, such as H_2^+.

Even so, "higher resolution" would be in the direction of future developments. With higher resolution of both incident photons and threshold electron analyzer, it might be possible to develop several interesting new areas, such as intracluster ion–molecule reactions with size selection. If the size-dependent ionization potentials of a molecule in clusters are distinguished, the effect of various sizes of third body in ion–molecule reactions could be studied.

Extention of our technique to the reactions of ions of unstable species would be another future problem. Extention to the study of state-selected neutral reactions is also possible by utilizing photodetached electrons from negative ions. In the latter case, however, the detection of reaction products poses a severe problem. All these future developments certainly await a breakthrough in highly sensitive detection technique as well as in high-intensity, highly monochromatic photon sources. In this context, the use of synchrotron radiation and lasers (both single-photon and multiphoton ionization) will become increasingly popular, as already has been initiated.[81–85]

Acknowledgments

The work reviewed here was performed while the authors were at the Institute for Molecular Science. The authors would like to thank their many collaborators who joined us and contributed greatly in various phases of this continuing project: Tatsuhisa Kato, Shinzo Suzuki, Takashi Imamura, Kenji Honma, Jean Durup, Paul-Marie Guyon, Zdenek Herman, Teruhiko Ogata, and Shinji Tomoda. In particular, the contributions of T. K. to the materials given in Sections III and IV and of S. S. to the materials given in section VI were enormous.

References

1. P. R. Brooks and E. F. Hayes (eds.), *State-to-state Chemistry*, American Chemical Society, Washington, DC, 1977.
2. R. D. Levine and R. B. Bernstein, *Molecular Reaction Dynamics and Chemical Reactivity*, Oxford University, New York, 1987.
3. A. H. Zewail, *Science* **242**, 1645 (1988); A. H. Zewail and R. B. Bernstein, *Chem. Eng. News* **66**, 24 (1988).
4. R. B. Metz, T. Kitsopoulos, A. Weaver, and D. M. Neumark, *J. Chem. Phys.* **88**, 1463 (1988): A. Weaver, R. B. Metz, S. E. Bradforth, and D. M. Neumark, *J. Phys. Chem.* **92**, 5558 (1988); R. B. Metz, A Weaver, S. E. Bradforth, T. N. Kitsopoulos, and D. M. Neumark, *ibid.* **94**, 1377 (1990);S. E. Bradforth, A. Weaver, D. W. Arnold, R. B. Metz, and D. M. Neumark *J. Chem. Phys.* **92**, 7205 (1990); A. Weaver, R. B. Metz, S. E. Bradforth, and D. M. Neumark, *ibid.* **93**, 5352 (1990).
5. C. Y. Ng, in *Techniques for the Study of Ion-Molecule Reactions, Techniques of Chemistry*, J. M. Farrar and W. H. Saunders, Jr. (eds.), Wiley, New York, 1988, Vol. XX, p. 430.
6. K. Tanaka and I. Koyano, *J. Chem. Phys.* **69**, 3422 (1978).
7. S. R. Leone, *Ann. Rev. Phy. Chem.* **35**, 109 (1984).
8. J. H. Futrell, in *Gaseous Ion Chemistry and Mass Spectrometry*, J. H. Futrell, ed., Wiley, New York, 1986, p. 207.
9. J. H. D Eland, *Photoelectron Spectroscopy, Butterworths*, London, 1984, p. 221.
10. T. Baer, in *Gas Phase Ion Chemistry*, M. T. Bowers, ed., Academic, New York, 1979, Vol. 1, p. 153.

11. L. Squires and T. Baer, *J. Chem. Phys.* **65**, 4001 (1976).
12. T. Baer, L. Squires, and A. S. Werner, *Chem. Phys.* **6,** 325 (1974).
13. I. Koyano and K. Tanaka, *J. Chem. Phys.* **72**, 4858 (1980).
14. W. B. Peatman, G. B. Kasting, and D. J. Wilson, *J. Electron Spectrosc.* **7**, 233 (1975).
15. K. Tanaka, T. Kato, and I. Koyano, *J. Chem. Phys.* **84**, 750 (1986).
16. T. Kato, K. Tanaka, and I. Koyano, *J. Chem. Phys.* **77**, 834 (1982).
17. D. W. Turner, C. Baker, A. D. Baker, and C. R. Brundle, *Molecular Photoelectron Spectroscopy*, Wiley, London, 1970, p. 46: K. Kimura, S. Katsumata, Y. Achiba, T. Yamazaki, and S. Iwata, *Handbook of HeI Photoelectron Spectra of Fundamental Organic Moelcules,* Japan Scientific Societies Press, Tokyo, 1981, p. 25.
18. R. F. Holland and W. B. Maier II, *J. Chem. Phys.* **56**, 5229 (1972).
19. K. Honma, T. Kato, K. Tanaka, and I. Koyano, *J. Chem. Phys.* **81**, 5666 (1984).
20. I. Koyano, K. Tanaka, and T. Kato, in *Electronic and Atomic Collisions*, D. C. Lorents, W. E. Meyerhof, and J. R. Peterson, eds., Elesvier, Amsterdam, 1986, p. 529.
21. J. E. Pollard, D. J. Trevor, J. E. Reutt, Y. T. Lee, and D. A. Shirely, *J. Chem. Phys.* **81**, 5302 (1984).
22. Z. Herman, K. Tanaka, T. Kato, and I. Koyano, *J. Chem. Phys.* **85**, 5705 (1986).
23. S. Suzuki, S. Nagaoka, I. Koyano, K. Tanaka, and T. Kato, *Z. Phys. D* **4**, 111 (1986).
24. K. Tanaka, J. Durup, T. Kato, and I. Koyano, *J. Chem. Phys.* **73**, 586 (1980); **74**, 5561 (1981).
25. K. Tanaka, T. Kato, and I. Koyano, *J. Chem. Phys.* **75**, 4941 (1981).
26. C.-L. Liao, R. Xu, G. D. Flesch, M. Baer, and C. Y. Ng, *J. Chem. Phys.* **93**, 4818 (1990).
27. C.-L. Liao, R. Xu, S. Nourbakhsh, G. D. Flesch, M. Baer, and C. Y. Ng, *J. Chem. Phys.* **93**, 4832 (1990).
28. S. Chapman and R. K. Preston, *J. Chem. Phys.* **60**, 650 (1974); S. Chapman, *J. Chem. Phys.* **82**, 4033 (1985).
29. M. Baer and J. A. Beswick, *Chem. Phys. Lett.* **51**, 360 (1977); *Phys. Rev. A* **19**, 1559 (1979); M. Baer, *Molec. Phys.* **35**, 1637 (1978).
30. M. Baer and H. Nakamura, *J. Chem. Phys.* **87**, 4651 (1987); *J. Phys. Chem.* **91**, 5503 (1987); M. Baer, H. Nakamura, and A. Ohsaki, *Chem. Phys. Lett.* **131**, 468 (1986).
31. M. Baer, C.-L. Liao, R. Xu, G. D. Flesch, S. Nourbakhsh, and C. Y. Ng, *J. Chem. Phys.* **93**, 4845 (1990).
32. T. Kato, *J. Chem. Phys.* **80**, 6105 (1984).
33. G. D. Flesch, S. Nourbakhsh, and C. Y. Ng. *J. Chem. Phys.* **92**, 3590 (1990).
34. S. Scherbarth and D. Gerlich, *J. Chem. Phys.* **90**, 1610 (1989).
35. T. Kato, K. Tanaka, and I. Koyano, *J. Chem. Phys.* **79**, 5969 (1983).
36. T. Kato, K. Tanaka, and I. Koyano, *J. Chem. Phys.* **77**, 337 (1982).
37. S. Suzuki, K. Tanaka, and I. Koyano (unpublished).
38. T. R. Govers, P. M. Guyon, T. Baer, K. Cole, H. Frohlich, and M. Lavoliee, *Chem. Phys.* **87**, 373 (1984).
39. C.-L. Liao, R. Xu, and C. Y. Ng, *J. Chem. Phys.* **85**, 7136 (1986); J.-D. Shao, Y.-G. Li, G. D. Flesch, and C. Y. Ng, *J. Chem. Phys.* **86**, 170 (1987).
40. M. R. Spalburg and E. A. Gislason, *Chem. Phys.* **94**, 339 (1985).
41. G. Parlant and E. A. Gislason, *Chem. Phys.* **101**, 227 (1986).
42. D. Rapp and W. E. Francis, *J. Chem. Phys.* **37**, 2631 (1962).

43. C.-L. Liao, R. Xu, and C. Y. Ng, *J. Chem. Phys.* **84**, 1948 (1986).
44. P. M. Guyon, T. R. Govers, and T. Baer, *Z. Phys. D* **4**, 89 (1986).
45. G. Parlant and E. A. Gislason, *J. Chem. Phys.* **88**, 1633 (1988).
46. T. Imamura, S. Suzuki, T. Imajo, and I. Koyano (unpublished).
47. M. Baer, S. Suzuki, K. Tanaka, I. Koyano, H. Nakamura, Z. Herman, and D. J. Kouri, *Phys. Rev. A* **34**, 1748 (1986).
48. Z. Herman and I. Koyano, *J. Chem. Soc. Faraday Trans. 2*, **83**, 127 (1987).
49. W. A. Chupka, M. E. Russell, and K. Refaey, *J. Chem. Phys.* **48**, 1518 (1968).
50. S. L. Anderson, F. A. Houle, D. Gerlich, and Y. T. Lee, *J. Chem. Phys.* **75**, 2153 (1981); S. L. Anderson. T. Tunner, B. H. Mahan, and Y. T. Lee, *ibid.* **77**, 1842 (1982); T. Turner, O. Dutuit, and Y. T. Lee, *ibid.* **81**, 3475 (1985).
51. C. L. Liao, C. X. Liao, and C. Y. Ng, Chem. *Phys. Lett.* **103**, 418 (1984); *J. Chem. Phys.* **81**, 5672 (1984); J. D. Shao and C. Y. Ng, *J. Chem. Phys.* **84**, 4317 (1986).
52. F. M. Campbell, R. Browning, and C. J. Latimer, *J. Phys. B* **14**, 3493 (1981).
53. D. van Pijkeren, J. van Eck, and A. Niehaus, *Chem. Phys. Lett.* **96**, 20 (1983); D. van Pijkeren, E. Boltjes, J. van Eck, and A. Niehaus, *Chem. Phys.* **91**, 293 (1984).
54. I. Koyano, K. Tanaka, T. Kato. and S. Suzuki, *Faraday Discuss. Chem. Soc.* **84**, 265 (1987).
55. K. Tanaka, T. Kato, P. M. Guyon, and I. Koyano, *J. Chem. Phys.* **77**, 4441 (1982); **79**, 4302 (1983).
56. D. C. Moule and A. D. Walsh, *Chem. Rev.* **75**, 67 (1975).
57. R. W. Nichols, *J. Res. Natl. Bur. Std. A* **65**, 451 (1961).
58. S. Suzuki, K. Tanaka, T. Kato, and I. Koyano (unpublished).
59. I. Koyano and K. Tanaka, in *Electronic and Atomic Collisions,* N. Oda and K. Takayanagi (eds.), North-Holland, Amsterdam, 1980, p. 547.
60. I. Dotan, F. C. Fehsenfeld, and D. L. Albritton, *J. Chem. Phys.* **68**, 5665 (1978).
61. E. Alge, H. Villinger, and W. Lindinger, *Plasma Chem. Plasma Proccess.* **1**, 65 (1981).
62. M. Durup-Ferguson, H. Böhringer, D. W. Fahey, F. C. Fehsenfeld, and E. E. Ferguson, *J. Chem. Phys.* **81**, 2657 (1984).
63. B. R. Rowe, G. Dupeyrat, J. B. Marquette, D. Smith, N. G. Adams, and E. E. Ferguson, *J. Chem. Phys.* **80**, 241 (1984).
64. J. M. Van Doren, S. E. Barlow, C. H. Depuy, V. M. Bierbaum, I. Dotan, and E. E. Ferguson, *J. Phys. Chem.* **90**, 2772 (1986); S. E. Barlow, J. M. Van Doren, C. H. Depuy, V. M. Bierbaum, I. Dotan, E. E. Ferguson, N. G. Adams, D. Smith, B. R. Rowe, J. B. Marquette, G. Dupeyrat, and M. Durup-Ferguson, *J. Chem. Phys.* **85**, 3851 (1986).
65. Z. Herman, A. Lee. and R. Wolfgang, *J. Chem. Phys.* **51**, 452 (1969).
66. K. Honma, K. Tanaka, and I. Koyano, *J. Chem. Phys.* **86**, 688 (1987).
67. T. Ogata, S. Suzuki, and I. Koyano, in *Chemistry and Spectroscopy of Interstellar Molecules,* S. Saito, N. Kaifu, D. K. Bohme, and E. Herbst (eds.), Univ. of Tokyo Press, Tokyo, 1991.
68. Z. Herman, M. Henchman, and B. Friedrich, *J. Chem. Phys.* **93**, 4916 (1990).
69. A. Henglein, *Advan. Chem. Ser.* **58**, 63 (1966).
70. A. J. Mason, P. F. Fennelly, and M. J. Henchman, *Adv. Mass Spectrom.* **5**, 20, 207 (1971).
71. S. Suzuki and I. Koyano, *Int. J. Mass Spectrom. Ion Process.* **80**, 187 (1987).
72. S. Suzuki and I. Koyano, *Radiochim. Acta* **43**, 115 (1988).
73. S. Suzuki, *J. Chem. Phys.* **93**, 4102 (1990).

74. B. P. Tsai, A. S. Werner, and T. Baer, *J. Chem. Phys.* **63**, 4384 (1975).

75. T. Baer, P. T. Murray, and L. Squires, *J. Chem. Phys.* **68**, 4901 (1978).

76. T. Baer and P. T. Murray, *J. Chem. Phys.* **75**, 4477 (1981).

77. A. B. Lees and P. K. Rol, *J. Chem. Phys.* **61,** 4444 (1974); C. H. Douglas, D. J. McClure, and W. R. Gentry, *ibid.* **67**, 4931 (1977).

78. J. R. Krenos, K. K. Lehman, J. C. Tully, P. M. Hierl, and G. P. Smith, *Chem. Phys.* **16**, 109 (1976).

79. S. Suzuki and I. Koyano (unpublished).

80. S. Tomoda, S. Suzuki, and I. Koyano, *J. Chem. Phys.* **89**, 7268 (1988).

81. S. Suzuki, T. Imamura, I. Koyano, and K. Okuno, *Rev. Sci. Instr.* **60**, 2186 (1989).

82. P. M. Guyon and E. A. Gislason, in *Synchrotron Radiation in Chemistry and Biology III, Topics in Current Chemistry*, E. Mandelkow (ed.), Springer, Berlin, 1989, Vol.151, p. 161.

83. S. T. Pratt, P. M. Dehmer, and J. L. Dehmer, *J. Chem. Phys.* **80**, 1706 (1984).

84. W. E. Conaway, R. J. S. Morrison, and R. N. Zare, *Chem. Phys. Lett.* **113**, 429 (1985); R. J. S. Morrison, W. E. Conaway, and R. N. Zare, *ibid.* **113**, 435 (1985); R. J. S. Morrisaon, W. E. Conaway, T. Ebata, and R. N. Zare, *J. Chem. Phys.* **84**, 5527 (1986); W. E. Conaway, T. Ebata, and R. N. Zare, *ibid.* **87**, 3447, 3453 (1987).

85. T. M. Orlando and S. L. Anderson, *J. Chem. Phys.* **87**, 852 (1987); B. Yang, M. H. Eslami, and S. L. Anderson, *ibid.* **89**, 5527 (1988); T. M. Orlando, B. Yang, and S. L. Anderson, *ibid.* **90**, 1577 (1989); T. M. Orlando, B. Yang, Y.-H. Chiu, and S. L. Anderson, *ibid.* **92**,7356 (1990).

MULTICOINCIDENCE DETECTION IN BEAM STUDIES OF ION-MOLECULE REACTIONS: TECHNIQUE AND APPLICATION TO $X^- + H_2$ REACTIONS

JEAN-CLAUDE BRENOT

LCAM, Université Paris-Sud, Orsay, France

MARIE DURUP-FERGUSON

LPCR, Université Paris-Sud, Orsay, France

CONTENTS

State-Selected and State-to-State Ion–Molecule Reaction Dynamics, Part 1: Experiment, Edited by Cheuk-Yiu Ng and Michael Baer. Advances in Chemical Physics Series, Vol. LXXXII.
ISBN 0-471-53258-4

I. INTRODUCTION

Parallel processing is becoming increasingly important for theoretical studies of physical and chemical processes. The way in which current problems are tackled is significantly modified by the use of vector processors: classical treatments are more global and accurate and new approaches that correspond more closely to the physical structure of these computers are now being introduced.

A similar evolution is occurring in experimental studies: many new experiments are based on parallel-data-acquisition techniques and extensive data processing. This "parallelization" is likely to be of great use for measuring well-defined cross sections in the state-to-state chemistry: the purpose of this chapter is to describe the coincidence method and its "vectorized" version, the multicoincidence method, as one of the new and powerful tools of collision physics.

A. Coincidences and Multicoincidences

A coincidence experiment provides microscopic information about collisional systems for any channel leading to at least two detectable product particles.

Particles that arise from the same single collision are precisely correlated in time: valuable information about this interaction may thus be extracted in the laboratory frame from measurement of the time correlation between impacts on two detectors of products resulting from this interaction.

This basic principle implies three main characteristics of a coincidence set-up:

1. A coinncidence analysis is a time analysis between at least two sets of particles: these particles have to be detected individually by some multiplying devices and the resulting signals are time analyzed.

2. The observed time delay has to be accurately related to the collision process; this condition requires a crossed-beam configuration that ensures a well-defined collision region and thus well-defined flight paths to the detectors. Generally, at least one of the crossed beams is a molecular beam; such a target is usually obtained from a supersonic expansion that reduces the velocity spreads and cools the internal degrees of freedom of these molecular species; such a choice is especially pertinent in the present case since the high directivity of these jets ensures geometrically well-defined collision volumes and convenient target densities.

3. The interactions between the particles of these macroscopically continuous crossed beams are uncontrolled and fortuitous correlations may happen: the coincidence method provides information that can be accounted for in a statistical analysis; a first accumulation step is needed and accidental coincidences are removed by an off-line analysis. In a coincidence experiment the stream of data is thus necessarily processed in a digital way because the successively observed time delays have to be stored in memory: the time range in which physically acceptable delays may be obtained is divided into adjacent intervals; the time delay is digitized and the corresponding code may be directly memorized or only "summarized" in an histogram (hereafter

called a coincidence spectrum). This usual procedure of statistical analysis is easily performed in a digital memory area where the memory location assigned to each digitized time delay is incremented at each occurrence of this time delay.

The basic aspects of the time correlation are developed in Section II.

A basic multicoincidence experiment performs simultaneously the acquisition of the coincidence spectra, which are obtained by a sequential scanning in conventional experiments. The multicoincidence method is thus a practical improvement of the "monocoincidence" method leading to a doubly differential analysis. This improvement is both quantitative and qualitative:

1. The overall analysis time of a collisional system is evidently reduced if a given statistical accuracy is required. The total accumulation time may be managed at the convenience of the experimentalist in order to get the data either more statistically accurate or more detailed; furthermore, it is also possible to explore regions or systems where cross sections are too small to be studied reasonably by scanning the relevant experimental parameter.
2. All the spectra obtained within the same time interval are intrinsically consistent and no strict beam monitoring is needed.
3. The "mean statistical accuracy" is enhanced: this (secondary) point is noticeable every time that large discrepancies in the count rates are observed within the analysis range, the large accumulation time needed for some channels provide fairly high S/N ratios for the favored ones.

Ideal conditions are fulfilled when the two detectors collect the two particles over their entire dynamics, for example, the full collision cone for scattered particles: a single run provides a complete and coherent analysis of the collisional system even under unstable conditions. This complete information is also obtained if symmetry considerations may be invoked, for example the azimuthal independence for randomly oriented and aligned incident particles.

This parallelized accumulation procedure implies two practical consequences:

1. Each coincidence spectrum has to be labeled in a convenient way during the accumulation time.
2. The main analyzing parameter is still the time but the second one is necessarily working as a spatial parameter. (If a direct determination of energy, velocity, or wavelenght is needed, the focal plane of the spectrometer has to be analyzed spatially.)

The time code in which the physical delay is converted has to be associated with a "spatial" code: the range within which the physically pertinent parameter is varied in conventional scanning methods has to be quantized and this quantization is a digital position analysis. The technical aspects of this dissection are presented in Section III.

The multicoincidence procedure may be extended to the analysis of more and more "dimensions" in which collision processes provide information. The first historical applications[1–4] of coincidence methods in atomic physics were limited to sort the few cases produced by a spectroscopically well-defined process from a stream of scattered particles. The operating field of the multicoincidence methods is much larger now: the two detectors may be two dimensional and the ubiquitous microchannel plates (MCP) provide the ideal tool for this purpose. Furthermore the "analyzing dimensions" may be purely spatial[5,6] or mixed when a suitable analyser displays, in orthogonal directions, preserved spatial information and velocity-dispersed spectra.[7]

The various ways the correlation may be taken into account on the hardware and software levels are described in Section IV. In Section V a brief review of the main results previously obtained on the anion–molecule systems introduces the description of the advances made in Orsay on the X^{-}–H_2 systems using a multicoincidence method. In the concluding section a tentative summary of possible applications is proposed.

B. Preliminary Remarks and Comments

Some preliminary remarks and comments have to be made before proceeding.

1. A Multicoincidence Set-Up May Be Used for Lower-Level Analyses. A convenient multicoincidence set-up has to be versatile enough to allow the included time and position digitizers to be used in various configurations. After all, the natural way to undertake the study of a new system is just the progressive activation of more and more refined experimental procedures:

The simple spatial analysis of scattered particles provides a first look at the complete spatial map of total cross sections.

The coupled position and time-of-flight (TOF) analyses provide differential cross sections in a second step. This procedure may be used of course for both particles.

The operating conditions being well known by these preliminary measurements, the coincidence measurements successfully provide the ultimate information.

2. A Multicoincidence Set-Up May Be Used for Higher-Level Analysis. More efficient checking methods, alternate analyses or even new fields of investigation become possible when a proper initial design allows the basic

elements to be interlaced in various configurations. Some examples will be given later.

3. The Basic Tool of These Procedures is the Time: This choice is necessary:

Time is used to mark the synchronism of two events, and thus is a false/true (boolean) variable. (In many cases in nuclear physics, simple coincidence gates are used for this purpose.)

Time is used for an analogical analysis: the "analyzer" is a simple free-flight path supplanting a magnetic or electrostatic analyzer.

Time allows the multicoincidence: these "analyzers" only need free space to be "parallelized" and then to perform simultaneous measurements

This choice is convenient:

A free-flight-path, as analyzer, works for charged particles and also for neutral species; it "looks at" the total collision volume directly, that is, without any discriminating optical system. When conventional spectrometers are involved in a coincidence measurement, a major problem arises from the partial overlap of the two fields of view: such a handicap is avoided when only flight times are used.

Time measurements are intercorrelation measurements and these methods intrinsically provide good signal-to-noise ratios: interferometry in optics, lock-in techniques in the electronic filtering of noisy analog signal, and Dirac or pseudorandom TOF measurements in collision physics.

In spite of these features, conventional (electrostatic or magnetic) analyzers are often preferred to TOF analysers. This choice is based on two arguments:

Conventional analyzers act on steady beams, and would be faster.

TOF spectrometers are not as accurate as high-resolution spectrometers.

In most cases these considerations are questionable and the following remarks may be useful in this present introduction to a time-based method.

The first comment is intentionally crude in that "forgetting" the fact that time, energy, and velocity parameters are nonlinearly related and lead to dissimilar spectra: the effective accumulation time per channel is generally defined by the expected resolution and not by the procedure of analysis. In TOF experiments, the channels are (virtually) simultaneously opened during the beam pulses, whereas, in a conventional scanning, the channels are opened sequentially. In both cases the time sharing is similarly determined by the number of analyzed channels and the accumulation time per channel by the filtering bandwidth. In TOF experiments the simultaneity provides intrinsically coherent spectra, whereas only high-level multiscaler analyzers perform time-saving scannings (i.e., omitting "dead zones" and adapting the analysis time to the count rate[8].)

The second comment is related to the resolving power of spectrometers: high resolution is always obtained by decelerating the scattered particles, that is, at low energy. If the decelerator is stigmatic, a TOF method also leads to better resolutions.

In other words and as a last argument favoring the promotion of the time analysis in chemical physics, time, as pointed out by Pollard et al.,[9] is a free dimension in which information may be "written" and "read" in a much more simple and versatile way than if only spatial discrimination is used (as in conventional analyzers).

4. The goal of coincidence methods is to provide valuable information on quantal systems from measurements performed on two parts of these systems. Recent tests of the Bell's inequalities[10] have clearly demonstrated that such a procedure ought to be formulated within the basic concepts of quantum mechanics. The actual state-of-art in chemical physics, presumably, does not yet provide situations in which discrepancies can be found between classical and quantal interpretations of the same measurement: the wave function is projected just after the collision and the products fly apart as classical particles.

5. For the sake of clarity, the technical aspects of the method, which are closely related to analog and digital signal processing, will be analyzed within the usual electronic formalism.

6. In this introduction, a kind of "laboratory measurement space" (LMS) has been implicitly introduced. This picture is convenient for visualizing the operating field of a multicoincidence experiment and thus is well suited for qualitative descriptions (but of course not for doing any algebra). The "dimension" of the LMS is the number of discriminated variables and the LMS is partitioned into cells corresponding to the smallest unit available of this space, and to one memory location in the most detailed histogram of the experimental results. For example, when the two dimensions are determined for the two detectors, the LMS is five dimensional and is divided into 2^{60} cells if every time and spatial variable is encoded in 12-bit words. The "projection" on the time dimension provides the time histogram of all the observed events whatever the positions are; the projection on the two-dimension space of the detector A for a single time channel, independently of the position on the second detector B, provides a spatial mapping on A of all processes characterized by a well-defined time delay. A convenient software handling has to allow any of these arbitrary "cuts" in the LMS (in fact, these dimensions correspond to multidimensional arrays of integer or real variables).

II. TIME CORRELATION

A. Ideal Correlation

The coincidence concept has been introduced intuitively in the preceding section, but a short mathematical introduction is convenient for elucidating some aspects of the method.

Consider the simple conceptual coincidence set-up in Fig. 1 in which particles scattered from the crossing point of two macroscopically continuous incident beams hit two detectors A and B. This experiment generates a random sequence of successive collision times, but these collisions are "viewed" by two detectors, which both have limited geometrical acceptances and limited efficiencies. Therefore (binary) collisions are divided into four exclusive sets: a first set groups collisions producing product particles that are both detected, two sets correspond to collisions observed by only one detector, and a fourth set groups collisions that are not viewed at all in the laboratory frame. The random flux of particles impinging a detector and effectively detected may be described as the sum $s(t)$ of Dirac "functions" depicting these events. In the present case the two functions $s_A(t)$ and $s_B(t)$ describing the discriminated outputs of A and B may be written as the sums of two components

$$s_A(t) = c_A(t) + r_A(t) \quad \text{and} \quad s_B(t) = c_B(t) + r_B(t)$$

where $c_A(t)$ and $c_B(t)$ represent the fluxes on A and B of the related particles and $r_A(t)$ and $r_B(t)$ correspond to collisions exclusively detected on A or on B.

The cross correlation function

$$\mathrm{CC}_{BA}(\tau) = \int_{-\infty}^{+\infty} s_B(t) s_A(t-\tau)\,dt = s_B(t) \otimes s_A(t)$$

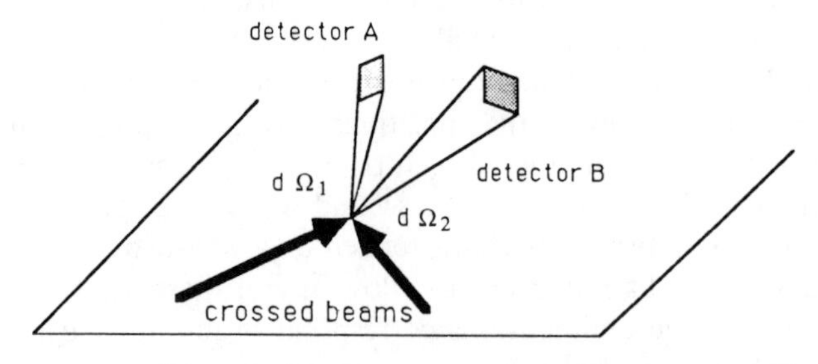

Figure 1. Monocoincidence configuration.

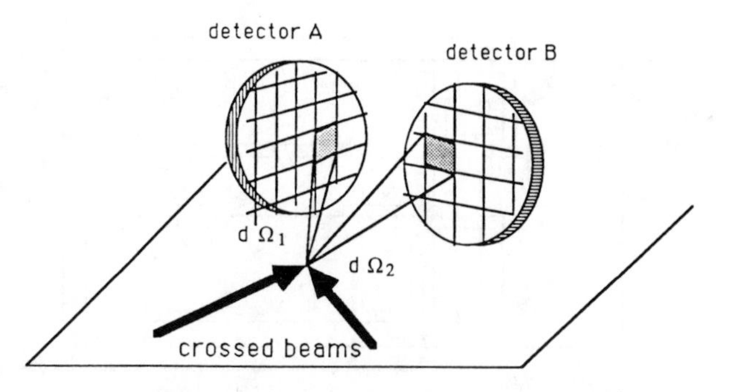

Figure 2. Multicoincidence configuration.

is the sum of four terms, $c_B(t) \otimes r_A(t)$, $r_B(t) \otimes c_A(t)$ and $r_B(t) \otimes r_A(t)$: these cross correlation functions of unrelated fluxes are flat if the incident beams are unstructured; $c_B(t) \otimes c_A(t)$: if the efficiencies of both detectors are constant for the energy range of the hitting particles, this function is the time histogram of the effectively time-related pairs (upon another flat background).

The function $CC_{BA}(\tau)$ then provides, when the flat backgrounds are removed, the relative probabilities of processes leading to equal time delays without any distortion.

In the typical multicoincidence experiment of Fig. 2, these considerations hold for each pair of pixels. (This affirmation will be discussed later.) Note that the randomicity (in fact the abbreviated form of "with a constant probability law") of the collision times is not a necessary condition to get valuable information from the cross correlation function. Steady beams provide easily "readable" spectra, but the method still holds if the probability law is no longer flat but is given by the product of two time-modulated beam densities. An interesting particular case is pointed to in Section IV E 3; when one of the beams is chopped to produce periodic short deltalike pulses in the collision region, TOF and multicoincidences methods are coupled in a powerful measuring tool.

B. Practical Correlation

To our knowledge, no hardware assembly is able to perform on-line an exact determination of this cross correlation function: the conceptual "full correlator" schematized in Fig. 3 is too "large" for a concrete realization from discrete digital circuits, but would certainly be easily integrated in a VLSI device.

A full correlation may of course be done off-line: the events on the two (or more) detectors are separately memorized for an off-line treatment in

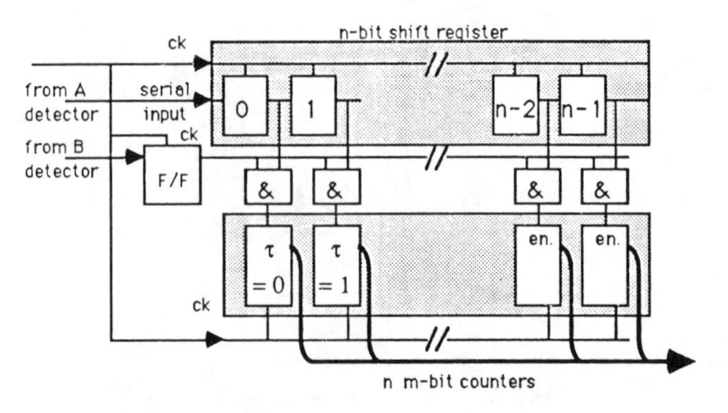

Figure 3. Principle of a full digital correlator. This apparatus is a real time digital n-integrator. The A input signal is delayed n times; when the B input is true, the true A signals memorized in the convenient flip-flops (F/F) of the shift register command the increment of the corresponding counter (purely schematic).

which the integrating counters of Fig. 3 are replaced by memory locations (and then easily "multiplied").

Two attempts to a "quasifull" on-line correlation are briefly described in Section IV C but, practically, the "correlator" is usually a simple unique time digitizer: this device is triggered by detector A start signals and stopped by detector B output (eventually conveniently delayed if any time overlap may occur). The model developed in Appendix A shows that this method leads to the true cross correlation function only for vanishing count rates. This point will be discussed later in relation to the results of the next section.

C. Efficiency

For the sake of simplicity this analysis of the intrinsic resolving power of an ideal coincidence experiment is performed on the illustrative example of a single correlating counter for very small count rates; therefore, a complete decoupling of false and true coincidence events is assumed.

1. *Qualitative Analysis*

Each start signal triggers a single time digitizer. A stop signal appearing within the convenient i_0-$i_0 + \Delta i$ time window at the ith δt time interval led to increment the corresponding i channel. If the two signals are uncorrelated, the counter is still running at the ith time interval with the probability $P(i)$. This probability is given by $P(i) = P(i-1)\,(1 - \dot{N}_B\delta t)$, where $\dot{N}_B$ is the mean count rate of the stop signal. Such a recurrence law leads to $P(i) = (1 - \dot{N}_B\delta t)^i = \exp[i \ln(1 - \dot{N}_B\delta t)]$ and, if $\dot{N}_B\delta t \ll 1$, to the usual exponential law $P(i) = \exp(-\dot{N}_B i\delta t)$, Therefore, after N_A counts have been received at the start port, the mean content of the ith channel is given by $N_A \exp(-\dot{N}_B\, i\delta t)\dot{N}_B\delta t$.

On this exponential background true coincidences appear, if any, as peaks; if $P(i)$ is the (cross-section-related) probability to get a stop signal at channel i, the content of the peak i is given by $N_A p(i)$.

The "contrast," that is, the peak-to-background ratio

$$C = \frac{p(i)\exp(\dot{N}_B i\delta t)}{\dot{N}_B \delta t} \cong \frac{p(i)}{\dot{N}_B \delta t}$$

may be done arbitrarily large by a proper choice of the time unit and the true $p(i)$ values are easily obtained.

This formula merely illustrates the power of coincidence measurements: related events are obtained at their own characteristic time delay whatever δt is, whereas unrelated ones can be "scattered" on so many channels that they become negligible in each of these channels.

2. *Quantitative Analysis*

Nevertheless, the preceding formula cannot be a guide for an evaluation of the limiting factors of the method and a more refined analysis[11] is required in order to determine the conditions for the proper evaluation of a given process.

Let $\overset{\frown}{\sigma}_k(\theta_1, \varphi_1, \theta_2, \varphi_2)\, d\Omega_1\, d\Omega_2$ be some "coincidence cross section" to observe the k process on both small detectors A and B, that is, to detect within the solid angles $d\Omega_1$ and $d\Omega_2$ the particles 1 and 2 issued from the process k (a more precise definition of this cross section will be given later).

The parts of the collision volume V effectively seen from the detectors A and B are v_A and v_B and their intersection is v_{AB}; the detector efficiencies are ε_A and ε_B and σ_T is the total cross section producing the $\dot{N}$ collision rate.

In these conditions the preceding $\dot{N}_A$ and $\dot{N}_B$ values may be evaluated explicitly

$$\dot{N}_A = \varepsilon_A \frac{v_A}{V} \frac{\dot{N}}{\sigma_T} \sum_k \int_2 \overset{\frown}{\sigma}_k d\Omega_2\, d\Omega_1$$

and

$$\dot{N}_B = \varepsilon_B \frac{v_B}{V} \frac{\dot{N}}{\sigma_T} \sum_k \int_1 \overset{\frown}{\sigma}_k d\Omega_1\, d\Omega_2$$

(Note that the dark noise of the detectors could easily be included if necessary.) The number of coincidences for an integration time T is given by

$$N_C = T\varepsilon_A \varepsilon_B \frac{v_{AB}}{V} \frac{\dot{N}}{\sigma_T} \overset{\frown}{\sigma}_k d\Omega_1\, d\Omega_2$$

A convenient measure of the "intrinsic" uncertainty of coincidence methods owing to the unrecoverable false coincidence background is the signal to noise ratio $S/N = N_C/N_F^{1/2}$ where N_F represents the mean noise in a δt wide channel—$N_F = T\dot{N}_A\dot{N}_B\delta t$—

$$S/N = [\varepsilon_A\varepsilon_B]^{1/2}\frac{v_{AB}}{[v_Av_B]^{1/2}}\left(\frac{\overset{\frown}{\sigma}_k d\Omega_1 d\Omega_2}{d\Omega_1\sum_k\int_2 \overset{\frown}{\sigma}_k d\Omega_2}\right)^{1/2} \times \left(\frac{\overset{\frown}{\sigma}_k d\Omega_1 d\Omega_2}{d\Omega_2\sum_k\int_1 \sigma_k d\Omega_1}\right)^{1/2}\left(\frac{T}{\delta t}\right)^{1/2} \qquad \text{(II-1)}$$

The S/N ratio is independent of the total count rate. This feature, which is related to the fact that the background is a quadratic function of $\dot{N}$, has to be pointed out: in a coincidence experiment, the beam intensities may be kept low, at least when the product $T\dot{N}$ is high enough to provide the desired usual statistical uncertainty $S^{1/2}$. The influence of the yields is clear. The geometrical factor that may be dramatically small with analyzers is equal to unity in the present case.

The time factor $T/\delta t$, which may be quite high, about 10^{12} in typical experiments (e.g., $T = 10^4$s, $\delta t = 10$ ns), has to be compared with the two cross-section terms. These terms represent the ratios of the desired cross section to the total cross sections integrated on each detector: this formula would predict, for example, a S/N ratio equal to 10 for a process participating for only 0.01% in the total signals received by both 10% efficient detectors.

This result has of course to be qualified:

1. It supposes that the number of observed true coincidences is high enough, for example, greater than 100 in this example, in order to be limited mainly by the coincidence uncertainty. [The definition of this "intrinsic" S/N ratio is questionable but is justified by the striking result of Eq. II-1].
2. An arbitrarily small δt time resolution is meaningless, since the intrinsic time resolution is determined by the geometrical extent of the interaction zone. This point is developed in Section II D in a more general context.

A more detailed discussion is necessary for the various possible coincidence configurations:

1. Coincidences between massive particles issued from a binary collision[3,5,6]:

The observed time delay, the position of at least one impact, and the momentum conservation rules may be used to fully determine the kinematics of the collision (see Sections II D and IV E 6 and Appendix E)

$\overset{\frown}{\sigma}_k$ reduces to the usual cross section σ_k for the effectively correlated positions and vanishes elsewhere.

These points fail if more than two particles are involved: this case will be discussed in Section IV E 3.

2. Coincidences with electrons[2,6] in ionization or detachment processes:

The TOF of the coincident massive particle is quite close to the observed delay.

$\overset{\frown}{\sigma}_k$ is the product of the usual cross section by some steric function describing the emission and the collection of the electron. This function may be chosen almost arbitrarily whether the definition and/or the determination of the ejection angle is possible and expected, or not.

The case of a complete collection of the electrons is particularly interesting: if no autoionizing state is excited, a weak extracting field in the collision region transforms a coincidence experiment into a "self-signed" TOF experiment:

This "TOF" experiment is "signed," that is, works only on the electron-generating processes.

This "TOF" experiment is "self-chopped," hence is working on steady beams, that is, with a 100% duty cycle.

3. Coincidences with photons.[1,4] Similar observations hold:

The observed TOF spectra result from a convolution of the true one with two time functions owing to the finite collision volume and to the lifetime of the excited state (eventually perturbed by cascading, which can then be controlled).

Photons may be spatially analyzed or simply collected over a large solid angle using a parabolic mirror[1] or a large aperture lens[4] to get spectroscopically defined differential cross sections.

The case of the spatial and/or polarization analysis has to be particularly outlined in this volume devoted to state-to-state chemistry, since the most precise information on the electronically excited output states can be deduced[12] from the measurements of their alignment and orientation as a function of the scattering angle[4,13]. A limiting factor of these experiments arises from the photon analysis, which practically is performed with interference filters.

A practical conclusion including the results of Appendix A is that a coincidence experiment works well with moderate count rates: higher count rates do not increase the resolving power of the coincidence procedure but induce undesirable consequences such as spectrum distortion, increased dead time, and wasted memory. Convenient conditions are merely fulfilled in multicoincidence experiments by a careful collimation of the crossing beams: a large set of coherent spectra is obtained for a reasonable overall accumulation time and each spectrum is precise and directly significant. Furthermore, the attainment of a given S/N "in the laboratory coordinates" may not lead to an early data reduction, since the final results are generally presented in the more physical and essential CM frame: if the full resolution is not wasted before the final lab–CM transformation, a "natural" smoothing occurs and even poorly defined individual spectra will lead to a large overall S/N ratio if symmetry considerations such as Φ-independence "multiply" the accumulation time by allowing some summation of many equivalent spectra.

D. *Redundant Variables as a Tool for Noise Removal*

A further comment may be given now about the "scattering" of the background outlined in the qualitative analysis of Section II C 1. (The term "dispersion" would be less confusing in the present context, but this word does not indicate in a satisfactory way the basic power of the multicoincidence experiments.) The key of a multicoincidence measurement lies in the fact that the signal and the noise have quite different distributions in the LMS (defined at the end of the introduction). The different processes are condensed only in some few specific cells of the LMS, which are closely defined by conservation laws, and this localization is only broadened by the experimental uncertainties (geometric and kinematic spreads and quantizing noise). At variance, the background noise is (almost) uniformly distributed along the time axis and smoothly shared out on the other dimensions: the background "density" is given by the product of the "global profiles" produced by all the collision events on each of these dimensions (or couples of dimensions for two-dimensional detectors: the factorization is possible for unrelated particles; the impact coordinates of a single particle are obviously tied). This remark introduces in a natural way the idea of redundant measurements. In a versatile coincident set-up the analysis of the collisions may be extended to variables that are not strictly indispensable for a proper data reduction in the CM frame. Such a procedure is not only a test for fun of the versatility of the apparatus, since two improvements of the basic methods may be expected:

1. A more precise experimental description of the collisions could be used to check the system or even to get more accurate results; the difficult problem of taking into account these redundant measurements is discussed in Section IV E 6.

2. A more evident enhancement is obtained for the S/N ratio, since the false coincidence background is spread along the "cross-section" profile of such an "additional" variable. The data reduction routine ought at least to use this variable as a discriminator in order to reject patented false events. (Note that such a routine requires the performance of some simulation calculations in order to come to convenient decisions.)

Looking back at the first comments of Eq. II-1, an arbitrarily large time accuracy of course does not sweep out the "degeneracy" between false and true events in the restricted space of the only necessary variables, but a fair "splitting" of these events is possible if the collisions show themselves in a more complete LMS.

III. POSITION-SENSITIVE DETECTION

A. The Physical Layer: The Microchannel Plate Detector

The Microchannel plate (MCP) is the ideal interface between a single particle impact and electronic devices. Now extensively used for civilian purposes (e.g., high-performance CRT tubes, image-intensifying or X-rays imaging), they are commercially available under a lot of shapes and dimensions.

Basic information on the MCPs may be found in the literature[14,15] or in the manufacturers' data sheets: the principle and the main characteristics of the MCPs are supposed to be known and only some particular aspects are pointed out:

Gain: two cascaded MCPs (in a chevron mounting in order to limit the ion feedback) provide picocoulomb range charges in subnanosecond pulses; these signals are easily processed as well by charge-sensitive amplifiers as by fast (therefore matched) voltage amplifiers. Unfortunately, the statistical nature of the secondary emission mechanism usually provides an unsuitable charge distribution if no further mechanism is occurring; this quasiexponential distribution is always troublesome: for timing applications, the choice of the discriminator thresholds is hazardous and the practical efficiency is lowered; the problems become worse when a further division of this widely varying charge is used to sense the impact position (Sections III B 3 and 4). In the "standard picture," the regulating phenomenon is the lowering of the emission yield when a too large space charge limits the kinetic energy of electrons impinging the channel walls. (The mechanism is not very clear and is still discussed; the maximum charge per channel is seemingly about 1 femtocoulomb; a pulse of fwhm 100 ps and a mean velocity of 10^6 m/s provide a rough evaluation of the radial electric field produced in the neighborhood of the channel wall. The orders of magnitude of this radial field and of the longitudinal external one are similar: the secondary electrons are then closely reflected toward the wall and hit it prematurely. They are then poorly accelerated by the external field and a dynamic equilibrium occurs when the mean velocity of these "sliding" electrons corresponds to a mean effective emission yield of one.)

This saturation effect is obtained by increasing the accelerating voltage or, more conveniently, by stacking three (or more) MCPs in a "Z-assembly," that is, with alternate inclinations of the channel axes. A proper MCP spacing allows the saturation of some hundreds channels providing, for usual channel diameters, overall gains about 10^7–10^8. Note that in these "Z-assemblies" the voltage drop per MCP, and thus the number of spurious field emitted electrons, are limited.

Efficiency: A recent compilation by Seah[16] of the published data on the "efficiency" of the parent channeltron to electron impact, illustrates the difficulty of such a question: for MCPs worse conditions are encountered:

1. This efficiency is position-dependent.
2. The pulse width is smaller: a proper matching is required on a larger frequency range to get a convenient pulse discrimination. (Some strange remarks are found in the literature on this particular point and some "practical" recipes seem to be far away from the usual "theory" of transmission lines: when gigahertz frequencies are involved, and in fact down to the DC limit, the reduction of a coaxial line to its "capacitance," as well as to its "inductance," independently of the terminating charge would frighten the author of the Smith chart.)
3. For every position-determination technique, many hardware and/or software thresholds are simultaneously limiting the number of accepted data (for example, the two-dimension rise-time method involves at least five discriminators).
4. Pollution and aging problems are more crucial than for channeltrons. Too many parameters, quantitative, such as the bandwidth(s) and the threshold(s) of the electronic equipment, or more qualitative, such as the mechanical description of the pulse path, ought to be given for a valuable information. Therefore no numerical data are given in the present paper but the further remarks have to be made.
5. A significant increase (20%[17]) of the observed efficiency is obtained by applying an inverse electric field above the first MCP: secondary electrons emitted on the 40% "blind" interchannel area are more efficiently collected by an adjacent channel; furthermore, the timing performances are enhanced.[18]
6. Cesium may be evaporated on the front face of the first MCP in order to enhance the sensitivity to low neutral particles.

7. A crude variation of the efficiency is observed when the particle trajectory is close to the channel axis.

Local and global maximum count rates. This important parameter is also quite depended on the operating conditions: distribution of the impinging particles and resistance and decoupling of the biasing network. An efficient checking of the overall amplification system is obtained by an on-line control of the histogram of the total charge.

The standard mounting[19] used in Orsay is represented Fig. 4: the resistor chain is installed in the vacuum close to the detector in order to minimize reflections; the resistances are high enough to provide a low heat dissipation and an efficient protection against an accidental runaway but allow a generally convenient total count rate without important gain loss. The time pick-up system is also shown. The successive couplings to the resistor-matched amplifier are designed in order to avoid as much as possible any discontinuity in the guiding of the electromagnetic wave.

Note that the commercially available image intensifiers use an electrostatic lens in order to provide, on the MCP, a magnified image of the photocathode output; such a lens may also be used between the first two stages of amplification. A convincing spectrum obtained by Gao et al.[20] shows that the channels of the first MCP may be individually identified.

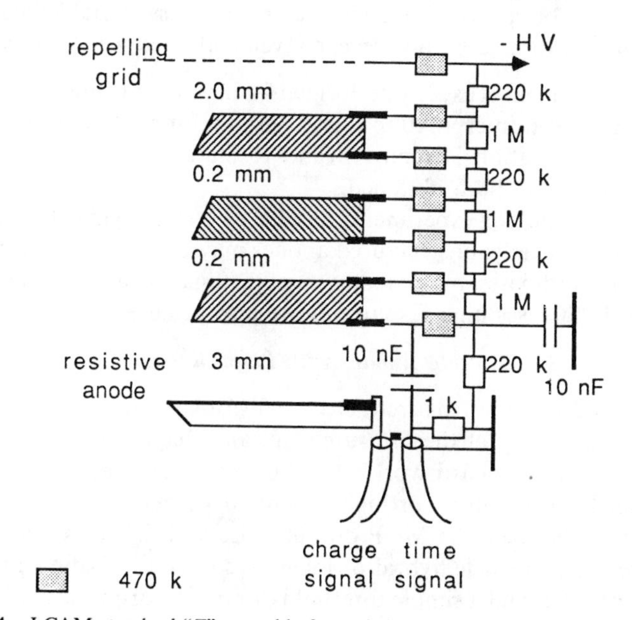

Figure 4. LCAM standard "Z" assembly for resistive anode (charge division encoding).

B. The Encoding Layer: Position-Detection Methods

1. *Brief General Considerations*

Many methods have been proposed and utilized. These methods can be classified into various classes according to

The number of resolved dimensions.

The nature of the discriminating anode.

The balance between analogic and digital processing.

The ability to be used for time-resolved experiments.

The following discussion is limited to the MCP array detector for low-energy particles, but the various position-determination methods are examined within a larger framework than the time resolved one.

Users take advantage of the MCP's large operating area but, in some cases, the corresponding collector may be designed in structured anodes that fit the intrinsic symmetries of the experiment; when possible, such a method is quite valuable, since

The hardware is simpler and the data flow is reduced by half if this collector presents in fact a one-dimension structure.

The data-processing time is lowered since the spatial filtering as well as the integration procedure over the meaningless variable are automatically performed by a simple, conveniently designed collector.

A fully digital treatment is usually performed for high-count-rate applications, but an analogic linear interpolation is also possible for moderate count rates: the spatial quantification is removed and resolutions down to some few channel spacings have been obtained.

These methods fail for experiments in which no simplifying structure is *a priori* possible, or steady; unstructured anodes may be used to determine the two cartesian coordinates, but in many cases cartesian-structured anodes are preferred. In both cases high resolutions are also achieved.

2. *Discrete Anodes with Logic Readout*

The collector is divided into discrete, electrically isolated areas that determine directly the granularity of the measurement; any shape (Fig. 5) is possible. A double-sided epoxy board works well but more elaborate collectors are preferable for high-resolution or high-vacuum experiments.

The interface between the MCP output pulse and the succeeding digital circuits is generally a single hybrid or integrated circuit including a more or less charge amplifier and a simple threshold discriminator-monostable. Some are commercially available by chip or as 16-chip printed boards, but devices

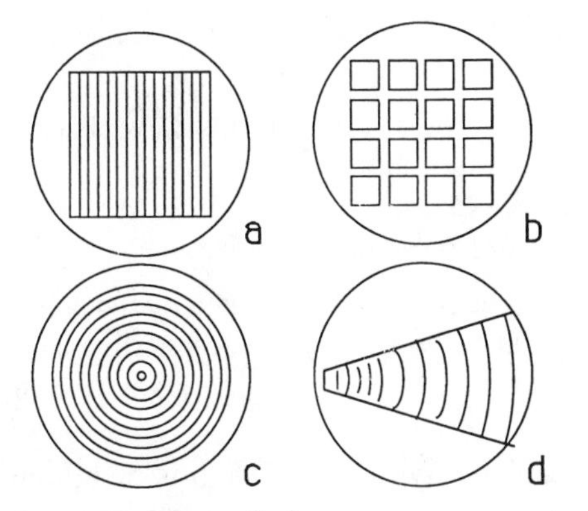

Figure 5. Some discrete collectors for circular MCPs (very schematic): (*a*, *b*) Basic one/two-dimension cartesian patterns. (*c*) Circular-shaped anodes for Φ-independent scattering. A quite efficient tool for photodissociation studies is obtained when the two halves are separately detected.[5] A Faraday cup or holed MCPs are necessary for direct reactions. (*d*) Similar pattern for large angular range. The incident beam does not impinge the MCPs. The output pulses are guided by microstrips (not represented). The width of these microstrips formed with the grounded conducting backplane is chosen to get the characteristic impedance of the next transmission lines. (Note that these guides are rather striplines when the collector is close to the MCPs for high resolution.)

designed for multiwire chambers are also quite convenient at a lower price (the signals are much faster but their larger amplitude allows an efficient detection). The input bandwidth of these ECL devices is some hundred megahertz and particular attention is required for a proper mounting to prevent crosstalk or general oscillation. The inputs are generally differential allowing the use of twisted pairs in "congested" systems; the unsaturated ECL logic is "quiet," but a reliable system needs a close decoupling of the analog and digital subsets.

The ECL (or NIM if AC-coupled) outputs can be used directly, for example, on individual scalers[21–23] or encoded (latches prevent any problem in the case of shortened monostable output). This reduction, which makes the further data handling in multicoincidence experiments easier, is easily performed using ECL priority encoders and multiplexers; less than 50 ns are needed for 64 channels[3]. Two improvements of this basic technique have to be outlined:

1. A sophisticated collector allows a basically two-dimension dissection using only $(n + m)$ amplifiers for a $(n \times m)$ pixels array with a 50μm resolution.[24]
2. Adding and subtracting the results of two opposite encodings[8] allow

the determination of the midpoint of the electron cloud and the rejection of double pulses.

In some small systems alternate methods have been used, for example, position encoding through individual delay lines[4] or direct correlation determination between 25 collector pairs using AND coincidence gates.[25]

The major features of these methods are

1. Fast processing: the count rate is limited by the pulse pair discrimination (typically 50–100 ns) and/or by the access time to the following layers; furthermore, the immediate encoding leads to simplified handling of spatial data.
2. Good integral linearity: the resolution is determined by the collector design and can be almost down to the channel spacing limiting value (see for instance the review of Richter[14]).

The main shortcoming is in fact the poor uniformity in gain: the independent treatment of each channel introduces another random component to the intrinsic inhomogeneity of the MCP, especially when the MCP assembly does not work in a saturated mode. If no independent total cross-section data are available, the correcting factors have to be evaluated for the same operating conditions. Note that this difficult operation is avoided if every channel of the detector is used to accumulate data for each part of the final spectrum.[8,26]

In these methods problems due to the electronic noise are generally negligible with respect to the MCP dark noise or to crosstalk; however, "electronically cooled" terminations[27] may be useful for small signals.

In a last method a first analog stage is used in order to simplify the digital encoding (binary charge division[28,29] or coincidence method[15] or "summation encoding"). This ingenious method needs only n amplifiers for 2^n pixels for "weighting" simultaneously the bits of the impinged pixel code; unfortunately no reliable apparatus has ever been made (to our knowledge). A possible arrangement is suggested on Fig. 6; it also needs only $n+1$ amplifiers and uses the well-known Gray encoding of mechanical position sensors and an adaptation level detector. More attractive solutions could be found by adjusting the code distance to the maximum size of the electron cloud. (The distance of a code is the minimum number of bits in which two valid codes differ.)

3. *Discrete Anodes and Analog Encoding*

These methods can be seen as the ideal balance between analog and digital techniques: the collectors may still be shaped for convenience and a fairly high accuracy is obtained by a linear interpolation with relatively simple electronic devices.

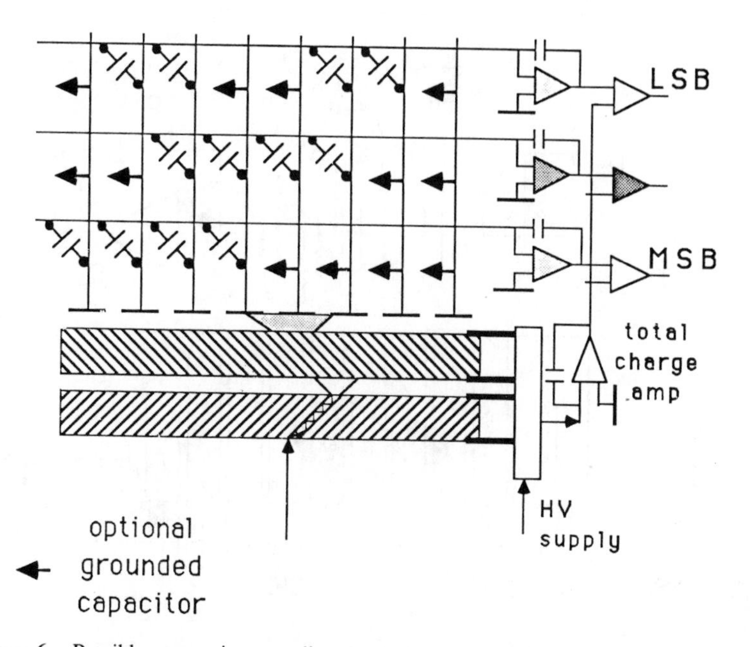

Figure 6. Possible summation encoding: 2^n pixels are encoded using $n + 1$ amplifiers. The Gray encoding prevents false codes when two adjacent strips are involved. The total charge information is given by integrating the current pulse induced in the supply line when a particle is detected.

One or two dimensions may be coded using capacitive or spatial charge division. In both cases the position is deduced from an arithmetic treatment of the charge-sensitive-amplifier outputs. In the following, typical set-ups are first briefly described and are than discussed and compared.

a. The "FOM" Method. This old faithful method has been used as a standard in the FOM Institute for a long time[28–30]. The principle is illustrated on Fig. 7: the incoming charge is shared between the two adjacent capacitors in a linear position-dependent way and "propagated" along the capacitor chain to two terminating charge amplifiers. This linearity still works when parallel resistors limit the charge integration (the gain of a charge amplifier is inversely proportional to its input impedance and then proportional to the number n of serialized capacitor/resistor parallel assemblies); furthermore, this linearity allows the direct evaluation of the center of gravity of a diffuse electron cloud. This feature is of course fully used and leads to a fairly continuous and accurate position determination when this cloud is empirically adjusted to be about five collectors wide.

When large collectors are used, the long capacitor chain may be divided into adjacent smaller ones in which nanofarad-range coupling capacitances

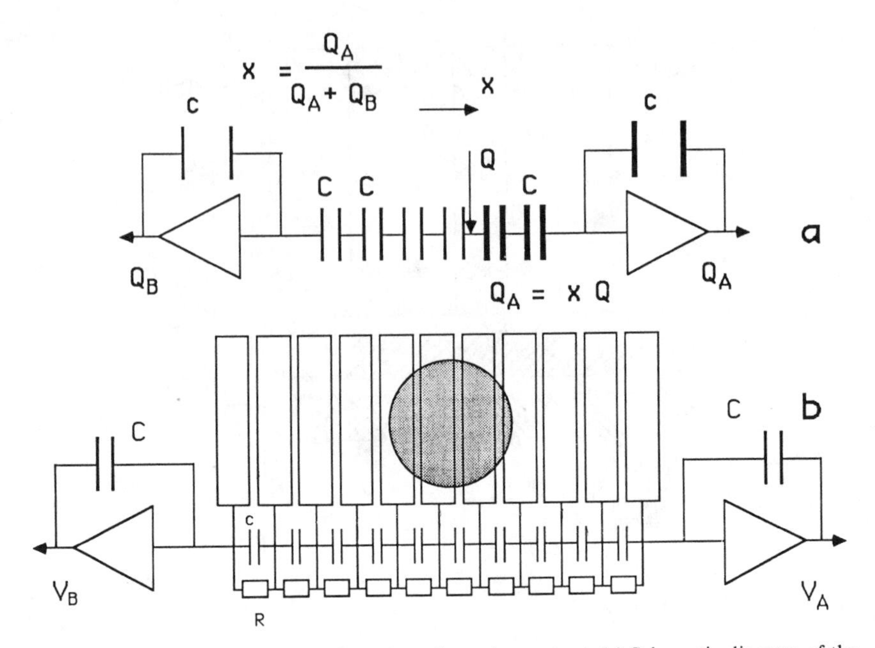

Figure 7. The FOM method 28–30 (one-dimension system): (*a*) Schematic diagram of the charge division. (*b*) Centroid determination. The (hardware or software) discriminators needed for safe operating conditions are not shown.

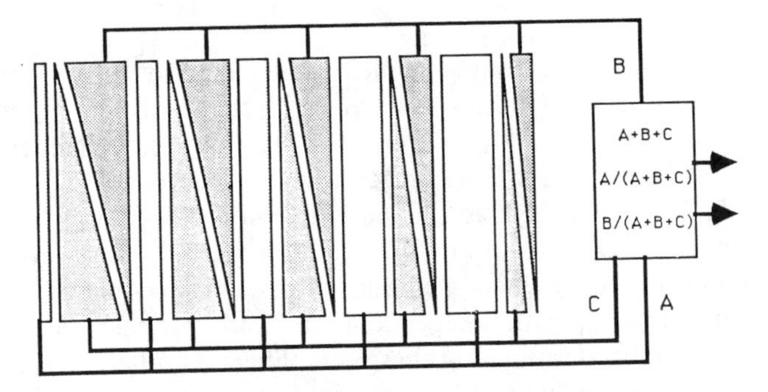

Figure 8. Possible wedge-and-strip anodes.[31,32] The (hardware or software) discriminators needed for safe operating conditions are not shown.

are still used. The nonlinearities due to the stray capacitances are then limited. This method may be extended to a two-dimension dissection[28] using two independent orthogonal systems.

b. The Wedge-and Strip Method (WS). This method uses the spatial charge division on conveniently shaped collectors (Fig. 8). The original design allowed the determination of the centroid coordinates through two independent X and Y systems of linear charge partition. Alternate patterns using three electrodes have been proposed for different symmetries[31,32]. The total charge is used for both dimensions and the periodicity of the unit cell can be decreased; these "triplets" allow a finer division of the expected gaussian charge profile, which reduces the distortion. A proper determination of the centroid is obtained when the cloud overlaps more than one triplet. The linearity and resolution will be given later, but note that the coupling capacitance between electrodes may be important for large and high "spatial frequency" collectors for which a slightly different method may be applied[33].

c. The "Backgammon" Method. This method (Fig. 9) is obviously the superposition of the two previous techniques working with the same charges but separately devoted to each dimension (see, for example, Ref. 34).

d. Signal Processing. Once more, many techniques can be used to perform the summations and divisions:

1. Chains of summing, log and antilog operational amplifiers provide analog representations of the expected variables; only the ADC last stage is digital.

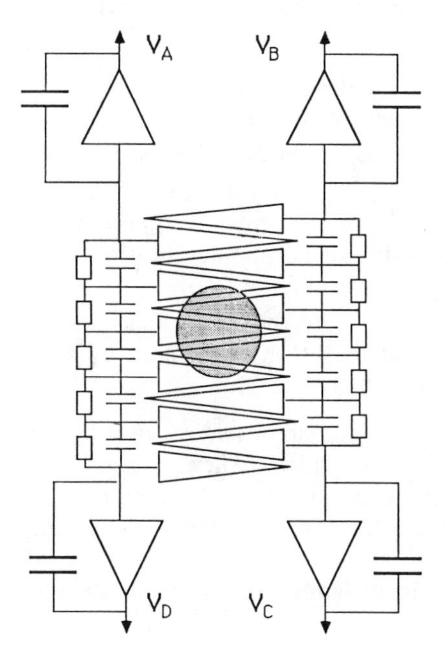

Figure 9. Possible backgammon anodes.[34] The (hardware or software) discriminators needed for safe operating conditions are now shown.

2. The voltages are analogically summed; the division is then performed by position encoders used in nuclear physics or by simple ADCs. In the first method, the rsult is the coded time needed for a divider-controlled current to cancel a charge proportional to the dividend; in the second method, the dividend is directly weighted to the divider used as reference input of successive approximation ADCs or flash converters.
3. Each voltage is individually encoded; the numerical treatment is done by the host computer or by a fast coprocessor[35].

Analogic treatments are generally less expensive but suffer the usual offset, temperature, and noise problems; furthermore, a parallel control of saturation is necessary. Most of these problems are avoided in the last digital case, which allows a full software control, in particular, the on-line checking of the total charge may be valuable. It appears that 12-bit ADCs are convenient: the quantification noise is generally negligible and the dynamical range large enough to prevent too frequent overflows and rounding errors.

e. Performances. Most of the reported results show that these methods are fairly accurate: the order of magnitude of the spatial resolution is about the channel pitch and the integral linearity is better than 0.05% over the full range. (The differential linearity is included in the MCP spatial efficiency.) Detailed analyses may be found in the literature and the main limiting points are briefly summarized. The effects of the first ones are small or may be minimized.

1. The channel pitch and the asymmetry in the spreading of the electrons[29] are inherent but practically lead to negligible limitations.
2. The electron cloud spreading induces a "partition" noise,[31] which becomes negligible when the gain is sufficiently high.
3. The accuracy of the capacity chain[29], or the geometrical accuracy of the charge dividing collectors[31], and the equalization of the gain of two or four channels have to be checked carefully.
4. The DC offset in amplifiers or ADC[29] and the stray capacitances[29] introduce specific nonlinearities that may then be reduced by some proper adjustments.

The noise of the amplifiers is a more important factor and requires the use of high MCP gains and high-quality charge amplifiers. [The equivalent input charges range typically from 100 electrons rms (spectroscopic amplifiers) to 1 fC (simple charge sensitive amplifiers) for low-capacitance source and microsecond shaping.]

The last problem of the pile-up is certainly the most effective for multicoincidence experiments: an increasing count rate gradually spreads well-defined structures and in fact convolutes the whole image with a gaussian (?) but unpredictable function; this effect is reduced if the shaping amplifier "restores" efficiently the baseline and decouples the successive (randomly occurring) charge pulses.

In conclusion one should note that resolution over large operating areas, efficiency, and fast processing are contradictory features, even in the ideal conditions of well-saturated MCP regimes.

Three points have to be kept in mind:

1. The randomness is a quite perturbing factor: the time performances have to be oversized.
2. The limited range of linear data processing: if high sensitivities are expected, the incidence of nonlinearities in analog systems or the number of unresolvable events for digital systems increases and leads to position-dependent resolution or efficiency.
3. The rapid degradation of the "static" performances of fast processing.

4. *Continuous Anodes*

Continuous anode techniques are becoming increasingly popular especially for static imaging applications; the electronics may be quite simple and inexpensive or is included in a commercially available ready-for-use system[36] which makes this solution quite attractive, at least for standard applications.

The basic tool of these methods is a sheet of resistive material on which the electron cloud produced by a MCP assembly is received. Two methods are used for evaluating the impact cartesian coordinates:

1. The charge division method: the sharing of the initial charge is now resistive.
2. The risetime method: the sheet is associated with a parallel conducting plane to form a low-phase-velocity propagating medium: the positions result from time measurements.

These methods have been introduced for nuclear physics detector readouts and therefore extensively described and analyzed. Complete information may be found, in particular, in the basic papers of the Leicester group[37–43]. A short presentation of the principles is proposed in Appendix B for a one-dimension configuration.

a. The Charge-Division Method (Fig. 10). Numerical calculations have been performed at the DC limit for square[41] and circular[42] anodes (the intensities sinked at the four ports are deduced through the reciprocity theorem from solutions of Laplace's equation): the images are always distorted, but an

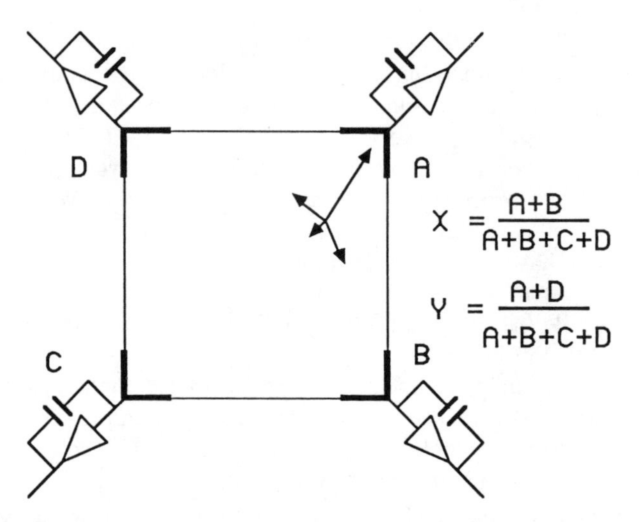

Figure 10. Possible configuration for charge-division method using a square resistive anode. Note that stray capacitances have to be kept as low as possible in order to reduce nonlinearities. More sophisticated pick-up electrodes provide "distortion-free" anodes.[36] The (hardware or software) discriminators needed for safe operating conditions are not shown.

optimization is possible about the pincushion-to-barrel transition that is observed when the pick-up electrode geometry is varied.

These defects are certainly modified by the stray capacitances and only two solutions are realistic:

1. To get a distortion-free anode (commercially available by unity in various shapes[36]).
2. To determine experimentally the code-to-position function.

When the linearity problem is solved, this method is similar to the two previous analogous ones except for two points:

1. The position determination is now intrinsically continuous.
2. The contribution of the series resistance r to the total noise: the corresponding equivalent charge may be roughly estimated from[44] $Q = (4kT\tau/r)^{1/2}$ through the usual thermal noise formula (τ is the shaping time constant). The corresponding value is comparable to the amplifiers' contribution and their incidence may be lowered by increasing the MCP gain.

Nevertheless, the observed performances are quite convincing, since 50 μm resolutions are obtained with standard MCP for 10 μs dead time. A new commercial position analyzer is available[36] in which flash ADCs allow a 8-bit spatial digitization with a 400 ns dead time; a 70 ns pulse-pair rejection is also provided, but an important flat-field correction is needed.

b. The Risetime Method. The basic two-dimension set-up is shown Fig. 11. The initial sharp charge pulse is shared among the four electrodes. The charge peaks that are broadened along the transit length in the dispersive *RC* line are amplified and shaped by a double differentiating and single integrating circuit (cr-cr-rc) into a bipolar signal; the zero-crossing time provides the position information.

A complete theoretical evaluation[43] of the linearity as a function of electrodes shapes and time constants allows a significant optimization of these parameters. A further empirical study may be useful.

This method has some interesting features:

1. The *R–C* characteristics of the anode may be largely varied and chosen to fit the MCP dimensions and/or provide a given maximum count rate. High *RC* anodes provide small sensitivities (e.g., 10 mm/μs for a 100 kHz central frequency), which allows low-frequency shaping amplifiers and a direct position quantization with TTL counters. The electronic equipment is simple and cheap but the count rate is limited by the pile-up of the long impulse responses. "Quick" surfaces allow higher count rates, but need more elaborate time digitizers. (The measure of the time difference between opposite electrodes doubles the sensitivity and reduces the nonlinearity.)
2. The resolution may be comparable to previous methods[39,45].

Nondissipative propagating media may also be used, for example, zigzag microstrips printed on a simple epoxy board[46] have successfully been used with MCP (resolution 70 μm FWHM with constant fraction discriminators).

5. *Photodiode Arrays*

The capability of the photodiode array for a parallel integration of a spatial information given by a MCP assembly has led to interesting applications[26], but this integrating function is in opposition to the principles of conventional coincidence experiments. However, CCD photodiode arrays could be of interest to record the results of a time-to-position conversion for high-time-resolution applications; in this case the usual procedure using a phosphor screen and an optical fiber coupling might be replaced by a direct electron-hole production within the collecting diode's electric field. This procedure necessitates a partial removal of the passivation SiO_2 layer, but provides an important amplification if some tens of kilovolts of acceleration voltage is applied to the impinging electrons.

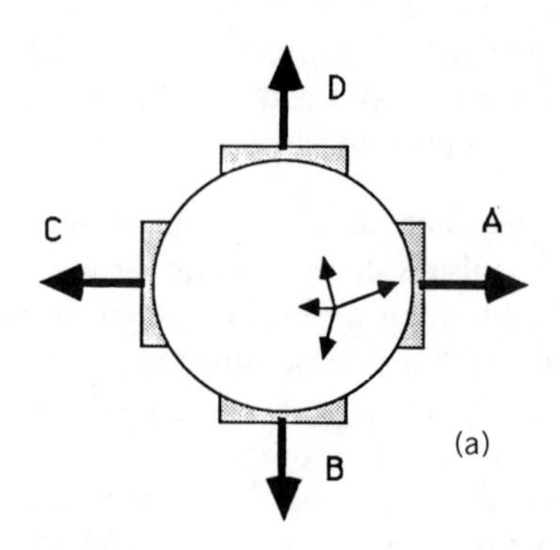
D
C
A
B

(a)

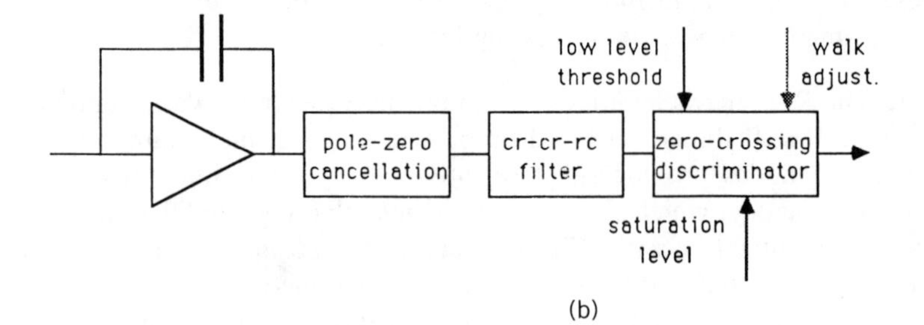
low level
threshold
walk
adjust.
pole-zero
cancellation
cr-cr-rc
filter
zero-crossing
discriminator
saturation
level

(b)

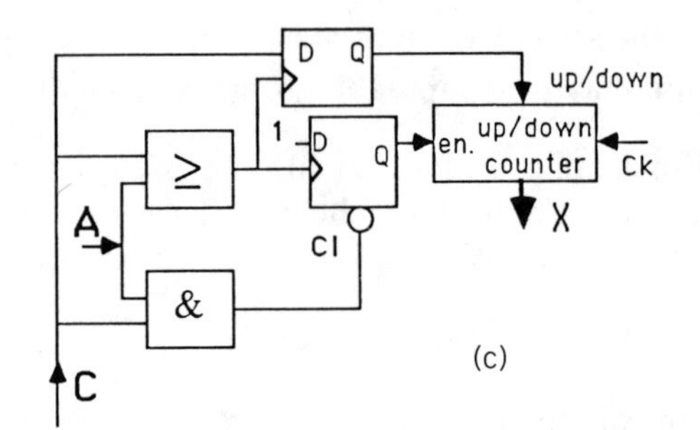
D Q
up/down
1
D
Q
en.
up/down
counter
Ck
≥
A
Cl
X
&
C

(c)

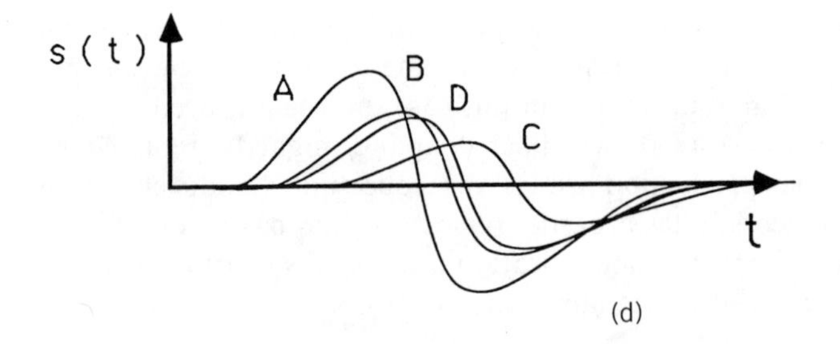
s (t)
A
B
D
C
t

(d)

IV. THE MULTICOINCIDENCE LAYER

A schematic view of a complete multicoincidence experiment is given in Fig. 12. The experiment of Fig. 2 is reproduced with the successive layers up to the host computer. Each position encoder provides a fast timing signal and two position information (Section III). For the sake of simplicity, this information is supposed to be the true position codes, that is, two logical words in which the two spatial parameters are encoded.

The multicoincidence layer processes these "signals" into a LMS "vector" describing each successive coincidence event. In the case of Fig. 12 this vector is five-dimensional: a time delay code and the two sets of two-dimension codes describing the impact locations on detectors A and B.

Two units are included in this layer:

1. A time digitizer that performs the time correlation (Section II).
2. A digital module that links time- and position-sensitive modules and aggregates the resulting codes into a LMS vector.

Figure 13 shows one of the simplest systems ensuring the coherence of the associated variables. A standard time-to-amplitude converter (TAC) is the coincidence supervisor: the two fast timing signals determine the time conversion, and the true start and true stop outputs trigger the position encoders at the convenient times, that is, ensure that time and position determinations are logically consistent. The SCA output, which becomes valid for successful time determination, enables the storage of data of interest by the host computer.

All the necessary functions are performed by this system but the data flow is obviously almost uncontrolled and more constraining systems have to be planned to ensure fail-safe operations in the basically random context of a coincidence experiment.

Figure 11. Risetime method. (*a*) Possible configuration for circular resistive anode. Note that this system is based on propagation phenomena: the backplane is equipotential and stray capacitances have to preserve the structure of the propagating medium. (*b*) Typical analog-to-time treatment of each output. The zero-crossing detector needs the same two adjustments as constant fraction discriminators. The threshold adjustment determines the minimum pulse height enabling the zero-crossing detector. The walk adjustment allows a fine determination of the time pick-up level. A third discriminator disabling the zero-crossing detector for saturated signals would be convenient. Note that safe operation needs eight conditions to be fulfilled for successful two-dimension encoding (a ninth one occurs on the time signal for time-dependent applications). (*c*) Schematic block diagram for the x-determination in Fig. 11*a*. A simple up/down counter provides a two's complement code (reset signals are not shown). (*d*) Schematic waveforms at the cr-cr-rc filter outputs for the electron cloud impact of Fig. 11*a*.

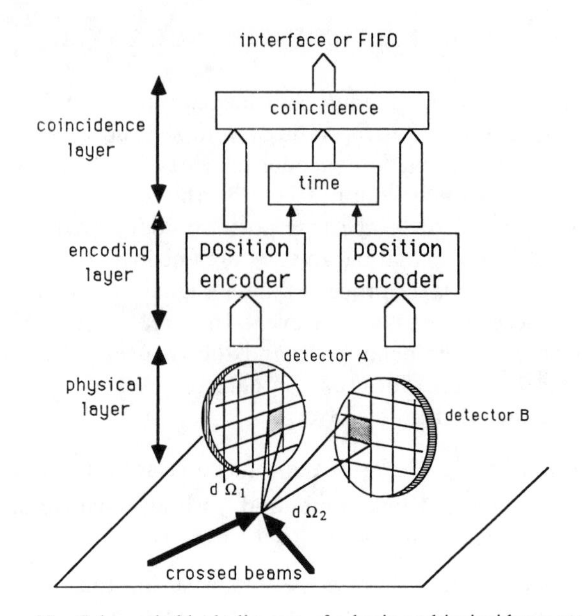

Figure 12. Schematic block diagram of a basic multicoincidence set-up.

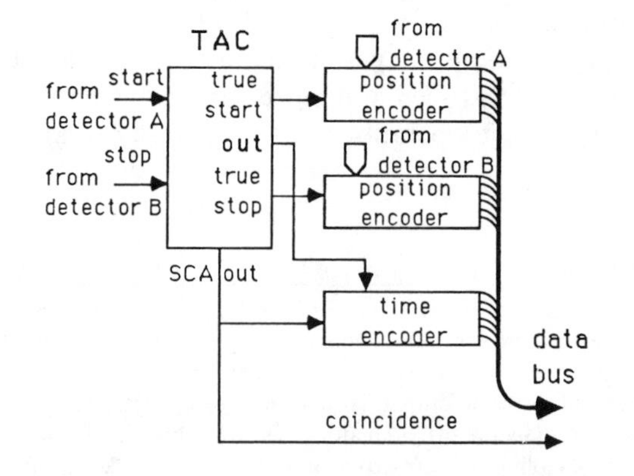

Figure 13. The simplest multicoincidence set-up. The start and stop inputs receive randomly occurring pulses from the timing amplifiers/discriminators of the two detectors. The TAC generates a synchronous true start at the beginning of each time measurement and a true stop signal only for valid operation. No control signal depending on the "black box" internal logic is displayed.

Many hardware solutions are possible depending of the energy range, the expected resolution, and, more crudely, of the technical and financial assistance, and these quite variable conditions prevent the definition of a standard multicoincidence set-up. General considerations are thus restricted to two basic points, namely, the time determination and the data handling. These principles are illustrated in Appendix C in which an example is described in some technical detail; this "multipurpose" set-up, which fits the prescriptions developed in this paper, is homemade and thus requires technical expertise. A similar system could be built using standard CAMAC modules and full software handling. Intermediate solutions may be considered depending on the relative weights of hardware and software treatments: software solutions use standard hardware and feature (generally) short development time but customized hardware solutions allow faster data acquisition.

A. Analog or digital time determination?

Two methods are generally used for time-to-digital conversions:

1. The "analogic" method takes advantage of a TAC to provide a fairly good ultimate absolute accuracy (about some tens of picoseconds), but the relative accuracy is limited by the TAC and the analog-to-digital converter (ADC) to about 10^{-3} of full scale.
2. The digital method uses a simple gated counter. This method presents for single measurements an intrinsic resolution equal to the clock period, but large time intervals are measured with a fairly high relative accuracy.

The specifications of the two methods are then opposite and the criteria for a proper choice are discovered by comparison of TOF with coincidence methods. Both methods need accurate time determinations, but the specifications are opposite.

Consider first the absolute accuracy. In TOF experiments a fine time resolution is useful because, in spite of the finite collision volume, a close relationship is observed between the overall TOF and the energy loss: this circumstance is a simple consequence of the principle of least action, which "dictates" the convenient elastic loss. The cancellation of the geometrical effects on the total TOF from the chopper to the detector does not work for inelastic channels, but the time spreading is still limited

At variance, coincidence spectra present a "natural" time width because in most cases the time delay depends on the localization of the collision: the experimental spectra result of a convolution with an experiment-dependent time window, but nanosecond resolutions are good enough taking into account the usual velocities and beam collimation in chemical physics experiments.

Consider now the relative accuracy. The corresponding problems are generally overcome by differential methods, namely, for time measurements by delaying the start event. In TOF experiments, this time offset is easily obtained from the quartz-controlled chopping clock and the problem is avoided. At variance, in coincidence experiments, this problem is critical because the difference between the TOF of the two particles from the interaction point to their respective detectors determines the time range of the measuring apparatus. If the analogic method is chosen, the time resolution will always be about 10^{-3} of full scale, because this delay is necessarily obtained through analogic delay lines (unfortunately, TACs and analogic delay lines present the same poor relative accuracy inherent in their common principle).

The TACs are then inadequate for coincidence measurements if the time delays are greater than some microseconds, that is, for most of the chemical physics experiments involving heavy and low-energy particles. Note also that the predefined time ranges of the TACs provide further practical problems. [The same remarks hold for the commercially available Time-to-Digital Converters (TDC) that have been especially designed for TOF measurements.]

Conversely, digital circuits seem specifically intended for coincidence experiments.

1. Typical resolutions of 10 ns (fast TTL) or 3 ns (basic ECL) are easily obtained on printed circuits boards for any time scale and allow a generally convenient sampling of the natural width of ion–molecule crossed-beam experiments. Higher performances[47] have been obtained or can be expected in the near future.[48] Furthermore, suppose a coincidence experiment in which only one time interval is observed. If n measurements are performed with an asynchronous clock (and then in a digital way), it is well known that the true time uncertainty is no more the time period δt, but the significantly lower value $n^{-1/2}\,\delta t$. This effect (used in all universal counter/timers) cannot be used of course when time structures are overlapping, but sharp structures are less convoluted by asynchronous determinations than by TAC synchronous measurements.

2. The velocity measurements are effectively enhanced if the flight paths are lengthened: the velocity accuracy $\delta V/V$ is directly related to the flight length uncertainty $\delta L/L$, where δL is determined by the extension of the collision volume and the direction of both flight paths. In other words, if the time unit is of the order of magnitude of the transit time in the collision region, a digital time measurement will always provide the best available accuracy.

3. Counters may be included into the correlation supervisor (see an example in Appendix C).

4. The resulting time code is of course directly available and ready-for-use for the storage layer.

B. Multicoincidences with a Single Time Digitizer

1. *Histogram Mode*

For small systems or dedicated applications, coincidence measurements may be handled by conventional multichannel analyzers or even directly by the host computer through a convenient parallel interface. In this case, a time-consuming "handshake" (or "look at me") procedure has to be used between the microcomputer and the correlator in order to handle by software not only the data transfers but also the data acquisition; this interaction between the computer and the measuring tool is directly possible for standard bus systems such as the well-known CAMAC system.

The time code is merged with the position code(s) to provide a pointer to the storage memory and the multiple histogram is simply obtained by incrementing the corresponding memory location. An example may be found in the nice experiment built at the FOM Institute for the study of dissociative processes in fast beams[5]: in this case a TAC is needed to take a full advantage of the "velocity amplification" of this "translational spectroscopy" method and provides a simultaneous energy and angular analysis of the dissociation processes.

2. *Stack Mode*

For multicoincidence experiments in which several parameters are simultaneously and accurately measured, too large memories and too many diskettes would be necessary; moreover, owing to the spread of information in the LMS, the memory cells would be underemployed. No data compression is *a priori* conceivable; therefore, the only solution is to stack the coordinates of the successive coincidences in files, which are stored in this form into diskettes. This "serial" procedure fits well the versatility of a multipurpose set-up, but is time consuming (in the example of Fig. 12 and for charge-division techniques, five or nine read cycles are necessary whether the positions are preprocessed or not). However, for self-governing correlators, the computer handles only successful coincidence events; in this case and for this serial structure, temporary registers as FIFOs are particularly well suited to derandomize the data flow and the total acquisition rate is mainly limited by the acquisition routine dead time. An illustration of these principles is given in Appendix C.

C. Enhanced Hardware Methods

The preceding solutions are not convenient when the coincidence time delays are not much smaller than the mean time interval between particles in both channels. This situation is unavoidable when high-power pumping lasers, pulsed ion sources, or supersonic beams produce high-peaked count rates (see Appendix A). This situation may be intentional when interesting low cross-section channels are buried in a large total cross section. If the S/N ratio is large enough (see Section II C), the experiment is possible but high continuous count rate is required to get a convenient $S^{1/2}$ factor.

In both cases the standard procedure induces large distortions and may even be "idle" in low-repetition-rate pulsed experiments. More sophisticated hardware solutions are necessary. Four solutions are proposed. The first two are hardware solutions for a better intercorrelation determination. The last two use a software correlation from memorized time data and apply to sharply peaked count rates.

1. Spatial Multiplexing (or Parallel Handling). The problem is the sharing of a limited set of time digitizers among a large number of detectors and even, for a same detector, between successive events. It is then similar to a telecommunication problem or, in a more convenient time scale, to the memory or resources sharing in a multiprocessor environment. A set-up is described in Ref. 49 in a nuclear physics context, but simpler designs using some fast programmable arrays or bus arbitrator circuits are certainly possible. In some cases, commercially available multistop digitizers in a bus structure may provide an almost ready-for-use solution. (Note that this solution is called in reliability textbooks "active redundancy" for which all systems are switched on in the same time. A "stand-by redundant" system would be a chain of time digitizers in which after a first start, each stop halts the running device and starts the following one; a similar procedure could also be applied to the start signals.)

2. Time Multiplexing (or Sequential Handling). The time interval may also be measured in three steps using a type of double vernier. Only two time digitizers and two digital delay lines are needed. The justification and an illustration of this method are given in Appendix D in which the first LCAM multicoincidence set-up is described.

3. Logic Analyzers. The new generation of logic analyzers may also provide valuable solutions for extreme experimental conditions as fragmentation induced by short laser pulses. Very high sampling rates are now available with versatile modular systems, for example, a complete registration of the evolution of 160 channels for 4 k successive times is possible with a 0.5 ns resolution (maximum configuration of the 92HS8 Tektronix logic analyzer).

4. Time to Position Conversion. The fast sweeping of photoelectrons allows a picosecond range time analysis (streak cameras); a coupling with integrating CCD devices could provide a quite powerful analyzer.

D. Data Processing

The dominant recurring themes of this chapter are the versatility of a multicoincidence set-up and that software problems are known to be highly dependent on subjective feelings or available computing power or software assistance. No general recipe may thus be given in this section, but some important points have to be outlined:

A computerized treatment is necessary when hundreds, thousands, or more spectra are involved.

A computerized treatment may be a simple "translation" in a computer language of "manual" treatments or may include more sophisticated data-reduction algorithms.

The number of "cells" in which the experimental measurements are scattered in a multicoincidence experiment is comparable or greater than the number of available memory cells in a current microcomputer (in 1990).

Some aspects of these problems are discussed in Section IV E 6. The subject is now restricted to the extension of handy treatments in a multicoincidence framework.

In a manual procedure, the treatment is only a bijective transformation between two spaces of similar structure (with some precautions with multiform functions),[50] and redundant variables in the LMS are ignored. In the histogram mode, the data processing is strictly "conventional": the background is conveniently removed and the remaining signal is transformed to the suitable coordinate system. The same procedure holds in the stack mode for poorly resolved experiments. The final result is obtained by successive additions of off-line built histogram channels. This method fails if too many sets of histograms are needed; in this case, another method has to be used, in particular, noise is removed after the treatment. These general considerations are illustrated in Appendix D by the description of the data-reduction algorithms used for the $X^- + H_2$ experiment.

E. Miscellaneous Remarks and Dreams

1. *Spatial and Time Correlations*

The conceptual and practical aspects of the coincidence method have deliberately been described in a conventional framework: the time is the two-faced main tool and the position determination is mainly a convenient but secondary feature. In fact, when the crossed beams are poorly defined, the correlations between companion particles observed in the laboratory frame are spread. Such a circumstance occurs when a relatively fast particle impinges an uncooled target of similar mass (which may be the case for expensive or rare species); the recoil particle is mainly sideways scattered (preserving the angular correlation), but its small velocity is widely perturbed by the thermal effect. Fayeton et al.[13] have clearly shown (the experimental set-up is described briefly in Appendix D) that valuable energy information could still be obtained from the angular dependence of dramatically spreaded coincidence spectra.

2. *Coalescing Coincidence Spectra*

Preliminary note: in this section we are dreaming of some ideal but possible experiment. An experimentalist always enjoys to "look at" the main characteristics of the system under study on the experimental spectra. This serene situation holds when, in spite of perspective effects, the main features of the "velocity vector diagram sphere" are still perceptible in the laboratory frame, in other words, when at least one of the multiple LMS variables' is so closely related to a "physical" CM variable that physically equivalent events are grouped along this variable. One should remark that, in fact, the serenity of the experimentalist is a secondary feature and that these advantageous situations determine nothing less than the feasibility of an experiment, since a limited and efficient scanning over well-chosen "laboratory variables" provides valuable global information.

In chemical physics measurements, these fair conditions are approximatively fulfilled at high collision energy or, at low energy, under quite asymmetric conditions: most of the differential cross-section measurements have then been done in these favorable configurations. However, some circumstances are existing, or may be created, in which the multicoincidence technique allows one to overcome these limitations. This point may be illustrated on the example of an ion pair formation experiment at thermal energies.[51] Suppose the endothermic reaction $CO + N_2 \rightarrow CN^- + NO^+$ ($\Delta H = 12.1$ eV) and the experimental set-up (Fig. 14*a*) in which two seeded and heated supersonic beams are crossed at various angles in order to get differential cross sections for some CM energies.

Here a multicoincidence detection is possible, since the two particles may be detected, but under two unpleasant conditions: as easily seen on the Newton diagram (Figs. 14*b* and 14*c*), the ions may be scattered in the laboratory frame in any direction and with (almost) any velocity.

The time and position measurement then have to be performed over very large ranges: if the count rate is relatively high (the total cross section is low but two supersonic beams are crossed), the time spectra will be distorted and the spread true coincidences buried into the false coincidence background.

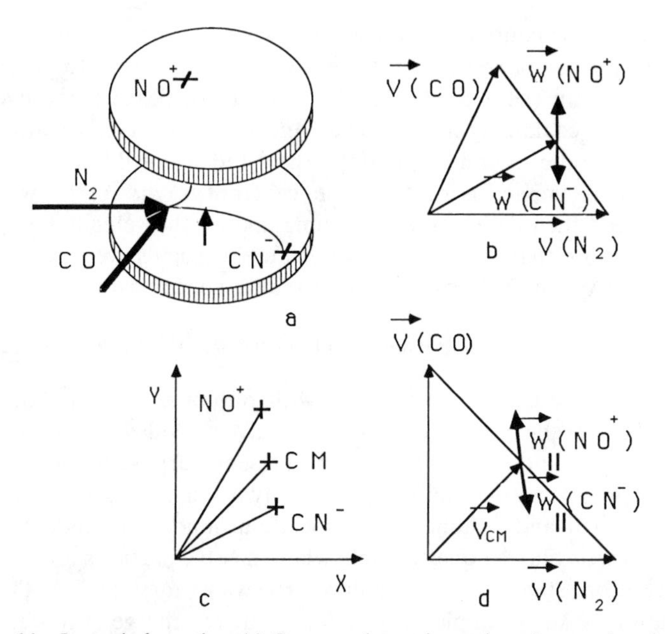

Figure 14. Ion pair formation. (*a*) Conceptual experimental set (perspective view). Ionic species produced at the crossing point of the molecular beams are pulled to two MCPs parallel to the horizontal macroscopic collision plane by a vertical uniform electric field. (*b*) Perspective view of the Newton diagram of an ion pair formation. (*c*, *d*) Projections on an horizontal plane of parts *a* and *b*. Part *d* is the projected Newton diagram. Only horizontal (or parallel) velocity components are represented. The two molecular beams are orthogonal for drawing convenience. The CM collision energy may be varied by heating the supersonic jets and by rotating one of the incident beam. Part *c* displays the projection of ion impacts on an horizontal plane. The origin is the projection of the scattering center; the axes are defined by the two-dimension position analysis (and coincide in the present case with the axes of the two incident beams).

A solution could be to put the collision region in a weak electric field that separates the two ionic species and allows an overall collection by two MCP assemblies (Fig. 14*a*). The velocity information is not lost in a multicoincidence experiment, since the time delay is measured. If the masses are known, this time delay, the conservation rules, and the localization of the impacts allow the full restoration of the initial conditions of the flights of the two particles. The multicoincidence method then allows a full spatial collection as in a total cross-section measurement but without any information loss. Is such an experiment feasible and what about the resolution? The problem of time-spread histograms is still pending, but an experimental parameter is still free: the ratio between the distances of the two detectors to the macroscopic collision plane. Any student easily proves, using the momentum conservation along the electric field direction, that the flight times of the companion ions to the two detectors are equal (whatever these velocity components are) if the distances are inversely proportional to the masses. In these conditions the coincidence peaks for a given mass ratio are then coalescing at zero delay and kinematical information is still available. In the second question the student will prove that the "center-of-mass impact" is located at the "center of gravity" of the two impacts (Fig. 14*d*) and then deduce the flight time and all the velocity components. Note once more that this exercise is not purely speculative since the Orsay experiment (Appendix E) uses a partial spatial concentration.

3. *Multicoincidence for Three-Body Systems*

Dissociations induced by interaction with photons, electrons, or massive particles have been studied extensively and have provided a lot of nice results on mechanisms and structures. For the present purpose we will only note that these processes are usually analyzed as two-step processes: the molecules are first excited and assumed to dissociate around their undeflected CM. This hypothesis may be questionable when massive particles are involved in the excitation step and is certainly irrelevant for "really" three-body dissociations. An example is the dissociative charge transfer (DCT): $X^- + H_2 \rightarrow X + H + H^-$, where the exchanged electron is the signature of such a complex process. (This channel has been observed in TSH calculations by Sizun et al. for the $Cl^- + H_2$ system at high collision energy.)

The multicoincidence provides two possibilities for a complete kinematical analysis:

1. The three particles are detectable: triple coincidence. A recent $(e, 3e)$ experiment[52] proved the feasibility of this method. The three-way coincidence set-up is simple: the same starting particle triggers two parallel two-way coincidence systems independently stopped by the two other particles. The AND function of these systems provides three-way coincident events; two OR functions provide two sets of two-way coincident events. The true triple coincidence events appear in a bidimensional histogram of the two time delays as structures superposed on a false coincidence background. This background is itself structured as a biexponential surface corresponding to really false coincidences crossed by three "walls"; these walls correspond to measurement sequences in which only two particles issued from a given interaction have been detected. Recently, Roncin et al. have observed triple coincidences between two ions and one photon for the following electron-capture reaction:[7,34]

$$\begin{aligned} B^{3+}(1s^2) + He &\rightarrow B^{2+}(1s^2\,2p) + He^+ \\ &\rightarrow B^{2+}(1s^2\,2s) + h\nu \text{ (206 nm)}. \end{aligned}$$

Note that the B^{2+} ion is energy and angle analyzed.[7] One should remark that a multicoincidence experiment involving an energy analyzer is also a self-chopped TOF experiment (cf Section II). In this case the collision time is also given by the photon: this redundancy sweeps out most of the false coincidence.

2. Only two particles are detectable: double coincident TOF. In this situation (of course more frequent in chemical physics) the lacking information could be obtained through a TOF method. The usual crossed-beam structure is unchanged, but the charged beam is swept back and forth over a slit by a square wave electric field; this slit allows only short packets of ions to interact with the target beam at well-defined times. In the DCT example some hundreds X^- ions concentrated in a short pulse are thus colliding with the H_2 beam; when a collision occurs, the negative species is first detected. Such an event triggers a first two-way coincidence system; if a neutral particle impinges in a convenient time window the corresponding detector, the first coincidence system is stopped to provide the time delay as in a usual coincidence measurement and a second time measurement is done to determine, as in an usual TOF experiment, the TOF of the neutral. Such a measurement is done by a second coincidence system started by the accepted neutral and stopped by the beam chopping clock. The true TOF of both particles are then determined. Could this experiment provide the expected result, that is, the characterization of the unambiguously isolated DCT process?

The beam modulation in TOF experiments normally prevents coincidence measurements, since a structured beam artificially induces a time correlation. The observation of a bidimensional plot of the two TOF spectra would confirm this fact: the usual (almost flat) biexponential structure is replaced by a convolution of the two TOF spectra except for the convenient delay. In this case, coincident particles are concentrated in enhanced structures superposed to the physically unrelated (but) modulated background and allow a first filtering. (This point has been outlined in a more mathematical way in Section II.)

The angular characteristics of a "true" three-body system are experimentally unknown and their determination could be puzzling if large and smooth time and position structures were dominant as allowed by the less stringent conservation rules in this three-body configuration. A "signature" of the DCT process is then needed. This signature may be given by a convenient data treatment: the angular and velocity information obtained by the TOF analysis of the two detected particles completely determine the velocity vector of the third particle and then the total CM final kinetic energy. The histogram of this released energy shows unambiguously the DCT process and allows, by a second (energy) filtering, the determination of its properties. The calculated released energy is then used as an experimentally determined test variable for the DCT process. One should remark that this procedure may easily be checked: the released energy histogram must present two structures, the sharp DCT peak and the reactive charge transfer profile. For this profile the calculated H velocity vector has to be compatible with the HX formation, that is, has to be found equal to the measured velocity of the thereby known neutral particle.

This experiment would use seven modules of the new system described in Appendix C.

4. *Multicoincidence for Many-Body Systems*

As yet these problems are only considered in high-energy physics; if the orders of magnitude (energy and cost) are for a while forgotten, two solutions matching the reduced scale of the chemical physics experiments may be considered.

The first one is an extension of the preceding three-body procedure. Bus-oriented systems such as the Orsay system (Appendix C) may be convenient if only few (coincidence) "ways" have to be opened. In fact, if a possible application is the analysis of some explosion mechanisms in fast beams of polyatomic molecules or clusters, the experimental set-up is similar to a photodissociation analyzer: all species impinge the same MCP and discrete anodes are necessary. In this case, the number of simultaneous two-way coincidence systems between n collectors increases as $n(n-1)/2$: a qualitative improvement of the techniques is necessary.

The second solution using a logic analyzer (cf. Section IV C) may be the basic tool of such a step. Every "way" is equally handled and the correlation needs only computer memory as hardware and a lot of software "components."

5. *Multicoincidences and High-Energy Resolution*

As pointed out by Doverspike [53], time may be used as a premonochromator: a pulsed source of energy-spread particles produces a time-spread pulse of energy-defined particles in the collision region: the collision energy is time dependent but is well known if the collision time is known. This time is easily determined for collisions leading to electron or photon emission or may be recovered with a double-coincident TOF or a conventional analyzer.

6. *Hardware and Software*

Returning to the data-processing sections (IV D Appendix E), one can be concerned by the gap between a sophisticated hardware and a rather poor software: experimental data are processed in a "manual" way and their redundancy is in fact not really used to enhance the accuracy of the contour map determination. In other words and using the example of Appendix E, three independent equations obtained through the three independent measurements (the time delay and two positions y and z) provide the three components of the CM velocity vector of the neutral particle. A fourth datum is obtained (position x on the underlying collector) leading to a fourth dependent equation: Is it possible to take advantage of this redundant equation?

The deconvolution method is essentially determinist and provides only a practical procedure to recover detailed data that have been blurred (linearly) by experimental conditions. This method does not seem to be useful in multicoincidence measurements:

1. Highly resolved experimental spectra are noisy; it is well known that a Fourier deconvolution that involves the division of frequency spectra generates, in this case, spurious structures.
2. Fourier deconvolution could be applied on CM reduced data, but in this space the evaluation of the "apparatus function" would be hazardous.
3. A three-dimension Fourier deconvolution is possible, but time and memory consuming.

The "softer" framework of fitting methods is *a priori* more suitable. This subject may be introduced by describing the first determinations of the reactive charge transfer (R) in the Orsay experiment (Appendix E).

As will be seen in Section V, the R process is sharply peaked; successive Monte Carlo simulations of peaked R processes were performed for various values and widths of both the deflection angle and the endothermicity until a convenient "visual" fitting was obtained among all the experimental and simulated redundant histograms. (These simulations included many experimental random parameters such as the definition of the beams.) Therefore redundancy may be taken into account but some general comments have to be made.

1. Fitting or maximum likelihood methods are the extensions for "dirty conditions" of the projection function. "Dirty conditions" mean that these "projections" are performed in a noisy environment and on nonorthogonal bases. In these cases the results are dependent of the chosen criteria (the determinism of a least square fit cannot hide this fact).
2. The "manual" data treatment is a convenient safeguard and may provide a useful starting point.
3. The history of the measurements of the speed of light has proved that

the overlap of the error bars is not always the best measure, even for fundamental constants. Is a sophisticated "computer-aided" data reduction possible using the present computers?

An improvement of the manual treatment might be obtained utilizing a random or gradient optimization using bounds defined by this preliminary treatment. The random optimization locates the main minimum of the error function, which is then obtained by the iterative gradient optimization. Two problems arise: all the processes have to be determined in the same procedure and convenient optimization criteria have to be chosen. This second problem, namely, the choice of a proper error function for this many-dimensional LMS (including sensitivities to the different time and position LMS variables), is not yet solved and conditions the next step, which would be a fully automatic data reduction. If this problem were solved, practical problems would still remain, for example, a reduced set of try spectra has to be constructed to provide efficient optimization for reasonable treatment times.

The main problem of multicoincidence experiments is then the lack of efficient algorithms that could take full advantage of the large amount of information provided by the hardware. Conventional data-reduction methods apparently do not work, and some new concepts are needed. Solutions may be expected from two directions. On one side is computer science: expert systems or even artificial intelligence concepts could help us to "dress" our too straight procedures with some "physical feeling." On the other side, we may think about some relatively new mathematical concepts such as fractals, which are now more and more used for the analysis of complex systems and the "dilute medium" of the LMS is certainly a complex one.

This emphasis put on unsolved problems is intentional. The development of such techniques involving high-level "hardware" information should generate parallel development of high-level software. Experimental papers generally escape comments on this "scientific instrument" (which also could be reviewed).

7. *State-to-State Chemistry and Shannon's Theorem*

State-to-state chemistry involves well-defined initial and final states, and differential state-to-state chemistry will led to measurements of the angular dependence of well-defined cross sections. In some cases these processes result if only some few coupling and pure, that is, not averaged, structures may appear (such as Stueckelberg oscillations in atomic physics).

These possible spatial features are observed on the quantized basis of a MCP or sampled by a conventional analyzer; in both cases, the low count rate and the dispersion of fine structures over a presumably large angular range may lead to improper sampling. The main result of digital-signal-processing theory is unfortunately that data obtained through an improper sampling are not poor but wrong. When usual time signals are involved, the sampling is performed after an initial low-pass filtering has removed any frequency component higher than half the sampling frequency: the method to ensure in collision physics that a spatial spectrum does not contain "spatial frequencies" higher than half the inverse resolution is not obvious. In some conditions a multicoincidence technique may overcome these difficulties. As was seen Section IV E 2, a multicoincidence method may allow one to collect charged particles on a high-resolution spatial detector without information loss.

V. APPLICATION TO $X^- + H_2$ REACTIONS

Negative ions colliding with molecules undergo a complex variety of reactions owing to the existence of ionic and detachment channels.

In a simple colliding system such as $A^- + BC$, the purely ionic channels are

Charge transfer, $A + BC^-$.

Atom abstraction, $AB^- + C$.

Reactive charge transfer, $AB + C^-$.

Similar processes are involved in positive ion reactions; they are here restricted to the ones in which stable species as regard to detachment are produced.

The detachment channels are

Simple detachment, $A + BC + e^-$.

Reactive detachment, $AB + C + e^-$.

Associative detachment, $ABC + e^-$ if ABC (and/or BAC) are stable molecules.

Many reactions producing ions have been studied (see the review of Franklin[55]). A few examples are:

1. Reactions involving charge transfer:

Simple charge transfer $A^-(BC)BC^-$

$Br^- + O_2 \rightarrow O_2^- + Br$ [Ref. 56].

Ion transfer $AB^-(CD)BCD^-$

$NO_2^- + H \rightarrow NO + OH^-$[Ref. 57].

Reactive charge transfer or positive ion abstraction $A^-(BC)C^-$

$OH^- + CH_3Cl \rightarrow CH_3OH + Cl^- +$ [Ref. 58].

2. Reactions involving atoms (or group of atoms), abstraction $A^-(BC)AB^-$ or transfer $AB^-(CD)A^-$:

$O^- + H_2 \rightarrow OH^- + H$ [Ref. 59]
$Br^- + CH_4 \rightarrow CH_2Br^- + H_2$ [Ref. 60]
$SF^{-6} + H \rightarrow SF_5^- + HF$ [Ref. 61]

Detachment processes, other than associative detachment (AD), have been studied less:

Simple detachment $A^-(BC)A + e^-$
$X^- + H_2 \rightarrow X + H_2 + e^-$[Ref. 6, 62–65].
Associative detachment $A^-(BC)ABC + e^-$
$O^- + CO \rightarrow CO_2 + e^-$ [Ref. 66]
and
$A^-(BC)BAC + + e^-$
$O^- + H_2 \rightarrow H_2O + e^-$ [Ref. 67].
Reactive detachment $A^-(BC)AB + e^-$
$Cl^- + H_2 \rightarrow ClH + H + e^-$ [Ref. 6, 62–65].

A crucial point for the understanding of the dynamics of anion systems is the interplay between the two families of reactions. Only a dynamical study of all the channels can provide information on the interdependence of these competing processes. The dearth of studies on the dynamics of anion reactions may reflect the difficulty of a complete analysis of a given system. It will be shown here that the multicoincidence method partly solves that problem.

Outside the remarkable studies performed in Leone's group on neutral-product-state analysis, there is no information on product-state internal energy distribution in anion–molecule reactions. The probing of product ions by infrared multiphoton dissociation or photodetached electron spectroscopy has seldom been used. The lack of techniques for preparing selected internal energy states may explain the fact that no anion reaction has been studied in the framework of state-to-state chemistry.

A. Ion-Formation Reactions

As for positive-ion reactions, the flowing-afterglow method has led to a wealth of data in exothermic anion reactions in the temperature range 80–900 K (0.01–0.1 eV). These successes have been reviewed in some detail by Ferguson.[68, 69] The channeling of reaction exothermicity into vibrational and rotational levels of products has been amenable to study by coupling flowing-afterglow techniques with optical-detection methods of infrared luminescence and laser-induced fluorescence (LIF). In $Y^- + HX \rightarrow HY + X^-$

reactions (with $Y \equiv$ halogen,[70,71] O,[72] or CN,[73] and $X \equiv$ halogen) the studies demonstrate that these reactions are direct and that a substantial amount of the available energy is deposited in product vibration. The highest vibrational level allowed by the exothermicity is always observed.[74]

A wider range of energy, 0.01–3 eV, difficult to investigate by other techniques, is offered by the flow drift tube.[75,76] Different types of energy dependence were found for exothermic reactions and have been discussed in the framework of reaction models.[77]

At higher collision energies, the total cross sections have been measured by tandem-mass-spectrometry techniques for many systems and over wide energy ranges.[78,79] A sharp decrease of the total cross section with increasing energy is observed for exothermal reactions, whereas, in endothermic reactions, the cross section rises steeply above the threshold energy, reaches a maximum value and then falls more or less rapidly. Such studies yield information on electron affinity and binding energies.[60] In the case of S_N2 (which denotes second-order nucleophilic substitution) reactions of the form $X^- + CH_3Y \rightarrow Y^- + CH_3X$, the observation of an additional barrier above the reaction enthalpy is tentatively explained by the assumption that these reactions proceed via a transition state with C_s symmetry.[80]

The use of isotopically labeled reactants has led to a deeper insight into anion–molecule reactions and to reaction models. As an example, the exothermic reaction $^{16}O^- + D_2{}^{18}O$ has received a lot of interest.[81–84] The evolution of the branching ratio of $^{18}O^-$,$^{16}OD^-$ and $^{18}OD^-$ [83,85] has given indication that the reaction would proceed below 1 eV collision energy through a long-lived complex, whereas a stripping mechanism would apply at energies above 5 eV. However, only the velocity angular distribution would provide a more detailed insight into the involved mechanisms. Such an experiment performed by Karnett above 3 eV, shows that the stripping of D obeys a spectator stripping model.[86]

Very few dynamical studies are available. The first one, performed by Doverspike[87] on the $O^- + D_2 \rightarrow OD^- + D$ exothermic reaction, indicates a stripping behavior from 3.6 to about 7 eV, whereas at lower energy the reaction proceeds through an intermediate long-lived complex.[88]

As a general rule the only reactions that have been angularly analyzed are atom abstraction $A^-(BC)AB^-$: OD^- formed on $O^- + D_2$,[87,89] $O^- + D_2O$,[86] and $O^- + C_3H_4$.[90] The reason is simple: the atom abstraction produces a fast ion forward scattered in the laboratory frame, whereas the recoil ion C^- resulting from a reactive charge transfer is scattered through all laboratory angles and therefore is more difficult to analyze in classical experiments.

Recently, the energy analysis of the recoil ion has been performed by Esaulov et al.[91] for the $O^- + H_2 \rightarrow OH + H^-$ reaction in the 0–10 eV energy

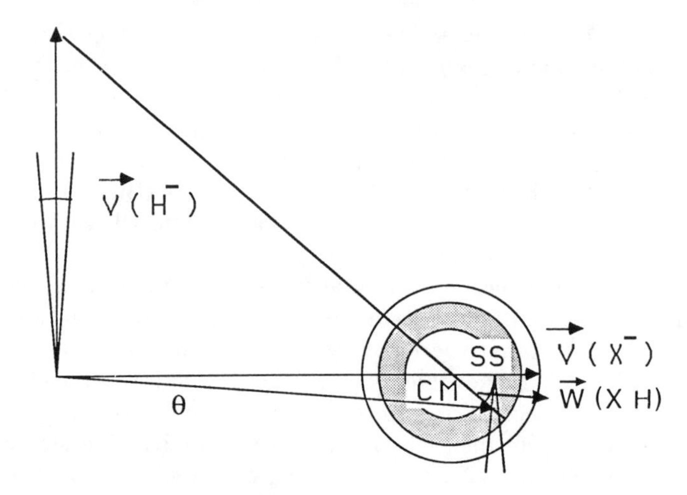

Figure 15. Representation in the CM frame of the angular velocity **W** (XH) distributions corresponding to the H^- collected perpendicularly to the fast incident beam.[91]

range. Only the H^- ions ejected in a perpendicular direction are detected, therefore, just a small portion of the velocity and angular distributions is observed (see Fig. 15). The vibrational excitation of the OH fragment deduced from the structured H^- distribution is very small at the lowest energies ($v = 0$ for $E = 2\,\text{eV}$) and increases with the collision energy ($v = 1,2,3$ at 5.5 eV).The H^- energy distribution is found to be quite similar to the spectra resulting from dissociative attachment in $e^- + H_2O$ scattering, which raises the question of the relevance of negative ion resonance states in heavy-ion collisions.

B. Electron-Detachment Reactions

The detachment processes were studied only through the energy behavior of the global detachment cross sections. When the ABC or BAC molecule exists, the associative detachment is exothermic; at thermal energy only associative detachment reaction is possible and is then studied alone. This intriguing ion–molecule process was first observed in flowing-afterglow experiments.[67] Information on the vibrational-state distribution of the molecules thus formed is now available from the flowing-afterglow infrared chemiluminescence technique developed by Leone's group.[92] The vibrational distribution provides important information concerning the dynamics of this detachment process. The associative detachment (AD) has largely been studied in drift-tube experiments as a function of collision energy. (See the review of Fehsenfeld[93].) Because of its low mass, the electron can take away only an extremely small part of the orbital angular momentum of the reactants. Its very small kinetic energy (< 0.1 eV) has been measured by Mauer and Schulz[94] and recently confirmed by Esaulov et al.[91] The energy of the electron slighly increases with collision energy when other detachment processes are also involved. Comer and Schulz [95] were able to investigate, in a beam experiment with a gas-filled chamber, the total detachment cross section (TDCS) of O^- colliding with various molecules (N_2, CO_2, NO, C_2H_2, C_2H_4, CO, O_2, and H_2) in the energy range 0.5–20 eV, which contains the threshold of all the other detachment channels. The TDCS has also been measured in the same energy range by tandem mass spectrometry. This was made possible by addition of SF_6 to the collision chamber.[83] SF_6 ions are monitored as a measure of the AD process, since the resonant-electron-attachment cross section of SF_6 is very sharp and peaks quite close to 0 eV. All these complementary techniques give results in surprisingly good agreement. Outside of some insertion reactions, the typical behavior of the AD process is a sharp decrease of the cross section when the collision energy increases.[67,83,96–98]

General experimental conclusions and theoretical descriptions about AD have been drawn by Fehsenfeld.[93] In the special case of insertion reactions $A^- + BC \rightarrow BAC + e^-$, a kinematic two-step mechanism has been proposed,[97] which predicts whether or not the reaction takes place at thermal energy:

$$A^- + BC \xrightarrow{(1)} [AB^- + C] \xrightarrow{(2)} BAC + e^-.$$

An initial capture of B by A^- is followed by a detachment recombination of AB^- and C before they separate. According to this, the insertion reaction is observed at thermal energy if the first step is exothermic ($O^- + H_2$, $S^- + O_2$, and $O_2^- + N$); it is not observed at thermal energy and is enhanced by collision energy if the first step is endothermic ($C^- + H_2$, $C^- + O_2$, $S^- + H_2$, $S^- + O_2$, and $OH^- + N$). When the energy is large enough, simple detachment and reactive detachment become possible, and the two contributions are overlapping in the energy-dependent global detachment cross section.

At keV collision energy when the velocity of the active electron matches the collisional velocity, quasiresonant charge exchange to a shape resonance of the molecular target is observed through the detached energy electron spectra; this subject has been recently reviewed by Esaulov.[99] Such a process cannot be observed in the low-energy range of interest in which reactive detachment occurs.

In an attempt to see different behaviors for collisional detachment and reactive detachment, the Williamsburg group has made systematic studies of various atomic ions colliding with H_2, HD, D_2 in the energy range 0–30 eV: halogen anions,[100] H^-,[101] O^-, S^-,[102] and Na^-, K^-.[103] In these experiments the released electrons are axially confined in the collision chamber by a magnetic field and attracted to a collecting plate. Since the energy-dependent profile of the H^- total cross section independently measured usually displays a peak for small energies, the existence of a similar structure in the electron cross-section profile is believed to be associated with a reactive detachment.

In some cases ($F^- + H_2$ at high energy) a strong isotope effect is observed, indicating that detachment is induced by nuclear motion of collision partners and therefore does not necessarily have to involve a surface crossing and is more likely due to a direct coupling with the continuum. A theoretical model of this process has been discussed by Gauyacq[104,105] for collisional detachment of H^- by Ne and He using the zero-range-potential model proposed by Demkov.

In most cases the isotopic behavior is rather complicated; it is inferred that some peculiarities such as $\sigma(H_2) > \sigma(D_2) > \sigma(HD)$ or $\sigma(HD) > \sigma(H_2) > \sigma(D_2)$ indicate the presence of reactive detachment due to surface crossing. Similarities are observed between the isotopic behavior $O^- + H_2 \rightarrow OH + H^-$ and the dissociative attachment $e^- + H_2O \rightarrow H^- + OH$, that lead these authors[102] to speculate about the relevance of intermediate molecular negative resonant states in these heavy-ion reactive collisions.

These studies raise many questions and show that the underlying chemistry is very complex. It is clear that the velocity analysis of the neutral products is a necessary step in the discrimination of the two reactive and nonreactive channels. Until recently the experiment of Cheung and Datz[106] has been the only attempt in that direction. In this experiment the H_2 or D_2 target effusing from a cold multichannel array crosses the ion beam at right angles. The TOFs of the fast neutrals issued from the collisional process are measured. The observation is done at a laboratory angle of 1^0; therefore, just a part of the total angular distribution is collected. Four different energy-loss channels were observed in the CM energy range of 7–50 eV. Two of them are clearly assigned, respectively, to a collisional detachment, which, whatever the collisional energy, leaves the H_2 molecule in the $v = 0$ vibrational level and a dissociative one through a sort of spectator kinematics. However, this energy analysis does not separate the neutral products resulting from detachment and charge-transfer reactions.

C. Competition between Ion-Formation and Electron-Detachment Reactions in the $X^- + H_2$ Systems

An unambiguous analysis of these anion–molecule systems implies the simultaneous determinations of

the family of the process (either detachment or charge-transfer reaction)

the velocities

the corresponding angular distributions.

In the Orsay apparatus described in Appendix E determination of these items is ensured by a multicoincidence method. The doubly differential cross sections accounting for the out of plane scattering that this method provides are plotted as contours on the velocity vector diagrams. The relative contribution to the total cross section of each competing channel (except atom abstraction) appears with its individual angular distribution and energy disposal.

This study of the $X^- + H_2$ systems extends from the simpler case where X is an halogen atom in which the competition at low collision energy is limited to the three channels:[6,62-64]

Reactive charge transfer, $XH + H^-$ (R)
Reactive detachment, $XH + H + e^-$ (RD)
Simple detachment, $X + H_2 + e^-$ (SD)

to the more complex case where X is S[65] in which two more channels enter into the competition:

H atom abstraction, $XH^- + H$ (HA)
Associative detachment, $XH_2 + e^-$ (AD).

At higher energies two dissociative channels are opened:

Dissociative charge transfer, $X + H + H^-$ (DCT)
Dissociative detachment, $X + H + H + e^-$ (DD).

The various endothermicities of each channel are reported in Table 1.

The evolution of the velocity vector diagrams has been studied for each system in the energy range extending from 20 eV to the lowest collision energy giving detectable neutral particles: below the first threshold for S^- and about 3 eV above threshold in halide systems. Each diagram exhibits the various competing channels: for instance in Fig. 16 for the Br^- system

TABLE I
Enthalpy of the Various Processes

ΔH_{eV}	S	F	Cl	Br	I
AD	- 0.99	—	—	—	—
R	2.12	3.28	2.91	3.73	1.28
RD	2.87	4.08	3.66	4.48	2.03
SD	2.07	3.36	3.60	3.06	3.40
DD	6.55	7.83	8.05	7.51	7.87
DCT	5.90	7.08	7.30	6.76	7.12
HA	0.68	—	—	—	—

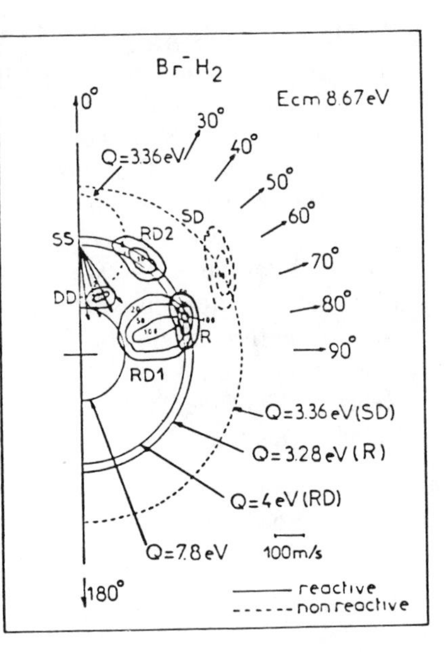

Figure 16. Contour diagram for system $Br^- + H_2$. Reprinted from Ref. 63 with permission of Elsevier Science Publishers.

at 8.67 eV, the various structures correspond to simple detachment, reactive detachment in two different dynamical behaviors RD1 and RD2, reactive charge transfer, and a dissociative-detachment process.

Some general rules emerge from the evolution of such maps as a function of the collision energy and remarkable features appear as characteristics of the various channels. All the systems exhibit two different regimes as a function of the collision energy, which appear through two different dynamical behaviors called RD1 and SD1 at low energy and RD2 and SD2 at high energy. The limit between the two regimes is about 10 eV for halide systems and about 6 eV in the S^- system.

1. *Total-Cross-Section Energy Profiles*

Absolute cross sections are calibrated using the total cross sections obtained by the Williamsburg group (Figs. 17, 18).[100,101]

1. In the low-energy regime, R, RD1, and SD1 cross sections exhibit peak structure. The R and RD1 have the same magnitude for all the systems with two exceptions: the RD1 channel is not populated in iodine and the R channel is 10 times higher in fluorine. For all systems, R appears at its enthalpy threshold, whereas RD1 has an onset above its reaction enthalpy. At the lowest energy the associative detachment (AD) is observed in the S^- system up to its dissociation limit in SH + H. At this energy RD is observed in

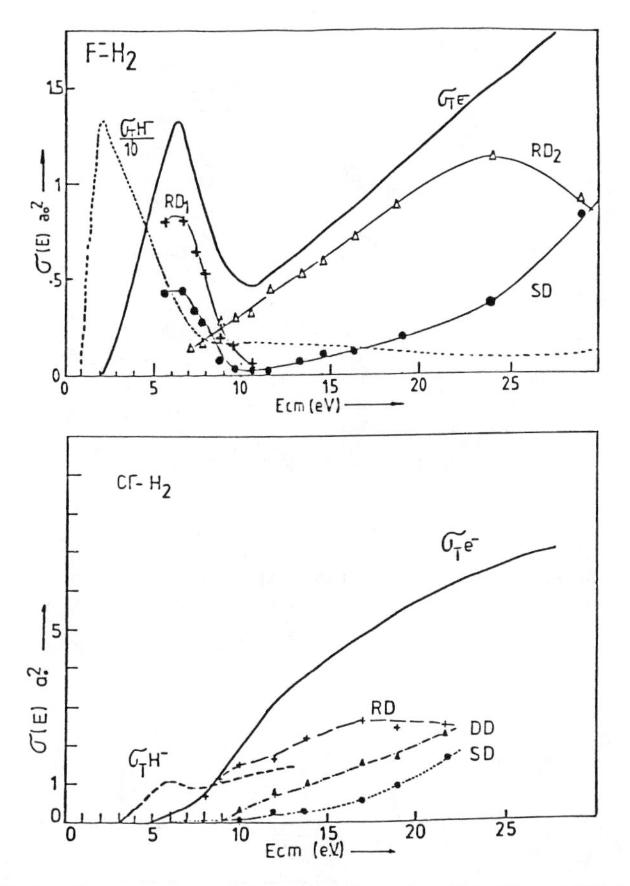

Figure 17. Total cross sections of the various channels in $F^- + H_2$ and $Cl^- + H_2$ systems normalized to the total cross sections measured by Huq et al.[100] Reprinted from Ref. 64 with permission of Elsevier Science Publishers.

an ephemeral dynamical behavior called RD0, which disappears at 4 eV (Fig. 17).[65]

2. In the high-energy regime RD2 and SD2 have increasing cross sections accompanied by an important DD process, which does not show up for F^-.

2. *Velocity Diagrams*

At the lowest energies a particular case is of course observed for the $S^- - H_2$ system. Below the SD threshold the AD process is observed; the SD appears at threshold in a forward/backward distribution, and when the RD channel is opened, the AD flux turns into the side-scattered distribution called RD0

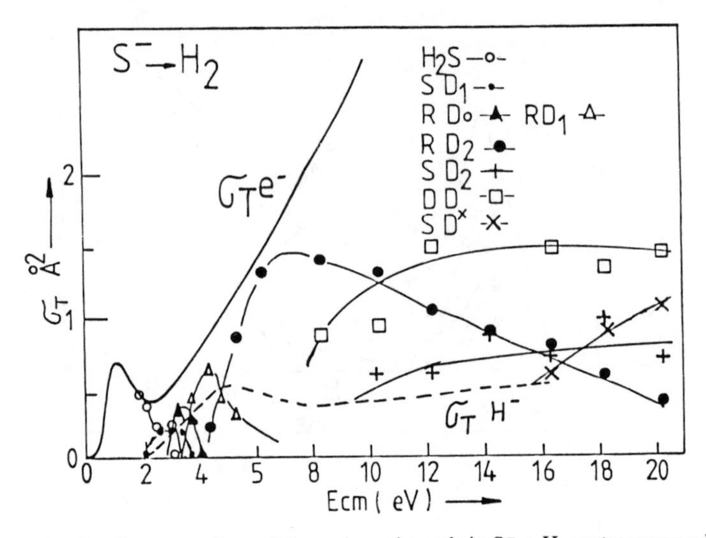

Figure 18. Total cross sections of the various channels in $S^- + H_2$ systems normalized to the total cross sections measured by Huq et al.[101] At the lowest energy, the total cross section corresponds to the AD reaction.[93,95] Reprinted from Ref.[65] with permission of Elsevier Science Publishers.

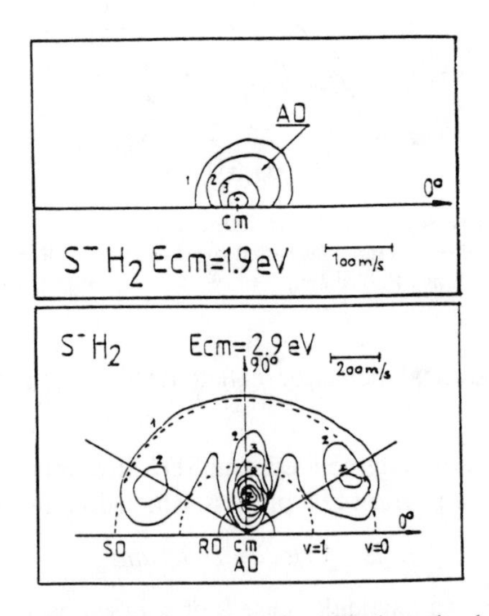

Figure 19. Contour diagrams for $S^- + H_2$ system. The intensity decreases by a factor of 8 between two successive contours. The various circles are the loci of maximum velocity corresponding to the endothermic threshold of each process. Partly reprinted from Ref.[65] with permission of Elsevier Science Publishers.

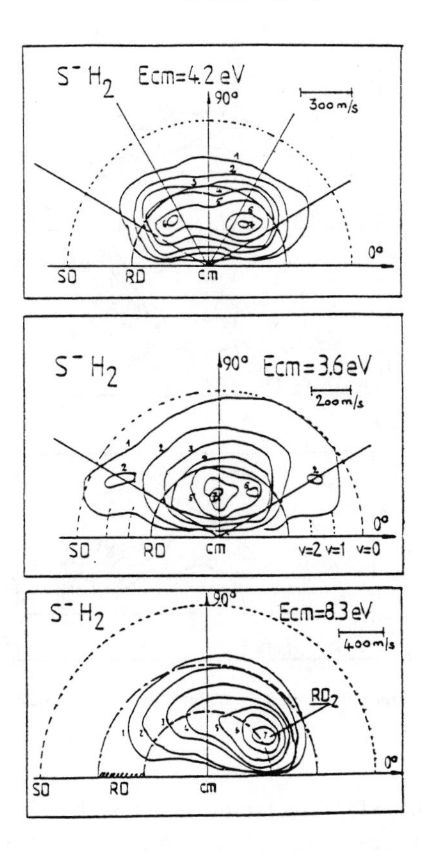

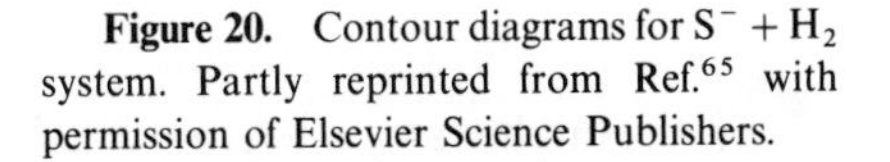

Figure 20. Contour diagrams for $S^- + H_2$ system. Partly reprinted from Ref.[65] with permission of Elsevier Science Publishers.

(Fig. 19). This dynamical behavior is usually recognized as resulting from an intermediate long-lived complex in an oblate symmetric top geometry.[107,108] The RD_1 channel exhibits a similar distribution than SD_1 (Fig. 20): for both channels the forward/backward ratio increases with collision energy, the lower observed limit being 1.3. This feature indicates the existence of an osculating intermediate complex.[108]

Outside these peculiarities, the velocity diagrams display features that emerge as common characteristics for all systems:

RDI and R have the same angular distribution and the internal energy of XH in R is always limited ($< 2\,eV$).

RDI appears in the continuity of R but, depending on the system, the internal energy can be either limited (F^-, S^-) (Figs. 20 and 21) or extended to the dissociation limit (Cl^-, Br^-) (Fig. 22). The continuity of this structure in the contour map beyond the dissociation limit in XH may be seen as an indication that the dissociation leading to DD process occurs in the reactive channel.

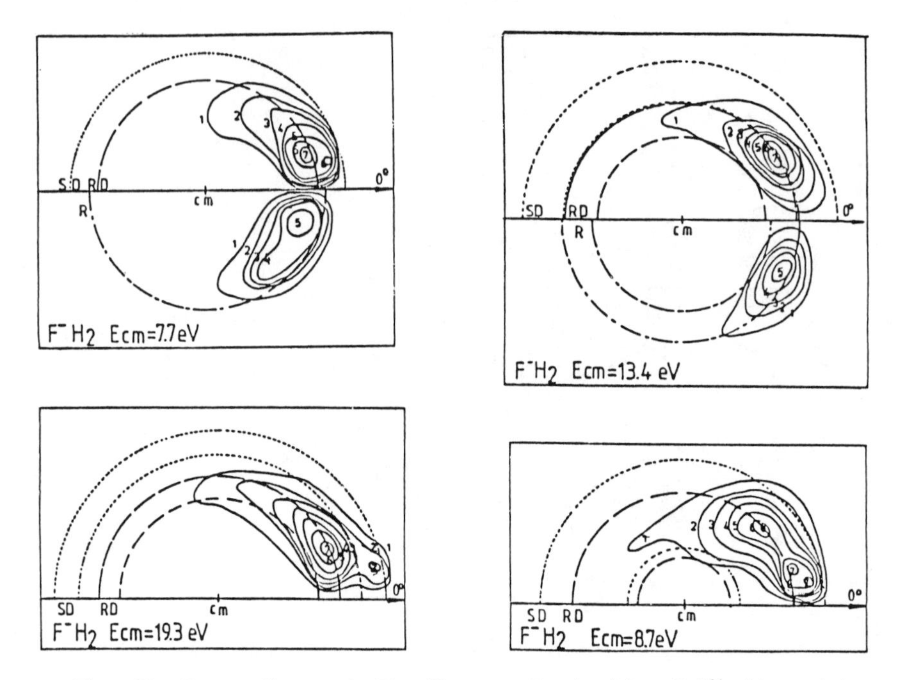

Figure 21. Contour diagrams for $F^- + H_2$ system. Reprinted from Ref.[64] with permission of Elsevier Science Publishers.

In the high-energy regime, RD2 and SD2 are for all systems sharply forward χ peaked (Figs. 21 and 23). The energy disposal is such that the XH or H_2 molecules have a limited internal energy except in the S^- system; in this case, the internal excitation of SH extends beyond the dissociation limit giving rise to the DD process (Fig. 23).

An intriguing feature of the Cl^- and Br^- systems is that R and the ridge of RD1 and DD processes closely follow the circle of elastic velocity of X^- on a unique free H atom of the H_2 molecule. (This circle, called C, has its center at the velocity of the CM of X^- and one H). This experimental observation suggests that the overall correlated reactions R, RD1, and DD are dominated by an impulsive interaction with one H atom. Independent of collision energy, the R channel that produces XH in the lowest vibrational levels is located at the intersection of the C circle with the circle $Q = \Delta H(\mathrm{R})$, whereas the various XH_v products appear at the intersection with more and more internal Q circles for larger internal energies of XH.

The overall reactive process (R, RD1, DD) acts as a prism scattering the whole spectrum of internal energies of XH and also beyond the continuum of dissociation: this spectrum appears on the frontal detector as concentric

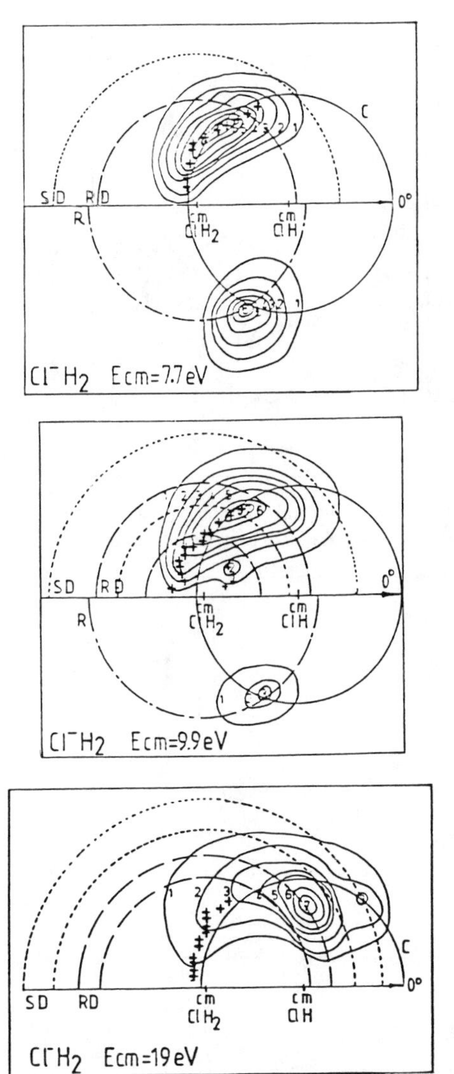

Figure 22. Contour diagrams for $Cl^- + H_2$ system. The C circle corresponds to elastic scattering of Cl on H. Crosses indicate the ridge of the contours. Reprinted from Ref.[64] with permission of Elsevier Science Publishers.

"circles" of equal "internal" energy. It may be interesting to notice that the calculated locus of H^- and H partners is a circle centered at the lab scattering center with a radius equal to the lab velocity of X^-.

In the halide systems a key parameter seems to be the endoergicity of the R channel. When low, as for F^-–H_2 (1.29 eV), the whole molecule is involved in the collision process, the R cross section is large ($15a_0^2$ [100], and the scattering

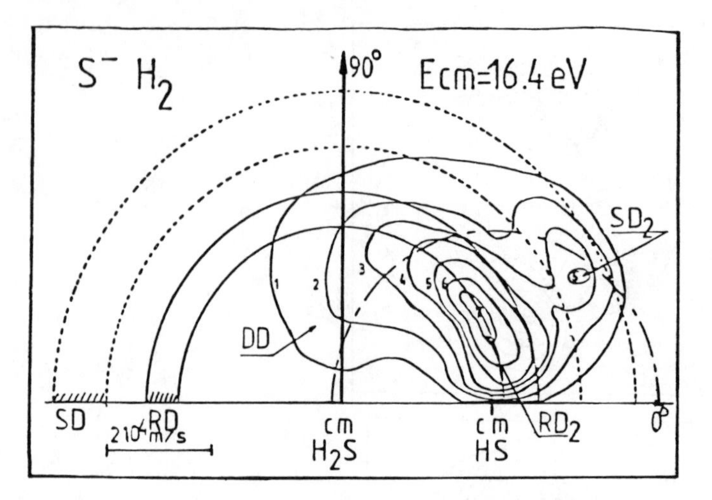

Figure 23. Contour diagram for $S^- + H_2$ system. The stability zone for RD and SD are indicated as dashed areas. Reprinted from Ref.[65] with permission of Elsevier Science Publishers.

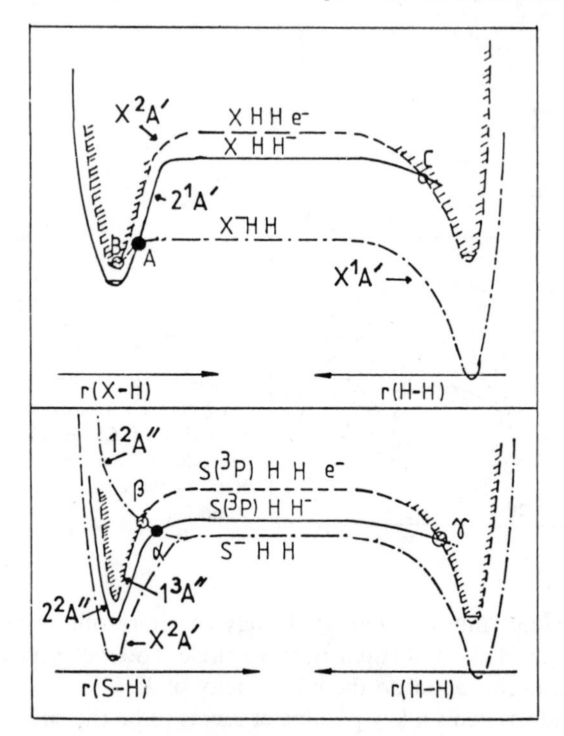

Figure 24. Schematic cut of the potential energy surfaces at infinite distance in the reactants and products valleys: (*a*) for $Cl^- + H_2$ system; (*b*) for $S^- + H_2$ system. Reprinted from Ref.[65] with permission of Elsevier Science Publishers.

angle is small for R, RD1, SD: large impact parameters are involved. When the endothermicity is large (2.91 eV in Cl^-–H_2), the R cross section is small (1a_0^2[100] and the scattering angle is large (from 45° to backward in RD1 and DD). This indicates that, in this last case, the reaction selects smaller impact parameters, which is in favor of an impulsive interaction with only one atom of the H_2 molecule. In S^-–H_2 system, the small cross sections of R and RD1 indicate that very small impact parameters are involved; however, the impulsive interaction with one H atom is not observed. The reason may be that the range of energy for the RD1 channel is too low (< 5 eV) for such a behavior to occur. Then one can infer that this impulsive behavior is observed when small impact parameters are involved, for collision energies larger than the H_2 molecule bond energy.

3. *Potential Energy Surfaces*

Outside the angular distribution on the *C* circle, which seems to be relevant from a mechanistic model, most of the experimental data related to the competition between R, RD, and SD have been rationalized in the framework of the interaction between two ionic surfaces that are diabatic with respect to the outer electron and their interactions with the detachment continua. These surfaces are referred to as $[X^-H_2]$ and $[XH_2^-]$, respectively, indicating to which core the active electron preferentially belongs:

$[X^-H_2]$ correlates $X^- + H_2$ on the reactant side to $XH^- + H$ on the product side.

$[XH_2^-]$ correlates $X + H_2^-$ on the reactant side to $XH + H^-$ on the product side.

These two surfaces have a crossing seam, which is the locus where the active electron changes core $[X^-H_2] \leftrightarrow [XH_2^-]$. The topology of the surfaces at this crossing seam is an important parameter that influences the dynamics.

In the case of halide ion system, a very simple picture of the potential energy surfaces (PES) has been given by drawing their cuts at infinite distance between X and H_2 on the reactant side and between XH and H on the product side. The apex of the crossing seam appears on the product side cut as the point *A* located at an energy value equal to the dissociation limit of H_2 (about 4.5 eV) (Fig. 24).

The trajectories generating the R channel have to go through this seam and the observation of this channel at its endothermic threshold $\Delta H(R)$ indicates that the energy position of the seam at short internuclear distances is at most equal to $\Delta H(R)$, therefore, much lower than its apex *A*. In the exit XH-H^- valley the ion-induced dipole interaction can hold the two products together in an orbiting intermediate structure for a limited energy range. In such conditions the velocity vector diagrams of the products should present a forward-backward distribution. The two ionic PES also have other seams with the neutral surface. HX^- (except HI^-) and H_2^- become unstable with respect to autodetachment as internuclear separation decreases[109]: in the asymptotic cut, the $^2\Sigma(HCl)^-$ curve enters in HCl continuum around the B crossing[110] and vanishes as a virtual state inside the continuum. Similarly the $(H_2)^-$ $^2\Sigma_u$ state couples with the continuum around the C crossing.[111]

The drawing of the ionic seam appears on the diabatic surfaces calculated by Sizun et al.[112] by a diatomic in molecule (DIM) procedure for the Cl^-–H_2 system (Fig. 25). At short Cl–H and H–H distances it is located at the corner; it approximately bisects the angle between the reactant and the product valley and runs along an energy ridge in the product valley for the large H–H distances. In the product valley, the seam along which the continuum interacts with the ionic diabatic $[Cl^- H_2]$ surface is parallel to the ionic one (its apex is B). On the $[ClH_2^-]$ diabatic ionic surface the neutral seam cuts the angle and runs parallel to the H–Cl coordinate (its apex is C).

It is thus clear that the trajectories giving any product (not only by R channel but also by RD and SD channels) have to go through the ionic seam: the topology of this ionic seam is the key feature influencing the dynamic of all the competitive channels.

On the DIM surfaces calculated by Sizun et al., most of the endothermic barrier occurs just before the reactants reach the crossing seam. It is unfortunate that the DIM surfaces are less accurate at the corner; however, they allow a qualitative understanding of the salient features of the experimental data.

The results of trajectory surface hopping of Sizun et al. carried out at 9.7 eV collision energy reproduce reasonably well experimental total and differential cross sections. An *adiabatic* behavior favored at low collision energies produces the R channel and at higher energy a *diabatic* behavior induces the RD channel. The experimental results have shown that at low energy a first *adiabatic* behavior also produces an RD channel (called here RD1): as a matter of fact, after a first adiabatic crossing at the corner, vibrationally hot trajectories can again reach the ionic seam in the exit valley and thus emerge on the upper diabatic ionic surface by an electronic transition and then reach the neutral surface giving rise to RD1. The close relationship between trajectories leading to R and RD1 through the corner and further to the second crossing causes the continuity of R and RD1 angular behavior. The second crossing of the ionic seam samples the trajectories giving rise to RD1 and thus cuts the vibrational population of R. The energy level of the second crossing fixes the effective threshold for RD1. From Sizun et al.[92] in the Cl^-–H_2 system, the recrossing of the seam can happen 2.0 eV above the ground state of $HCl + H^-$, which is the onset (about 5 eV) experimentally observed by Huq et al.[100] in the total cross section of electron detachment, although the endothermicity of RD is 3.6 eV. In the F^-–H_2 system the onset

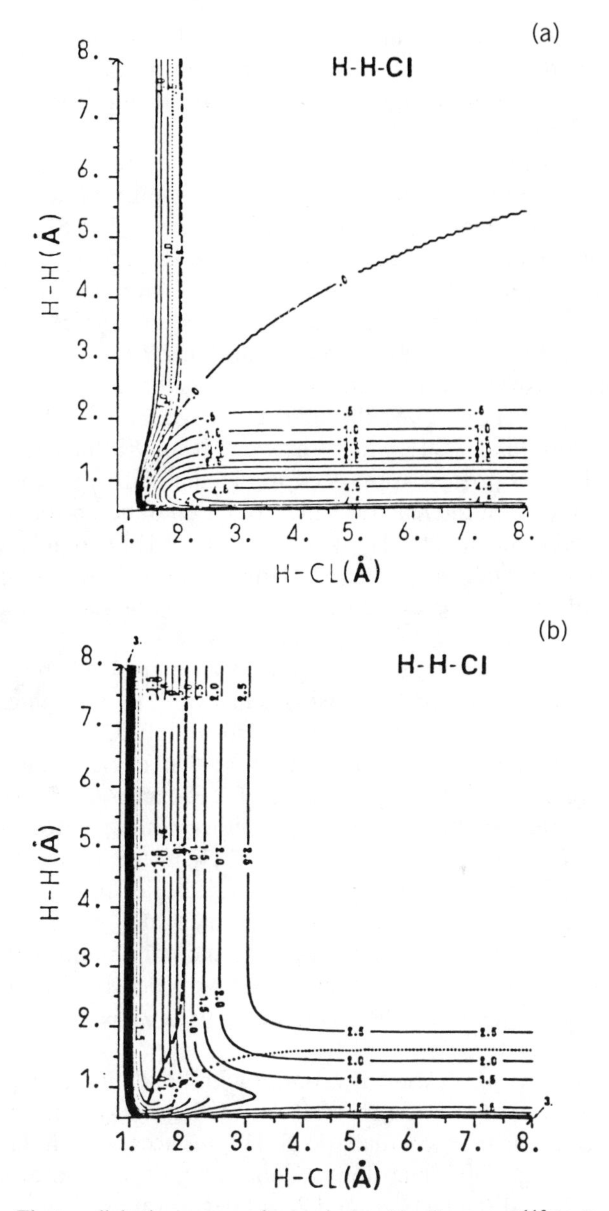

Figure 25. The two diabatic energy surfaces calculated by Sizun et al.[112] for linear (ClHH). The zero of energy corresponds to $Cl^- + H + H$. The dashed curve shows the crossing between the two surfaces; the dotted curve shows the crossing with the neutral ClH_2 surface. (*a*) Corresponds to the electronic configuration $Cl^- + H + H$. (*b*) Corresponds to the electronic configuration $Cl + H + H^-$ and $Cl + H^- + H$. Reprinted from Ref.[112] with permission of Elsevier Science Publishers.

of electron detachment is only 0.5 eV above the endothermic threshold of RD.[100] For the S^-–H_2 system, this onset is only 0.3 eV[101]: it depends for each system of the topology of the ionic seam in the reactive valley along the H–H coordinate.

More vibrational energy is stored in the R channel in F^- than in Cl^- system, which may explain that the branching ratio between RD1 and R (RD1/R) is much smaller in F^- than in Cl^- system. The SD channel is also induced by an adiabatic behavior at the ionic seam followed by a rebound on the repulsive wall. This is consistent with the more important SD1 in F^- than in Cl^- system: the access to the lower adiabatic surface is easier [ΔH(R) = 1.29 eV for F, 2.91 eV for Cl, and 2.12 eV for S], since, as was said previously, it is the endothermicity of R that fixes the energy of the crossing.

In the low-energy regime the total cross section of SD also presents a peak structure (like RD1), which therefore seems to indicate an adiabatic crossing. The SD process appears clearly mediated by the reactive crossing, it involves a path in the reactive valley and a rebound on the repulsive wall. It is remarkable that in $S^- - H_2$ system R, RD_1 and SD_1 angular distributions indicate the existence of the same intermediate osculating complex.

The adiabatic crossing of the ionic seam is the gate leading to R, RD1, SD1 channels.

In the same low-energy range, Huq et al. concluded from the absence of isotopic effects that surface crossing is involved. In the high-energy regime (above 10 eV) RD1 is no longer observed and RD2 takes place. RD2 is characterized by an angular peaking at smaller angles than RD1 and by a limited amount of internal excitation of XH except for S^- system. In the high-energy regime the DD process is either resulting from RD1 (Cl, Br systems) or from RD2 (S system). The RD2 channel is induced by a diabatic behavior at the seam: at the corner the trajectories penetrate directly into the continuum. For $F^- + H_2$ system, Huq et al.[100] concluded from the isotopic effect that a direct coupling with the continuum with no reactive process occurring. It is seen here that a reactive process (RD2) exists, however the diabatic behavior at the seam leads to a direct coupling with the continuum.

$S^- + H_2$ Particular Case. At variance with halide ion systems, which are well described using two ionic surfaces, in the S^- system, because of the open-shell character, two diabatic S–H_2 surfaces Σ and Π have to be considered (Fig. 24). The lowest one ($X\,^2A'$) correlates $S^- + H_2$ to $SH^-(^1\Sigma) + H$ and the $\Pi(1\,^2A''$ and $2\,^2A')$ degenerate state in $C_{\infty v}$ correlates $S^- + H_2$ to the $H + SH^-$ state dissociating in $S^- + H$. Only this Π state has a crossing seam with the $2\,^2A''$ (SHH^-) diabatic surface. This seam, named α in Fig. 24, plays the same role as the A seam in halide systems and corresponds to the electron exchange between the two cores S and H.

Three continua have to be considered. The lowest one correlates $S(^1D) + H_2$ to $SH + H$; it is the only one that is more stable in a bent configuration, which corresponds in C_{2v} to the X^1A' state of SH_2 molecule. The two upper ones $^3A''$ and $^1A''$ correlate $S(^3P) + H_2$ and $S(^1D) + H_2$ to $SH + H$; they are more stable in a linear configuration.

The same scenario as in halide ion systems is used: an adiabatic behavior at the α seam leads to 2 $^2A''$ diabatic surface and R process. Recrossing of the α seam samples trajectories that favor the jump of the active electron and lead to the 1 $^2A''$ surface, which penetrates in the continuum $^3A''$ at β seam leading to RD1. The adiabatic crossing of α seam also leads to SD1; the indication of an osculating complex for the three R, RD_1 and SD_1 channels strongly reinforces this point.

Passing the α seam diabatically gives rise to RD2 and at higher energy to SD2 by direct interaction of 2 $^2A'$ with the continuum 1 $^3A''$. An alternative interpretation of RD2 could be the autodetachment of SH^- product. Dynamical information on the $SH^- + H$ channel, which is the only one that this technique cannot analyze, could check this hypothesis. However, in a similar system OD^- has been found remarkably stable as regard to detachment.[87–89]

The AD and RD0 are not mediated by this α crossing and cannot be represented on the picture of the oversimplified surfaces at their infinite cuts. The lowest diabatic surface $[S^- -H_2]$ is more likely involved in the AD formation. This 1 $^2A'$ surface in C_s configuration is the only ionic surface that is more stable for a bent configuration. A probable representation of this surface is given in Fig. 25. Most of the energy gain is arbitrarily put at the corner between the two valleys of reactants $S^- + H_2$ and products $SH^- + H$. The S–H internuclear distance being about the same in SH and SH_2, it seems probable that, for the H–H internuclear distance corresponding to SH_2, the ionic surface, possibly in the configuration of the resonance $^2A_1$ state, has a seam with the continuum X $^1A'$ in C_{2v} corresponding to SH_2 molecule. It is assumed here that, on the way to the abstraction reaction, some trajectories corresponding to the adequate geometry enter the continuum leading to SH_2 formation.

This process, consistent with the observation of a threshold for the exothermic SH_2 formation, is in agreement with the two-step insertion mechanism suggested by the Boulder group.[97] Since the released electron carries away a very small amount of energy, the SH_2 molecule is formed with an internal energy equal to $\Delta H_{AD} + E_{CM}$. When this amount is equal to the dissociation energy either in S $(^1D) + H_2$ or in $SH + H$, the SD and RD channels should appear, at threshold, in an angular distribution corresponding to the dissociation of an oblate symmetric top (in SH_2 molecule, the largest moment of inertia is about the principal axis, the two

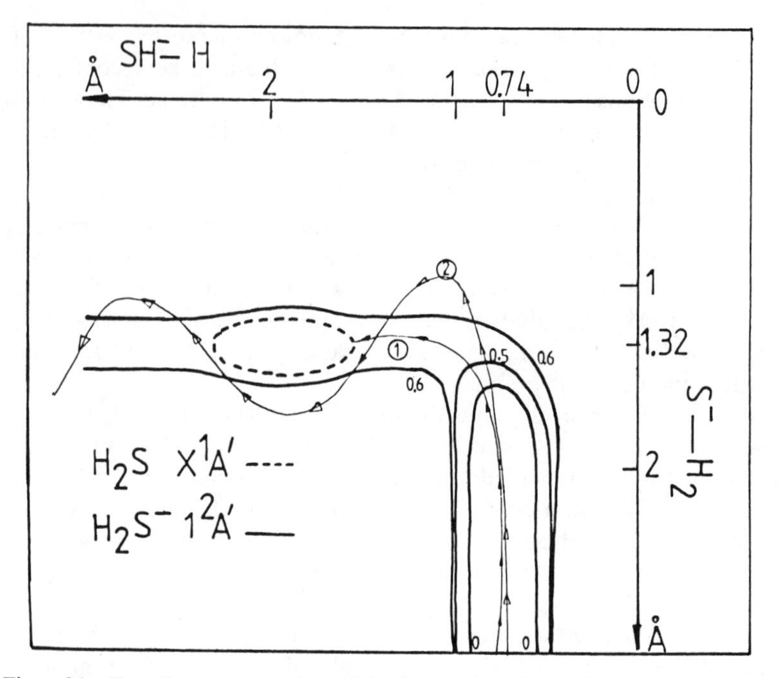

Figure 26. Tentative representation of the lowest $S^- + H_2$ surface in C_s configuration. Trajectory 1 leads to $SH_2 + e^-$; trajectory 2 leads to $SH^- + H$.

others being almost equal).[113] This is observed in RD0 and not in SD, which clearly shows up at threshold through another direct mechanism. The reason may be found in Fig. 26, which indicates that such a process occurs in the lowest reactive valley on the way to SH–H separation.

4. *Impulse Model*

The strange angular distribution of R, SD, RD1, and DD following the circle C of elastic velocity of X^- on one H atom of the H_2 molecule appears, in $X^- + H_2$ systems, limited at energies larger than 5 eV to reactions characterized by small impact parameters that result from trajectories evolving on the lower adiabatic surface. This surface, dissociative in $X^- + H + H$, is rather flat[112] and thus may allow a decoupling in the movement of the atoms. This typical distribution had also been observed by Gillen, Mahan, and Winn[114] in $O^+ + D_2 \rightarrow O^+ + D + D$. It is even more spectacular in $O^+ +$ HD in which two ridges are observed: one corresponding to the scattering on H and the second on D. It was suggested that much of this elastic scattering is described by a spectator or impulse model. The reactive scattering also indicates that, for that system, a decoupling in the motion of the atoms exists in some regions of the PES: spectator stripping is observed[115] at low energy (as for all the exoergic H-transfer reactions so far investigated except $Kr^+ + H_2$[116]) and is replaced at higher energy (> 10 eV), when the internal excitation of OD^+ makes it unstable to dissociation, by another impulse process that obeys the Bates sequential impulse model (SIM).[117]

In the SIM, the *AB* formation in *A*(*BC*)*AB* reaction is viewed as an event in which *A* hits *B* impulsively and elastically, *B* hits *C* in a like manner, and *A* combines with *B* (or *C*) if the appropriate energy of relative motion is less than the dissociation energy of the product molecule. The SIM, first postulated by Bates, Cooks, and Smith,[118] has been elaborated by Suplinskas[119] and George and Suplinskas[120,121] who showed that it can reproduce the major features of $Ar^+ + D_2$ reactive and non reactive scattering. Gillen, Mahan, and Winn[117] have found that a version of this model in which atoms interact via hard-sphere potential is consistent with the product angular distribution of OD^+ in $O^+ + D_2$, HD, H_2 at high relative energy. Armentrout[122] and Safron[123–126] have proposed extensions of this model for endothermic reactions.

By a vectorial analysis, Mahan has shown that the product velocity vector diagram generated by the SIM for an athermal reaction[127] (Fig. 27) is delimited by a cardioid, which has its cusp at the CM of the *AB* system (the *SS* point). It is interesting to notice that the position of the *AB* molecule

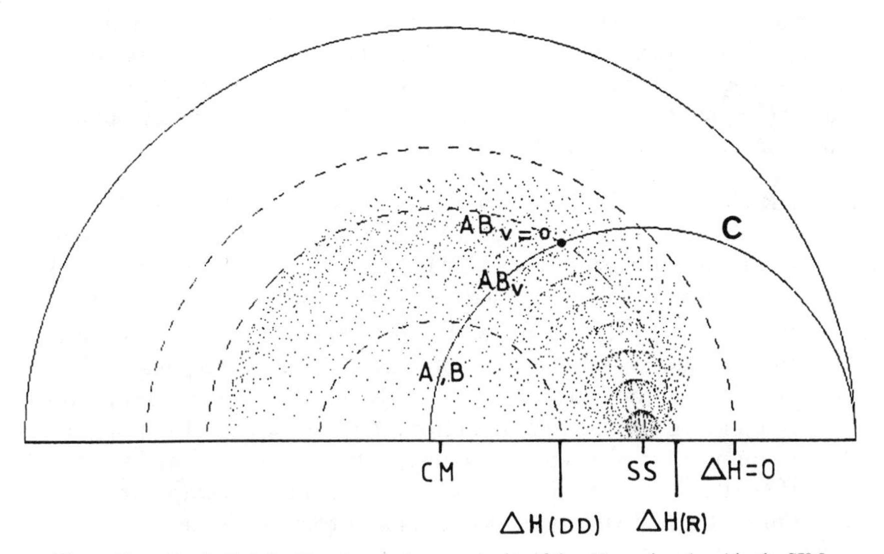

Figure 27. Cardioid delimiting the maximum velocity of the *AB* product (w_{AB}) in the SIM.

formed without internal energy is concentrated at a unique point located at the intersection of the C circle with the circle centered at the CM with a radius corresponding to $Q = 0$, which is also at the tangency with the wings of the cardioid. As a matter of fact, in such a model, the product AB formed without internal energy is obligatorily on the circle C since it results from the recombination of A and B having identical velocity vectors leading to $\mathbf{W}_{AB}(v = 0) = \mathbf{W}_A$ (**W** denotes velocity in the CM).

In the present study, the observation of the vibrationally cold HCl and HBr produced by the R and RD1 channels on the C circle leads us to imply that the first step of the collisional process is still an elastic binary collision, thus "memorized" by the halogen nucleus. In the other extreme situation, when dissociation is observed (DD process), the velocity of X after the first binary collision is also unchanged, since this dissociative process, in the frame of the SIM, results from collisions in which X and H cannot recombine.

For the vibrationally excited XH in the RD process, it is *a priori* difficult to figure out what sort of constraints will force the velocity vector of the products to stay on the C circle. However, in the framework of the SIM, limitations on the impact parameter and angle of attack in one or both binary collisions do force the distribution in the vicinity of the C circle. This is compatible with the finding that overall small impact parameters are obligatorily involved when this distribution is observed.

5. *Conclusion*

This method, which allows a simultaneous dynamical analysis of the various competing channels, reveals that, in the low-energy regime, the competition between them (except AD) is mediated by the jump of the active electron between two extreme positions on X and H cores. Although naive, this pictorial representation, which ignores details of the surfaces, allows a rationalization of the experimental results.

The R channel, which as a whole corresonds to a proton transfer and is usually named that way, does not in fact correspond to a proton transfer as an elementary step: it is a kind of electron harpoon transfer from X^- to the farther H followed by a recombination of X and the middle H.

In the low-energy regime the reactive detachment is initiated by reactive charge transfer but the RD1 process does not corespond to just a detachment from $X\text{H} + \text{H}^-$. The active electron experiences a rather complicated story: a first jump from X to the farther H followed by a recombination of X and H leading to $X\text{H–H}^-$ then a jump back if XH is stretched by vibrational motion, and finally an escape from $X\text{H}^-$ during the half vibration when X and H become closer.

The so-called simple detachment is not such a simple process and implies after a first jump of the electron a "U-turn" in the reactive region.

The associative detachment that occurs in systems for which HX^- ion is a stable species does not involve an electron jump. The AD takes place in the path to HX^-; the electron detaches for the geometrical configuration of the nuclei in the SH_2 molecule.

In the particular systems for which the low-energy regime extends up to 10 eV and for which there is evidence that small impacts parameters are involved, the atoms behave as decoupled in kinematics obeying an impulse model for R, RD, and associated DD. In the high-energy regime the electron does not change core, direct interaction with the detachment continuum is involved in the SD process, and the reactive detachment may be seen as a stripping of H producing an unstable HX^- species which detaches.

VI. CONCLUDING REMARKS AND SUGGESTIONS

The multicoincidence detection combines the angular resolution of the crossed-beam technique and the sensitivity of coincidence measurements. Doubly differential cross sections are obtained for any collisional process producing two detectable products such as ions, fast neutrals, metastables, electrons, and photons.

This technique has been applied in both atomic physics and physical chemistry experiments. First applications in atom–atom collisions gave information on excitation of magnetic sublevels through the angular correlation between the scattered neutral and the emitted photon analyzed either angularly[13] or through its circular polarization.[4] In chemical physics the application to the ion–molecule $X^- + H_2$ systems provides the dynamical analysis of all the ionic and detachment competing channels giving a deep insight into the competition mechanism itself.

The same principles and the same hardware and software "components" applied in convenient geometrical experimental arrangements extend the operating field of dynamical studies to physical chemistry processes such as chemiionization or three-body dissociation.

The versatility of this technique allows "low-level" analyses using only one particle detection. In this case the coupling with a time-of-flight method provides an angular and energetical analysis of a lot of scattering processes including mass analysis, surface scattering, photochemistry, and so on.[128]

All these aspects have been detailed in this chapter. Now, new generation of experiments in state-to-state chemistry should arise from the coupling of multicoincidence detection with optical techniques. The sensitivity of the coincidence detection and the efficiency of the "parallel" acquisition is quite adapted to the low intensity beams of fast positive ions or neutrals prepared in a selected state. This coupling is straightforward and the capabilities, which have been outlined in this chapter, of analyzing rearrangement reactions would easily be extended to state-selected ones.

The final step of the state-to-state chemistry that would provide the angular distribution of individual internal quantum states could be achieved through the coupling with the laser-induced fluorescence method (LIF). This coupling appears to be as well an enhancement of LIF methods providing an easy angular analysis of state-analyzed particles as an enhancement of the multicoincidence method providing a test particle for processes leading to only one detectable particle. In this latter definition the laser-induced-fluorescence-multicoincidence analysis is the natural extension of the previous spontaneous-fluorescence-multicoincidence method.

APPENDIX A—PRACTICAL CORRELATION WITH HIGH COUNT RATES

The basic hypotheses of the simple model used in Section II 3 for the determination the time dependence of a coincidence spectrum, that is, the probability that a true coincidence be hidden by a spurious stop is negligible and the false coincidence background is unperturbed by the correlated events, have obviously to be revisited in the framework of a multicoincidence experiment where most of the collision products impinge their respective detectors—the received fluxes are *a priori* large and highly correlated—and multiple simultaneous experiments share a single measuring tool. This appendix is a tentative approach to the nonlinearities introduced by high counting rates in a coincidence experiment when a single time digitizer (hereafter named counter) is used. A model is developed in a monocoincidence context and briefly extended to multicoincidence experiments.

For the sake of simplicity, the model is introduced using Fig. 2 and notations of Section II 3, namely, the count rates observed on each pixel are still given by $\dot{N}_A$ and $\dot{N}_B$. Furthermore, the time quantization in δt intervals (e.g., 10 ns) is supposed to be directly obtained through digital circuits.

A. Qualitative Analysis

A coincidence experiment may be seen as successive independent sequences. The counter is first reset and waits for a start signal on detector A. When started, the counter is free-running until a defined count number is obtained, the delay being determined by the minimum actual time delay between two particles resulting from the same collision. A discriminated output signal of detector B is then allowed to stop the counter during a second physically defined time interval. The counter state obtained for the first stop signal (if any) is summarized in an histogram. In any case the system is reset, initiating the next sequence.

The problem is twofold:

1. If the counter is started by a convenient particle, that is, produced in such a collision that the second particle will impinge the second detector B, the measurement procedure fails either when an "alien" particle is first occurring or when the companion particle is "missed" by the detector: the detection yield of a "true" pair depends on the effective delay and introduces nonlinearities into the coincidence "apparatus function"; furthermore, a correlated but complicated structure may be expected for the "false" pairs collection law.

2. If the companion of the starting particle cannot be received on the second detector, the sequence is unprofitable but still contributes to the scrambling of the data.

These qualitative arguments are readily confirmed by Monte Carlo simulations. The illustrative example given Fig. 28 has been obtained under the following conditions: the total collision rate producing impacts on detector A is 0.1 per clock period δt; 20% are uncorrelated and only two processes expected with equal 40% probabilities may be observed on both detectors for $15\delta t$ and $20\delta t$ delays (the time determination is synchronous). Detector B receives uncorrelated particles, with a 10% per δt probability, and the correlated ones. The efficiency of both detectors is 80% and a $15\delta t$ recovery time is assumed for valid sequences and for unsuccessful ones after a $30\delta t$ delay. The one-dimensional histogram of the i time counts is self-explaining: level and slope discontinuities are observed and the nonlinear

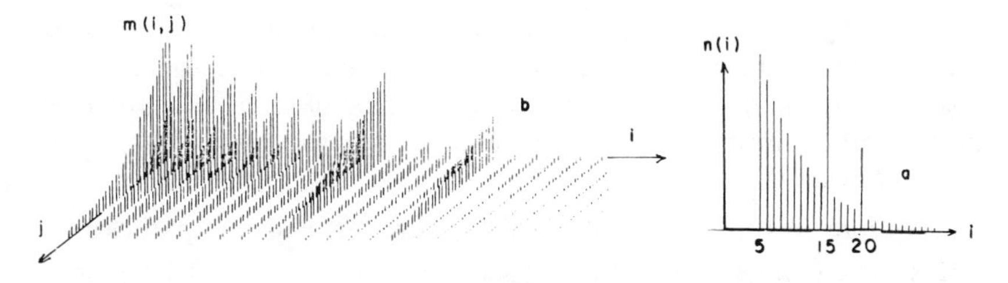

Figure 28. Monte Carlo simulation of a coincidence experiment (see text for the operating conditions). (*a*) Histogram of the count number $n(i)$ obtained for each possible time code (arbitrary units). The counter is enabled for $i = 5$. A first sharp nonexponential decrease is observed; each correlated peak at $i = 15$ and $i = 20$ introduces level and slope discontinuities on the background. Note that, whatever the evaluation of the false coincidence background is, the ratio between the two true coincidence peaks is quite different from the 1:1 simulated one. (*b*) Perspective view of the two-dimensional histogram of the count numbers $m(i,j)$ (see text). Remark the j profile for different i values. The exponential profile observed for $i = 5$ disappears progressively. Partial histograms may be obtained for any type of start and stop particles in order to enlighten the following study.

detection of the equally probable channels is evident. In the two-dimensional histogram the count number in each i channel is displayed as a function of a second variable j representing the latency time $j\delta t$ of the counter between the reset and the first detected start signal. The analysis of these phenomena is therefore needed in order to explain the observed structures.

The key problem is in fact the reliability of the counter against spurious stop signals, which may abort the correlation determination prematurely: a convenient approach may thus be founded on reliability concepts, which are briefly summarized in the next section.

B. Reliability

The reliability $R(t)$ of a system is the survival probability of this system as a function of time t. This function is either experimentally determined from the decay rate of a large number of similar and simultaneously started systems or calculated from the reliabilities of each of its constitutive elements.

The first derivative $\dot{R}(t)$ is the probability per time unit for the system to decay at time t. The failure rate (FR) $\lambda(t)$ is defined as the opposite of the log derivative $\dot{R}(t)/R(t)$ and specifies the probability for a system that is still running at time t to fail before $t + dt$. This function is the basic tool of reliability problems: in many cases this FR function is (or may be considered as) a constant and gives rise to exponentially decreasing populations (e.g., for many physical phenomena such as spontaneous radiative deexcitation or radioactive decays).

The extension to digitized time intervals δt is straightforward.

C. Monocoincidence Experiment Involving a Single Process

The probability to get two events within the same δt interval is assumed to be negligible, namely, $\dot{N}_A, \dot{N}_B \ll 2\delta t^{-1}$ (for example, some megahertz for a 10 ns clock period). Figure 29 presents a mapping of the FR$\lambda(i, j)$ of a time digitizer for a coincidence experiment involving a single process characterized by a $i_k\delta t$ delay.

$i_0\delta t$ represents the minimum delay between start and stop particles.

i is the current count and j represents the latency time $j\delta t$ between the resetting and triggering signals.

ε_A and ε_B are random variables depicting the efficiency of both detectors. Only the 0 and 1 values are allowed and the mathematical expectation is the quantum yield (the name of such a random variable is also varying).

p_k is the probability for a collision to occur within the time slice δt.

Several values of λ are possible: ε_B for the expected $i = i_k$ channel, $(1 - \varepsilon_A)\varepsilon_B p_k$ or $\varepsilon_B\, p_k$ whether the corresponding start particle could or could not be taken

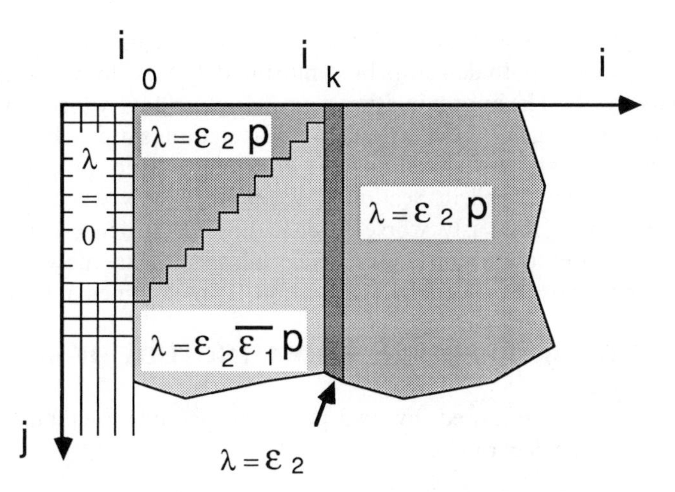

Figure 29. Failure rate mapping for a single coincidence channel.

into account. The light-shaded area would of course be "safe" for a 100% efficiency of the first detector.

The reliability of the counter $P(I,j) = \prod_{i=i_0}^{i=I-1}[1-\lambda(i,j)]$ is then j-dependent and the probability $P(I,j)$ to observe an I count in the counter started after j wait states is given by $P(I,j) = \lambda(I,j)\ R(I,j)$.

The shape of the coincidence spectrum is obtained by a proper weighting on j with $\varepsilon_A p_k(1-\varepsilon_A p_k)^j$:

$$C(I) = \varepsilon_A p_k \sum_{j=0}^{\infty} (1-\varepsilon_A p_k)^j P(I,j)$$

D. Monocoincidence Experiment Involving Several Processes

This model may be extended to real experiments: the p_k values correspond to differential cross sections of correlated events, whereas q_A and q_B account for unrelated ones received on A and B. The failure rate is now given in the Poisson approximation by

$$\lambda(i,j) = \varepsilon_B\left[\delta_{ik} + q_B + \left(1 - \varepsilon_A \sum_{k=i+1}^{k=i+j} p_k \Big/ \sum_{k=0}^{\infty} p_k\right) \sum_{k=0}^{\infty} p_k\right]$$

where $\delta_{ik} = 1$ only for a true coincidence configuration.

The previous procedure still holds: we calculate $P(I,j)$ for each start particle (including the unrelated one), average over j using the ε_A (q_A or p_k) $[1-\varepsilon_A\ (q_A + \sum_{k=0}^{\infty} p_k)]^j$ weighting factor; the total histogram is obtained after a final averaging with q_A and the different p_k.

These structures contain information, in particular on the effective quantum yields of both detectors but unfortunately these formulas cannot be inverted directly. The approximation $\varepsilon_A(q_A + \sum_{k=0}^{\infty} p_k) \ll 1$ which removes the j-dependance of $\lambda(i,j)$ makes the predictive calculations simpler, but the partial time spectra are still correlated and the linear dependence postulated in the simple model developed in Section II is obtained only when $\sum_{k=0}^{\infty} p_k \ll 1$. The situation is obviously worse when a multicoincidence experiment is performed under these conditions, since all spectra are now coupled. A solution is proposed in Appendix E Section B for intermediate cases.

APPENDIX B—FREQUENCY DOMAIN STUDY OF *RC* LINES

Resistive anodes are used by two-position encoding methods that are generally studied separately:

The charge division method is introduced in the static context of the Laplace equation and Ohm's law.

The risetime method is studied in the "time domain" through the Fourier heat conduction equation.

The following short introduction of a one-dimension dissection system is performed in the "frequency domain," which is particularly suitable to provide a qualitative but unified approach of both methods. (A similar study using the Laplace transform has been given by Borkowski et al.[129,130] in the risetime method framework.)

The arrival of an electron cloud on a resistive layer ($R\,\Omega$/m) isolated from a parallel conductive backplane (C F/m) can be considered as a deltalike current pulse exciting a distributed R-C line (Fig. 30). Low-frequency equations are valid in the infinitesimal dz interval (Fig. 31).

$$\frac{\partial v}{\partial z} = -Ri \quad \text{and} \quad \frac{\partial i}{\partial z} = -C\frac{\partial v}{\partial t}$$

and their decoupling provides the well-known propagation equation in dissipative media $\partial^2 v/\partial z^2 = RC\,\partial v/\partial t$ or in harmonic regime $\partial^2 v/\partial z^2 = jRC\omega v (j^2 = -1)$; the corresponding solutions are proportional to

$$\exp(\pm z/\delta)\exp[j(\omega t \pm z/\delta)] = \exp(j\omega t \pm \gamma z), \tag{B1}$$

δ being the skin-depth $(2/RC\omega)^{1/2}$ and $\gamma = (1+j)/\delta$. When both terminals are matched by the $R\delta/(1+j)$ characteristic impedance, only two wavelets are involved; the impulse response is obtained by Fourier transform or by the derivation of the step response [known to be proportional to the

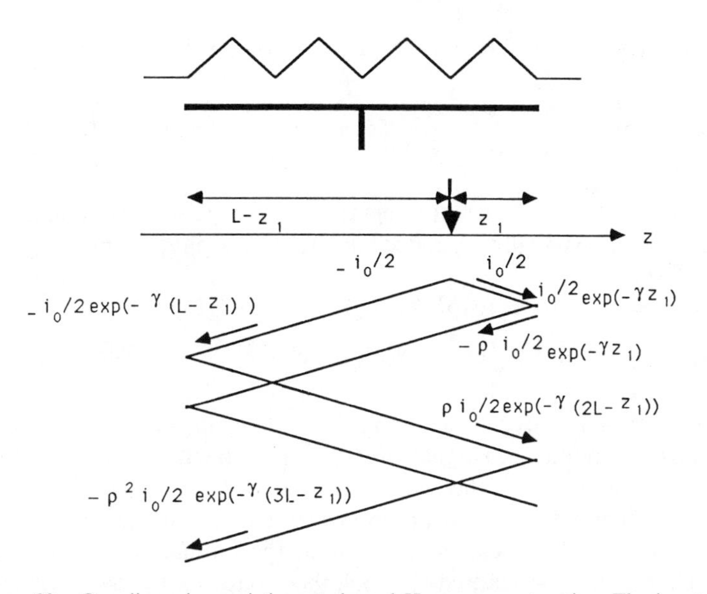

Figure 30. One-dimension resistive anode and Huygens construction. The input current pulse initiates two infinite sets of "traveling" waves corresponding to the flat continuous frequency spectrum of the deltalike excitation. The wavelets corresponding to a given small frequency interval are given by their voltage and intensity components as functions of the related intensity i_0. $(Z_0 i_0/2, i_0/2)$ propagates to the right side and $(Z_0 i_0/2, -i_0/2)$ to the left side. These two wavelets generate two independent infinite sets of wavelets resulting of successive reflections on both terminals. The intensity components of some first wavelets are displayed along the vertical time axis.

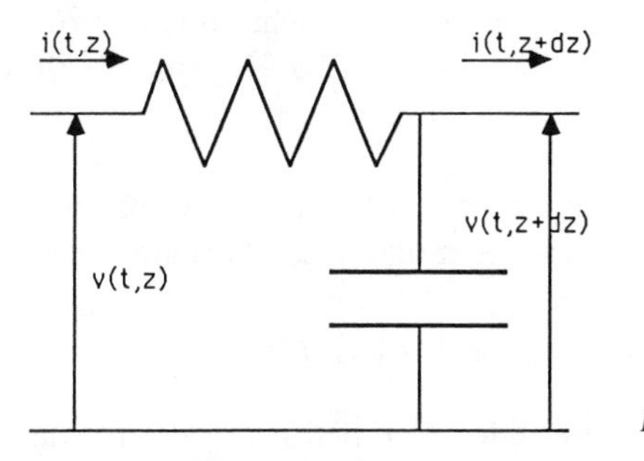

Figure 31. Model for a small portion of the *RC* line using localized elements.

complementary integral Gauss function of $z(RC/2t)^{1/2}$ [131])]. Otherwise, an infinite series of wavelets are observed. The resulting waves are easily obtained using a one-dimension Huygens construction. This diagram is given in Fig. 30 for a (sinusoidal) i_0 current injection on a L-long RC line at z_1 from its right side and equal voltage reflection coefficient ρ at both terminals. These wavelets interfere and give, for example, at the right side the output current

$$i_1 = 0.5 \cdot i_0 \frac{1-\rho}{1-\rho^2 \exp(-2\gamma L)} [\exp(-\gamma z_1) + \rho \exp(-\gamma(2L - z_1))].$$

In the practical case of current integration in charge-sensitive amplifiers, the terminations are virtually grounded and the two output currents are given by

$$\begin{aligned} i_1 &= 2 i_0 \sinh(\gamma z_2) \exp(-\gamma L)/[1 - \exp(-2\gamma L)] = I \sinh(\gamma z_2) \\ i_2 &= 2 i_0 \sinh(\gamma z_1) \exp(-\gamma L)/[1 - \exp(-2\gamma L)] = I \sinh(\gamma z_1) \end{aligned} \tag{B2}$$

(The grounding is only "virtual"; the input impedance of a charge amplifier is resistive for frequencies higher than the op-amp cut-off[27] and is supposed to be much smaller than the characteristic impedance $Z_0 = R\delta$.)

The DC limit[41,42] leads at first order to the current sharing $i_1/i_2 = z_2/z_1$ between two parallel resistors in the static approximation and corresponds to the charge-division method in which the capacitance must be vanishingly low: the backplane is removed, but the "stray" capacitance with the rear face of the last MCP still remains.

The general case corresponds to the so-called risetime method: the position is deduced from the dephasing of the output(s) as suggested by the linear position-to-phase dependence in Eq. (B1).

1. The Charge-division Method. The ratio of the charges integrated on both sides is clearly representative of the position, independent of the total shared charge, but the dependance cannot be linear; furthermore, the departure from linearity is depending of the bandwidth of the subsequent electronic treatment.

2. The Risetime Method. The position-dependent dephasings of sinusoidal components received at 1 and 2 detector terminals are the arguments of

$$\exp(z_{2/1}/\delta) \exp(j z_{2/1}/\delta) + \exp(-z_{2/1}/\delta) \exp(-j z_{2/1}/\delta) \tag{B3}$$

These expressions are once more not strictly linear functions of the impact location but an approximate determination might be obtained by measuring the dephasing of one harmonic component with the initial cosine obtained through the time pick-up. But this procedure, which supposes a dramatic filtering of the continuous frequency spectra, presents a poor energetic yield, a low count rate limited by the large transient time, and even a problematic noise rejection if the phase measurements are not averaged. These nasty consequences are fortunately avoided by the use of the universal cr-cr-rc shaping device. This band-pass filter exhibits a roughly 1:1 bandwidth-to-central frequency ratio: a larger part of the Fourier spectrum is then used, the transient response is a single bipolar output signal, and the zero crossing of this short signal appears to be a good measure of the position. (This zero-crossing corresponds to some "frequency-averaged" $\pi/2$ dephasing and is therefore less noise sensitive.) The order of magnitude of the spatial sensitivity may be obtained for a single output from the dephasing formula (B3): $S(\mathrm{m/s}) = (2\omega_0/RC)^{1/2}$, where ω_0 is the central circular frequency. Generally the relative dephasing between the two output signals is measured: in this case, the sensitivity is doubled and the linearity is enhanced.

Open-circuit RC line may also be used the formulas are similar to (B2) with hyperbolic cosines. The charge-division method does not work (of course!), but the risetime one does; in spite of many favorable arguments,[40,132] short-circuit RC lines seem to be generally preferred. Note also that the preamplifiers are usually charge sensitive and that a pole-zero cancellation is necessary to restore the frequency information before the final pulse shaping.

APPENDIX C—EXAMPLE OF MULTIPURPOSE SET-UP

This appendix presents the basic features of a multipurpose analyzer recently developed in Orsay.[133] The prototype is now running on the TOF apparatus Elvire[134] on which the position-sensitive detection is particularly useful to test the propensity rules in laser-assisted collisions.[135]

This modular and bus-oriented system is packaged in a VME standard rack and allows the processing of event vectors of any dimension with a 12-bit resolution in moderate count rate contexts. This system is basically designed for charge-division methods and supports as well digital or analogic time determinations. Only three types of modules are needed: the coincidence module that handles time signals, the ADC module that converts charges or TAC outputs, and the master module that links logically these modules to the computer. These modules are schematically drawn in Fig. 32.

The master module (Fig. 32*a*) includes two simple bipolar microprogram (Fig. 33) controllers allowing the simultaneous handling of coincidence data acquisition and of data storage. This feature is obtained through a systematic pipelining of all data registers enclosed in the time and ADC modules; a one-level pipeline regulates the data flow and allows significantly higher correlation rates.

The main controller (Fig. 33*a*) activates the correlation module and waits for its "busy" output signal; when this signal is negated, the controller waits for the end of the last (possible) conversion in the ADC modules. The

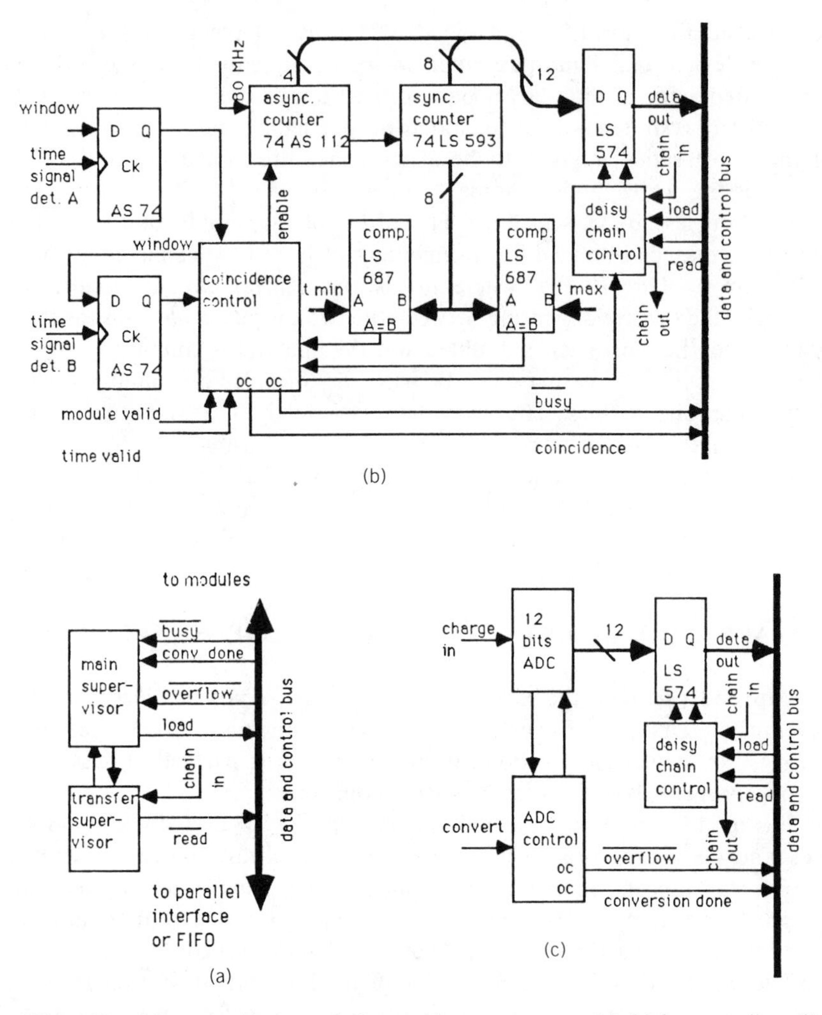

Figure 32. Schematic diagram of the multipurpose set-up. (*a*) Main controller. (*b*) coincidence unit. (*c*) ADC units.

controller then aborts the sequence if no valid data are available; if a stop signal is observed in the convenient time window and if all conversions are valid, the controller waits for the pipeline empty signal. When this signal is true, it loads the pipeline, activates the transfer controller, and resets the coincidence unit. Note that the hardware handling of unsuccessful measuring sequences allows high count rates and that the rate of successful sequences is mainly limited by the software.

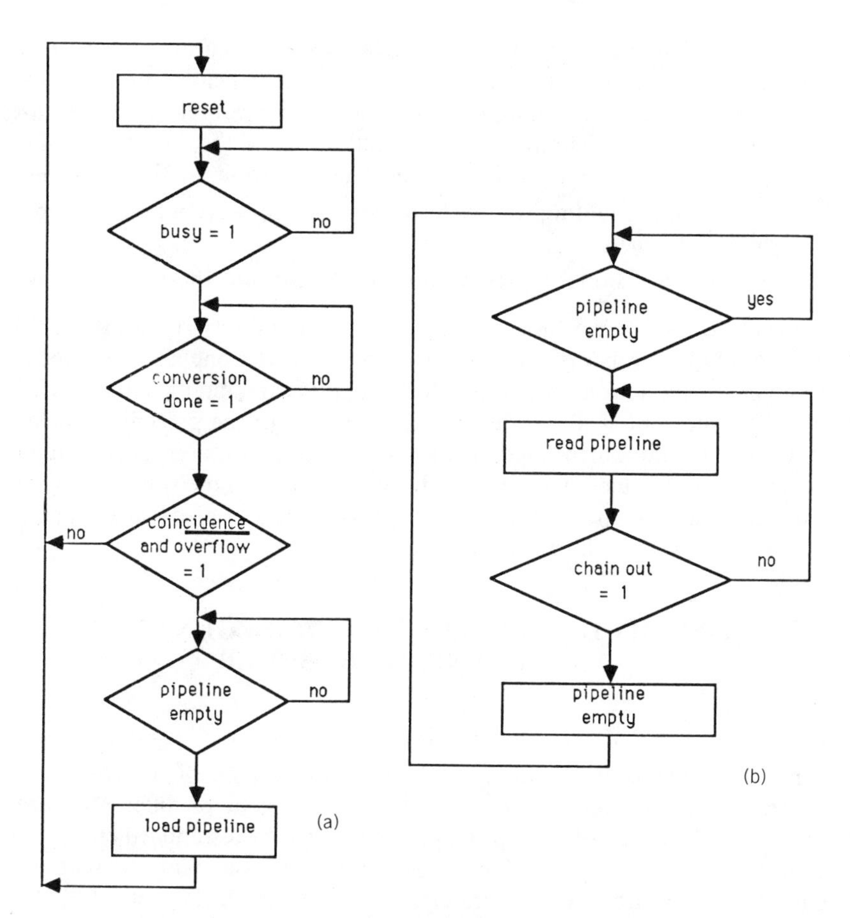

Figure 33. Flowcharts of (*a*) the main controller and (*b*) the transfer controller.

The linked-transfer controller (Fig. 33*b*) then stacks the stored data in the next layer. The versatility of this hardware structure is based on the daisy-chaining of the pipeline registers. The coincidence unit (Fig. 32*b*) uses fast flip-flops to get single true start and true stop signals, which command a 12-bit counter; the first asynchronous stages allow a 12.5 ns resolution, the maximum delay is then 51.2 μs and any time window may be chosen within 256 intervals of 400 ns.

The opto-isolated Datel converters of the ADC units (Fig. 32*c*) encode in 1.4 μs the output voltages of charge-sensitive amplifiers (and/or TAC outputs) in 12-bit words. Any encoding unit may be logically disconnected by a single switch to provide any working configuration. This feature is fully supported by the software, which performs until now three main functions:

1. The real time acquisition of stacked data; any configuration is supported by a fully reconfigurable routine, but fast dedicated programs are available for the standard accumulation procedures. In both cases on-line partial histograms (i.e., projections on all the dimensions of the present LMS) are provided for checking the accumulation procedure.
2. The off-line building of any histogram sets in a reconfigurable space of 2^{18} channels.
3. A two-dimensional representation of any cut into these histograms.

These routines are working on four successive subsets of primary LMS defined by the number of measured variable. The first one is defined for the real time accumulation procedure. The histograms are obtained in the second one. The third one is a two-dimension projection of the preceding one and is reduced to the fourth one when a zooming is performed. This system is potentially unlimited, since any addition may be handled as well by the hardware and software components. (This system is of course respecting most of the principles developed in this paper.)

APPENDIX D—EXAMPLE OF A HIGH COUNT RATE MULTICOINCIDENCE SET-UP

This apparatus has been built in Orsay some 15 years ago[3,13] and its principle would be enlighten by recalling that the main purpose of the coincidence experiment of Baudon's group was the measurement of differential cross sections of the processes that could not be isolated by a standard energy-loss analysis. The angular dependence was obtained by sweeping the scattering cone and the process was accurately assigned through the optical analysis of the emitted light.[1] These experiments were time consuming and then practically limited to a few model systems [for example, the state-analyzed $He^+ + He \rightarrow He^*(3\,^1P) + He^+$ channel has been studied in the keV range using the $3\,^3P \rightarrow 2\,^3S$ decay and the fast-scattered He^+ ion].

A larger application of this method could be obtained through parallel acquisition on large angular ranges. However, this procedure was incompatible with the energy preselection systematically performed on the scattered He^+ in order to reduce the false coincidence rate. The basic goal of this apparatus has been to prove that this limiting condition could be escaped if a convenient electronic system was used, in other words that a convenient signal-to-noise ratio might also be obtained with large signals if a high-bandwidth set-up prevents distortion and large dead times. In a pedagogical stretching of the usual rule, let us first describe the test experiment of the resulting fast multicoincidence set-up.

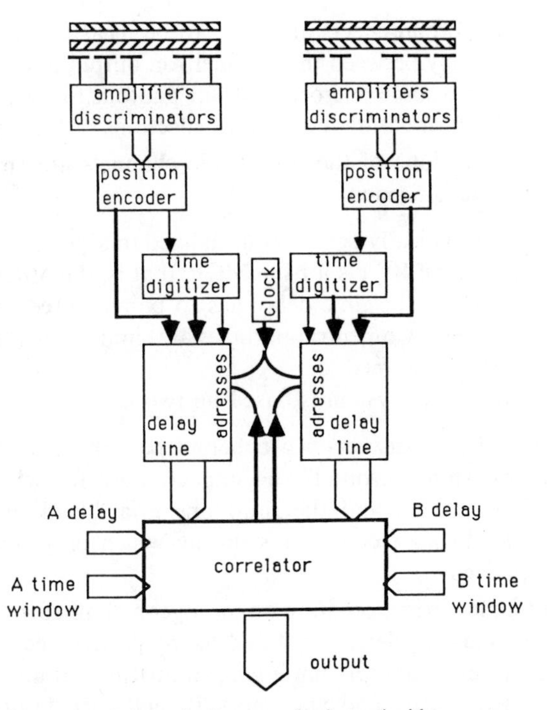

Figure 34. Schematic diagram of a fast coincidence set-up.

The simple and previously studied [136] He-He system is studied in the keV range in a crossed beam apparatus (Fig. 34). Fast helium scattered in a large solid angle by an helium effusive jet are received on a MCP and individually collected on 64 concentric annular rings for scattering angles ranging from 1° to 8°, on a 36° azimuthal range. (The width of the collectors increases with the scattering angle: the angular acceptance increases with the angular thermal spreading and the collecting area is decreased for small scattering angles.) The second detector is a channeltron, which rotates around the collision region and collects recoil metastable He and UV photons emitted by excited projectiles or targets. (The physical aspects are detailed in Ref. 13 and are not essential for the present purpose.)

Three negative circumstances hold for checking the efficiency of the electronic set-up:

1. The inelastic channels are deeply embedded at small angles into the elastic channel.
2. The TOF of the photons and recoil particles of the processes are quite different; furthermore, the TOF of the metastable particles are

spread by thermal effects and have to be measured over a large time range: the delay between the channeltron output and the fast particles varies between some microseconds for photons to "negative" values for metastables formed for the lowest deflections of the fast neutral.

3. The cross sections of the observable channels are small and a high collision rate is needed.

The count rate is typically some thousands counts per second (c/s) for the channeltron and some 10^5 c/s for the MCP, that is, the orders of magnitude of the range in which the time delay has to be measured and of the mean time interval between two neutral particles are similar: the (deliberate) worse conditions are then obtained.

The fast coincidence system is based on two facts:

1. The time delay between two coherent particles may be divided into three steps: two steps "around" the impact times for which an accurate time digitizer is needed and the third one is a "dead time" determined by velocity and distance constraints during which only the memorization function is required

2. A digital memorization allows one to get a "half correlation": in a full correlation between the detectors A and B, any particle received on A has to be correlated to each particle impinging B, at least within the "coherence" time of these two fluxes, and vice versa. If each particle hitting A, can be correlated to several particles received on B, the correlation tends to be perfect if the correlation of particles received on B, cannot practically be obtained for more than one particle impinging A. In the present case a large ratio between the coherence time of the stream of particles (some microseconds) and the mean time interval on channel A (more than hundred microseconds) ensures a quasi-full correlation. This result is obtained independently to the true flight times if the channels A and B, are, respectively, the photon/metastable channel and the high-count-rate channel.

In the resulting set-up (illustrated on Fig. 35), time and position data obtained at the start and stop channels are handled independently, and then symmetrically, until the last correlating layer in which the coincidence is resolved in the most advantageous way. Any start and/or stop event occurring within a 200 ns period of the master clock is immediately characterized by a logic work, which includes a status bit and the time and position codes. The 6-bit time code resulting from a homemade ECL digitizer determines the "fine" position in the period with a 3.2 ns resolution; the position codes are obtained by an ECL priority encoding of individual amplifiers–discriminators.

These start and stop words are written into a bipolar memory during the

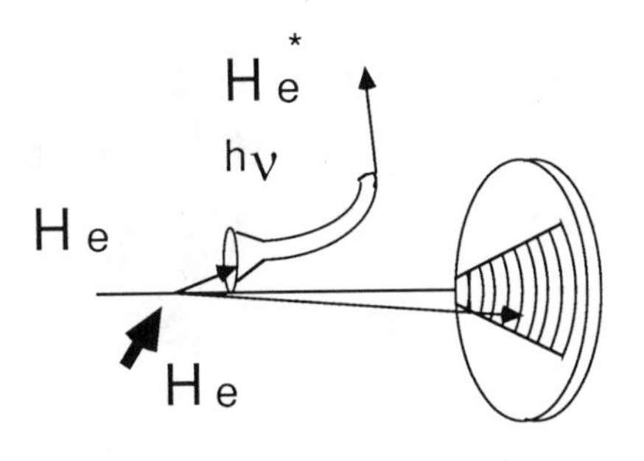

Figure 35. Experimental set-up.

first half of each period; the second half is used by the correlating layer to examine at convenient offset addresses these delayed data. Any physical start or stop channel may then be chosen as logical "start" channel and the more convenient start channel is the channel of lowest count rate: its sparse valid events are extracted from the digital delay line during a convenient time interval during which they may be correlated to several events occurring in the high-count-rate channel. This system is then presenting for the high-count-rate channel maximum and minimum dead times of 300 and 100 ns independently of the physical delay between the correlated particles. For the low-count-rate channel, the dead time is equal to the time window.

APPENDIX E—$X^- + H_2$ REACTIONS: THE ORSAY APPARATUS

The Orsay apparatus is designed to provide differential cross sections of reactive charge transfer and, in the case of anion–molecule reactions, of the competing detachment processes. A crossed-beam technique makes possible the angular analysis of energetically defined reactions, and this analysis is performed by a time correlating set-up: the coincidence technique allows an unambiguous separation between ionic and detachment processes. The multicoincidence technique provides complete angular and velocity information on these selected events.

A. Description of the Apparatus

A schematic view of the apparatus is given Fig. 36 for a $X^- + H_2$ system. The projectile ion beam, analyzed through a Wien filter, then decelerated and collimated, crosses at right angle a Campargue's supersonic beam[137] of H_2. The produced electrons and H^- are extracted from the collision region by a weak electric field (0.1–1 V/mm applied by two symmetric grids perpendicularly to the macroscopic collision plane) and then accelerated to an underlying detector. The scattered neutral products are detected on a frontal detector. The distance from this detector to the scattering center can

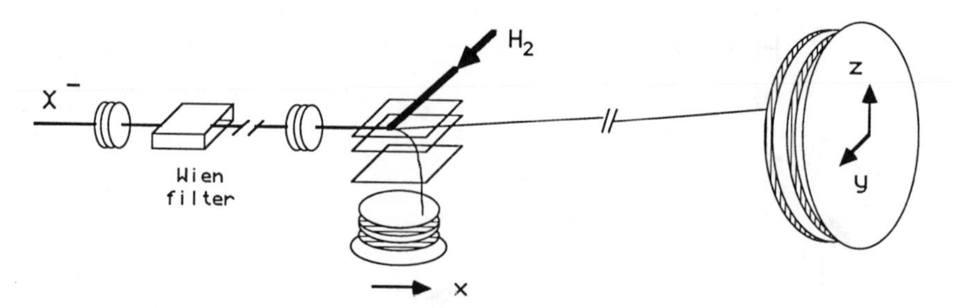

Figure 36. Experimental set-up.

be varied depending on the X/H mass ratio in order to collect the whole scattering cone. Both detectors are MCP assemblies with resistive anodes. On the electron and H^- detector only the position along the axis parallel to the incident beam (x) is detected, whereas y–z coordinates are detected on the frontal one. The charged particles start a 40-MHz counter, which may be stopped by the neutral ones within a programmable time window.

The position of each impinging event is determined by the risetime method: the characteristics of the resistive anode ($R = 50\,k\Omega\,\mathrm{cm}^2, C = 3.5\,\mathrm{pF/cm}^2$) and the central frequency of the shaping amplifiers (100 kHz) provide a sensitivity of about 10^4 m/s; four TTL 25-MHz counters are started by the corresponding timing signal and stopped by the four (or two) zero-crossing detectors. Seven 7-bit words are then stored in a intermediate Z-8000-controlled memory[138] for successful coincidences. These stacked data are transferred to the BFM86 host computer and processed to a four-dimension (t, x, y, z) vector of 7-bit time and position data. These data are then stored in a 1.2 MO diskette. This capacity of about 3×10^5 events is generally sufficient for an experiment, that is, one system at one energy. This system is then working in the "stack mode" (see Section IV B) and no histogram is integrated on-line except the four "projections" of the LMS on x, y, z, t axes and a fifth projection on the y, z LMS "plane" corresponding to the frontal detector. These projections are periodically refreshed during the accumulation time and provide a permanent control of the experiment.

In the collision region the hydrogen beam is about 1 mm wide; the maximum energy resolution in the laboratory frame is then 0.5% for a 400 mm flight path and the "natural" width of the coincidence peak is generally larger than the time resolution.

The negative incident ion is generally only 10–100 times faster than the hydrogen target and the Φ-symmetry in the CM frame is lost in the laboratory frame, that is, on the frontal detector; similarly, no structure may be predefined for the underlying detector: a cartesian position determination is then necessary on both detectors. The risetime method, which has been chosen for its simplicity, provides about 80 channels on each dimension, corresponding to 0.5 mm resolution. The nonlinearities are taken into account through an experimentally fitted parabolic function.

B. Data Reduction

Three phases are then needed to get the physical information from the stored data.

1. *Determination of the Fast-Beam Axis*

The velocity of the hydrogen molecule is known but the direction of the X^- axis in the collision region is not directly available since the extracting field is applied on both sides of this region. This information is obtained from histograms. These histograms are obtained off-line and observed with a convenient labeling by lab and CM data as deflection angle and endothermicity. Any projection of the LMS on a two-dimension space may be obtained. For the present purpose the visual analysis of successive cuts "parallel" to the y–z plane along the t axis allows the determination of the CM velocity vector.

2. *Determination of the Detachment Cross Sections*

The electron transit time is obviously negligible and the coincidence experiment works as a TOF experiment: the velocity of the neutral particle is directly obtained using the coordinates of the impinged pixel and the time delay; the transformation to the CM frame is straightforward.

In the fully automatic routine, time histograms are systematically constructed using the full resolution (eight successive readings are necessary to treat 2^{21} histograms on 128 time channels corresponding to each pixel of the front MCP). For each pixel, a mean noise value is obtained from event-free channels and subtracted from the content of each time channel of interest; these true coincidence are then added to the corresponding channels in the CM frame. Some remarks now have to be made.

The CM velocity space is digitized within a cube centered at the CM; the sides are equal to twice the incident X^-CM velocity (this cube includes the elastic sphere, or contour sphere). The parallel and perpendicular velocities of the neutral particle are coded on 64 and 32 values; the azimuthal angle is also encoded on 16 levels in order to check the validity of the incident axis determination. These cylindrical coordinates have been chosen to make easier the representation of the contour map (i.e., a planar representation in the fast neutral CM velocity space; equal magnitude of counts obtained in these cells are plotted for any Φ range; this point will be discussed later).

Coincidences with H^- lead to smaller apparent TOF's; if the corresponding TOF channels are obtained as superelastic, these events are automatically removed; if not, they appear in an asymmetric structure in the forward scattering and easily rejected too.

In this experiment a direct calculation is thought to be more convenient than the commonly used jacobian transformation. This method replaces the determination of a rather complicated analytical function by a simple vectorial analysis and, furthermore, allows a physically acceptable smoothing. The problem of the transfomation of digital data between two digitized three-dimension coordinate systems is never simple; in this case, the LMS system (y, z, t) is almost cartesian and the CM system is spherical or cylindrical along an axis which slightly diverges from the previous t-axis (at least due to the H_2 velocity). Spurious enhancements, or coalescences, occur in the "regions" where the structures are quite different as along the symmetry axis of the CM system. The same nasty effects are appearing in other parts of the contour "sphere" when roughly similar structures induce beats; these unphysical peaks are "smoothed" in this case by a linear interpolation (on four points) on the time channels, that is, in the LMS frame and using reasonable hypotheses. This method may also be seen as using partially the properties of the time interval asynchronous determination (cf. Section IV A).

3. *Determination of the Reactive Charge-Transfer Cross Sections*

The procedure is similar but, in this case, the measured time is the difference of two unknown flight times. In some circumstances,[5] lab measurements are directly related to CM data but in this case the symmetries in the CM frame are completely broken in the lab frame by the transverse electric field and by the molecular beam velocity, since

The CM velocity is not perpendicular to the Frontal MCP.

The two detectors are not in the same plane.

The anion crosses four successive uniform electric fields before impinging the underlying MCP.

A direct transformation is impossible and the following inversion procedure is performed: for each pixel of the frontal MCP, the unit vector of the HX lab velocity and some limiting values of its module are determined; these limiting values are obtained by the intersection of the velocity axis with the elastic sphere. This interval is sampled on 200 values and for each of these try points the CM velocity vector is known and a direct calculation of the two flight times and of the expected time delay is obvious. The quantization of the time delay giving the corresponding time channel number provides a pointer to the histogram of these delays and to four arrays in which the parallel and perpendicular CM velocities, the azimuthal angle, and the x value are accumulated. When this sweeping is over, the mean values of these variables are determined for each time delay channel; these correspondence tables are then used to add the content of each channel of the experimental time spectrum to the convenient cells of the contour map and to the histogram of the position (x) of the H^- impacts. The histogram of the redundant x variable allows the checking of the validity of the description of the extracting electric field and consequently of the CM data. Coincidences with electrons are also easily removed.

4. *Direct Determination of Cross Sections*

Most of the published results have been obtained by the preceding algorithms; as noted elsewhere, such a procedure would fail if the number of histograms is too large (e.g., with the new set-up described in Appendix C); furthermore, a complete contour map is obtained only after eight sweeps on the rough data diskette(s). An alternate algorithm has then been developed that "continuously" builds the contour map, and then allows the permanent control of its "growing." Two main points may be outlined:

1. The computer Time is Kept Low by A Linear Interpolation. The CM variables (and x) are first calculated for a convenient sampling on nine values of the lab variables (t, y, z); by this way each event in the lab frame is converted to the CM coordinates through three interpolations.

2. Noise is Removed "in the CM Frame." As already seen, no criterion allows one to reject a single event if its characteristics are compatible with the conservation laws. Each presumably true LMS event is then transformed into a CM event and the final contour map is the superposition of the "true" contour map and of a "false" one corresponding to the false coincidence background. A reasonable evaluation of this false contour map may be obtained as follows.

Events obtained for each pixel on the outside of the energetically allowed time interval are representative of the false coincidence rate for this pixel and their occurrence may be used to "simulate" a false coincidence within the allowed interval. A second "false" contour map is thus constructed in parallel with the true one by assigning to each "bad" event some "good" time delays. A conveniently weighted subtraction of the two contour maps after this treatment provides the noise-free data.

This procedure works well when the false coincidence background is flat and is even more valuable since the noise removal is performed on statistically well-defined numbers. (The subtraction of the mean noise value necessarily leads to "negative" channel contents: in this case, they are zeroed in the final controlled step. In the usual method this subtraction operates on smaller numbers, is then introducing larger relative errors, and is furthermore uncontrolled.)

The procedure also works well for noisy experiments, but negatively: the effect of the structured background appears quite clearly even if this high-count-rate regime is not evident on the individual spectra. (In fact, this method checks the validity of the direct "blind" noise removal on these histograms.)

If the operating conditions or the observed spectra allow one to assume a simple exponential structure, the corresponding time constant is estimated (or calculated from the count rate on the stop channel) and used to simulate this time varying structure. This simulation is easily performed from the standard uniformly distributed random function.

Some recent measurements obtained at low energy on the $O^- + H_2$ system under very unfavorable conditions (low MCP efficiency and high detached electron background) cannot be interpreted satisfactorily by the preceding algorithms. This failure has stimulated the deeper analysis proposed in Appendix A, which unfortunately shows that no analytic solution is *a priori* possible for a convenient data processing. The basic (but not yet experienced) idea is to use a linear predictive algorithm in order to unscramble the data in the same way they have been scrambled by this tricky time; in other words the (three-dimension) contour sphere is swept along the "experimental time axis" and, for each step, the signal level is deduced from the expected background and its value is used for the evaluation of the noise for the next step.

C. Comments

Some defects or failures of this experimental set-up may be outlined:

1. Only charged particle–fast neutral pairs can be detected and processes such as atom abstraction cannot be studied. Since classical methods allow the analysis of the channels leading to fast and then slightly deflected ions, only the first restriction is important, that is, the possibility of detecting the fast neutral: the momentum conservation implies that the projectile ion has to be energetic but chemical reactions are observed only if the energy in the CM is small (from threshold to about 20 eV). Most of the kinetic energy has then to appear as kinetic energy of the CM: only heavy-projectile–light-target systems may then been studied. This configuration is advantageous.

2. The fast neutrals are concentrated in a small solid angle in the lab frame allowing their complete collection.

3. The endoergic reaction can be studied at threshold but it requires a

supersonic target beam in order to cool the target gas and avoid energy and angular Doppler broadening.[139] (A short discussion on this point is given in Section IV E 1) Collisions involving expensive species (as HD for a fruitful study of isotopic effects) or vibrationally excited molecules cannot be studied.

4. Some assignations are ambiguous. The contour diagrams are more complex than the usual ones because all the competing channels appear. The reactive and detachment channels are separated by the coincidence but, among the detachment channels, the TOF-like method has not the capability of deciding whether XH or X is detected, that is, whether the reaction is reactive detachment (RD) or simple detachment (SD). An unambiguous attribution is possible only when, in the contour diagram, the structure is located inside the circle of maximum velocity of X but outside the circle of maximum velocity of XH. Any structure inside the RD circle may be as well a RD or a SD process (beyond the dissociation thresholds). However, the sharp angular anisotropy of all the observed structures and their continuous evolution as a function of the collision energy led to unambiguous attributions.

5. The absolute cross-section measurements are always difficult; in this case, two efficiencies have to be known precisely. An independent measurement is presumably preferable. When such data are available (as for $X^- + H_2$[101,102]), it is possible, by integration of each structure taking advantage of the fact that they are well individualized, to draw the energy profile of each detachment process. These data are very important for the understanding of the evolution of competing processes as a function of the collision energy. The contribution to the total electron detachment of the RD which was deduced in classical experiments from the isotopic behavior is here clearly evidenced.

6. Dissociative channels cannot of course be completely characterized. This point is discussed in Section IV E 3 b.

7. A constant problem in collision physics is the visual representation of "cross section." At least three definitions are commonly used, namely, σ_χ, σ_ω, and I; advantages and drawbacks of these representations are extensively discussed in textbooks and papers.[140–144] In this experiment another definition is used and has thus to be justified.

The transformation of the LMS vectors (using $y, z, \Delta t$ variables) into CM velocity data is obtained without geometrical correcting factors. All the particles involved in detachment channels are collected: the analysis is thus complete, that is, performed as well in the "macroscopic" collision plane as out of this plane. This is not obtained for the reaction charge transfer because the ions are only partially collected depending of the extracting electric field.

In this case, the azimuthal angular range within which the detection is effective is determined and considered as representative of the total contour map.

The treatment and the representation of the results are inferred by this acquisition procedure. The "natural" way is the performance of the data reduction directly in the CM velocity frame and not in the mixed spatial-velocity "intensity" context: the probability to get events in a given cell of the contour sphere results from counting the occurrences of velocity vectors of the neutral products in this cell. In this paper the unit cells are defined in the cylindrical reference system ($w_{\perp}, w_{\parallel}, \Phi$) defined with respect to the initial CM velocity vectors and their elementary volume is given by the usual term $w_{\perp} dw_{\perp} dw_{\parallel} d\Phi$.

When no azimuthal anisotropy is observed, a summation is performed over Φ and the new elementary volume is $2\pi w_{\perp} dw_{\perp} dw_{\parallel}$. Two representations are then possible. If a direct representation of the probability to get counts in the elementary torus is chosen, the function $\sigma_{\parallel\perp}$ is defined and normalized to the total cross section using $\sigma_{\text{tot}} = \iint \sigma_{\parallel\perp} dw_{\perp} dw_{\parallel}$. This representation is therefore different from the usual σ_{χ} and σ_{ω} and the correspondence is obtained through the total cross section $\sigma_{\text{tot}} = \iint \sigma_{\chi} dw\, d\chi = 2\pi \iint \sigma_{\omega} \sin \chi\, dw d\chi$.

These problems arise from the fact that all these representations are too closely related to the coordinate system. Everybody knows that the physical meaning of the divergence or curl operators do not appear clearly in their expressions in spherical coordinates but that students do "understand" their "intrinsic" definitions. An intrinsic representation of a collision process is obtained if $\sigma_{\parallel\perp}$ is replaced by a density $d_{\parallel\perp} = \sigma_{\parallel\perp}/2\pi w_{\perp} dw_{\perp} dw_{\parallel}$. This function provides, whatever the coordinate system is, a "local" evaluation of the probability to get events in small areas (of any shape) of the velocity diagram. A similar representation is possible for anisotropic scattering with small volume elements in the contour sphere.

The full determination of the impact positions of negative species (until now not allowed by mechanical constraints) would lead to a more complete experimental picture.

8. Analysis of anions. The kinematics of the reactive collisions is obtained from a time delay and a position analysis on the frontal MCP; the same procedure using the positions on the underlying one could be ambiguous; on the other hand, a multi-TOF experiment using these positions would obviously provide full information on this channel, and then could be used at lower energy, since the vertical component of the anion is a monotonic function of the TOF. (The peak of detached electrons on the time spectra provides the zero time of the anion flight.)

9. Analysis of electrons. The energy of the detached electrons has been

measured in other experiments[91,95] as inferior to 0.2 eV in the type of $X^- + H_2$ reactions studied here. This limit is compatible with the spread of the impacts around the projection of the intersection between the two beams. A two-dimension analysis of this spot in coincidence with the heavy X or XH partner would certainly be interesting, even without a complete analysis of the released electrons. At higher energy, the mapping of the electron impacts correlated to the measured or calculated velocities of X/XH and H/H_2 could provide some "propensity" rules for the electron ejection.

Acknowledgments

The authors feel very indebted to all the colleagues who participate in all these multicoincidence experiments: M. Barat, J. A. Fayeton, K. Gourdjil, J. C. Houver, M. Lecas, and J. B. Ozenne. They would like to thank J. Durup, V. Esaulov, J. P. Gauyacq, E. Gislason, V. Sidis, and M. Sizun for enlightening discussions and again V. Esaulov who suggested the application of this technique to Cl^-–H_2,

References

1. G. Rahmat, G. Vassilev, J. Baudon, and M. Barat, *Phys. Rev. Lett.* **26**, 1411–1413 (1971).
2. M. Eminyan, K. B. MacAdam, J. Slevin, and M. Kleinpoppen, *Phys. Rev. Lett.* **31**, 576–579 (1973).
3. J. C. Brenot, J. Fayeton, and J. C. Houver, *Rev. Sci. Instrum.* **51**, 1623–1629 (1980).
4. J. Herman, B. Menner, E. Reisacher, L. Zehnle, and V. Kempter, *J. Phys. B: At. Mol Phys.* **13**, L165–168 (1980).
5. D. P. de Bruijn and J. Los, *Rev Sci. Instrum.* **53**, 1020–1026 (1982).
6. M. Bharat, J. C. Brenot, J. A. Fayeton, J. C. Houver, J. B. Ozenne, R. S. Berry, and M. Durup-Ferguson, *Chem. Phys.* **97**, 165–177 (1985).
7. P. Roncin, C. Adjouri, M. N. Gaboriaud, L. Guillemot, M. Barat, and N. Andersen, *Phys. Rev. Lett.* **65**, 3261–3264 (1991).
8. D. Dhuicq, J. C. Brenot, and V. Sidis, *J. Phys. B: At. Mol. Phys.* **18**, 1395–1407 (1985).
9. J. E. Pollard, D. A. Lichtin, S. W. Janson, and R. B. Cohen, *Rev. Sci. Instrum.* **60**, 3171–3180 (1989).
10. A. Aspect, P. Granger, and, G. Roger, *Phys. Rev. Lett.* **49**, 91–94 (1982).
11. J. C. Houver and J. C. Brenot, LCAM internal seminar (1980).
12. J. Macek and D. H. Jaecks, *Phys. Rev., A* **4**, 2288–2300 (1971).
13. J. A. Fayeton, J. C. Houver, J. C. Brenot, and M. Barat, *J. Phys. B: At. Mol. Phys.* **14**, 2599 (1981).
14. J. L. Wiza, *Nucl. Instr. and Meth.* **162**, 587—601 (1979).
15. L. J. Richter and W. Ho, *Rev. Sci. Instrum.* **57**, 1469–1482 (1986).
16. M. P. Seah, *J. Electron Spectrosc. Relat. Phenom.* **50**, 137–157 (1990).
17. R. S. Gao, P. S. Gibner, J. H. Newman, K. A. Smith, and R. F. Stebbings, *Rev. Sci. Instrum.* **55**, 1756–1759 (1984).
18. I. Yamazaki, N. Tamai, H. Kume, H. Tsuchiya, and K. Oba, *Rev. Sci. Instrum.* **56**, 1187–1194 (1985).
19. J. B. Ozenne. (private communication).

20. R. S. Gao, W. E. Robert, G. J. Smith, K. A. Smith, and R. F. Stebbings, *Rev. Sci. Instrum.* **59**, 1954–1956 (1988).
21. R. K. Richards, H. H. Moos, and S. L. Allen, *Rev. Sci. Instrum.* **51**, 1–7 (1980).
22. C. J. Armentrout, *Rev. Sci. Instrum.* **56**, 1179–1186 (1985).
23. L. J. Richter, W. D. Miehler, L. J. Whitman, W. A. Noonan, and W. Ho., *Rev. Sci. Instrum.* **60**, 12–16 (1988).
24. J. G. Timothy and R. L. Bybee, *Rev. Sci. Instrum.* **46**, 1615–1623 (1975).
25. T. J. Skillman, J. R., E. D. Brooks, III, and M. A. Coplan, *Nucl. Instr. and Meth.* **155**, 267–272 (1978).
26. P. J. Hicks, S. Daviel, B. Wallbank, and J. Comer, *J. Phys. E: Sci. Instrum.* **13**, 713–715 (1980).
27. V. Radeka *IEEE Trans. Nucl. Sci.* **NS-21**, 51–63 (1974).
28. R. W. Wijnaendts van Resandt, H. C. den Harink, and J. Los, *J. Phys. E: Sci. Instrum.* **9**, 503–509 (1976).
29. D. P. de Bruijn and M. Spalburg, *Microchannel Plate Report* (FOM) H. Kerstein, ed. Chap. 3.
30. R. Gott, W. Parkes, and K. A. Pounds, *IEEE Trans. Nucl. Sci.* **17**, 367–373 (1970).
31. C. Martin. P. Jelinski, M. Lampton, R. F. Malina, and H. O. Anger, *Rev. Sci. Instrum.* **52**, 1067–1074 (1981).
32. A. Cerezo, T. J. Godfrey, and G. D. W. Smith, *Rev. Sci. Instrum.* **59**, 862–866 (1988).
33. C. L. C. M. Knibbeler, G. J. A. Hellings, H. J. Maaskamp, H. Ottevanger, and H. H. Brongersma, *Rev. Sci. Instrum.* **58**, 125–126 (1986).
34. P. Roncin, H. Laurent, and M. Barat, *J. Phys. E: Sci. Instrum.* **19**, 37–40 (1986).
35. P. Roncin (private communication): a simple hardware system using a 2910 microprogram controller and a 16-bit 74381-based ALU provides an overall treatment of CAMAC data in less than 3 μs.
36. Quantar Technology Incorporated, 3004 Mission Street Santa Cruz, CA, 95060.
37. E. Mathieson, *Nucl. Instr. and Meth.* **97**, 171–176 (1971).
38. E. Mathieson, K. D. Evans, W. Parkes, and P. F. Christie, *Nucl. Instr. and Meth.* **121**, 139–149 (1974).
39. W. Parkes, K. D. Evans, and A. Mathieson, *Nucl. Instr. and Meth.* **121**, 151–159 (1974).
40. E. Mathieson. K. D. Evans, K. Cole, and R. Everett, *Nucl. Instr. and Meth.* **126**, 199–204 (1975).
41. G. W. Fraser and E. Mathieson, *Nucl. Instr. and Meth.* **179**, 591 (1981).
42. G. W. Fraser and E. Mathieson, *Nucl. Instr. and Meth.* **184**, 537–542 (1981).
43. G. W. Fraser, E. Mathieson, M. Lewis, and M. Barstow, *Nucl. Instr. and Meth.* **190**, 53–65 (1981).
44. M. Lampton and F. Paresce, *Rev. Sci. Instrum.* **45**, 1098–1105 (1974).
45. J. B. Ozenne, M. Lavollée, J. C. Brenot, and M. Lemonnier, *Proc. Image Detection and Quality Symposium,* Paris, July 1986 pp. 63–66.
46. M. Lampton, O. Siegmund, and R. Raffanti, *Rev. Sci. Instrum.* **58**, 2298–2305 (1987).
47. C. Legrêle and J. C. Lugol, *IEEE Trans. Nucl. Sci.* **NS-30**, 297–301 (1983).
48. *Physics Today* (February, 1990).
49. J. Braunsfurth and K. Geske, *Nucl. Instr. and Meth.* **134**, 379–386 (1976).
50. D. Taupin *Probabilities, Data Reduction and Error Analysis in the Physical Sciences,* Les

Editions de Physique, Avenue du Hoggar, Zone industrielle de Courtaboeuf, B.P. 112, F-91944 Les Ulis Cedex, France (1988).

51. M. Barat, J. C. Brenot, M. Durup, J. Durup, J. C. Houver, J. B. Ozenne, G. Parlant, S. Berry, and S. Goursaud, report CNRS 1980.
52. A. Lahmam-Bennani, C. Dupré, and A. Duguet, *Phys. Rev. Lett.* **63**, 1582–1585 (1989).
53. L. D. Doverspike (private communication) during a mushroom gathering in the Evreux forest (France).
54. P. R. Bevington, *Data Reduction and Error Analysis for the Physical Sciences*, McGraw-Hill, New York, (1969).
55. J. L. Franklin, *Ann. Rev. Phys. Chem.* **25**, 485–526 (1974).
56. D. Vogt, W. Dreves, and J. Mischke, *Inter. J. Mass Spectrom. Ion Phys.* **24**, 285–296 (1977).
57. F. C. Fehsenfeld, and E. E. Ferguson, *J. Chem. Phys.: Planet. Space Sci.* **20**, 295 (1972).
58. D. K. Bohme and L. B. Young, *J. Am. Chem. Soc.* **92**, 7354 (1970).
59. D. A. Parkes, *J. Chem. Soc. Far. Tras.* **168**, 313 (1972).
60. M. Kashihira, E. Vietzke, and G. Kellerman, *Chem. Phys. Lett.* **39**, 316–319 (1976).
61. C. E. Hamilton, V. M. Bierbaum, and S. R. Leone, *J. Chem. Phys.* **80**, 1831 (1984).
62. M. Durup-Ferguson, J. C. Brenot, J. A. Fayeton, K. Provost, and M. Barat, *J. Chem. Phys.* **114**, 389 (1987).
63. M. Durup-Ferguson, J. A. Fayeton. J. C. Brenot, and M. Barat, *Intern. J. Mass Spectrom. Ion Processes* **80**, 211 (1987).
64. J. A. Fayeton, J. C. Brenot, M. Durup-Ferguson, and M. Barat, *Chem. Phys.* **133**, 259 (1989).
65. J. C. Brenot, M. Durup-Ferguson, J. A. Fayeton, K. Goudjil, Z. Herman, and M. Barat *Chem. Phys.* **146**, 263–272 (1990). Results presented in this paper for low collision energies are distorded by an expermental artefact. A correcting note will be sent to the same publisher.
66. V. M. Bierbaum, G. B. Ellison, J. H. Futrell, and S. R. Leone, *J. Chem. Phys.* **67**, 2375–2376 (1977).
67. F. C. Fehsenfeld, A. L. Schmeltekopf, H. I. Schiff, and E. E. Ferguson, *J. Chem. Phys.* **45**, 1844 (1966).
68. E. E. Ferguson, F. C. Fehsenfeld, and A. L. Schmeltekopf, *Advances in At. and Mol. Phys.* 1–56 (1969).
69. E. E. Ferguson, in *Ion Molecule Reactions*, J. Franklin, ed., Plenum Press, New York, 1972 p. 363.
70. J. C. Weisshaar, T. S. Zwier, and S. R. Leone, *J. Chem. Phys.* **75**, 4873 (1981).
71. T. S. Zwier, V. B. Bierbaum, J. B. Ellison, and R. S. Leone, *J. Chem. Phys.* **72**, 5426 (1980).
72. C. E. Hamilton, M. A. Duncan, T. S. Zwier, J. C. Weisshaar, G .B. Ellison, V. M. Bierbaum, and S. R. Leone, *Chem. Phys. Lett.* **94**, 4 (1983).
73. M. M. Maricq. M. A. Smith, C. J. S. M. Simpson, and J. B. Ellison, *J. Chem. Phys.* **74**, 6154 (1981).
74. V. M. Bierbaum, G. B. Ellison, and S. R. Leone, in *Gas Phase Ion Chemistry,* M. Bowers ed., Academic Press, New York, 1984, Vol. 3, Chap. 17, pp. 1–39.
75. M. A. Biondi, *Comments At. Mol. Phys.* **6**, 159–161 (1977).
76. D. L. Albritton, I. Dotan, W. Lindinger, M. McFarland, J. Tellinghuisen, and F. C. Fehsenfeld, *J. Chem. Phys.* **66**, 410–421 (1977).
77. W. Lindinger, D. L. Albritton, F. C. Fehsenfeld, and E. E. Ferguson, *J. Chem. Phys.* **63**, 3238–3242 (1975).

78. J. H. Futrell and T. O. Tiernan, in *Ion-Molecule Reactions* J. Franklin, ed., Plenum Press, New York, 1972, Vol. 2, Chap. 8, pp. 485–551.
79. J. F. Paulson, in *Ion-Molecule Reactions* J. Franklin, ed., Plenum Press, New York, 1972, Vol. 1, Chap. 4, pp. 77–100.
80. G. Zellermann and E. Vietzke, *Radiochimica Acta* **50**, 107–115 (1990).
81. C. E. Melton and J. A. Neece, *J. Am. Chem. Soc.* **93**, 6757 (1971).
82. D. Vogt, *Inter. J. Mass Spectrom. Ion Phys.* **3**, 81 (1969).
83. T. O. Tiernan, in *Interactions Between Ions and Molecules*, P. Ausloos, ed., Plenum Press, New York, 1974, pp. 353–385.
84. J. F. Paulson and J. P. Gale, *Advan. Mass Spectrom.* **7 A**, 263 (1978).
85. C. Lifschitz *J. Phys. Chem.* **86**, 3634–3637 (1982).
86. M. P. Karnett and R. J. Ross, *Chem. Phys. Lett.* **82**, 277–280 (1981).
87. L. D. Doverspike, R. L. Champion, and S. K. Lam, *J. Chem. Phys.* **58**, 1248–1250 (1973).
88. S. G. Johnson, L. N. Kremer, C. J. Metral, and R. J. Cross, *J. Chem. Phys.* **68**, 1444–1447 (1978).
89. E. Herbst, L. G. Payne, R. L. Champion, and L. D. Doverspike, *Chem. Phys.* **42**, 413–421 (1979).
90. M. P. Karnett and R. J. Ross, *Chem. Phys. Lett.* **84**, 501–503 (1981).
91. V. A. Esaulov, R. L. Champion, J. P. Grouard, R. I. Hall, J. L. Montmagnon, and F. Penent, *J. Chem. Phys.* **92**, 2305–2309 (1990).
92. V. M. Bierbaum. G. B. Ellison, J. H. Futtrel, and S. R. Leone, *J. Chem. Phys.* **67**, 2375 (1977).
93. F. C. Fehsenfeld, in *Interactions between Ions and Molecules*, P. Ausloos, ed., Plenum Press, New York, 1974, pp. 387–412.
94. J. L. Mauer and G. J. Schulz, *Phys. Rev. A* **7**, 593 (1973).
95. J. Comer and G. J. Schulz, *Phys. Rev. A* **10**, 2100–2106 (1974).
96. J. L. Moruzzi and A. V. Phelps, *J. Chem. Phys.* **45**, 4617 (1966).
97. M. McFarland, D. L. Albritton, F. C. Fehsenfeld, E. E. Ferguson, and A. L. Schmeltekopf, *J. Chem. Phys.* **59**, 6629 (1973).
98. J. Comer and G. J. Schulz, *J. Phys. B: At. and Mol. Phys.* **7**, 1249 (1974).
99. V. A. Esaulov, *Ann. Phys. Fr.* **11**, 493–592 (1988).
100. M. S. Huq, D. S. Fraedrich, L. D. Doverspike, R. L. Champion, and V. A. Esaulov, *J. Chem. Phys.* **76**, 4952–4960 (1982).
101. M. S. Huq, L. D. Doverspike, and R. L. Champion, *Phys. Rev. A* **27**, 785 (1983).
102. M. S. Huq, D. Scott, R. L. Champion, and L. D. Doverspike, *J. Chem. Phys.* **82**, 3118–3122 (1985).
103. D. Scott, M. S. Huq, R. L. Champion, and L. D. Doverspike, *Phys. Rev. A* **33**, 170 (1986).
104. J. P. Gauyacq, *J. Phys. B* **13**, L 501 (1980).
105. J. P. Gauyacq, *J. Phys. B* **13**, 4417 (1980).
106. J. T. Cheung and S. Datz, *J. Chem. Phys.* **73**, 3159–3165 (1980).
107. W. B. Miller, S. A. Safron, and D. R. Herschbach, *Discuss. Faraday Soc.* **44**, 108 (1967).
108. S. Stolte, A. E. Proctor, and R. B. Bernstein, *J. Chem. Phys.* **65**, 4990–5008 (1976).
109. G. A. Segal and K. Wolf, *J. Phys. B* **14**, 2291 (1981).
110. R. Azria, Y. Le Coat, D. Simon, and M. Tronc, *J. Phys. B* **13**, 1909 (1980).
111. J. N. Bardsley, A. Herzenberg, and F. Mandl, *Proc. Phys. Soc. Lond.* **98**, 321 (1966).

112. M. Sizun, E. A. Gislason, and G. Parlant, *Chem. Phys.* **107**, 311 (1986).
113. G. Herzberg, in *Electronic Spectra and Electronic Structure of Polyatomic Molecules*, D. Van Nostrand Company, Inc., Princeton New Jersey (1966), p. 586.
114. K. T. Gillen, B. H. Mahan, and J. S. Winn, *Chem. Phys. Lett.* **22**, 344–347 (1973).
115. K. T. Gillen, B. H. Mahan, and J. S. Winn, *J. Chem. Phys.* **55**, 5373–5384 (1973).
116. A. Henglein, *J. Phys. Chem.* **76**, 3883 (1972).
117. K. T. Gillen, B. H. Mahan, and J. S. Winn, *J. Chem. Phys.* **59**, 6380–6396 (1973).
118. D. R. Bates, C. J. Cook, and F. J. Smith, *Proc. Phys. Soc. Lond.* **83**, 49–57 (1964).
119. R. J. Suplinskas, *J. Chem. Phys.* **49**, 5046 (1968).
120. T. F. George and R. J. Suplinskas, *J. Chem. Phys.* **54**, 1037 (1971).
121. T. F. George and R. J. Suplinskas, *J. Chem. Phys.* **54**, 1046 (1971).
122. P. B. Armentrout and J. L. Beauchamp, *Chem. Phys.* **48**, 315–320 (1980).
123. S. A. Safron, G. W. Coppenger, and V. F. Smith, *J. Chem. Phys.* **80**, 1929–1936 (1984).
124. S. A. Safron and G. W. Coppenger, *J. Chem. Phys.* **80**, 4907–4914 (1984).
125. S. A. Safron, *J. Chem. Phys.* **89**, 5513–5519 (1985).
126. S. A. Safron, *J. Chem. Phys.* **89**, 5719–5722 (1985).
127. B. H. Mahan, W. E. W. Ruska, and J. S. Winn, *J. Chem. Phys.*, **65**, 3888–3896 (1976).
128. R. W. Wijnaendts van Resandt and J. Los, *Invited Papers and Progress Reports, XI ICPEAC, Kyoto*, (1979), N. Oda and K. Takayanagi, eds., North Holland, Amsterdam, 1979.
129. C. J. Borkowski and M. K. Kopp, *Rev. Sci. Instrum.* **39**, 1515–1522 (1968).
130. C. J. Borkowski and M. K. Kopp, *Rev. Sci. Instrum.* **46**, 951 (1975).
131. P. G. Fontolliet Traité d'électricité (Ecole polytechnique fédérale de Lausanne Vol. XVIII, Georgi, Lausanne CH (1983), p. 78.
132. H. Thyssen, N. Shenhav, H. P. Eulenberg, and A. Arbel, *Nucl. Instr. and Meth.* **165**, 265–278 (1979).
133. J. B. Ozenne, M. Lecas, and J. C. Brenot, (unpublished).
134. D. Dowek, J. C. Houver, J. Pommier, C. Richter, T. Royer, N. Andersen, and B. Palsdottir, *Phys. Rev. Let.* **64**, 1713–1716 (1990).
135. S. E. Nielsen and N. Andersen, *Z. Phys. D Atoms, Molecules and Clusters* **5**, 312–329 (1987).
136. J. C. Brenot, D. Dhuicq, J. P. Gauyacq, J. Pommier, V. Sidis, M. Barat, and E. Pollack, *Phys. Rev. A* **11**, 1933–1945 (1975).
137. R. Campargue, in Books of Abstracts, VI th International Symposium on Molecular Beams, Noordwijkerhout 1977, p. 247.
138. J. Agusti I Cullel, M. Lecas, J. C. Brenot, and J. C. Houver, *Rev. Sci. Instr.* **51**, 1623 (1985).
139. P. J. Chantry, *J. Chem. Phys.* **55**, 2746 (1971).
140. F. A. Morse and R. B. Bernstein, *J. Chem. Phys.* **37**, 2019 (1962).
141. K. T. Gillen, A. M. Rulis, and R. B. Bernstein, *J. Chem. Phys.* **54**, 2831 (1971).
143. B. Friedrich and Z. Herman, *Collect. Czech. Chem. Commun.* **49**, 570 (1984).
144. J. M. Farrar, *Techniques of Chemistry* **XX**, 325–416 (1988).

STATE-SELECTED AND STATE-TO-STATE ION-MOLECULAR REACTION DYNAMICS BY PHOTOIONIZATION AND DIFFERENTIAL REACTIVITY METHODS

CHEUK-YIU NG

Ames Laboratory, U.S. Department of Energy and Department of Chemistry
Iowa State University
Ames, Iowa

CONTENTS

State-Selected and State-to-State Ion–Molecule Reaction Dynamics, Part 1: Experiment, Edited by Cheuk-Yiu Ng and Michael Baer. Advances in Chemical Physics Series, Vol. LXXXII.
ISBN 0-471-53258-4

I. INTRODUCTION

The field of state-selected ion chemistry involving the reactivity studies of gaseous ions in their ground and excited rotational, vibrational, and electronic states was pioneered more than 20 years ago by Chupka and co-workers.[1-4] Their work, which showed the simplicity of preparing state-selected reactant ions by vacuum ultraviolet (VUV) photoionization, has inspired the younger generation of experimental molecular reaction dynamicists. The past decade has witnessed a rapid increase in activity in this field. Despite many difficulties, such as the low intensities of state-selected reactant ions obtainable from photoionization methods, absolute state-to-state cross section measurements for several simple ion–molecule reaction systems have been accomplished. Because of the development of photoelectron–photoion coincidence (PEPICO) techniques and intense VUV light sources, such as synchrotron radiations and VUV lasers, the field of state-selected ion–molecule reaction dynamics is expected to continue to grow at a rapid pace. The recent review on this subject emphasizes the technical developments achieved.[5]

The most important goal for experimental state-selected and state-to-state studies is to provide accurate cross-section data for comparisons with theoretical predictions. This chapter focuses on the discussion of state-selected and state-to-state cross-section measurements for several simple ion–molecule reaction systems. $Ar^+(^2P_{3/2,1/2}) + Ar, [Ar + N_2]^+, Ar^+(^2P_{3/2,1/2}) + CO(O_2), [Ar + H_2]^+, O^+(^4S, ^2D, ^2P) + N_2, O^+(^4S) + H_2$, and $H_2^+ + H_2$. These systems have been studied in great detail during the past few years in our and other laboratories. The importance of these systems stems from the fact that they are amenable to rigorous quantum-theoretical investigations. In cases where theoretical predictions exist, comparisons of the experimental and theoretical results are made. These comparisons, together with comparisons of experimental findings obtained from different laboratories, allow the evaluation of existing data sets, and they point to improvements in future measurements, which are necessary for critical testing of state-of-the-art theoretical calculations.

In Section II, the essential features of the crossed ion–neutral beam and triple-quadrupole–double-octople (TQDO) photoionization apparatuses built in our laboratory are described. The state-selected and state-to-state cross sections obtained using these apparatuses are discussed in Section III and compared to theoretical predictions and experimental results of other laboratories. The summary and anticipated future developments are given in Section IV.

II. EXPERIMENTAL CONSIDERATIONS AND PROCEDURES

Two specially designed ion–molecule reaction photoionization apparatuses have been constructed in our laboratory for state-selected and state-to-state cross-section measurements. Since the details of these apparatuses are discussed in the recent review,[5] only a brief account about their essential features and experimental procedures is given below.

A. Triple-Quadrupole–Double-Octopole Photoionization Apparatus

The detailed design and operation procedures of the TQDO photoionization apparatus (Fig. 1) have been described previously.[6–8] The apparatus consists of a 0.2 m VUV monochromator (McPherson 234), a discharge lamp, a tungsten photoelectric VUV light detector, three quadrupole mass filters (QMF), two radio-frequency (rf) octopole ion guide reaction gas cells, a supersonic-free-jet production system, and a scintillation ion detector. As shown in Fig. 1, the arrangement of the apparatus requires reactant ions formed in the photoionization region (1) to pass through, in sequential order, the reactant QMF (4), the lower rf octopole ion guide reaction gas cell (8), the middle QMF (11), the upper rf octopole ion guide reaction gas cell (13), and the product QMF (16) before their detection by the scintillation ion detector [(18) + (19) + (20)].

1. *Measurements of Absolute State-Selected Cross Sections*

The reactant ions in the ground state, or in a known mixture of ground and excited states, are prepared by photoionization of a free jet of the precursor molecules. The free jet is produced by supersonic expansion through a 5-μm-diameter quartz nozzle (2) at a stagnation pressure of 50–150 Torr. Under these expansion conditions, the photoionization chamber is maintained at a pressure of $1–3 \times 10^{-5}$ Torr. The free-jet arrangement allows the efficient trapping and the maintenance of a relatively thin volume of high number density precursor molecules in the photoionization region. The reactant ions formed in the photoionization region are extracted perpendicular to the free jet and are mass-selected by the reactant QMF before reacting with the neutral reactant molecules of interest in the upper rf octopole ion guide reaction gas cell. The neutral reactant gas cell pressure is monitored with a Baratron manometer and a gas cell pressure of $(0.1–3) \times 10^{-4}$ Torr is used. The reactant ions, and product ions formed in the upper rf octopole ion guide reaction gas cell, are mass-selected by the product QMF and detected by the scintillation ion detector. During the measurements of absolute cross sections, the lower octopole rf ion guide. (9) and the middle QMF are used as ion lenses to transmit ions of all masses.

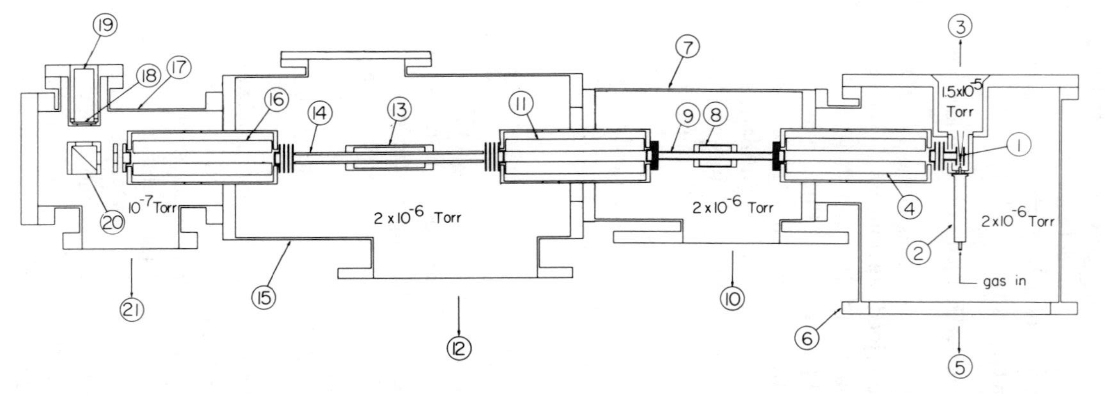

Figure 1. Cross-sectional view of the TQDO photoionization apparatus. (1) Photoionization region, (2) nozzle, (3) to freon-trapped 6 in. diffusion pump (DP), (4) reactant quadrupole mass filter (QMF), (5) to liquid-nitrogen (LN_2)-trapped 6 in. DP, (6) the reactant QMF chamber, (7) the lower rf octopole ion guide chamber, (8) the lower reaction gas cell, (9) the lower rf octopole ion guide, (10) to LN_2-trapped 6 in. DP, (11) the middle QMF, (12) to LN_2-trapped 4 in. DP, (13) the upper reaction gas cell, (14) the upper rf octopole ion guide, (15) the upper rf octopole ion guide chamber, (16) the product QMF, (17) detector chamber, (18) plastic scintillator window, (19) photomultiplier tube, (20) aluminum ion target, (21) to LN_2-trapped 2 in. DP.[3–6]

a. Calibration of Reactant Ion Beam Energies. The reactant-ion-beam energy is measured by the retarding-potential-energy method.[7,9–11] The retarding dc voltage is applied to the ion lenses at the exit of the rf octopole ion guide. Armentrout and co-workers use a retarding-field method in which the dc potential of the octopole ion guide is swept through the nominal ion energy zero.[12] The reactant-ion-beam energy is also calibrated using a time-of-flight (TOF) technique.[13] This method involves measuring the difference in flight times of the reactant-ion pulses created at the entrance and exit of the rf octopole ion guide. From the difference in flight times and the length of the octopole ion guide, the mean velocity and the energy spread of the reactant-ion beam are calculated. Figure 2 illustrates the results of such an exercise in measuring the kinetic energy of an Ar^+ reactant-ion beam. The first peak, at approximately 20 μs, is a measure of the flight time of reactant Ar^+ from the exit of the upper rf octopole ion guide to the ion detector. The TOF peaks in the range of 80–280 μs measure the flight times of Ar^+ reactant ions from the entrance of the upper rf octopole ion guide to the ion detector. The nominal laboratory kinetic energies (E_{lab}), which are equal to the difference of dc potentials between the photoionization region and the upper rf octopole ion guide reaction gas cell, are in the range of 0.3–5.0 eV. Using the TOF method, a reactant ion beam with $E_{lab} \leqslant 0.05$ eV can be measured with accuracy. Gerlich has reported an energy-calibration

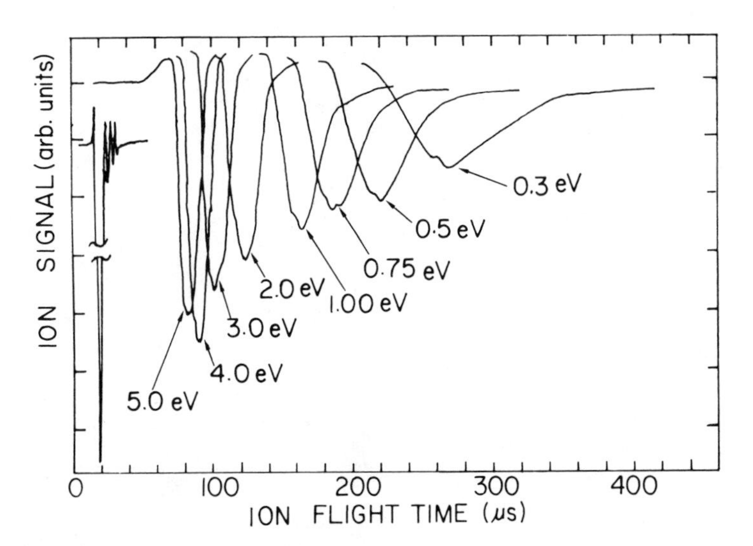

Figure 2. Time-of-flight (TOF) spectra for the reactant Ar^+ ion beam, which are a measure of the flight times of Ar^+ ion pulses from the exit and entrance of the upper rf octopole ion guide to the ion detector.[13]

method[14] that allows the kinetic energy of the reactant ion beam to be measured down to 6 meV. for $E_{lab} \geqslant 5$ eV, the mean reactant-ion-beam energies determined by the TOF and retarding-potential-energy methods are in agreement.

b. Ion Transmission Through the Mass Spectrometer and Octopole Ion Guide. At each E_{lab}, it is important to maximize the product-ion collection efficiencies by optimizing carefully the transmission of the ions through the ion lenses, product QMF, and upper rf octopole ion guide. Since the ion transmission through a QMF depends on the ion mass, it is often necessary to lower the product QMF resolution to minimize the transmission effect. As an example, the mass spectrum for reactant Ar^+ and product O_2^+, O^+, and ArO^+ ions formed in the reaction of $Ar^+(^2P_{3/2}) + O_2$ at $E_{lab} = 10$ eV is shown in Fig. 3. At this QMF resolution, the mass peaks exhibit flat-top peak structures and the ion intensity measurements are insensitive to minor

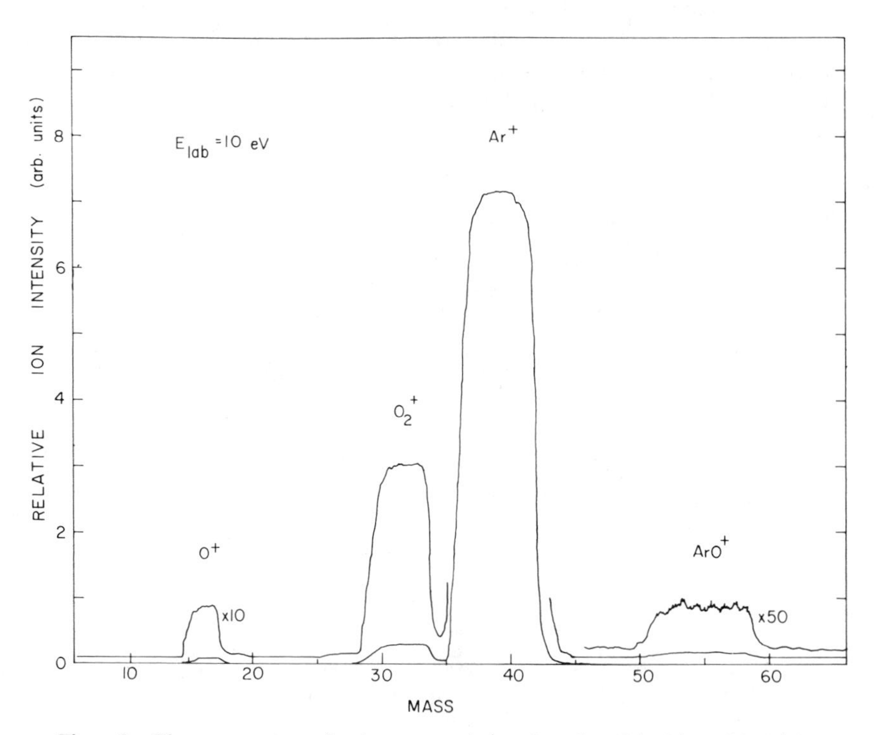

Figure 3. The mass spectrum for the reactant Ar^+ and product O_2^+, O^+, and ArO^+ formed by the reaction of $Ar^+(^2P_{3/2}) + O_2$ at $E_{lab} = 10$ eV. This mass spectrum illustrates the typical mass resolution used in the absolute cross section measurements.[13]

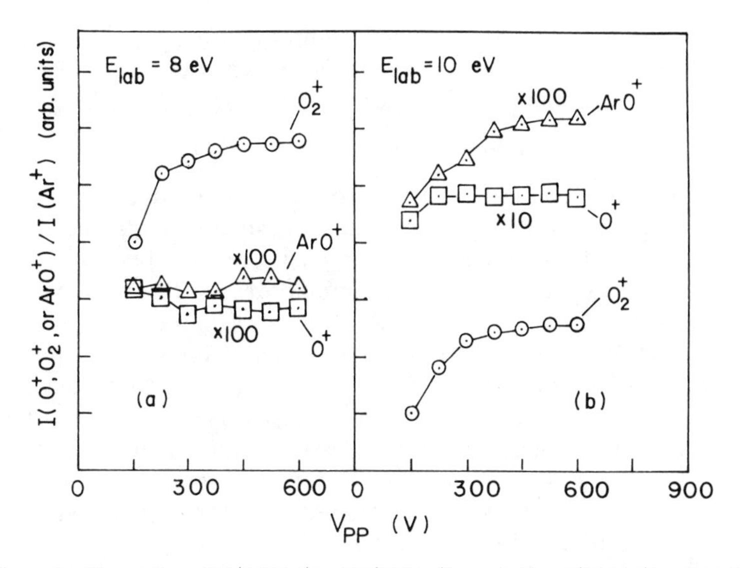

Figure 4. The ratios, $I(O_2^+)/I(Ar^+)$, $I(O^+)/I(Ar^+)$, and $I(ArO^+)/I(Ar^+)$, plotted as a function of the peak-to-peak rf voltage (V_{pp}) applied to the upper octopole ion guide. Here, $I(O_2^+)$, $I(O^+)$, $I(ArO^+)$, and $I(Ar^+)$ represent the intensities for the product O_2^+, O^+, and ArO^+ and reactant $Ar^+(^2P_{3/2})$, respectively. O_2^+, O^+, and ArO^+ are formed by the reaction $Ar^+(^2P_{3/2}) + O_2$ at (*a*) $E_{lab} = 8$ eV and (*b*) $E_{lab} = 15$ eV.[13]

mass drifting effects. Further lowering of the QMF resolution does not affect the relative intensities of these mass peaks.

Figures 4(*a*) and 4(*b*) show the plots of $I(O_2^+)/I(Ar^+)$, $I(O^+)/I(Ar^+)$, and $I(ArO^+)/I(Ar^+)$ as a function of peak-to-peak rf voltage (V_{pp}) applied to the upper rf octopole ion guide at $E_{lab} = 8$ and 10 eV, respectively. Here $I(O_2^+)$, $I(Ar^+)$, $I(O^+)$, and $I(ArO^+)$ represent the intensities for O_2^+, Ar^+, O^+, and ArO^+ formed in the $Ar^+(^2P_{3/2}) + O_2$ reaction. At sufficiently high rf voltages, these ratios remain essentially constant as the rf voltage is further increased, indicating that the trapping efficiencies for the product ions are maximized. The rf voltage required to maximize the trapping efficiency for a specific ion depends on E_{lab}. In order to obtain reliable absolute cross section data, it is important to examine the trapping curves of product ions to ensure that trapping efficiencies have been maximized.

From the attenuation of the reactant-ion beam measured at a known neutral reactant gas cell pressure, the upper limit for the sum of total cross sections for all reaction channels can be calculated. Consistency between the sum of the cross sections for all reactions and the upper limit for the total cross sections deduced by the attenuation measurement is an important criterion for showing that the absolute cross-section data are reliable. Taking into account the uncertainties in the length of the reaction gas cell and in the absolute pressure measurement of the neutral reactant gas, the accuracy of absolute cross sections can be determined to better than $\pm 25\%$.

2. *Retarding-Potential-Energy Analyses of Product Ions*

Information about the kinetic energy distributions of product ions along the octopole axis can also be obtained by the retarding-potential method. The retarding-potential-energy curves for product ions are obtained using the same experimental arrangement and procedures as those employed for measuring the retarding-potential-energy curve of the reactant ion. Using the product QMF to select the product ion of interest, the retarding-potential-energy curves for product ions at a given E_{lab} are recorded individually. The retarding potential curve for a product ion provides only a rough estimate of the product angular distribution, but when combined with other information, the estimate may bring out important mechanistic and dynamical details of the reaction. Examples of retarding potential analyses of product ions are given in Section III.

3. *Detection of Product-Ion State by the Differential Reactivity Method*

The utilization of the reactivity of an ion with a specific neutral reactant has been used to monitor the ion concentration in flow-tube experiments.[15] The detection of $O_2^+(X\,^2\Pi_g, \tilde{a}\,^4\Pi_u)$ ions formed in the charge transfer reaction $Ar^+(^2P_{3/2}) + O_2$ illustrates this method.

The ionization energy (IE) for $Ar^+(^2P_{3/2})$ is lower than that for $O_2^+(\tilde{a}\,^4\Pi_u, v' = 0)$ by 0.34 eV.[16] At a sufficiently low center-of-mass energy ($E_{c.m.}$), the cross section for the exothermic charge-transfer reaction,

$$O_2^+(\tilde{a}\,^4\Pi_u, v') + Ar \rightarrow O_2 + Ar^+, \tag{1}$$

is expected to be significantly higher than that for

$$O_2^+(\tilde{X}\,^2\Pi_g, v') + Ar \rightarrow O_2 + Ar^+, \tag{2}$$

because the $O_2^+(\tilde{X}\,^2\Pi_g, v' = 0) + Ar$ charge-transfer reaction is endothermic by 3.7 eV. The state selected study using the threshold PEPICO method[17] shows that at $E_{c.m.} \leqslant 5.8$ eV the cross sections for reaction (2) with $v' = 19$ and 20 are negligibly small compared to those for reaction (1) with $v' = 0$–7, in spite of the fact that the $O_2^+(\tilde{a}\,^4\Pi_u, v' = 0)$ state lies only 0.17 eV above the $O_2^+(\tilde{X}\,^2\Pi_g, v' = 20)$ level.[18,19] The negligibly low cross sections for charge transfer between $O_2^+(\tilde{X}\,^2\Pi_g, v' = 20)$ and Ar are attributed to the very low Franck–Condon factor for the ionization transition $O_2(X, v'') \rightarrow O_2^+(\tilde{X}\,^2\Pi_g, v' = 20)$.

Figures 5(*a*) and 5(*b*) compare the photoionization efficiency (PIE) spectrum for O_2^+, prepared by photoionization of O_2, with those O^+ and Ar^+ formed in the upper rf octopole ion guide reaction gas cell by the reaction between Ar and O_2^+ at $E_{lab} = 5$ and 2 eV, respectively. The positions for the O_2^+ ($\tilde{a}\,^4\Pi_u, \tilde{A}\,^2\Pi_u, \tilde{b}'\,^4\Pi_u, \tilde{b}\,^4\Sigma_g^-$) thresholds[19] are indicated in the figures. At $E_{lab} = 2$ eV, the PIEs for Ar^+ at photon energies below the O_2^+ ($\tilde{a}\,^4\Pi_u, v' = 0$) threshold are essentially zero, indicating that the cross sections for reaction (2) with $v' = 0$–20 are negligible. The PIE for product Ar^+ increases rapidly as the photon energy is increased above the O_2^+ ($\tilde{a}\,^4\Pi_u, v' = 0$) threshold. The PIE spectrum for Ar^+ formed at $E_{lab} = 5$ eV, which is not shown here, is essentially identical to that observed at $E_{lab} = 2$ eV.

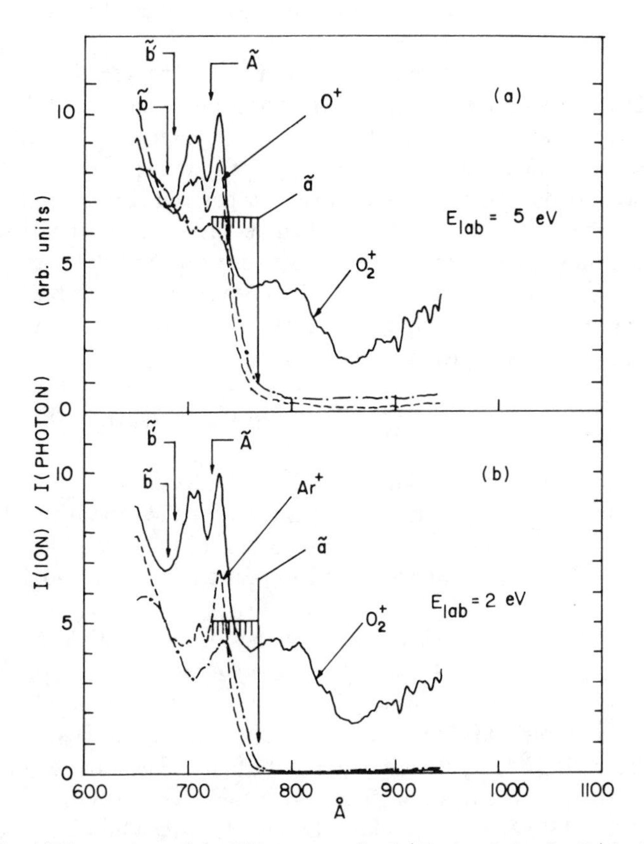

Figure 5. (*a*) Comparison of the PIE spectrum for O_2^+ (—) and that for O^+(– –) formed in the reaction $O_2^+ + Ar$ at $E_{lab} = 5$ eV; (–·–) ratio for PIE(O^+)/PIE(O_2^+). (*b*) Comparison of the PIE spectrum for O_2^+(—) and that for Ar^+(– –) formed in the reaction $O_2^+ + Ar$ at $E_{lab} = 2$ eV; (–·–) ratio for PIE(Ar^+)/PIE(O_2^+).[13]

The ratios PIE(O^+)/PIE(O_2^+) and PIE(Ar^+)/PIE(O_2^+), which are proportional to the cross sections for O^+ and Ar^+ formed in the $O_2^+ + Ar$ reaction, are also plotted in Fig. 5(*a*) and 5(*b*). These plots clearly show that at $E_{lab} = 2$–$5\,eV$ the cross sections for O^+ and Ar^+ due to $O_2^+(\tilde{a}\,^4\Pi_u, v')$ are substantially higher than those associated with $O_2^+(\tilde{X}\,^2\Pi_g, v' = 0$–$20)$. The thresholds of the vibrational levels $v' = 0$–8 of the $O_2^+(\tilde{a}\,^4\Pi_u)$ state are indicated in Figs. 5(*a*) and 5(*b*). The cross section for reaction (2) appears to peak at $v' = 5$ and shows a strong dependence on the vibrational state, an observation consistent with the results of the previous threshold PEPICO study.[17] The curve for PIE(Ar^+)/PIE(O_2^+) exhibits a minimum in the photon energy region between the $O_2^+(\tilde{A}\,^2\Pi_g, v' = 0)$ and $O_2^+(\tilde{b}\,^4\Sigma_g^-)$ thresholds. This minimum can be ascribed partly to the formation of $O_2^+(\tilde{X}\,^2\Pi_g, v')$ by rapid radiative decay[18] of $O_2^+(\tilde{A}\,^2\Pi_u, v')$ ions initially produced by photoionization of O_2.

Because of the large differences observed in the cross sections of reactions (1) and (2), these reactions are appropriate probing reactions for the detection of $O_2^+(\tilde{a}\,^4\Pi_u)$ formed by the $Ar^+(^2P_{3/2,1/2}) + O_2$ charge-transfer reaction. In this experiment, the reaction of $Ar^+(^2P_{3/2}) + O_2$ at a specific E_{lab} takes place in the lower rf octopole ion guide reaction gas cell. Product O_2^+ ions are mass-selected by the middle QMF and guided into the upper rf octopole ion guide reaction gas cell where reaction between O_2^+ and the probing Ar gas at a fixed laboratory collision energy (E_{pr}) produces Ar^+ and O^+. Absolute cross sections for the charge transfer reaction $O_2^+ + Ar$ and the collision-induced dissociation process,

$$O_2^+(\tilde{X}\,^2\Pi_g, \tilde{a}\,^4\Pi_u) + Ar \rightarrow O^+ + O + Ar, \tag{3}$$

are measured using the product QMF and procedures similar to those described above. The cross sections measured for the formation of Ar^+ and O^+ contain information about the population of $O_2^+(\tilde{a}\,^4\Pi_u)$ formed by the $Ar^+(^2P_{3/2}) + O_2$ charge transfer reaction in the lower rf octopole ion guide gas cell.

Figure 6 shows the cross sections for the formation of Ar^+ and O^+ by the reaction $O_2^+ + Ar$ at $E_{pr} = 5\,eV$, where O_2^+ ions are produced by the $Ar^+(^2P_{3/2}) + O_2$ change-transfer reaction in the E_{lab} range of 0.1–300 eV. The energy thresholds for the formation of $O_2^+(\tilde{a}\,^4\Pi_u, v' = 0$–$8)$ and $O_2^+(\tilde{A}\,^2\Pi_u, \tilde{b}'\,^4\Pi_g, \tilde{b}\,^4\Sigma_g^-)$ are indicated in the figure. The cross section for Ar^+ formed in the probing reaction rises above the background level in the E_{lab} region of 0.5–1.0 eV. At E_{lab} below the energy threshold ($E_{lab} = 3.1\,eV$) for the formation of $O_2^+(\tilde{A}\,^2\Pi_u)$, only the $O_2^+(\tilde{X}\,^2\Pi_g$ and $\tilde{a}\,^4\Pi_u)$ can be populated. Neglecting the formation of Ar^+ by reaction (2), we attribute the increase of the Ar^+ cross section in the E_{lab} range of 0.9–3 eV to reaction (1). The trend

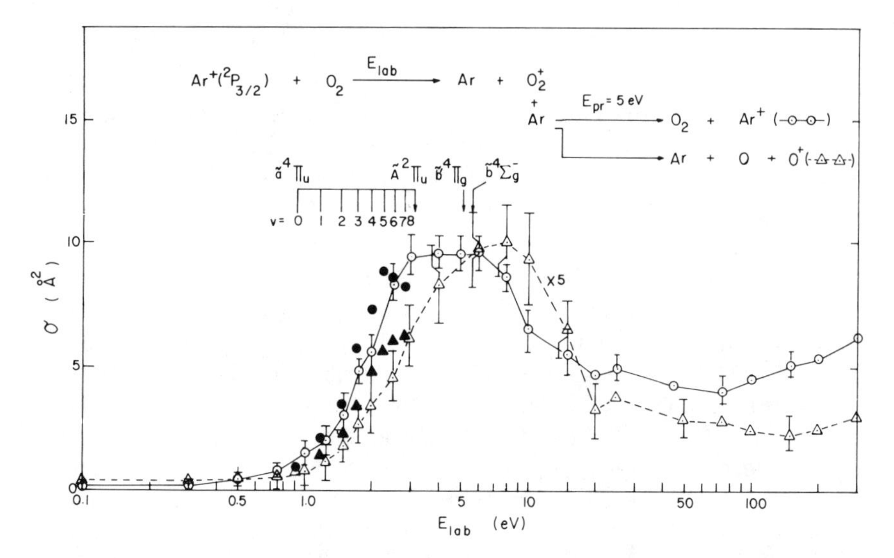

Figure 6. Absolute cross sections for Ar^+(○) and O^+(Δ) formed in the reaction $O_2^+ + Ar$ at $E_{pr} = 5$ eV. The reactant O_2^+ ions are produced in the reaction $Ar^+(^2P_{3/2}) + O_2$ at $E_{lab} = 0.1$–300 eV. (●) Ar^+ and (▲) O^+ cross sections due to O_2^+ formed by photoionization of O_2.[13]

for the increase of the Ar^+ cross section observed in this E_{lab} range is similar to that for the ratio $I(Ar^+)/I(O_2^+)$ plotted in Fig. 5(*b*) in the photon energy region corresponding to the formation of $O_2^+(\tilde{a}\,^4\Pi_u, v' = 0$–$8)$. This qualitative observation suggests that $O_2^+(\tilde{a}\,^4\Pi_u, v')$ ions are formed in the charge-transfer collisions of $Ar^+(^2P_{3/2}) + O_2$ as soon as the reaction becomes energetically allowed.

The absolute cross sections for the formation of Ar^+ and O^+ at $E_{lab} = 5$ eV by the reaction between Ar and O_2^+, produced in the photoionization of O_2 at the wavelength region (728–770 Å) between the $O_2^+(\tilde{a}\,^4\Pi_u, v' = 0)$ and $O_2^+(\tilde{A}\,^2\Pi_u, v' = 0)$ thresholds have also been measured. These absolute cross sections for Ar^+ and O^+ are compared with those due to O_2^+ formed in the charge-transfer collisions of $Ar^+ + O_2$ in Fig. 6. The good agreement observed between these two sets of absolute total cross-section data strongly supports the conclusion that the vibrational distributions for $O_2^+(\tilde{a}\,^4\Pi_u, v')$ formed in the photoionization of O_2 and in the $Ar^+(^2P_{3/2}) + O_2$ charge-transfer reaction are similar. It is known from photoionization experiments[20,21] that $O_2^+(\tilde{a}\,^4\Pi_u)$ ions in vibrational levels $v' = 0$–10 are produced in the photoionization of O_2. Therefore, this experiment provides strong evidence that vibrationally excited $O_2^+(\tilde{a}\,^4\Pi_u, v' = 0$–$10)$ ions are produced by the $Ar^+(^2P_{3/2}) + O_2$ charge-transfer reaction as soon as it is allowed energetically.

4. *Measurements of Absolute Spin–Orbit State-Transition Cross Sections*

The application of the TQDO photoionization apparatus for the measurements of absolute direct spin-orbit state excitation [$\sigma_{3/2\to1/2}$(M)] and relaxation [$\sigma_{1/2\to3/2}$(M)] cross sections in the collisions of $Ar^+(^2P_{3/2,1/2})$ + M has been reported.[5,7,10,22]

a. Spin–Orbit State Excitation Cross Sections. In this experiment, the reactant Ar^+ ion beam in the pure $^2P_{3/2}$ state is prepared by photoionization of Ar at 786 Å. The collisions between $Ar^+(^2P_{3/2})$ and the neutral gas molecules (M) of interest at a specific E_{lab} take place in the lower rf octopole ion guide reaction gas cell. After the collisions, the $Ar^+(^2P_J)$ ions in a mixture of $^2P_{3/2}$ and $^2P_{1/2}$ states are selected by the middle QMF and undergo further charge-transfer collisions with the H_2 at a specific E_{pr} in the upper rf octopole ion guide reaction gas cell. From the cross section measured for the formation of H_2^+ due to $Ar^+(^2P_J)$, along with the known spin–orbit state-selected cross sections $\sigma_{3/2,1/2}(H_2)$ for the $Ar^+(P_{3/2,1/2}) + H_2$ charge-transfer cross sections, the fraction of $Ar^+(^2P_J)$ ions in the $^2P_{1/2}$ state ($X_{1/2}$) is calculated.[23] Under thin target conditions, $\sigma_{3/2\to1/2}$(M) is related to $X_{1/2}$, the number density (n) for M, and the length (l) of the lower rf octopole ion guide reaction gas cell by the relation[7]

$$\sigma_{3/2\to1/2}(M) = X_{1/2}/nl. \tag{4}$$

For the cross-section measurement of a symmetric system such as $Ar^+(^2P_{3/2})$ + Ar, the mass spectrometer cannot distinguish the product $Ar^+(^2P_{1/2})$ ions due to direct excitation and those formed by charge transfer. However, the charge-transfer product $Ar^+(^2P_{1/2})$ ions are mostly ions with near thermal energies, while those associated with direct excitation have kinetic energy similar to the reactant $Ar^+(^2P_{3/2})$ ion beam. It is possible to prevent the slow charge-transfer product $Ar^+(^2P_{1/2})$ ions from getting into the upper reaction gas cell by applying proper dc potentials at the photoionization region, lower rf octopole reaction gas cell, and middle QMF.[7]

b. Spin–Orbit State Relaxation Cross Sections. In the measurements of $\sigma_{1/2\to3/2}$(M), the reactant Ar^+ ions are produced in a 2:1 mixture of $^2P_{3/2}$ and $^2P_{1/2}$ states by photoionization of Ar at 769 Å, which corresponds to a photon energy above the IE for $Ar^+(^2P_{1/2})$. After the collisions in the lower rf octopole ion guide reaction gas cell, the emerging Ar^+ ions consist of $Ar^+(^2P_{3/2})$ and $Ar^+(^2P_{1/2})$ with the intensities equal $I_{3/2}$ and $I_{1/2}$, respectively:

$$I_{3/2} = I_{\text{om}}\{\tfrac{2}{3}\exp(-nl\sigma_{3/2})\exp(-nl\sigma_{3/2\to 1/2}(\text{M})) + \tfrac{1}{3}\exp(-nl\sigma_{1/2})[1-\exp(-nl\sigma_{1/2\to 3/2}(\text{M}))]\}, \tag{5}$$

$$I_{1/2} = I_{\text{om}}\{\tfrac{1}{3}\exp(-nl\sigma_{1/2})\exp(-nl\sigma_{1/2\to 3/2}(\text{M})) + \tfrac{2}{3}\exp(-nl\sigma_{3/2})[1-\exp(-nl\sigma_{3/2\to 1/2}(\text{M}))]\}. \tag{6}$$

Here I_{om} is the measured intensity for the Ar^+ reactant beam when the lower and upper rf octopole ion guide reaction gas cells are empty; and $\sigma_{3/2}$ and $\sigma_{1/2}$ represent the total charge-transfer and reactive cross sections due to the collisions of $\text{Ar}^+({}^2P_{3/2,1/2}) + \text{M}$. Since the charge-transfer and reactive processes deplete the reactant Ar^+ ion beam, the total intensity of Ar^+ emerging from the lower rf octopole ion guide reaction gas cell, $I_{\text{m}} = I_{3/2} + I_{1/2}$, is less than I_{om}:

$$I_{\text{m}} = I_{3/2} + I_{1/2} = I_{\text{om}}[\tfrac{2}{3}\exp(-nl\sigma_{3/2}) + \tfrac{1}{3}\exp(-nl\sigma_{1/2})] \tag{7}$$

or

$$I_{\text{m}}/I_{\text{om}} = \tfrac{2}{3}\exp(-nl\sigma_{3/2}) + \tfrac{1}{3}\exp(-nl\sigma_{1/2}). \tag{8}$$

We note that

$$\exp(-nl\sigma_{3/2}) = I_{3/2}(786\,\text{Å})/I_{\text{o}3/2}(786\,\text{Å}), \tag{9}$$

where $I_{\text{o}3/2}(786\,\text{Å})$ and $I_{3/2}(786\,\text{Å})$ are the intensities for the reactant $\text{Ar}^+(P_{3/2})$ ion beam measured before and after the reaction with H_2, respectively, in the lower rf octopole ion guide reaction gas cell. The reactant Ar^+ ions in the pure ${}^2P_{3/2}$ state are prepared by photoionization of Ar at 786 Å, which corresponds to a photon energy below the IE for $\text{Ar}^+({}^2P_{1/2})$. Combining Eqs. (8) and (9), we obtain the equation

$$\exp(-nl\sigma_{1/2}) = 3(I_{\text{m}}/I_{\text{om}}) - 2[I_{3/2}(786\,\text{Å})I_{\text{o}3/2}(786\,\text{Å})]. \tag{10}$$

By substituting Eqs. (9) and (10) into (6) and dividing Eq. (6) by I_{m}, we can show that

$$\sigma_{1/2\to 3/2}(\text{M}) = [-1/(nl)]\ln \text{K}, \tag{11}$$

where

$$K = \frac{\{X_{1/2} - \tfrac{2}{3}(I_{\text{om}}/I_{\text{m}})[1-\exp(-nl\sigma_{3/2\to 1/2}(\text{M}))]\}}{1-\tfrac{2}{3}(I_{\text{om}}/I_{\text{m}})(I_{3/2}/I_{\text{o}3/2})}. \tag{12}$$

Therefore $\sigma_{1/2\rightarrow3/2}(M)$ can be determined from the measurements of n, l, and K.

B. Crossed Ion–Neutral Beam Photoionization Apparatus

The schematic diagram of the crossed ion–neutral beam photoionization apparatus is shown in Fig. 7.[5,6,23–25] The apparatus consists of a 3-m near-normal-incidence VUV monochromator (McPherson 2253M), a discharge lamp, a VUV light detector, two supersonic-beam production systems, a vertical QMF (10) for monitoring the intensity of the reactant ion, two horizontal QMFs [(11) + (12)] for detection of the charge-transfer product ions, and an rf octopole ion guide reaction gas cell (14). The electron energy analyzer shown in the diagram is not used in the experiments described here.

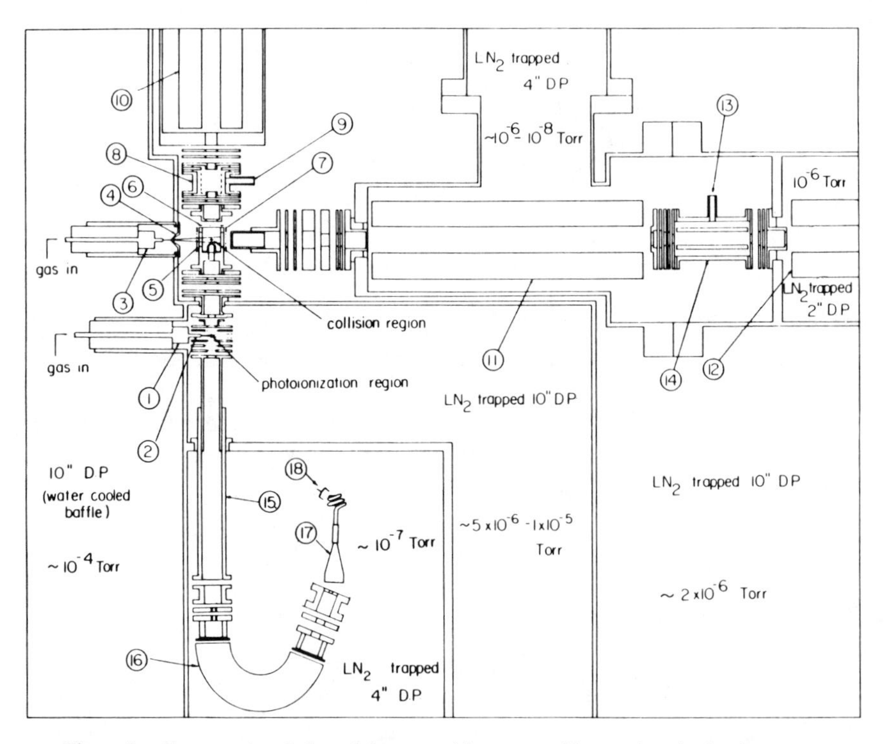

Figure 7. Cross-sectional view of the crossed ion–neutral beam photoionization apparatus: (1) lower supersonic nozzle; (2) quartz nozzle tip; (3) upper supersonic nozzle; (4) skimmer; (5), (6), (7) 90% transmission gold grids; (8) vertical gas cell; (9), (13) to Baratron manometer; (10) vertical QMF; (11) front horizontal QMF; (12) back horizontal QMF; (14) rf octopole ion guide reaction gas cell; (15) steradiancy electron energy analyzer; (16) sector electron energy analyzer; (17) channeltron detetor; (18) collector.[23–25]

Charge-transfer product H_2^+ ions initially formed by the reactions $H_2^+ + H_2$ and $Ar^+ + H_2$ may be converted efficiently into H_3^+ by secondary collisions within the gas cell. The crossed ion-neutral beam photoionization apparatus is most appropriate for the vibrational distribution measurements of charge-transfer product H_2^+.[22,24,26] In the crossed ion–neutral beam arrangement, the destruction of charge-transfer product H_2^+ due to secondary collisions is minimized and the nascent internal-state distribution of product H_2^+ is preserved for detection by the differential reactvity method. Furthermore, the crossed ion-neutral beam arrangement also makes possible the identification of reactant and product ions in symmetric charge-transfer reactions, which are difficult to study using the TQDO apparatus.

The measurement of the vibrational-state distributions of charge-transfer product $H_2^+(\tilde{X}, v')$ formed in the collisions of $Ar^+(^2P_{3/2,1/2}) + H_2$ is described subsequently to illustrate the differential reactivity method. The $Ar^+(^2P_{3/2})$ reactant ions formed in the photoionization region by photoionization of an Ar free jet are extracted and focused onto the neutral H_2 supersonic beam at an intersecting angle of 90°. Charge-transfer product $H_2^+(\tilde{X}, v')$ ions formed at the collision region are collected and mass-selected by the horizontal QMF before entering the rf octopole ion guide reaction gas cell. The N_2^+, Ar^+, and CO^+ formed there according to the charge-transfer processes,

$$H_2^+(\tilde{X}, v') + N_2 \rightarrow H_2 + N_2^+, \tag{13}$$

$$H_2^+(\tilde{X}, v') + Ar \rightarrow H_2 + Ar^+, \tag{14}$$

$$H_2^+(\tilde{X}, v') + CO \rightarrow H_2 + CO^+, \tag{15}$$

are measured at $E_{pr} = 10\,eV$. At sufficiently high collision energies, the charge-transfer product $H_2^+(\tilde{X}, v')$ ions scatter mostly in the direction of the H_2 neutral beam. Therefore, the collection of charge-transfer product ions along the neutral reactant beam direction is efficient. Figure 8 shows the relative vibrational-state-selected cross sections for reactions (13)–(15) at $E_{pr} = 20\,eV$. The fact that the charge-transfer cross sections for reactions (13)–(15) vary differently with v' makes them an ideal set of probing reactions for the determination of the $H_2^+(\tilde{X}, v')$ vibrational distribution. Assuming that $H_2^+(\tilde{X}, v')$ ions formed by the $Ar^+(^2P_{3/2}) + H_2$ charge-transfer reaction consist only of $H_2^+(\tilde{X})$ in the $v' = 0, 1, 2,$ and 3 states, the fractions $(X_{v'})$ of product $H_2^+(\tilde{X})$ ions in $v' = 0, 1, ,2,$ and 3 can be determined by solving the set of linear equations (16)–(19):

$$X_0 + X_1 + X_2 + X_3 = 1, \tag{16}$$

$$X_0 nl\sigma_0(N_2^+) + X_1 nl\sigma_1(N_2^+) + X_2 nl\sigma_2(N_2^+) + X_3 nl\sigma_3(N_2^+) = nl\sigma_m(N_2^+), \tag{17}$$

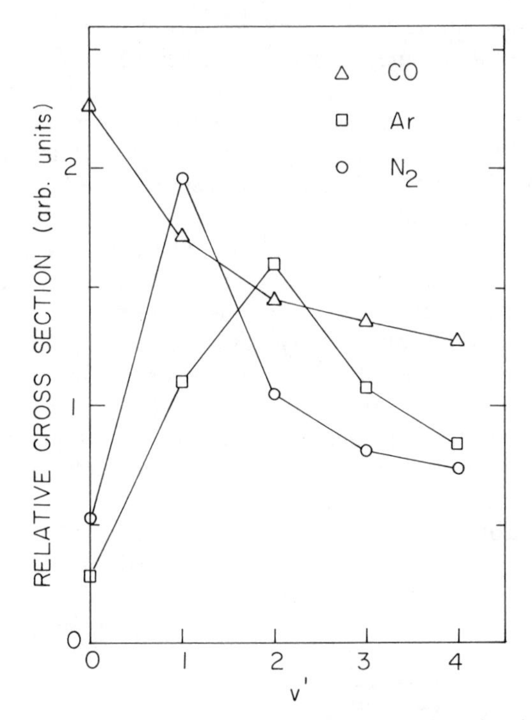

Figure 8. Relative total vibrational-state-selected cross sections for the charge-transfer reactions $H_2^+(\tilde{X}, v' = 0\text{–}4) + Ar(CO, N_2)$; (□) Ar; (Δ) CO; (⊙)$N_2$.[22]

$$X_0 nl\sigma_0(Ar^+) + X_1 nl\sigma_1(Ar^+) + X_2 nl\sigma_2(Ar^+) + X_3 nl\sigma_3(Ar^+) = nl\sigma_m(Ar^+), \tag{18}$$

$$X_0 nl\sigma_0(CO^+) + X_1 nl\sigma_1(CO^+) + X_2 nl\sigma_2(CO^+) + X_3 nl\sigma_3(CO^+) = nl\sigma_m(CO^+), \tag{19}$$

Here $\sigma_{v'}(N_2^+), \sigma_{v'}(Ar^+)$, and $\sigma_{v'}(CO^+), v' = 0\text{–}3$, are the vibrational-state-selected cross sections for reactions (13)–(15); $\sigma_m(N_2^+), \sigma_m(Ar^+)$, and $\sigma_m(CO^+)$ represent the cross sections for reactions (13)–(15) characteristic of $H_2^+(v')$ ions formed by the $Ar^+(^2P_{3/2}) + H_2$ charge-transfer reaction; n is the density of N_2(Ar or CO); and l is the effective length of the rf octopole ion guide reaction gas cell. The calculations of $X_{v'}, v' = 0\text{–}3$, need not involve the determination of absolute values for $nl\sigma_{v'}, (N_2^+), nl\sigma_{v'}(Ar^+)$, and $nl\sigma_{v'}(CO^+)$, $v' = 0\text{–}3$, $nl\sigma_m(N_2^+), nl\sigma_m(Ar^+)$, and $nl\sigma_m(CO^+)$, provided that these values are determined in the same gas cell with the same value n. The vibrational distributions of $H_2^+(v')$ determined in this experiment are shown in Section III.

The crossed ion–neutral beam apparatus may also be used to estimate the relative spin–orbit state selected cross sections for a charge-transfer reaction such as $Ar^+(^2P_{3/2,1/2}) + H_2$. The collection of charge-transfer product $H_2^+(v')$ ions along the neutral reactant H_2 direction may favor the near energy resonance channel, such as the formation of $H_2^+(v'=2)+Ar$ from $Ar^+(^2P_{1/2})+H_2$, which is endothermic by only 16 meV. Experimental evidence shown in Section II indicates that the value for the ratio of the charge-transfer cross sections due to $Ar^+(^2P_{1/2})$ and $Ar^+(^2P_{3/2})$, $\sigma_{1/2}(H_2^+)/\sigma_{3/2}(H_2^+)$, measured by the crossed ion–neutral beam apparatus at low $E_{c.m.}$, may be too high.

III. EXPERIMENTAL RESULTS AND DISCUSSION

A. Atom–Atom System

1. $[Ar + Ar]^+$

a. $Ar^+(^2P_{3/2,1/2})+Ar$. The trend of symmetric charge-transfer cross sections observed for atomic systems has been shown to agree qualitatively with the prediction of the two-state model.[27] The charge-transfer reaction between Ar^+ and Ar is among the most studied atomic systems.[28–43] Because of the small spin–orbit splitting (0.178 eV) for $Ar^+(^2P)$, the cross sections for the direct and charge-transfer excitation and relaxation channels in the collisions of $Ar^+(^2P_{3/2,1/2})+Ar$ are expected to be large at relatively low collisional energies. Yet, the separation of the $Ar^+(^2P_{3/2})$ and $Ar^+(^2P_{1/2})$ states is sufficiently large for state selection of these ions by photoionization with a moderate wavelength resolution. Furthermore, the product spin–orbit states can be identified by time-of-flight techniques. These considerations make the $Ar^+(^2P_{3/2,1/2})+Ar$ reaction an ideal atom–atom system for detailed state-to-state cross section measurements.

I. CHARGE TRANSFER. The theoretical calculation on this system,[44] which took into account the Σ and Π energy states of the Ar_2^+ quasimolecular ion, has provided state-to-state cross sections, $\sigma_{3/2\to3/2}(CT)$, $\sigma_{3/2\to1/2}(CT)$, $\sigma_{1/2\to3/2}(CT)$, and $\sigma_{1/2\to1/2}(CT)$, for the charge transfer processes (20)–(23), respectively:

$$Ar^+(^2P_{3/2})+Ar(^1S_0)\rightarrow Ar(^1S_0)+Ar^+(^2P_{3/2}), \tag{20}$$

$$\rightarrow Ar(^1S_0)+Ar^+(^2P_{1/2}), \tag{21}$$

$$Ar^+(^2P_{1/2})+Ar(^1S_0)\rightarrow Ar(^1S_0)+Ar^+(^2P_{1/2}), \tag{22}$$

$$\rightarrow Ar(^1S_0)+Ar^+(^2P_{3/2}), \tag{23}$$

McAfee et al.[37] have derived the probabilities for spin–orbit transitions at $E_{lab}=123$ eV by means of a transformation analysis of the energy and angular distributions of charge-transfer product Ar^+ ions. The ratios of the spin–orbit state-selected cross sections at selected E_{lab} have been obtained in a PEPICO study.[42] The relative state-to-state cross sections for reactions

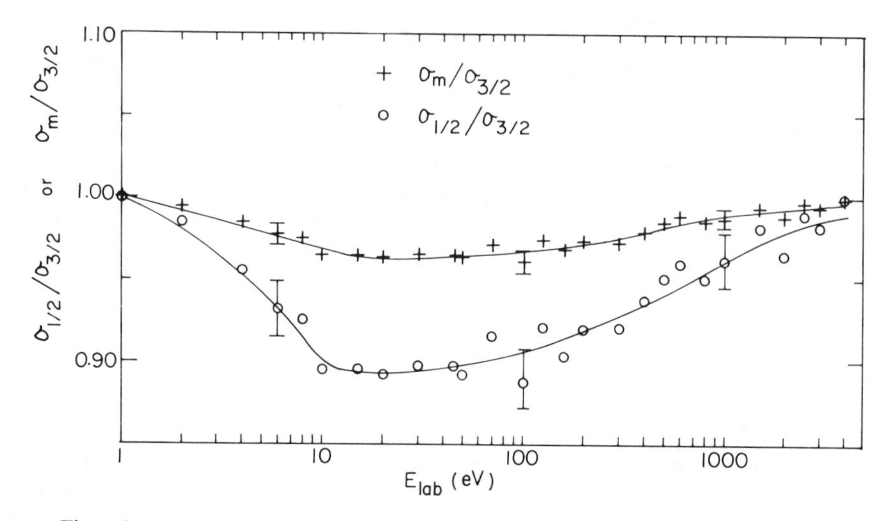

Figure 9. Values for $\sigma_{1/2}(\mathrm{CT})/\sigma_{3/2}(\mathrm{CT})$ (⊙) and $\sigma_{\mathrm{m}}(\mathrm{CT})/\sigma_{3/2}(\mathrm{CT})$ (+) plotted as a function of E_{lab} in the range of 1–4000 eV.[23]

(20)–(23) at $E_{\mathrm{lab}} = 1$–$4000\,\mathrm{eV}$ have been determined using the crossed ion-neutral beam photoionization apparatus.[23] The values for $\sigma_{3/2\to1/2}(\mathrm{CT})/\sigma_{3/2}(\mathrm{CT})$ at $E_{\mathrm{lab}} \leqslant 200\,\mathrm{eV}$ have also been measured in a crossed-beam experiment.[43]

Figure 9 shows the values for $\sigma_{\mathrm{m}}(\mathrm{CT})/\sigma_{3/2}(\mathrm{CT})$ and $\sigma_{1/2}(\mathrm{CT})/\sigma_{3/2}(\mathrm{CT})$ in the E_{lab} range of 1–4000 eV.[23] Here, $\sigma_{\mathrm{m}}(\mathrm{CT})$ is equal to $\frac{2}{3}\sigma_{3/2}(\mathrm{CT}) + \frac{1}{3}\sigma_{1/2}(\mathrm{CT})$. The experimental values for $\sigma_{1/2}(\mathrm{CT})/\sigma_{3/2}(\mathrm{CT})$ approach unity smoothly toward both low and high collision energies from a broad minimum of 0.89 at 10–50 eV. The values for $\sigma_{1/2}(\mathrm{CT})/\sigma_{3/2}(\mathrm{CT})$ being less than unity agree with the recent PEPICO measurements.[39] The impact parameter calculation[44] gives a value of 1.00 ± 0.05 for $\sigma_{1/2}(\mathrm{CT})/\sigma_{3/2}(\mathrm{CT})$ at $E_{\mathrm{lab}} \geqslant 5\,\mathrm{eV}$, which is higher than the experimental values. Although the calculation indicates that $\sigma_{1/2}(\mathrm{CT})/\sigma_{3/2}(\mathrm{CT})$ is essentially independent of E_{lab}, the profiles of the kinetic energy dependences for the ratios of the total charge-transfer cross sections for the $^2P_{3/2}$ and $^2P_{1/2}$ states obtained for the $\mathrm{Kr}^+ + \mathrm{Kr}$ and the $\mathrm{Xe}^+ + \mathrm{Xe}$ systems in similar calculations[45] resemble that observed for the $\mathrm{Ar}^+ + \mathrm{Ar}$ system shown in Fig. 9. The calculated total cross sections for Kr^+ and Xe^+ in the $^2P_{3/2}$ and $^2P_{1/2}$ states change their slopes going from the low energy region, where (J_a, J_b) coupling dominates, to the higher energy region, where (Λ, S) coupling dominates. The changes in slope occur roughly over the energy region where the velocity goes from a value equal to the spin–orbit splitting in atomic unit to half that value. This energy region also corresponds to the location of the broad minimum for the ratio of the total cross sections for the $^2P_{1/2}$ and $^2P_{3/2}$ states. Since the spin–orbit splitting for Ar^+ is substantially smaller than those for Kr^+ and Xe^+, the position of the broad minimum for $\sigma_{1/2}(CT)/\sigma_{3/2}(CT)$ is expected to be at a lower energy region for Ar. Based on the known spin–orbit splitting for $Ar^+(^2P)$, the E_{lab} region predicted for the location of the minimum for $\sigma_{1/2}(CT)/\sigma_{3/2}(CT)$ is ≈ 40 eV, in agreement with the experimental observation.

As E_{lab} is increased, the nonadiabatic transitions between a quasimolecular state derived from $Ar^+(^2P_{3/2})$ to that derived from $Ar^+(^2P_{1/2})$ become more important. These nonadiabatic transitions lead to the increase in $\sigma_{1/2}(CT)/\sigma_{3/2}(CT)$ as a function of collision energy at $E_{lab} > 50$ eV. At sufficiently high E_{lab}, the charge-transfer cross section is expected to be independent of the initial J state of the ion. Johnson pointed out that at high collision energies the value for $\sigma_{1/2}(CT)/\sigma_{3/2}(CT)$ depends strictly on the IEs for the $^2P_{3/2}$ and $^2P_{1/2}$ states and would be predicted correctly by the two-state model. The simple two-state model of Rapp and Francis[27] gives a value of 0.98 for $\sigma_{1/2}(CT)/\sigma_{3/2}(CT)$, which is in agreement with experimental values measured at $E_{lab} \geqslant 1500$ eV. The conditions of the collision become more adiabatic as E_{lab} is decreased from ≈ 20 eV. At very low E_{lab}, the cross section should be determined solely by the polarizability of the neutral Ar atom[42] and E_{lab}, and thus should be independent of the ionic state. The observation that the value $\sigma_{1/2}(CT)/\sigma_{3/2}(CT)$ approaches unity at low E_{lab} is in accord with this theoretical expectation.

The values for $\sigma_{3/2\to1/2}(CT)/\sigma_{3/2}(CT)$, $\sigma_{3/2\to3/2}(CT)/\sigma_{3/2}(CT)$, $\sigma_{1/2\to3/2}(CT)/\sigma_{3/2}(CT)$, and $\sigma_{1/2\to1/2}(CT)/\sigma_{3/2}(CT)$ in the E_{lab} range of 1–4000 eV have been reported by Liao et al.[23] The experimental and theoretical values[44] for $\sigma_{3/2\to1/2}(CT)/\sigma_{3/2}(CT)$ are compared in Fig. 10. The experimental values for $\sigma_{3/2\to1/2}(CT)/\sigma_{3/2}(CT)$ are approximately 40% lower than those predicted by the calculation with the inclusion of rotational coupling (RC), and slightly higher than the theoretical results obtained by neglecting rotational couplings (NR). Nevertheless, the theoretical prediction of the profile for the probability of fine-structure transitions as a function of E_{lab} is in good accord with the experimental results. Both the experimental and RC theoretical results show that $\sigma_{3/2\to1/2}(CT)/\sigma_{3/2}(CT)$ reaches a maximum at $E_{lab} \approx 2000$–3000 eV. Johnson points out that the greatest uncertainty in the calculation of the cross sections is due to the potentials. An overall uncertainty of 20% in the potential for the important internuclear distances can give rise to a shift in the calculated excitation cross sections in the $\log_{10} E_{lab}$ scale by ± 0.2. In addition to uncertainties due to the possibility of being shifted in the E_{lab} scale, the magnitudes the calculated transition cross sections in the vicinity of the maximum are estimated to have an uncertainty of about $\pm 20\%$. The change in momentum of the captured electron and polarization effects have also been neglected in the theoretical calculation. As mentioned previously, the

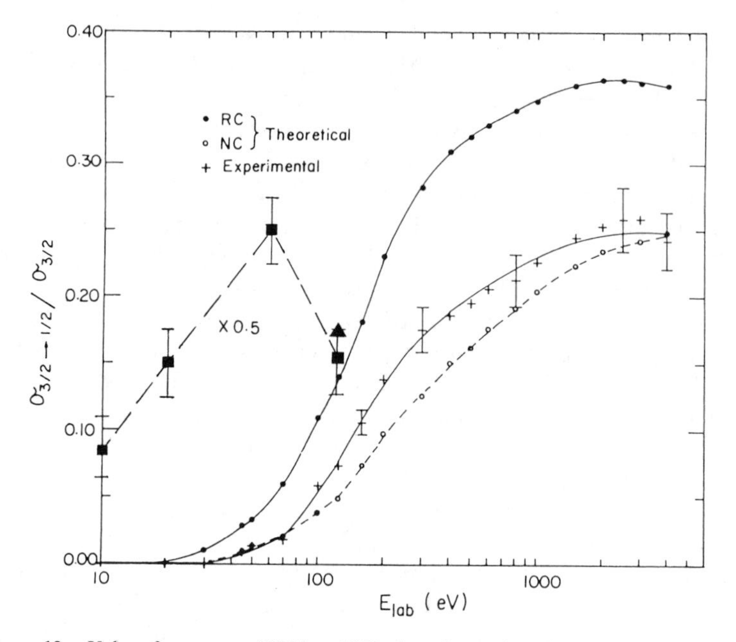

Figure 10. Values for $\sigma_{3/2\rightarrow1/2}(\mathrm{CT})/\sigma_{3/2}(\mathrm{CT})$ plotted as a function of E_{lab} in the range of 1–4000 eV. Experimental values: (+) Ref. 23, (Δ) Ref. 37, (■) Ref. 43. Theoretical values: (●) with rotational coupling and (⊙) without rotational coupling, Ref. 44.

collection of the charge-transfer Ar^+ product ions using the crossed ion-neutral beam photoionization apparatus is likely to discriminate against inelastic charge-transfer product Ar^+ ions scattered at wide angles with respect to the Ar neutral-reactant-beam direction. The discriminative effects are expected to be more important at low E_{lab}. Thus, the experimental results for $\sigma_{3/2\rightarrow1/2}(\mathrm{CT})/\sigma_{3/2}(\mathrm{CT})$ are considered as lower limits.

The values for the ratio $\sigma_{3/2\rightarrow1/2}(\mathrm{CT})/\sigma_{3/2}(\mathrm{CT})$ in the E_{lab} range 4–200 eV measured recently in a crossed-beam experiment by Howard et al.[43] reveal that the ratio first increases with increasing E_{lab} to a maximum of about 0.5 at 60 eV and then decreases at higher E_{lab}. Their measurements are included in Fig. 10. The probabilities of the fine-structure excitation channel at low energies observed in the crossed-beam study are substantially greater than the experimental results observed by Liao, Liao, and Ng and the theoretical predictions obtained by Johnson. Howard et al. suggest that a curve-crossing mechanism not previously considered and the perturbation of trajectories at low energies resulting from the strongly bound Ar_2^+ intermediate may be responsible for the unexpected observation. In similar studies[46] of the $Kr^+(^2P_{3/2}) + Kr$ and the $Xe^+(^2P_{3/2}) + Xe$ charge-transfer reactions, Futrell and co-workers have found the thresholds and the maxima in cross sections for the excitation channels at collision energies well below those predicted by theoretical calculations. Since accurate *ab initio* potential energy curves for Ar_2^+ are now available, a calculation using the more accurate potentials should shed light on the discrepancy in the experimental observations.

II. DIRECT SPIN–ORBIT STATE EXCITATION AND RELAXATION. The theoretical calculation of Johnson[44] also provides predictions for the cross sections of the direct excitation reaction (24) $[\sigma_{3/2\to 1/2}(\mathrm{Ar})]$ and relaxation reaction (25) $[\sigma_{1/2\to 3/2}(\mathrm{Ar})]$:

$$\mathrm{Ar}^+(^2P_{3/2}) + \mathrm{Ar}(^1S_0) \to \mathrm{Ar}^+(^2P_{1/2}) + \mathrm{Ar}(^1S_0), \tag{24}$$

$$\mathrm{Ar}^+(^2P_{1/2}) + \mathrm{Ar}(^1S_0) \to \mathrm{Ar}^+(^2P_{3/2}) + \mathrm{Ar}(^1S_0), \tag{25}$$

The values for $\sigma_{3/2\to 1/2}(\mathrm{Ar})$ and $\sigma_{1/2\to 3/2}$ (Ar) have been measured by Itoh, Kobayashi, and Kaneko[47] using the energy–loss spectroscopic method. The values for $\sigma_{3/2\to 1/2}(\mathrm{Ar})$, which are measured by observing the energy loss of Ar^+ along the reactant Ar^+ beam direction with an detector acceptance angle of 0.45°, are lower than the RC theoretical predictions by a factor of 2. Since in their experiment the angular distribution of the inelastically scattered ions is slightly broader than the acceptance angle of the detector, their values for $\sigma_{3/2\to 1/2}(\mathrm{Ar})$ should be considered as lower limits. Absolute cross sections for process (24) in the E_{lab} range of 5–400 eV have also been determined using the TQDO apparatus, and the results of such measurements[7,18] are compared to those of the energy loss experiment in Fig. 11(*d*). The two sets of experimental results are in general agreement, indicating that $\sigma_{3/2\to 1/2}(\mathrm{Ar})$ increases nearly linearly as E_{lab} is increased. The slightly higher values for the excitation cross sections obtained using the TQDO photoionization apparatus are attributed to higher collection efficiencies for inelastically scattered $\mathrm{Ar}^+(^2P_{1/2})$ ions. The RC theoretical predictions for $\sigma_{3/2\to 1/2}(\mathrm{Ar})$ and $\sigma_{3/2\to 1/2}(\mathrm{CT})$ at $E_{\mathrm{lab}} < 600$ eV are essentially identical. The observation that experimental values for both $\sigma_{3/2\to 1/2}(\mathrm{Ar})$ and $\sigma_{3/2\to 1/2}(\mathrm{CT})/\sigma_{3/2}(\mathrm{CT})$ are lower than the RC theoretical predictions by a similar amount shows that the experimental cross sections for the charge transfer and direct excitations are self-consistent.

B. Atom–Diatom Systems

1. $[Ar + N_2]^+$

Charge transfer is the major channel in the collisions of $\mathrm{Ar}^+(^2P_{3/2,1/2}) + N_2$ and $N_2^+(\tilde{X}, \tilde{A}, v') + \mathrm{Ar}$. Owing to the mutual stimulation between experimental and theoretical studies, the $[\mathrm{Ar} + N_2]^+$ charge-transfer system

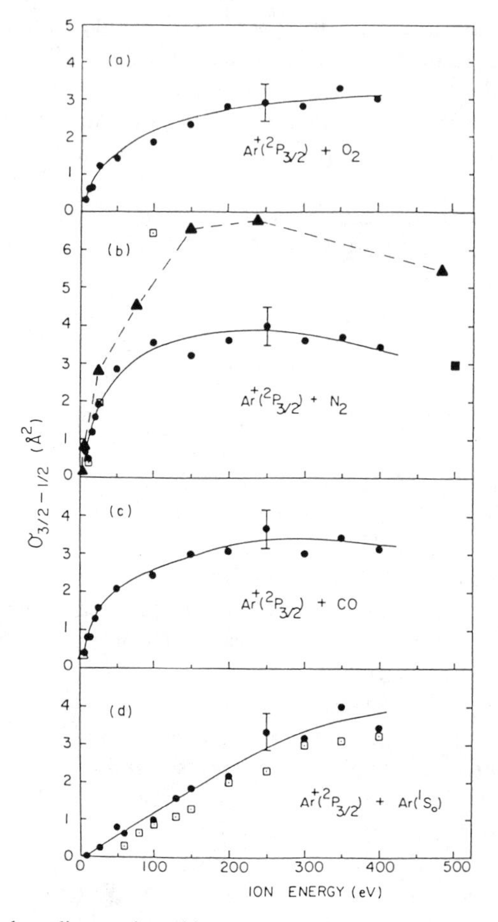

Figure 11. Absolute direct spin–orbit state excitation cross sections $[\sigma_{3/2\to 1/2}(M)]$ for the process, $Ar^+(^2P_{3/2}) + M \to Ar^+(^2P_{1/2}) + M$. (*a*) $M = O_2$, experimental: (●) Ref. 13. (*b*) $M = N_2$, experimental: (●) Refs. 7 and 72, (■) Ref. 71; theoretical: (Δ) Ref. 63, (□) Ref. 59; (*c*) $M = CO$, experimental: (●) Ref. 72; theoretical: (Δ) Ref. 89; (*d*) $M = Ar$, experimental: (●) Refs. 7 and 71, (□) Ref. 47.

has become the best-studies ion–molecule reaction, both experimentally[6–8,25,48–58] and theorecically.[59–66]

The experimental state-selected and state-to-state cross sections for the reactions[6]

$$Ar^+(^2P_{3/2,1/2}) + N_2(X, v''=0) \to Ar(^1S_0) + N_2{}^+(\tilde{X}, \tilde{A}, v'), \qquad (26)$$

and the vibration-state selected cross sections for the reaction,[8,46,47]

$$N_2^+(\tilde{X}, \tilde{A}, v') + Ar(^1S_0) \rightarrow N_2(v'') + Ar^+(^2P_{3/2,1/2}) \quad (27)$$

are in general accord with theoretical predictions obtained using the vibronic semiclassical method[59-62] and *ab initio* surfaces[65] of the $[Ar + N_2]^+$ system. Experimental values for the total state-to-state cross sections of reactions (26) and (27) measured at selected $E_{c.m.}$ are consistent with microscopic reversibility.[6,8,25] Since these charge-transfer reactions have been the subject of several recent reviews,[5,51,67-69] only an update on the most recent experimental and theoretical investigations on the $[Ar + N_2]^+$ system is given.

a. $Ar^+(^2P_{3/2,1/2}) + N_2$

I. CHARGE TRANSFER. The most controversial experimental findings on this system concern the ratio of the spin–orbit state-selected cross sections for reaction (26) $[\sigma_{3/2,1/2}(N_2^+)]$. Values for $\sigma_{1/2}(N_2^+)/\sigma_{3/2}(N_2^+)$ have been reported by several laboratories[6,48-51,70] and the data are widely scattered. The difficulty stems partly from the efficient relaxation of $Ar^+(^2P_{1/2})$ reactant ions by collision with background molecules along the ion flight path before entering the reaction region. To avoid such an effect, it is important to maintain low background pressure between the photoionization and reaction regions as well as thin target conditions in the reaction region.[6] The data set for $\sigma_{1/2}(N_2^+)/\sigma_{3/2}(N_2^+)$ obtained using the crossed ion–neutral beam photoionization apparatus,[48] which is lower than that measured using the TQDO photoionization apparatus,[6] is more likely to suffer from the discriminative effect of N_2^+ collection efficiencies mentioned in Section II. B. Taking into account the experimental uncertainties, the data set obtained using the TQDO photoionization apparatus agrees with the more recent experimental results of Lindsay and Latimer[57] and Guyon.[70] The agreement between these experimental results and the theoretical predictions can only be considered as fair.

II. SPIN–ORBIT STATE EXCITATION AND RELAXATION. Experimental measurements[7,71,72] and theoretical calculations[63] indicate that fine-structure transitions [reactions (28) and (29)] and direct vibrational excitation of N_2[reactions (30) and (31)] have unexpectedly high cross sections allowed by impulsive collisions:

$$Ar^+(^2P_{3/2}) + N_2(v'' = 0) \rightarrow Ar^+(^2P_{1/2}) + N_2(v') \quad (28)$$

$$Ar^+(^2P_{1/2}) + N_2(v'' = 0) \rightarrow Ar^+(^2P_{3/2}) + N_2(v') \quad (29)$$

$$Ar^+(^2P_{3/2}) + N_2(v'' = 0) \rightarrow Ar^+(^2P_{3/2}) + N_2(v') \qquad (30)$$

$$Ar^+(^2P_{1/2}) + N_2(v'' = 0) \rightarrow Ar^+(^2P_{1/2}) + N_2(v') \qquad (31)$$

In the vibronic semiclassical calulations, there is no direct coupling in the Hamiltonian matrix between the $Ar^+(^2P_{3/2}) + N_2$ and $Ar^+(^2P_{1/2}) + N_2$ states. Therefore, fine-structure transitions in this system occur only because of the presence of the $N_2^+(\tilde{X}, \tilde{A}, v') + Ar$ charge-transfer states. Since the *ab initio* diabatic potential surfaces used have no dependence on the vibrational coordinate of N_2, impulsive momentum transfer between Ar^+ and N_2 cannot occur. Direct vibrational excitation processes are also due to the existence of the intermediate charge-transfer states $N_2^+(\tilde{X}, \tilde{A}, v') + Ar$. That is, reactions (28)–(31) may be considered the results of two stepwise charge-transfer processes.

The state-to-state cross sections for processes (28)–(31) in the $E_{c.m.}$ range of 200–800 eV have been measured using the ion-energy-loss spectroscopy method.[71] Assuming that the theoretical and experimental angular distributions for inelastically scattered Ar^+ ions are identical, Parlant and Gislason[63] have scaled down the theoretical cross sections based on the acceptance angle of the detector used in the energy-loss experiment.[71] The corrected theoretical cross sections are in good agreement with the experimental results.

Abolute state-selected cross sections for reactions (28) and (29) in the E_{lab} range of 5–400 eV have been measured by the TQDO photoionization apparatus.[7,72] The cross section for reaction (28) at 164 eV is consistent with those determined in the ion-energy-loss study. The experimental and theoretical spin–orbit state excitation cross sections are shown in Fig. 11(*b*). It is clear that the theoretical cross sections[63] are significantly higher than the cross sections obtained using ion-energy-loss spectroscopy[71] and the TQDO photoionization apparatus.[72]

III. DISSOCIATION AND REACTION. The previous theoretical calculations[59–66] on the $Ar^+(^2P_{3/2,1/2}) + N_2$ charge-transfer and inelastic collisions do not take into account the dissociative and reactive channels,

$$Ar^+(^2P_{3/2,1/2}) + N_2 \rightarrow N^+ + N + Ar, \qquad (32)$$

$$\rightarrow ArN^+ + N. \qquad (33)$$

Owing to the strong bond dissociation energy of N_2^+, reactions (32) and (33) are highly endothermic (≈ 8.5 eV),[16] and the neglect of the formation of N^+ and ArN^+ in the theoretical calculations seems to be justified. Contrary to this expectation, both channels have been observed.[10] In a related experiment, C^+, O^+, and ArC^+ are also identified as product channels in the collisions of $Ar^+(^2P_{3/2,1/2}) + CO$.[9,72]

Absolute total cross sections for the formation of N_2^+ and N^+ [$\sigma_{3/2}(N^+)$] from the reaction of $Ar^+(^2P_{3/2}) + N_2$ in the E_{lab} range of 3–300 eV are compared in Fig. 12(*a*). The profile of the collision energy dependence of $\sigma_{3/2}(N_2^+)$ is explained nicely by the vibronic semiclassical calculation.[62] The broad peak centered at $E_{lab} = 4$–7 eV is due to the formation of $N_2^+(\tilde{X}, v' = 1)$. The rise in $\sigma_{3/2}(N_2^+)$ at E_{lab} above 70 eV is the result of the increasing population of the $N_2^+(\tilde{X}, v' = 0)$ and $N_2^+(\tilde{A}, v' = 0$–$2)$ states. The

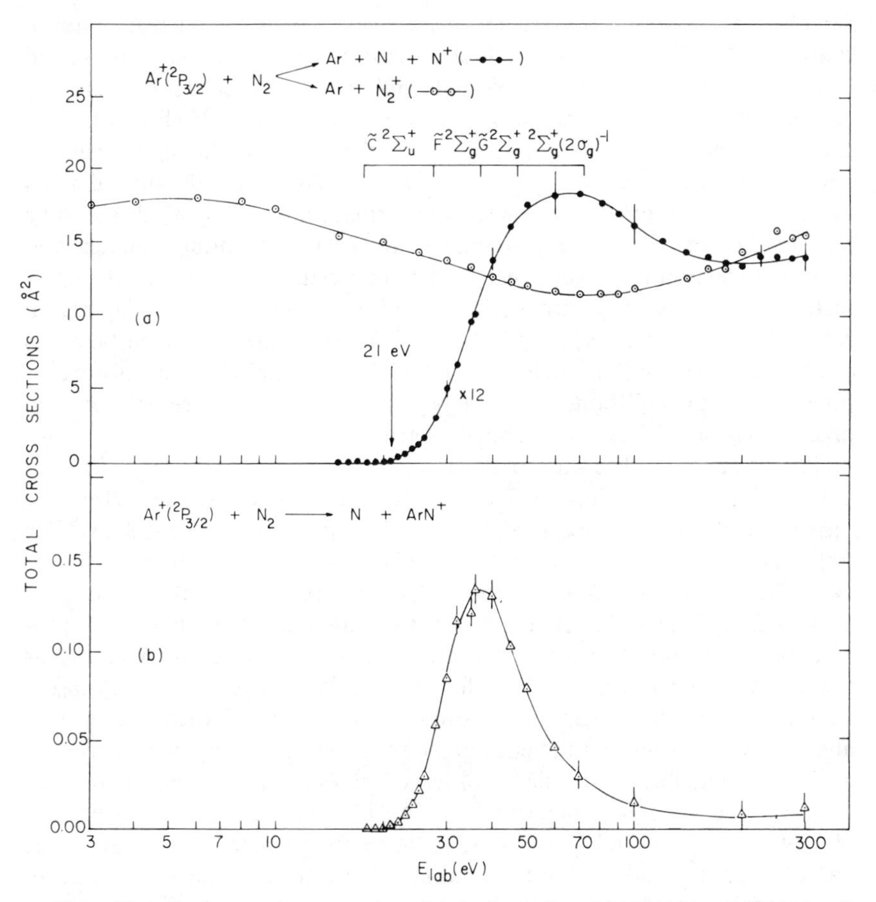

Figure 12. Absolute total cross sections for the formation of (*a*) N^+(●) and N_2^+(⊙) and (*b*) ArN^+(△) in the collisions of $Ar^+(^2P_{1/2})$ with N_2 at $E_{lab} = 3$–300 eV.[10]

crossover point for the formation of $N_2^+(\tilde{X}, v'=1)$ and $N_2^+(\tilde{X}, v'=0)$ marks the minimum at $E_{lab} \approx 70$ eV in the N_2^+ cross-section curve.

The cross section curve for N^+ exhibits a maximum at $E_{lab} = 50$–80 eV. At the maximum, $\sigma_{3/2}(N^+)$ is slightly more than 10% of that for $\sigma_{3/2}(N_2^+)$. The appearance energy (AE) for N^+ [$E_{lab} = 21.0 \pm 0.5$ eV ($E_{c.m.} = 8.65 \pm 0.21$ eV)] is in excellent agreement with the thermochemical threshold for reaction (32). The cross sections for the dissociative photoionization process

$$N_2 + h\nu \rightarrow N^+ + N + e^- \tag{34}$$

have been measured in the photon energy range of 24–70 eV.[73–77] In this energy range four broad electronic bands of N_2^+, $\tilde{C}\,^2\Sigma_u^+$, $\tilde{F}\,^2\Sigma_g^+$, $\tilde{G}\,^2\Sigma_g^+$, and $^2\Sigma_g^+(2\sigma_g)^{-1}$, are identified in the photoelectron spectrum obtained at a photon energy of 50.1 eV.[78] These are multielectron excitation states that have mixed configuration of the $2\sigma_g$ single-hole state and the outer valence states, $2\sigma_u$, $1\pi_u$, and $3\sigma_g$. Previous experimental and theoretical studies[76–83] show that N^+ can be formed by predissociation of all the four electronic bands. The positions of the $\tilde{C}\,^2\Sigma_u^+$, $\tilde{F}\,^2\Sigma_g^+$, $\tilde{G}\,^2\Sigma_g^+$, and $^2\Sigma_g^+(2\sigma_g)^{-1}$ bands are indicated in Fig. 12(*a*). The cross sections for the formation of N^+ from dissociative photoionization of N_2 are approximately an order of magnitude smaller than those for reaction (32). Nevertheless, it is interesting to note that the cross sections for N^+ from processes (32) and (34) both peak in the energy region corresponding to the $^2\Sigma_g^+(2\sigma_g)^{-1}$ band. This observation may be taken as evidence supporting the conclusion that the four multielectron states are also populated in the collisions of $Ar^+ + N_2$, and that N^+ ions are produced by predissociation of these N_2^+ excited electronic states.

The measured absolute total cross sections for ArN^+ [$\sigma_{3/2}(ArN^+)$] produced by the reaction $Ar^+(^2P_{3/2}) + N_2$ are depicted in Fig. 12(*b*). The formation of ArN^+ is observed only in the E_{lab} range of 20–100 eV. The ArN^+ cross section at the peak ($E_{lab} \approx 35$ eV) is approximately 1% that for N_2^+. The AE for ArN^+ is determined to be $E_{lab} = 20.0 \pm 0.5$ eV ($E_{c.m.} = 8.24 \pm 0.21$ eV). The difference between this value and the AE for N^+ provides a lower bound of 0.5 eV for the dissociation energy of ArN^+. This lower bound substantially below the values of > 2 eV for the ArN^+ dissociation energy obtained in the theoretical calculation of Frenking et al.[84] and in the recent photofragment kinetic spectroscopic study of Brostrom et al.[85] The negligible ArN^+ signal observed at $E_{lab} \geqslant 100$ eV may be due to high internal energies deposited in ArN^+ causing the fragmentation of ArN^+ into $N^+ + Ar$. Figure 13 shows the retarding-potential-energy curves for N^+, N_2^+, Ar^+, and ArN^+ at a nominal laboratory reactant Ar^+ ion-beam energy of 36.5 eV. To facilitate the discussion concerning the interpretation of the retarding-potential-energy curves, the Newton diagram for the collision

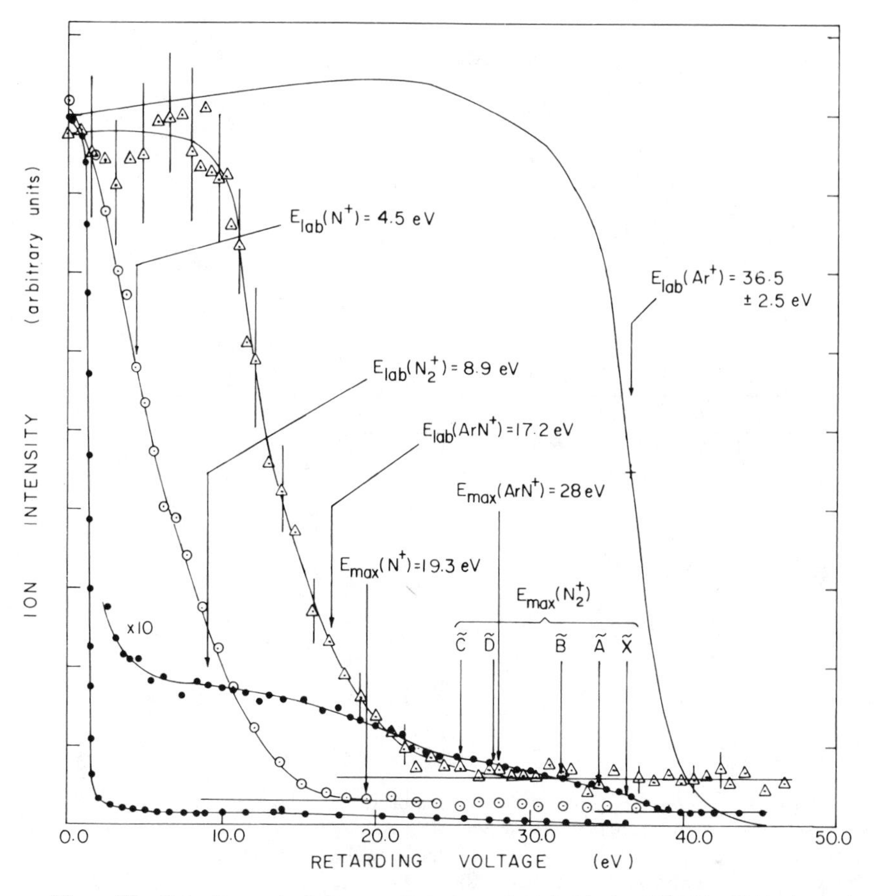

Figure 13. Retarding-potential energy curves for Ar^+(—), N_2^+(●), N^+(⊙), and ArN^+(△) at a nominal $E_{lab} = 35$ eV. $E_{max}(N^+)$ and $E_{max}(ArN^+)$ are the maximum laboratory energies for N^+ and ArN^+ allowed by energy conservation.[10]

of Ar^+ with N_2 at $E_{lab} = 36.5$ eV is depicted in Fig. 14. The z axis coincides with the central axis of the octopole ion guide. The scattering of product ions is expected to have a cylindrical symmetry around the z axis. The retarding-potential-energy analysis provides information about the velocity components for the product ions projected along the z axis. Here, $\mathbf{v}_{c.m.}$, $\mathbf{v}_{c.m.}(N_2)$, and $\mathbf{v}_{c.m.}(Ar^+)$ are the center-of-mass velocities of the $Ar^+ + N_2$ system, reactant N_2, and reactant Ar^+, respectively. The possible recoil velocities of N_2^+ corresponding to the formation of $N_2^+(\tilde{X}, \tilde{A}, \tilde{B}, \tilde{C}, \tilde{D})$ are shown by cirlces in the figure.[16] The circle marked $\mathbf{v}_{D.T.}(N_2^+)$ defines the c.m. velocity for charge-transfer product N_2^+ at the threshold for the

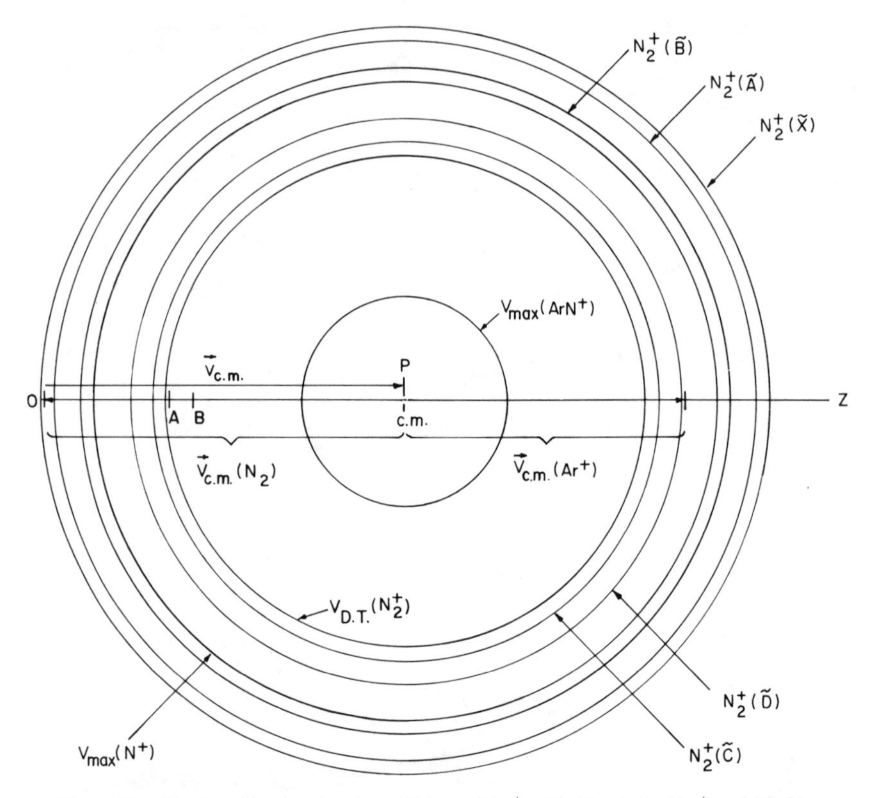

Figure 14. Newton diagram for the collision of Ar^+ with N_2 at $E_{lab}(Ar^+) = 36.5$ eV. $\mathbf{v}_{c.m.}$, $\mathbf{v}_{c.m.}(N_2)$, and $\mathbf{v}_{c.m.}(Ar^+)$ are the c.m. velocities of the $Ar^+ + N_2$ system, reactant N_2, and reactant Ar^+, respectively. The z axis coincides with the central axis of the octopole ion guide. The maximum c.m. velocities for N_2^+ corresponding to the formation of $N_2^+(\tilde{X}, \tilde{A}, \tilde{B}, \tilde{C}, \tilde{D})$ are shown by circles. The circles marked $\mathbf{v}_{max}(N^+)$ and $\mathbf{v}_{max}(ArN^+)$ define the maximum c.m. velocities for N^+ and ArN^+ allowed by energy conservation. At the threshold for the dissociation reaction (32), the c.m. velocity for N_2^+ is $\mathbf{v}_{D.T.}$.[10]

dissociation reaction (32). For c.m. velocities of N_2^+ below $v_{D.T.}(N_2^+)$, the internal energies of charge-transfer product N_2^+ are expected to be greater than the dissociation energy of N_2^+. At $E_{lab} = 36.5$ eV, the maximum energy available for distribution to the translational energies of the fragments $ArN + N^+$ is 6.6 eV, which corresponds to the maximum c.m. recoil velocity of 8.5×10^5 cm/s and 2.2×10^5 cm/s for N^+ [$\mathbf{v}_{max}(N^+)$] and ArN^+ [$\mathbf{v}_{max}$ (ArN^+)], respectively, as shown in Fig. 14.

The maximum laboratory velocities are equal to $\mathbf{v}_{c.m.} + \mathbf{v}_{max}(N^+)$ for N^+ and $\mathbf{v}_{c.m.} + \mathbf{v}_{max}(ArN^+)$ for ArN^+. The maximum laboratory energies (E_{max})

for N^+ and ArN^+ corresponding to these maximum laboratory velocities are 19.3 and 28 eV, respectively. These are the maximum E_{lab} values for N^+ and ArN^+ allowed by energy conservation. As shown in Fig. 13, the retarding-potential-energy curves for N^+ and ArN^+ are in accord with the constraint of energy conservation, that is, the N^+ and ArN^+ intensities drop to the noise level when the retarding voltages are greater than $E_{max}(N^+)$ and $E_{max}(ArN^+)$. If N^+, N_2^+ and ArN^+ have the same laboratory velocities as $\mathbf{v}_{c.m.}$ of the system, the values for $E_{lab}(N^+)$ $E_{lab}(N_2^+)$, and $E_{lab}(ArN^+)$ are predicted to be 4.5, 8.9, and 17.2 eV, respectively (see Fig. 13). Depending on whether the laboratory translational energies for the product ions are higher or lower than these E_{lab} values, the product ions are scattered in the forward or backward hemispheres with respect to $\mathbf{v}_{c.m.}(Ar^+)$.

At $E_{lab}(Ar^+) = 36.5$ eV, $\sigma_{3/2}(ArN^+)$ is near the maximum. The retarding-potential-energy curve shows that the intensity for ArN^+ drops nearly 80% as the retarding voltage increases to 17.2 eV. This observation indicates that the majority of ArN^+ ions are backward scattered with respect to $\mathbf{v}_{c.m.}(Ar^+)$. At collision energies close to the threshold of reaction (33), the most effective collision geometry to convert kinetic energy into internal energies of N_2, and to induce the formation of $ArN^+ + N$, is the collinear Ar^+–N–N configuration. The breaking of the ArN^+–N bond in a nearly collinear collision configuration results in the scattering of ArN^+ in the backward hemisphere with respect to $\mathbf{v}_{c.m.}(Ar^+)$, a prediction consistent with the experimental observation.

The intensity for N_2^+ decreases sharply at the retarding voltage of 1.3 ± 0.2 eV, indicating that the overwhelming majority of N_2^+ are slow ions with $E_{lab} \leqslant 1.3$ eV. The magnitudes of the laboratory velocities corresponding to $E_{lab} = 1.1$ and 1.5 eV are measured as OA and OB in Fig. 14. The result of the retarding-potential-energy analysis for N_2^+ is in accord with the interpretation that $\geqslant 95\%$ of charge-transfer product N_2^+ ions are formed in the $\tilde{X}, \tilde{A}, \tilde{B}, \tilde{C}$, and/or $\tilde{D}$ states and are scattered in the backward hemisphere with respect to $\mathbf{v}_{c.m.}(Ar^+)$. Within the time scale of this experiment, charge-transfer product N_2^+ formed in the excited $N_2^+(\tilde{A}, \tilde{B}, \tilde{C}, \tilde{D})$ states may be stabilized by radiative decay.[79,80] The positions of $E_{max}(N_2^+)$ for the formation of $N_2^+(\tilde{X}, \tilde{A}, \tilde{B}, \tilde{C}, \tilde{D})$ are marked in Fig. 13. The retarding-potential-energy curve for N_2^+ shows that about 3% of the total N_2^+ ions are scattered in the forward hemisphere with respect to $\mathbf{v}_{c.m.}(Ar^+)$ with E_{lab} in the range of 10–36 eV.

The retarding-potential-energy data for N_2^+ show that the intensity for charge-transfer N_2^+ scattered in the forward hemisphere with respect to $\mathbf{v}_{c.m.}(Ar^+)$ decreases from about 8% to less than 1% of the total N_2^+ intensity as $E_{lab}(Ar^+)$ is increased from 26.5 to 205 eV. The fact that the overwhelming proportion of N_2^+ are slow ions with $E_{lab} \leqslant 1.5$ eV supports the long-range electron jump mechanism for charge transfer, by which little momentum is transferred from Ar^+ to N_2^+.

At $E_{lab}(Ar^+) = 26.5$ and $36.5\,eV$, it appears that the intensity for N^+ scattered in the forward hemisphere is substantially greater than that scattered in the backward hemisphere. In order to rationalize the scattering patterns of ArN^+ and N^+ observed at $E_{lab}(Ar^+)$ near the threshold of reaction (33), Flesch and Ng[10] propose a nearly collinear charge-transfer predissociation mechanism for the formation of ArN^+ and N^+ near their thresholds.

$$Ar^+ + N\text{–}N \xrightarrow{(a)} Ar\cdots(N - N)^{*+} \xrightarrow{(b)} Ar\cdots N^+ \cdots N + Ar\cdots N\cdots N^+, \quad (35)$$

$$Ar\cdots N\cdots N^+ \rightarrow N^+ + ArN\,(\text{or } Ar + N), \quad (36)$$

$$Ar\cdots N^+ \cdots N \xrightarrow{(a)} ArN^+ + N \xrightarrow{(b)} N^+ + Ar + N. \quad (37)$$

Process 35(a) represents a charge-transfer process that takes place in a nearly collinear configuration to produce a near-collinear collision intermediate $Ar\cdots(N - N)^{*+}$. The N_2^{*+} moiety of the collision intermediate may be formed in a predissociative electronic state. Process 35(b) shows that as the dissociation of N_2^{*+} proceeds, the charge may reside on the N atom adjacent to or away from Ar, giving the reaction intermediates with the linear structures, $Ar\cdots N^+\cdots N$ and $Ar\cdots N\cdots N^+$, respectively. The dissociation of $Ar\cdots N\cdots N^+$ yields N^+ and ArN(or Ar + N), as indicated in process (36), while ArN^+ and N are produced in process 37(a) by the decomposition of $Ar\cdots N^+\cdots N$. The ArN^+ ion may further fragment into $Ar + N^+$ [process 37(b)] when the internal energy of ArN^+ exceeds its dissociation energy of about 0.5 eV.

As $E_{lab}(Ar^+)$ is increased from 26.5 to 205 eV, N^+ ions are scattered increasingly in the backward direction with respect to $\mathbf{v}_{c.m.}(Ar^+)$. At $E_{lab}(Ar^+) = 205\,eV$, the portion of forward scattered N^+ comprises $\leqslant 7\%$ of the total N^+ intensity. The increase of N^+ intensity in the backward hemisphere can be attributed partly to the further dissociation of ArN^+ at higher $E_{lab}(Ar^+)$. At a sufficiently high $E_{lab}(Ar^+)$, reaction (32) may proceed via other collision configurations in addition to the collinear geometry. The observed scattering patterns of N^+ at higher $E_{lab}(Ar^+)$ values reflect the formation of N^+ via different collision geometries. The results of this experiment suggest that N^+ ions are produced mainly by the predissociation of excited N_2^+ formed in the charge-transfer collisions of $Ar^+(^2P_{3/2,1/2})$ with N_2. The predissociative $N_2^+\,[\tilde{C}, \tilde{F}, \tilde{G}, (2\sigma_g)^{-1}]$ states responsible for the formation of N^+ by the photoionization process (34) are likely to play a role in reactions (32) and (33). In order to obtain accurate predictions of the cross sections for the $[Ar + N_2]^+$ charge-transfer system, it is necessary to consider the dissociative and reactive channels and to include many energetically accessible electronic states in the theoretical calculation. Since the previous theoretical studies have not included the reactive channels, the cross sections of which are $\approx 10\%$ that of the charge transfer, it is unlikely that the theoretical predictions for the charge-transfer cross sections at $E_{c.m.} > 8.5$ eV can be more accurate than 10%.

b. $N_2^+(\tilde{X}, \tilde{A}, v') + Ar$

I. CHARGE TRANSFER. Recently, Parlant and Gislason[64] have reported state-selected and state-to-state cross sections for reaction (27). Although the theoretical state-selected cross sections $[\sigma_{v'}(Ar^+)\ (v' = 0\text{–}2)]$ for the $N_2^+(\tilde{X}, v') + Ar$ charge-transfer reaction in the $E_{c.m.}$ range of 1–320 eV are consistent with experiment findings,[8,48] the theoretical branching ratios for producing the two product $Ar^+(^2P_{3/2})$ and $Ar^+(^2P_{1/2})$ are substantially different from the experimental results. Figure 15 compares the experimental and theoretical values for $\sigma_{v' \to 1/2}(Ar^+)/\sigma_{v'}(Ar^+)$, indicating that the theoretical values in some cases are greater than the experimental measurements by more than a factor of two. Here. $\sigma_{v' \to 1/2}(Ar^+)$ represents

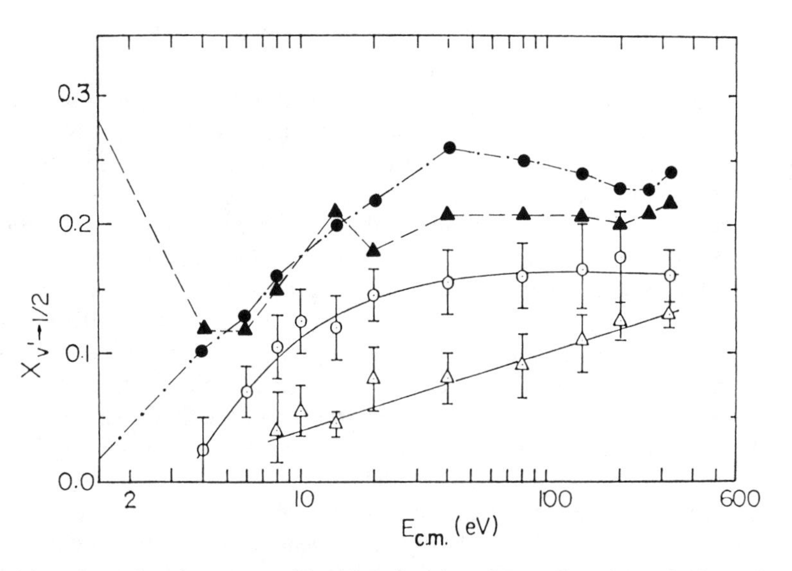

Figure 15. Values for $X_{v' \to 1/2}$ plotted as a function of $E_{c.m.}$. Experimental: (Δ) $v' = 0$ and ($\odot$) $v' = 1$, Ref. 25; theoretical: ($\blacktriangle$) $v' = 0$ and ($\bullet$) $v' = 1$, Ref. 64.

the total partial state-to-state cross section for the $N_2^+(\tilde{X}, v') + Ar$ charge-transfer reaction. Since the formation of $Ar^+(^2P_{1/2}) + N_2(v' = 0)$ is endothermic by 0.36 eV, as compared to 0.18 eV for the formation of $Ar^+(^2P_{3/2}) + N_2(v' = 0)$, it is puzzling that the theoretical value for $\sigma_{v' \to 1/2}(Ar^+)/\sigma_{v'}(Ar^+)$ at $E_{c.m.} = 1.2$ eV is 0.32, a value higher than that at $E_{c.m.} \geqslant 4.1$ eV. Other than this value, the general trend for the kinetic energy dependence of $\sigma_{v' \to 1/2}(Ar^+)/\sigma_{v'}(Ar^+)$ measured by Liao, Xu, and Ng[25] is consistent with that found by the vibronic semiclassical calculation.

In spite of the success of the vibronic semiclassical method, many aspects of the theoretical scheme can be improved. The *ab initio* surfaces used are calculated with the N–N distance fixed at the equilibrium value of N_2. The orientation angle of N_2 relative to Ar is also assumed to be fixed during the collision. Furthermore, the theoretical cross sections are based only on calculations at two orientations of the colliding pairs, namely, a linear and a perpendicular configuration. The resolution of this discrepancy requires further experimental and theoretical investigations in the future.

2. $[Ar + CO]^+$

a. $Ar^+(^2P_{3/2,1/2}) + CO$ Reactions

I. CHARGE TRANSFER. The charge transfer reactions

$$Ar^+(^2P_{3/2,1/2}) + CO(X, v'' = 0) \to Ar(^1S_0) + CO^+(\tilde{X}, \tilde{A}, v') \tag{38}$$

have been the subject of many experimental[49,86–90] and theoretical studies.[61,91] In contrast to the strongly coupled $[Ar + N_2]^+$ system, the $[Ar + CO]^+$ system is characterized by very weak effective charge-transfer couplings for the near-resonance channels $Ar + CO^+(\tilde{X}, v' = 6, 7)$ owing to the very small Franck–Condon factors for the ionization transitions $CO(X, v'' = 0) \to CO^+(\tilde{X}, v' = 6, 7)$. As a result of this, charge transfer in the $[Ar + CO]^+$ system likely results from short-range interactions. The accurate prediction of the dynamics of this system may require a theoretical model that allows the reorientation of the colliding pair. It will also be necessary to use detailed potential surfaces for the interactions of $Ar + CO^+$ and $Ar^+ + CO$ which take into account the variation of the C–O bond distance. The $[Ar + CO]^+$ charge-transfer system is believed to be a more stringent test for the vibronic semiclassical model.[61,91]

Recently, absolute total cross sections for reaction (38) $[\sigma_{3/2,1/2}(CO^+)]$ at the E_{lab} range of 0.1–300 eV have been measured by Flesch, Nourbakhsh, and Ng.[9,92] As shown in Fig. 16(*a*), the kinetic energy dependence of $\sigma_{3/2}(CO^+)$ exhibits a pronounced minimum at $E_{lab} \approx 3$–5 eV. A second minimum is

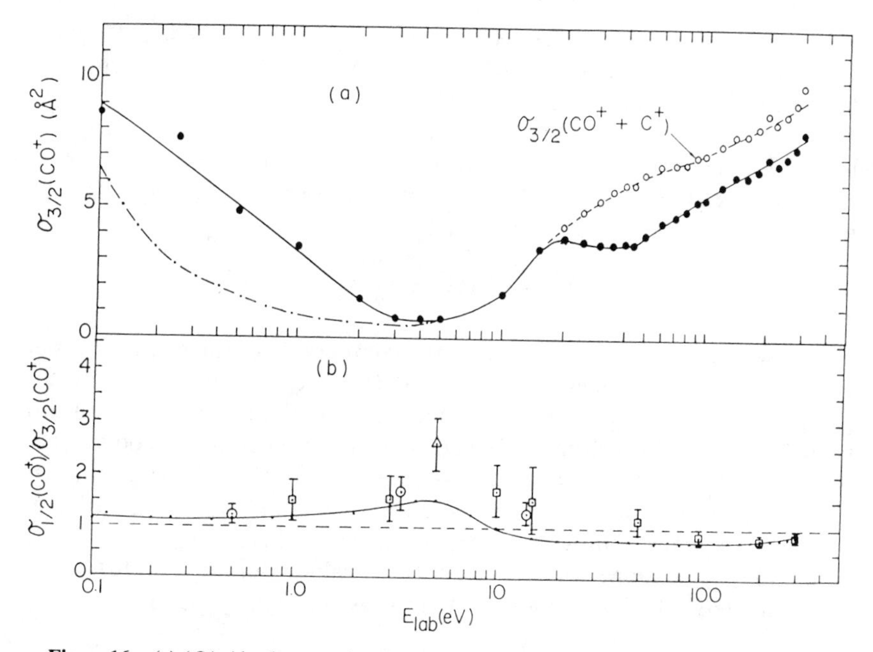

Figure 16. (*a*) (●) Absolute total cross sections for CO^+ [$\sigma_{3/2}(CO^+)$] and (⊙) sum of absolute total cross sections for CO^+ and C^+ [$\sigma_{3/2}(CO^+ + C^+)$] formed in the reaction of $Ar^+(^2P_{3/2}) + CO$ at $E_{lab} = 0.1$–300 eV; (–·–) effective total cross sections converted from rate constants, Ref. 90. (*b*) Values for the ratio $\sigma_{1/2}(CO^+)/\sigma_{3/2}(CO^+)$ plotted versus E_{lab} in the range of 0.1–300 eV; experimental: (· — ·) Ref. 92, (⊙) Ref. 49, (□) Ref. 57; theoretical: (△) Ref. 91.[92]

found at $E_{lab} \approx 20$–40 eV. The effective cross sections σ in the E_{lab} range of 0.1–5 eV, converted from experimental rate constants[90] (k) using the approximate relation, $k \approx \sigma v$, are also plotted in the figure. Here v is the average relative velocity of the colliding pair. Since electron impact ionization was used to prepare reactant Ar^+ in the rate constant measurement, it is most likely that the σ values represent the cross sections due to a mixture of $Ar^+(^2P_{3/2})$ and $Ar^+(^2P_{1/2})$. The values for $\sigma_{3/2}(CO^+)$ are higher than the σ values by about a factor of 2. The observation of a rapid increase in $\sigma_{3/2}(CO^+)$ as E_{lab} goes from 3 eV toward thermal energy is in qualitative accord with the Langevin–Gioumousis–Stevenson (LGS) orbiting complex model.[93,94]

The ratios $\sigma_{1/2}(CO^+)/\sigma_{3/2}(CO^+)$ are compared to previous experimental[49,57] and theoretical[91] values in Fig. 16(*b*). The experimental results are in general agreement, indicating that $\sigma_{1/2}(CO^+)/\sigma_{3/2}(CO^+)$ is greater than one at $E_{lab} < 10$ eV, and becomes lower than unity at higher E_{lab}. The maximum for the ratio $\sigma_{1/2}(CO^+)/\sigma_{3/2}(CO^+)$ coincides with the minimum for $\sigma_{3/2}(CO^+)$ at $E_{lab} \approx 3$–$6\,eV$. The value for the ratio predicted by the vibronic semiclassical calculation[91] at $E_{lab} = 5\,eV$ is more than a factor of 2 higher than the experimental value.

II. SPIN–ORBIT STATE EXCITATION AND RELAXATION. The spin–orbit excitation cross sections $[\sigma_{3/2\to1/2}(CO)]$ in the E_{lab} range of 5–400 eV for the process

$$Ar^+(^2P_{3/2}) + CO \to Ar^+(^2P_{1/2}) + CO \tag{39}$$

are shown in Fig. 11(*c*).[92] Similar to the comparison of the charge-transfer cross sections, the values for $[\sigma_{3/2\to1/2}(CO)]$ are smaller than those for $[\sigma_{3/2\to1/2}(N_2)]$. At $E_{lab} < 200\,eV$, the spin–orbit excitation cross sections for process (39) are also substantially higher than those due to the collisions of $Ar^+(^2P_{3/2}) + Ar$. This observation supports the stepwise charge-transfer mechanism for reaction (39). The theoretical prediction[91] for $\sigma_{3/2\to1/2}(CO)$ at $E_{lab} = 5\,eV$ is consistent with the experimental result.

In the E_{lab} range of 5–400 eV, the values for $\sigma_{1/2\to3/2}(CO)$ are approximately a factor of 2 greater than those for $\sigma_{3/2\to1/2}(CO)$.[92]

III. DISSOCIATION AND REACTION. The dissociative and reactive processes,

$$Ar^+(^2P_{3/2,1/2}) + CO \to C^+(^2P) + O(^3P) + Ar, \tag{40}$$

$$\to O^+(^4S) + C(^3P) + Ar, \tag{41}$$

$$\to ArC^+ + O(^3P), \tag{42}$$

have also been observed by Flesch et al.[9,92] The measured total cross sections for C^+, O^+, and ArC^+ formed in the $Ar^+(^2P_{3/2}) + CO$ reaction over the E_{lab} range of 10–300 eV are shown in Fig. 17(*a*). The cross sections for C^+, O^+, and ArC^+ produced in the $Ar^+(^2P_{1/2}) + CO$ reaction are ≈ 5–30% lower than those for the $Ar^+(^2P_{3/2}) + CO$ reaction. The highest cross sections for O^+ and ArC^+ are < 0.05 that for C^+. The AEs for C^+ and O^+ are $E_{lab} = 16 \pm 1\,eV$ ($E_{c.m.} = 6.6 \pm 0.4\,eV$) and $21 \pm 1\,eV$ ($8.6 \pm 0.4\,eV$) and are in agreement with the thermochemical threshold[16,19] for reactions (40) and (41), respectively. The cross sections for C^+ at $E_{lab} = 35$–$55\,eV$ are $> 50\%$ those for CO^+. The formation of ArC^+ is observed only in the E_{lab} range of 15–70 eV. The observed AE for ArC^+ yields a lower bound of 0.5 eV for the ArC^+ bond dissociation energy. This compares to a theoretical[84] and an experimental[98] value of $\approx 0.9\,eV$.

At energies above the threshold for reaction (40), five electronic bands of

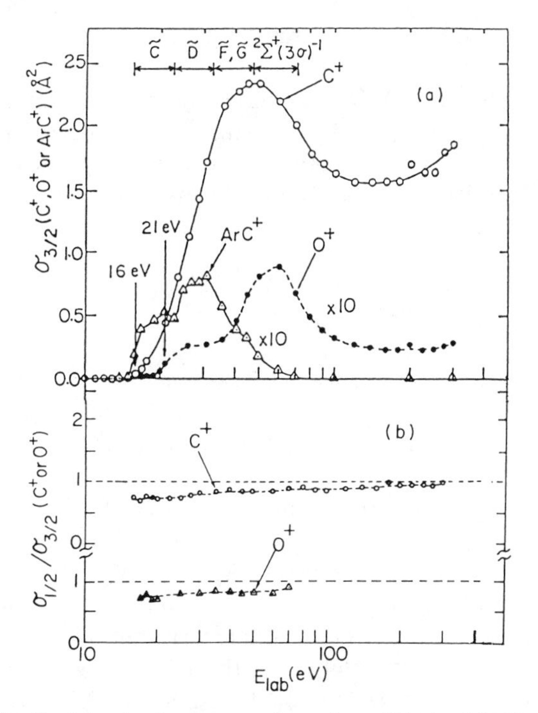

Figure 17. (*a*) Absolute total cross sections for C^+(⊙), O^+(●), and ArC^+(△) [$\sigma_{3/2}(C^+, O^+$, and $ArC^+)$] formed in the reaction of $Ar^+(^2P_{3/2}) + CO$ in the E_{lab} range of 10–300 eV. (*b*) (⊙) Values for the ratios $\sigma_{1/2}(C^+)/\sigma_{3/2}(C^+)$ and (△) $\sigma_{1/2}(O^+)/\sigma_{3/2}(O^+)$.[9,92]

CO^+, $\tilde{C}\,^2\Sigma^+, \tilde{D}\,^2\Pi, \tilde{F}\,^2\Sigma, \tilde{G}\,^2\Sigma^+$, and $^2\Sigma^+(3\sigma)^{-1}$ have been identified in the photoelectron spectrum obtained at a photon energy of 50.1 eV.[95] These multielectron excitation states have mixed configurations of the 3σ single-hole state and outer-valence shell states. The $\tilde{C}\,^2\Sigma^+$ band predissociates into $C^+ + O$.[96,97] The $\tilde{D}\,^2\Pi, \tilde{F}\,^2\Sigma, \tilde{G}\,^2\Sigma^+$, and $^2\Sigma^+(3\sigma)^{-1}$ bands predissociate into both $C^+ + O$ and $O^+ + C$.[74] The positions of these bands are indicated in Fig. 17(*a*). The sum of $\sigma_{3/2}(CO^+)$ and $\sigma_{3/2}(C^+)$ is plotted as a function of E_{lab} in Fig. 16(*a*). The comparison of the kinetic energy dependences for $\sigma_{3/2}(CO^+)$ and the sum of $\sigma_{3/2}(CO^+ + C^+)$ leads to the conclusion that the minimum for $\sigma_{3/2}(CO^+)$ observed at $E_{lab} \approx 20$–40 eV is the result of dissociation of excited CO^+ formed in the predissociation states, $\tilde{C}\,^2\Sigma^+, \tilde{D}\,^2\Pi, \tilde{F}\,^2\Sigma, \tilde{G}\,^2\Sigma^+$, and $^2\Sigma^+(3\sigma)^{-1}$, by charge-transfer collisions of $Ar^+(^2P_{3/2}) + CO$. The ratios $\sigma_{1/2}(C^+)/\sigma_{3/2}(C^+)$ and $\sigma_{1/2}(O^+)/\sigma_{3/2}(O^+)$ [see Fig. 17(*b*)], which vary from a value of ≈ 0.75 at $E_{lab} = 16$ eV to slightly less than one at $E_{lab} = 300$ eV, are similar to the values found for $\sigma_{1/2}(CO^+)/\sigma_{3/2}(CO^+)$. This observation is also in accord with the charge-transfer predissociation mechanism for the formation of C^+ and O^+.

It is interesting that ArO^+ ions are not observed in the reaction of $Ar^+ + CO$. A plausible explanation is that the predissociation of CO^+ favors overwhelmingly the formation of C^+. The substantially lower intensity for O^+ compared to that for C^+ found in the experiment is in agreement with this rationalization.

3. $[Ar + O_2]^+$

a. $Ar^+(^2P_{3/2,1/2}) + O_2$ *Reactions*

I. CHARGE TRANSFER. The charge-transfer reactions

$$Ar^+(^2P_{3/2,1/2}) + O_2 \rightarrow Ar + O_2^+ \tag{43}$$

have been studied extensively in the past.[17,90,99–109] Rate coefficients[90,99,103–107] and absolute total cross sections[109] for reaction (43) have been reported from thermal energies to $E_{c.m.} = 4\,eV$. The spin–orbit effect of reaction (43) has been examined at $E_{c.m.} = 1.4$ and $5.8\,eV$ using the threshold PEPICO technique.[17] The internal-state distributions for charge-transfer product O_2^+ have been measured by the time-of-flight (TOF),[102,109] the chemical reactivity,[99] and the laser photofragmentation[109] methods. All these studies are in general agreement that O_2^+ ions are formed preferably in the $O_2^+(\tilde{a}\,^4\Pi_u)$ metastable state as $E_{c.m.}$ is raised above the threshold for the formation of $O_2^+(\tilde{a}\,^4\Pi_u)$. The recent laser photofragmantation experiment of Scherbarth and Gerlich[109] provides rovibrational-state distributions of $O_2^+(\tilde{a}\,^4\Pi_u)$ formed by the $Ar^+ + O_2$ charge-transfer reaction at $E_{c.m.} = 0.5$–$1.4\,eV$.

Absolute spin–orbit state-selected cross sections for the charge-transfer reactions (43) $[\sigma_{3/2,1/2}(O_2^+)]$ in the E_{lab} range of 0.1–300 eV, measured using the TQDO photoionization apparatus,[13] are shown in Fig. 18(*a*). The $\sigma_{3/2}(O_2^+)$ curve exhibits considerable structure. The value for $\sigma_{3/2}(O_2^+)$ decreases from $11\,Å^2$ at $E_{lab} = 0.1\,eV$ to a minimum of $2.6\,Å^2$ at $E_{lab} \simeq 2\,eV$, followed by a maximum of $4.5\,Å^2$ at $E_{lab} \approx 8\,eV$ as E_{lab} is increased. The guided-ion-beam cross sections[109] and effective cross sections converted from flow-tube reaction rates[88,99] for the $Ar^+(^2P_J) + O_2$ charge-transfer reaction in the E_{lab} range of 0.04–7 eV are in reasonable agreement. Although these cross sections are not spin–orbit state-selected results, they are also included in Fig. 18(*a*). The minimum at $E_{lab} \approx 2\,eV$ observed in the TQDO experiment compares to minima at $E_{lab} \approx 0.7$ and 1.2 eV found in the guided-ion-beam and flow-tube studies, respectively. The differences are attributed party to the fact that the Ar^+ reactant ions prepared in the latter studies may consist of

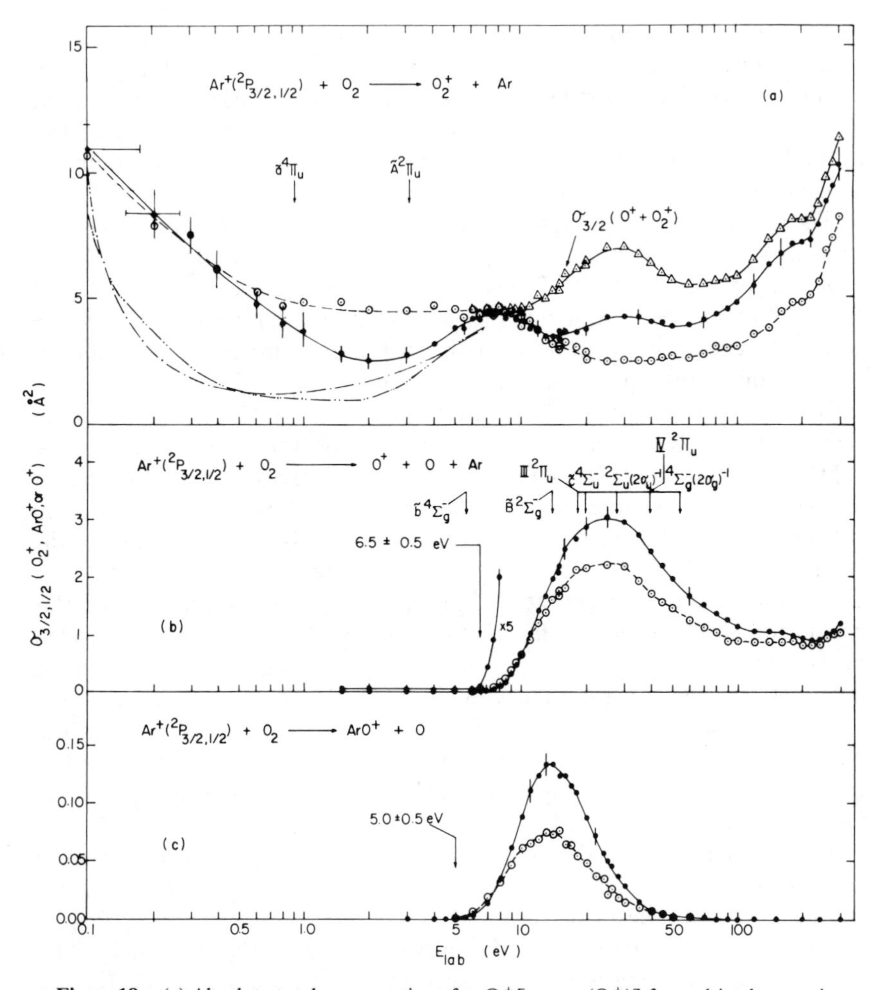

Figure 18. (*a*) Absolute total cross sections for $O_2^+[\sigma_{3/2,1/2}(O_2^+)]$ formed in the reactions of $Ar^+(^2P_{3/2}) + O_2(\bullet)$ and $Ar^+(^2P_{1/2}) + O_2(\odot)$ at $E_{lab} = 0.1$–300 eV. ($\triangle$) Sum of absolute total cross sections for O_2^+ and $O^+(\sigma_{3/2})$ formed in the reaction $Ar^+(^2P_{3/2}) + O_2$. (— · —) Absolute total cross sections for O_2^+ reported in Ref. 109. (— · · —) Effective total cross sections for O_2^+ based on rate constants reported in Refs. 90 and 99. (*b*) Absolute total cross sections for $O^+[\sigma_{3/2,1/2}(O^+)]$ formed in the reactions of $Ar^+(^2P_{3/2}) + O_2(\bullet)$ and $Ar^+(^2P_{1/2}) + O_2$ ($\odot$) at $E_{lab} = 15$–300 eV. (*c*) Absolute total cross sections for $ArO^+[\sigma_{3/2,1/2}(ArO^+)]$ formed in the reactions of $Ar^+(^2P_{3/2}) + O_2(\bullet)$ and $Ar^+(^2P_{1/2}) + O_2$ ($\odot$) at $E_{lab} = 3$–300 eV.[13]

a mixture of $^2P_{3/2}$ and $^2P_{1/2}$ states. Taking into account the experimental uncertainties, the cross sections for O_2^+ at $E_{lab} = 0.1$ eV and $E_{lab} \geqslant 3$ eV obtained in the TQDO study are in agreement with those obtained in the guided-ion-beam and flow-tube studies. However, the cross sections for O_2^+ in the E_{lab} range of 0.1–1.5 eV measured in these studies are lower than the TQDO values for $\sigma_{3/2}(O_2^+)$ by more than a factor of 2.

The observation of the rapid increase in O_2^+ cross section as E_{lab} decreases from 2 eV toward thermal energy is in qualitative accord with the LGS orbiting complex model.[93,94] The values for $\sigma_{3/2}(O_2^+)$, and $\sigma_{1/2}(O_2^+)$ at $E_{lab} < 0.6$ eV are identical within the experimental uncertainties. This finding is also consistent with the conclusion that the reaction proceeds via a collision complex mechanism at low collision energies.

In the E_{lab} range of 0.6–6 eV, the values for $\sigma_{1/2}(O_2^+)$ are greater than those for $\sigma_{3/2}(O_2^+)$. The energy thresholds for the formation of $O_2^+(\tilde{a}\,^4\Pi_u, \tilde{A}\,^2\Pi_u)$ are indicated in Fig. 18(*a*). The increase of the cross section for the $Ar^+ + O_2$ charge-transfer reaction observed as E_{lab} is increased in the region of 0.9–7.0 eV ($E_{c.m.} = 0.4$–3.0 eV) has been attributed to the formation of $O_2^+(\tilde{a}\,^4\Pi_u, v')$.[99,102,109] As described in Section II A 3, the differential reactivity experiment[13] provides very strong support for this conclusion. The preference for the formation of $O_2^+(\tilde{a}\,^4\Pi_u)$ is predicted because the $Ar^+(^2P_{3/2,1/2}) + O_2$ state is in closer energy resonance with the $O_2^+(\tilde{a}\,^4\Pi_u, v' = 0) + Ar$ state than with the $O_2^+(\tilde{X}\,^2\Pi_g, v' = 0) + Ar$ state in their asymptotic limits. Furthermore, the experiment of Flesch, Nourbakhsh, and Ng[13] also suggests that the populations of $O_2^+(\tilde{a}\,^4\Pi_u, v')$ ions produced in the charge-transfer collisions of $Ar^+(^2P_{3/2}) + O_2$ and by photoionization of O_2 are similar.

If energy resonance is the dominant factor controlling the charge-transfer cross section, $\sigma_{1/2}(O_2^+)$ is predicted to be greater than $\sigma_{3/2}(O_2^+)$ because the endothermicity for the formation of $O_2^+(\tilde{a}\,^4\Pi_u, v' = 0) + Ar$ from $Ar^+(^2P_{1/2}) + O_2$ is 0.22 eV compared to a value of 0.4 eV from $Ar^+(^2P_{3/2}) + O_2$. This prediction is consistent with the experimental observation that $\sigma_{3/2}(O_2^+)$ is less then $\sigma_{1/2}(O_2^+)$ at $E_{lab} = 0.6$–6 eV. The $O_2^+(\tilde{a}\,^4\Pi_u)$ and $O_2^+(\tilde{A}\,^2\Pi_u)$ states are derived from the same molecular-orbital configuration. The previous TOF experiment[102] provides evidence suggesting that $O_2^+(\tilde{A}\,^2\Pi_u)$ ions are also formed in reaction (43) at E_{lab} above the $O_2^+(\tilde{A}\,^2\Pi_u, v' = 0)$ threshold.

In the E_{lab} range of 13–300 eV, the values for $\sigma_{3/2}(O_2^+)$ are higher than those for $\sigma_{1/2}(O_2^+)$, contrary to the order of values for $\sigma_{3/2}(O_2^+)$ and $\sigma_{1/2}(O_2^+)$ observed at $E_{lab} = 0.6$–6 eV. At high E_{lab}, the energy difference between the $^2P_{3/2}$ and $^2P_{1/2}$ states of Ar^+ is expected to have less effect on the charge-transfer cross section. This expectation is consistent with the finding that $\sigma_{3/2}(O_2^+)/\sigma_{1/2}(O_2^+)$ decreases from 1.68 at $E_{lab} = 30$ eV to 1.26 at $E_{lab} = 300$ eV. At $E_{lab} \approx 6$–13 eV, the value for $\sigma_{3/2}(O_2^+)/\sigma_{1/2}(O_2^+)$ is essentially unity. The variation of $\sigma_{3/2}(CO^+)/\sigma_{1/2}(CO^+)$ as a function of E_{lab} observed in the $Ar^+(^2P_{3/2,1/2}) + CO$ charge-transfer reaction is similar to the variation of the kinetic energy dependence of $\sigma_{3/2}(O_2^+)/\sigma_{1/2}(O_2^+)$ found for reaction (43). It is interesting to note that, for both $\sigma_{3/2}(O_2^+)$ and $\sigma_{3/2}(CO^+)$, the pronounced minimum is located at the E_{lab} value where the ratio of the charge-transfer cross section due to $Ar^+(^2P_{1/2})$ to that associated with $Ar^+(^2P_{3/2})$ reaches the maximum.

The retarding-potential curves for charge transfer O_2^+, as well as those for O^+ and ArO^+, formed in the collisions of $Ar^+(^2P_{3/2}) + O_2$ at $E_{lab} =$ 8–104 eV are shown in Figs. 19(*a*)–19(*f*). As in the retarding-potential analysis of product ions formed in the $Ar^+(^2P_{3/2}) + N_2$ reaction, the E_{max} values are the maximum laboratory energies allowed by energy conservation. The E_{lab} values indicated in Figs. 19(*a*)–19(*f*) are the calculated laboratory energies that divide the forward- and backward-scattered product ions. The retarding-potential curves indicate that the intensities of O_2^+ decrease sharply at the retarding potential of 1.3 ± 0.2 eV. At E_{lab} $(Ar^+) = 8$ eV, the intensities for the forward and backward scattered O_2^+ are nearly comparable. As $E_{lab}(Ar^+)$ is increased, the fraction of forward scattered O_2^+ decreases rapidly. At $E_{lab}(Ar^+) \geqslant 31.6$ eV, the retarding-potential-energy curves for O_2^+ reveal that the overwhelming proportion ($\geqslant 98\%$) of O_2^+ are slow ions with $E_{lab}(O_2^+) \leqslant 1.5$ eV and are scattered in the backward hemisphere with respect to $\mathbf{v}_{c.m.}(Ar^+)$. This observation indicates that the long-range electron jump mechanism is the major pathway for change transfer $E_{lab}(Ar^+) > 32$ eV. Depending on $E_{lab}(Ar^+)$, O_2^+ may be formed in different excited states and subsequently stabilized by radiative decay. It is interesting to note that 1–2% of O_2^+ ions formed at $E_{lab}(Ar^+) = 31.6, 53$ and 104 eV have E_{lab} in the range of 25–35, 45–56, and 60–107 eV, respectively.

II. SPIN–ORBIT STATE EXCITATION AND RELAXATION. The absolute partial state-to-state total cross sections $[\sigma_{3/2\to1/2}(O_2)]$ for the excitation process,

$$Ar^+(^2P_{3/2}) + O_2 \to Ar^+(^2P_{1/2}) + O_2, \tag{44}$$

at $E_{lab} = 5$–400 eV are depicted in Fig. 11(*a*).[13] Similar to the observations found in the comparison of $\sigma_{3/2\to1/2}(N_2, CO)$ and $\sigma_{3/2\to1/2}(Ar)$, the values for $\sigma_{3/2\to1/2}(O_2)$ and $\sigma_{3/2\to1/2}(Ar)$ are essentially identical at $E_{lab} \approx 250$–400 eV after taking into account experimental uncertainties. However, at E_{lab} below 250 eV, values for $\sigma_{3/2\to1/2}(O_2)$ are greater than those for $\sigma_{3/2\to1/2}(Ar)$. The significantly higher values observed for $\sigma_{3/2\to1/2}(O_2)$, compared to $\sigma_{3/2\to1/2}(Ar)$ at lower E_{lab}, may be ascribed to the existence of intermediate $Ar + O_2^+$ charge-transfer states in the $[Ar + O_2]^+$ systems.[13] That is, the spin–orbit state excitation process, such as reaction (44), may be viewed to

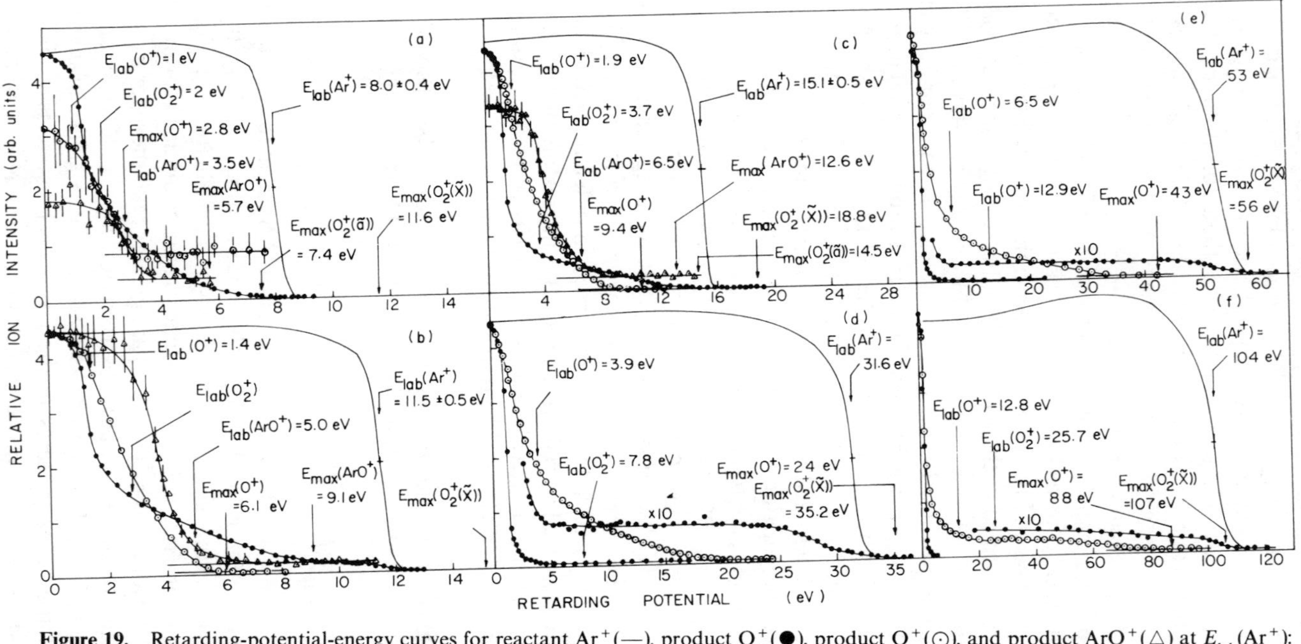

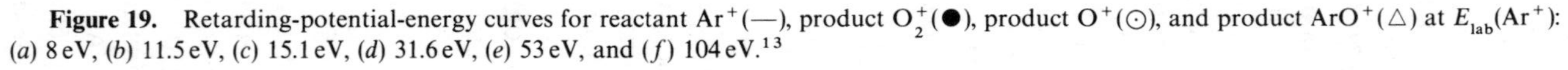

Figure 19. Retarding-potential-energy curves for reactant Ar^+(—), product O_2^+(●), product O^+(⊙), and product ArO^+(△) at $E_{lab}(Ar^+)$: (*a*) 8 eV, (*b*) 11.5 eV, (*c*) 15.1 eV, (*d*) 31.6 eV, (*e*) 53 eV, and (*f*) 104 eV.[13]

proceed according to processes 45(*a*) and 45(*b*).

$$Ar^+(^2P_{3/2}) + O_2 \xrightarrow{(a)} O_2^+ + Ar \xrightarrow{(b)} Ar^+(^2P_{1/2}) + O_2. \quad (45)$$

The involvement of intermediate $O_2^+ + Ar$ charge-transfer states should enhance indirectly the coupling between the reactant $Ar^+(^2P_{3/2}) + O_2$ and the product $Ar^+(^2P_{1/2}) + O_2$ states.

Rough estimates have also been obtained for the spin–orbit relaxation cross sections $[\sigma_{1/2\to3/2}(O_2)]$ in the collisions of $Ar^+(^2P_{1/2}) + O_2$. The values for $\sigma_{1/2\to3/2}(O_2)$ are approximately a factor of 2 greater than those for $\sigma_{3/2\to1/2}(O_2)$ at corresponding E_{lab}.

III. DISSOCIATION AND REACTIONS. Absolute total cross sections for the dissociative and reactive processes,[13]

$$Ar^+(^2P_{3/2,1/2}) + O_2 \to O^+ + O + Ar, \quad (46)$$

$$\to ArO^+ + O, \quad (47)$$

are shown in Figs. 18(*b*) and 18(*c*). The formations of O^+ and ArO^+ by the reaction $Ar^+(^2P_{3/2}) + O_2$ are endothermic[16,110] by 2.97 and 2.29 eV, respectively. The observation of ArO^+ has also been reported by Scherbarth and Gerlich.[109]

The $\sigma_{3/2}(O^+)$ and $\sigma_{1/2}(O^+)$ curves shown in Fig. 18(*b*) exhibit a maximum at $E_{lab} \approx 20$–30 eV. In this E_{lab} range, the cross sections for O_2^+ and O^+ are comparable in magnitude. The AE for the formation of O^+ from the reaction of $Ar^+(^2P_{3/2}) + O_2$ is determined to be $E_{lab} = 6.5 \pm 0.5$ eV ($E_{c.m.} = 2.89 \pm 0.22$ eV), in good agreement with the thermochemical threshold.

The kinetic energy dependence of the ratio $\sigma_{3/2}(O^+)/\sigma_{1/2}(O^+)$ follows closely with that for $\sigma_{3/2}(O^+)/\sigma_{1/2}(O^+)$. Namely, $\sigma_{3/2}(O^+) \approx \sigma_{1/2}(O^+)$ at $E_{lab} = 6$–13 eV and $\sigma_{3/2}(O^+) > \sigma_{1/2}(O^+)$ at $E_{lab} > 13$ eV.

The thresholds for the formation of $O_2^+[\tilde{b}\,^4\Sigma_g^-, \tilde{B}\,^2\Sigma_g^-, \mathrm{III}\,^2\Pi_u, \tilde{c}\,^4\Sigma_u^-, ^2\Sigma_u^-(2\sigma_u)^{-1}, \mathrm{IV}\,^2\Pi_u$, and $^4\Sigma_g^-(2\sigma_g)^{-1}]$ are indicated in Fig. 18(*b*).[111–116] The (pre)dissociation of the $O_2^+(\tilde{b}\,^4\Sigma_g^-, \tilde{B}\,^2\Sigma_g^-, \mathrm{III}\,^2\Pi_u$, and $\tilde{c}\,^4\Sigma_u^-)$ states has been the subject of many previous PIE,[111,117–119] PEPICO,[120–127] and laser photofragmentation[128–133] studies. The weakly bound $O_2^+(\tilde{b}'\,^4\Pi_g)$ state, which is about 0.2 eV lower than the $O_2^+(\tilde{b}\,^4\Sigma_g^-)$ state, is also found to be predissociative at higher vibrational states by laser photofragmentation experiments.[129] Both the $O_2^+(\tilde{b}'\,^4\Pi_g)$ and $O_2^+(\tilde{b}\,^4\Sigma_g^-)$ states dissociate to the first dissociation limit $O^+(^4S) + O(^3P)$. Previous PEPICO experiments show that the $O_2^+(\tilde{B}\,^2\Sigma_g^-, v')$ and $O_2^+(\tilde{c}\,^4\Sigma_u^-, v = 0, 1)$ states may predissociate to the first as well as the second dissociation limit

$O^+(^4S) + O(^1D)$.[43,121–126] The $O_2^+(\text{III}\ ^2\Pi_u)$ state dissociates into the $O^+(^2D) + O(^3P)$ and the $O^+(^2P) + O(^3P)$ dissociation limits.[123] The $O_2^+[^2\Sigma_u^-(^2\sigma_u)^{-1}, \text{III}\ ^2\Pi_u, ^4\Sigma_g^-(2\sigma_g)^{-1}]$ are multielectron states that have mixed configurations of the $2\sigma_u$ and $2\sigma_g$ single-hole states and outer valence states,[133,134] and have been shown to be dissociative in previous experiments.[111,116,117,135]

The PIE spectrum for fragment O^+ produced in the photoionization of O_2 in the photon energy range of 15–110 eV shows two unresolved broad peaks.[111,117] The first peak of the PIE spectrum of O^+ from O_2 and the maximum of the O^+ cross-section curves shown in Fig. 18(*b*) are nearly in the energy region covering the $O_2^+[\tilde{B}\ ^2\Sigma_g^-, \text{III}\ ^2\Pi_u, \tilde{c}\ ^4\Sigma_u^-, ^2\Sigma_u^-(2\sigma_u)^1, \text{IV}\ ^2\Pi_u]$ states. The second broad peak observed in the O^+ PIE spectrum is approximately in the energy region of the $O_2^+[^2\Sigma_g^-(2\sigma_g)^{-1}]$ state, but no corresponding peak is found in the O^+ cross-section curve. The thermochemical thresholds for the processes,

$$O_2 + h\nu \rightarrow O^+(^4S) + O^+(^4S) + 2e^{-1}, \quad (48)$$

$$\rightarrow O_2^{2+}[\tilde{X}\ ^1\Sigma_g^+(1\pi_g)^{-2}] + 2e^{-1}, \quad (49)$$

are 32.4 and 36.7 eV,[136,137] respectively. The energy thresholds for the formation of $2O^+(^4S)$ and $O_2^{2+}[\tilde{X}\ ^1\Sigma_g^+(1\pi_g)^{-2}]$ from the reactions,

$$Ar^+(^2P_{3/2}) + O_2 \rightarrow O^+(^4S) + O^+(^4S) + Ar + e^{-1}, \quad (50)$$

$$\rightarrow O_2^{2+}[\tilde{X}\ ^1\Sigma_g^-(1\pi_g)^{-2}] + 2e^{-1}, \quad (51)$$

are $E_{\text{c.m.}} = 16.6$ eV ($E_{\text{lab}} = 37.4$ eV) and $E_{\text{c.m.}} = 20.94$ eV ($E_{\text{lab}} = 47.2$ eV), respectively. Assuming that the second broad peak observed in the O^+ PIE spectrum in the photon energy range of 35–50 eV arises from double ionization processes, one may conclude that the cross section for O^+ formed by reaction (46) and the PIE for O^+ formed in the dissociative photoionization process,

$$O_2 + h\nu \rightarrow O^+ + O + e^{-1}, \quad (52)$$

are both peaked in the same energy range. The fact that a second peak in the E_{lab} region of approximately 40–80 eV is not found in the O^+ cross-section curves shown in Fig. 18(*b*) is evidence that reactions (50) and (51) are not important in that E_{lab} range.

By monitoring the formation of NO^+ in the collisions of O^+ and N_2, the overwhelming majority of O^+ ions formed by reaction (46) are shown to be in the $O^+(^4S)$ ground state.

The production of ArO^+ by reaction (47) is observed only in the E_{lab} range of 5–60 eV ($E_{c.m.} = 2.2$–26.6 eV) and peaks at $E_{lab} \approx 14$ eV. The profile of the cross section curves for ArO^+ observed here are similar to those for ArC^+ (ArN^+) formed in the reactions of $Ar^+(^2P_{3/2,1/2}) + CO(N_2)$.[9,10,92] The negligible ArO^+ signal observed at $E_{lab} \geqslant 60$ eV is attributed to high internal energies deposited in ArO^+, which cause its fragmentation into $O^+ + Ar$. The measured values for $\sigma_{3/2,1/2}(ArO^+)$ at the peak are approximately 5% of the maximum values for $\sigma_{3/2,1/2}(O^+)$. The AE for ArO^+ formed in the reaction of $Ar^+(^2P_{3/2}) + O_2$ is determined to be 5.0 ± 0.5 eV ($E_{c.m.} = 2.2 \pm 0.2$ eV), a value consistent with that observed by Scherbarth and Gerlich.[109] The difference between the AEs of O^+ and ArO^+ give an estimate of 0.7 ± 0.3 eV for the binding energy of Ar–O^+. This value is in good agreement with the well depth for the Ar–O^+ potential measured using the elastic differential scattering technique.[110]

At $E_{lab} = 6$–15 eV, the decrease in $\sigma_{3/2}(O_2^+)$ coincides with the increase in $\sigma_{3/2}(O^+)$ and $\sigma_{3/2}(ArO^+)$. The sums $[\sigma_{3/2}(O_2^+ + O^+)]$ of the values for $\sigma_{3/2}(O_2^+)$ and $\sigma_{3/2}(O^+)$ in the E_{lab} range of 6–300 eV are plotted in Fig. 18(*a*). The variation of $\sigma_{3/2}(O_2^+ + O^+)$ in the E_{lab} range of 6–15 eV is quite smooth. This observation supports the predissociation mechanism for the formation of O^+. The increase of $\sigma_{3/2}(O_2^+)$ as E_{lab} increases from 2 eV toward higher E_{lab} is attributed to the formation of O_2^+ [$\tilde{a}\,^4\Pi_u$, $\tilde{A}\,^2\Pi_u$, $\tilde{b}'\,^4\Pi_g$, $\tilde{b}\,^4\Sigma_g^-$, $\tilde{B}\,^2\Sigma_g^-$, III $^2\Pi_u$, $\tilde{c}\,^4\Sigma_u^-$, $^2\Sigma_u^-(2\sigma_u)^{-1}$, IV $^2\Pi_u$, and $^4\Sigma_g^-(2\sigma_g)^{-1}$, etc.]. As E_{lab} is increased above the thermochemical threshold for reaction (46), the (pre)-dissociation of these O_2^+ excited states may give $O^+ + O$. Within the time scale of the experiment, charge-transfer product O_2^+ ions formed in these excited states may also be stabilized by radiative decay processes.

As shown by the retarding-potential-energy curves in Fig. 19(*a*), nearly 100% of the ArO^+ formed at $E_{lab}(Ar^+) = 8$ eV are scattered in the backward hemisphere with respect to $\mathbf{v}_{c.m.}(Ar^+)$. At $E_{lab}(Ar^+) = 11.5$ and 15.1 eV, it appears that the fractions of ArO^+ ions scattered into the forward hemisphere are greater compared to that at $E_{lab} = 8$ eV. Nevertheless, more than 95% of the ArO^+ ions remain backward scattered at these collision energies. Contrary to the retarding-potential-energy curves for ArO^+ formed at $E_{lab}(Ar^+) = 8$, 11.5, and 15.1 eV, which show that ArO^+ ions are scattered in the backward hemisphere with respect to $\mathbf{v}_{c.m.}(Ar^+)$, the retarding-potential-energy curves for O^+ measured at these collision energies reveal that product O^+ ions are scattered predominantly in the forward hemisphere. The scattering patterns of ArO^+ and O^+ formed by reactions (46) and (47) at collision energies near the thresholds of these reactions indicate that ArO^+ and O^+ are produced via the collinear charge-transfer predissociation mechanism, similar to the production of ArN^+ and N^+ by the reaction of $Ar^+(^2P_{3/2}) + N_2$.[7] According to the charge-transfer predissociation mechanism, the production of ArO^+ and O^+ follows the charge transfer between Ar^+ and O_2. Thus, the charge-transfer predissociation mechanism also accounts for the observation that the kinetic energy dependences for the ratios $\sigma_{3/2}(O^+)/\sigma_{1/2}(O^+)$ and $\sigma_{3/2}(ArO^+)/\sigma_{1/2}(ArO^+)$ are similar.

As $E_{lab}(Ar^+)$ is increased from 15 to 104 eV, the portion of forward-scattered product O^+ ions decreases. At $E_{lab}(Ar^+) = 31.6$ eV, the intensities for the forward and backward scattered O^+ are about equal, while at $E_{lab}(Ar^+) = 104$ eV the portion of O^+ scattered in the forward hemisphere comprises only approximately 12% of the total O^+ intensity. The increase of O^+ intensity in the backward hemisphere reflects that the formation of O^+ proceeds via other collision geometries in addition to the collinear geometry.

4. $[Ar + H_2]^+$

The $[Ar + H_2]^+$ ion–molecule reaction system, which involves competition between charge transfer and chemical reactions, has been considered a prototype system for detailed theoretical study.[138–145]

a. $H_2^+(\tilde{X}, v') + Ar$. The $H_2^+ + Ar$ ion–molecule reaction has been the subject of studies using photoionization[1,146–148] and crossed-beam[149,150] techniques. These studies show that at sufficiently low $E_{c.m.}$, the collisions of H_2^+ with Ar give rise to product Ar^+ and ArH^+ corresponding to the charge-transfer reaction (53) and the proton-transfer reaction (54):

$$H_2^+(\tilde{X},v') + Ar \rightarrow H_2(X,v'') + Ar^+(^2P_{3/2,1/2}), \tag{53}$$

$$\rightarrow ArH^+ + H, \tag{54}$$

It has long been speculated that at a specific $E_{c.m.}$ the most important product channel of a charge-transfer reaction yields a product state in energy resonance with the reactant state.[151] The most interesting feature of the $[Ar + H_2]^+$ ion–molecule reaction system is that the energetics for the $H_2^+(\tilde{X}, v' = 2) + Ar(^1S_0)$ and $H_2(X, v'' = 0) + Ar^+(^2P_{1/2})$ states at their asymptotic limits differ only by 16 meV (Fig. 20).[19] Thus the $[Ar + H_2]^+$ charge-transfer system is an ideal model system for illustrating the energy resonance effect on the cross section of a charge-transfer reaction.

The previous photoionization studies[1,146–148] of the charge-transfer reactions $H_2^+(\tilde{X}, v' = 0–4) + Ar$ reveal that the cross sections at low $E_{c.m.}$ are peaked at $v' = 2$, which seems to support the energy resonance argument. Although there is general agreement that the energy resonance behavior is manifested in the vibrational-state-selected charge-transfer cross sections, quantitative agreement is not observed.[147,148] The inconsistency of

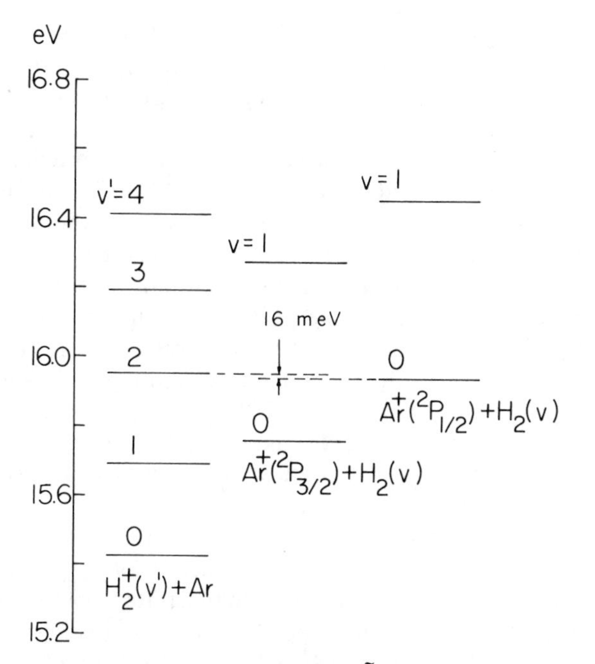

Figure 20. Energy diagram for the reactant $H_2^+(\tilde{X}, v') + Ar$ and charge-transfer product $Ar^+(^2P_{3/2,1/2}) + H_2(x, v'')$ states.[152]

experimental state-selected cross sections for reaction (53) and its reverse reaction, based on the consideration of microscopic reversibility, has been pointed out recently by Baer and Nakamura.[138]

Reaction (54) is exothermic by 1.30 eV.[16,19] The vibrational excitation of reactant H_2^+ has less effect on the cross section for reaction (54) compared to that for reaction (53).[146–149] Similar to the measurements of the cross section for the charge-transfer channel, quantitative agreement is also lacking between the vibrational-state-selected cross sections for reaction (58) reported in previous experiments.[147,148]

In order to shed light on the disagreements of previous experimental measurements and to provide accurate cross sections of this system for assisting the development of quantum-theoretical models, Liao et al.[152] have obtained detailed state-selected and state-to-state cross sections for reactions (53) and (54) and the collision-induced dissociation process

$$H_2^+(\tilde{X}, v') + Ar \rightarrow H^+ + H + Ar. \qquad (55)$$

I. CHARGE TRANSFER AND REACTION. Absolute vibrational-state-selected total cross sections for the charge-transfer reaction (53) $[\sigma_{v'}(Ar^+), v' = 0\text{–}4]$

and for the proton-transfer reaction (54) [$\sigma_{v'}(ArH^+), v' = 0$–4] measured in the $E_{c.m.}$ range of 0.48–95.3 eV are summarized in Tables I and II, respectively.[152] Experimental results reported previously[147,148] are also included in the tables.

Tanaka et al.[147] have measured vibrational-state-selected cross sections for the formation of Ar^+ and ArH^+ at $E_{c.m.} = 0.48, 0.77$, and 1.24 eV using the threshold PEPICO method. Taking into account the uncertainties of both studies, their values for $\sigma_{v'}(Ar^+)$, $\sigma_{v'}(ArH^+)$, and $\sigma_{v'}(Ar^+ + ArH^+), v' =$ 0–4, are in very good agreement with those shown in Table I and II. Absolute state-selected cross sections for reactions (53) and (54) have also been examined by Houle et al.[148] in the $E_{c.m.}$ range of 1–9 eV using the photoionization and rf octopole ion guide techniques. Their values for $\sigma_{v'}(Ar^+), \sigma_{v'}(ArH^+)$, and $\sigma_{v'}(Ar^+ + ArH^+), v' = 1$–4, are nearly twice those observed by Liao et al.[152] and Tanaka et al.[147] The relative values for $\sigma_{v'}(Ar^+), v' = 0$–5, at $E_{c.m.} = 20$ eV have been measured using the HeI PEPICO technique.[146]

The values for $\sigma_{v'}(Ar^+), \sigma_{v'}(ArH^+)$, and $\sigma_{v'}(Ar^+ + ArH^+), v' = 0$–4, obtained by Liao et al. are plotted as a function of $E_{c.m.}$ in the range of 0.48–95.2 eV in Figs. 21(*a*)–21(*e*). In order to illustrate the effect of vibrational excitation, the vibrational-state dependences for $\sigma_{v'}(Ar^+), \sigma_{v'}(ArH^+)$, and $\sigma_{v'}(Ar^+ + ArH^+)$ at $E_{c.m.} = 0.48, 0.95, 2.86, 8.57, 19.0$, and 47.6 eV are shown in Figs. 22(*a*)–22(*f*). The values for $\sigma_0(Ar^+)$ and $\sigma_1(Ar^+)$ are relatively low at $E_{c.m.} \leqslant 2$ eV, but both increase as $E_{c.m.}$ is increased. This behavior is attributed to the fact that the formation of Ar^+ by the charge-transfer reaction $H_2^+(\tilde{X}, v' = 0, 1) + Ar$ is endothermic. Because of the endothermicity of the $H_2^+(\tilde{X}, v' = 0) + Ar$ charge-transfer reaction, the values for $\sigma_0(Ar^+)$ remain lower than those for $\sigma_{v'}(Ar^+), v' \geqslant 1$, at all collision energies. The values for $\sigma_3(Ar^+)$ and $\sigma_4(Ar^+)$ are only weakly dependent on $E_{c.m.}$, while $\sigma_2(Ar^+)$ decreases monotonically with increasing $E_{c.m.}$. At $E_{c.m.} \leqslant 47.6$ eV, the values for $\sigma_2(Ar^+)$ are higher than those for other $\sigma_{v'}(Ar^+)$. The high values for $\sigma_2(Ar^+)$ are due to the close energy resonance of the $H_2^+(\tilde{X}, v = 2) + Ar$ and $Ar^+(^2P_{1/2}) + H_2(X, v'' = 0)$ states. The gradual decrease of $\sigma_2(Ar^+)$ with increasing $E_{c.m.}$ reflects the decreasing importance of the near resonance factor in governing the charge-transfer reaction at higher $E_{c.m.}$. At $E_{c.m.} \geqslant 47.6$ eV, the values for all $\sigma_{v'}(Ar^+), v' = 0$–4, converge to a range of 10–15 Å^2, and the energy resonance effect is no longer evident in the state-selected charge-transfer cross sections.

As shown in Figs. 21(*a*)–21(*e*), the collision energy dependences for $\sigma_{v'}(ArH^+), v' = 0$–4, are essentially the same. Namely, the values for $\sigma_{v'}(ArH^+), v' = 0$–4, are high at low $E_{c.m.}$ and they decrease rapidly as $E_{c.m.}$ is increased. This behavior for $\sigma_{v'}(ArH^+)$ is consistent with qualitative predictions based on the LGS orbiting model.[93,94] At $E_{c.m.} \geqslant 19$ eV,

TABLE I
Absolute Vibrational-State-Selected Cross Sections [$\sigma_{v'}(Ar^+)$]($\AA^2$) for the Charge-Transfer Reaction $H_2^+(\tilde{X}, v'=0\text{–}4) + Ar(^1S_0) \rightarrow H_2 + Ar^+$ in the $E_{c.m.}$ Range of 0.48–95.2 eV[a]

$E_{c.m.}$(eV)	$\sigma_0(Ar^+)$	$\sigma_1(Ar^+)$	$\sigma_2(Ar^+)$	$\sigma_3(Ar^+)$	$\sigma_4(Ar^+)$
0.48	2.0	5.3	27.5	14.6	13.6
	0.0±2.8[b]	3.3±2.8[b]	25.9±2.8[b]	10.8±2.8[b]	6.9±2.8[b]
	2.6[c]	24.0[c]	—	—	—
	0.2[g]	3.3[g]	31.2[g]	7.7[g]	12.0[g]
0.77	0.3±2.6[b]	3.6±2.8[b]	25.8±2.6[b]	14.2±2.6[b]	9.6±2.6[b]
	0.7[g]	4.2[g]	40.5[g]	8.3[g]	—
0.95	2.0	5.9	27.0	15.5	11.7
	3.1[d]	17.7[d]	47.2[d]	28.4[d]	24.4[d]
	1.9±0.5[e]	28.6±3.1[e]	22.4±3.6[e]	—	21.1±3.6[e]
	0.6[g]	6.6[g]	44.0[g]	—	—
1.24	1.0±3.4[b]	6.4±3.4[b]	28.4±3.4[b]	16.9±3.4[b]	9.4±3.4[b]
	0.8[g]	7.1[g]	—	—	—
1.90	1.9	5.9	23.4	13.9	12.1
2.86	2.4	8.0	23.1	14.0	14.0
	1.64[d]	20.6[d]	44.9[d]	30.0[d]	24.4[d]
	1.4±0.6[e]	25.3±2.1[e]	24.1±3.7[e]	—	—
3.81	2.9	9.2	23.0	14.7	12.8
5.71	3.6	11.8	21.2	15.8	13.6
	1.51[d]	24.8[d]	38.9[d]	30.1[d]	27.8[d]
	2.0±0.8[e]	30.3±4.2[e]	17.5±2.9[e]	—	—
8.57	4.4	14.4	21.4	17.2	13.1
	2.1[d]	26.3	38.3[d]	29.9[d]	22.8[d]
	2.1±0.8[e]	—	—	—	—
9.52	4.6	15.3	21.4	16.9	14.5
19.3	7.4	17.3	19.8	16.1	15.0
	1.6[f]	14.5[f]	26.8[f]	20.6	16.1[f]
47.6	9.1	15.3	15.6	13.9	13.2
	3.6[f]	13.2[f]	21.5[f]	19.6[f]	14.4[f]
95.2	8.5	14.7	13.8	12.9	10.8

[a]Reference 152. The uncertainties for absolute cross sections are estimated to be ⩽25%.
[b]Reference 147. Experimental cross sections obtained using the threshold PEPICO method.
[c]Reference 138. Theoretical cross sections calculated using the RIOSA.
[d]RSeference 148. Experimental cross sections obtained using the photoionization and rf octopole ion guide techniques. The data sets listed here with $E_{c.m.}$ = 0.95, 2.86, 5.71, and 8.57 eV correspond to $E_{c.m.}$ = 1.0, 3.0, 6.0, and 9.0 eV, respectively.
[e]Reference 141. Theoretical cross sections calculated using the TSH model. The data sets listed here with $E_{c.m.}$ = 0.95, 2.86, 5.71, and 8.57 eV correspond to $E_{c.m.}$ = 1.0, 3.0, 6.0, and 9.0 eV, respectively.
[f]Reference 152. Theoretical cross sections calculated using the NRIOSA.
[g]Reference 190.

TABLE II
Absolute Vibrational-State-Selected Cross Sections $[\sigma_{v'}(ArH^+)](Å^2)$ for the reaction $H_2^+(\tilde{X}, v'=0\text{–}4) + Ar(^1S_0) \rightarrow ArH^+ + H$ in the $E_{c.m.}$ Range of 0.48–95.2 eV[a]

$E_{c.m.}$(eV)	$\sigma_0(ArH^+)$	$\sigma_1(ArH^+)$	$\sigma_2(ArH^+)$	$\sigma_3(ArH^+)$	$\sigma_4(ArH^+)$
0.48	27.0	31.0	22.5	26.5	28.0
	27.4 ± 4.0[b]	34.0 ± 4.0[b]	21.3 ± 4.0[b]	32.3 ± 4.0[b]	36.4 ± 4.0[b]
	8.8[c]	20.0[c]	—	—	—
	24.5[f]	32.8[f]	12.9[f]	27.4[f]	28.0[f]
0.77	22.6 ± 4.7[b]	24.5 ± 4.7[b]	28.0 ± 4.7[b]	26.6 ± 4.7[b]	32.8 ± 4.7[b]
	17.8[f]	26.4[f]	19.9[f]	29.7[f]	—
0.95	25.6	28.0	22.0	23.3	25.0
	51.8[d]	66.1[d]	69.1[d]	67.7[d]	73.4[d]
	15.0 ± 1.3[e]	28.6 ± 3.1[e]	26.2 ± 2.9[e]	—	22.5 ± 2.6[e]
	17.8[f]	23.1[f]	22.0[f]	—	—
1.24	21.5 ± 5.4[b]	23.6 ± 5.4[b]	26.0 ± 5.4[b]	31.8 ± 5.4[b]	25.4 ± 5.4[b]
	16.4[f]	21.7[f]	—	—	—
1.90	21.0	23.0	17.0	20.5	21.0
2.86	16.0	19.0	15.0	18.0	16.0
	12.3[d]	30.4[d]	44.6[d]	38.9[d]	45.3[d]
	5.1 ± 0.8[e]	8.9 ± 1.3[e]	11.1 ± 1.7[e]	—	—
3.81	12.0	15.0	12.0	13.0	14.0
5.71	6.9	7.5	7.2	7.4	9.0
	5.0[d]	12.5[d]	18.2[d]	14.6[d]	14.0[d]
	4.1 ± 0.8[e]	4.4 ± 0.8[e]	4.1 ± 0.8[e]	—	—
8.57	3.6	2.3	2.4	2.1	2.1
	3.2[d]	8.9[d]	12.1[d]	10.8[d]	7.2[d]
	3.4 ± 0.7[e]	—	—	—	—
9.52	2.4	2.1	2.1	2.1	2.0
19.3	0.2	0.4	0.5	0.6	0.3
47.6	0.4	0.5	0.5	0.5	0.3
95.2	—	—	—	—	—

[a]Reference 151. The uncertainties for absolute cross sections are estimated to be ≤25%.
[b]Reference 147. Experimental cross sections obtained using the threshold PEPICO method.
[c]Reference 138. Theoretical cross sections calculated using the RIOSA.
[d]Reference 148. Experimental cross sections obtained using the photoionization and rf octopole ion guide techniques. The data sets listed here under $E_{c.m.} = 0.95$, 2.86, 5.71, and 8.57 eV correspond to $E_{c.m.} = 1.0$, 3.0, 6.0, and 9.0 eV, respectively.
[e]Reference 141. Theoretical cross sections calculated using the TSH model. The data sets listed here with $E_{c.m.} = 0.95$, 2.86, 5.71, and 8.57 eV correspond to $E_{c.m.} = 1.0$, 3.0, 6.0, and 9.0 eV, respectively.
[f]Reference 190.

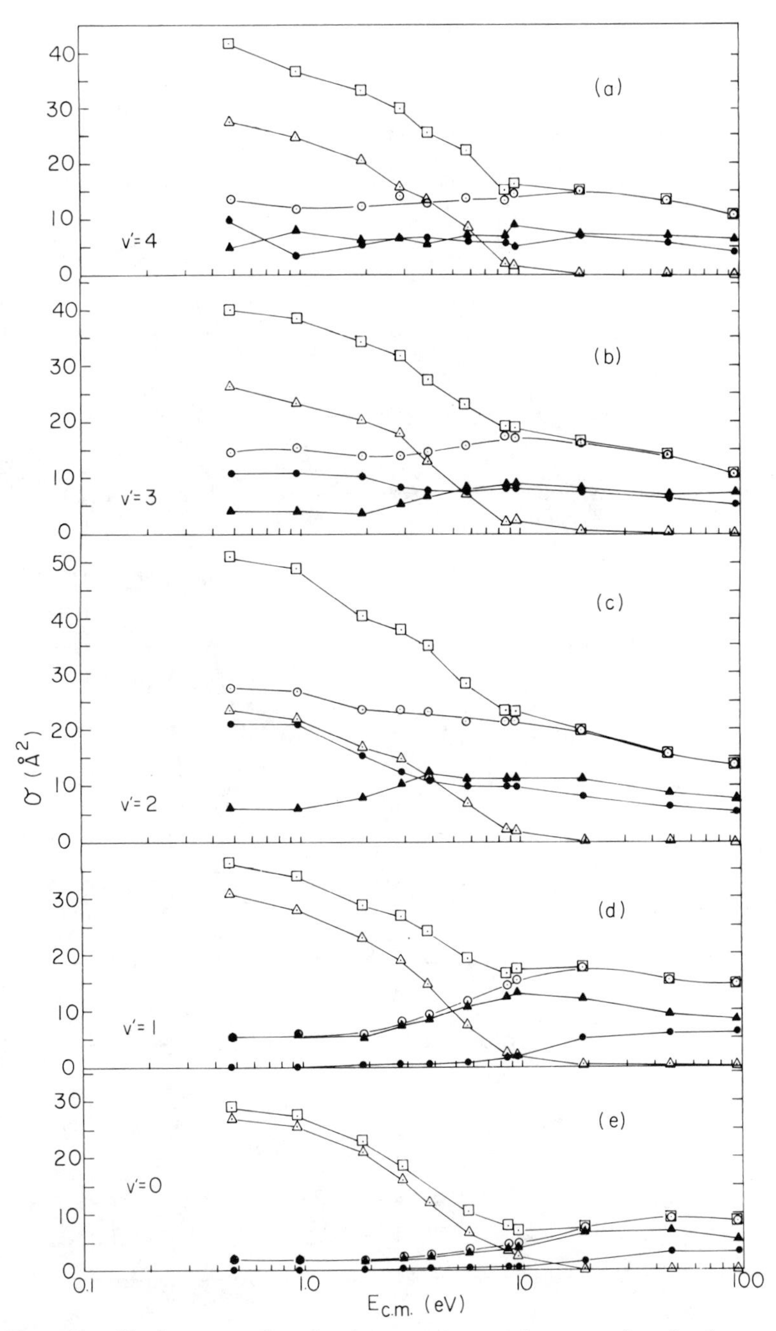

Figure 21. Absolute state-selected and state-to-state total cross sections for the reaction $H_2^+(\tilde{X}, v' = 0\text{–}4) + Ar$ plotted as a function of $E_{c.m.}$. (*a*) $v' = 4$; (*b*) $v' = 3$; (*c*) $v' = 2$; (*d*) $v' = 1$; and (*e*) $v' = 0$. (□) $\sigma_{v'}(ArH^+ + Ar^+)$; (△) $\sigma_{v'}(ArH^+)$; (⊙)$\sigma_{v'}(Ar^+)$; (●)$\sigma_{v' \to 1/2}$; (▲)$\sigma_{v' \to 3/2}$.[152]

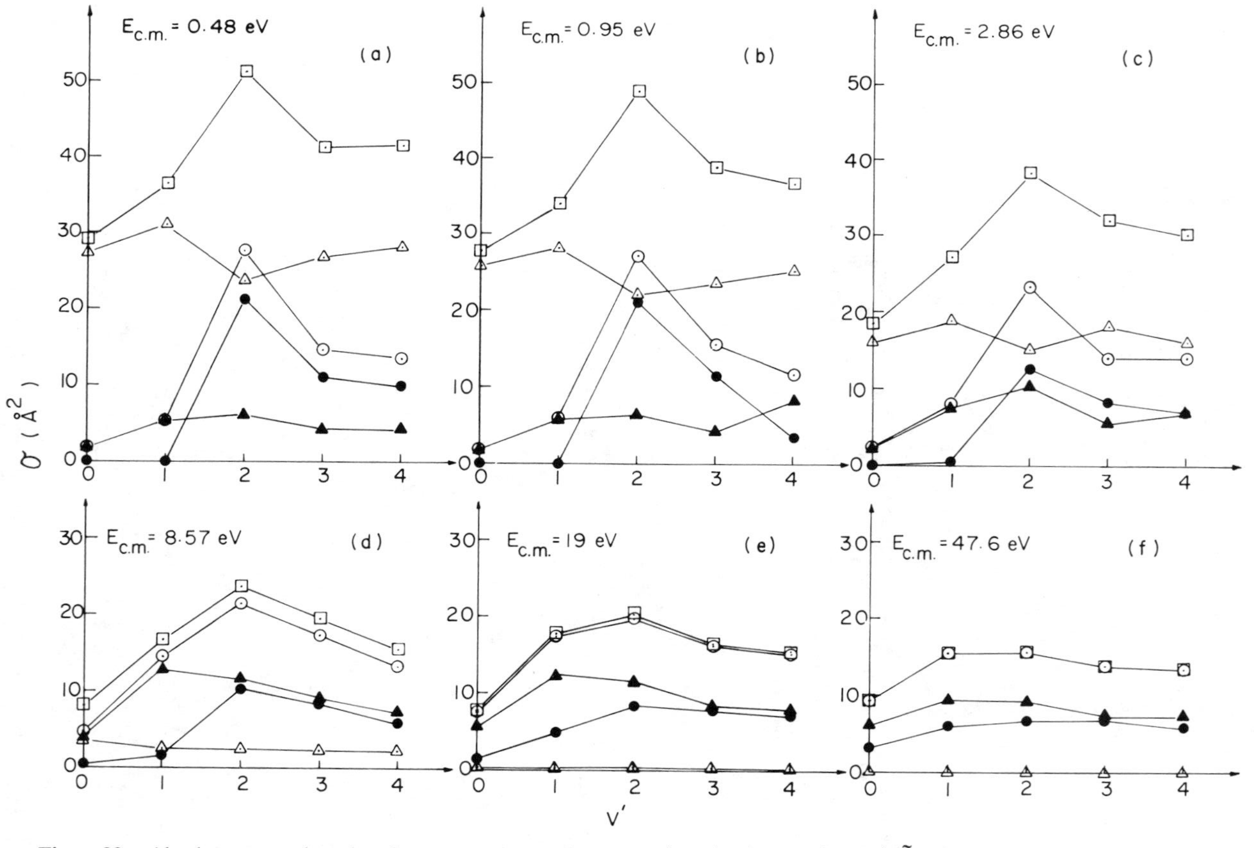

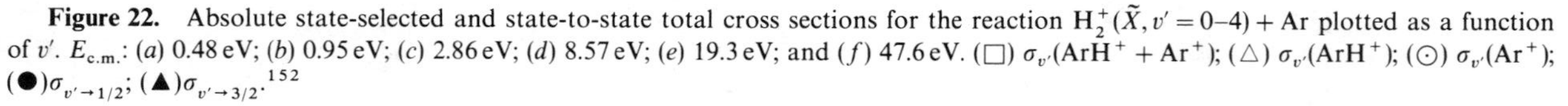

Figure 22. Absolute state-selected and state-to-state total cross sections for the reaction $H_2^+(\tilde{X}, v' = 0\text{–}4) + Ar$ plotted as a function of v'. $E_{c.m.}$: (*a*) 0.48 eV; (*b*) 0.95 eV; (*c*) 2.86 eV; (*d*) 8.57 eV; (*e*) 19.3 eV; and (*f*) 47.6 eV. (□) $\sigma_{v'}(ArH^+ + Ar^+)$; (△) $\sigma_{v'}(ArH^+)$; (⊙) $\sigma_{v'}(Ar^+)$; (●)$\sigma_{v' \to 1/2}$; (▲)$\sigma_{v' \to 3/2}$.[152]

$\sigma_{v'}(\mathrm{ArH}^+)$ are negligible compared to $\sigma_{v'}(\mathrm{Ar}^+)$. The plots in Figs. 22(*a*)–22(*f*) show that $\sigma_{v'}(\mathrm{ArH}^+)$ are only weakly dependent on vibrational excitation. Because of this, the vibrational-state dependences for $\sigma_{v'}(\mathrm{ArH}^+ + \mathrm{Ar}^+)$ have appearances similar to those for $\sigma_{v'}(\mathrm{Ar}^+)$, showing a maximum at $v' = 2$ in the $E_{\mathrm{c.m.}}$ range $\leqslant 19\,\mathrm{eV}$. However, the collision energy dependences for $\sigma_{v'}(\mathrm{ArH}^+ + \mathrm{Ar}^+)$ are dominated by the contributions of $\sigma_{v'}(\mathrm{ArH}^+)$.

At $E_{\mathrm{c.m.}} = 0.48\,\mathrm{eV}$, the vibrational-state dependence for $\sigma_{v'}(\mathrm{ArH}^+)$ dips at $v' = 2$, in agreement with the finding of the threshold PEPICO study.[147] However, contrary to the PEPICO measurements, Liao et al.[152] find that the dip at $v' = 2$ persists at $E_{\mathrm{c.m.}} \leqslant 2.86\,\mathrm{eV}$ [see Figs. 22(*a*)–22(*c*)]. Based on symmetry arguments, the ground $\mathrm{ArH}^+(^1\Sigma^+) + \mathrm{H}(^2S)$ product state correlates adiabatically with the reactant $\mathrm{H}_2^+(\tilde{X}\,^2\Sigma_g^+) + \mathrm{Ar}(^1S_0)$ state. At low $E_{\mathrm{c.m.}}$, the effect of long-range charge transfer from $\mathrm{H}_2^+(\tilde{X}\,^2\Sigma_g^+) + \mathrm{Ar}$ to form $\mathrm{Ar}^+ + \mathrm{H}_2(X\,^1\Sigma^+)$ is to decrease the cross section for reaction (54). Therefore, the enhancement of charge transfer for $v' = 2$ corresponds to the reduction in cross section for the formation of ArH^+. The finding that the vibrational-state dependences for $\sigma_{v'}(\mathrm{ArH}^+), v' = 2$–4, at $E_{\mathrm{c.m.}} = 0.48$ and $0.95\,\mathrm{eV}$ show the reverse trend compared to those for $\sigma_{v'}(\mathrm{Ar}^+), v' = 2$–4, is in accord with this interpretation.

The values for the fractions of charge transfer product Ar^+ formed in the $^2P_{1/2}$ state $(\mathrm{X}_{v' \to 1/2})$ at $E_{\mathrm{c.m.}} = 0.48$–$200\,\mathrm{eV}$ obtained using the crossed ion–neutral beam apparatus are depicted in Figs. 23(*a*)–23(*e*) for $v' = 0$–4. It is interesting that the profiles for $X_{v' \to 1/2}$ and $\sigma_{v'}(\mathrm{Ar}^+), v' = 0$–4, for the corresponding v' are similar. The values for $X_{v' \to 1/2}, v' = 0$ and 1, at $E_{\mathrm{c.m.}} < 1.9\,\mathrm{eV}$ are very small because the formation of $\mathrm{Ar}^+(^2P_{1/2}) + \mathrm{H}_2(X\,^1\Sigma_g^+, v'' = 0)$ by the charge-transfer reaction (53) is exothermic by $0.50\,\mathrm{eV}$ for $v' = 0$ and $0.23\,\mathrm{eV}$ for $v' = 1$. As $E_{\mathrm{c.m.}}$ is increased, the values for both $X_{0 \to 1/2}$ and $X_{1 \to 1/2}$ increase to ≈ 0.4 at $E_{\mathrm{c.m.}} \geqslant 95.2\,\mathrm{eV}$. The most interesting results are those for $v' = 2$ where close energy resonance exists between the $\mathrm{H}_2^+(\tilde{X}, v' = 2) + \mathrm{Ar}$ and $\mathrm{Ar}^+(^2P_{1/2}) + \mathrm{H}_2(X, v'' = 0)$ vibronic states and the value for $X_{2 \to 1/2}$ is expected to be higher than 0.5. Liao et al. find that at low $E_{\mathrm{c.m.}}$ ($\leqslant 1.9\,\mathrm{eV}$), the formation of $\mathrm{Ar}^+(^2P_{1/2})$ is indeed greater than that of $\mathrm{Ar}^+(^2P_{3/2})$. However, the value for $X_{2 \to 1/2}$ decreases from about 0.8 at $E_{\mathrm{c.m.}} = 0.48$–$0.95\,\mathrm{eV}$ to the range of 0.41–0.47 at $E_{\mathrm{c.m.}} > 4\,\mathrm{eV}$. The measured values for $X_{2 \to 1/2}$ support the conclusion that close energy resonance is the dominant factor governing the charge-transfer product channel only at relatively low collision energies. As $E_{\mathrm{c.m.}}$ is increased to a value $> 5\,\mathrm{eV}$, the charge-transfer reaction seems to be governed by the energetics consideration, which favors the formation of $\mathrm{Ar}^+(^2P_{3/2})$ corresponding most likely to the lower-energy reaction product channel $\mathrm{Ar}^+(^2P_{3/2}) + \mathrm{H}_2(X, v'' = 0)$. Although the experimental uncertainties for $X_{3 \to 1/2}$ at $E_{\mathrm{c.m.}} \leqslant 1.9\,\mathrm{eV}$ are relatively large, the results indicate that the intensities for product $\mathrm{Ar}^+(^2P_{1/2})$ are greater

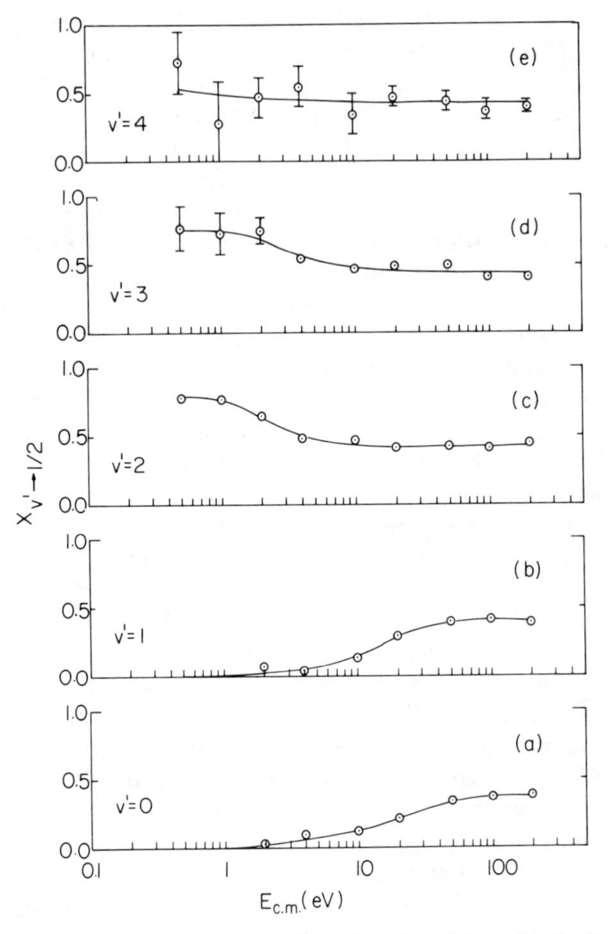

Figure 23. The values for $X_{v' \to 1/2}$ plotted as a function of $E_{c.m.}$. (*a*) $v' = 0$; (*b*) $v' = 1$; (*c*) $v' = 2$; (*d*) $v' = 3$; and (*e*) $v' = 4$. $X_{v' \to 1/2}$ is defined to be $\sigma_{v' \to 1/2}/(\sigma_{v' \to 3/2} + \sigma_{v' \to 1/2})$.[152]

than those for product $Ar^+(^2P_{3/2})$ at these $E_{c.m.}$. Similar to the collision energy dependence for $X_{2 \to 1/2}$, the value for $X_{3 \to 1/2}$ drops below 0.5 at $E_{c.m.} > 4$ eV. At $E_{c.m.} = 19$–119 eV, the values for $X_{3 \to 1/2}$ vary in the range of 0.40–0.49. It appears that $X_{4 \to 1/2}$ is insensitive to $E_{c.m.}$ and has a value close to 0.5. Again, the values, for $X_{4 \to 1/2}$ at $E_{c.m.} > 4$ eV are approximately equal to 0.4.

By definition, $X_{v' \to 1/2}$ is $\sigma_{v' \to 1/2}/(\sigma_{v' \to 3/2} + \sigma_{v' \to 1/2})$ or $\sigma_{v' \to 1/2}/\sigma_{v'}(Ar^+)$, where $\sigma_{v' \to 3/2}(= \Sigma\sigma_{v' \to 3/2v''})$ amd $\sigma_{v' \to 1/2}(= \Sigma\sigma_{v' \to 1/2v''})$ are partial state-to-state cross sections for reaction (53). Using the known values for

TABLE III

Absolute Partial State-to-State Total Cross Sections $[\sigma_{v'\to J}(=\sum\sigma_{v'\to Jv''}), J=3/2, 1/2]$ for the Charge-Transfer Reaction $H_2^+(\tilde{X}, v'=0\text{–}4) + Ar(^1S_0) \to H_2(X, v'') + Ar^+(^2P_{3/2,1/2})$ in the $E_{c.m.}$ Range of 0.48–95.2 eV

	Absolute Partial State-to-State Cross Sections (Å^2)[a]				
$E_{c.m.}$(eV)	$\sigma_{0\to3/2,1/2}$	$\sigma_{1\to3/2,1/2}$	$\sigma_{2\to3/2,1/2}$	$\sigma_{3\to3/2,1/2}$	$\sigma_{4\to3/2,1/2}$
0.48	2.0(0.2)[b]	5.3(3.1)[b]	6.0(1.0)[b]	4±3(16)[b]	4±3(5.4)[b]
	0.0(—)[c]	0.0(0.2)[b]	21.5(30.6)[b]	11±3(6.1)[b]	10±3(5.0)[b]
0.95	2.0(0.6)[b]	5.9(6.2)[b]	6.3(3.0)[b]	4±1.5	8.2±3.5
	0.0(0.1)[b]	0.0(0.4)[b]	21.3(40.8)[b]	11.5±1.5	3.5±3.5
1.90	1.8	5.5	8.2	3.6±1.3	6.3±1.9
	0.1	0.4	15.2	10.3±1.3	5.8±1.9
2.86	2.3	7.6	10.4	5.6	7.0±2.1
	0.1	0.4	12.7	8.4	7.0±2.1
3.81	2.6	8.7	12.1	6.9	5.9±1.9
	0.3	0.5	10.9	7.8	6.9±1.9
5.71	3.2	11.0	11.2	7.7	7.5±2.0
	0.4	0.8	10.0	8.1	6.1±2.0
8.57	3.9	12.6	11.3	8.9	7.2±2.0
	0.5	1.8	10.1	8.3	5.9±2.0
9.52	4.0	13.3	11.3	9.0	9.4±2.2
	0.6	2.0	10.1	7.9	5.1±2.2
19.3	5.8(1.2)[d]	12.3(11.9)[d]	11.5(19.8)[d]	8.3(12.2)[d]	7.9(9.5)[d]
	1.6(0.5)[d]	5.0(2.6)[d]	8.3(7.1)[d]	7.8(8.4)[d]	7.1(6.6)[d]
47.6	6.0(2.8)[d]	9.3(10.3)[d]	9.0(15.0)[d]	7.1(12.7)[d]	7.4(8.1)[d]
	3.1(0.85)[d]	6.0(3.1)[d]	6.6(6.2)[d]	6.8(7.3)[d]	5.8(6.7)[d]
95.2	5.4	8.6	8.1	7.6	6.8
	3.1	6.1	5.7	5.3	4.0

[a]Reference 152. At each $E_{c.m.}$, values for $\sigma_{v'\to3/2}$ and $\sigma_{v'\to1/2}$ are listed in the first and second rows, respectively. The overall uncertainties are estimated conservatively to be approximately ≤25% for $v'=0$, 1, and 2 and ≤30% for $v'=4$.

[b]Reference 190. Theoretical cross sections calculate using the complex RIOSA.

[c]Population of the product $Ar^+(^2P_{1/2})$ state is not allowed due to energy constraint.

[d]Reference 152. Theoretical cross sections calculated using the NRIOSA.

$X_{v'\to1/2}$ and $\sigma_{v'}(Ar^+)$, Liao et al. have calcuated the values for $\sigma_{v'\to3/2}$ and $\sigma_{v'\to1/2}$, as summarized in Table III. The plots for $\sigma_{v'\to3/2,1/2}, v'=0\text{–}4$, as a function of $E_{c.m.}$ shown in Figs. 21(*a*)–21(*e*) and the collision energy dependence for $X_{v'\to1/2}$ depicted in Figs. 23(*a*)–23(*e*) contain similar information. The values for $\sigma_{v'\to3/2,1/2}, v'=0\text{–}4$, at $E_{c.m.}=0.48, 0.95, 2.86, 8.57, 19$, and 47.6 eV are also plotted as a function of v' in Figs. 22(*a*)–22(*f*). It is clear from these figures that $\sigma_{v'\to3/2}$ is only weakly dependent on vibrational excitation, and that at $E_{c.m.} \leqslant 2.86\,\text{eV}$ the prominent structure observed for $\sigma_{v'}(\text{Ar}^+)$ are mainly due to $\sigma_{v' \to 1/2}$.

The structure showing that $\sigma_{v' \to 1/2}$ is peaked at $v' = 2$ is discernible at $E_{c.m.} = 0.48$–$19\,\text{eV}$. This observation seems to indicate that the factor of close energy resonance has an effect on the charge-transfer reaction (54) for $E_{c.m.}$ up to 19 eV. Such a conclusion probably is not justified because the low cross sections for $v' = 0$ and 1 are mainly due to the endothermic nature of the charge-transfer reactions between $\text{H}_2^+(\tilde{X}, v' = 0, 1)$ and Ar. Although the values for $\sigma_{v' \to 1/2}$ are slightly higher than those for $\sigma_{v' \to 1/2}$, $v' = 3, 4$, at $E_{c.m.} = 3$–$19\,\text{eV}$, the lower values for $\sigma_{v' \to 1/2}$, $v' = 3, 4$, may be the consequence of higher cross sections for the collision-induced dissociation of $\text{H}_2^+(\tilde{X}, v' = 3, 4)$ compared to that of $\text{H}_2^+(\tilde{X}, v' = 2)$.

II. DISSOCIATION. Figure 24 shows the absolute total cross sections for the formation of $\text{H}^+[\sigma_{v'}(\text{H}^+)]$ in the collisions of $\text{H}_2^+(\tilde{X}, v' = 0$–$4)$ with Ar. The thermochemical thresholds for the collision-induced dissociation of $\text{H}_2^+(\tilde{X}, v')$, $v' = 0, 1, 2, 3$, and 4, are 2.65, 2.36, 2.07, 1.79, and 1.50 eV, respectively.[16,19] The AEs observed for $\sigma_{v'}(\text{H}^+)$ agree with these values. The $\sigma_{v'}(\text{H}^+)$ curves

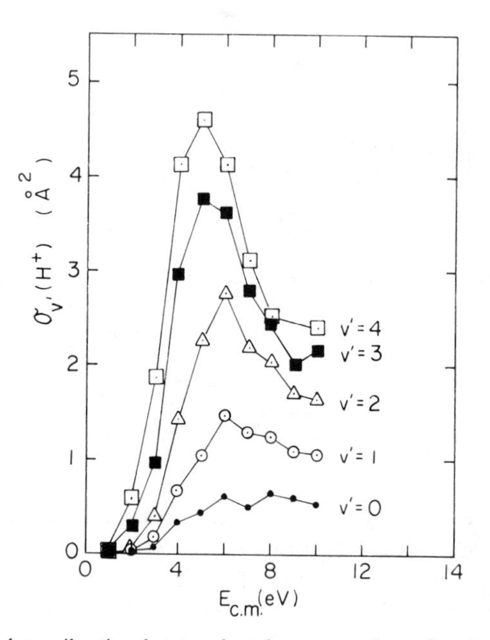

Figure 24. Absolute vibrational-state-selected cross sections for the collision-induced dissociation process $\text{H}_2^+(\tilde{X}, v' = 0$–$4) + \text{Ar} \to \text{H}^+ + \text{H} + \text{Ar}$ in the $E_{c.m.}$ range of 1–10 eV. (●) $v' = 0$; (⊙) $v' = 1$; (△) $v' = 2$; (■) $v' = 3$, (□) $v' = 4$.[152]

exhibit a maximum in the $E_{c.m.}$ range of 4–8 eV. The $E_{c.m.}$ values corresponding to these maxima decrease as v' is increased. As expected, the value for $\sigma_{v'}(H^+)$ increases successively as v' is varied from 0 to 4. At the maxima of the $\sigma_{v'}(H^+)$ curves, the values for $\sigma_{v'}(H^+)$ are $\approx 20\%$ of those for $\sigma_{v'}(H_2^+)$. The decrease of $\sigma_{v'}(H^+)$ at higher $E_{c.m.}$ is anticipated due to the shortening of the interaction time for the colliding pair. It appears that all $\sigma_{v'}(H^+)$ become nearly constant at $E_{c.m.} = 8$–10 eV with values ranging fromapproximately 2.5 Å^2 for $v' = 4$ to 0.5 Å^2 for $v' = 0$.

b. $Ar^+(^2P_{3/2,1/2}) + H_2$. The ion–molecule reactions,

$$Ar^+(^2P_{3/2,1/2}) + H_2(X, v'' = 0) \rightarrow Ar(^1S_0) + H_2^+(\tilde{X}, v'), \quad (56)$$

$$\rightarrow ArH^+ + H, \quad (57)$$

have been the focus of numerous experimental studies[1,12,17,57,90,149,153–187] in the past 30 years. The photoionization[1] and PEPICO[17,57,183,184,186] studies provide valuable information about the spin–orbit effects on the cross sections for reactions (56) and (57). The recent guided-ion-beam mass spectrometric experiments[12,184] have determined detailed absolute total cross sections for reaction (57) in the $E_{c.m.}$ range of 0.01–30 eV.

The spin–orbit excitation and relaxation processes induced by the collisions of $Ar^+(^2P_{3/2,1/2})$ with H_2, reactions (58) and (59), respectively, have been examined previously by the energy-loss spectroscopic method[185]:

$$Ar^+(^2P_{3/2}) + H_2(X, v'' = 0) \rightarrow Ar^+(^2P_{1/2}) + H_2(X, v'') \quad (58)$$

$$Ar^+(^2P_{1/2}) + H_2(X, v'' = 0) \rightarrow Ar^+(^2P_{3/2}) + H_2(X, v'') \quad (59)$$

Detailed state-selected and state-to-state cross sections for reactions (56), (57), (58), and (59), as well as the dissociation process,

$$Ar^+(^2P_{3/2,1/2}) + H_2(X, v'' = 0) \rightarrow H^+ + H + Ar, \quad (60)$$

have been obtained most recently by Liao et al.[22]

I. CHARGE TRANSFER AND REACTION. Values for the relative spin–orbit state-selected cross sections for the $Ar^+(^2P_{3/2,1/2}) + H_2$ charge-transfer reactions have been reported by several group.[17,57,183,184,186] The values measured for the ratios, $\sigma_{1/2}(H_2^+)/\sigma_{3/2}(H_2^+)$, using the TQDO and crossed ion–neutral beam photoionization apparatuses,[22] and the results of Tanaka et al.[183] and Latimer and co-workers[57,186] in the $E_{c.m.}$ range of 0.1–19.1 eV are compared in Fig. 25.

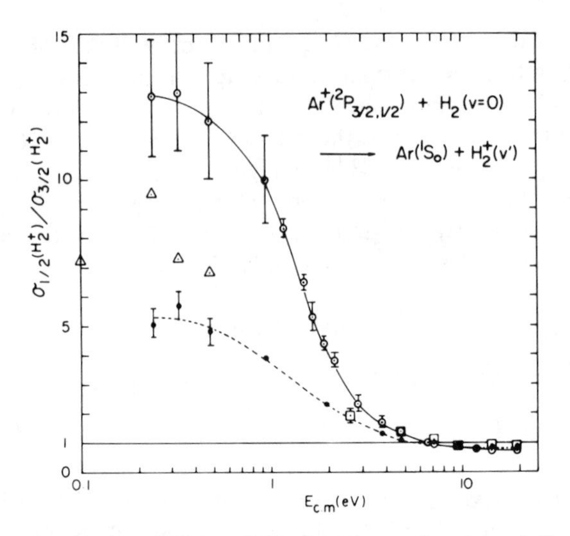

Figure 25. Values for $\sigma_{1/2}(H_2^+)/\sigma_{3/2}(H_2^+)$ plotted as a function of $E_{c.m.}$. (●) Values measured using the TQDO photoionization apparatus and (⊙) values measured using the crossed ion–neutral beam apparatus, Ref. 22; (△) Refs. 57 and 186.

At $E_{c.m.} = 0.24$–$0.48\,eV$, the values for the ratio $\sigma_{1/2}(H_2^+)/\sigma_{3/2}(H_2^+)$ measured using the TQDO photoionization apparatus are in the range of 4.8–5.7.[22] The values for the ratio reported for a similar $E_{c.m.}$ range by Tanaka et al.[183] and by Henri et al.[184] are higher than these values. After taking into account the experimental uncertainties, the values reported by Tanaka et al. agree with the results of the measurements using the TQDO apparatuses. The high values for the ratio at $E_{c.m.} \leqslant 5\,eV$, can be attributed to the energy resonance effect of the $Ar^+(^2P_{1/2}) + H_2(X, v''=0)$ and $H_2^+(\tilde{X}, v'=2) + Ar$ vibronic states. As $E_{c.m.}$ is increased, the couplings of the reactant $Ar^+(^2P_{1/2}) + H_2(X, v''=0)$ states with charge-transfer states other than the near-resonance states become more important. This may account for the drop in the value for $\sigma_{1/2}(H_2^+)/\sigma_{3/2}(H_2^+)$ with increasing $E_{c.m.}$. At $E_{c.m.} \geqslant 6\,eV$, the values obtained by the TQDO photoionization apparatuses agree with those reported by Latimer and co-workers.[57,186] The ratio essentially levels off with a value of 0.8 ± 0.1 at $E_{c.m.} \geqslant 9\,eV$. Owing to the higher collection efficiency for H_2^+ formed in near-resonance charge-transfer channels, the values measured at low $E_{c.m.}$ using the crossed ion–neutral beam apparatus represent upper limits.

Absolute total spin–orbit state-selected cross sections for reactions (56) measured using the TQDO apparatus[22] and those obtained previously by Tanaka et al.[180] and Henri et al.[181] are compared in Table IV. The absolute

TABLE IV
Absolute Spin–Orbit State-Selected Cross Sections [$\sigma_{3/2,1/2}(H_2^+)$] ($Å^2$) for the Charge-Transfer Reaction $Ar^+(^2P_{3/2,1/2})+H_2(X, v''=0) \rightarrow H_2^+ + Ar$ in the $E_{c.m.}$ Range of 0.05–19.3 eV[a]

$E_{c.m.}$(eV)	$\sigma_{3/2}(H_2^+)$	$\sigma_{1/2}(H_2^+)$	$\sigma_{1/2}(H_2^+)/\sigma_{3/2}(H_2^+)$
0.05	0.8±0.45[b]	5.0±2.5[b]	6.2[b]
0.10	0.5[b]	3.6[b]	7.2[b]
0.24	3.0	15.3	5.1
	0.4[b]	3.8[b]	9.5[b]
	—	—	10.0[c]
0.33	2.9	16.5	5.7
	0.4[b]	2.9[b]	7.3[b]
0.48	3.4	16.2	4.8
	0.4[b]	2.7[b]	6.8[b]
	2.9[d]	21.0[d]	7.1[d]
0.69	4.8[d]	28.5[d]	5.9[d]
0.95	4.0	15.6	3.9
1.50	—	—	6.0[c]
1.90	6.9	16.0	2.3
2.86	—	—	1.9±0.3[e]
3.81	9.2	12.1	1.3
4.76	9.3	9.7	1.04
	—	—	1.3[e]
7.14	—	—	1.10[e]
9.52	10.7	9.1	0.85
	—	—	0.87[e]
11.9	10.7	8.0	0.75
14.3	10.7	8.6	0.81
	—	—	0.87[e]
19.3	11.6	9.6	0.83
	—	—	0.82[e]
	22.1[f]	21.6[f]	0.98[f]

[a]Reference 22. The overall uncertainties for absolute cross sections are estimated to be ⩽25%.
[b]Reference 183. Typical uncertainties for absolute cross sections are similar to those indicated for cross sections at $E_{c.m.} = 0.05$ eV.
[c]Reference 17.
[d]Reference 184.
[e]Reference 54 and 186.
[f]Reference 22. Theoretical cross sections predicted by the NRIOSA.

cross sections reported by Tanaka et al. are substantially lower than those obtained by Liao et al. and Henri et al.

The absolute total spin–orbit state-selected cross sections determined using the TQDO apparatus are plotted in Fig. 26 as a function of $E_{c.m.}$ in the range of 0.24–19.3 eV. The value for $\sigma_{3/2}(H_2^+)$ is equal to about 3 $Å^2$ at

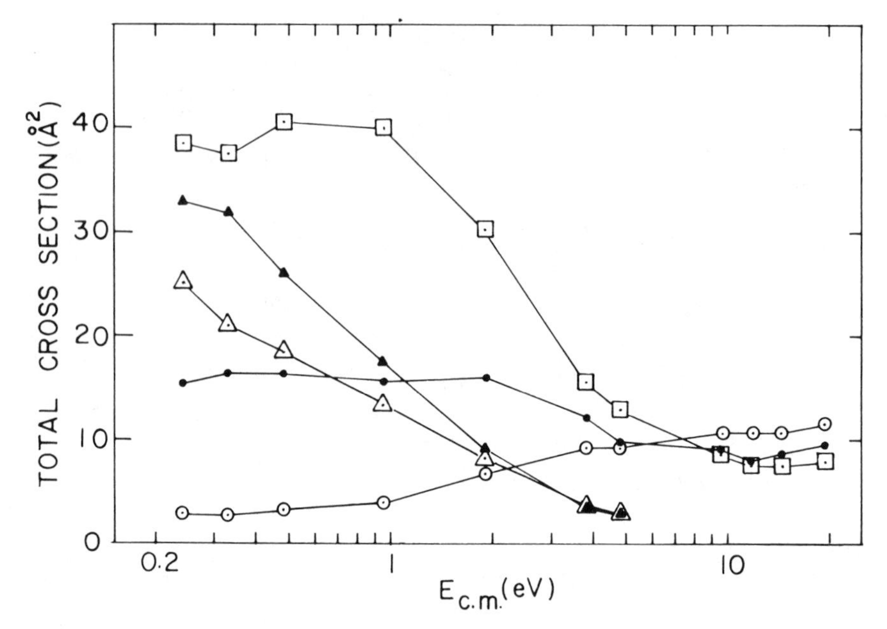

Figure 26. Total spin–orbit state-selected cross sections. (●) $\sigma_{3/2}(H_2^+)$; (⊙) $\sigma_{3/2}(H_2^+)$; (▲) $\sigma_{3/2}(ArH^+)$; (△) $\sigma_{1/2}(ArH^+)$; (□) upper limits for $\sigma_{1/2}(H_2^+)$.[22]

$E_{c.m.} = 0.24$–0.48 eV and increases gradually to 10–11 $\mathrm{\AA}^2$ at $E_{c.m.} \geqslant 4$ eV. The value for $\sigma_{1/2}(H_2^+)$ is 15–16 $\mathrm{\AA}^2$ at $E_{c.m.} = 0.24$–2 eV and decreases to 8–10 $\mathrm{\AA}^2$ at $E_{c.m.} \geqslant 5$ eV. The estimates for $\sigma_{1/2}(H_2^+)$, calculated using values for $\sigma_{1/2}(H_2^+)/\sigma_{3/2}(H_2^+)$ measured with the crossed ion–neutral beam apparatus and the absolute values for $\sigma_{3/2}(H_2^+)$, are also shown in the figure. These estimates corresponding to $E_{c.m.} \leqslant 2$ eV should be considered as upper limits for $\sigma_{1/2}(H_2^+)$.

The fractions of H_2^+ ions formed in the v' state by the reaction $Ar^+(^2P_J) + H_2(X, v''=0)$, $X_{J0\to v'}(=X_{J\to v'}) = \sigma_{J0\to v'}/\sigma_{J0}(H_2^+)\ [=\sigma_{J\to v'}/\sigma_J(H_2^+)]$, measured at $E_{c.m.} = 0.16$–19.3 eV are shown in Figs. 27(*a*)–27(*h*) for $J = 3/2$ and Figs. 28(a)–28(*h*) for $J = 1/2$. At low $E_{c.m.}$ ($\leqslant 0.54$ eV), $X_{3/2\to v'}$ has a maximum at $v' = 0$. However, the values for $X_{3/2\to v'}$, $v' = 0$ and 1, are essentially the same after taking into account the experimental uncertainties. In the $E_{c.m.}$ range of 2.0–19.3 eV, the vibrational distributions for $X_{3/2\to v'}$ are quite flat with a slight maximum at about $v' = 1$–2. Contrary to the observation of $X_{3/2\to v'}$ the vibrational distributions for $X_{1/2\to v'}$ peak sharply at $v' = 2$, except that for $E_{c.m.} = 19.3$ eV, where $X_{1/2\to v'}$ depends only mildly on v'. The strong peaking of $X_{1/2\to v'}$ at $v' = 2$ shows unambiguously that charge-transfer product H_2^+ of the reaction $Ar^+(^2P_{1/2}) + H_2(X, v''=0)$ are

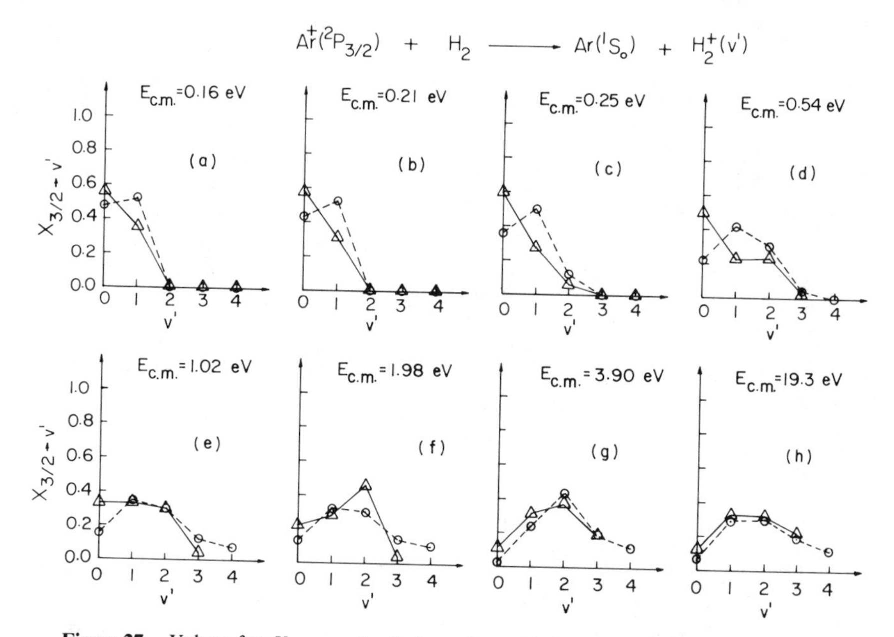

Figure 27. Values for $X_{3/2\to v'}$, $v' = 0$–4, at $E_{\text{c.m.}}$: (*a*) 0.16 eV; (*b*) 0.21 eV; (*c*) 0.25 eV; (*d*) 0.54 eV; (*e*) 1.02 eV; (*f*) 1.98 eV; (*g*) 3.9 eV; (*h*) 19.3 eV. (△) Experimental values; (⊙) predicted values using the microscopic reversibility relation [Eq. (67)]. See text.[22]

formed preferentially in the closest energy resonance channel, namely $H_2^+(\tilde{X}, v' = 2) + Ar$.

Table VI lists the absolute values for $\sigma_{3/2\to v'}$ and $\sigma_{1/2\to v'}$, $v' = 0$–3, calculated from the values for $X_{3/2\to v'}$, $X_{1/2\to v'}$, $\sigma_{3/2}(H_2^+)$, and $\sigma_{1/2}(H_2^+)$. With the exception of $\sigma_{1/2\to 2}$, which decreases rapidly as $E_{\text{c.m.}}$ is increased beyond 2 eV, weak collision energy dependences are observed for $\sigma_{3/2\to v'}$ and $\sigma_{1/2\to v'}$. At $E_{\text{c.m.}} \leqslant 2$ eV, $\sigma_{1/2\to 2}(H_2^+)$ is clearly the predominant component of $\sigma_{1/2}(H_2^+)$.

The total spin–orbit state-selected absolute cross sections for the formation of ArH^+ by reaction (57) [$\sigma_{3/2,1/2}(ArH^+)$] obtained using the TQDO photoionization apparatus in the $E_{\text{c.m.}}$ range of 0.24–4.76 eV, and those reported by Tanaka et al.,[183] are listed in Table V. The absolute values for $\sigma_{3/2,1/2}(ArH^+)$ at $E_{\text{c.m.}} = 0.24$–0.48 eV determined by Tanaka et al. are greater than those obtained by Liao et al.[22] The values for $\sigma_{1/2}(ArH^+)/\sigma_{3/2}(H_2^+)$ at $E_{\text{c.m.}} = 0.24$–0.48 eV found by Liao et al. are in the range of 1.34–1.54, as compared to the range of 1.54–1.71 obtained by Tanaka et al. In a single gas cell photoionization experiment, Chupka and Russell have measured a value of 1.3 for the ratio at nominally zero kinetic energy.[1]

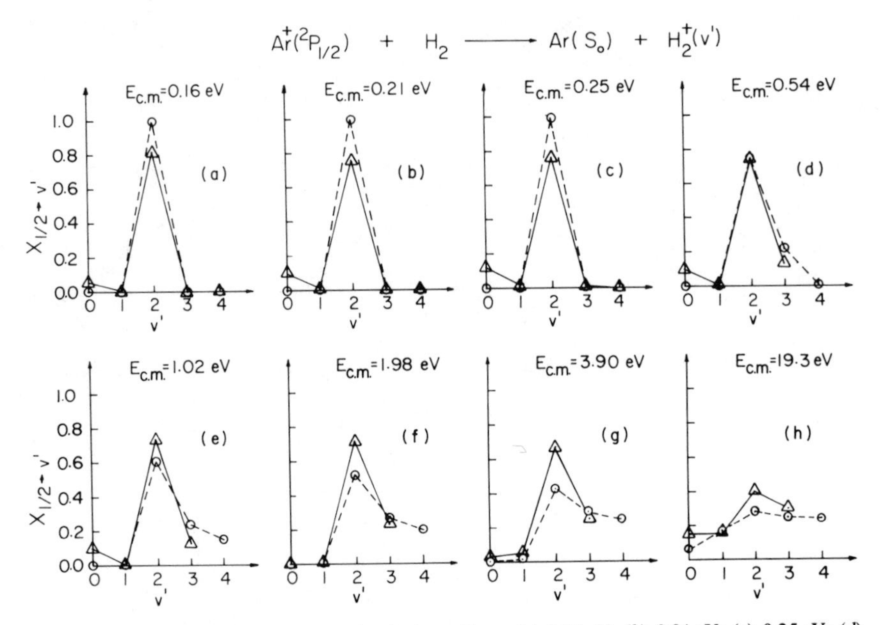

Figure 28. Values for $X_{1/2\to v'}$, $v' = 0$–4, at $E_{\text{c.m.}}$: (*a*) 0.16 eV; (*b*) 0.21 eV; (*c*) 0.25 eV; (*d*) 0.54 eV; (*e*) 1.02 eV; (*f*) 1.98 eV; (*g*) 3.9 eV; (*h*) 19.3 eV. (△) Experimental values; (⊙) predicted values using the microscopic reversibility relation [Eq. (67)]. See text.[22]

TABLE V

Absolute Spin–Orbit State-Selected Cross Sections $[\sigma_{3/2,1/2}(ArH^+)]$ ($Å^2$) for the Reaction $Ar^+(^2P_{3/2,1/2}) + H_2(X, v''=0) \to ArH^+ + H$ in the $E_{\text{c.m.}}$ Range of 0.24–4.76 eV[a]

$E_{\text{c.m.}}$(eV)	$\sigma_{3/2}(ArH^+)$	$\sigma_{1/2}(ArH^+)$	$\sigma_{1/2}(ArH^+)/\sigma_{3/2}(ArH^+)$
0.24	25	33	1.34
	33[b]	55[b]	1.67[b]
0.33	21	32	1.54
	22.5[b]	38.5[b]	1.71[b]
0.48	18.4	26	1.41
	22[b]	34[b]	1.54[b]
0.95	13.5	17.6	1.30
1.90	8.2	9.0	1.09
3.81	3.9	3.7	0.95
4.76	3.0	2.9	0.97

[a]Reference 22. The overall uncertainties for absolute cross sections are estimated to be ≤ 25%.

[b]Reference 183.

TABLE VI
Absolute Partial State-to-State Total Cross Sections ($\sigma_{3/2\to v'}$, $\sigma_{1/2\to v'}$) ($Å^2$) for the reactions $Ar^+(^2P_{3/2,1/2}) + H_2(X, v''=0) \to Ar + H_2^+(\tilde{X}, v')$ in the $E_{c.m.}$ Range of 0.21–19.3 eV[a]

$E_{c.m.}$(eV)	$\sigma_{3/2,1/2\to 0}$	$\sigma_{3/2,1/2\to 1}$	$\sigma_{3/2,1/2\to 2}$	$\sigma_{3/2,1/2\to 3}$
0.21	2.0	1.0	—[b]	—[b]
	1.9	0.3	13.1	—[b]
0.25	2.0	0.9	0.2	—[b]
	2.2	0.5	12.6	0.0
0.54	1.7	0.8	0.8	0.1
	1.6	0.3	12.0	2.3
1.02	1.3	1.3	1.2	0.2
	1.6	0.5	11.5	2.0
1.98	1.5	1.9	3.2	0.3
	0.3	0.5	11.5	3.7
3.90	1.0	2.9	3.4	2.1
	0.4	0.6	7.5	3.8
19.3	1.4(0.4)[c]	3.8(5.1)[c]	3.7(9.0)[c]	2.7(4.0)[c]
	1.4(0.2)[c]	1.6(1.8)[c]	3.8(5.6)[c]	2.9(6.2)[c]

[a]Reference 22. At each $E_{c.m.}$, the values for $\sigma_{3/2\to v'}$ and $\sigma_{1/2\to v'}$ are listed in the first and second rows, respectively. The uncertainties for $\sigma_{3/2\to v'}$ at $E_{c.m.} = 0.21$–1.02 eV are estimated to be $\leqslant 50\%$. The uncertainties for all other values are estimated to be $\leqslant 35\%$.

[b]Population of these H_2^+ vibrational states are not allowed due to energy constraint.

[c]Reference 22. Theoretical cross section obtained using the NRIOSA.

The TQDO values for $\sigma_{3/2,1/2}(ArH^+)$ at $E_{c.m.} = 0.24$–4.76 are also shown in Fig. 26. Both $\sigma_{3/2}(ArH^+)$ and $\sigma_{1/2}(ArH^+)$ decrease rapidly with increasing $E_{c.m.}$. At $E_{c.m.} \geqslant 2$ eV, they are essentially identical. In the $E_{c.m.}$ range of 0.05–0.5 eV, Tanaka et al.[183] find that $\sigma_{3/2,1/2}(ArH^+)$ has a $(E_{c.m.})^{-1/2.5}$ collision energy dependence, which is milder than that predicted by the LGS orbiting model.[93,94] This observation is consistent with the later finding of Ervin and Armentrout[12] obtained without spin–orbit state selection. In a recent study using high resolution in $E_{c.m.}$, Tosi et al.[187] have observed a dip in the cross-section curve centered at $E_{c.m.} \approx 0.05$ eV. The nature of this dip is not clear at the present. It may be the result of the avoided crossing of the vibronic curves.

II. DISSOCIATION. In Fig. 29, the values measured for $\sigma_{3/2}(H^+)/\sigma_{3/2}(H_2^+)$ and for $\sigma_{3/2}(H^+)$ are plotted as a function of $E_{c.m.}$ in the range of 2.2–11 eV. The values for $\sigma_{3/2}(H^+)$ are $\leqslant 3\%$ of those for $\sigma_{3/2}(H_2^+)$. The AE determined for the formation of H^+ is 2.5 ± 0.15 eV, in good agreement with the

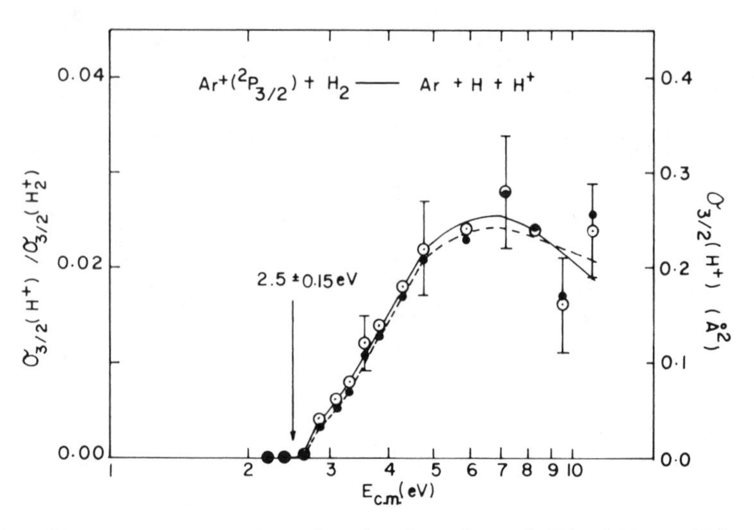

Figure 29. Total cross sections for the formation of H^+. (●) $\sigma_{3/2}(H^+)$; (⊙) $\sigma_{3/2}(H^+)/\sigma_{3/2}(H_2^+)$.[22]

thermochemical threshold[16,29] of 2.39 eV for reaction (60) with reactant Ar^+ in the ${}^2P_{3/2}$ state. The cross section for the formation of H^+ exhibits a maximum at $E_{c.m.} \approx 6$–8 eV.

II. SPIN–ORBIT EXCITATION AND RELAXATION. Total absolute partial state-to-state cross sections, $\sigma_{3/2\to1/2}(H_2)$ and $\sigma_{1/2\to3/2}(H_2)$, for reactions (58) and (59), respectively, measured using the TQDO apparatus in the $E_{c.m.}$ range of 0.24–19.3 eV, are depicted in Fig. 30. Both cross sections increase as $E_{c.m.}$ is increased from 0.24 eV. The values for $\sigma_{3/2\to1/2}(H_2)$ level off at about 3 Å^2 for $E_{c.m.} \geqslant 10$ eV. The values for $\sigma_{1/2\to3/2}(H_2)$ are nearly twice those for $\sigma_{1/2\to3/2}(H_2)$ at the corresponding $E_{c.m.}$.

The values for $\sigma_{3/2\to1/2}(H_2)$ and $\sigma_{1/2\to3/2}(H_2)$ obtained using the energy-loss spectroscopic method[185] are also plotted in Fig. 30. The general profiles for the excitation and relaxation cross-section curves observed by the Liao et al.[22] are in accord with those found by Nakamura, Kobayashi, and Kaneko.[185] However, the absolute values for $\sigma_{3/2\to1/2}(H_2)$ and $\sigma_{1/2\to3/2}(H_2)$ determined using the TQDO photoionization apparatus are nearly a factor of 2 greater than those measured by the energy-loss spectroscopic method. As was pointed out in previous studies,[5,7] the higher values for the excitation and relaxation cross sections observed here, compared to those measured in the energy-loss experiment, are attributed to higher collection efficiencies achieved in this experiment for inelastically scattered Ar^+ ions.

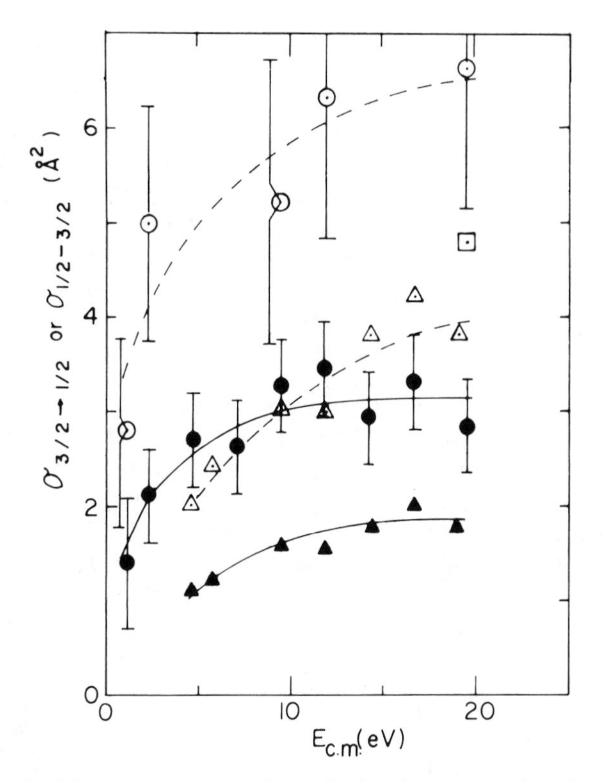

Figure 30. Partial state-to-state excitation and relaxation cross sections. (▲) $\sigma_{3/2\to1/2}$ and (●) $\sigma_{1/2\to3/2}$, Ref. 185; (△) $\sigma_{3/2\to1/2}$ and (⊙) $\sigma_{1/2\to3/2}$, Ref. 22; (□) theoretical values for $\sigma_{3/2\to1/2}$ calculated using the NRIOSA, Ref. 22 (the corresponding value for $\sigma_{1/2\to3/2}$ is 16.9 $\mathring{A}^2$).

At relatively low $E_{c.m.}$($\leqslant 50$ eV), the values for $\sigma_{3/2\to1/2}(H_2)$ are significantly greater than those for $\sigma_{3/2\to1/2}$(Ar).[13,22] This observation is consistent with the charge-transfer mechanism for fine-structure transitions.[63] That is, the spin–orbit state transition such as process (58) may be considered to involve a two-step charge-transfer process[185]:

$$Ar^+(^2P_{3/2}) + H_2 \xrightarrow{(a)} H_2^+(\tilde{X}, v' = 2) + Ar \xrightarrow{(b)} Ar^+(^2P_{1/2}) + H_2. \quad (61)$$

In favourable cases, the energy defect for each of the steps [processes 61(a) and 61(b)] is smaller, compared to the energy difference of the spin–orbit states of Ar^+, making the two-step mechanism more favorable than the direct excitation process indicated by reaction (58). Applying the argument of microscopic reversibility, we expect that the two-step charge-transfer mechanism also enhances the spin–orbit state relaxation process in the collisions of $Ar^+(^2P_{1/2}) + H_2$.

c. Comparison of Charge-Transfer Cross Sections from the Consideration of Microscopic Reversibility. Liao et al.[22] use an approximated scheme based on microscopic reversibility to show that values for $X_{J \to v'}$ can be calculated from the ratios $\sigma_{v' \to J}(Ar^+)/\sigma_{v'' \to J}(Ar^+)$, where $v' \neq v''$.

From the consideration of microscopic reversibility,[188]

$$E_{v' \to Jv''} g_1 \sigma_{v' \to Jv''}(E_{v' \to Jv''}) = E_{Jv'' \to v'} g_2 \sigma_{Jv'' \to v'}(E_{Jv'' \to v'}), \tag{62}$$

where $E_{v' \to Jv''}$ and $E_{Jv'' \to v'}$ are the asymptotic relative kinetic energies for the reactant $H_2^+(\tilde{X}, v' = 0\text{–}4) + Ar(^1S_0)$ and the product $H_2(X, v'') + Ar^+(^2P_{3/2,1/2})$ pairs, respectively; g_1 and g_2 are the degeneracies for the reactant and product states, respectively; and $\sigma_{v' \to Jv''}(E_{v' \to Jv''})$ and $\sigma_{Jv'' \to v'}(E_{Jv'' \to v'})$ are the state-to-state cross sections for reaction (53) and its reverse at $E_{v' \to Jv''}$ and $E_{Jv'' \to v'}$, respectively. If the degeneracies due to the rotational states of H_2 and H_2^+ are ignored, g, equals the degeneracy for the $H_2^+(\tilde{X}\ ^2\Sigma_g^+)$ electronic state and g_2 equals the degeneracy for the $Ar^+(^2P_J)$ spin–orbit state. Therefore, applying Eq. (62) to the $[Ar + H_2]^+$ charge-transfer system, we have

$$\frac{\sigma_{Jv'' \to v'}}{\sigma_{v' \to Jv''}} = \frac{2E_{v' \to Jv''}}{(2J+1)E_{Jv'' \to v'}}, \tag{63}$$

$$\frac{1}{X_{Jv'' \to v'}} = \frac{\sigma_{Jv''}}{\sigma_{Jv'' \to v'}} = \sum_{v} \frac{(E_{v \to Jv''}/E_{Jv'' \to v'})\sigma_{v \to Jv''}(E_{v \to Jv''})}{(E_{v' \to Jv''}/E_{Jv'' \to v'})\sigma_{v' \to Jv''}(E_{v' \to Jv''})}. \tag{64}$$

We note that $E_{Jv'' \to v'} = E_{Jv'' \to v}$. Therefore

$$\frac{1}{X_{Jv'' \to v'}} = 1 + \sum_{v \neq v'} \frac{E_{v \to Jv''}\sigma_{v' \to Jv''}(E_{v \to Jv''})}{E_{v' \to Jv''}\sigma_{v' \to Jv''}(E_{v' \to Jv''})}. \tag{65}$$

The absolute values for $\Sigma_{v''}\sigma_{v' \to Jv''}(E_{v' \to Jv''})$ have been measured . For $E_{c.m.} \leqslant 19\,eV$ and $v' \leqslant 3$, the sum $\Sigma_{v''}\sigma_{v' \to Jv''}(E_{v' \to Jv''}) = \sigma_{v' \to J}(E_{v' \to J})$ is expected to be dominated by $\sigma_{v' \to J0}(E_{v' \to J0})$. This expectation is supported by the theoretical state-to-state cross sections.[152] Assuming that

$$\frac{\sigma_{v \to Jv''}}{\sigma_{v' \to Jv''}} = \frac{\sigma_{v \to J}}{\sigma_{v' \to J}}, \tag{66}$$

we arrive at the relation,

$$\frac{1}{X_{J0\to v'}} \approx 1 + \sum_{v \neq v'} \frac{E_{v\to J0}\sigma_{v\to J}(E_{v\to J0})}{E_{v'\to J0}\sigma_{v'\to J}(E_{v'\to J0})}. \tag{67}$$

The values for $X_{J0\to v'}(=X_{J\to v'})$, $J = 3/2$ and $1/2$, calculated using Eq. (67) are compared to experimental values for $X_{3/2\to v'}$ and $X_{1/2\to v'}$ in Figs. 27 and 28, respectively. At $E_{\text{c.m.}} = 0.16$–0.54 eV, the predicted values of $X_{3/2\to v'}$ show that the vibrational distributions for the charge-transfer product H_2^+ ions peak at $v' = 1$ instead of $v' = 0$, as observed by experiment. Nevertheless, the experimental and predicted values for $X_{3/2\to v'}$, $v' = 0$–3, are in good agreement after taking into account the experimental uncertainties. As shown in Fig. 28, excellent agreement between the experimental and predicted values is found for all $E_{\text{c.m.}}$ in the range of 0.16–19.3 eV, showing the preferential formation of H_2^+ in $v' = 2$.

Liao et al.[22] have also compared the absolute experimental values for $\sigma_{J0\to v'}, v' = 1, 2$ and $J = 3/2, 1/2$, with those calculated using the microscopic reversibility relation given in Eq. (63). The overall agreement between the experimental and predicted values is fair when the experimental uncertainties are taken into account. However, the comparison reveals that the experimental values for $\sigma_{J0\to v'}$ may be too small, or those for $\sigma_{v'\to J}$ values may be too large.

d. Comparison of Experimental and Theoretical Cross Sections. Previous theoretical values for $\sigma_{v'}(Ar^+)$ and $\sigma_{v'}(ArH^+)$ calculated using the trajectory-surface-hoping (TSH) model[141] or the reactive infinite-order sudden approximation (RIOSA)[138] are compared to the experimental values in Tables I–II. These calculations do not include the spin–orbit splitting of $Ar^+(^2P)$. Since the factor of close energy resonance between the $H_2^+(\tilde{X}, v' = 2) + Ar$ and $Ar^+(^2P_{1/2}) + H_2(X, v' = 0)$ states, which is most important in governing the structure of the cross sections for the $[Ar + H_2]^+$ reaction system, is not taken into account in the theoretical models, we expect that these calculations do not always yield the correct cross sections. For instance, the TSH calculation predicts that $\sigma_{v'}(Ar^+)$ and $\sigma_{v'}(Ar^+ + ArH^+)$ have their maximum values at $v' = 1$, contrary to the experimental observation. With the exception of this discrepancy, the TSH cross sections in most cases fit reasonably well with experimental results. The results of the previous RIOSA treatment are in poor agreement with experimental cross sections.

Theoretical cross sections at $E_{\text{c.m.}} = 19.3$ and 47.6 eV for the $H_2^+(\tilde{X}, v') + Ar$ and the $Ar^+(^2P_{3/2,1/2}) + H_2$ charge-transfer reactions calculated by Liao et al.[22,152] using the nonreactive infinite-order approximation (NRIOSA) are also included in Table I, III, IV, and VI. The spin–orbit splitting of $Ar^+(^2P)$

is taken into account in these calculations. Fair agreement is observed between experimental and theoretical cross sections for the $H_2^+(\tilde{X}, v') + Ar$ charge-transfer reaction, whereas the theoretical values for $\sigma_{3/2,1/2}(H_2^+)$ at $E_{c.m.} = 19.3$ eV are about twice the experimental values. Since the theoretical cross sections for the $H_2^+(\tilde{X}, v') + Ar$ and $Ar^+(^2P_{3/2,1/2}) + H_2$ charge-transfer reactions are in accord with the principle of detailed balance, this comparison again points to the possibility that the experimental values for $\sigma_{3/2,1/2}(H_2^+)$ may be too low.

Using a new version of the RIOSA based on the complex potential idea,[189–191] Baer and co-workers have computed state-selected and state-to-state cross sections for charge transfer, proton and hydrogen transfer, and fine-structure transitions of the $[Ar + H_2]^+$ reaction system. By incorporating the three potential energy surfaces for $Ar^+(^2P_{3/2}) + H_2$, $Ar^+(^2P_{1/2}) + H_2$, and $H_2^+(\tilde{X}) + Ar$ into the new calculation, they find the fit between the theoretical predictions and experimental results encouraging. Some of the theoretical cross sections for the $H_2^+(\tilde{X}, v') + Ar$ reactions are summarized in Tables I–III. The application of this new RIOSA theoretical scheme to the $[Ar + H_2]^+$ reaction system is a subject of discussion in the chapter by Baer in the theoretical review volume.

5. $[O + N_2]^+$

a. $O^+(^4S) + N_2$. The O^+ ion produced by solar ultraviolet photoionization of atomic oxygen is the major ionic species in the ionosphere.[192–196] Since the rates of radiative recombination of O^+ with electrons are several orders of magnitude smaller than those for dissociative recombination of N_2^+, O_2^+, and NO^+ with electrons,[197,198] the reactions which convert O^+ into N_2^+, O_2^+, and NO^+ constitute the most important ion-molecule reactions in the ionosphere. The relevance of the reactions,

$$O^+(^4S) + N_2 \rightarrow NO^+(\tilde{X}\,^1\Sigma^+) + N(^4S), \quad (68)$$

$$\rightarrow N_2^+(\tilde{X}\,^2\Sigma_g^+) + O(^3P), \quad (69)$$

to the ion chemistry of the ionosphere has provided much impetus for the detailed experimental[199–204] and theoretical[205–209] investigations of the reaction dynamics of this system. In a guided-ion-beam mass spectrometric study, Burley et al.[200] have measured the absolute cross sections for reactions (68) and (69) in $E_{c.m.}$ range from near thermal to ≈ 20 eV using a high-pressure electron impact ion source to produce $O^+(^4S)$. They find that the cross section for the formation of charge transfer N_2^+ has an onset that coincides with the thermochemical threshold of reaction (69) at $E_{c.m.} = 1.96$ eV. Their experimental also reveals that the cross section for N_2^+ increases from the onset and reaches a maximum at $E_{c.m.} \approx 10$ eV before declining rapidly toward higher $E_{c.m.}$, an observation in fair agreement with the prediction of the phase-space calculation of Wolf.[209] The rapid decline in N_2^+ cross section occurs at a collision energy close to the thermochemical threshold ($E_{c.m.} = 10.7$ eV) for the reaction,

$$O^+(^4S) + N_2 \rightarrow N^+(^3P) + N(^4S) + O(^3P). \tag{70}$$

Recently, Flesch and Ng[210] have undertaken a state-selected study of the $O^+(^4S) + N_2$ reaction to search for the formation of N^+. In their experiment, the $O^+(^4S)$ reactant ions are produced by the dissociaive photoionization of O_2,

$$O_2 + h\nu \rightarrow O^+(^4S) + O(^3P) + e^-, \tag{71}$$

at 640 Å. Figure 31 shows the absolute total cross sections for NO^+, N_2^+, and N^+ formed by reactions (68)–(70) in the E_{lab} range of 1–300 eV. The cross sections for NO^+ are in general accord with those reported previously,[200,203,204] indicating that the high-pressure $O^+(^4S)$ ion sources

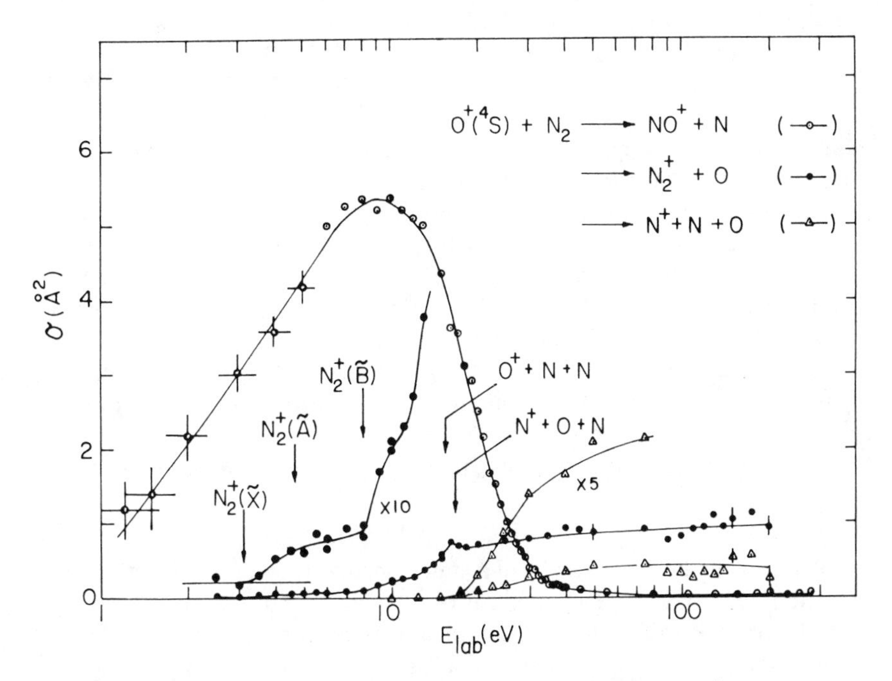

Figure 31. Absolute total cross sections for NO^+ (⊙), N_2^+ (●), and N^+ (△) formed in collisions of $O^+(^4S)$ with N_2 in the E_{lab} range of 1–300 eV.[210]

used in previous experiments have little contamination from metastable $O^+(^2D, ^2P)$ ions. The measured cross sections for N_2^+ at $E_{lab} \leqslant 16.8$ eV ($E_{c.m.} = 10.7$ eV), the onset for reaction (70), are also in good agreement with those obtained by Burley et al.[200] The increases in N_2^+ cross section at $E_{lab} \approx 2$ and 8 eV correlate with the thresholds for the formation of $N_2^+(\tilde{X})$ and $N_2^+(\tilde{B})$, respectively. This observation supports the view that the formation of N_2^+ in the collision of $O^+(^4S)$ and N_2 proceeds preferentially by the attack of $O^+(^4S)$ along the N–N bond direction via a near linear $(O–N–N)^+$ intermediate. Contrary to the previous observation[200] that the N_2^+ cross section decreases abruptly at $E_{lab} \geqslant 16.8$ eV, Flesch and Ng[210] find that it increases slowly from 0.7Å at $E_{lab} = 16$ eV to ~ 1 Å^2 at $E_{lab} = 200$ eV. As expected, the steplike feature observed at ~ 16 eV in the N_2^+ cross section marks the onset for the $N^+ + N + O$ channel. Taking into account the experimental uncertainties, the onset of the formation of N^+ is found to be consistent with the thermochemical threshold of reaction (70). The N^+ cross section increases from the onset and becomes nearly constant at ~ 0.4 Å^2 in the E_{lab} range of 50–200 eV.

b. $O^+(^2D, ^2P) + N_2$. The $O^+(^2D)$ and $O^+(^2P)$ atomic ions are also produced by ultraviolet photoionization of atomic oxygen in the ionosphere.[192] The branching ratios for $O^+(^4S)$:$O^+(^2D)$:$O^+(^2P)$ observed in the photoionization of oxygen atoms are roughly 0.43:0.29:0.28.[194–196] The $O^+(^2D)$ and $O^+(^3P)$ states are metastable with radiative lifetimes of 3.6 hr and 4.57 s, respectively. Thermal rate constants for many reactions involving these metastable ions have been determined from analaysis of the data obtained from the satellite Atmospheric Explorer.[211–215] Electron impact ionization of O_2 and dissociative charge transfer of $He^+ + O_2$ have been utilized as the source of excited $(O^+(^2D, ^2P)$ reactant ions for the reactivity studies of $O^+(^2D, ^2P) + N_2$.[201,203,216–218] The exact proportion of the two excited states is not well characterized in thes experiments. The results of these studies show that the quenching [process (72)] and charge-transfer [process (73)] channels are dominant in the collisions of $O^+(^2D, ^2P) + N_2$:

$$O^+(^2D, ^2P) + N_2 \rightarrow O^+(^4S) + N_2 \tag{72}$$

$$\rightarrow O + N_2^+ \tag{73}$$

Recently, preliminary results of a state-selected study of the charge-transfer reaction (73) have been reported by Lavollee and Henri[219] using the threshold PEPICO technique. In their experiment, $O^+(^2D)$ and $O^+(^2P)$ atomic ions are prepared by the dissociative photoionization process

$$O_2 + h\nu \rightarrow O_2^+(\mathrm{III}\ ^2\Pi_u) \rightarrow O^+(^2D, ^2P) + O + e^-. \tag{74}$$

The $O^+(^2D)$ and $O^+(^2P)$ ions are formed via predissociation of $O_2^+(\text{III}\,^2\Pi_u)$ initially prepared by photoionization at $h\nu = 22.06$ and 23.75 eV, respectively. The coincidence with a threshold electron does not ensure the selective dissociation of $O_2^+(\text{III}\,^2\Pi_u)$ according to reaction (74). Nevertheless, a previous study has shown that the dissociation is strongly selective.[123] Furthermore, production of O^+ in the ^4S state corresponds to fast O^+ ions, which can be distinguished from the slow O^+ ions in the ^2D and ^2P metastable states by the TOF method. One of the major shortcomings of utilizing dissociative photoionization processes such as process (74) for the preparation of excited atomic ions is the very low efficiency of threshold electron production.

In the E_{lab} range of ≈ 7–20 eV, the ratio of the charge-transfer cross section for $O^+(^2P)$ to that associated with $O^+(^2D)$ is found to be 0.5–0.8.[219] The absolute charge-transfer cross section for $O^+(^2D)$ is estimated to be 25 Å^2, which is in agreement with the previous measurement of Rutherford and Vroom.[203] The large cross sections for reaction (73) have been attributed to the close resonances of the $O^+(^2D) + N_2$ and $N_2^+(\tilde{A}, v' = 1) + O(^3P)$ states and the $O^+(^2P) + N_2$ and $N_2^+(\tilde{A}, v' = 0) + O(^1D)$ states.

6. $[O + H_2]^+$

a. $O^+(^4S) + H_2$. The ion–molecule reaction

$$O^+(^4S) + H_2(X\,^1\Sigma_g^+) \rightarrow OH^+(\tilde{X}\,^3\Sigma^-) + H(^2S) \qquad (75)$$

is important in H_2-rich interstellar clouds where product OH^+ ions react further with H_2 to yield H_2O^+ and H_3O^+ molecular ions.[220–222] This, together with the relative simplicity of this system, has attracted many experimental[223–229] and theoretical[230–242] investigations. At $E_{\text{c.m.}}$ below ≈ 0.3 eV, experimental rate constants[225–227] and total cross sections[228] for reaction (75) are found to be in excellent agreement with predictions based on the LGS model.[93,94] Experimental studies at $E_{\text{c.m.}} = 0.8$–8.0 eV, using crossed ion–neutral beam[223] and ion beam gas cell[224] arrangements, show that the dynamics of reaction (75) are dominated by the spectator stripping mechanism.[243] Correlation diagrams[223,242] constructed for collinear and C_{2v} configurations for the $[O + H_2]^+$ system indicate that reaction (75) proceeds via a near collinear $O^+\cdots H\text{–}H$ geometry on the ground potential energy surface, whereas the reaction in a C_{2v} approach has a high potential barrier. Detailed *ab initio* calculations[231–239] support this picture. The most interesting observation of this charge-transfer system concerns the formation of OH^+ and OD^+ in the collisions of $O^+(^4S) + HD$.[228] The latter reaction has been reviewed recently.[224] Results of many previous experiments[219,220,223] suggest that product OH^+ formed at sufficiently high $E_{c.m.}$ may contain enough internal energy for it to dissociate into $H^+ + O$.

Absolute total cross sections for the charge-transfer reaction (76) and the dissociative charge-transfer reaction (77) have been measured by Flesch and Ng[245]:

$$O^+(^4S) + H_2 \rightarrow O + H_2^+ \quad \Delta E = 1.80\,\text{eV}, \tag{76}$$

$$\rightarrow O + H + H^+, \quad \Delta E = 4.48\,\text{eV}. \tag{77}$$

Figure 32 depicts the absolute total cross sections for OH^+, H_2^+, and H^+ [$\sigma(OH^+, H_2^+$, and $H^+)$] formed by reactions (75)–(77) in the $E_{c.m.}$ range of 1.33–22.22 eV. Owing to the worsening resolution of the product QMF as the laboratory O^+ ion-beam energy is increased beyond 60 eV ($E_{c.m.} \simeq 7$ eV), the values for $\sigma(H^+)$ at $E_{c.m.} > 7$ eV have not been measured. The values for $\sigma(OH^+)$ are taken from Ref. 228. As shown in the figure, the

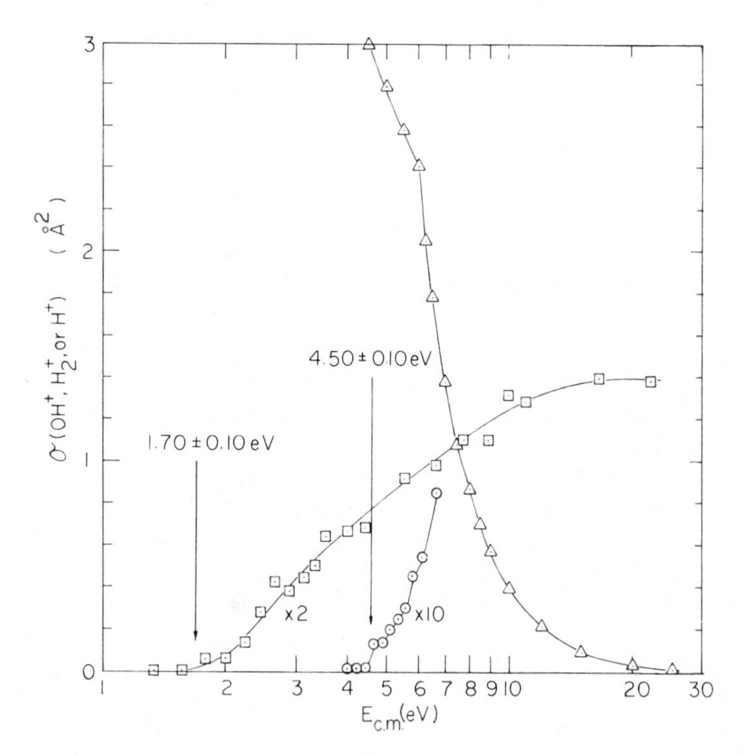

Figure 32. Absolute total cross sections for OH^+ (△), H_2^+ (□), and H^+(⊙) formed in the reaction $O^+(^4S) + H_2$.[245]

value for $\sigma(OH^+)$ drops steeply as $E_{c.m.}$ is increased above 6.5 eV. This feature has been interpreted to be the result of further dissociation of excited product OH^+. When values for $\sigma(OH^+)$ in the fall-off region are analyzed with a fitting routine, an empirical onset of $E_{c.m.} = 6.25 \pm 0.50$ eV is deduced for reaction (77).[228] This value compares to the dissociation threshold of 8.6 eV based on the spectator stripping model.[243] Since the empirical onset is well above the thermochemical threshold of 4.48 eV, [19,246,247] Burley et al. conclude that some of the energy available to the products of reaction (75) is placed preferentially in translation.[228] The fact that the AE of $E_{c.m.} = 4.50 \pm 0.10$ eV for H^+ observed in the experiment of Flesch and Ng[245] is in excellent accord with the thermochemical threshold for reaction (75) does not support that conclusion. Very weak H^+ signals are still observable at $E_{c.m.}$ below 4.5 eV, but these are likely the results of imperfect background corrections stemming from the procedures adopted in the experiment. Nevertheless, it is possible that these weak signals arise from the near thermoneutral[19,246,247] reaction.

$$O^+(^4S) + H_2(X\,^1\Sigma_g^+) \rightarrow OH(X\,^2\Pi) + H^+(^1S), \qquad \Delta E = 0.06\,\text{eV} \qquad (78)$$

These negligibly small H^+ signals are consistent with the expectation that reaction (78) is not important because it violates the spin conservation rule.

The experimental AE of $E_{c.m.} = 1.70 \pm 0.10$ eV found for the formation of H_2^+ is in agreement with the thermochemical threshold[19,246,247] for reaction (76). The values for $\sigma(H_2^+)$ at $E_{c.m.} < 7$ eV are substantially greater than those for $\sigma(H^+)$. The sum of $\sigma(H_2^+)$ and $\sigma(H^+)$ at $E_{c.m.} = 6$ eV is approximately 20% that for $\sigma(OH^+)$. At $E_{c.m.} > 9$ eV, $\sigma(H_2^+)$ becomes higher than $\sigma(OH^+)$. Recently, predictions for $\sigma(OH^+)$ in the $E_{c.m.} = 0.25$–6.3 eV have been obtained in a quasiclassical trajectory calculation.[236] Comparison of results from the trajectory calculation and experimental data show good accord, except at $E_{c.m.} > 1.0$ eV. The experimental results of Flesch and Ng[245] show that it is necessary to include the charge-transfer and dissociative charge-transfer channels at $E_{c.m.} > 2$ eV in order to obtain accurate comparisons of theoretical and experimental cross section data for the $O^+(^4S) + H_2$ reaction.

C. Diatom–Diatom System

1. $[H_2 + H_2]^+$

a. $H_2^+(\tilde{X}, v_0' = 0) + H_2(X, v_0'' = 0)$. The simplicity of the reaction between $H_2^+ + H_2$ makes it a system of fundamental importance to molecular reaction dynamics. The $H_2^+ + H_2$ reaction and its isotopic variations have been investigated[2,3,11,24,26,153–155,248–291] by nearly all accessible experimental techniques. In spite of its apparent simplicity, the reaction is rich in chemistry. Combining the results of previous experimental and theoretical[292-304] studies, a qualitative picture of the $H_2^+ + H_2$ reaction has emerged. At low $E_{c.m.}$, the major product channel is the formation of $H_3^+ + H$, which has cross sections close to the predictions of the LGS model.[93,94] In principle, the H_3^+ formation can result from the proton- and the hydrogen-atom-transfer mechanisms. The isotopic-substitution experiments[254,256,257] show that the cross sections for nominal proton- and hydrogen-atom-transfer reactions are similar. This is interpreted to be the result of rapid charge equilibrium between the reactants H_2^+ and H_2 prior to H_3^+ formation. The $H_3^+ + H$ channel is believed to proceed with a direct mechanism. As $E_{c.m.}$ is increased, the cross section for H_3^+ formation decreases rapidly with concomitant increase in the cross section for charge transfer. At $E_{c.m.} > 5\,eV$, charge transfer becomes the dominant product channel. The formation of $H^+ + H + H_2$ is also a viable process when $E_{c.m.}$ is greater than the dissociation energy of H_2^+.[250,252,274]

Recently, several theoretical calculations[293-295,303-305] on the $H_2^+ + H_2$ reaction system have been reported. Applying the TSH method[306] to H_4^+ valence bond DIM potential energy surfaces,[307,308] Stine and Muckerman[295,296] and Eaker and Schatz[293,304] have performed extensive calculations on the $H_2^+ + H_2$ reaction. Combining the quasiclassical trajectory approach and a Fourier transform method,[293] Eaker and Schatz have calculated cross sections for the $H_3^+ + H$ channel at $E_{c.m.} < 1\,eV$, with an emphasis on the H_3^+ product rotational and vibrational distributions. State-to-state cross sections have been obtained by Lee and DePristo[294] for $E_{c.m.} \geqslant 8\,eV$ using the semiclassical-energy-conserving-trajectory (SCECT) formulation. The interaction potentials used are derived from a simple one-active-electron model[309] and agree with the *ab initio* potential-energy curves for the $H_2^+ + H_2$ system.[310] Since their calculation incorporates the effects of molecular orientations in the charge-exchange dynamics, it is expected to be more accurate than previous theoretical studies. Cross sections for the $H_3^+ + H$ and the charge-transfer channels have also been computed by Baer and Ng using the RIOSA approach which incorporates the complex-potential method.[303]

Recent state-selected and state-to-state cross-section measurements for the processes,

$$H_2^+(\tilde{X}, v_0') + H_2(X, v_0'' = 0) \rightarrow H_2(X, v') + H_2^+(\tilde{X}, v''), \tag{79}$$

$$\rightarrow H_3^+ + H, \tag{80}$$

$$\rightarrow H^+ + H + H_2, \tag{81}$$

are summarized subsequently.

I. CHARGE TRANSFER. The relative total cross section for charge transfer of H_2^+ with H_2 as a function of H_2^+ vibrational state was first examined by Chupka at $E_{c.m.} = 215$ eV.[3] Combining the photoionization and the guided-beam technique, Anderson et al.[252] have measured state-selected charge-transfer cross sections for several isotopic $H_2^+ + H_2$ reactions for $v'_0 = 0$–4 over the $E_{c.m.}$ range of 0.23–6.1 eV. Vibrational state-selected cross sections for the symmetric charge transfer process (79) have been also reported by Campbell et al.[251] at $E_{c.m.} = 8,78$, 200, and 500 eV and by Guyon and co-workers[250] at $E_{c.m.} = 4$, 8, 12, 16 eV using the PEPICO methods. Recently, Liao et al.[24,26] have measured state-selected and state-to-state cross sections for reaction (79) in the $E_{c.m.}$ range of 2–200 eV.

The experiment of Liao et al.[24,26] was performed using the crossed ion–neutral beam photoionization apparatus in which the reactant $H_2^+(\tilde{X}, v'_0)$ and product $H_2^+(\tilde{X}, v'')$ are detected by the vertical and horizontal QMF, respectively. Figure 33 compares the PIE spectra for product H_2^+ observed at $E_{lab} = 8, 16, 45$, and 400 eV and the PIE spectrum for the reactant H_2^+ ion, recorded with a wavelength resolution of 1.4 Å. The PIEs of the product and reactant H_2^+ ion spectra in the region of 790–805 Å, which corresponds to the formation of H_2^+ in the $v'_0 = 0$ state, were normalized to the same value. From the comparison of these spectra, it is obvious that the charge-transfer cross section for reaction (79) $[\sigma_{vo'}(H_2^+)]$ depends on both the vibrational and kinetic energies of the H_2^+ reactant ions. At $E_{lab} = 8$ eV [Fig. 33(*a*)], the PIEs for the product H_2^+ ions in vibrational state $v'_0 \geqslant 1$ are $\approx 50\%$ higher than those for the reactant H_2^+ ions, an observation consistent with the conclusion that $\sigma_{vo'}(H_2^+)$ increases with vibrational excitation of the reactant H_2^+ ion. The vibrational enhancement for $\sigma_{vo'}(H_2^+)$, manifested by the spectra at $E_{lab} = 16$ eV [Fig. 33(*b*)], is reduced considerably. As E_{lab} is increased to 45 eV, Fig. 33(*c*) shows that $\sigma_{vo'}(H_2^+)$ is inhibited by vibrational excitations of the reactant H_2^+ ions. At $E_{lab} = 400$ eV, the vibrational enhancement for $\sigma_{vo'}(H_2^+)$ is again evident from the comparison depicted in Fig. 33(*d*). The vibrational effect on $\sigma_{vo'}(H_2^+)$ at $E_{lab} = 400$ eV revealed in the comparison in Fig. 33(*d*) is consistent with that observed at $E_{lab} = 430$ eV by Chupka.[3]

By comparing the relative heights for reactant and product H_2^+ produced at selected autoionization peaks at a wavelength resolution of 0.3 Å (FWHM), Liao, et al.[24] have determined relative values for $\sigma_{vo'}(H_2^+), v'_0 = 0$–4, at $E_{c.m.} = 2$–200 eV. The derived values for $\sigma_{vo'}(H_2^+)/\sigma_{vo'=0}(H_2^+)$ at $E_{c.m.} = 2$, 4, 8, 16, 22.5, and 200 eV are plotted as a function of v'_0 in Figs. 34(*a*)–34(*f*) to compare with the theoretical predictions of Lee and DePristo[24,305] and Eaker and Schatz[304] and the experimental results of Campbell, Browning, and Latimer.[251] The best agreement between the experimental results and the SCECT theoretical calculations was observed

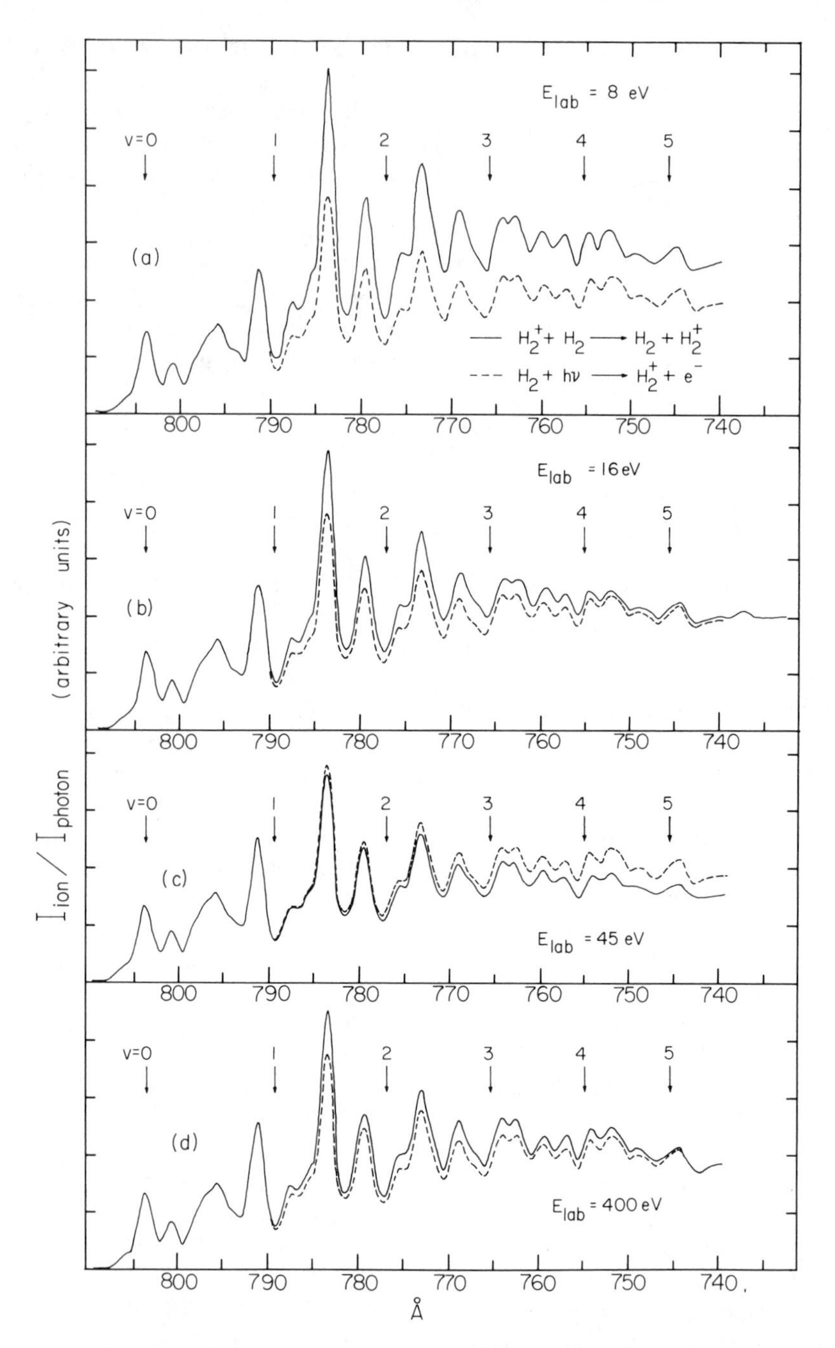

Figure 33. The comparisons of PIE spectra for the H_2^+ product ions (—) formed at E_{lab} equal to (*a*) 8 eV, (*b*) 16 eV, (*c*) 45 eV, and (*d*) 400 eV with that for the H_2^+ reactant ions (——) in the regions of 730–810 Å. The PIEs for the product and reactant ions are normalized to the same values in the region of 790–810 Å, which corresponds to the formation of $H_2^+(v_0' = 0)$.[24]

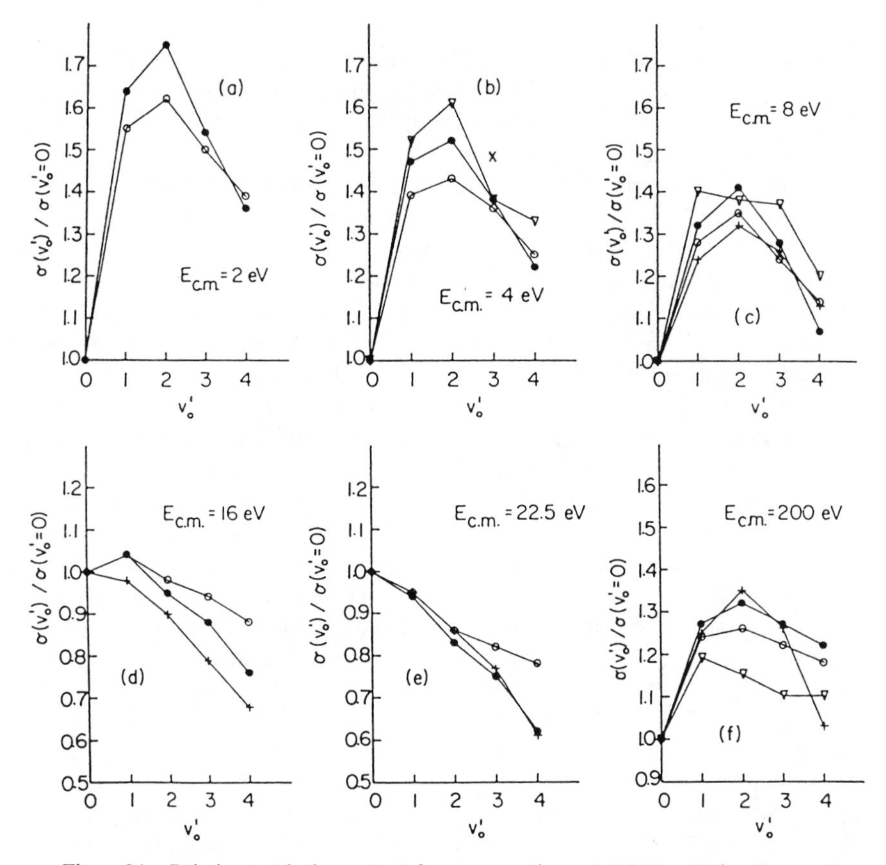

Figure 34. Relative total charge transfer cross sections $\sigma(v_0')/\sigma(v_0'=0)$ for the reaction $H_2^+(v_0') + H_2(v_0'')$ plotted as a function of v_0'. $E_{c.m.}$: (*a*) 2 eV, (*b*) 4 eV, (*c*) 8 eV, (*d*) 16 eV, (*e*) 22.5 eV, (*f*) 200 eV. Experimental: (●) Ref. 24; (⊙) values obtained without corrections of direct photoionization, Ref. 24; (∇) Ref. 251. Theoretical: (+) Ref. 305; (×) Ref. 304.

at $E_{c.m.} = 22.5$ eV, where $\sigma_{vo'}(H_2^+)$ decreases monotonically with v_0'. The experimental values for $\sigma_{vo'}(H_2^+)/\sigma_{vo'=0}(H_2^+)$, $v_0' = 1$–4, at $E_{c.m.} = 16$ eV are higher than the corresponding calculated values by ≈ 5–10%. At $E_{c.m.} = 8$ eV, the value for $\sigma_{vo'=0}(H_2^+)$ is lower than those for $\sigma_{vo'}(H_2^+)$, $v_0' = 1$–4. The experimental profile for the vibrational dependence of $\sigma_{vo'}(H_2^+)$ is consistent with the theoretical prediction. Unlike the results of the PEPICO experiment,[251] which show a maximum for $\sigma_{vo'=1}(H_2^+)$, both the SCECT calculations and the experimental results indicate that the value for $\sigma_{vo'}(H_2^+)$ peaks at $v_0' = 2$ at $E_{c.m.} = 8$ eV. Taking into account the error limits, the results of Liao, Liao, and Ng[24] and Campbell et al.[251] are in agreement at

$E_{c.m.} = 4\,eV$. The greatest vibrational enhancement for reaction (79) is observed at $E_{c.m.} = 2\,eV$. No SCECT theoretical results are available for $E_{c.m.} < 8\,eV$. Since the SCECT calculation does not include the $H_3^+ + H$ channel, which is a major product channel at low $E_{c.m.}$, its results are not expected to be accurate at low $E_{c.m.}$.

Previous calculations[257,296] of the H_4^+ potential surface by the DIM method provide insight into the charge-transfer process for the $H_2^+ + H_2$ system. Resonance charge transfer essentially involves the crossing of a barrier of the ground-state adiabatic surface, which divides the two charge-transfer states. The barrier height depends on the $H_2^+ + H_2$ distance. At a H_2^+–H_2 distance of $\geqslant 4.2\,Å$, the calculations show that the reactant pair at zero collision energy cannot cross the barrier. For shorter H_2^+–H_2 distances, the barrier is reduced as a result of a larger splitting of the ground and the first excited surfaces, and the reactants, even in their ground vibrational states, can traverse the barrier. Since vibrational excitations of the reactants afford the crossing of the barrier, vibrational enhancement for reaction (79) is expected at low $E_{c.m.}$. The experimental observation of Liao et al.,[24] which indicates the difference between $\sigma_{vo'=0}(H_2^+)$ and $\sigma_{vo'=1}(H_2^+)$ increases as $E_{c.m.}$, is decreased in the range of $E_{c.m.} = 2$–$8\,eV$ is in accord with this picture. However, their experiment also shows that this difference and vibrational enhancement becomes less pronounced as $E_{c.m.}$ is decreased in the range of 0.38–2 eV. This is contrary to the prediction of the TSH calculation that the ratio $\sigma_{vo'=3}(H_2^+)/\sigma_{vo'=0}(H_2^+)$ continues to increase as $E_{c.m.}$ decreases in the range of 0.25–3.0 eV. The values of 1.48 and 1.24 for $\sigma_{vo'=3}(H_2^+)/\sigma_{vo'=0}(H_2^+)$ at $E_{c.m.} = 4$ and 8 eV, obtained by Eaker and Schatz[304] in a recent TSH calculation, are in reasonable agreement with the SCECT results [see Fig. 34(*b*) and 34(*c*)]. It is interesting to note that the vibrational dependences of $\sigma_{vo'}(H_2^+)$ measured at $E_{c.m.} = 2, 4$, and 8 are in qualitative agreement with the Franck–Condon factors for the transitions from $H_2(X, v_0'' = 0)$ to $H_2^+(\tilde{X}, v_0' = 0$–$4)$.[311]

The magnitude of vibrational enhancement observed at $E_{c.m.} = 200\,eV$ is comparable to that found at $E_{c.m.} = 8\,eV$, except that the falloff of $\sigma_{vo'}(H_2^+), v_0' = 3$ and 4, with respect to $\sigma_{vo'=2} = (H_2^+)$ is less dramatic than that at $E_{c.m.} = 8\,eV$. The measured relative values for $\sigma_{vo'}(H_2^+), v_0' = 0$–$3$, are in excellent agreement with theoretical predictions. The slow reduction of $\sigma_{vo'}(H_2^+), v_0' = 3$ and 4, with respect to $\sigma_{vo'=2}(H_2^+)$ observed at $E_{c.m.} = 200\,eV$ is probably due to higher cross sections for nonresonant charge-transfer channels. The calculation indeed shows that the cross sections of many off-resonance product channels at $E_{c.m.} = 200\,eV$ are as large as those of the resonant channels. The vibrational enhancements observed at $E_{c.m.} = 200\,eV$ in the PEPICO study are less than those found by Liao et al.[24]

The variation of the relative values for $\sigma_{vo'}(H_2^+)$ as a function of v_0' shown in

Figs. 34(*a*)–34(*f*) suggests that the kinetic energy dependence of $\sigma_{vo'}(H_2^+)$ may exhibit structure in the $E_{c.m.}$ region of 2–200 eV. The previous measurements of the absolute total charge-transfer cross sections for the $H_2^+ + H_2$ system, which used low-energy electron ionization to prepare the reactant H_2^+ ions, have been reviewed by Barnett et al.[291] Within the E_{lab} range of 8–400 eV, essentially all these measurements show that the total charge-transfer cross section decreases monotonically as E_{lab} increases. The recommended values for the absolute total charge transfer cross sections for the $H_2^+ + H_2$ reaction

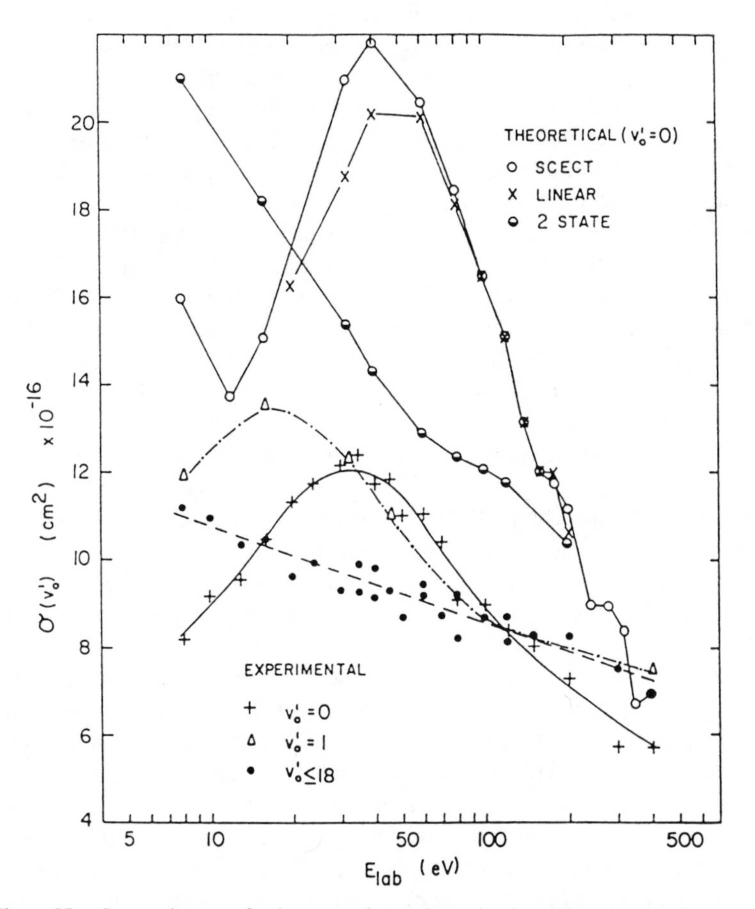

Figure 35. Comparisons of the experimental and theoretical state-selected total charge-transfer cross sections $\sigma(v_0')$ for the reaction $H_2^+(v_0') + H_2(v_0'' = 0)$. Experimental: (— —) electron impact values recommended by Ref. 291; (+) $\sigma(v_0' = 0)$, (Δ) $\sigma(v_0' = 1)$, and (●) $\sigma(v_0' \leqslant 18)$, Ref. 24; Theoretical: (⊙) $\sigma(v_0' = 0)$ SCECT, (×) $\sigma(v_0' = 0)$ linear trajectory, and (⊖) $\sigma(v_0' = 0)$ two state, Ref. 305.

in the range of $E_{lab} = 8$–$400\,eV$ are shown as the dashed line in Fig. 35. Most of the previous electron ionization studies used electron energies greater than 18 eV. The vibrational distribution of the H_2^+ reactant ions produced in electron impact ionization can be characterized by the Franck–Condon factors for the ionization transition $H_2(X, v_0' = 0) \rightarrow H_2^+(\tilde{X}, v')$. Assuming that the vibrational distribution of H_2^+ formed by photoionization of H_2 at 688 Å is similar to that prepared by electron ionization, absolute values for $\sigma_{vo'=0}(H_2^+)$ at $E_{lab} = 8$–$400\,eV$ have been determined by measuring the relative cross sections at 688 and 792 Å and by calibrating the relative cross sections at 688 Å to the absolute values observed in electron ionization studies.[291] At 792 Å, all reactant H_2^+ are formed in $v_0' = 0$, while at 688 Å H_2^+ ions are expected to have a broad vibrational distribution with $v_0' = 0$–18, characterized by the Franck–Condon factors for the ionization transitions.

The absolute experimental values for $\sigma_{vo'}(H_2^+)$ thus determined reveal a broad peak centered at $E_{lab} \approx 35\,eV$, as shown in Fig. 35. The nature of this broad peak has been explored by DePristo and co-workers.[24,305] The SCECT theoretical values for $\sigma_{vo'=0}(H_2^+)$ also show a broad peak at $E_{lab} \approx 40\,eV$, in agreement with the experimental obseravation. The absolute theoretical values are about a factor of 2 greater than the experimental results. However, after taking into account the error in the absolute magnitude of the electron impact data ($\approx \pm 40\%$) and the uncertainty of the theoretical values ($\approx \pm 30\%$), the magnitude of the experimental and theoretical results are in fair agreement. The SCECT calculation shows that the initial rise of $\sigma_{vo'=0}(H_2^+)$ is due to the increase of both the resonant and nonresonant cross sections, while the decrease of $\sigma_{vo'=0}(H_2^+)$ at $E_{lab} > 40\,eV$ is the result of the falloff the resonant channel. Since the $H_3^+ + H$ channel is the dominant product channel at low E_{lab}, it is logical to conclude that the decrease of $\sigma_{vo'=0}(H_2^+)$ at low E_{lab} is partly due to the competition of reaction (80). The theoretical results for $\sigma_{vo'=0}(H_2^+)$ based on the calculations for a rectilinear trajectory are similar to those obtained using the SCECT method. This indicates that the use of Ehrenfest's theorem and the direct–direct and exchange–exchange interactions in the SCECT formulation to generate curved trajectories is not a significant factor in predicting the kinetic energy dependence of $\sigma_{vo'=0}(H_2^+)$. Interestingly, when the calculations are restricted to the two channels, resonant direct ($v_0' = 0, v_0'' = 0$) and exchange ($v' = 0, v'' = 0$), the calculated total charge-transfer cross sections decrease monotonically as a function of E_{lab} (Fig. 35). The theoretical analysis clearly shows that the broad peak of the $\sigma_{vo'=0}(H_2^+)$ curve observed in Fig. 35 arises from the strongly coupled multistate nature of the dynamics of reaction (79). A complementary interpretation could be made in terms of the transition from diabatic to adiabatic vibrational dynamics as the kinetic energy decreases.

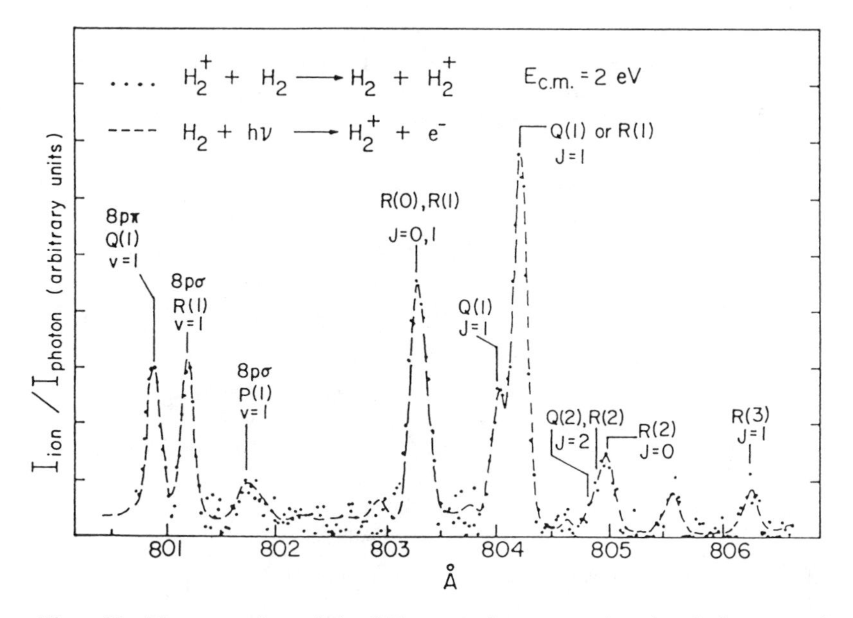

Figure 36. The comparison of the PIE spectra for reactant (— —) and charge-transfer product (···) H_2^+ formed at $E_{c.m.} = 2$ eV in the wavelength region of 800.5–806.5 Å [wavelength resolution = 0.14 Å (FWHM)].[24]

As a consequence of autoionization selection rules and energy constraint, Chupka and co-workers[2] have shown that $H_2^+(v_0')$ with $J = 0$–2 can be formed by photoionization of H_2 in high resolution.[312,313] Liao, Liao, and Ng[24] have examined the rotational energy effect on reaction (79) at $E_{c.m.} = 2$ eV. Figure 36 compares the PIE spectrum for product H_2^+ at $E_{c.m.} = 2$ eV with that for reactant H_2^+ obtained using a wavelength resolution of 0.14 Å. After normalizing the heights of the autoionization peaks at 804.12 Å, the PIEspectra for the reactant and product H_2^+ are found to be superimposable, indicating that changing the rotational quantum number of $H_2^+(v_0' = 0)$ from $J = 0$ to $J = 2$ has no observable effect on reaction (79) at $E_{c.m.} = 2$ eV. In order for the rotational energy of $H_2^+(v_0')$ to play a role in the charge-transfer reaction $H_2^+(v_0', J) + H_2$, the time for molecular interaction between $H_2^+ + H_2$ in the collision should be of the same order of magnitude as the rotational period of $H_2^+(\approx 5 \times 10^{-13}$s). Assuming an effective distance for charge exchange of ≈ 7 Å, a collision time of 5×10^{-13}s corresponds to a $E_{c.m.}$ value much less than 1 eV. Therefore, the lack of rotational energy effect for reaction (79) at $E_{c.m.} = 2$ eV is not surprising.

Using the differential reactivity technique, Liao and Ng[26] have also determined the vibrational distributions of product H_2^+ ions formed by

TABLE VII
Vibrational-State Distributions of Product $H_2^+(v')$ formed by the Charge-Transfer Reaction $H_2^+(\tilde{X}, v'_0=0$ or $1)+H_2(X, v'_0=0)\rightarrow H_2(X, v')+H_2^+(\tilde{X}, v'')$ in the $E_{c.m.}$ Range of 2–16 eV[a,b]

$E_{c.m.}$(eV)	v'_0	X_0	X_1	X_2
2	0	1.00	0.00	0.00
	1	0.21	0.79	0.00
4	0	0.91	0.04	0.05
	1	0.25	0.66	0.09
6	0	0.88	0.09	0.03
	1	0.31	0.54	0.15
8	0	0.86(0.92)	0.10(0.07)	0.04(0.01)
	1	0.34(0.17)	0.57(0.76)	0.09(0.07)
12	0	0.83	0.14	0.03
	1	0.40	0.53	0.07
16	0	0.82(0.87)	0.17(0.12)	0.01(0.01)
	1	0.46(0.39)	0.43(0.50)	0.11(0.11)

[a] X_0, X_1, and X_2 are the fractions of product H_2^+ formed in the $v''=0$, 1, and 2 states.
[b] Theoretical values are in parentheses. Reference 305

reaction (79) at $E_{c.m.}=2$–$16\,eV$. Table VII summarizes the experimental values for the fractions of product $H_2^+(v'')$, $X_{v''}(v''=0$–$2)$, formed in the charge-transfer reaction $H_2^+(v'_0=0,1)+H_2(v''_2=0)$ at $E_{c.m.}=2$–$16\,eV$. The SCECT theoretical values obtained by Lee and DePristo at $E_{c.m.}=8$ and 16 eV are also included in the table.

When the reactant H_2^+ ions are in $v'_0=0$, nearly all the charge-transfer product H_2^+ ions formed at $E_{c.m.}=2\,eV$ are in the $v'_0=0$ state, indicating that resonance charge-transfer is the dominant process. As $E_{c.m.}$ is increased from 2 to 16 eV, the value for X_1 increases steadily from 0.0 to 0.17, while values for X_2 remain small. The SCECT predictions for $X_{v''}, v''=0$–2 at $E_{c.m.}=8$ and 16 eV are in fair agreement with the experimental findings.

For reactant H_2^+ prepared in $v'_0=1$, the inelastic relaxation channel forming $H_2^+(v''=0)$ is significant at all $E_{c.m.}$. The degree of relaxation is found to increase as $E_{c.m.}$ is increased. The most interesting observation is that the extent of charge-transfer relaxation is substantially greater than that of charge-transfer excitation. Although SCECT calculations also predict such a trend, the theoretical predictions underestimate the degree of relaxation. A better agreement between experiment and theory is found at $E_{c.m.}=16\,eV$. The observed efficient vibrational relaxation of $H_2^+(v'_0=1)$ via the symmetric charge-transfer process (79) is consistent with the interpretation that charge transfer involves mainly the long-range electron jump mechanism, by which the vibrational energy of the reactant H_2^+ ion is efficiently distributed between the product H_2^+ and H_2.

Relative cross sections for reaction (79) with $v_0' = 0$–10 and $E_{c.m.} = 4, 6, 8,$ and 16 eV have also been reported by Guyon et al.[250] By calibrating their relative cross section with the average absolute charge-transfer cross sections for $v_0' = 0$ and 1 obtained by Liao et al.[24,305] they have placed their relative cross sections on an absolute basis as shown in Figs. 37(*a*)–37(*d*). With the exception of the vibrational energy dependence of $\sigma_{vo'}(H_2^+)$ at $E_{c.m.} = 16$ eV, the trends of $\sigma_{vo'}(H_2^+)$ for other $E_{c.m.}$ are not in agreement with the experimental results of Liao et al.[24] and Campbell et al.[251] Since time-of-flight was used to distinguish the reactant H_2^+ and product H_2^+ and H_3^+ ions in their experiment, product H_2^+ ions with flight times overlapping the reactant

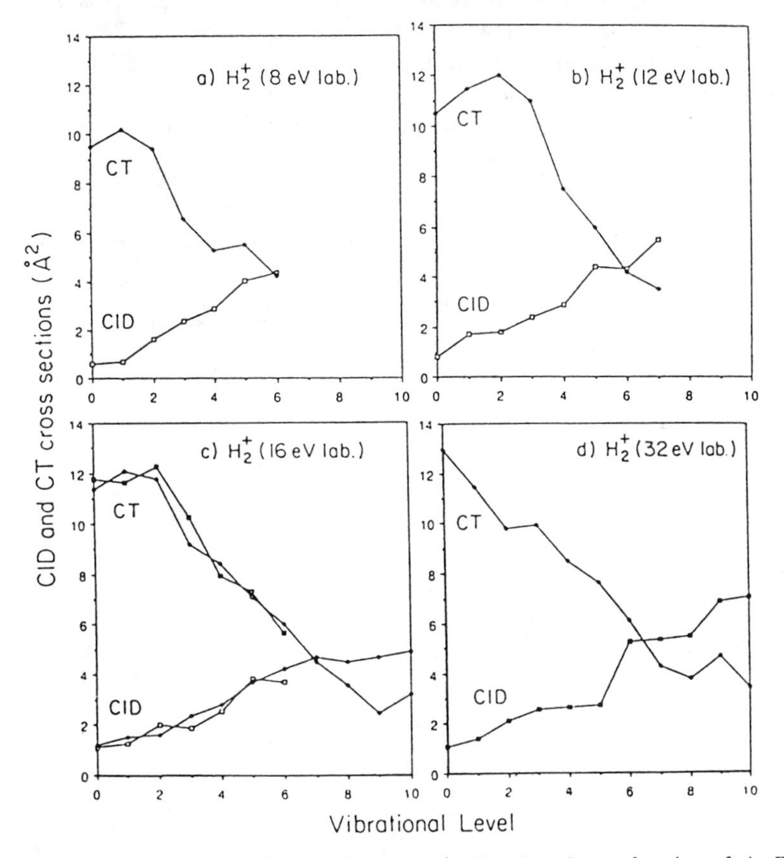

Figure 37. Values for $\sigma_{vo'}(H^+)$ (□) and $\sigma_{vo'}(H_2^+)$ (●) plotted as a function of v_0'. E_{lab}: (*a*) 8 eV, (*b*) 12 eV, (*c*) 16 eV, (*d*) 32 eV.[250]

and product H_3^+ peaks may not be identified. Furthermore, it is possible that some slow product H_2^+ ions may be converted into H_2^+ as they travel from the reaction region to the ion detector.

II. DISSOCIATION AND REACTION. The first state-selected experiment on reaction (80) was performed by Chupka and co-workers.[2] Vibrational-state-selected cross sections for reaction (80) at $E_{c.m.} < 1\,eV$ have also been measured by Koyano and Tanaka[253] and van Pijkeren et al.[248] using the PEPICO methods. In all these experiments, photoionization and subsequent reactions take place in a single gas cell. Owing to a continuous potential drop across the gas cell, the collision energy is ill-defined.

Shao and Ng[11] have measured absolute total cross sections for reaction (80) using a tandem photoionization mass psectrometer equipped with an rf octopole reaction cas cell. Figure 38(*a*)–38(*f*) show the relative cross sections, $\sigma_{vo'}(H_3^+)/\sigma_{vo'=0}(H_3^+)$, $v_0' = 0$–4, at $E_{c.m.} = 0.04, 0.25, 0.46, 0.5, 0.75$, and $1.0\,eV$, respectively. The experimental results obtained by Koyano and Tanaka and van Pikjeren et al. are included in Figs. 38(*a*), 38(*c*), and 38(*f*). The relative cross sections at thermal energy reported by van Pikjeren et al. are in good

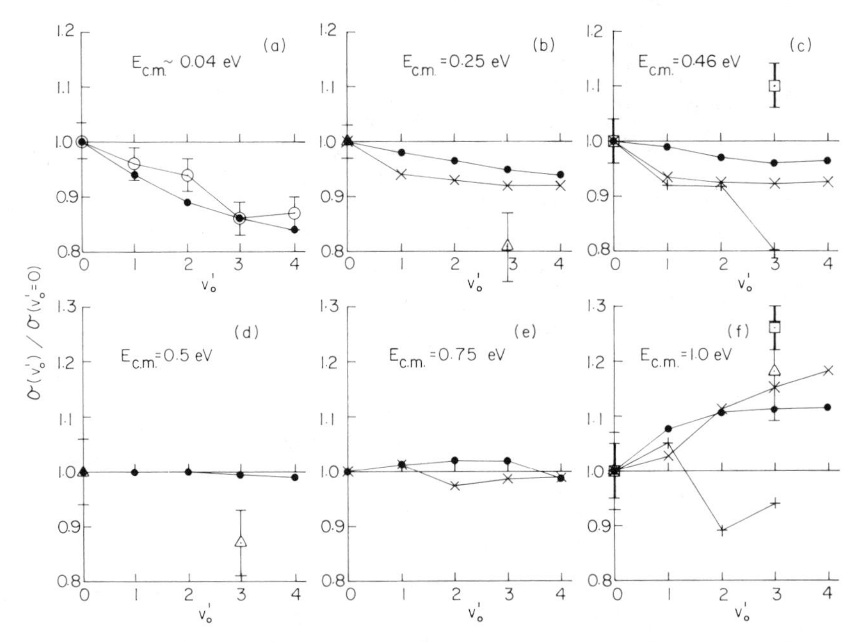

Figure 38. Relative total cross sections $\sigma(v_0')/\sigma(v_0' = 0)$ for reaction (80) at $E_{c.m.} = 0.04$–$1\,eV$ plotted as a function of v_0'. Experimental: (●) Ref. 11, (⊙) Ref. 248, (×) mean values of proton and atom transfers, Ref. 252; theoretical: (Δ) Ref. 295. (□) Ref. 293.

agreement with those determined by Shao and Ng[Fig. 38(*a*)]. Although the values obtained at $E_{c.m.} = 0.46\,eV$ for $\sigma_{vo'}(H_3^+)/\sigma_{vo'=0}(H_2^+)$, $v'_0 = 1$–3, by Koyano and Tanaka are lower than those determined by Shao and Ng, the results are within exprimental uncertainties, showing vibrational inhibition for $\sigma_{vo'}(H_3^+)$. In a single-chamber experiment,[2,253] the uncertainty in $E_{c.m.}$ becomes greater as $E_{c.m.}$ is increased. In Fig. 38(*f*), the experimental results obtained by Koyano and Tanaka at $E_{c.m.} = 0.93\,eV$ and those observed by Shao and Ng at $E_{c.m.} = 1.0\,eV$ are in poor agreement.

Absolute total cross sections for the reactions

$$H_2^+(v'_0) + D_2 \rightarrow D_2H^+ + H, \tag{82}$$

$$D_2^+(v'_0) + H_2 \rightarrow D_2H^+ + H, \tag{83}$$

have been measured by Anderson et al.[252] The formation of D_2H^+ by reactions (82) and (83) are referred to as the nominal proton and atom-transfer reactions, respectively. If an electron hops repeatedly between H_2^+ and H_2 in the entrance channel prior to the formation of H_3^+, the value for $\sigma_{vo'}(H_3^+)$ should be in accord with the mean value of the total cross sections for reactions (82) and (83). Shao and Ng have calculated the average cross sections for reaction (82) and (83) at $E_{c.m.} = 0.23$, 0.43, 0.77, and 1.1 eV based on the experimental results of Anderson et al. As shown in the Figs. 38(*b*), 38(*c*), 38(*e*), and 38(*f*), the relative values for the average cross sections are in good agreement with those determined for reaction (79). This observation gives strong support for the rapid charge hopping interpretation.

For $E_{c.m.} \leqslant 1\,eV$, the results of two theoretical calculations are available to compare with the experimental data. Trajectory surface hopping calculations by Stine and Muckerman[295] and by Eaker and Schatz[293] provide only $\sigma_{vo'}(H_3^+)$, $v'_0 = 0$ and 3. Their results are shown in Figs. 38(*b*), 38(*c*), 38(*d*), and 38(*f*). In the quasiclassical calculation of Eaker and Schatz, the surface hopping part of the dynamics is treated by using an approximation to the usual TSH model. They assume that the trajectories follow the diabatic surfaces up to a particular separation and the ground adiabatic surfaces thereafter. The DIM surface used by Eaker and Schatz is similar to that used by Stine and Muckerman but has a different parametrization of the potentials. The fact that only a finite number of trajectories were sampled causes large uncertainties in the theoretical results. Taking into acount the uncertainties of the calculated and experimental values, the theoretical predictions of Stine and Muckerman at $E_{c.m.} = 0.25$, 0.5, and 1.0 eV are in agreement with the experimental measurements shown in Fig. 38. At $E_{c.m.} = 0.46\,eV$, the calculations of Eaker and Schatz predict vibrational enhancement for $\sigma_{vo'}(H_3^+)$, whereas the experimental values show the opposite trend.

The experimental values for $\sigma_{vo'}(H_3^+)/\sigma_{vo'=0}(H_3^+)$, $v_0' = 0$–4, measured by Shao and Ng over the $E_{c.m.}$ range of 2–15 eV are plotted in Figs. 39(*a*)–38(*i*). Vibrational enhancement for $\sigma_{vo'}(H_3^+)$ reaches a maximum near $E_{c.m.} = 3$ eV. As $E_{c.m.}$ is increased in the range of 3–15eV, the values for $\sigma_{vo'>0}(H_3^+)$ decrease steadily with respect to $\sigma_{vo'=0}(H_3^+)$. At $E_{c.m.} = 12$ and 15 eV, $\sigma_{vo'}(H_3^+)$ decreases monotonically as a function of v_0'. The relative values for $\sigma_{vo'}(H_3^+)$, $v_0' = 0$ and 3, at $E_{c.m.} = 3$ and 5 eV obtained by the TSH calculations of Stine and Muckerman also agree with the experimental findings [Figs. 39(*b*) and 39(*d*)].

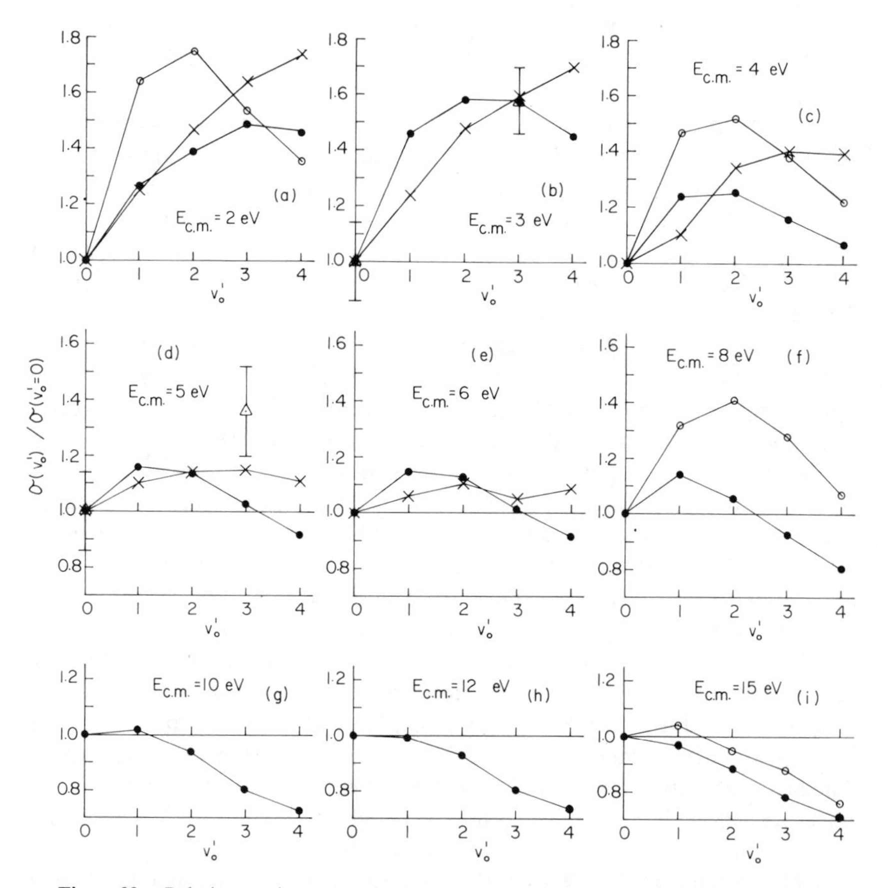

Figure 39. Relative total cross sections $\sigma(v_0')/\sigma(v_0' = 0)$ for reaction (80) at $E_{c.m.} = 2$–15 eV plotted as a function of v_0'. Experimental: (●) Ref. 11, (×) mean values of proton and atom transfers, Ref. 252; theoretical: (Δ) Ref. 295; (⊙) relative total cross sections for the $H_2^+ + H_2$ charge-transfer reaction, Ref. 24.

The mean values for $\sigma_{vo'}(H_3^+)/\sigma_{vo'=0}(H_3^+)$ of proton and atom transfers at $E_{c.m.} = 2.1$, 2.8, 4.1, 5.1, and 6.1 eV observed by Anderson et al. are also included in Figs. 39(*a*)–39(*e*). The relative cross sections for reaction (79) and the mean cross sections exhibit similar functional dependences on v_0'. Nevertheless, the two sets of data show better agreement at $E_{c.m.} = 0.25$–1 eV than at $E_{c.m.} = 2$–6 eV. Since a finite length of time is needed for an electron to jump from H_2 to H_2^+, the shorter collision time at high $E_{c.m.}$ limits the number of electron jumps between H_2^+ and H_2 at the entrance channel. If charge equilibrium cannot be achieved at high $E_{c.m.}$ at the entrance channel prior to the formation of H_3^+, the vibrational dependences for $\sigma_{vo'}(H_3^+)/\sigma_{vo'=0}(H_3^+)$ and the relative mean cross sections for reactions (82) and (83) will be different from these at low $E_{c.m.}$. The resluts of Anderson et al. show that the cross sections for reaction (82) and (83) have similar vibrational depedences at low $E_{c.m.}$, whereas at high $E_{c.m.}$, the cross sections for atom and proton transfers are quite different.

The relative vibrational-state-selected total cross sections for reaction (79) at $E_{c.m.} = 2$, 4, 8, and 16 eV obtained by Liao et al.[24] are included in Figs. 39(*a*), 39(*c*), 39(*f*), and 39(*i*), respectively. The vibrational enhancements for the cross sections of symmetric charge transfer are greater than those observed for the $H_3^+ + H$ channel. It is interesting that both charge transfer and H_3^+ formation decrease with respecct to that at $v_0' = 0$ as $E_{c.m.}$ is increased in the $E_{c.m.}$ range of 4–15 eV.

The TSH calculation reveals that the critical impact parameter charge transfer is larger than that for H_3^+ formation and is relatively independent of $E_{c.m.}$. The impact parameter for H_3^+ formation decreases with increasing $E_{c.m.}$. Eaker and Schatz give a rationalization for the observed vibrational dependence of $\sigma_{vo'}(H_3^+)$ in the low and intermediate $E_{c.m.}$ (< 3 eV) range. They note that, from their trajectory results, the probability of nonreactive charge transfer increases as v_0' is increased, but at the same time the maximum impact parameter leading to H_3^+ formation is larger. The vibrational inhibition for $\sigma_{vo'}(H_3^+)$ observed at $E_{c.m.} \leqslant 0.5$ eV indicates that the first effect is more important. In the intermediate $E_{c.m.}$ range of 1–3 eV, the vibrational enhancements for $\sigma_{vo'}(H_3^+)$ are attributed to the dominance of the second effect. An increase in reagent vibrational excitation is expected to favor H_3^+ formation by providing more translational energy for the products to overcome the product centrifugal barrier. For a direct proton-transfer reaction, the vibrational excitations of the reactant H_2^+ ions should facilitate the H_2^+ bond breaking and enhance the formation of H_3^+.

At $E_{c.m.} \geqslant 2$ eV, collision-induced dissociation becomes energetically allowed. These processes have the major contribution from low-impact-parameter collisions. The quasiclassical trajectory calculation of Eaker and Muzyka[292] suggests that the fall off of vibrational enhancement for H_3^+

formation from 4 eV to higher $E_{c.m.}$ is due to the increase in further dissociation of highly rotationally excited H_3^+.

The absolute values for $\sigma_{vo'}(H_3^+)$, $v'_0 = 0$ and 3, measured by Shao and Ng over the $E_{c.m.}$ range of 0.25–15 eV are plotted in Fig. 40. Table VIII summarizes the experimental and theoretical values for $\sigma_{vo'}(H_3^+)$, $v'_0 = 0$ and 3, at $E_{c.m.} = 0.25$–5 eV. The experimental and theoretical cross sections at $E_{c.m.} = 0.5$ and 1.0 eV are in good agreement. With the exception of $\sigma_{vo'=0}(H_3^+)$ at $E_{c.m.} = 0.25$ eV, the TSH cross sections of Stine and Muckerman are consistent with the experimental results (Fig. 40). The results of the recent TSH calculation by Eaker and Schatz,[314] which are in exellent agreement with those of Stine and Muckerman, are also listed in Table VIII.

Although it has been shown that the previous quasiclassical trajectory calculation accounts for the general features of the experimental cross-sectional data, the detailed features are predicted only by the TSH model, which includes nonadiabatic surface hopping throughout the reaction. In practice, two methods[296,315,316] are available for the calculation of surface hopping trajectories. It is still unsettled[315] which method is more correct

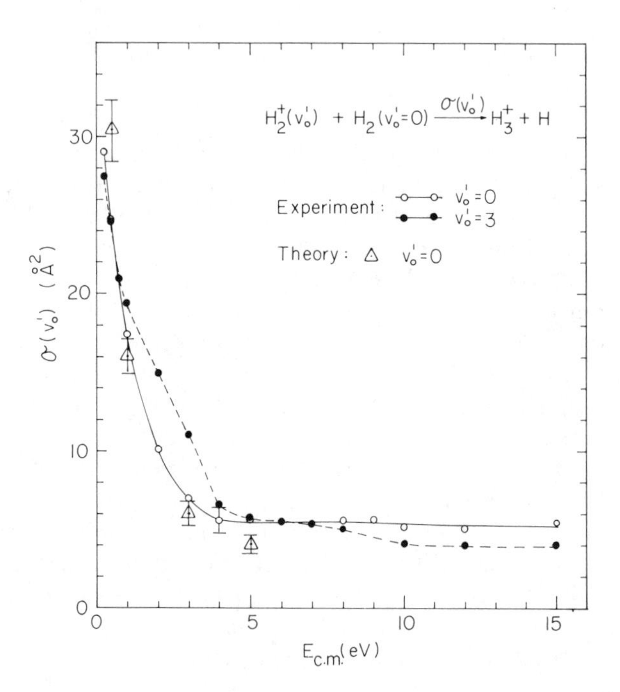

Figure 40. Absolute vibrational-state-selected total cross sections $\sigma(v'_0)$, $v'_0 = 0$ and 3, for reaction (80) plotted as a function of $E_{c.m.}$. Experimental: (⊙) $\sigma(v'_0 = 0)$ and (●) $\sigma(v'_0 = 3)$, Ref. 11; theoretical: (Δ) $\sigma(v'_0 = 0)$, Ref. 295.

TABLE VIII
Total Cross Sections $[\sigma_{v0'}(H_3^+)]$ for the reaction $H_2^+(v_0'=0,3)+H_2(v_0''=0)\rightarrow H_3^+ + H$ at $E_{c.m.}=0.25$–5 eV

		Experimental (Å^2)		Theoretical (Å^2)			
$E_{c.m.}$(eV)	v_0'	SN[a]	KT[b]	SM[c]	ES[d]	ES(TSH)[e]	LGS[f]
0.25	0	34.0±4.0	···	47.8±1.4	···	46.4±2.0	30.5
	3	32.3±4.0	···	38.8±2.2	···	37.4±2.0	30.5
0.5	0	26.6±2.0	24.3±1.5	30.6±1.7	30.6±1.4	30.6±1.4	21.5
	3	26.3±2.0	19.5±2.0	26.7±1.7	33.7±1.4	26.4±2.0	21.5
1.0	0	$17.4^{+1.5}_{-2.5}$	16.9±1.5	16.0±1.1	18.3±0.8	16.6±1.1	15.2
	3	$19.3^{+1.5}_{-2.5}$	16.2±1.5	18.8±1.7	23.0±0.8	18.5±1.4	15.2
3.0	0	$7.5^{+0.8}_{-1.6}$	···	5.8±0.8	···	6.5±0.6	8.8
	3	$11.8^{+0.8}_{-1.6}$	···	9.3±1.1	···	9.6±0.8	8.8
5.0	0	$6.1^{+0.6}_{-2.0}$	···	3.9±0.6	···	3.7±0.3	6.8
	3	$6.3^{+0.6}_{-2.0}$	···	5.3±0.8	···	5.1±0.6	6.8

[a]Reference 11.
[b]Reference 253.
[c]Reference 295.
[d]Results of quasiclassical trajectory calculation (Ref. 293)
[e]Results of TSH calculation obtained by Eaker and Schatz (Private communication).
[f]Values predicted by the Langevin–Gioumousis–Stevens model (Refs. 93 and 94).

and practical. The method of Stine and Muckerman is more stable.[317] The application of the Stine and Muckerman surface hopping procedures is suitable for the $H_2^+ + H_2$ system.[296,304] As shown in the table, the predictions by the LGS model are in fair agreement with the experimental results, suggesting that the dynamics for reaction (75) are governed mainly by the ion-induced dipole interaction.

Most recently, Baer and Ng[303] have reported preliminary results for a three-dimensional quantum-mechanical study of the $H_2^+ + H_2$ system using the RIOSA and complex-potential approach. Absolute cross sections for reactions (79) and (80) for $v_0' = 0$ and $E_{c.m.} = 0.25$ and 0.52 eV are obtained from their calculations. Their cross sections for charge transfer are higher than those observed in the TSH calculation, while those for H_3^+ formation are lower than the TSH results. The RIOSA values of 28 and 19 Å^2 are slightly lower than the experimental cross sections of 34 ± 4 and 26.6 ± 2.0 Å for $\sigma_{vo'=0}(H_3^+)$ at $E_{c.m.} = 0.25$ and 0.52 eV, respectively. The application of the RIOSA calculation to the $H_2^+ + H_2$ system is discussed in the review of Baer in the theoretical volume.

The cross sections $[\sigma_{vo'}(H^+)]$ for the collision-induced dissociation process (81) at $E_{lab} = 8$, 12, 16, and 32 eV obtained by Guyon et al.[250] are shown in

Figs. 37(*a*)–37(*d*). The endothermicity of reaction (81) is 2.65 eV and, as expected, $\sigma_{vo'}(H^+)$ increases nearly linearly with v'_0. The values for $\sigma_{vo'}(H^+)$ at high v'_0 are comparable to those for $\sigma_{vo'}(H_2^+)$. An interesting observation is that $\sigma_{vo'}(H^+)$ depends only very weakly on $E_{c.m.}$. The H^+ product ion TOF distribution splits into a low- and a high-energy peak and becomes skewed toward the higher energy peak with increasing v'_0. The slow and fast groups of H^+ are associated with the target and projectile molecules, respectively, based on the observed TOF distributions of D^+ and H^+ resulting from the collisions of $D_2^+ + H_2$. Guyon et al. have considered two models for the production of H^+. The first is the direct excitation mechanism by which H_2^+ ion are excited directly above the dissociation limit to formed $H^+ + H$. The second mechanism involves the formation of H_3^+ prior to the dissociation to yield $H^+ + H_2$. Arguing that excitation to high vibrational state is unlikely in collisions of $H_2^+ + H_2$ in the $E_{c.m.}$ range of 4–16 eV, Guyon et al. conclude that the H^+ product ion TOF distributions observed in their experiment are consistent with a direct dissociation model via excitation to the repulsive $^2\Sigma_u^+$ surface. The fast H^+ peak arises from the dissociation of the projectile, while the slow H^+ peak is attributed to dissociation of the target after charge transfer.

The experimental work of Guyon et al. stimulated a theoretical TSH study by Eaker and Schatz.[304] In the TSH calculation, a second excited-state DIM surface, corresponding to the excited $H_2^+(^2\Sigma_u^+) + H_2(X\,^1\Sigma_g^+)$ state, has been included and surface hopping is allowed for all geometries. When the trajectories are followed for a sufficiently long time (6×10^{-12}s), Eaker and Schatz find that a major pathway for producing H^+ at low $E_{c.m.}$ and low v'_0 is attributable to the further dissociation of excited H_3^+ intermediates. As the vibrational energy of reactant H_2^+ increases and as $E_{c.m.}$ increases, the direct mechanism becomes more important. The TSH study shows that the direct dissociation pathway via excitation to the $H_2^+(^2\Sigma_u^+) + H^2(X\,^1\Sigma_g^+)$ surface suggested by Guyon et al. plays only a minor role in reaction (81).

IV. CONCLUSIONS AND FUTURE DEVELOPMENTS

The series of reactions discussed in this review are considered important model systems for accurate comparisons of experimental results and theoretical calculations. Detailed state-selected and state-to-state cross sections for these systems obtained using the photoionization and differential reactivity methods have provided critical tests for state-of-art dynamical calculations. The agreement observed between experimental results and theoretical predictions for these systems represents a great achievement and an important step in the application of the theoretical models to more complex reaction systems.

When reactive and dissociative channels are not important, the semiclassical vibronic models work well in predicting total cross sections of charge-transfer reactions that proceed via the long-range electron jump mechanism. Further improvements of the semiclassical models to allow reorientation of the colliding pair during the reaction, together with the use of more accurate potential energy surfaces, are needed for accurate predictions of short-range charge-transfer processes. The TSH calculations yield accurate cross sections for reactive and dissociative channels of simple ion–molecule reactions. Since the TSH model does not take into account quantum-resonance effects, it is not expected to work well for charge-transfer reactions. How well the TSH model will work for charge transfer remains to be tested. The quantum-mechanical RIOSA, which incorporates the complex-potential approach, has been applied with some success to predict both charge- and atom-transfer channels in ion–molecule collisions.

Besides providing accurate cross sections for assisting the development of theoretical models, important physical insights have also been gained from the experiment results. Energy resonance has been shown to be an important factor governing charge-transfer processes at low collision energies. As $E_{\text{c.m.}}$ is increased, the resonance effect becomes less important. Evidence has been found that product molecular ion states formed by endothermic charge transfer involving atomic ion and molecule collisions are similar to those produced by photoionization. Furthermore, endothermic charge-transfer processes may proceed with high cross sections. The analyses of experimental results of the $Ar^+(^2P_{3/2,1/2}) + N_2$ (CO, O_2) reactions indicate that the $N^+(C^+, O^+)$ formations proceed via the charge-transfer predissociation mechanism. The "true" threshold for a dissociative charge-transfer process can be determined with accuracy provided that predissociative excited states of the molecular ion exist at the threshold region of the dissociative process.

Resonant charge transfer is an efficient vibrational-relaxation mechansim. The high cross sections observed for direct inelastic ion–molecule reactions are consistent with the interpretation that the existence of charge-transfer states enhances the coupling between reactant and inelastic product states. That is, inelastic excitation and relaxation may be viewed as a two-stepwise charge-transfer process. Since the time scale for electron jump is much shorter than that for nuclear rearrangements, and electron jump is a long-range process, it is possible for charge switching to occur in the entrance channel prior to chemical reaction. This picture suggests that inelastic, dissociative, and reactive channels of ion–molecule reactions are all subject to the influence of charge transfer during the ion–molecule collision. The dynamics of ion–molecule reactions are known to be governed by long-range ion–multipole interactions. Other than this, the charge switching mechanism

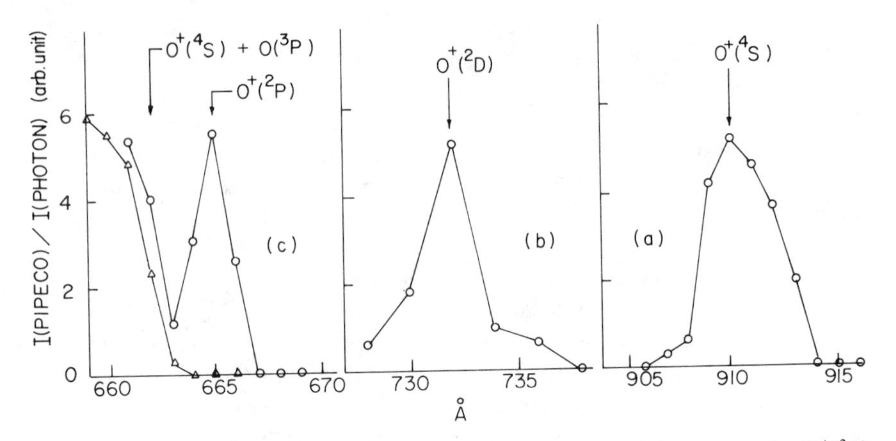

Figure 41. (*a*) PEPICO spectrum for $O^+(^4S)$ from O; (*b*) PEPICO spectrum for $O^+(^2D)$ from O; (*c*) (⊙) PEPICO spectrum for $O^+(^2P)$ from O and $O^+(^4S)$ from O_2 obtained with microwave power "on," (Δ) PEPICO spectrum for $O^+(^4S)$ from O_2 obtained with microwave power "off."

is probably the most important characteristic distinguishing ion–molecule and neutral–neutral reactions.

It is obvious that the most powerful photoionization technique for investigating state-selected ion–molecule processes is the PEPICO method, which is the subject of the review chapter by Koyano in this volume. Using the PEPICO method, the total cross-section measurements of most vibrational- and electronic-state-selected ion–molecule reactions involving reactant ions formed from stable precursor molecules can now be made routinely in the laboratory. Many important ion–molecule reactions involve excited state ions, such as the $O^+(^2D, ^2P)$ described in this review, are difficult to prepare in abundance by photoionization of stable precursors. However, it is possible to prepare state-selected $O^+(^2D, ^2P)$ reactant ions straightforwardly using the PEPICO method in the photoionization of O atoms. As an example, Figs. 41(*a*)–41(*c*) show the PEPICO spectra for $O^+(^4S, ^2D, ^2P)$ formed by the photoionization of $O(^3P)$.[318] The $O(^3P)$ atoms are produced by a microwave discharge of O_2. The O^+ PEPICO spectra obtained with the microwave discharge "on" and "off" are compared in Fig. 41(*c*). The $O^+(^2P)$ PEPICO peak is clearly resolved. The increase in O^+ PEPICO intensity at wavelength < 663 Å are due to the dissociative photoionization process (71). This scheme for generating state-selected radical ions will certainly become easier when the available VUV light intensities increase in future VUV sources.

Using the specially designed guided-beam apparatus of the Freiburg group, information about product ion angular distributions can be obtained. The incoporation of the PEPICO technique into such a guided-beam apparatus is expected to be very powerful. Angular distributions of scattered product ions resulting from state-selected ion–neutral collisions may also be obtained by coupling the PEPICO technique with a position-sensitive coincidence scheme similar to that discussed in the experimental chapter by Brenot and Durup-Ferguson.

Better control of vibrational modes and rotational states of reactant ions can be achieved by laser-ionization methods. Further efforts in detailed product state detections will rely also on the use of laser techniques.

It is fair to say that the field of state-selected and state-to-state ion–moleucle reaction dynamics has developed to a stage that we are ready to extend state-selected and state-to-state measurments to more complex and chemically interesting ion–molecule reactions.

Acknowledgments

This work is supported by the National Science Foundation Grant CHE 8913283 and ATM 8914356. Acknowledgment is also made to the donors of the Petroleum Research Fund, administered by the American Chemical Society, for the partial support of this research. Part of this review was prepared while the author was visiting the Institute for Molecular Science (IMS) in Okazaki, Japan, as a foreign invited scholar supported by the Ministry of Education, Science, and Culture of Japan. The author wishes to thank his hosts, Profs. Shobatake and Kimura, for a stimulating and memorable stay at IMS. The author is grateful to Mr. G. D. Flesch for reading the manuscript.

References

1. W. A. Chupka and M. E. Russell, *J. Chem. Phys.* **48**, 1527 (1968); **49**, 5426 (1968).
2. W. A. Chupka, M. E. Russell, K. Refaey, *J. Chem. Phys.* **48**, 1518 (1968).
3. W. A. Chupka, in *Ion–Molecule Reactions*, J. L. Franklin, Ed., Plenum, New York, 1972, p. 33.
4. W. A. Chupka, in *Chemical Spectroscopy and Photochemistry in the Vacuum Ultraviolet*, NATO–Advanced Study Institutes Series C, C. Sandorfy, P. Ausloos, and M. B. Robin, Ed., Reidel, Boston, 1973, Vol. 8, p. 433.
5. C. Y. Ng, State-Selected and State-to-State Ion–Molecule Reaction Dynamics by Photoionization Methods, in *Techniques for the Study of Gas-Phase Ion–Molecule Reactions*, J. M. Farrar and W. H. Saunder, Jr., Ed., Wiley, New York, 1988, p. 417 and references therein.
6. C.-L Liao, J.-D. Shao, R. Xu, G. D. Flesch, Y.-G. Li, and C. Y. Ng, *J. Chem. Phys.* **85**, 3874 (1986).
7. J.-D. Shao, Y.-G. Li, G. D. Flesch, and C. Y. Ng, *Chem. Phys. Lett.* **132**, 58 (1986).
8. J.-D. Shao, Y.-G. Li, G. D. Flesch, and C. Y. Ng, *J. Chem. Phys.* **86**, 170 (1987).
9. G. D. Flesch and C. Y. Ng, *J. Chem. Phys.* **89**, 3381 (1988).
10. G. D. Flesch and C. Y. Ng, *J. Chem. Phys.* **92**, 2876 (1990).
11. J.-D. Shao and C. Y. Ng, *J. Chem. Phys.* **84**, 4317 (1986).
12. K. M. Ervin and P. B. Armentrout, *J. Chem. Phys.* **83**, 166 (1985).

13. G. D. Flesch, S. Nourbakhsh, and C. Y. Ng, *J. Chem. Phys.* **92**, 3590 (1990).
14. D. Gerlich, in *Electronic and Atomic Collisions*, D. C. Lorentz, W. E. Meyerhof, and J. R. Peterson, Eds., North-Holland, Amsterdam, 1986, p. 541.
15. E. E. Ferguson, *J. Phys. Chem.* **90**, 731 (1986).
16. H. M. Rosenstock, K. Draxl, B. W. Steiner, and J. T. Herron, *J. Phys. Chem. Ref. Data Suppl.* **6** (1977).
17. T. Kato, *J. Chem. Phys.* **80**, 6105 (1984).
18. K. Yoshino and Y. Tanaka, *J. Chem. Phys.* **48**, 4859 (1968).
19. K. P. Huber and G. Herzberg, *Constant of Diatomic Molecules*, Van Nostrand Reinhold, New York, 1979.
20. J. M. Ajello, K. D. Pang, and K. M. Monahan, *J. Chem. Phys.* **61**, 3152 (1974).
21. P. M. Dehmer and W. A. Chupka, *J. Chem. Phys.* **62**, 2228 (1975).
22. C.-L. Liao, R. Xu, J.-D. Shao, S. Nourbakhsh, G. D. Flesch, M. Baer, and C. Y. Ng, *J. Chem. Phys.* **93**, 4832 (1990).
23. C.-L. Liao, C.-X. Liao, and C. Y. Ng, *J. Chem. Phys.* **82**, 5489 (1985).
24. C.-L. Liao, C.-X. Liao, and C. Y. Ng, *J. Chem. Phys.* **81**, 5672 (1984); *Chem. Phys. Lett.* **103**, 418 (1984).
25. C.-L. Liao, R. Xu, and C. Y. Ng, *J. Chem. Phys.* **85**, 7136 (1986).
26. C.-L. Liao and C. Y. Ng, *J. Chem. Phys.* **84**, 197 (1986).
27. D. Rapp and W. E. Francis, *J. Chem. Phys.* **37**, 2631 (1962).
28. R. C. Amme and H. C. Hayden *J. Chem. Phys.* **42**, 2011 (1965).
29. W. H. Cramer, *J. Chem. Phys.* **30**, 641 (1959).
30. I. P. Flaks and E. S. Solov'ev, *Sov. Phys.-Tech. Phys.* **3**, 564 (1958).
31. J. B. Hasted, *Proc. R. Soc. London Ser. A.* **205**, 421 (1951).
32. R. M. Kushmir, B. M. Polynkh, and L. A. Sena, *Bul. Acad. Sci. USSR, Phys. Ser.* **23**, 995 (1959).
33. P. Mahadevan and G. D. Magnuson, *Phys. Rev.* **171**, 103 (1968).
34. R. F. Potter, *J. Chem. Phys.* **22**, 974 (1954).
35. B. Ziegler, *Z. Phys.* **136**, 108 (1953).
36. R. H. Neynaber, S. M. Trujillo, and E. W. Rothe, *Phys. Rev.* **157**, 101 (1967).
37. K. B. McAfee, Jr., W. E. Falconer, R. S. Hozack, and D. J. McClure, *Phys. Rev. A* **21**, 827 (1980).
38. Y. Itoh, N. Kobayashi, and Y. Kaneko, *J. Phys. Soc. Jpn.* **50**, 3541 (1981).
39. K. B. McAfeee, Jr., R. S. Hozack, and R. E. Johnsom, *Phys. Rev. Lett.* **44**, 1247 (1980).
40. N. Hishinuma, *J. Phys. Soc. Jpn.* **32**, 227 (1972).
41. E. W. Kaiser, A. Crowe, and W. E. Falconer, *J. Chem. Phys.* **61**, 2720 (1974).
42. F. M. Campbell, R. Browning, and C. J. Latimer *J. Phys. B* **14**, 1183 (1881).
43. S. L. Howard, A. L. Rockwood, S. G. Anderson, and J. H. Futrell, *J. Chem Phys.* **91**, 2922 (1989).
44. R. E. Johnson, *J. Phys. B* **3**, 539 (1970).
45. R. E. Johnson, *J. Phys. Soc. Jpn.* **32**, 1612 (1972).
46. A. L. Rockwood, S. L. Howard, W.-H. Du, P. Tosi, W. Lindinger, and J. H. Futrell, *Chem. Phys. Lett.* **114**, 486 (1987); *Can J. Phys.* **65**, 1077 (1987).

47. Y. Itoh, N. Kobayashi, and Kaneko, *J. Phys. Soc. Jpn.* **50**, 3541 (1981).
48. C.-L. Liao, R. Xu, and C. Y. Ng, *J. Chem. Phys.* **84**, 1948 (1986).
49. T. Kato, K. Tanaka, and I. Koyano, *J. Chem. Phys.* **77**, 337 (1982); 770, 834 (1982).
50. T. R. Govers, P. -M. Guyon, T. Baer, K. Cole, H. Fröhlich, and M. Lavollée *Chem. Phys.* **87**, 373 (1984).
51. P. M. Guyon, T. R. Grovers, and T. Baer, *Z. Phys. D* **4**, 89 (1986).
52. L. Hüwel, D. R. Guyer, G. H. Lin, and S. R. Leone, *J. Chem. Phys.* **81**, 3520 (1984).
53. B. Friedrich, W. Trafton, A. L. Rockwood, S. L. Howard, and J. H. Futrell *J. Chem. Phys.* **8**, 2537 (1984).
54. A. L. Rockwood, S. L. Howard, W.-H. Du, P. Toshi, W. Lindinger, and J. H. Futrell, *Chem. Phys. Lett.* **114**, 486 (1985).
55. J. H. Futrell, *Int. J. Quant. Chem.* **31**, 133 (1987).
56. D. M. Sonnenfroh and S. R. Leone, *J. Chem. Phys.* **90**, 1677 (1989).
57. B. G. Lindsay and C. J. Latimer, *J. Phys. B* **21**, 1617 (1988).
58. A. A. Viggiano, J. M. Van Doren, R. A. Morris, and J. F. Paulson, *J. Chem. Phys.* **93**, 4761 (1990).
59. M. R. Spalburg and E. A. Gislason, *Chem. Phys.* **94**, 339 (1985).
60. G. Parlant and E. A. Gislason, *Chem. Phys.* **101**, 227 (1986).
61. G. Parlant and E. A. Gislason, *J. Chem. Phys.* **86**, 6183 (1987); G. Parlant and E. A. Gislason, in *Electronic and Atomic Collisions*, H. B. Gilbody, W. R. Newell, F. H. Read, and A. C. H. Smith, Eds., North-Hollands, Amsterdam, 1988, p. 357.
62. G. Parlant and E. A. Gislason, *J. Chem. Phys.* **86**, 6183 (1987).
63. G. Parlant and E. A. Gislason, *J. Chem. Phys.* **88**, 1633 (1988).
64. G. Parlant and E. A. Gislason, *J. Chem. Phys.* **91**, 5359 (1989).
65. P. Archirel and B. Levy, *Chem. Phys.* **106**, 51 (1986).
66. E. E. Nikitin, M. Y. Ovchinnikova, and D. V. Shalashilin, *Chem. Phys.* **111**, 313 (1987).
67. E. A. Gislason and G. Parlant, *Comments At. Mol. Phys.* **19**, 157 (1987).
68. P. M. Guyon and E. A. Gislason, *Topics in Current Chem.* **151**, 161 (1989).
69. V. Sidis, *Adv. At. Mol. Phys.* **26**, 161 (1989).
70. P. M. Guyon (private communication).
71. T. Nakamura, N. Kobayashi, and Y. Kaneko, *J. Phys. Soc. Jpn.* **55**, 3831 (1986).
72. G. D. Flesch and C. Y. Ng, *J. Chem. Phys.* (to be published).
73. J. W. Gallagher, C. E. Brion, J. A. R. Samson, and P. W. Langhoff, *J. Phys. Chem. Ref. Data* **17**, 9 (1988).
74. G. R. Wight, M. J. Van der Wiel, and C. E. Brion, *J. Phys. B* **9**, 675 (1976).
75. J. Fryar and R. Browning, *Planet. Space Sci.* **21**, 709 (1973).
76. T. Masuoka and H. Fujikawa, *J. Chem. Phys.* **84**, 3771 (1986).
77. J. L. Gardner and J. A. R. Samson, *J. Chem. Phys.* **62**, 1447 (1975).
78. S. Krummacher, V. Schmidt, and F. Wuilleumier, *J. Phys. B* **13**, 3993 (1980)
79. C. A. van de Runstraat, F. J. de Heer, and T. R. Govers, *Chem. Phys.* **3**, 431 (1974).
80. P. Erman, *Phys. Scr.* **14**, 51 (1976).
81. J. C. Lorquet and M. Desouter, *Chem. Phys. Lett.* **16**, 136 (1972).

82. A. J. Lorquet and J. C. Lorquet, *Chem. Phys. Lett.* **26**, 138 (1974).
83. A. L. Roche and H. Lefebvre–Brion, *Chem. Phys. Lett.* **32**, 155 (1975).
84. G. Frenking, W. Koch, D. Cremer, J. Gauss, and J. F. Liebman, *J. Phys. Chem.* **93**, 3410 (1989).
85. L. Broström, M. Larrson, S. Mannervick, and D. Sonnek, *J. Chem. Phys.* **94**, 2734 (1991).
86. J. Danon and R. Marx, *Chem. Phys.* **68**, 255 (1982).
87. R. Marx, G. Mauclaire, and R. Derai, *Int. J. Mass Spectrom. Ion Phys.* **47**, 155 (1983).
88. G.-H. Lin, J. Maier, and S. R. Leone, *J. Chem. Phys.* **82**, 5527 (1985).
89. C. E. Hamilton, V. M. Bierbaum, and S. R. Leone, *J. Chem. Phys.* **83**, 2284 (1985).
90. I. Dotan and W. Lindinger, *J. Chem. Phys.* **76**, 4972 (1982).
91. E. A. Gislason, G. Parlant, P. Archirel, and M. Sizun, *Faraday Discuss. Chem. Soc.* **84,** 325 (1987).
92. G. D. Flesch, S. Nourbakhsh, and C. Y. Ng, *J. Chem. Phys.* **95**, 3381 (1991).
93. P. Langevin, *Ann. Chim. Phys.* **5**, 245 (1905).
94. G. Gioumousis and D. P. Stevenson, *J. Chem. Phys.* **29**, 294 (1958).
95. S. Krummacher, V. Schmidt, F. Wuilleumier, J. M. Bizau, and D. Ederer, *J. Phys. B* **16**, 1733 (1983).
96. P. W. Langhoff, S. R. Langhoff, T. N. Rescigno, J. Schirmer, L. S. Cederbaum, W. Domcke, and W. Von Niessen, *Chem. Phys.* **58**, 71 (1981).
97. J. Schirmer and O. Walter, *Chem. Phys.* **78**, 201 (1983).
98. I. H. Hillier, M. F. Guest, A. Ding, J. Karlau, and J. Weise, *J. Chem. Phys.* **70**, 864 (1979).
99. W. Lindinger, H. Villinger, and F. Howorka, *Int. J. Mass Spectrom.* Ion Phys. **41**, 89 (1981).
100. N. Kobayashi, *J. Phys. Soc. Jpn.* **36**, 259 (1974).
101. G. Mauclaire, R. Derai, S. Fenistein, and R. Marx, *J. Chem. Phys.* **70**, 4017 (1979).
102. T. Matsuo, N. Kobayashi, and Y. Kaneko, *J. Phys. Soc. Jpn.* **55**, 3045 (1986).
103. N. G. Adams, D. K. Bohme, D. B. Dunkin, and F. C. Fehsenfeld, *J. Chem. Phys.* **52**, 1951 (1970).
104. D. Smith, C. V. Goodall, N. G. Adams, and A. G. Dean, *J. Phys. B* **30**, 34 (1970).
105. J. B. Laudenslager, W. T. Huntress, Jr., and M. T. Bowers, *J. Chem. Phys.* **61**, 4600 (1974).
106. N. G. Adams and D. Smith, *Int. J. Mass Spectrom. Ion Phys.* **21**, 349 (1976).
107. B. L. Upschulte, R. J. Shul, R. Passarella, R. G. Keesee, and A. W. Castleman, Jr., *Int. J. Mass Spectrom. Ion Phys.* **75**, 27 (1987).
108. M. Sizun and P. J. Kuntz, in Abstracts of Contributed Papers, 2'ECAMP, A. E. de Vries and M. J. van der Wiel, eds., EPS, Amsterdam, 1985.
109. S. Scherbarth and D. Gerlich, *J. Chem. Phys.* **90**, 1610 (1989).
110. A. Ding, J. Karlau, and J. Weise, *Chem. Phys. Lett.* **45**, 92 (1977).
111. C. E. Brion, K. H. Tan, M. J. Van der Wiel, and Ph. E. Van der Leeuw, *J. Electron Spectrosc. and Related Phenomena* **17**, 101 (1979).
112. T. Gustafsson, *Chem. Phys. Lett.* **75**, 505 (1980).
113. P. Morin, I. Nenner, M. Y. Adam, M. J. Hubin–Franskin, J. Delwiche, H. Lefebvre-Brion, and A. Giusti-Suzor, *Chem. Phys. Lett.* **92**, 609 (1982).

114. O. Edqvist, E. Lindholm, L. E. Selin, and L. Åsbrink, *Phys. Scr.* **1**, 25 (1970).
115. P. M. Guyon and I Nenner, *Appl. Opt.* **19**, 4068 (1980).
116. J. L. Gardner and J. A. R. Samson, *J. Chem. Phys.* **62**, 4460 (1975).
117. J. A. R. Samson, G. H. Rayborn, and P. N. Pareek, *J. Chem. Phys.* **76**, 393 (1982).
118. P. M. Dehmer and W. A. Chupka, *J. Chem. Phys.* **62**, 4525 (1975).
119. T. Masuoka, *Z. Phys.* **D4**, 43 (1986).
120. C. J. Danby and J. H. D. Eland, *Int. J. Mass Spectrom. Ion Phys.* **8**, 153 (1972).
121. P. M. Guyon, T. Baer, L. F. A. Ferreira, I. Nenner, A. Tabche-Fouhaile, R. Botter, and T. R. Govers, *J. Phys. B* **11**, L141 (1978).
122. L. J. Frasinski, K. J. Randall, and K. Codling, *J. Phys. B* **18**, L129 (1985).
123. M. Richard-Viard, O. Dutuit, M. Lavollée, T. R. Govers, P. M. Guyon, and J. Durup, *J. Chem. Phys.* **82**, 4054 (1985).
124. R. G. C. Blyth, I. Powis, and C. J. Danby, *Chem. Phys. Lett.* **84**, 272 (1981).
125. R. Bombach, A. Schmelzer, and J. P. Stadelmann, Int. J. Mass Spectrom. Ion Phys. 43, 211 (1982).
126. T. Akahori, Y. Morioka, M. Watanabe, T. Hayaishi, K. Ito, and M. Nakamura, *J. Phys. B* **18**, 2219 (1985).
127. K. Codling, L. J. Frasinski, and K. J. Randall, *J. Phys. B* **18**, L251 (1985).
128. J. T. Moseley, P. C. Cosby, J. B. Ozenne, and J. Durup, *J. Chem. Phys.* **70**, 1474 (1979).
129. P. C. Cosby, J. B. Ozenne, J. T. Moseley, and D. L. Albritton, *J. Mol. Spectrosc.* **79**, 203 (1980).
130. J. C. Hansen, M. M. Graff, J. T. Moseley, and P. C. Cosby, *J. Chem. Phys.* **74**, 2195 (1981).
131. J. C. Hansen, J. T. Moseley, A. L. Roche, and P. C. Cosby, *J. Chem. Phys.* **77**, 1206 (1982).
132. A. Carrington, P. G. Roberts, and P. J. Sarre, *Mol. Phys.*, **35**, 1523 (1978).
133. M. Carre, M. Druetta, M. L. Gaillard, H. H. Bukow, M. Horani, A. L. Roche, and M. Velghe, *Mol. Phys.* **40**, 1453 (1980).
134. N. H. F. Beebe, E. W. Thulstrup, and A. Andersen, *J. Chem. Phys.* **64**, 2080 (1976).
135. A. Gerwer, C. Asaro, B. V. McKoy, and P. W. Langhoff, *J. Chem. Phys.* **72**, 713 (1980).
136. K. Siegbahn, C. Nordling, G. Johansson, J. Hedman, P. F. Heden, K. Hamrin, U. Gelius, T. Bergmark, L. O. Werme, R. Manne, and Y. Baer, *ESCA Applied To Free Molecules*, North-Holland, Amsterdam, 1969.
137. W. E. Moddeman, T. A. Carlson, M. O. Krause, B. P. Pullen, W. E. Bull, and G. K. Schweitzer, *J. Chem. Phys.* **55**, 2317 (1971).
138. M. Baer and H. Nakamura, *J. Chem. Phys.* **87**, 4651 (1987).
139. P. J. Kuntz and A. C. Roach, *J. Chem. Soc. Faraday Trans.* II **68**, 259 (1972).
140. S. Chapman and R. K. Preston, *J. Chem. Phys.* **60**, 650 (1974).
141. S. Chapman, *J. Chem. Phys.* **82**, 4033 (1985).
142. M. Baer and J. A. Beswick, *Chem Phys. Lett.* **51**, 360 (1977).
143. M. Baer and J. A. Beswick, *Phys. Rev. A* **19**, 1559 (1979).
144. M. Baer, H. Nakamura, and A. Ohsaki, *Chem. Phys. Lett. 131*, **468** (1986).
145. M. Baer and H. Nakamura, *J. Phys. Chem.* **91**, 5503 (1987).

146. F. M. Campbell, R. Browning, and C. J. Latimer, *J. Phys. B* **13**, 4257 (1980).
147. K. Tanaka, T. Kato, and I. Koyano, *J. Chem. Phys.* **75**, 4941 (1981).
148. F. A. Houle, S. L. Anderson, D. Gerlich, T. Turner, and Y. T. Lee, *J. Chem. Phys.* **77**, 748 (1982); *Chem. Phys. Lett.* **82**, 392 (1981).
149. P. M. Hierl, V. Pacak, and Z. Herman, *J. Chem. Phys.* **67**, 2678 (1977). Bowers, Ed., Academic, New York, 1979, Vol. 1, p. 45.
150. R. M. Bilotta, F. N. Preuninger, and J. M. Farrar, *Chem. Phys. Lett.* **74**, 95 (1980); *J. Chem. Phys.* **73**, 1637 (1980); **74**, 1699 (1981).
151. E. Lindholm, *Adv. Chem. Ser.* **58**, 1 (1966).
152. C.-L. Liao, R. Xu, G. D. Flesch, M. Baer, and C. Y. Ng, *J. Chem. Phys.* **93**, 4818 (1990).
153. H. Gutbier, *Z. Naturforsch.* **12a**, 499 (1957).
154. D. P. Stevenson and D. O. Schissler, *J. Chem. Phys.* **29**, 282 (1958).
155. C. F. Giese and W. B. Maier, II, *J. Chem. Phys.* **39**, 739 (1963).
156. Z. Herman and V. Cermak, *Collection Czech. Chem. Commun.* **28**, 799 (1963).
157. F. S. Klein and L. Friedman, *J. Chem. Phys.* **41**, 1789 (1964).
158. J. B. Homer, R. S. Lehrle, J. C. Robb, and D. W. Thomas, Nature **202**, 795 (1964).
159. J. H. Green and D. M. Pinkerton, *J. Phys. Chem.* **68**, 1107 (1964).
160. V. Aquilanti, A Galli, A. Giardini–Guidoni, and G. G. Volpi, *J. Chem. Phys.* **43**, 1969 (1965).
161. A. G. Harrison and J. J. Myher, *J. Chem. Phys.* **46**, 3276 (1967).
162. A. Henglein, K. Lacmann, and B. Knoll, *J. Chem. Phys.* **43**, 1048 (1965).
163. K. Lacmann and A. Henglein, *Ber. Bunsenges. Phys. Chem.* **69**, 286 (1965).
164. L. D. Doverspike, R. L. Champion, and T. L. Bailey, *J. Chem. Phys.* **45**, 4385 (1966).
165. R. C. Amme and J. F. McIlwain, *J. Chem. Phys.* **45**, 1224 (1966).
166. A. Ding, K. Lacmann, and A. Henglein, *Ber. Bunsenges. Phys. Chem.* **71**, 596 (1967).
167. Z. Herman, K. Kerstetter, T. Rose, and R. Wolfgang, *Discuss. Faraday Soc.* **44**, 123 (1967).
168. R. D. Fink and J. S. King Jr., *J. Chem. Phys.* **47**, 1857 (1967).
169. D. Hyatt and K. Lacmann, *Z. Naturforsch.* **23a**, 2080 (1968).
170. T. F. Moran and P. C. Cosby, *J. Chem. Phys.* **51**, 5724 (1969).
171. M. T. Bowers and D. D. Elleman, *J. Chem. Phys.* **51**, 4606 (1969).
172. N. G. Adams, D. K. Bohme, D. B. Dunkin, and F. C. Fehsenfeld, *J. Chem. Phys.* **52**, 1951 (1970).
173. R. P. Clow and J. H. Futrell, *Int. J. Mass Spectrom. Ion Phys.* **4**, 165 (1970).
174. M. Chiang, E. A. Gislason, B. H. Mahan, C. W. Tsao, and A. S. Werner, *J. Chem. Phys.* **52**, 2698 (1970).
175. K. R. Ryan and I. G. Graham, *J. Chem. Phys.* **59**, 4260 (1973).
176. R. D. Smith, D. L. Smith, and J. H. Futrell, *Chem. Phys. Lett.* **32**, 513 (1975); *Int. J. Mass Spectrom. Ion Phys.* **19**, 395 (1976).
177. E. Teloy and D. Gerlich, *Chem. Phys.* **4**, 417 (1974).
178. W. L. Hodge, Jr., A. L. Goldberger, M. Vedder, and E. Pollack, Phys. Rev. A **16**, 2360 (1977).
179. A. B. Rakshit and P. Warneck, *J. Chem. Phys.* **73**, 2673 (1980).
180. P. R. Kemper and M. T. Bowers, *Int. J. Mass Spectrom. Ion Phys.* **52**, 1 (1983).
181. W. Lindinger, E. Alge, H. Störi, M. Pahl, and R. N. Varney, *J. Chem. Phys.* **67**, 3495 (1977).

182. M. Hamdan, K. Birkinshaw, and N. D. Twiddy, *Int. J. Mass Spectrom. Ion Proc.* **62**, 297 (1984).

183. K. Tanaka, J. Durup, T. Kato, and I. Koyano, *J. Chem. Phys.* **74**, 5561 (1981).

184. G. Henri, M. Lavollée, O. Dutuit, J. B. Ozenne, P. M. Guyon, and E. A. Gislason, *J. Chem. Phys.* **88**, 6381 (1988)

185. T. Nakamura, N. Kobayashi, and Y. Kaneko, *J. Phys. Soc. Jpn.* **55**, 3831 (1986).

186. C. J. Latimer and F. M. Campbell, *J. Phys. B* **15**, 1765 (1982).

187 P. Tosi, F. Boldo, F. Eccher, M. Filippi, and D. Bassi, *Chem. Phys. Lett.* **164**, 471 (1989).

188. J. Ross, J. Light, and K. E. Schuler, in *Kinetic Processes in Gases and Plasma*, A. R. Hochstim, Ed., Academic, New York, 1969.

189. M. Baer, C. Y. Ng, and D. Neuhauser, *Chem. Phys. Lett.* **169**, 534 (1990).

190. M. Baer, C.-L. Liao, R. Xu, S. Nourbakhsh, G. D. Flesch, C. Y. Ng, and D. Neuhauser, *J. Chem. Phys.* **93**, 4845 (1990).

191. D. Neuhauser and M. Baer, *J. Chem. Phys.* **90**, 4351 (1989); *J. Phys. Chem.* **94**, 185 (1990); *J. Chem. Phys.* **91**, 4651 (1989).

192. E. E. Ferguson, F. C. Fehsenfeld, and D. L. Albritton, in *Gas Phase Ion Chemistry*, M. T. Bowers, Ed., Academic, New York, 1979, Vol. 1, p. 45.

193. M. R. Torr and D. G. Torr, *Rev. Geophys. Space Phys.* **20**, 91 (1982).

194. A. Dalgarno and M. B. McElroy, *Planet. Space Sci.* **13**, 947 (1965).

195. R. J. W. Henry, *Astrophys. J.* **161**, 1153 (1970).

196. J. L. Kohl, G. P. Lafyatis, H. P. Palenius, and W. H. Parkinson, *Phys. Rev. A* **18**, 571 (1978).

197. A. Dalgarno, *Adv. At. Mol. Phys.* **15**, 37 (1979).

198. D. R. Bates and H. S. Massey, *Proc. R. Soc. London Ser. A* **192**, 1 (1947).

199. M. McFarland, D. L. Albritton, F. C. Fehsenfeld, E. E. Ferguson, and A, L. Schmeltekopf, *J. Chem. Phys.* **59**, 6610, 6620, 6629 (1973).

200. J. D. Burley, K. M. Ervin, and P. B. Armentrout, *J. Chem. Phys.* **86**, 1944 (1987).

201. B. R. Turner, J. A. Rutherford, and D. M. J. Compton, *J. Chem. Phys.* **48**, 1602 (1968).

202. J. J. Leventhal, *J. Chem. Phys.* **54**, 5102 (1971).

203. J. A. Rutherford and D. A. Vroom, *J. Chem. Phys.* **55**, 5622 (1971).

204. D. L. Albritton, I. Dotan, W. Lindinger, M. McFarland, J. Tellinghuisen, and F. C. Fehsenfeld, *J. Chem. Phys.* **66**, 410 (1977).

205. J. J. Kaufman and W. S. Koski, *J. Chem. Phys.* **50**, 1942 (1969).

206. T. F. O'Malley, *J. Chem. Phys.* **52**, 3269 (1970).

207. A. Pipano and J. J. Kaufman, *J. Chem. Phys.* **56**, 5258 (1972).

208. D. G. Hopper, *Chem. Phys. Lett.* **31**, 446 (1975); *J. Am. Chem. Soc.* **100**, 1019 (1978); *J. Chem. Phys.* **72**, 3679, 4676 (1980); **76**, 1068 (1982); **77**, 314 (1982).

209. F. A. Wolf, *J. Chem. Phys.* **44**, 1619 (1966).

210. G. D. Flesch and C. Y. Ng, *J. Chem. Phys.* **92**, 3235 (1990).

211. M. Oppenheimer, A. Dalgarno, and H. C. Brinton, *J. Geophys. Res.* **81**, 3762 (1976).

212. D. W. Rusch, D. G. Torr, P. B. Hays, and J. C. G. Walker, *J. Geophys. Res.* **82**, 719 (1977).

213. D. G. Torr and N. Orsini, *Planet. Space Sci.* **25**, 1171 (1977).

214. N. Orsini, D. G. Torr, M. R. Torr, H. C. Brinton, L. H. Brace, A. O. Nier, and J. C. Walker, *J. Geophys. Res.* **82**, 4829 (1977).

215. D. G. Torr, K. Donahue, D. W. Rusch, M. R. Torr, A. O. Nier, D. Kayser, W. B. Hanson, and J. H. Hoffman, *J. Geophys. Res.* **84**, 387 (1979).
216. J. Glosik, A. B. Rakshit, N. D. Twiddy, N. G. Adams, and D. Smith, *J. Phys. B* **11**, 3365 (1978).
217. R. Johnsen and M. A. Biondi, *J. Chem. Phys.* **73**, 190 (1980).
218. B. R. Rowe, D. W. Fahey, F. C. Fehsenfeld, and D. L. Albritton, *J. Chem. Phys.* **73**, 194 (1980).
219. M. Lavollee and G. Henri, *J. Phys. B* **22**, 2019 (1989).
220. M. Mendillo, G. S. Hawkins, and J. A. Klobuchar, *J. Geophys. Res.* **80**, 2217 (1975).
221. W. D. Watson, *Accounts Chem. Res.* **10**, 221 (1977).
222. W. W. Duley and D. A. Williams, *Interstellar Chemistry* Academic Press, New York, 1984.
223. K. T. Gillen, B. H. Mahan, and J. S. Winn, *J. Chem. Phys.* **58**, 5373 (1973); **59**, 6380 (1973).
224. H. H. Harris and J. J. Leventhal, *J. Chem. Phys.* **58**, 2333 (1973); **64**, 3185 (1976).
225. F. C. Fehsenfeld, A. L. Schmeltekopf, and E. E. Ferguson, *J. Chem. Phys.* **46**, 2802 (1967).
226. J. K. Kim, L. P. Theard, and W. T. Huntress, Jr., *J. Chem. Phys.* **62**, 45 (1975).
227. D. H. Smith, N. G. Adams, and T. M. Miller, *J. Chem. Phys.* **69**, 308 (1978).
228. J. D. Burley, K. M. Ervin, and P. B. Armentrout, *Int. J. Mass Spectrom. Ion Proc.* **80**, 153 (1987).
229. L. S. Sunderlin and P. B. Armentrout, *Chem. Phys. Lett.* **167**, 188 (1990).
230. R. A. Rouse, *J. Chem. Phys.* **64**, 1244 (1976).
231. G. Chambaud, Ph. Millie, and B. Levy, *J. Phys. B* **11**, L211 (1978).
232. M. Gerard-Ain, *J. Phys. B* **13**, L131 (1980).
233. C. F. Jackels, *J. Chem. Phys.* **72**, 4873 (1980).
234. D. M. Hirst, *J. Phys. B* **17**, L505 (1984).
235. M. González, A. Aguilar, and J. Virgili, *Chem. Phys. Lett.* **113**, 179 (1985).
236. M. González, A. Aguilar, and M. Gilibert, *Chem. Phys.* **131**, 335 (1989); **131**, 347 (1989).
237. J. C. Leclerc, J. A. Horsley, and J. C. Lorquet, *Chem. Phys.* **4**, 337 (1974).
238. A. J. Lorquet and J. C. Lorquet, *Chem. Phys.* **4**, 353 (1974).
239. J. A. Smith, P. Jorgensen, and Y. Öhrn, *J. Chem. Phys.* **62**, 1285 (1975).
240. F. O. Ellison and M. L. McCandlish, *J. Phys. B* **15**, L229 (1982).
241. C. E. Dateo and D. C. Clary, *J. Chem. Soc. Faraday Trans. II* **85**, 1685 (1989).
242. F. Fiquet–Fayard and P.-M. Guyon *Mol. Phys.* **11**, 17 (1966).
243. A. Henglein, in *Molecular Beams and Reaction Kinetics*, Ch. Schlier, Ed., Academic Press, New York, 1970, p. 154.
244. P. B. Armentrout, *Comments At. Mol. Phys.* **22**, 133 (1988).
245. G. D. Flesch and C. Y. Ng, *J. Chem. Phys.* **94**, 2372 (1991).
246. K. E. McCulloh, *Int. J. Mass Sectrom. Ion Phys.* **21**, 333 (1976).
247. M. W. Chase, Jr., C. A. Davies, J. R. Downey, Jr., D. J. Frurip, R. A. McDonald, and A. N. Syverud, *J. Phys. Chem. Ref. Data*, **14**, Suppl. 1 (1985).
248. D. Van Pijkeren, E. Boltjes, J. Van Eck, and A. Niehaus, *Chem. Phys.* **91**, 293 (1984).
249. A. Carrington and R. A. Kennedy. *J. Chem. Phys.* **81**, 91 (1984).
250. S. K. Cole, T. Baer, P.-M. Guyon, and T. R. Govers, *Chem. Phys. Lett.* **109**, 285 (1984); P.-M. Guyon, T. Baer, S. K. cole, and T. R. Govers, *Chem. Phys.* **119**, 145 (1988).

251. F. M. Campbell, R. Browning, and C. J. Latimer, *J. Phys B* **14**, 3493 (1981).
252. S. L. Anderson, F. A. Houle, D. Gerlich, and Y. T. Lee, *J. Chem. Phys.* **75**, 2153 (1981).
253. I. Koyano and Tanaka, *J. Chem. Phys.* **72**, 4858 (1980).
254. P. M. Hierl and Z. Herman, *Chem. Phys.* **50**, 249 (1980).
255. A. Ding, *Faraday Discuss. Chem. Soc.* **67**, 353 (1979).
256. C. H. Douglass, D. J. McClure, and W. R. Gentry, *J. Chem. Phys.* **67**, 4931 (1977).
257. J. R. Krenos, K. K. Lehmann, J. C. Tully, P. M. Hierl, and G. P. Smith, *Chem. Phys.* **16**, 109 (1976).
258. A. B. Lees and P. K. Rol, *J. Chem. Phys.* **61**, 4444 (1974).
259. R. N. Stocker and H. Neumann, *J. Chem. Phys.* **61**, 3852 (1974).
260. L. P. Theard and W. T. Huntress Jr., *J. Chem. Phys.* **60**, 2840 (1974).
261. A. Henglein, *J. Phys. Chem.* **76**, 3883 (1972).
262. W. T. Huntress Jr., D. D. Elleman, and M. T. Bowers. *J. Chem. Phys.* **55**, 5413 (1971).
263. J. Krenos, P. M. Hierl, J. C. Tully, Z. Herman, and R. Wolfgang, *Adv. Mass Spectrom.* **5**, 213 (1971).
264. L. V. Sumin and M. V. Gurev, *Dokl. Akad. Nauk SSSR* **193**, 858 (1970).
266. M. T. Bowers, D. D. Elleman, and J. King, Jr., *J. Chem. Phys.* **50**, 4787 (1969).
266. T. F. Moran and J. R. Roberts, *J. Chem. Phys.* **49**, 3411 (1968).
267. A. G. Harrison and J. C. J. Thynne, *Trans. Faraday Soc.* **64**, 945 (1968).
268. R. H. Neynaber and S. M. Trujillo, *Phys. Rev.* **167**, 63 (1968); **171**, 282E (1968).
269. M. Yamane, *J. Chem. Phys.* **49**, 4624 (1968).
270. J. J. Leventhal, T. F. Moran, and L. Friedman, *J. Chem. Phys.* **46**, 4666 (1967).
271. L. Matus, I. Opauszky, D. Hyatt, A. J. Masson, K. Birkinshaw, and M. J. Henchman, *Discuss. Faraday Soc.* **44**, 146 (1967).
272. L. D. Doverspike and R. L. Champion, *J. Chem. Phys.* **46**, 4718 (1967).
273. P. Warneck, *J. Chem. Phys.* **46**, 502 (1967).
274. D. W. Vance and T. L. Bailey, *J. Chem. Phys.* **44**, 486 (1966).
275. J. H. Futrell and F. P. Abramson, *Adv. Chem. Ser.* **58**, 123 (1966).
276. A. G. Harrison, A. Ivko, and T. W. Shannon, *Can. J. Chem.* **44**, 1351 (1966).
277. M. Saporoschenko, *J. Chem. Phys.* **42**, 2760 (1965).
278. A. Weingartshofer and E. M. Clarke, *Phys. Rev. Lett.* **12**, 591 (1964).
279. B. G. Reuben and L. Friedman, *J. Chem. Phys.* **37**, 1636 (1962).
280. V. L. Talroze, *Izv. Akad. Nauk SSSR Ser. Fiz.* **24**, 1001 (1960).
281. C. J. Latimer, R. Browning, and H. B. Gilbody, *J. Phys. B* **2**, 1055 (1969).
282. W. H. Cramer, *J. Chem. Phys.* **35**, 836 (1961).
283. H. L. Rothwell, B. V. Zyl, and R. C. Amme, *J. Chem. Phys.* **61**, 3851 (1974).
284. F. Wolf, *Ann. Phys.* **29**, 33 (1937).
285. J. H. Simons, C. M. Fontana, H. T. Francis, and L. G. Unger, *J. Chem. Phys.* **11**, 312 (1943).
286. J. H. Simons, C. M. Fontana, Muschlitz, and S. R. Jackson, *J. Chem. Phys.* **11**, 307 (1943).
287. J. B. H. Stedeford and J. B. Hasted, *Proc. R. Soc. London Ser. A* **227**, 466 (1955).
288. W. H. Cramer and A. B. Marcus, *J. Chem. Phys.* **32**, 186 (1960).
289. O. Hollricher, *Z. Phys. 187*, 41 (1965).
290. D. W. Koopman, *Phys. Rev.* **154**, 79 (1967).

291. C. F. Barnett, J. A. Ray, E. Ricci, M. I. Wilker, E. W. McDaniel, E. W. Thomas, and H. B. Gilbody, Oak Ridge National Laboratory Report 5206 (1977).
292. C. W. Eaker and J. L. Muzyka, *Chem. Phys. Lett.* **119**, 169 (1985).
293. C. W. Eaker and G. C. Schatz, *J. Phys. Chem.* **89**, 2612 (1985).
294. C.-Y. Lee and A. E. DePristo, *J. Chem. Phys.* **80**, 1116 (1984).
295. J. T. Muckerman, in *Theoretical Chemistry*, D. Henderson, Ed., Academic, New York, 1981.
296. J. R. Stine and J. T. Muckerman, *J. Chem. Phys.* **68**, 185 (1978); **65**, 3975 (1976).
297. T. F. Moran, K. J. McCann, M. R. Flannery, and D. L. Albritton, *J. Chem. Phys.* **65**, 3172 (1976).
298. M. R. Flannery, J. V. Hornstein, and T. F. Moran, *Chem. Phys. Lett.* **32**, 455 (1975).
299. T. F. Moran, M. R. Flannery, and D. L. Albritton, *J. Chem. Phys.* **62**, 2869 (1975).
300. F. A. Wolf and J. L. Haller, *J. Chem. Phys.* **52**, 5910 (1970).
301. D. R. Bates and R. H. G. Reid, *Proc. R. Soc. London Ser. A* **1**, 310 (1969).
302. H. Eyring, J. O. Hirschfelder, and H. S. Taylor, *J. Chem. Phys.* **4**, 479 (1936).
303. M. Baer and C. Y. Ng, *J. Chem. Phys.* **93**, 7787 (1990).
304. C. W. Eaker and G. C. Schatz, *J. Chem. Phys.* **89**, 6713 (1988).
305. C.-Y. Lee, A. E. DePristo, C.-L. Liao, C.-X. Liao, and C. Y. Ng, *Chem. Phys. Lett.* **116**, 534 (1985); S. K. Cole and A. E. DePristo, *J. Chem. Phys.* **85**, 1389 (1986).
306. J. C. Tully and R. K. Preston, *J. Chem. Phys.* **55**, 562 (1971).
307. F. O. Ellison, *J. Am. Chem. Soc.* **85**, 3540 (1963).
308. P. J. Kuntz, in *Theory of Chemical Reaction Dynamics*, M. Bear, Ed., CRC Press, Boca Raton, 1985, Vol. I, Chapter 2.
309. C.-Y. Lee and A. E. DePristo, *J. Am. Chem. Soc.* **105**, 6775 (1983).
310. R. F. Borkman and M. Cobb, *J. Chem. Phys.* **74**, 2920 (1981).
311. R. W. Nicholls, *J. Phys. B* **1**, 1192 (1969).
312. W. A. Chupka and J. Berkowitz, *J. Chem. Phys.* **51**, 4244 (1969).
313. P. M. Dehmer and W. A. Chupka, *J. Chem. Phys.* **65**, 2243 (1976).
314. C. W. Eaker and G. C. Schatz (private communication).
315. N. C. Blasis and D. G. Truhlar, *J. Chem. Phys.* **79**, 1334 (1983).
316. C. A. Mead and D. G. Truhlar, *J. Chem. Phys.* **84**, 1055 (1986).
317. K. Yamashita and K. Morokuma, *J. Chem. Phys.* **91**, 7477 (1989).
318. K. Norwood and C. Y. Ng (unpublished results).

CROSSED-MOLECULAR BEAM STUDIES OF STATE-TO-STATE REACTION DYNAMICS

JEAN H. FUTRELL

Department of Chemistry and Biochemistry
University of Delaware
Newark, Delaware

CONTENTS

I. INTRODUCTION

As is amply demonstrated elsewhere in this volume, the investigation of the state-to-state reaction dynamics of ion–molecule reactions has developed rapidly over the past several years, paralleling in many respects similar developments concerning the kinematics of neutral reactions. The main goal of these studies is the elucidation of the microscopic mechanism of reaction that describes precisely how reactants evolve into products. Intimately connected with this question is the disposal of energy—internal, translation, and reaction exo- or endothermicity—in the reaction products. We shall discuss in this chapter some particular insights that were obtained using the combined angle- and energy-resolved measurement capabilities of the crossed-molecular-beam method as practiced in our laboratory.

Most of the crossed-beam data demonstrating state-specific, angular-specific reactive scattering were obtained in the study of charge-transfer

State-Selected and State-to-State Ion–Molecule Reaction Dynamics, Part 1: Experiment, Edited by Cheuk-Yiu Ng and Michael Baer. Advances in Chemical Physics Series, Vol. LXXXII.
ISBN 0-471-53258-4

reaction dynamics. Since this necessarily involves electronic transitions and, for polyatomic systems, nonadiabatic transitions involving the breakdown of the Born–Oppenheimer approximation, it is plausible that this is an appropriate place to search for interesting dynamical effects. Accordingly, this chapter will review selected data mainly involving charge transfer in atomic and small molecule systems. We shall also describe some new results for a quite different class of problems, the collision-induced dissociation of polyatomic ions. For these systems we find quite unexpectedely efficient coupling of translational and electronic energy in low-energy collisions. This is demonstrated in quite specific reactive scattering, which is strongly dependent on collision energy, just as has been observed in charge transfer.

Most of the results will be presented as cartesian probability contour diagrams, frequently referred to simply as scattering diagrams (or Newton diagrams), showing the relative intensity of the product ion at a particular scattering angle (defined with respect to the reactant-ion-beam direction as zero angle) and at a particular velocity (relative to the center-of-mass velocity). Both scattering angle and relative velocity are defined in the center-of-mass (CM) reference frame. Except where otherwise stated, the CM velocity is calculated using the measured average velocities of the reactant ion and neutral beams.

II. CHARGE TRANSFER

A. Fine-Structure Transitions in the Rare Gases Ar, Kr, and Xe

Charge-transfer reactions for simple ion–atom systems are well suited for investigating details of energy-conversion processes in single-collision chemical reactions. The precision with which translational-energy changes are readily measured permits such changes to be interpreted as changes in specific internal quantum states as the reactants evolve into products. Several studies have used state-selected reactant ions to probe these processes in considerable detail.[1–8] Total cross sections for these processes as a function of ion internal state and translational energy have been reported.[1–3] Fine-structure transitions in the reaction $Ar^+(Ar, Ar)Ar^+$ have been investigated using crossed beams by McAfee et al., at a collision energy of 61.5 eV.[4] Liao, Liao, and Ng have measured total cross sections for state-selected reactant ions with state analysis of products over the collision energy range 0.5–1000 eV for the same reaction.[3] We have measured differential cross sections at lower energy for the ion–atom pairs $Ar^+(Ar, Ar)Ar^+$,[7] $Kr^+(Kr, Kr)Kr^+$,[5,6] and $Xe^+(Xe, Xe)Xe^+$[8] using state-selected reactant ions.

The state-to-state analysis of charge transfer in our experiments is based on simple expressions derived from the conservation of energy. The total energy, E_{tot}, is the sum of the relative translational energy of the reactants, T, and the internal energy, U.

$$E_{tot} = T + U = T' + U' \tag{1}$$

The unprimed quantities refer to reactants and the primes to products. It therefore follows that

$$\Delta T = T' - T = (U' - U) \tag{2}$$

Measurement of translational-energy changes can therefore be used to investigate internal-energy changes in charge-transfer reactions. For low-energy collisions in the homonuclear atomic ion–atom cases considered here, ΔT is precisely correlated with the endothermicity of the fine-structure ${}^2P_{3/2} \rightarrow {}^2P_{1/2}$ transition.

Figure 1 is an experimental scattering diagram which demonstrates that both the resonant charge-transfer reaction [reaction [3a)] and the endoergic fine-structure reaction [reaction (3b)] are observed

$$\mathrm{Kr}^+({}^2P_{3/2}) + \mathrm{Kr}({}^1S_0) \rightarrow \mathrm{Kr}({}^1S_0) + \mathrm{Kr}^+({}^2P_{3/2}) \tag{3a}$$

$$\mathrm{Kr}^+({}^2P_{3/2}) + \mathrm{Kr}({}^1S_0) \rightarrow \mathrm{Kr}({}^1S_0) + \mathrm{Kr}^+({}^2P_{1/2}) \tag{3b}$$

at 9.3 eV collision energy. The dashed lines indicate the loci of velocity vectors for the products $\mathrm{Kr}^+({}^2P_{3/2})$ and $\mathrm{Kr}^+({}^2P_{1/2})$. The exactly resonant reaction (3a) product is located on the relative velocity vector with essentially the same velocity distribution as the reactant neutral Kr beam prior to collision (measured experimentally using electron beam ionization at the collision center). As will be discussed, we suggest that a direct mechanism involving electron exchange occurring predominantly from large-impact-parameter collisions correctly describes this reaction at 9 eV. However, reaction (3b) exhibits large-angle scattering typically associated with small-impact-parameter collisions. We suggest that this transition may occur via a curve-crossing mechanism in the repulsive part of the Kr_2^+ potential energy curve in a region where the shapes of the curves are strongly influenced by spin–orbit coupling.

The results of a similar scattering experiment designed to investigate the analogous reactions

$$\mathrm{Ar}^+({}^2P_{3/2}) + \mathrm{Ar}({}^1S_0) \rightarrow \mathrm{Ar}({}^1S_0) + \mathrm{Ar}^+({}^2P_{3/2}) \tag{4a}$$

$$\mathrm{Ar}^+({}^2P_{3/2}) + \mathrm{Ar}({}^1S_0) \rightarrow \mathrm{Ar}({}^1S_0) + \mathrm{Ar}^+({}^2P_{1/2}) \tag{4b}$$

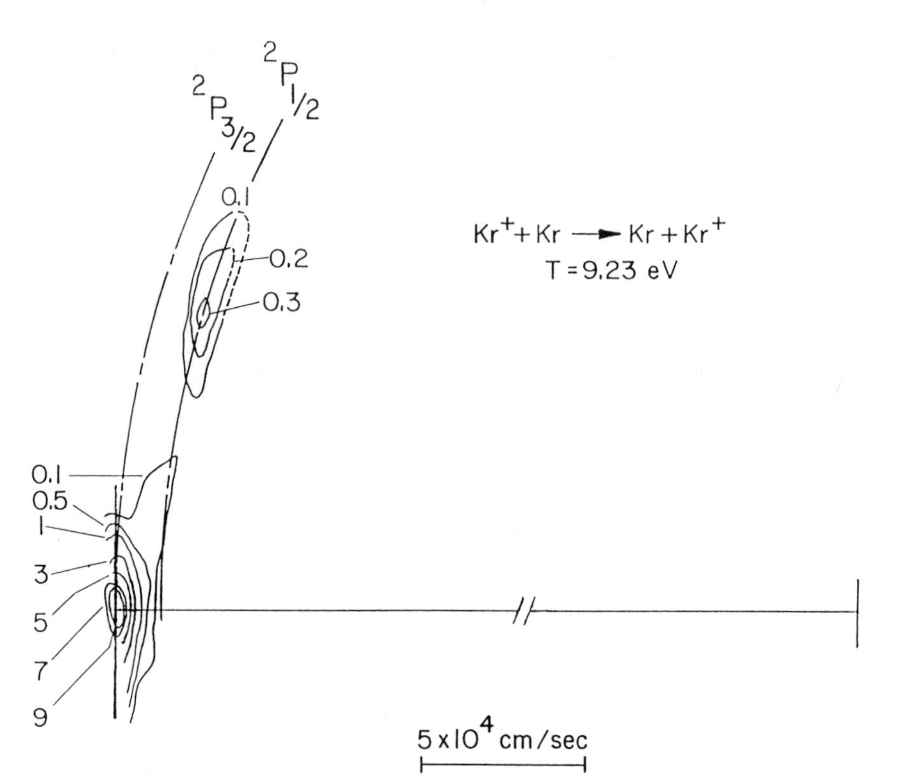

Figure 1. Scattering contour diagram for the charge-transfer reaction $Kr^+(^2P_{3/2})(Kr, Kr)Kr^-$ at 9.23 eV collision energy. The vertical lines mark the center-of-mass (CM) velocity and the CM velocity of Kr prior to collision. The break in line indicates that the velocity vector difference between the Kr neutral and the CM velocity vector is not to scale. Contours define the cartesian probability densities of product velocities and illustrate that a direct mechanism populates a resonant charge-transfer channel and that the slightly endoergic fine-structure transition producing $Kr^+(^2P_{3/2})$. Reprinted from Ref. 5 with the kind permission of the Canadian Institute of Physics.

at the same CM energy are presented in Fig. 2. In contrast with Kr^+/Kr, the exactly resonant $Ar^+(^2P_{3/2})$ charge-transfer product from reaction (4a) masks the intensity of the fine-structure transition $Ar^+(^2P_{1/2})$ product at or near the relative velocity vector in Fig. 2. The experimental evidence that $Ar^+(^2P_{1/2})$ is formed is broadening of the charge-transfer peak as an incompletely resolved shoulder located at the proper angle and energy for this endothermic channel. The data require deconvolution to remove the dominant resonant charge-transfer product in order to expose the fine-structure channel relative intensity.

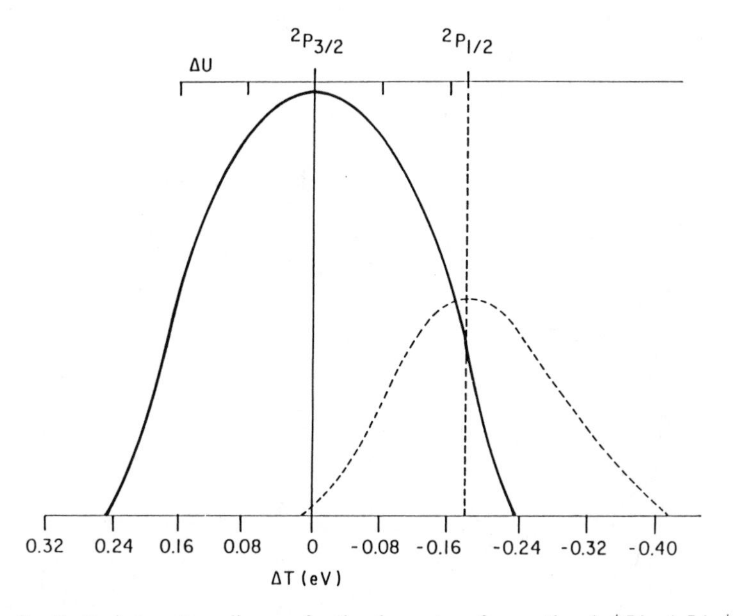

Figure 2. Scattering contour diagram for the charge-transfer reaction Ar^+ [Ar, Ar] Ar^+ at 9.28 eV collision energy. The vertical lines mark the center-of-mass (CM) velocity of Ar prior to collision. Contours define the probability densities of product velocities and illustrate that a long-range electron-transfer mechanism populates a resonant charge-transfer channel; peak broadening suggests that the slightly endoergic reaction fine-structure transition may also be populated at this collision energy. Reprinted from Ref. 7 with the kind permission of the American Institute of Physics.

Our deconvolution procedure has been described elsewhere[5]; it is based on the assumption that the resonant charge-transfer channel involves no momentum change, a hyopothesis partly validated by Fig. 1. The velocity distribution of this product is therefore assumed to be identical to the neutral-beam velocity vector distribution. Accordingly, the measured energy and angular profiles of the neutral beam (ionized by an off-axis electron gun located at the collision center) were normalized to the observed intensity maximum in Fig. 2 and subtracted from the observed reaction product intensities. The resultant difference contour diagram in which the resonant channel reaction (3) suppressed is shown in Fig. 3. The original normalization of the contour diagram at maximum peak intensity equal to 10 in Fig. 2 is preserved in Fig. 3.

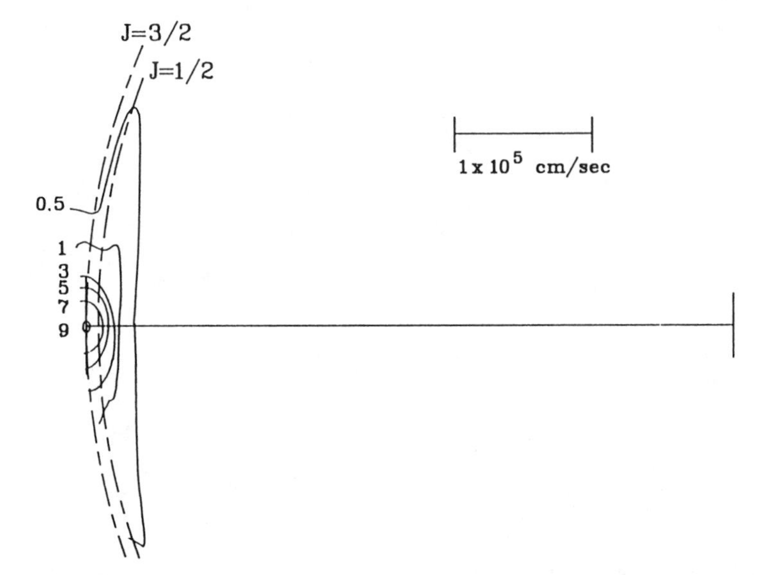

Figure 3. Scattering diagram for the fine-structure reaction channel present in the charge-transfer reaction Ar^{+} [Ar, Ar] Ar^{+} at 9.28 eV collision energy. The crosses mark the CM velocity and the CM velocity of Ar prior to collision. Contours define the probability densities of product velocities and domonstrate that after subtracting the exactly resonant reaction in Fig. 2 that the fine-structure transition channel contributes significantly to the overall cross section. Reprinted from Ref. 7 with the kind permission of the American Institute of Physics.

Figure 3 demonstrates that the fine-structure transition channel exhibits modest but somewhat larger angular scattering than the resonant channel in Fig. 2. In contrast with our observations for Kr^{+}/Kr[5,6] and Xe^{+}/Xe[8] in which the fine-structure transition channel exhibits well-resolved and energy-dependent angular scattering (in the same low-collision-energy range), the maxima for both channels are found on the relative velocity vector for Ar^{+}/Ar at all energies investigated. Very likely this reflects the much closer energy spacing of the $^{2}P_{3/2}$ and $^{2}P_{1/2}$ levels in Ar^{+} (0.178 eV) rather than any fundamental difference in mechanism. In all three systems the fine-structure transition mechanism exhibits larger angular scattering than the exactly resonant charge-transfer product.

While scattering diagrams provide the maximum information about mechanism, they are misleading with regard to the relative cross sections for the two processes. This results from the fact that the scattering diagram displays the relative probability of events in the plane of the collision. Integration over scattering angles and velocities is required to deduce the relative importance of the two processes, which are defined by their characteristic CM velocity vectors. The appropriate transformation is[9]

$$P(T') = u \int P_c(u, \Theta) \sin \Theta \, d\Theta \tag{5}$$

where $P(T')$ is the total probability that a product is formed with kinetic energy T', u and Θ are the CM velocity and scattering angle for the detected product, and $P_c(u, \Theta)$ is the probability density plotted in the contour diagrams described above.

Figure 4 illustrates the results of integrating the original and deconvoluted scattering diagrams in Figs. 2 and 3 using Eq. (5). The integration confirms that the endoergic product matches the energy shift for the fine-structure channel, reaction (4b). We can also deduce from this figure that the ratio, $\sigma_{3/2 \to 1/2}/\sigma_T$, is 0.17 ± 0.05 at 9.3 eV collision energy. Data for this ratio from a series of experiments are plotted as a function of collision energy in Fig. 5, which also includes information for Kr^+/Kr[5,6] and Xe^+/Xe.[8] This figure compares the experimentally determined values of the ratio of these cross sections with Johnson's theoretical calculations.[10,11] The solid line theoretical curve labeled Ar in Fig. 5 indicates that the predicted onset energy for argon to have an observable threshold is about 16 eV collision energy and that the maximum should occur in the neighborhood of 700 eV for argon (and at much higher values for krypton and xenon). The theoretical curves plotted are the relative probability calculated for the fine-structure transition occurring in the charge-transfer channel. In the original references[10,11] these curves are the calculations that include rotational coupling and are labeled CT (for charge transfer; the direct, noncharge, transfer channel also calculated by Johnston is not measurable in crossed beam experiments).

Experimental results by Liao et al.[3] for argon charge transfer are included in Fig. 5 along with the single data point of McAfee et al.[4] (at a laboratory collision energy of 123 eV, which corresponds to a center-of-mass collision energy of 61.5 eV). The results of Liao et al.[3] are in good agreement with the theoretical curve for Ar^+/Ar demonstrating that the ratio of the cross sections is well described by the Johnson[10,11]/Nikitin[12] theory at high collision energy. However, this theory does not account for our results for charge transfer at lower collision energy. For example, we find the onset of the fine-structure transition for Ar^+/Ar is below 1.94 eV rather than the predicted value of 16 eV. The maximum in the cross-section ratio occurs at about 30 eV in our experiments and drops significantly at energies below the rise in the

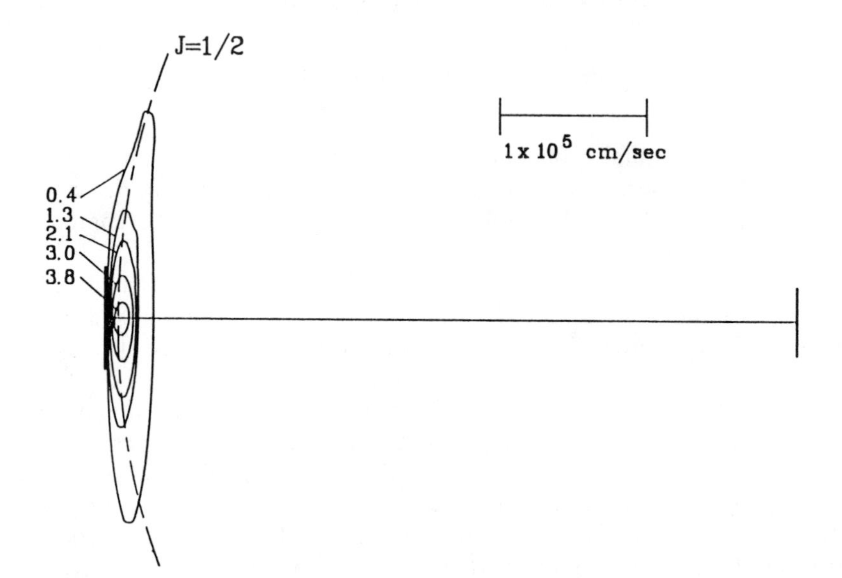

Figure 4. Relative translational energy distributions of the products expressed in terms of the translational exoergicity observed for the two channels in the reaction Ar^+ [Ar, Ar] Ar^+ at 9.28 eV collision energy. The solid curve shows the contribution of the resonant reaction, and the dashed line shows the contribution of the fine-structure reaction. The correlation with Δu is shown with a superposed scale for the $J = 1/2$ and $3/2$ states of Ar^+ product. The relative cross sections for $J = 1/2$ and $3/2$ are proportional to the areas of these curves. Reprinted from Ref. 7 with the kind permission of the American Institute of Physics.

excitation function for the higher-energy mechanism predicted by theory. The situation for Kr^+/Kr and Xe^+/Xe is even more dramatic. Our results show these reactions to occur in the same low-energy regime, while theory predicts a shift to kilovolt energies for observing the analogous fine-structure transitions.

Clearly the mechanism that is operative in the crossed-beam investigations at low energy is not the one described theoretically by Johnson.[10,11] As he pointed out, the principal uncertainties in the calculations involve the detailed shapes of the six ground-state potential curves for the rare-gas dimer ions when spin–orbit coupling is included. Johnson's treatment emphasizes the long-range potential and does not include the repulsive region and the

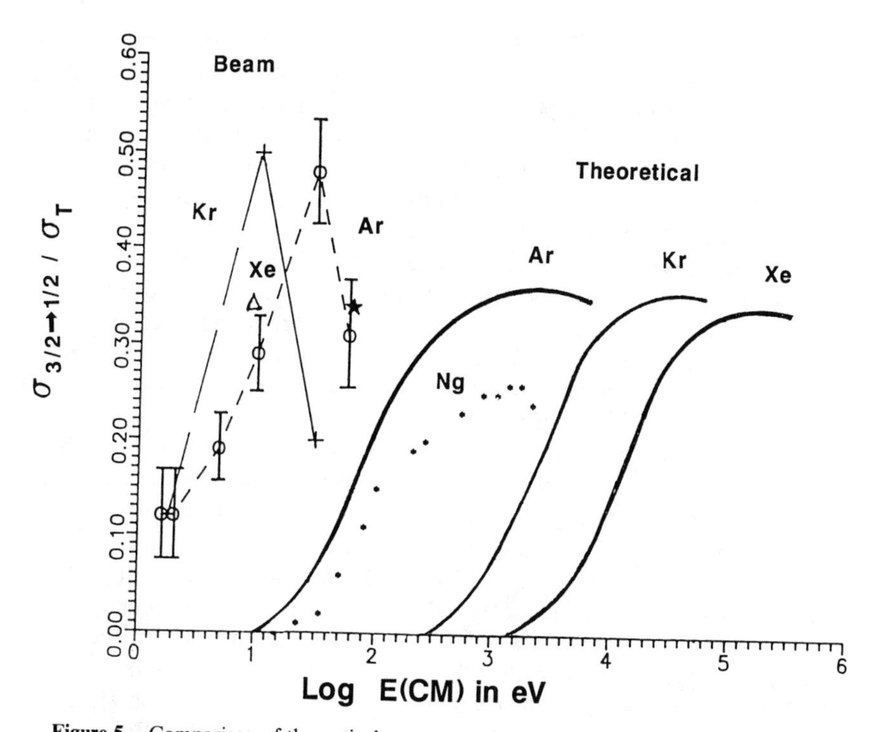

Figure 5. Comparison of theoretical curves as calculated by Johnson (Refs. 10 and 11) for the fine-structure transitions in argon and krypton and xenon with experimenta data. Data of Ng and co-workers (Ref. 3) for argon are shown by filled circles and the experimental data for argon, krypton, and xenon from our work are shown by circles, crosses, and a triangle, respectively. The argon data point of McAfee et al. (Ref. 4) is shown as a star *.

potential-well region, both of which are likely important in our low-energy experiments. More accurate recent calculations provide improved potential curves for the rare-gas dimer ions carried out by Wadt[13] and by Stace and co-workers.[14] It is quite possible that trajectory calculations utilizing these potentials would explain the low-energy mechanism satisfactorily.

We also note that the Johnson theory is an impact-parameter method for linear trajectories. In his treatment, the exchange potentials were estimated by using a one-electron Coulomb wave function as in the Rapp and Francis model for symmetric charge-transfer reactions.[15] A first-order approximation of the exchange interaction was obtained using the Firsov integration method.[16] In this approximation only the lead terms in $1/R$ were used (R is the internuclear separation of Ar_2^+) for the interaction potential and rectilinear trajectories for a range of impact parameters were numerically integrated. Such factors as the high binding energy of the rare-gas dimer ions (1.34 eV[13] for the $^2\Sigma_u^+$ state of Ar_2^+) and the use of more accurate interaction potential curves that properly incorporate spin–orbit coupling effects would substantially perturb collision trajectories, particularly at low energy. Since the detailed shapes of the curves in the avoided crossing region determine transition probabilities, it is not surprising that theory does not account properly for reaction probabilities at low energy (which we surmise occur primarily in small R, intimate collisions).

No evidence was found for a low-energy mechanism for reaction (4b) in the elegant photoionization study by Liao, Liao, and Ng,[3] whose data are also shown in Fig. 5. These workers determined state-to-state reaction cross sections using a photoionization ion–neutral beam apparatus. They utilized a chemical-detector method that relied on the quite different cross sections for the ion–molecule reactions of $Ar^+(^2P_{1/2})$ and $Ar^+(^2P_{3/2})$ with H_2 to determine the fine-structure population of product ions. The good agreement between their results and theory in Fig. 5 serves to validate that the theoretical description is an essentially correct description of charge transfer in rare-gas systems at intermediate- and high-translational energy. However, as we have demonstrated, the ability to scan products in angle as well as energy greatly enhances the probability that scattered product ions can be detected. Indeed, it is the angular measurement capability of the crossed-beam method that enabled McAfee et al.[4] to detect the fine-structure transitions in argon at 123 eV laboratory energy. Their value for $\sigma_{3/2\to 1/2}/\sigma_T$ at this energy is 0.35 versus 0.07 as reported by Liao et al.[3] This value is plotted in Fig. 5 and is in excellent agreement with our value at approximately the same energy.

Liao et al.[3] were well aware that angular scattering associated with reactions (4a, b) could affect their total and partial cross-section measurements. In their discussion of this point, they analyze the laboratory angular scattering expected for 5 eV $Ar^+(^2P_{3/2})$ ions colliding with a supersonic jet of Ar at 90°. They estimate that product ions could be scattered by up to 15° from the mean neutral-beam velocity vector if the center-of-mass scattering angle is zero. Since their beam is restricted experimentally to a divergence angle of 10° and the acceptance angle of the detector is estimated as ±25°, they conclude that all products should be collected by their apparatus. At higher collision energies their experimental procedure allows the use of higher extraction voltages, which collapse the laboratory collection angle about the neutral velocity vector and there is little doubt that the high-energy data correctly represent the relative cross sections. However, our discovery that *nonzero* center-of-mass scattering angles are associated with reaction (4b) at energies below the theoretically predicted threshold implies that a significant fraction of $Ar^+(^2P_{1/2})$ product may be lost at low energies in their experiments. We suggest this tentative explanation for both their failure to detect products from the low-energy mechanism below 45 eV laboratory collision energy (approximately 22 eV center-of-mass); this also rationalizes the discrepancy between their value of 0.07 for this ratio at 61 eV collision energy and 0.35 obtained in the two crossed-beam studies, both of which have angular scanning capabilities.

It is clear that the fine-structure transitions detected in our differential cross-section apparatus and plotted at the low-energy end of Fig 5 are quite different phenomenologically from those occurring at high energy. They all appear in the same low energy range (ca. 0–15 eV CM) and do not scale with the $^2P_{3/2}$–$^2P_{1/2}$ energy spacing as theoretically predicted. Significant angular momentum exchange accompanies the transition (see Fig. 1, for example), and this mechanism peaks at low energy and declines with increasing collision energy well below the onset of the "normal" transition treated by the Johnson theory. More accurate potential curves[13,14] demonstrate that the I$(1/2)_u$ curve [which has $Ar^+(^2P_{3/2})$ as its asymptotic limit approaches very closely the II$(1/2)_u$ curve [which has $Ar^+(^2P_{1/2})$ as its asymptotic limit] in the repulsive region, and we suggest curve crossing between these curves may be responsible for the low-energy angular deflection mechanism. Subtleties in the shapes of the six potential curves in the bound and repulsive region revealed in the most accurate calculation to date[14] are likely to be important at low energy where the transition from adiabatic to diabatic behavior occurs.

Our crossed-beam scattering results for all three systems are summarized in Fig. 5, which also shows theoretical curves for relative cross sections for the fine-structure transitions calculated by Johnson.[10,11] Theory predicts an observable "threshold" for the endothermic, fine-structure transition occurring in charge-transfer collisions at collision energies of about 16 eV for Ar, 700 eV for Kr, and 2000 eV for Xe. Broad maxima are predicted (and found for Ar) at 1500 eV, 5000 eV, and 35,000 eV, respectively. In contrast our data demonstrate that these reactions occur with much lower threshold energies lower energy maxima—thresholds below 1, 9 and 10 eV for Ar,[7] Kr, and Xe[8] and have maxima well below the theoretically predicted *thresholds.*

In conclusion, this study illustrates the importance of the angular scanning capability intrinsic to crossed-beam measurements for characterizing state-specific reactions in charge transfer. From the viewpoint of translational spectroscopy, the small energy change in the $Ar^+(^2P_{3/2}) \rightarrow Ar^+(^2P_{1/2})$ transition is resolved in our experiments mainly because of its angular scattering characteristics rather than its energy shift. The capability to measure both kinetic energy and angle has permitted the resolution of the dynamics features discussed in this section.

Previous experimental and theoretical studies of rare gases have suggested that this relatively simple example of charge-transfer reactions is completely understood. As shown in Fig. 5, this supposition is clearly not the case. Qualitative explanations for these low-energy processes, which differ dramatically in their dynamical characteristics from those treated by the theory, were suggested. A true understanding of these phenomena for each respective rare gas will require trajectory calculations using accurate potential energy curves for at least the six ground-state potential curves for R_2^+ having $R^+(^2P_{3/2})/R(^1S_0)$ and $R^+(^2P_{1/2})/R(^1S_0)$ as their limits. Accurate potential curves have just been computed for Ar_2^+ [14] and could be calculated, in principle, for Kr_2^+ with equal accuracy.

B. Charge Transfer of N_2^+ with N_2

Symmetrically resonant charge exchange in molecular systems differs from the atomic systems just discussed in that the exact symmetry is broken by molecular orientation with respect to the collision axis. Thus, the description of the collision intermediate in terms of the internuclear distance as the only important dimension is no longer possible. Vibrational and rotational energy transfer in molecular systems further complicates the dynamics of charge-exchange collisions.

The nitrogen system has extensively been investigated as a simple example of resonant charge transfer for molecular systems. Total charge transfer cross sections were measured by Ghosh and Sheridan,[17] Gustafsson and Lindholm,[18] Stebbings, Turner, and Smith,[19] Flannery, Cosby, and Moran,[20] and Utterbach and Miller.[21] Theoretical studies relating to charge transfer in this system have been carried out by Flannery et al.,[20,22] by deCastro, Shaefer, and Pitzer,[23] and by McAfee, Szmanda, Hozack, and Johnson.[24] One dynamics investigation of the system at the relatively high energy of 55.5 eV impact CM has been carried out by McAfee et al.[24,25] Binding energies for the N_4^+ intermediate have been measured by Linn, Ono, and Ng,[26] by Stefan, Märk, Futrell, and Helm,[27] and by Norwood, Luo and Ng.[28] State-to-state studies of nitrogen charge transfer have been carried out by Mahan, Martner, and O'Keefe[29] in an ion trap using the laser-induced fluorescence method (LIF).

The low-energy-limit reaction dynamics of this system have been characterized using the laser-induced fluorescence method to measure the nascent vibrational, rotational, and (by Doppler widths) translational distributions for N_2^+ formed by electron impact and these distributions after one to three collisions.[29] The "effective vibrational temperature" of several thousand degrees Kelvin for the nascent distribution is determined largely by the Franck–Condon factors for vertical ionization. The translational temperature of the ions was also several thousand degrees Kelvin in the electrostatic ion trap used for these experiments, reflecting the formation of ions in a potential gradient in the trap cell. In contrast, the rotational distribution was characterized by a temperature of the order of 300 K. This much cooler distribution results from the fact that the ionizing electron imparts very little angular momentum to the heavy molecular species; the resultant distribution mirrors that of the room-temperature neutral precursor.

These characteristic temperatures changed dramatically after an average of one N_2^+/N_2 collision had occurred.[29] Nearly complete quenching of $N_2^+(X\,^2\Sigma_g, v=1)$ is observed, resulting from the nearly resonant charge-transfer reaction

$$N_2^+(X\,^2\Sigma_g^+, v=1) + N_2^+(X\,^1\Sigma_g, v=0) \rightarrow N_2(X\,^1\Sigma_g, v=1) + N_2^+(X\,^2\Sigma_g, v=0), \tag{6}$$

which is endothermic by only $150\,\mathrm{cm}^{-1}$. This quenching reaction occurs with about unit efficiency, much too rapidly for conventional $V-T$ energy relaxation. Translational "cooling" and rotational "heating" were also shown to occur as a result of ion–neutral collisions, which rapidly brought the translational, vibrational, and rotational temperatures to their steady-state values.

Figure 6 is the CM probability contour plot measured for the charge-transfer reaction at 0.74 eV collision energy.[30] At this kinetic energy (approximately twice the translational energy investigated in the LIF experiment[29]), two mechanisms are observed with distinct characteristics

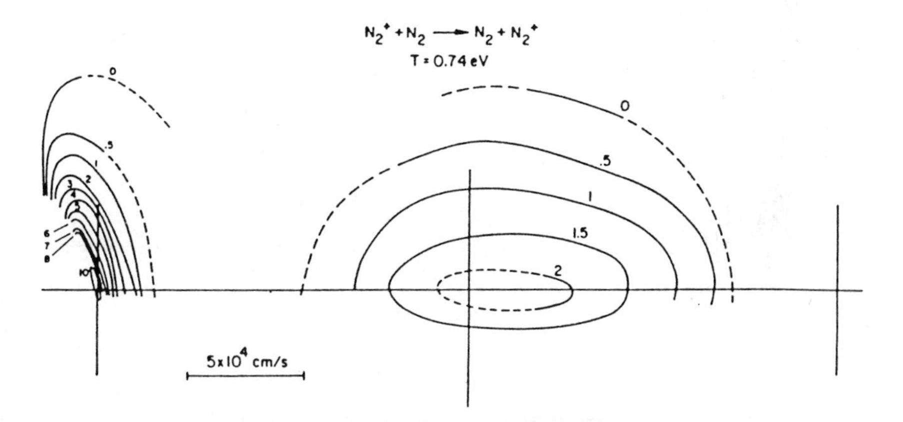

Figure 6. Scattering diagram for the charge-transfer reaction $N_2^+(N_2, N_2)N_2^+$ at 0.74 eV collision energy. The crosses mark the velocity of the center-of-mass of the system and the CM velocity of N_2 prior to collision. Contours define the relative probability densities of product velocities. Reprinted from Ref. 30 with the kind permission of Elsevier Science Publishers.

regarding disposal of angular momentum in the reaction. These reactions may be described as a direct, or impulsive, mechanism and one proceeding via a reaction complex. For the direct mechanism the locus of product vectors is essentially identical to that of the neutral-beam N_2 velocity vectors. Almost no angular momentum is exchanged, and electron transfer evidently occurs at large impact parameters. Since our experiments involved only ground-state N_2^+ ions [excited species having been relaxed by collisions inside the high-pressure ion source by reaction (6)], the reaction is exactly analogous to the symmetric resonant case discussed for atomic systems in Section A.

The second, or complex, mechanism is defined by complete symmetry, within experimental error, about the CM velocity vector in Fig. 6. This high degree of forward–backward symmetry for the reacction complex product implies a lifetime that exceeds the rotational period of the N_4^+ intermediate. The product ions from this mechanism leave the center of mass with greatly reduced kinetic energy, implying efficient energy transfer into internal modes of the reaction products.

Figure 7 is the scattering diagram at the significantly higher energy of 9.96 eV.[30] Under these conditions the collision complex mechanism has

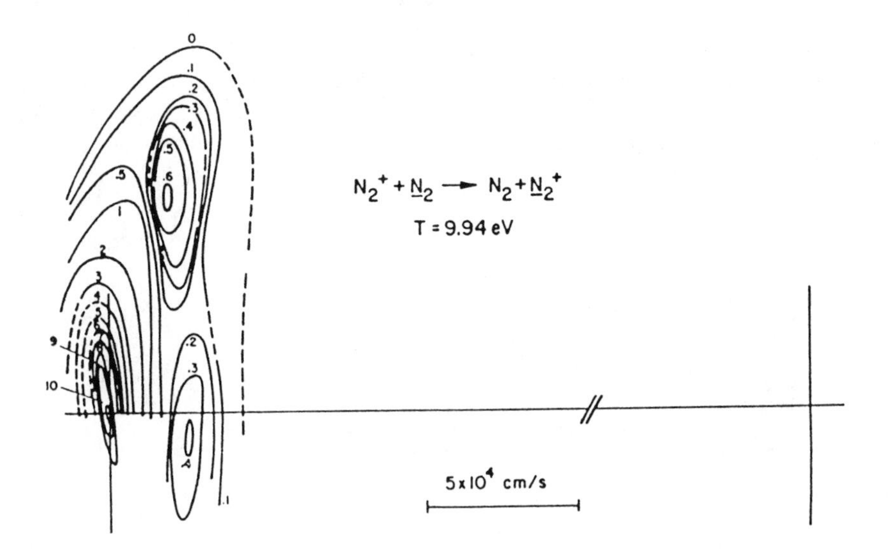

Figure 7. Scattering diagram for the charge-transfer reaction $N_2^+(N_2, N_2)N_2^+$ at 9.94 eV collision energy. The crosses mark the velocity of the center of mass of the system and the CM velocity of N_2 prior to collision. Contours define the relative probability densities of product velocities. The broken line indicates that the vector is not to scale: the CM velocity is equal to the distance between the crosses multiplied by 2.38. Reprinted from Ref. 30 with the kind permission of Elsevier Science Publishers.

completely disappeared. Although the collision complex mechanism is the dominant one at collision energies below 0.7 eV CM, at and above 10 eV only the direct mechanism is observed. This is consistent with the expected dramatic decrease in the lifetime of the N_4^+ intermediate with increasing total energy of the system. Under these conditions the available energy exceeds the binding energy of the N_4^+ complex by more than an order of magnitude, and the scattering diagram provides no evidence for a contribution from the persistent complex mechanism. The dominant direct product at high energy is the translationally resonant, $\Delta v = 0$, reaction product. However, we also note in this figure the formation of satellite peaks indicating significant exchange of angular momentum. These ridges of intensity are translationally endoergic processes whose maxima match approximately the vibrational spacings of N_2^+ and/or N_2 molecules.

This underlying vibrational structure can be resolved using the same deconvolution technique, which was developed for treating resonant charge transfer in atomic systems[5] and described in Section A. The results are shown in Figs. 8 and 9. Figure 8 shows the contour diagram after subtracting the intensity of the resonant charge-transfer channel corresponding to the formation of $N_2^+(X\,^2\Sigma_g, v = 0)$. The same normalization as the original

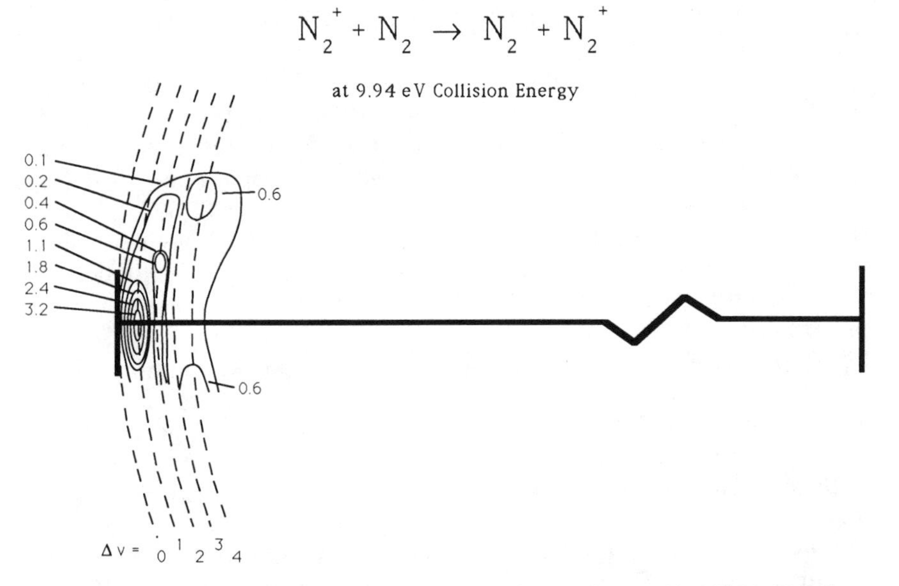

Figure 8. Deconvoluted scattering contour diagram for the reaction $N_2^+(N_2, N_2)N_2^+$ at 9.94 eV collision energy. After removal of the dominant $\Delta v = 0$ reaction, the channels that produce $\Delta v = 1, 2$, and 4 are readily evident. Contours retain the normalization of Fig. 7.

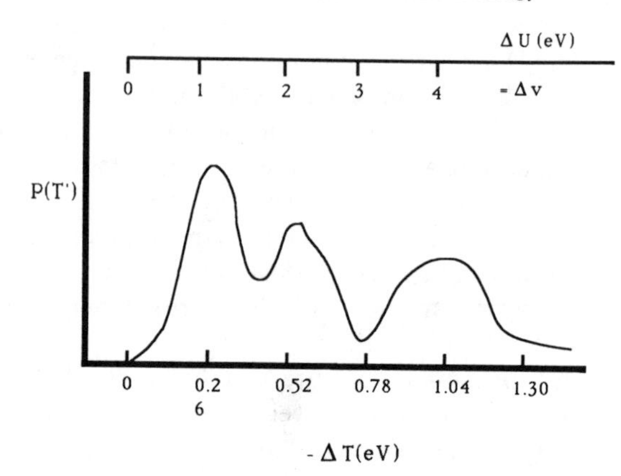

Figure 9. Product relative translational energy distribution at 9.94 eV collision energy for the reaction $N_2^+(N_2, N_2)N_2^+$ as shown in Fig. 8. The correlation with internal energy changes in N_2^+ (or N_2) is shown by the superimposed scale showing vibrational energy levels.

diagram, in which the maximum intensity for the resonant channel is assigned the value of 10, is preserved in Fig. 8.

Figure 8 shows substantial intensity that corresponds to the formation of $N_2^+(X\,^2\Sigma_g, v = 1)$ [or $N_2(X\,^1\Sigma_g, v = 1)$] with no angular deflection (located on the relative velocity vector). This structure was unresolved in the original data. Slightly lower intensity peaks correspond to the deposition of two and four quanta of vibrational, respectively, in the products with both of these channels accompanied by significant angular scattering. No detectable intensity is found on the dashed line circle corresponding to excitation of three vibrational quanta. Larger angle scattering associated with progressively higher vibrational excitation is plausibly interpreted as resulting from progressively smaller impact parameters and impulsive collisions. Since the vibrational spacings in N_2^+ and N_2 are nearly identical, it is impossible for us to distinguish in these experiments whether the ion, or the neutral, or both are vibrationally excited.

Figure 9 shows the integration of the differential cross section of Fig. 8 using Eq. (5). As discussed earlier, this integration gives the relative cross sections for generating products with one, two, and four quanta of internal vibrational energy. The total area of these peaks amounts to about 35% of total product intensity at this collision energy. As indicated in the figure, the cross-sectional area for formation of $\Delta v = 1, 2,$ and 4 are nearly the same.

Numerical integration of the peaks gives the relative intensities as 0.65, 0.12, 0.11, 0.12 for $\Delta v = 0$, 1, 2, and 4.[29] This demonstrates that collisions at 10 eV are beginning to exhibit the propensity rule pointed out by McAfee et al.[24] for forming products with $\Delta v = 0, 2$ and 4.

McAfee et al. demonstrated a remarkable property of the direct CT mechanism in N_2 at 55.5 eV CM. Figure 10, taken from their paper,[25] illustrates a pronounced alternative of relative intensities with preference for exchange of an even number of vibrational quanta. (We note

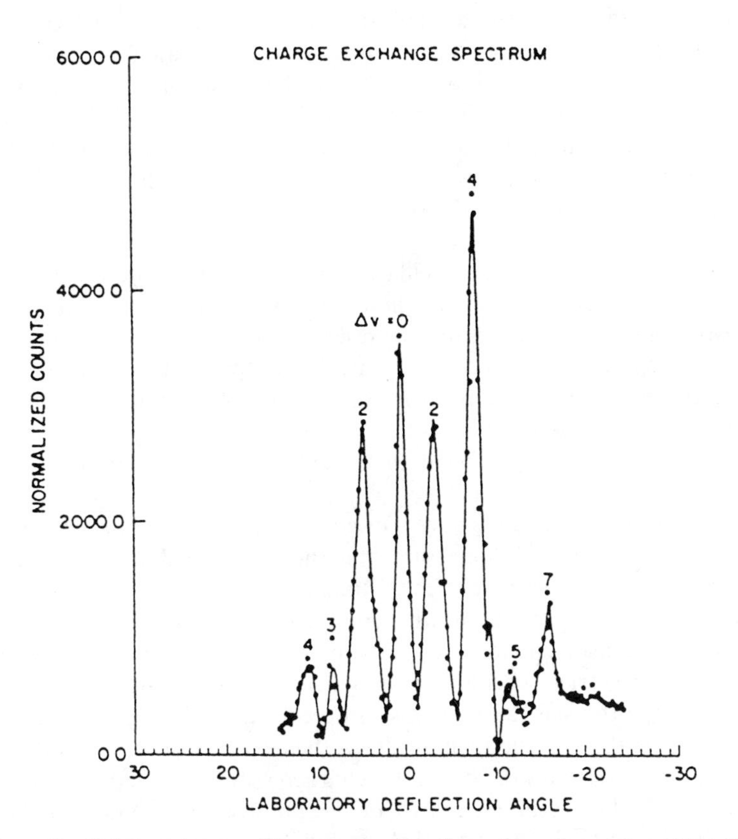

Figure 10. Relative intensity of low-energy scattered ions from the reaction of 111 eV N_2^+ with a neutral supersonic beam of N_2 (55.5 eV CM). Points are experimental the solid lines are Gaussian fits using a fifth-order spline function to locate the transition center positions designated. A broad background scattering contribution has been subtracted. Negative angles represent exothermic (superelastic) charge transfer of vibrationally excited N_2^{+*}; positive angles are the endothermic charge transfer reactions of $N_2^+(X\,^2\Sigma_g c = 0)$. The pronounced preference for transfer of even numbers of quanta is clearly evident in the figure. Reprinted from Ref. 25 with the kind permission of the American Institute of Physics.

parenthetically that the experimental data shown in this figure involve both ground-state and vibrationally excited N_2^+. Charge-transfer processes that are superelastic because of the transfer of vibrational energy into translation are responsible for the scattering into negative laboratory angles shown in the figure.) This result was rationalized on the basis of a Landau–Zener–Stueckelberg curve crossing model using approximate potential energy curves. The quadrupolelike character of the wave function accounted for an enhanced transition probability for $\Delta v = 2, 4, 6$. It was concluded that the coupling terms at large R between N_2^+ and N_2 are symmetric in the internal vibrational coordinates. This model also assumes that charge transfer with vibrational excitation occurs only in the incoming trajectory—for example, that no vibrational or charge transfer occurs in the exit channel.

An alternative, simpler explanation first suggested to us by Professor Douglas Ridge is that the collision of two identical harmonic oscillators should exhibit the resonance effect of a higher probability for depositing equal numbers of quanta in the two oscillators. This has been confirmed in a detailed trajectory calculation that first considered harmonic oscillator potentials, then unharmonic oscillators, frequency mismatch, and, finally, collisions proceeding on an *ab initio* N_4^+ potential energy surface.[31] A propensity for equal excitation is preserved, along with a preference for $\Delta v = 0, 1$ excitation that accounts for the experimental results at least qualitatively. An analogous propensity for $\Delta v = 2, 4$ excitation is also found in the accurate high-energy quantum and semiclassical results of Depristo for O_2^+/O_2 scattering,[32] which has also been found experimentally by McAfee et al.[25]

In conclusion, the nitrogen system constitutes a “textbook example” of the Rapp and Francis[15] conjecture that there is a fundamental change in the mechanism of charge transfer as kinetic energy is reduced to the range where formation of transient orbiting complexes is anticipated. At high energy a direct mechanism applies and electron transfer takes place largely from rectilinear trajectories. As collision energies are reduced, a significant number of the charge-transfer interactions proceed via ‘orbiting” or “capture” trajectories; in the low-energy limit this is the only mechanism operative and charge transfer assumes the dynamics characteristic of typical ion–molecule reactions. The high binding energy of the putative N_4^+ complex (which is estimated theoretically to be 1.4 eV[23] and has been measured in recent experiments to be about 1 eV[26–28]) is also an important parameter. This provides a potential well that can trap the collision partners if energy transfer to internal modes occurs during the collision. This trapping behind the centrifugal barrier extends the lifetime of the collision intermediate, further enhancing the probability for energy equilibration prior to dissociation.

C. Charge Transfer of $Ar^+(^2P_{3/2})$ with $NO(^2\Pi, \nu = 0)$

The charge-transfer reaction of $Ar^+(^2P_{3/2}, ^2P_{1/2})$ with NO was the first crossed-molecular-beam study that demonstrated the forward–backward symmetry "signature" of proceeding via an orbiting complex with a lifetime exceeding its rotational period.[32] In addition to addressing such questions as energy transfer and angular momentum exchange at low and intermediate energy, this study specifically addressed the apparent change in mechanism between low energy (below 1 eV) and high energy (above 2 eV) reported by Kobayashi[34,35] and Birkinshaw and Hasted[36] for this reaction in injected ion drift-tube experiments.

The experiments of Kobayashi,[34,35] in particular, demonstrated that several rare-gas–diatomic molecule systems exhibit a minimum in the energy dependence of the charge-transfer cross section. In the Ar^+–NO system, for example, the cross section obeyed an $E^{-1/2}$ dependence at low energy, consistent with the hypothesis that long-range ion-induced dipole forces between the reactants dominate the reaction. The cross section is a minimum at 1.3 eV and increases at higher relative energies in a manner described at least qualitatively by the rectilinear trajectory model of Rapp and Francis.[15] Of the several systems investigated, Ar^+–NO had the largest cross section (about 0.4 of the Langevin limit) and also exhibited a minimum at the highest relative energy (1.3 eV). Charge transfer appears to be the only reaction occurring in the Ar^+–NO system at about 1 eV; no competing ion–atom exchange reactions have been reported. Finally, potential energy curves of high accuracy are available for both ground and excited states of NO and NO^+.[37]

This reaction (and its inverse) have been further explored in selected ion flow drift tubes[38,39] by ion cyclotron resonance[40] and in a state-selected TESICO (threshold electron secondary ion coincidence) beam measurement by Kato.[41] Except for details, all are in essential agreement with the conclusions from the earlier beam experiment that a combination of energy resonance (energy defect) and Frank–Condon factors cause the reaction to be highly state specific, populating only low vibrational levels of the excited $a\,^3\Sigma$ state of NO^+. Kato's study[41] shows this most dramatically for the reverse reaction, while the drift-tube study[39] shows that the forward reaction gives more than 85% $NO^+(a\,^3\Sigma)$. Kato also carried out *ab initio* molecular-orbital calculations to investigate the effect of collision geometry, orientation, electronic state, and spin–orbit coupling effects on the charge-transfer reaction.[41] These results were used to rationalize the interesting experimental observation of a *resonance enhancement* of the cross section of $NO^+(a\,^3\Sigma, v = 2)$ for the reverse reaction, explained as the result of strong coupling of this state with $Ar^+(^2P_{1/2}) + NO$, and the *absence* of any dependence of the relative cross section on J for $Ar^+(^2P_{1/2})$ vs $Ar^+(^2P_{3/2})$ in the forward reaction of Ar^+ with NO. Kato reached the interesting conclusion that different mechanisms are traced in the forward and reverse directions, a small-impact-parameter Landau–Zener curve-crossing mechanism for Ar^+ reacting with NO and a long-range, Demkov-type direct mechanism for the reverse reaction of $NO^+(a\,^3\Sigma)$ with Ar.

These extensive new data prompted us to repeat the earlier beam experiment of Herman, Pacak, Yencha, and Futrell[33] utilizing state-selected $Ar^+(^2P_{3/2})$, a seeded supersonic jet to improve angular definition (and reduce energy spread in the reactants) and with improved energy and angular resolution. The use of a He-seeded NO beam produces rotationally cooled species that may enhance alignment during low-energy collisions. The experiment was also done at slightly lower and slightly higher collision energies.[42]

Our data for the charge transfer of $Ar^+(^2P_{3/2})$ with NO at the very low collision energy of 0.43 eV are shown in Fig. 11. [Our reactant ions are

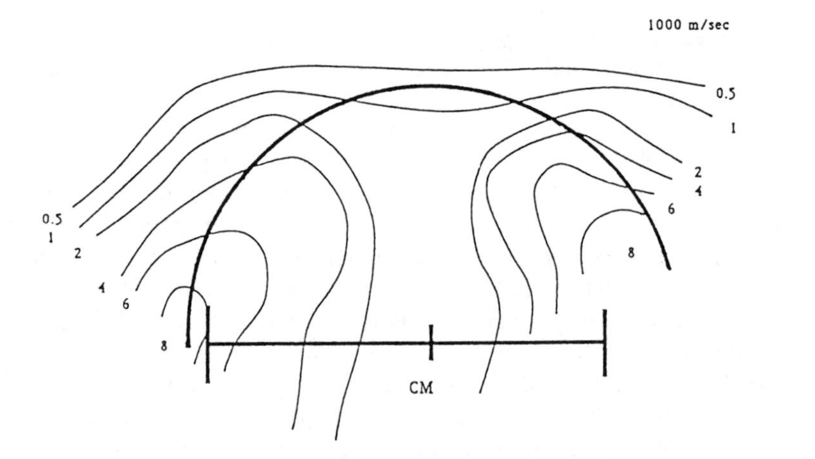

Figure 11. Center-of-mass scattering contour diagrams for the charge-transfer reaction of $Ar^+(^2P_{3/2})$ with NO at 0.43 eV collision energy. Velocity contours indicate relative intensity and vertical bars to the right of the CM denotes the tip of the Ar^+ velocity vector while the vertical bar to the left denotes the neutral velocity vector. The circle denotes the loci of product ion velocity vectors calculated for the state specific reaction $Ar^+(^2P_{3/2})$ and $NO(^2\Pi, v=0) \rightarrow Ar(^1S_0) + NO^+(a\,^3\Sigma^+, v=0)$.

generated in a plasma source that reduces the initial distribution to the $Ar^+(^2P_{3/2})$ level.[43]] The charge-transfer process is translationally slightly exothermic with the product ions leaving the collision center with slightly higher velocity than the initial NO neutral velocity. In Fig. 11 the velocity of the argon ion is indicated by the vertical bar to the right of the center of mass, while the velocity of the NO neutral reactant is indicated by the solid bar to the left of the center of mass. The circumscribed circle shows the velocity that the particles would have if the charge-transfer reaction is given by the following equation:

$$Ar^+(^2P_{3/2}) + NO(^2\Pi, v=0) \rightarrow Ar(^1S_0) + NO^+(a\,^3\Sigma^+, v=0) + 0.097\,\text{eV}. \quad (7)$$

We infer from the fact that the position of maximum intensity in the reactive scattering contour diagram matches the energetics of reaction (7) rather well that this is the reaction that actually occurs. It appears that NO^+ is formed in the nearly resonant transition producing the ground vibrational level of the first electronic excited state rather that the much more exothermic reaction to generate ground state NO^+:

$$Ar^+(^2P_{3/2}) + NO(^2\Pi, v=0) \rightarrow Ar^+(^1S_0) + NO^+(X\,^1\Sigma^+, v=0) + 6.399\,\text{eV}. \quad (8)$$

This general conclusion was reached both in our earlier beam study[33] and in drift-tube studies of the same reaction.[39] The interesting additional feature in Fig. 11 is the development of nearly complete backward–forward symmetry about the center of mass at very low collision energy. This is the characteristic dynamic signature of an orbiting complex. In contrast with our earlier study,[33] which involved a mixture of $Ar^+(^2P_{3/2})$ and $Ar^+(^2P_{1/2})$, the present experiment appears to generate $NO^+(a\,^3\Sigma, v=0)$ quite cleanly, with modest rotational excitation but no detectable population of $v=1$. The backward–forward peaking of intensity implies that collisional angular momentum is largely conserved, as is also found in the earlier study at 0.65 eV.[33] At the lower energy of 0.45 eV the development of backward–forward symmetry anticipated for an $[ArNO^+]$intermediate with a lifetime greater than a rotational period is essentially complete.

Figure 12 shows our data for this reaction at the higher translational energy of 4.75 eV. As before, the energetics correspond to reaction (7) rather than to reaction (8). The dramatic change in the scattering patterns with a tenfold increase in collision energy is the complete disappearance of backward–forward symmetry. Clearly the reaction mechanism is direct at this collision energy. We deduce from the similar state population and dramatically different angular scattering that the charge-transfer reaction

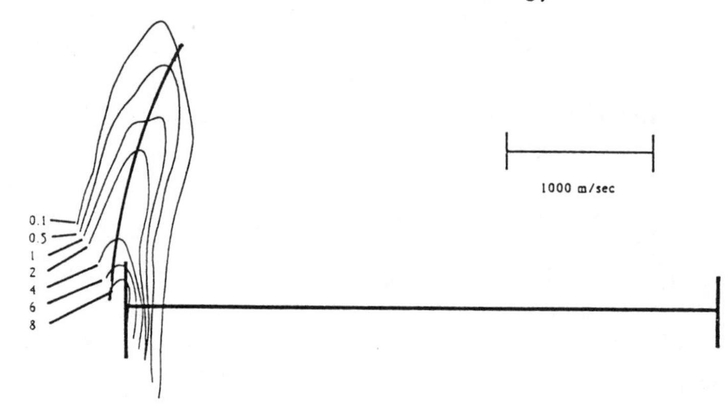

Figure 12. Center-of-mass scattering contour diagram for the charge-transfer reaction of $Ar^+(^2P_{3/2})$ with NO at 4.75 eV collision energy. Velocity contours indicate relative intensity and vertical bars to the right of the CM denotes the tip of the Ar^+ velocity vector while the vertical bar to the left denotes the neutral velocity vector. The circle denotes the loci of product ion velocity vectors calculated for the state specific reaction $Ar^+(^2P_{3/2})$ and NO $(^2\Pi, v = 0) \rightarrow Ar(^1S_0) + NO^+(a\,^3\Sigma^+, v = 0)$.

occurs at relatively large internuclear distances over this energy range with little exchange of angular momentum—an "electron jump" mechanism. Since the asymptotic energy levels of the reactants and products of reaction (7) are nearly energy resonant, we infer that a curve-crossing mechanism of the Demkov type occurs as the reactants approach each other. This likely occurs in the incoming reactant channel. The strongly attractive potential well of $ArNO^+$ (which is isoelectronic with NOCl) causes orbiting trajectories at low collision energies (Fig. 11). However, orbital and angular momentum exchange is modest in the collision complex (presumably relatively short-lived, a few rotational periods) and the products separate without energy randomization. At higher collision energy (Fig. 12) the transition is direct and the velocity distribution is essentially that of the neutral beam, displaced by the slight exothermicity of reaction (7).

From a molecular-orbital perspective, as described by Kato[41] and discussed briefly by Herman et al.,[33] only triplet potential surfaces need to be considered. Furthermore, only the $NO^+(a\,^3\Sigma^+)$ and $NO^+(b\,^3\Pi)[+Ar(^1S_0)]$ are located close energetically to $NO(X\,^2\Pi) + Ar_J^+$. The $a\,^3\Sigma^+$ state corresponds to transfer of a $^1\Pi$ electron to Ar^+, while the $b\,^3\Pi$ state is formed by transfer of a 5σ electron. Since these electrons extend perpendicular to and along the internuclear axis of NO, they lead to different exchange energies for broadside and linear approach trajectories by the reactant $Ar^+(^2P_{3/2})$. The former is a $^3A'$ state of the complex and the latter is $^3A''$. Kato's calculations show for the $^3A'$ state (perpendicular approach) that $Ar^+(^2P_{3/2})$ correlates asymptotically to the $NO^+(a\,^3\Sigma^+)$ state. For the $^3A''$ state a strongly bound Π-state intermediate is produced by the charge-transfer interaction, while the $NO^+(a\,^3\Sigma^+)$ (+ Ar) state is repulsive in this geometry. These two surfaces cross at small internuclear distances. Kato suggested that these two approach geometries were responsible for Demkov-type and Landau–Zener-type curve crossing mechanisms, respectively.[41] Herman et al.[33] used the isoelectronic principle (based on the NOC1 molecule) to argue that the bound state of $ArNO^+$ should have a bond angle of 120°, intermediate between the two cases discussed by Kato[41] and that the NO bond distance in the potential well should be perturbed only slightly from that of the isolated NO molecule.

Kato's molecular-orbital calculation (Figs. 12 and 13 of Ref. 41) suggests that Demkov-type curve crossing between $Ar^+(^2P_{3/2})[+NO(^2\Pi)]$ and $NO^+(a\,^3\Sigma^+)[+Ar(^1S_0)]$ occurs in broadside approach on the $^3A^1$ surface at internuclear separations of around 4 Å, while in collinear approach Landau–Zener curve crossing between the $^3A^1(^3\Sigma)$ and $^3A^{11}(^3\Pi)$ surfaces can occur at about 3.6 Å. Both mechanisms would therefore be direct and involve modest scattering when the diabatic curves are followed at moderate collision energies, as in Fig. 12. However, at the much lower energy of Fig. 11, it is plausible that adiabatic coupling into the strongly bound potential well would be favored, leading to a long-lived intermediate and nearly symmetric scattering of the NO^+ product.

D. Reactive and Unreactive Scattering of $Ar^+(^2P_{3/2})$ and $Ar^+(^2P_{1/2})$ by N_2

The charge-transfer reaction of argon ions with nitrogen and the reverse reaction of excited nitrogen ions with argon is one of the most thoroughly investigated reactions in ion chemistry.[9,29,44–80] We have investigated both the elastic and inelastic scattering of the $Ar^+(^2P_{1/2})$ and $Ar^+(^2P_{3/2})$ ions with N_2 and the reactive scattering of these ions with special emphasis on the low-energy regime. Our crossed-beam investigation of this system has revealed the very unusual feature that angular-specific and quantum-state-specific scattering are observed within a narrow energy range while a different state-specific mechanism is followed at both lower and higher energies.[9,67–69] This unexpected result, plus the recent theoretical investigations by Gislason and co-workers,[61–65] which include competition between reactive and quenching channels in this system, will be discussed in this section. Since all these processes are governed by the same potential-energy hypersurfaces, the present data offer a significant challenge for theoretical interpretations of charge and energy transfer in this triatomic system.

The following processes will be discussed:

$$Ar^+(^2P_{3/2}) + N_2(X\,^1\Sigma_g^+, v=0)$$

$$\rightarrow Ar^+(^2P_{3/2}) + N_2(X\,^1\Sigma_g^+, v'), \quad (9a)$$

$$\rightarrow Ar^+(^2P_{1/2}) + N_2(X\,^1\Sigma_g^+, v'), \quad (9b)$$

$$\rightarrow Ar(^1S_0) + N_2^+(X\,^2\Sigma_g^+, v'); \quad (9c)$$

$$Ar^+(^2P_{1/2}) + N_2(X\,^1\Sigma_g^+, v=0)$$

$$\rightarrow Ar^+(^2P_{1/2}) + N_2(X\,^1\Sigma_g^+, v'), \quad (10a)$$

$$\rightarrow Ar^+(^2P_{3/2}) + N_2(X\,^1\Sigma_g^+, v'), \quad (10b)$$

As described elsewhere,[43] a variation in ion-operating parameters was used to produce, in turn, in essentially pure $Ar^+(^2P_{3/2})$ beam or a beam of $Ar^+(^2P_{3/2})$ containing variable amounts of $Ar^+(^2P_{1/2})$/Ar to the statistical ratio of 1/2 for $Ar^+(^2P_{1/2})/Ar^+(^2P_{3/2})$ formed by 750 eV electron impact. The neutral-beam internal temperature is about 7 K and only the lowest rotational levels of the $v=0$ state of $N_2(^1\Sigma_g^+)$ are present. Since the rotational distribution of the reactant N_2 is very narrow and $Ar(^1S_0)$ is the only charge-transfer atomic product state, the internal energy of the N_2 or N_2^+ products can be determined approximately by careful analysis of the energy and angular distributions of the scattered ions.

It is now well established that the exothermic $N_2^+(v=0)$ product is dominant only at 80 K.[79] Using LIF analysis of products from a crossed-beam experiment, Leone and co-workers[57,58] have demonstrated that $N_2^+(X\,^2\Sigma^g$, $v=1)$ constitutes about 87% of the total product at 0.24 eV CM. The remainder is $N_2^+(X\,^2\Sigma_g, v=0)$. Leone et al. have also analyzed the rotational energy of $N_2^+(X\,^2\Sigma_g, v=1)$ and find, in their most recent study (which utilized a supersonic jet of N_2 at 0.24 and 0.40 eV CM collision energies[59] that there is a bimodal rotational distribution, with about 14% corresponding to a rotational temperature of 80 ± 10 K and 86% corresponding to 680 ± 30 K for the $v=1$ product at 0.24 eV CM.

Since the selective population of the endoergic product channel $N_2^+(X\,^2\Sigma_g^+, v'=1)$ in preference to the exoergic $N_2^+(X\,^2\Sigma_g^+, v=0)$ is not simply explained, the mechanism of this reaction has been the subject of several theoretical studies.[60-65,76,77] Our present understanding is best summarized in recent papers by Gislason et al.[63-65] and in the papers by Sonenfroh and

Leone[59] and by Clary and Sonenfroh.[60] Parlant and Gislason carried out classical trajectory time-dependent quantum-mechanical studies of this system using *ab initio* surfaces calculated for C_{2v} and linear geometries by Archirel and Levy.[76] Clary and Sonenfroh[60] carried out a three-dimensional quantum-mechanical rotation close-coupling calculation with the specific objective of calculating rotational distributions for the 0.24 eV LIF study. Parlant and Gislason's calculations[63–65] addressed the entire range of experiments on this benchmark system, and their conclusions are quite relevant to the present results.

The nonreactively scattered $Ar^+(^2P_{3/2})$ results [Eq. (9a)] are depicted in Fig. 13 and the quenched but nonreactively scattered $Ar^+(^2P_{1/2})$ [eq. (2b)] are shown in Fig. 15, respectively. Figure 13 is constructed directly for ion-source conditions generating about 99% $Ar^+(^2P_{3/2})$. Figure 14 is the same experiment conducted with an ion-beam mixture of $Ar^+(^2P_{3/2})$ and

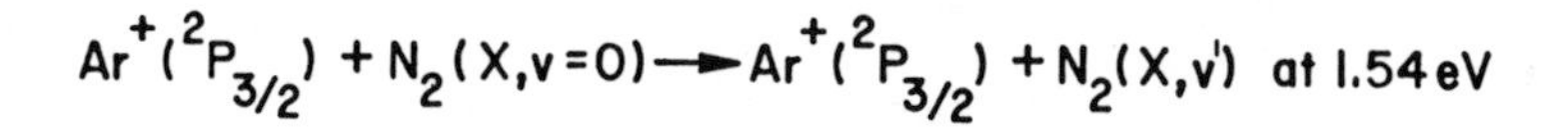

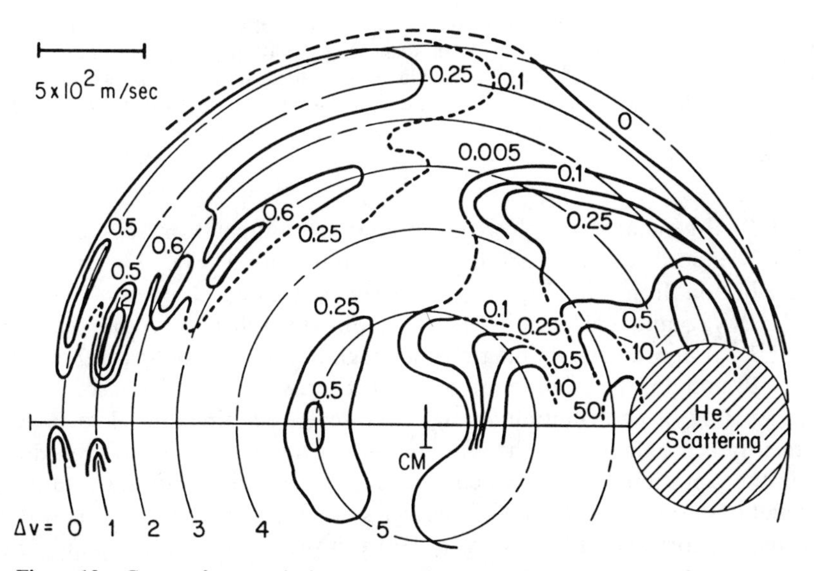

Figure 13. Center-of-mass velocity contour diagram for the reaction $Ar^+(^2P_{3/2}) + N_2(X, v = 0) \rightarrow Ar^+(^2P_{3/2}) + N_2(X, v')$ at 1.54 eV collision energy. Dashed curves indicate the uncertainty in the contours because of contributions from both $Ar^+(^2P_{1/2})$ and $Ar^+(^2P_{3/2})$. The circles indicate the expected location of the scattered ions when the N_2 product is in the vibrational state marked on the circle. The shaded circle shows the region where the scattering by He in the He-seeded supersonic beam. Reprinted from Ref. 69 with the kind permission of Elsevier Science Publishers.

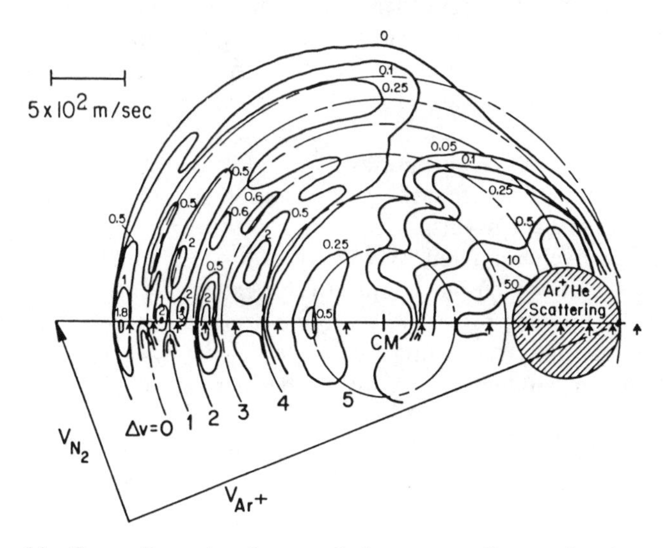

Figure 14. Composite center-of-mass velocity contour diagram for the scattering of $Ar^+(^2P_{3/2}$ and $^2P_{1/2})$ with N_2 at 1.54 eV collision energy. Source pressure was reduced so that both $Ar^+(^2P_{3/2})$ and $Ar^+(^2P_{1/2})$ were present in the ion source. Arrows on the relative velocity vector indicate the expected locations of the scattered ions from $Ar^+(^2P_{1/2})$ with the N_2 product is in different vibrational states. Reprinted from Ref. 69 with the kind permission of Elsevier Science Publishers.

$Ar^+(^2P_{1/2})$ generated in a low-pressure electron-impact ion source. Figure 15 is deduced by subtracting off the normalized Fig. 13 $Ar^+(^2P_{3/2})$ contribution from the Fig. 14 composite scattering diagram. Since the $Ar^+(^2P_{3/2})$component in Fig. 14 is readily identified and the scattered $Ar^+(^2P_{1/2})$ appears to dominate other regions of velocity space (compare Figs. 13 and 15) the subtraction appears to provide adequate resolution of the $Ar^+(^2P_{1/2})$ contribution in the back-scattered direction. The labeled circles indicate the loci of the scattered ions when the N_2 or N_2^+ "product" is in the vibrational state marked on the circle. Where significant rotational excitation also occurs, the kinetic energies of the separating products will be decreased, correspondingly the gaps between the contours corresponding to vibrational excitation are "filled in."

The structure of the contour diagrams for scattered Ar^+ in the backward direction is striking. The fortuitous result that the distribution of scattered Ar^+ has a different characteristic location for $Ar^+(^2P_{3/2})$ and $Ar^+(^2P_{1/2})$ is the key factor that allows us to resolve the structure shown in these diagrams. In the forward direction no clear separation of scattered ion distributions

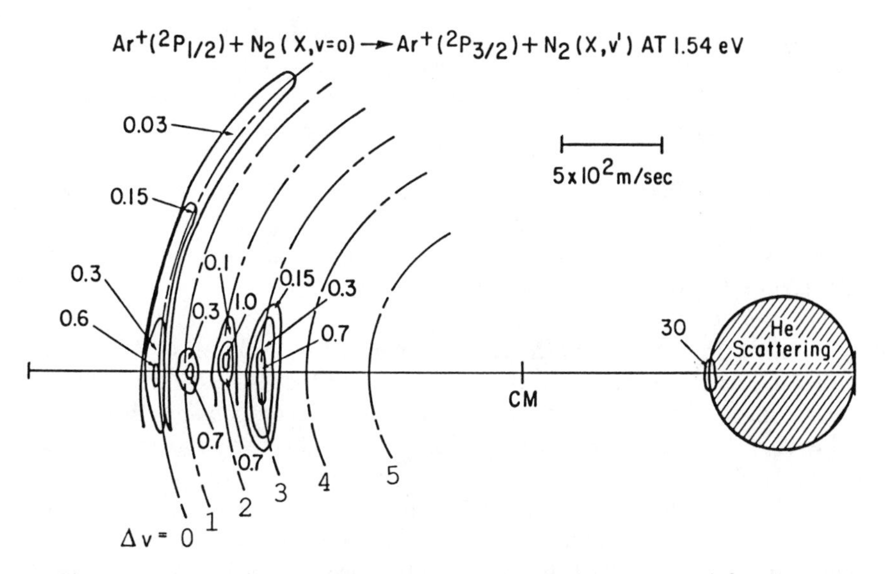

Figure 15. Center-of-mass velocity contour diagram for the reaction $Ar^+(^2P_{1/2}) + N_2(X, v=0) \rightarrow Ar^+(^2P_{3/2}) + N_2(X, v')$ at 1.54 eV collision energy. This contour diagram was obtained by first constructing a contour diagram when $Ar^+(^2P_{3/2})$ and $Ar^+(^2P_{1/2})$ were both present in the ion source and then subtracting the contribution from $Ar^+(^2P_{3/2})$ scattering (see text for explanation). Reprinted from Ref. 69 with the kind permission of Elsevier Science Publishers.

could be deduced; experiments in which $Ar^+(^2P_{1/2})$ was removed by changing ion-source operating parameters gave only small variations in the relative heights of peaks in the contour diagram. Dashed contours are shown in Fig. 15 where our experiments indicate contributions from both $Ar^+(^2P_{3/2})$ and $Ar^+(^2P_{1/2})$, and the identification of relative intensities as resulting from $Ar^+(^2P_{1/2})$ scattering is uncertain.

Parlant and Gislason[65] have shown that the fine-structure transitions in the $Ar^+(^2P_J)/N_2$ system have surprisingly large cross sections. The quenching cross section for $Ar^+(^2P_{1/2})$ actually exceeds the value for charge transfer above 10 eV. The excitation cross section for $Ar^+(^2P_{3/2})$, although appreciable, is much less than the charge-transfer cross section at all energies. Fine-structure transitions accompanied by vibrational excitation have appreciable cross sections for $\Delta v = +1$ and smaller values for higher levels. However, vibrational excitation occurs, in their model,[65] only via the intermediate charge-transfer state $N_2^+(A \rightarrow X) + Ar$. There is no term in their vibronic Hamiltonian allowing vibrational excitation in a single (diabatic) electronic state. Impulsive vibration is also excluded, and it is certain that cross sections for vibrational excitation are underestimated. Rotational excitation is not included in their calculational model.

Reactive scattering to form N_2^+, reaction (1c), is illustrated in Figs. 16–18 for 0.6, 0.8, and 1.5 eV collision energies. At all energies the angular distribution of the N_2^+ is strikingly asymmetric. At the lower (0.6 eV, Fig. 16) and higher collision energy (1.5 eV, Fig. 8) the scattering diagrams show clear evidence for formation of $N_2^+(X\,^2\Sigma_g^+, v=0, 1)$ with $v=1$ the dominant product. Although the angular range of the present apparatus is not sufficient to map completely the generation of $N_2^+(X\,^2\Sigma^+, v=0)$ in superelastic collisions, this product is clearly present in an amount consistent with the LIF results noted above by Leone et al.[57–59]

The precise analysis by Sonenfroh and Leone[59] of rotational distributions of $N_2^+(X\,^2\Sigma_g, g=0)$ and $N_2^+(X\,^2\Sigma_g, v=1)$ formed by reaction of $Ar^+(^2P_{3/2})$ with $N_2(X\,^1\Sigma_g, v=0)$ cooled to 10 K at 0.28 ± 0.22 and 0.40 ± 0.22 eV collision energy in supersonic jet expansion and their accompanying theoretical analysis of this problem provide valuable new insights regrading the detailed dynamics of this reaction. As they point out, for the low-energy range explored in their experiments and, *vide infra*, for our experiments reported in Figs. 16 and 17 as well, an adiabatic description is more appropriate to describe dynamics than the diabatic picture used by Gislason et al.[61,63–65] for most of their theoretical analysis. In an earlier paper Spalberg, Los, and Gislason[62] actually used the adiabatic framework, but did so with approximately flat potential functions prior to the availability of the Archirel and Levy *ab initio* surface.[76]

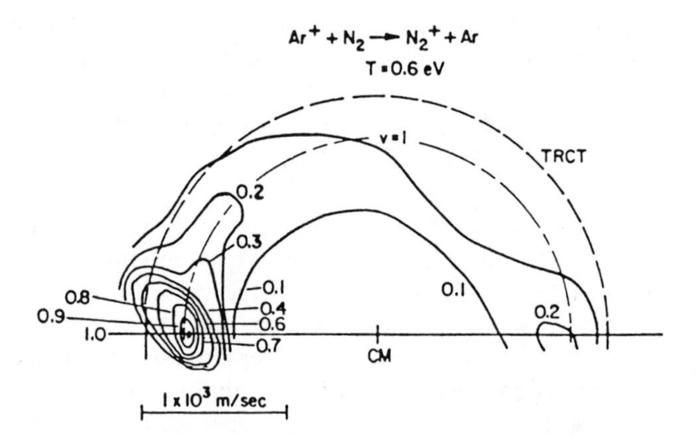

Figure 16. Scattering diagram for the charge-transfer reaction $Ar^+(N_2, Ar)N_2^+$ at 0.60 eV collision energy. The maximum intensity at 0° and the 20% contour at 180° (with respect to the initial N_2 velocity vector) both correspond to $-\Delta U = 0.1$ eV and match the endothermicity of the reaction $Ar^+(^2P_{3/2}) + N_2(X\,^1\Sigma_g^-, v=0) \rightarrow Ar(^1S) + N_2^-(X\,^2\Sigma_g^-, v=1)$. The symmetry beginning to develop in this diagram suggests that about 40% of all reactive collisions result in orbiting of the center of mass. Reprinted from Ref. 69 with the kind permission of Elsevier Science Publishers.

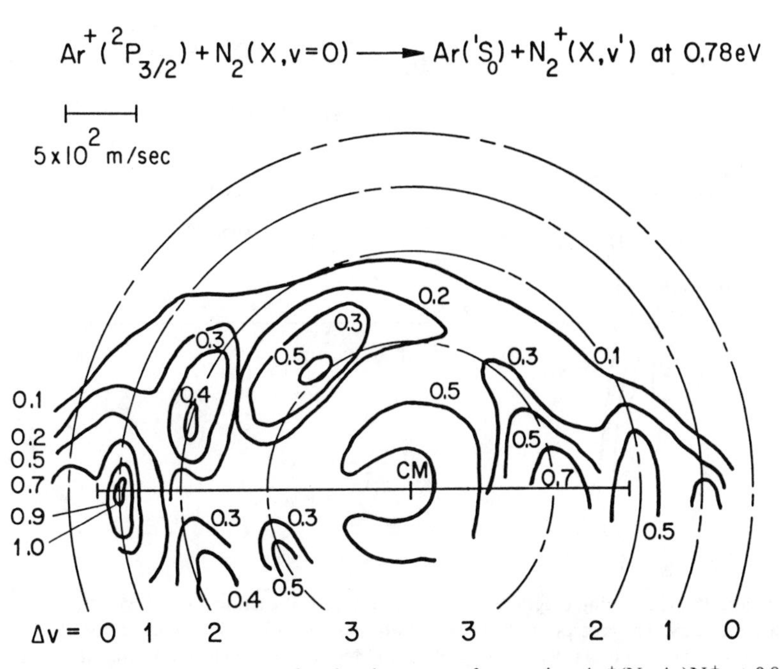

Figure 17. Scattering diagram for the charge-transfer reaction $Ar^+(N_2, Ar)N_2^+$ at 0.8 eV collision energy. The crosses mark the center-of-mass velocity and the CM velocity of N_2 prior to collision. Contours define the probability densities of product velocities (in Cartesian coordinates), whereas the inscribed circles define the loci of velocity vectors for N_2^+ products leaving the collision center with translation-to-internal energy conversion corresponding to the formation of $N_2^+(X\,^2\Sigma_g)$ with 0, 1, 2, 3, and 4 quanta of vibration with no energy deposited in rotation. Reprinted from Ref. 69 with the kind permission of Elsevier Science Publishers.

Sonnenfroh and Leone calculated adiabatic potentials for linear ($C_{\infty v}$) and C_{2v} configurations of ArN_2^+ using Parlant and Gislason's diabatic potential curves and couplings for the energetics of the six lowest energy states of the system[63,64] as a function of $[Ar{-}N_2]^+$ separation. These curves are fitted to the Archivel and Levy *ab initio* data[76] and represent the best information currently available on ArN_2^+. These potentials are highly anisotropic, the collinear approach ($C_{\infty v}$) being highly attractive and the perpendicular approach (C_{2v}) being essentially repulsive. The $C_{\infty v}$ curve with $[Ar^+(^2P_{3/2}) + N_2(X\,^1\Sigma_g, v = 0)]$ as its asymptote is bound by 0.6 eV, while the ground state of ArN_2^+ is with linear C_∞ configuration with $[N_2^+(X\,^2\Sigma_g, v = 0) + Ar(^1S_0)]$ asymptote and is bound by 1 eV. The latter value matches experimental binding energies measured by Teng and Conway[78] and by Ding, Futrell, and Cassidy.[81]

These highly anisotopic adiabatic potential surfaces are suggested by

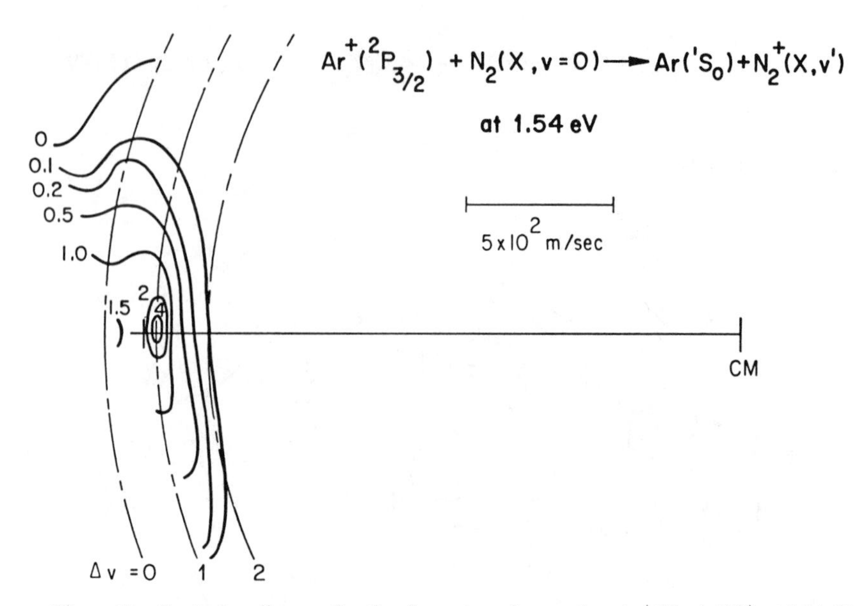

Figure 18. Scattering diagram for the charge-transfer reaction $Ar^+(N_2, Ar)N_2^+$ at 1.54 eV collision energy. The crosses mark the center-of-mass velocity and the CM vector of N_2 prior to collision. The contours define the probability densities of product velocities (in Cartesian coordinates), whereass the circular radii correspond to the translational-energy changes required to drive the reactions $Ar^+(^2P_{3/2}) + N_2(X, v = 0) = Ar(^1S_0) + N_2^-(X, v = 0, 1, 2)$ with no energy deposited in rotation. Reprinted from Ref. 69 with the kind permission of Elsevier Science Publishers.

Sonnenfroh and Leone as the explanation for the bimodal rotational distributions which they observe (and confirmed by the 3D closely coupled quantum calculations of Clary and Sonnenfroh[60]. Collinear approach samples both the strongly attractive well and a steeply repulsive potential at the classical turning point for trajectories in which charge transfer occurs as the products separate. Furthermore, as demonstrated by Fig. 16, at least one-third of the N_2^+ products formed at 0.6 eV are formed via an orbiting complex with a lifetime of the order of a rotational period. This mechanism, the formation of a transient complex, provides a ready alternative explanation for the high rotational energy distribution component.

The perpendicular approach of reactants (C_{2v} geometry) is essentially repulsive at all distances.[59,60] It neither samples the strongly attractive potential curve nor the steepest portion of the repulsive potential at the classical trajectory turning points. Charge transfer in this geometry occurs at larger internuclear distances following more closely a Demkov-type picture, while the collinear geometry approach favors a Landau–Zener model.

It is plausibly suggested that this dynamical route to products accounts for the low-rotational-energy part of the bimodal distribution.

Finally, it is argued by Sonnenfroh and Leone[59] that their use of a supersonic jet to generate the rotationally cold $N_2(X\,^1\Sigma_g, v=0$, low $N')$ reactants is an important feature of their dynamics study, for the reason that it allows the long range ion–quadrupole interaction of $C_{\infty v}$ geometry to align the reactants in collinear geometry. This overcomes the statistical preference for reactants to approach in C_{2v} rather than $C_{\infty v}$ geometry and the slightly higher CT probability predicted for perpendicular approach because of the slightly closer energy resonance. The suggested alignment effect also explains the fact that the bimodal rotational energy distribution was not detected in the earlier experiments of Leone et al.[57,58] which utilized an effusive beam of N_2 reactants rather than supersonic jet expansion.

The fact that these dynamical features are observed in the LIF study at 0.40 ± 0.22 eV collision energy[59] and that essentially the same reactant preparation of $Ar^+(^2P_{3/2})$ and seeded supersonic jet expansion of N_2 is followed in our Fig. 16 experiment at 0.6 ± 0.2 eV strongly suggests that the same dynamics considerations apply. The partial backward–forward symmetry of Fig. 16 demonstrates that at least one-third of the reaction proceeds by a complex surviving at least a rotational period. The width of the distribution easily accommodates the rotational branching into $\langle N'' \rangle = 2$ and $\langle N'' \rangle = 10$–14 measured by Sonnenfroh and Leone[59] and/or the higher distribution logically extrapolated from their model for our slightly higher energy experiment.

Consequently, we interpret our Fig. 16 experiment as the superposition of three classes of trajectories. The direct mechanism found on the relative velocity vector is primarily the product of trajectories with C_{2v} (and C_s) alignment with a second component representing collinear approach in which charge exchange occurs only in the incoming trajectory. The collision complex component, which is symmetric about the CM, results from CT occurring for those nearly collinear trajectories that sample the strongly attractive potential well. Energy transfer into internal modes leads to centrifugal trapping by reducing the rotational energy of the complex, extending its lifetime long enough for some degree of randomization of energy prior to dissociation.

This dynamics description is readily extended to Fig. 18, where evidence for an orbiting complex is no longer detected. At this energy collision times are significantly shortened and rotational alignment to the energetically preferred collinear configuration by long-range quadrupolar forces is less probable. Consequently, only direct mechanisms are possible. Accordingly, the diabatic description of Parlant and Gislason[64] becomes the preferred dynamics model for this collision energy and higher. Their description appears to provide a generally satisfactory theoretical framework for this reaction from about 2 to 4000 eV.

The scattering diagram for 0.78 eV collision energy (Fig. 17) is strickingly different. At this energy readily measurable N_2^+ ion signals could be detected at all scattering angles corresponding to the generation of all the energetically accessible vibrational states of $N_2^+(X\,^2\Sigma_g^+, v = 0, 1, 2, \text{and } 3)$ in reaction (9c). A much broader angular scattering pattern with the population of all energetically accessible vibrational states demonstrates a dramatic change in mechansim in the narrow energy range between 0.6 and 1.5 eV. In contrast to the dominance of the $N_2^+(^2\Sigma_g^+, v' = 1)$ peak at 0.6 and 1.5 eV collision energy (Fig.s 16 and 18), the $v' = 3$ peak is dominant at 0.78 eV [because of the $\sin\Theta$ factor in Eq. (5) is applied]. The intergration of Fig. 19 using Eq. (5) indicates that the normalized relative populations of $v' = 1$, 2 and 3 are about 0.2, 0.3, and 0.5, respectively.

This energy range where the mechanism is so different has been examined at collision energy intervals of about 0.2 eV. A pattern similar to Fig. 7 is observed from 0.8 to 1.2 eV and is absent at 1.4 eV.[82] Consideration of the

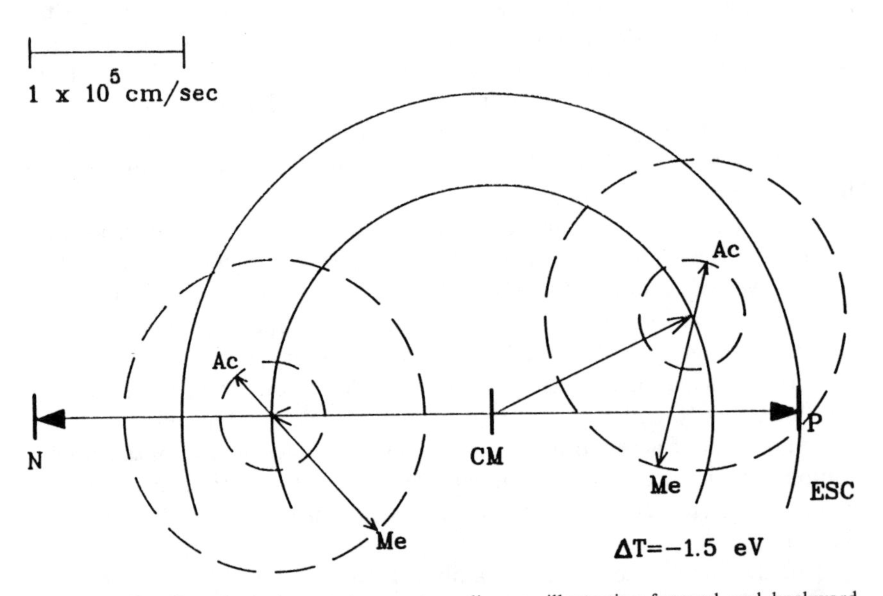

Figure 19. Hypothetical scattering contour diagram illustrating forward and backward scattering of acetone ion on collisional activation. The outer circle marked ESC is the elastic scattering circle and the inner circle marked $\Delta T = -1.5$ eV corresponds to transfer of 1.5 eV of translational energy into internal modes of acetone ions on collision. The dashed circles show the velocity vectors for the isotropic dissociation of acetone into acetyl ion and methyl radical with the assumed release of 0.1 eV energy on dissociation. Reprinted from Ref. 86 with the kind permission of American Institute of Physics.

energy widths in the ion and neutral beams allows us to place a lower limit of 0.8 eV and upper limit of 1.2 eV for the collision energy range where this mechanism is observed. This places the "window" at 1.0 ± 0.2 eV with the possibility that the energy interval where this mechanism is operative could be even narrower. This characteristic strongly suggests we are observing a scattering resonance, which has not been found in theoretical calculations carried out thus far.

The wealth of information on this system provides a very good overall understanding of the relevant reaction dynamics. This understanding is nicely summarized in the recent papers by Parlant and Gislason.[63–65] These quenching reactions proceed via charge-transfer intermediate states. Their calculations reproduced most of the reported observations for the forward reaction of $Ar^+(^2P_{3/2})$ with N_2 and for the back reaction of $N_2^+(X\,^2\Sigma_g^+, v)$ and $N_2^+(A\,^2\Pi_u, v)$ with $Ar(^1S_0)$. Their calculations also predict significant cross sections for quenching of $Ar^+(^2P_{3/2})$ accompanied by vibrational excitation, as observed in the present experiments. A comparison between these calculations and the present results shows the following.

1. The $Ar^+(^2P_{1/2})$ was detected in the present results by its quenching rather than its reaction with N_2. No features attributable to reaction, implying either no reaction occurs or the reactive scattering pattern for $Ar^+(^2P_{1/2})$ is indistinguishable from that of $Ar^+(^2P_{3/2})$. The high probability for quenching relative to reaction agrees with theoretical predictions.

2. The lowest collision energy calculated by Parlant and Gislason[63–65] is 1.2 eV and predicts $N_2^+(^2\Sigma_g, v = 1)$ is the only product. Our experimental results, strongly supported by other research cited previously, is that about 10% of $N_2^+(^2\Sigma_g^+, v' = 0)$ is also formed.

3. As discussed previously, theoretical calculations fail completely to account for the observed opening of new quantum-state product channels associated with particular scattering angles in the narrow energy "window" at 1 ± 0.2 eV.

4. The calculations indicate the $^2P_{1/2}$ quenching reaction (10b) with $N_2(v = 0)$ to the $v' = 1$ state to be the most probable even product at 1.2 and 4.1 eV collision energies. Experimentally we find no such preference for $v' = 1$ in the backward direction at 1.54 eV, but rather find an approximately equal distribution of intensity (within a factor of 2) between the $v' = 0$–3 products. This conclusion refers only to the back-scattered Ar^+ and the possibility remains that in the forward direction a different ratio of v' states for process (10) is masked in our experiments by the He scattering circle and by $Ar^+(^2P_{3/2})$ scattering.

In conclusion, it is clear that rectilinear trajectories on diabatic surfaces cannot explain the low-energy-scattering results or rotational distributions. Trajectory calculations on adiabatic surfaces addressing these issues would be extremely useful to aid our understanding of this bechmark system.

III. COLLISION-INDUCED DISSOCIATION OF POLYATOMIC IONS

We have recently completed a study of the collision-induced-dissociation (CID) reaction dynamics of the molecular ions of acetone, nitromethane, and propane.[83-90] The results for these three example of polyatomic ions are quite different. Propane is an archetypical example of RRKM-type ions for which the rapid intramolecular vibrational relaxation of internal energy precedes dissociation of the collisionally excited ion. A statistical model is the appropriate description for this ion. Our dynamics study of nitromethane and acetone ions provides strong evidence for isolated electronic states playing a major role in the CID of these non-RRKM-type ions. The acetone ion is the most striking example; at low energy (below 5 eV CM) a long-lived excited state dominates the CID of this species.

A. Acetone

Our recent publication[83] on the threshold energy collision-induced dissociation (CID) of acetone molecular ion with helium

$$CH_3COCH_3^+ \rightarrow CH_3CO^+ + CH_3 + He \tag{11}$$

provided conclusive evidence that a significant fraction of acetone ions formed by 70 eV electron impact is in the electronically excited A state, which survives more than 38 μs (required for the ion to travel from the ion source to the collision center in our crossed-beam apparatus). Low-energy collisions of these excited acetone ions convert their internal energy (electronic and possibly vibrational) into recoil kinetic energy of the acetone ion and neutral. The recoiling acetone ion rapidly dissociates into CH_3CO^+ and CH_3 on the ground-state potential surface. Fragment ions from the decomposition of superelastically scattered molecular ions are back scattered in the center-of-mass (CM) frame.

Reducing the energy of the ionizing electrons in the ions source to 12 eV completely suppressed the superelastic scattering mechanism, demonstrating the excited state(s) origin of this mechanism. The superelastic scattering mechanism is also suppressed at high collision energies[84,88] where forward-scattered (but nonzero scattering angle) highly inelastic (endothermic) processes dominate acetone ion CID. The observed dynamics are consistent with the hypothesis that the endothermic CID channel proceeds on both ground and electronically excited surfaces and primarily involves the efficient $T \rightarrow E$ excitation of acetone ions.

Figure 19 illustrates two hypothetical dynamics for acetone ion CID. In this postcollision diagram the reactant acetone ion, Ac^+, retreats from the CM toward the right and the neutral atom retreats from the CM toward the left. These two directions in the CM frame define the forward-scattering (direction of primary ion beam) and back-scattering (direction of primary neutral beam) hemispheres, respectively. After the collision, the excited ion and neutral collider recoil from the CM. Model (a) illustrates forward scattering (referenced to the primary ion direction) of the excited ion, while model (b) is back-scattered. The neutral-atom velocity vectors for models (a) and (b) represent recoil from the CM with momenta equal and opposite to those of the excited acetone molecular ions. The subsequent dissociation of the activated acetone ions into acetyl ions (and methyl radicals) originate with the translationally excited acetone as the CM for the dissociation step. Both models (a) and (b) assume 1.5 eV translational endothermicity, which is approximately the average energy transfer for the lowest-energy translationally endothermic mechanism actually observed for the acetone ion.[94,88] The 0.1 eV energy release in dissociation is the measured energy release in the PIPECO study by Baer et al.[91] for acetone ions excited by 1.5 eV photons.

Before discussing these models further, we must confront the unfortunate fact illsutrated in Fig. 19 that there is a fundamental indeterminancy in discussing the reaction dynamics of CID processes—namely, that two particles collide and a minimum of three leave the scattering center. Only the product-ion velocity vector is measured, and the scattering angles and velocities of the two neutral species can have all values allowed by conservation of energy and angular momentum. Consequently, the scattering diagram does not provide an unique determination of the energetics and angular properties of CID reactions. This dilemma is resolved by assuming that there is a time delay between the excitation processes and the decomposition of the excited ion. We think that this assumption is very well founded, especially for polyatomic ions, as the time delay need only be long enough for the neutral collider to escape the potential field of the ion. For typical collisions this will be less than 10^{-13}s. Since the decomposition step is preceded by at least one and typically by thousands to millions of vibrational oscillations, we postulate that acetone ion CID is a two-step process in our analysis.

Our two models (a) and (b) assume identical energy depositions and dissociation kinetics. The inner circle is offset from the ESC reference circle by a constant velocity shift corresponding to the translational to internal energy conversion excitation, $\Delta T'$ (illustrated in Fig. 19 by $\Delta T' = -1.5$ eV).

The zero-scattering-angle excitation model (a) corresponds dynamically to a large-impact-parameter glancing collision. The back-scattered mechanism shown as model (b) assumes that a small-impact-parameter impulsive collision is involved in the excitation step. For each mechanism, the centroid for the acetyl ion product velocities can be used to deduce $\Delta T'$, which defines the energy deposited in the acetone ion by the collision. The width of the distribution is a measure of kinetic energy release in the dissociation step. Our hypothetical models correspond to δ-function excitation and unique scattering angles. It follows that the range of excitation energies, scattering angles and energy release values actually anticipated for polyatomic ions will broaden features in CID scattering diagrams significantly in comparison with the hypothetical examples of Fig. 19.

Figure 20 illustrates the dissociation pattern actually observed for the acetone molecular ion CID by helium collisions at 1.3 eV collision energy. The inner circle marked ESC on this diagram is elastic scattering circle described in Fig. 19. The outer circle marked $\Delta T = 2.2$ eV passes through the maximum intensity contour of the acetyl product ion assumes the mechanism described previously, which releases 2.2 eV of internal energy of the ions into translational energy. The fact this peak maximum corresponds to a kinetic

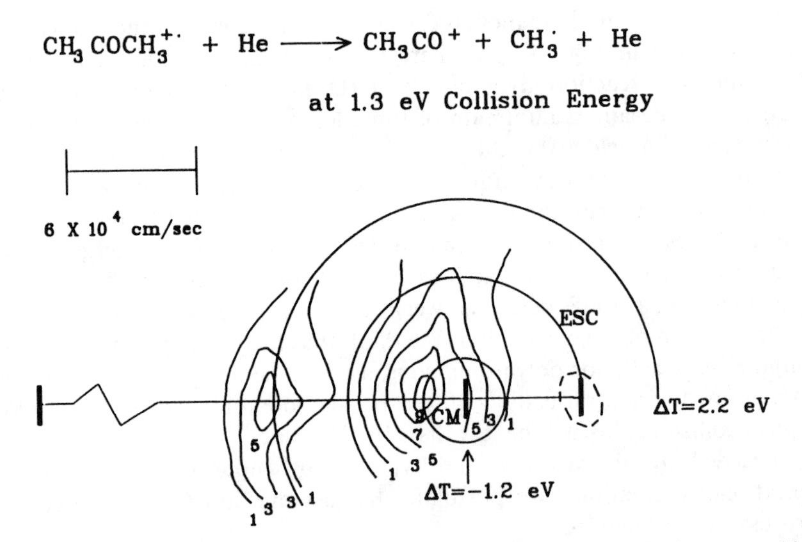

Figure 20. Scattering contour diagram for the CID of acetone molecular ion to acetyl ion on collision with helium neutrals at 1.3 eV collision energy using the in-place crossed-beam instrument. The small dashed-line oval at the right of the figure shows the FWHM profile on the primary ion beam at this energy. Reprinted from Ref. 86 with the kind permission of American Institute of Physics.

energy *gain* by the system of 2.2 eV demonstrates that a large excess of internal energy is converted into translation of the recoiling acetone ion and helium atom *before* the dissociation occurs. The acetone ion then dissociates to acetyl ions with a centroid located on this circle, which is drawn to correspond to recoil energy of 3.5 eV (2.2 + 1.3) with respect to the CM.

A "normal" endothermic product peak located within the ESC near the CM is also observed in Fig. 20. The maxima of both the superelastic (2.2 eV exothermic) and inelastic (1.1 eV endothermic) scattering mechanisms lie directly on the relative velocity vector. Consequently, we may identify these collisions as those occurring with near zero impact paramteres such that the excited acetone ions and neutral reverse their initial directions and recoil from the collision center. These "hard core" collisions are responsible both for triggering the decay of the long-lived (lifetime greater than 38 μs), electronically excited state acetone molecular ion into acetyl ion and methyl radical and also responsible for a "normal" inelastic scattering mechanism giving the same products.

The fact that the scattering contours are relatively tight also demonstrates that the dissociation of the excited acetone ion into acetyl ion and methyl radical occurs with relatively small release of kinetic energy in the dissociation step. In fact, the breadth of the distribution for the exothermic channel is only slightly greater than can be accounted for by the velocity distribution of the reactant ion beam. The ion-beam FWHM velocity and angular distribution is shown in Fig. 20 as dashed lines; convolution with the supersonic jet neutral broadens this only slightly in defining the CM distribution of reactants. About half of the actual distribution for the superelastic mechanism in Fig. 20 is accounted for by convolution of the distribution of the primary ion and neutral velocities. Consequently the combination of dispersion in $E \rightarrow T$ energy exchange and kinetic energy release taken together is estimated to be less than 0.1 eV. Clearly, energy exchange and angular scattering characteristics are extraordinarily well defined for this polyatomic ion CID mechanism.

The fundamental reasons for this extraordinary behavior are found in the photoelectron spectroscopy of acetone.[92,93] The adiabatic ionization potential for forming the A state of acetone is located is about 2.2 eV above the ground state. The vertical ionization is 2.9 eV above the X state. Therefore, the most probable ionization process generates ions with around 0.7 eV of vibrational excitation. This internal vibrational energy is sufficient for C–C bond scission on the ground-state surface but apparently not enough on the excited electronic state surface. Since the electronic transition between the A state and the ground state is optically forbidden, and since internal conversion is also very slow, this electronic energy can be stored for a relatively long period of time under collision free conditions.

We infer the following mechanism for CID of excited *A* state acetone ions in He and Ar collisions. Collision induces a curve crossing of the ion–neutral potential surface of excited acetone ion with that of ground-state acetone ion as suggested qualitatively in Fig. 21. Here the acetone ion is represented as a quasidiatomic potential curve and the ion–neutral potential is shown on an orthogonal axis, the *z* axis in this figure. The excited and ground states are shown as isolated systems, which are coupled by collision with the rare-gas atom. These collisions lead to deexcitation of the ion to the ground state releasing the excess energy to translational recoil of the acetone ion and neutral with release of the *adiabatic* energy difference. Vibrational excitation present in the ion is preserved as internal energy in the ground state and provides the energy required to drive the dissociation to acetyl ion and methyl radicals. The nearly exact energy balace is responsible for the very narrow peak in Fig. 20.

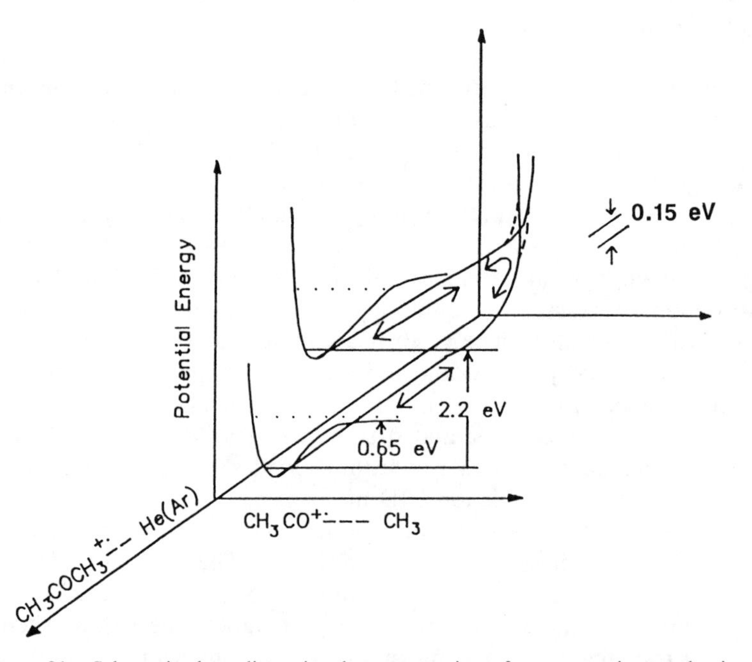

Figure 21. Schematic three-dimensional representation of curve-crossing mechanism for superelastic scattering of acetone ion proceeding a release of excess internal energy into translation prior to dissociation. Acetone ions in the upper, electronically excited state collide with He or Ar atoms and undergo curve-crossing deexcitation to the ground state with release of 2.2 eV into recoil kinetic energy of the ion and neutral atom. Those ions retaining more than 0.65 eV of internal energy dissociate into acetyl ions and methyl radicals on the gound-state hypersurface. As discussed in the text, we have determined the threshold energy of the postulated curve-crossing mechanism is about 0.15 eV.

We have further tested this mechanism by examining the microscopic inverse—excitation of ground-state ions to the excited-state surface. Figure 22 presents our data for the nondissociative, inelastic scattering of acetone ions, at 2.3 eV collision energy; the peak near the CM demonstrates a large cross section for a transition absorbing 2.2 ± 0.1 eV demonstrating a bound–bound transition ($X \rightarrow A$). At this energy we also find two dissociative channels displayed in Fig. 23, one of which is endothermic by 2.2 eV and one which is exothermic by 2.2 eV. The major endothermic scattering mechanism in Fig. 23 represents the collision-induced bound-unbound transition, which transfers a population of vibrationally excited acetone ions (near but below the dissociation limit) from the X state to the A state, again with the transfer of approximately 2.2 eV. This population of vibrationally excited ions retains sufficient vibrational energy during the transition to the excited-state surface to decompose into acetyl ion and methyl radical on the excited-state surface. The endothermic channel represents the excitation of vibrationally excited ground-state acetone ions above the dissociation limit of the excited-state hypersurface, while the translationally exothermic channel is the process discussed earlier. Clearly we are able to trace the reaction pathways of Fig. 21 in both directions.

The unexpectedly long-lived excited state of acetone has been further characterized in a series of experiments conducted in the Orsay tricyclotron

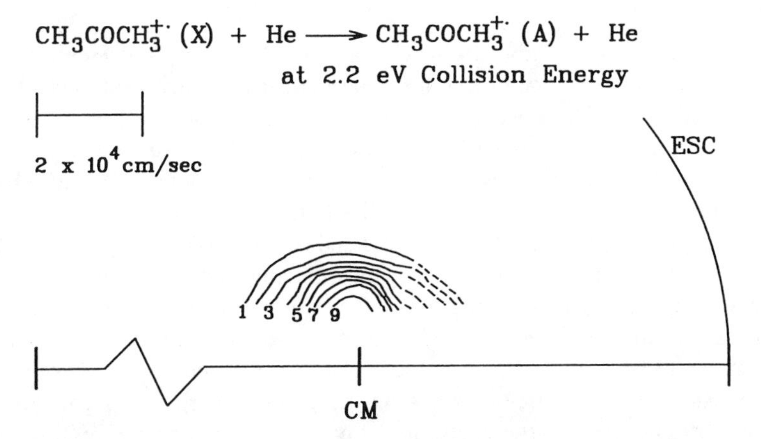

Figure 22. Velocity contour diagram for the inelastic collisions of acetone molecular ions with a supersonic beam of helium neutrals at 2.2 eV center-of-mass collision energy. CM and ESC denote center-of-mass and elastic scattering circle, respectively. The numbers marked on contours represent relative intensity of each contour. Location of product at the CM corresponds to conversion of 2.2 eV translational energy into internal energy. Reprinted from Ref. 90 with the kind permission of Elsevier Science Publishers.

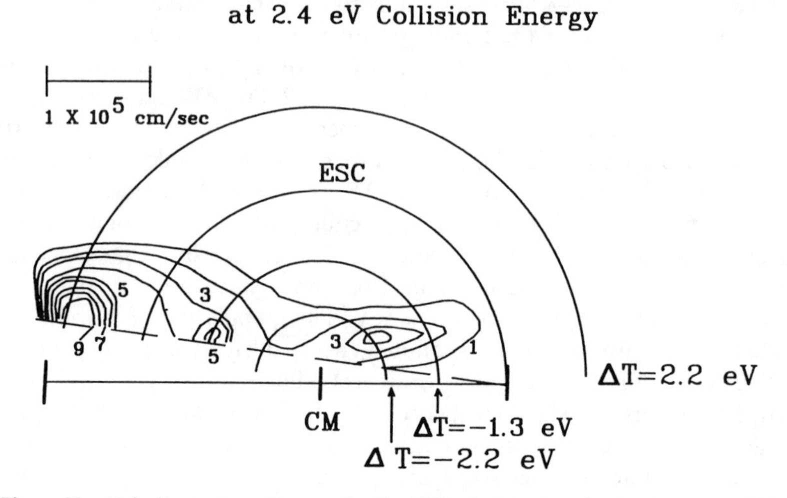

Figure 23. Velocity contour diagram for the CID of acetone molecular ions on collision with argon at 2.4 eV collision energy. Circles marked $\Delta T = -1.3$ and -2.2 eV correspond to inelastic energy transfer of 1.3 and 2.2 eV into internal excitation and the circle marked $\Delta T = 2.2$ eV corresponds to energy release of 2.2 eV from internal to translational modes of the dissociating acetone ion. Reprinted from Ref. 90 with the kind permission of Elsevier Science Publishers.

magnetic ion trap in collaboration with Marx, Mauclaire, and Heninger.[94] In these experiments acetone ions generated by electron impact were mass-selected, stored for a variable time period in a trapping cell, and reacted with oxygen (a monitor gas whose ionization potential is above the ground state of acetone and matches the A-state ionization energy) or caused to decompose by colliding them with helium in the reaction ion cyclotron resonance cell. A lifetime of 10 ± 4 ms was deduced from these measurements. We also measured the threshold energy for inducing excited acetone ion CID by varying the translational energy of the acetone ions prior to collision. The results of this experiment are shown in Fig. 24, which plots the fraction of acetone ions (aged for 1.2 ms) that decompose as a function of CM collision energy. As discussed elsewhere,[94] the threshold energy is significantly Doppler-broadened and the excitation function is plausibly represented as a step function with a threshold energy of 0.15 ± 0.05 eV. This places the barrier height in Fig. 21 for the collision-induced curve-crossing mechanism at about 0.15 eV, as marked on this figure.

With increasing collision energy the translationally exothermic CID reaction path closes and only translationally endothermic dissociation of

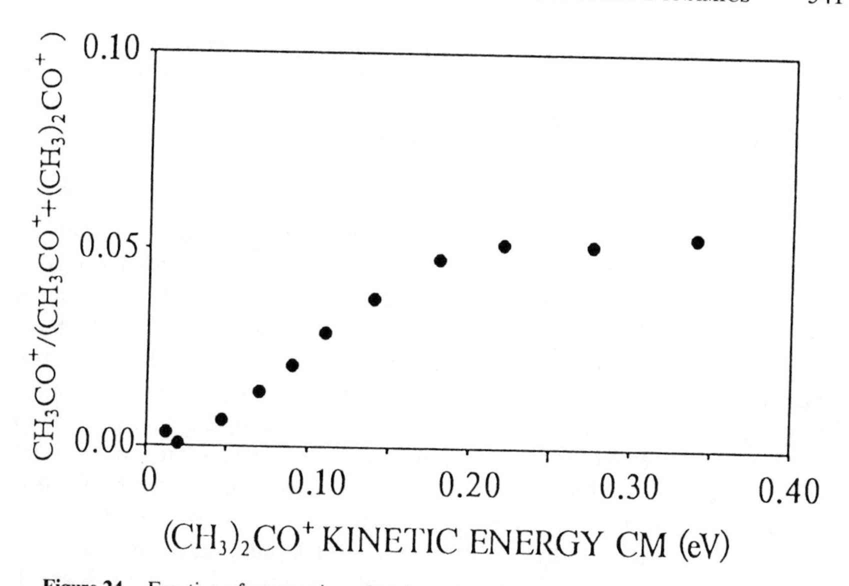

Figure 24. Fraction of acetone ions that decompose into acetyl ions and methyl radicals on collision with He atoms as a function of collision energy. The width of this excitation function is accounted for by the kinetic energy distributions of the ion and neutral collider in this trapped cell experiment, implying a step function threshold of 0.15 ± 0.05 eV.

acetone is observed. Above 10 eV CM all acetone to acetyl ion CID become forward-scattered, and the endothermicity increases smoothly to an average value of about 6 eV. Zero-angle scattering is never observed, even at KeV collision energies.[84] The high-energy mechanism, endothermic by more than an order of magnitude than the thermochemical requirement for dissociation to occur on the ground-state hypersurface, is taken as evidence that electronically excited states are readily reached in the CID of this non-RRKM-type ion. Nonzero scattering even at the highest energy probed is evidence that an impulsive-type mechanism is responsible for CID excitation of acetone. Our general conclusion is that efficient $T \leftrightarrow E$ excitation/deexcitation dominates acetone ion CID at both low and high collision energies. Only in the intermediate-energy regime do we find that excitation/dissociation collisional activation occurs on and dissociation occurs from the ground electronic hypersurface.

Finally, the disappearance of the long-lived excited-state exothermic CID mechansim with increasing collision energy is evidence that the curve-crossing $E \rightarrow T$ transition is no longer induced efficiently by impulsive collisions, probably because the collision time—determined by collision velocity—is too short. As shown in Fig. 24, the threshold energy to induce this mechanism is of the order of 0.15 eV, and this mechanism is no longer observed at collision energies 30 times higher. Alternatively, the opening of direct dissociation channels may simply compete very effectively with the $E \rightarrow T$ curve-crossing mechanisms.

B. Propane and Nitromethane Ion CID

Our investigation of CID dynamics of propane and nitromethane had the objective of defining the scattering characteristics over a broad range of collision energies for an ion ($C_3H_8^+$) that is an archetypical example of statistical (RRKM or QET) dissociation plus a second example ($CH_3NO_2^+$) whose dissociation does not follow this kind of description. The photoelectron spectrum of propane is broad, with unresolved features characteristic of overlapping vibrónic energy levels that are strongly coupled. This provides a mechanism for efficient relaxation of internal excitation to the ground electronic state.[93] In contrast, the nitromethane photoelectron spectrum is discontinuous with three overlapping electronic states separated by about 2 eV from a higher band of at least three overlapping electronic states,[95] analogous in these respects to acetone. The experimental breakdown graph for propane is in excellent agreement with RRKM/QET calculations,[96] while the nitromethane ion exhibits characteristics of isolated electronic states or rearragement structures, or both. The CID of both ions differs significantly from that of the acetone described in Section A.

Two decomposition channels for the propane ion

$$C_3H_8^+ + He(Ar) \rightarrow C_2H_4^+ + CH_4 \tag{12}$$

$$\rightarrow C_2H_5^+ + CH_3 \tag{13}$$

have been investigated previously at low every[97] and at higher energy, up to 1 KeV.[88] Figures 25 and 26 are scattering diagrams for these two reactions at 4.2 eV collision energy.[97] Qualitatively, the characteristics of (1) nonzero scattering angle (which correlates with reaction endothermicity) and (2) energy deposition that exceeds by only a few tenths of an electron volt the threshold energy for these reactions are the distinctive CID dynamics features for this RRKM-type ion. As collision energy is increased to 1 KeV only quantitative changes are observed.[88] The angular deflection closes to smaller values with increasing energy, and the mean energy shifts to slightly broader distributions with slightly higher mean values. These characteristics are those anticipated for the collisional excitation and dissociation of a polyatomic ion whose unimolecular decay is accurately described by a theoretical/experimental RRKM/QET model in which the energy deposition function follows the Massey adiabatic criterion dependence on primary ion/collision neutral relative velocity.[88]

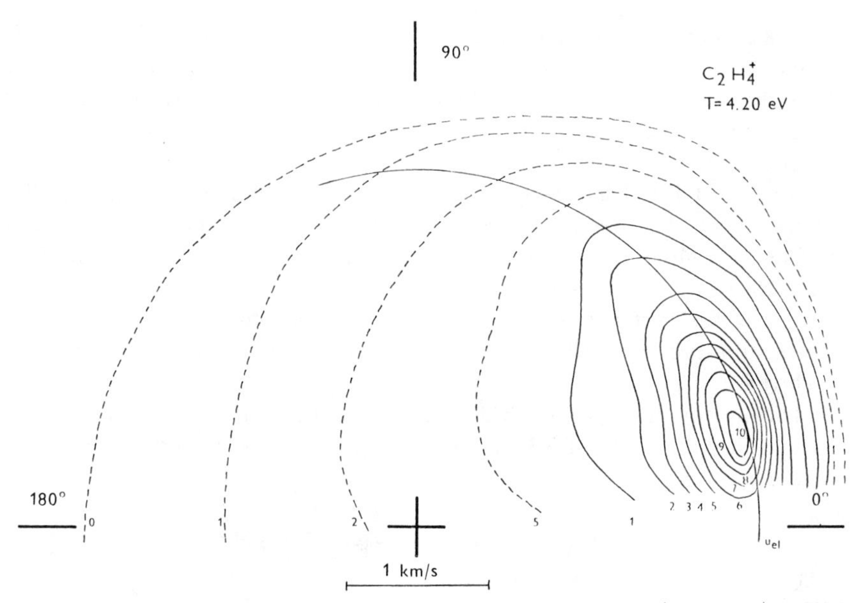

Figure 25. Velocity contour for collision-induced dissociation of $C_3H_8^+$ into $C_2H_4^+(+CH_4)$ on collision with Ne at 4.20 eV CM energy. Reprinted from Ref. 97 with the kind permission of Elsevier Publishers.

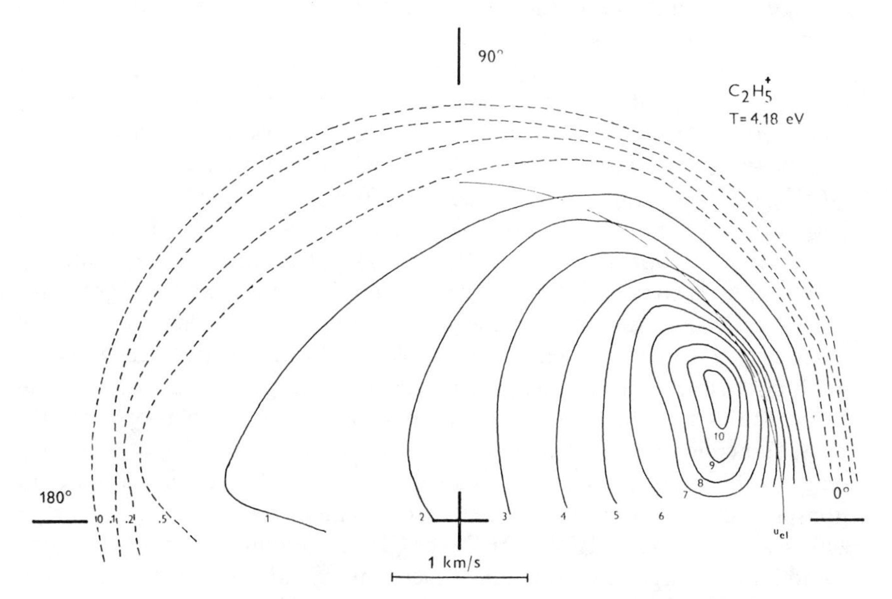

Figure 26. Velocity contour for collision-induced dissociation of $C_3H_8^+$ into $C_2H_4^+(+CH_3)$ on collision with Ne at 4.18 eV CM energy. Reprinted from Ref. 97 with the kind permission of Elsevier Publishers.

It is of interest to compare Figs. 26 and 23, which constrast the nominally similar C–C bond cleavage reactions (13) and (11) for propane and acetone at a similar collision energy. The highly structured scattering pattern of acetone in Fig. 23 and the opening and closing of reaction pathways as a function of collision energy (velocity) found for acetone are completely absent in propane CID. Tentatively, we use the information from propane CID dynamics as a baseline reference for statistical behavior and attribute major differences observed in acetone (and nitromethane) to non-RRKM characteristics. In particular, back-scattering mechanisms and highly structured scattering patterns are suggestive that decomposition from isolated states/structures and efficient $E \leftrightarrow T$ energy transfer mechanisms may be involved.

The nitromethane cation CID dynamics exhibits features that we interpret as evidence for non-RRKM/QET behavior. We have studied two CID reactions

$$CH_3NO_2^+ + He(Ar) \rightarrow NO^{2+} + CH_3 \quad (14)$$

$$\rightarrow NO^+ + CH_3O \quad (15)$$

over a collision energy range from 1.5–119 eV CM.[88] We did not find evidence at any energy for a long-lived excited state that would generate CID products superelastically (analogous to acetone). At low collision energy (below 10 eV CM) reactions (14) and (15) exhibit similarities and some differences from our expectations for RRKM/QET behavior. The NO_2^+ channel [reaction (14)], endothermic by 1 eV, is scattered at larger angles than reaction (15), endothermic by 0.6 eV, as expected for normal RRKM behavior. At threshold NO^+ [reaction (15)] exhibits both a forward-scattered, large-impact-parameter excitation mechanism and a back-scattered peak indicative of an impulsive excitation mechanism. Both exhibit very small energy shifts, as predicted from consideration of a RRKM/QET based breakdown graph for the lowest-energy dissociation mechanism for the molecular ion. At 3 eV reaction (14) is first detected as a back-scattered peak located near the CM. With increasing collision energy the back-scattered peaks shift forward and below 20 eV CM both reactions become dynamically similar to propane CID.

As shown in Fig. 27, at 18.6 eV collision energy a very different scattering diagram for reaction (15) is obtained, which we interpret, in part, as nonstatistical dissociation of the nitromethane cation. In this figure a zero-scattering-angle RRKM/QET-type CID is a dominant dynamics feature. In addition a large-scattering-angle mechanism with the large energy shift is easily detected. This represents the opening of a new reaction channel with increasing collision energy. Integration of this scattering pattern using

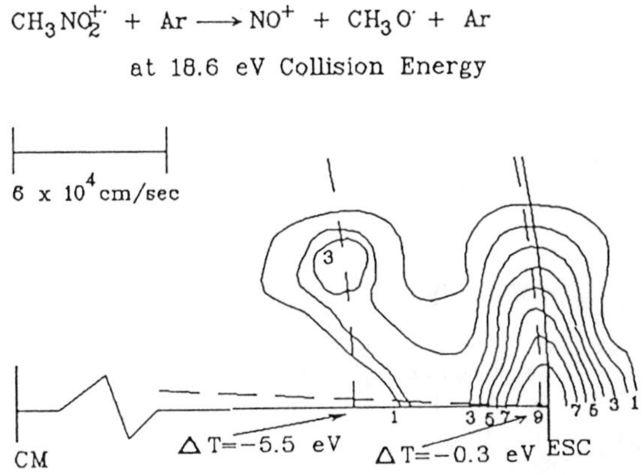

$CH_3NO_2^{+\cdot} + Ar \longrightarrow NO^+ + CH_3O^\cdot + Ar$

at 18.6 eV Collision Energy

6 x 10^4 cm/sec

CM

$\Delta T = -5.5$ eV

$\Delta T = -0.3$ eV

ESC

Figure 27. CM velocity contour map for the collision-induced dissociation of nitromethane ion to NO^+ on collision with Ar at 18.6 eV collision energy. The circle marked $\Delta T = -5.5$ eV corresponds to transfer of 5.5 eV energy into internal excitation of the nitromethane ions prior of dissociation.

Eq. (5) shows that this new mechanism accounts for about half of the total reaction probability at 18.6 eV.[87] At higher energy it becomes the dominant mechanism.[89]

The explanation for this behavior is found in the complex potential energy surface for nitromethane decomposition, which is still a controversial topic. A schematic view which shows relevant details is given in Fig. 28.[99] This figure shows that several potential wells corresponding to several stable structures, transition-state intermediates, and additional local minima (not shown) further complicate the interpretation of experiments on the unimolecular decay and CID of nitromethane and methyl nitrite cations. The existence of ion–dipole complexes and other stable structures are complicating additional features not depicted in this figure that may be important in unimolecular decay processes. It is probable that many of these structures are not reached in CID since the unimolecular decomposition following relatively high-energy excitation occurs too quickly for the entire hypersurface to be explored. Stated slightly differently, the subtleties found in PIPECO and metastable studies may occur in CID but are obscured by the dominant features of rapid dissociation following the deposit of a broad range and large amount of excitation energy.

The key feature of the hypersurface manifested in CID is the existence of (at least) two deep minima in the hypersurface, which represent the nitromethane and methyl nitrite cations, respectively. Our interpretation of our scattering results assumes electron impact ionization of nitromethane creates

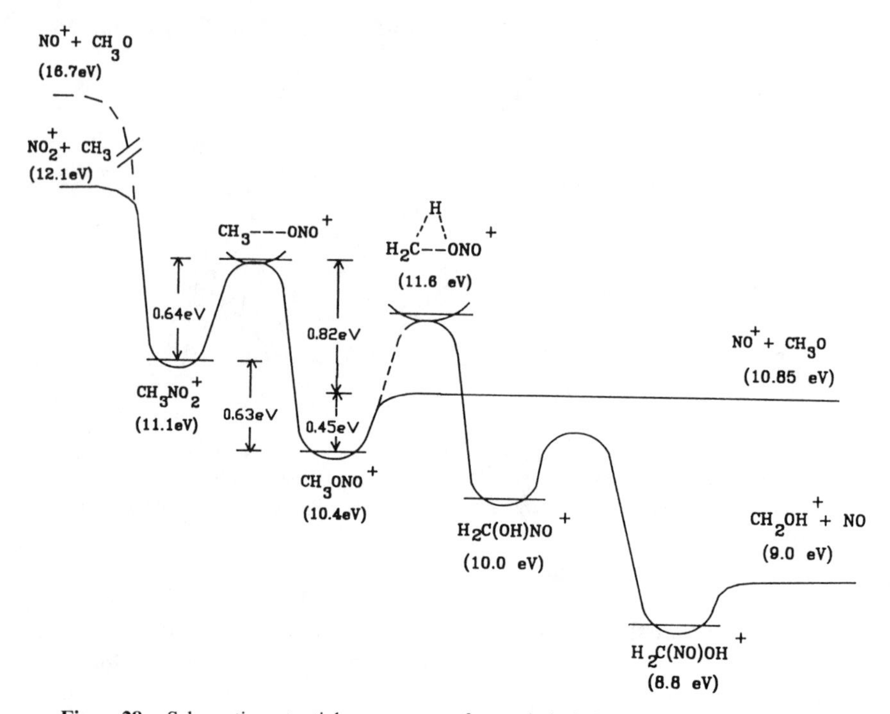

Figure 28. Schematic potential energy curve for methyl nitrite and nitromethane ions and their dissociation products.

a population of vibrationally-excited ions, some of which contain almost enough energy to rearrange to methyl nitrite ions. As shown schematically in Fig. 28, the highest levels lie near the transition state and have a configuration that can be described as highly excited methyl nitrite cations. Less excited ions nearer the nitromethane cation minimum retain the structure of the precursor neutral. Consequently, our ion beam contains a population of vibrationally excited "methyl nitrite" and "nitromethane" ions. The "methyl nitrite" cations are easily dissociated in large-impact-parameter collisions and are the forward-scattering mechansim of Fig. 27. This reaction mechanism follows the solid line path in Fig. 28 leading to NO^+ and CH_3O. The dashed line connecting the transition state for successive rearrangements leading eventually to the metastable ion product $CH_2OH^+ + NO$ is not traversed in CID.

The nitromethane structure cations can absorb 5.5 eV in an efficient $T \rightarrow E$ collisional excitation step and are the high-energy, large-scattering-angle, impulsive mechanism in Fig. 27. Supporting evidence for this hypothesis is found in the photoelectron spectra of nitromethane and methyl nitrite.[94] Nitromethane exhibits a large peak for the sixth ionization band, which has NO^+ as its major dissociation reaction.[96] No such feature is present in the PES of methyl nitrite.

More speculative, but plausible, is the hypothesis that the correllated low-energy, back-scattering mechanism for forming NO^+ [reaction (15)] and NO_2^+ [reaction (14)], which exhibit analogous energy/angle correlations as collision energy increases, suggesting they are competing reactions for the same species. If this is correct, CID excites nitromethane ions, which are in the potential well on the LHS of Fig. 28 and induces the competing reactions of rearrangement—going over the barrier to the right—and direct dissociation to NO_2^+ products on the left. Methyl nitrite cations formed by rearrangement of nitromethane in the ion source undergo the large-impact-parameter, small-scattering-excitation-characteristic RRKM/QET-type dissociation mechanism generating NO^+ via the low-energy pathway accessible to the right. The third mechanism, already discussed, is the high-energy, electronically excited hypersurface reaction path shown in the upper LHS of Fig. 28.

In conclusion, nitromethane CID is an interesting intermediate case exhibiting some features analogous to propane ion CID and some analogous to acetone ion CID. Clear evidence is obtained for rearrangements being an important feature of nitromethane ion CID. It is a second example for which efficient $T \rightarrow E$ collisional excitation appears to be an important CID energy transfer mechanism.

IV. FUTURE DEVELOPMENTS

It is clear that several examples now exist where experiment has outstripped the ability of existing theory to explain dynamics features that have been measured in crossed-beam measurements of charge-transfer and collision-induced dissociation reactions. In some cases more precise potential surfaces are needed. For example, a high quality surface for ground and excited states of the acetone cation will be required to understand both the reason for the extraordinary lifetime of the excited state and the coupling of these surfaces in repulsive regions, which is induced so efficiently by low-energy collisions. A higher-level calculation is also required to achieve better-than-qualitative understanding of the complex hypersurface explored in the CID and unimolecular decay of nitromethane and methyl nitrite cations. Better trajectory calculations using high-quality potential curves now available for Ar_2^+ are required to explain the unexpectedly high efficiency with which fine-structure transitions are induced in low-energy collisions. Better hypersurfaces *and* better trajectory calculations are likely required to understand the reactive-resonance behavior observed for ArN_2^+ in the neighborhood of 1 eV CM, the transition region between diabatic and adiabatic regions and the region where trajectories are severely perturbed by the *ca.* 1 eV potential well.

Experimentally there is much to be learned by continuing the study of charge transfer and energy transfer in atomic-ion/diatomic-molecule systems. For ArN_2^+ a crossed-beam study of the back reactions of N_2^+ in the ground and vibrationally-excited states with Ar is desirable to complement the theoretical and experimental studies of total cross sections for these reactions. Extending the studies to higher energy to define scattering dynamics in the region where new reaction channels open will also be of interest, as will extending these studies to other systems such as $ArCO^+$, which have been well explored by other techniques and which indicate the strong participation of excited states. Charge-transfer studies of molecular systems that better define the transition from complex formation to direct mechanisms continues to be an interesting research topic as well.

During the last year our attention has been riveted on the striking surprises encountered in our studies of polyatomic ion CID dynamics. These include the long-lived excited state of the acetone ion described in detail in Section III, the remarkably tightly defined dynamics in acetone and nitromethane, and the discovery that efficient $E \rightarrow T$ energy transfer can be collisionally induced at low kinetic energy. These issues require further exploration to learn whether they are general phenomena or interesting curiosities peculiar to the systems we chose for investigation.

Our continued investigation of this problem will focus on metal carbonyls, whose CID spectra have served as benchmarks for describing energy deposition in the CID of polyatomic ions. They are also prime candidates for exhibiting characteristics of isolated electronic states and bottlenecks in intramolecular relaxation. For example, the ground state of Cr^+ is $(3d)^5 4s$, while the ground state of $Cr(CO)_6$ is a singlet. Ionization and ligand loss processes necessarily involve extensive rearrangement in the electronic configuration of the metal, implying nonadiabatic processes that lead to energy barriers and excess energy in the fragments. Consequently, over 70% of Cr^+ generated by electron impact ionization of $Cr(CO)_6$ is electronically excited, with a radiative lifetime exceeding 2 s.[100] Jarrold, Misev, and Bowers[101] have reported metastable decays from $CrCO^+$ and $Cr(CO)_2^+$, which they explain as occurring via electronic predissociation involving spin-forbidden curve-crossing mechanisms. Meisels et al. have reported stable molecular ions for $Cr(CO)_6^+$ generated 3.2 eV above the ionization potential of the molecule.[102]

Despite these anomalies, a PEPICO study of $Cr(CO)_6$ has successfully generated a breakdown graph for the molecular ion that follows the QET formalism and ion energetics quite well, assuming successive CO loss is the decomposition mechanism.[102] Only the Cr^+ ion exhibits any evidence for energetically distinguishable mechanisms. Pronounced autoionization from Rydberg states was suggested to explain both significant ionization at photon energies in the PES window region and the appearance of undissociated parent ions at 11.4 eV. Thus, the possibility remains that metal carbonyls exhibit mainly normal QET-type behavior in these decomposition patterns, including CID. We suggest that a detailed dynamics of this system is likely to turn up interesting features of energy disposal and angular scattering.

Acknowledgments

The author gratefully acknowledges many helpful discussions of these processes with Professors Steve Leone, Paolo Tosi, Eric Gislason, Tomas Baer, and Douglas Ridge. I also thank graduate students Anderson, Biggerstaff, Howard, Qian, and Sohlberg and Postdoctoral Fellows Friedrich and Rockwood for carrying out most of the work described in this paper. I owe a special debt of gratitute to Dr. Anil Shukla for assistance in carrying out a substantial body of the work, developing one of our beam instruments and clarifying our interpretation through frequent discussions and debates. I am gratefull to the National Science Foundation for support of most of this work. Finally, I thank Betty Painter for her assistance in fitting the composition and typing of this manuscript into our busy schedules and thank the Editor, Cheuk Ng, for his patience and encouragement.

References

1. T. Kato, K. Tanaka, and I. Koyano, *J. Chem. Phys.* **77**, 337 (1982).
2. K. Tanaka, J. Durup, T. Kato, and I. Koyano, *J. Chem. Phys.* **73**, 586 (1980).
3. C.-L. Liao, X.-X. Liao, and C. Ng, *J. Chem. Phys.* **82**, 5489 (1985).
4. K. McAfee, Jr., W. Falconer, R. Hozack, and D. McClure, *Phys. Rev. A.* **21**, 827 (1980).
5. S. L. Howard, A. L. Rockwood, W. Trafton, B. Friedrich, S. G. Anderson, and J. H. Futrell, *Can. J. Phys.* **65**, 1077 (1987).
6. S. L. Howard, A. L. Rockwood, W. Trafton, B. Friedrich, S. G. Anderson, and J. H. Futrell, *Chem. Phys. Lett.* **140**, 385 (1987).
7. S. L. Howard, A. L. Rockwood, S. G. Anderson, and J. H. Futrell, *J. Chem. Phys.* **91**, 2922 (1989).
8. S. L. Howard, A. L. Rockwood, and J. H. Futrell, *Chem. Phys. Lett.* **170**, 99 (1990).
9. J. H. Futrell, *Int. J. Quantum Chem.* **31**, 131 (1987).
10. R. Johnson, *J. Phys. B.* **3**, 539 (1970). The ordinate in Figs. 5 and in this article have been corrected to log E rather than ln E, as explained in Fig. 1 and footnote 5 of Ref. 7.
11. R. Johanson, *J. Phys. Soc. Jpn.* **32**, 1612 (1962).
12. E. Nikitin, *Opt. Spectrosc. (Engl. Transl.)* **19**, 91 (1965).
13. W. Wadt, *J. Chem. Phys.* **68**, 402 (1978).
14. B. J. Whitaker, C. A. Woodward, P. J. Knowles, and A. J. Stace, *J. Chem. Phys.* **93**, 376 (1990).
15. D. Rapp and W. Francis, *J. Chem. Phys.* **37**, 2631 (1962).
16. O. Firsov, *Zh. Eksp. Theor. Fiz.* **21**, 1001 (1951).

17. N. Ghosh and W. F. Sheridan, *Indian J. Phys.* **31**, 337 (1957).
18. E. Gustafsson and E. Lindholm, *Ark. Fys.* **18**, 219 (1960).
19. R. F. Stebbings, B. R. Turner, and A. C. Smith, *J. Chem. Phys.* **38**, 2277 (1963).
20. M. R. Flannery, P. C. Cosby, and T. F. Moran, *J. Chem. Phys.* **59**, 5494 (1973).
21. N. G. Utterbach and G. H. Miller, *Rev. Sci. Instrum.* **32**, 1101 (1961).
22. K. J. McCann, M. R. Flannery, J. V. Hornstead, and T. F. Moran, *J. Chem. Phys.* **63**, 4998 (1975).
23. S. C. deCastro, H. F. Schaefer, III, and R. M. Pitzer, *J. Chem. Phys.* **74**, 550 (1981).
24. K. B. McAfee, Jr., C. R. Szmanda, and R. S. Hozack, *J. Phys. B* **14**, L243 (1981).
25. K. B. McAfee, Jr., C. R. Szmanda, R. S. Hozack, and R. E. Johanson, *J. Chem. Phys.* **77**, 2399 (1982).
26. S. H. Linn, Y. Ono, and C. Y. Ng, *J. Chem. Phys.* **74**, 3342 (1981).
27. K. Stephan, J. H. Futrell, H. Helm, and T. D. Märk, *J. Chem. Phys.* **80**, 3185 (1984).
28. K. Norwood, G. Luo, and C. Y. Ng, *J. Chem. Phys.* **91**, 849 (1989).
29. B. H. Mahan, C. Martner, and A. O'Keefe, *J. Chem. Phys.* **76**, 4433 (1982).
30. B. Friedrich, S. L. Howard, A. L. Rockwood, W. E. Trafton, Jr., De Wen-Hu, and J. H. Futrell, *Int. J. Mass Spectrum. Ion Proc.* **59**, 203 (1984).
31. K. Sohlberg, J. Futrell, and K. Szalewicz, *J. Chem. Phys.* **94**, 6500 (1991).
32. A. E. dePristo, *J. Chem. Phys.* **78**, 1237 (1983).
33. Z. Herman, V. Pacak, A. J. Yencha, and J. Futrell, *Chem. Phys. Lett.* **37**, 329 (1975).
34. N. Kobayashi, *J. Phys. Soc. Jpn.* **36**, 259 (1974).
35. N. Kobayashi, Y. Kaneko, F. Koike, and T. Watanabe, *J. Phys. Soc. Jpn.* **42**, 1701 (1977).
36. K. Birkinshaw and J. B. Hasted, *J. Phys. B.* **84**, 1711 (1971).
37. D. L. Albritton, A. L. Schmeltekopf, and R. N. Zare, *J. Chem. Phys.* **71**, 3271 (1979).
38. I. Dotan, F. C. Fehsenfeld, and D. L. Albritton, *J. Chem. Phys.* **71**, 3280 (1979).
39. I. Dotan, F. C. Fehsenfeld, and D. L. Albritton, *J. Chem. Phys.* **71**, 3289 (1979).
40. R. Marx, G. Mauclaire, and R. Derai, *Int. J. Mass Spectrom. Ion Phys.* **47**, 155 (1983).
41. T. Kato, *J. Chem. Phys.* **80**, 6105 (1984).
42. S. L. Howard and J. H. Futrell (unpublished).
43. K. Birkinsahw, A. K. Shukla, S. Howard, J. Biggerstaff, and J. H. Futell, *Int. J. Mass Spectrom. Ion Proc.* **84**, 283 (1988).
44. P. Warneck, *J. Chem. Phys.* **46**, 513 (1967).
45. Y. Kaneko, N. Kobayashi, and I. Kanomata. *J. Phys. Soc. Jpn.* **27**, 992 (1969).
46. N. G. Adams, D. K. Bohme, and E. E. Ferguson, *J. Chem. Phys.* **52**, 5101 (1970).
47. D. Hyatt and P. K. Knewstubb, *J. Chem. Soc. Faraday Trans. 2* **68**, 202 (1972).
48. N. G. Adams, A. G. Dean, and D. Smith, *Int. J. Mass Spectrom. Ion Phys.* **10**, 63 (1972).
49. J. B. Laudenslager, W. T. Huntress, and M. T. Bowers, *J. Chem Phys.* **61**, 4600 (1974).
50. F. Howorka, *J. Chem. Phys.* **68**, 804 (1978).
51. R. Thomas, A. Barassin, and R. R. Burke, *Int. J. Mass Spectrom. Ion. Phys.* **28**, 275 (1978).
52. W. Lindinger, F. Howorka, P. Lukac, S. Kuhn, H. Villinger, E. Alge, and H. Ramler, *Phys. Rev. A.* **23**, 2319 (1981).
53. D. Smith and N. G. Adams, *Phys. Rev. A.* **23**, 2327 (1981).
54. T. Kato, K. Tanaka, and I. Koyano, *J. Chem. Phys.* **77**, 337 (1982).

55. T. R. Govers, P.-M. Guyon, T. Baer, K. Cole, H. Frolich, and M. Lavolee, *Chem. Phys.* **87**, 373 (1984).
56. P. M. Guyon, T. R. Govers, and T. Baer, *Z. Phys. D.* **4**, 89 (1986).
57. L. Huwel, D. R. Guyer, G. H. Lin, and S. R. Leone, *J. Chem. Phys.* **81**, 3520 (1984).
58. G. H. Lin, J. Maier, and S. R. Leone, *Chem. Phys. Lett.* **125**, 557 (1986).
59. D. M. Sonnenfroh and S. R. Leone, *J. Chem. Phys.* **90**, 1677 (1989).
60. D. C. Clary and D. M. Sonnenfroh, *J. Chem. Phys.* **90**, 1686 (1989).
61. M. R. Spalburg and E. A. Gislason, *Chem. Phys.* **94**, 327 (1985).
62. M. R. Spalburg, J. Los, and E. A. Gislason, *Chem. Phys.* **94**, 327 (1985).
63. G. Parlant and E. A. Gislason, *Chem. Phys.* **101**, 227 (1986).
64. G. Parlant and E. A. Gislason, *J. Chem. Phys.* **86**, 6183 (1987).
65. G. Parlant and E. A. Gislason, *J. Chem. Phys.* **88**, 1633 (1988).
66. M. Hamdan, K. Birkinshaw, and N. D. Twiddy, *Int. J. Mass Spectrom. Ion Proc.* **57**, 225 (1984).
67. B. Friedrich, W. Trafton, A. L. Rockwood, S. Howard, and J. Futrell, *J. Chem. Phys.* **80**, 2537 (1984).
68. A. L. Rockwood, S. L. Howard, S.-H. Du, P. Tosi, W. Lindinger, and J. Futrell, *Chem. Phys. Lett.* **114**, 486 (1985).
69. K. Birkinshaw, A. Shukla, S. Howard, and J. H. Futell, *Chem. Phys.* **113**, 149 (1987).
70. J. D. Shao, Y. G. Li, G. D. Flesch, and C. Y. Ng, *Chem. Phys. Lett.* **132**, 58 (1986).
71. C.-L. Liao, J.-D. Shao, R. Xu, G. D. Flesch, Y.-G. Li, and C. Y. Ng, *J. Chem. Phys.* **85**, 3874 (1986).
72. C.-L. Liao, R. Xu, and C. Y. Ng, *J. Chem. Phys.* **85**, 7136 (1986).
73. C.-L. Liao, R. Xu, and C. Y. Ng, *J. Chem. Phys.* **84**, 1948 (1986).
74. J. D. Shao, Y. G. Li, G. D. Flesch, and C. Y. Ng, *J. Chem. Phys.* **86**, 170 (1987).
75. G. D. Flesch, and C. Y. Ng, *J. Chem. Phys.* **92**, 2876 (1990).
76. P. Archirel and Levy, *Chem. Phys.* **106**, 51 (1986).
77. E. E. Nitikin, M. Y. Ovchinnikova, and D. V. Shalashilin, *Chem. Phys.* **111**, 313 (1987).
78. H. H. Teng and D. C. Conway, *J. Chem. Phys.* **59**, 2316 (1973).
79. C. R. Brion, B. R. Rose, and J. B. Marquette, *J. Chem. Phys.* **91**, 6142 (1989).
80. A. A. Viggiano, J. M. Van Doren, R. A. Morris, and J. F. Paulson, *J. Chem. Phys.* **93**, 4761 (1990).
81. A. Ding, J. H. Futrell, R. A. Cassidy, and J. Hesslich, *Surf. Sci.* **156**, 282 (1985).
82. S. Howard, and J. H. Futrell (unpublished).
83. K. Qian, A. Shukla, S. Howard, S. Anderson, and J. Futrell, *J. Phys. Chem.* **93**, 3889 (1989)
84. A. K. Shukla, K. Qian, S. L. Howard, S. G. Anderson, K. W. Sohlberg, and J. H. Futrell, *Int. J. Mass Spectrom. & Ion Proc.* **92**, 147 (1989).
85. A. K. Shukla, K. Qian, S. Anderson, and J. H. Futrell, *J. Amer. Soc. Mass Spec.* **1**, 6 (1990)
86. K. Qian, A. Shukla, and J. Futrell, *J. Chem. Phys.* **92**, 5988 (1990).
87. K. Qian, A. Shukla, and J. Futrell, *Rapid Comm. Mass Spec.* **4**, 222 (1990).
88. K. Qian, in "Studies of Collision-Induced Dissociation Reactions of Polyatomic Ions by Crossed-Beam Tandem Mass Spectrometry, " University of Delaware, 1990.
89. K. Qian, A. Shukla, and J. H. Futrell (unpublished).

90. K. Qian, A. K. Shukla, and J. H. Futrell (unpublished).
91. D. M. Mintz and T. Baer, *Int. J. Mass Spectrom. Ion Phys.* **25**, 39 (1977).
92. R. Bombach, J. P. Stadelmann, and J. Vogt, *Chem. Phys.* **72**, 259 (1982).
93. K. Kimura, S, Katsumata, Y. Achiba, T. Yamazaki, and S. Iwata, in *Handbook of HeI Photoelectron Spectra of Fundamental Organic Molecules*, Halsted Press, New York, NY, 1981.
94. R. Marx, G. Mauclaire, M. Heninger, and J. H. Futrell (unpublished).
95. Y. Nirva, S. Tajima, T. and T. Tsuchiya, *Int. J. Mass Spectrom Ion Proc.* **40**, 287 (1981).
96. M. Brehm, J. H. D. Eland, R. Fry, and H. Schrelte, *Int. J. Mass Spectrom. Ion Phys.* **21**, 373 (1976).
97. Z. Herman, J. H. Futrell, and B. Friedrich, *Int. J. Mass Spec. and Ion Physics* **58**, 181 (1984).
98. H. S. W. Massey, *Rep. Pr. Phys.* **12**, 3279 (1949).
99. This figure is based, in part, on data for methyl nitrite decomposition pathways and relevant transition states communicated by Professor Tomas Baer.
100. W. D. Reents, Jr., F. Strobel, R. B. Freas, III, and D. P. Ridge, *J. Phys. Chem.* **89**, 5666(1985).
101. M. F. Jarrold, L. Mesev, and M. T. Bowers, *J. Chem. Phys.* **88**, 3928 (1984).
102. P. R. Das, T. Nishimura, and G. G. Meisels, *J. Phys. Chem.* **89**, 2808 (1985).

PROTON ENERGY LOSS SPECTROSCOPY AS A STATE-TO-STATE PROBE OF MOLECULAR DYNAMICS

GEREON NIEDNER-SCHATTEBURG*
and
J. PETER TOENNIES

Max-Planck-Institut für Strömungsforschung, Göttingen, F.R. Germany

CONTENTS

* *Present address*: Institut für Physikalische und Theoretische Chemie, Technische Universität München, Garching, F. R. Germany.

State-Selected and State-to-State Ion–Molecule Reaction Dynamics, Part 1: Experiment, Edited by Cheuk-Yiu Ng and Michael Baer. Advances in Chemical Physics Series, Vol. LXXXII.
ISBN 0-471-53258-4

I. INTRODUCTION

This chapter surveys recent high-resolution scattering experiments of protons from atoms, diatoms, and polyatomic molecules. These are of two types: inelastic scattering (process I, detection of the protons) and charge-transfer scattering (process II, detection of the H atoms produced):

$$H^+ + M(j_i, v_i, e_i) \rightarrow H^+ + M(J_f, v_f, e_f) \quad \text{(I)}$$

$$\rightarrow H + M^+(j_f, v_f, e_f) \quad \text{(II)}$$

Here M denotes the target and j, v, and e specify the rotational, vibrational, and electronic quantum states of the target before (index i) and after (index f) scattering. Collision energies between 1.9 and 120 eV are considered. The focus is on collision energies between 10 and 30 eV where extensive excitation of all of preceding degrees of freedom is observed. Only data from differential-cross-section measurements will be presented and discussed. State resolution is achieved in all cases by a high-resolution analysis of the energy distributions of the scattered protons or H atoms.

Proton interactions are of great importance in nature and in numerous different technical applications. For example, the solar wind consists mainly of protons with mean kinetic energies of 1–2 keV. In the earth's atmosphere so-called solar proton events can deplete the stratospheric ozone by up to 15%.[42,157] By various inelastic processes the protons lose energy and ultimately reach the energy range considered in this review. These processes are best understood for interaction of protons with H_2. Figure 1 shows the energy dependence of all the known processes occurring in proton H_2 collisions.[147,182] In addition, we remind the reader of the importance of protons in biological systems (proton-transfer reactions), in electrochemistry, and in the solid state. In many situations the reactants' proton affinity plays an important role. This has motivated extensive research on the proton affinities of various molecules.[4,47,67,144,145] Protons and their isotopes also find important technical applications in fusion reactors and in most plasma applications.

Fundamentally, the choice of protons as a scattering projectile is motivated by the fact that the proton is the simplest of all ions and has no internal degrees of freedom. Being devoid of any electrons, protons can penetrate much more deeply (≈ 1 Å) into the molecular orbitals of the target compound than, for example, Li^+ ions (≈ 3 Å). Finally, because of their low mass compared to all target molecules (except H_2) the kinematics are especially favorable and in many cases the laboratory and center-of-mass systems are nearly identical.

The experimental results obtained with protons have of course many

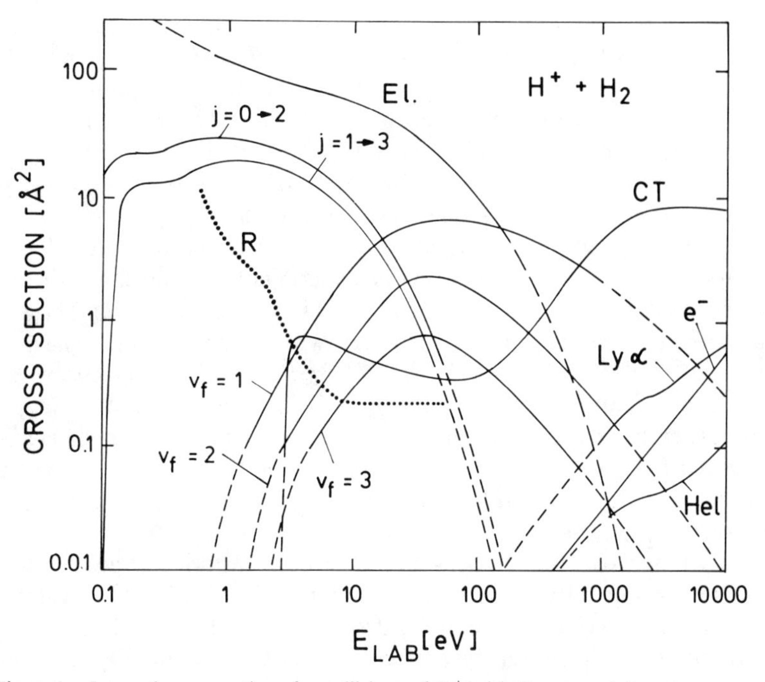

Figure 1. Integral cross sections for collisions of H^+ with H_2 versus laboratory energy of H^+ (H_2 at rest).[182] Cross sections are shown for elastic scattering (EI), rotational excitation ($j = 0 \rightarrow 2, j = 1 \rightarrow 3$), vibrational excitation ($v_f = 1, 2, 3$), charge transfer (CT), Lyman α and H α emission (Ly α, H α), and ionization (e^-) over a range of collision energies covering five decades. Reactive cross sections for the $H^+ + D_2$ system (R) are also indicated.[147]

features in common with other ion–molecule scattering processes. Experimental and theoretical studies of ion–molecule scattering have been reviewed many times. Ion–atom scattering has been extensively investigated and is now quite well understood.[41,178] Ion–molecule scattering is much more complicated largely as a result of the vibrational and rotational degrees of freedom. Their role in electronically adiabatic and nonadiabatic processes such as charge transfer is presently not well understood.

Early attempts to resolve vibrational excitation in ion-beam scattering focused on hydrogen molecule targets because of their wide level spacings. The first inelastic-energy-loss experiments with ions were reported simultaneously and independently by two groups.[46,200] The first scattering experiment to resolve energy losses to single vibrational levels used H^+ at $E_{cm} = 100$–600 eV[162] and Ar^+ at $E_{cm} = 14$ eV.[163] The first experiments to resolve individual rotational levels in energy loss were reported for the system $Li^+ + H_2$ at $E_{lab} = 2.7$ eV.[231] At about the same time extensive studies of angular and energy-loss distributions were reported for $H^+ + H_2$ at laboratory energies between 4 and 20 eV.[227,228] The study of ion–molecule reactions in scattering experiments date back to the same period and the first experiments on the reaction of H^+ with D_2 was reported in 1968.[129] Ion–molecule scattering studies have been extensively reviewed.[48,64,65,74,79,80,127,128,156] A recent book compares the various new methods of ion-beam scattering.[58] The previous experimental work on inelastic proton scattering as well as on the closely related lithium ion investigations has been part of several reviews.[59,61,219,220] Stimulated by the experimental progress the theoretical understanding of inelastic, nonadiabatic, and reactive elementary processes involving simple ions has grown steadily and has been reviewed.[8–10,66,76,106,111,120,150,166,177,192,217,225,237] One review article was devoted entirely to the selectivity of vibrational excitation in proton molecule collisions reflecting an early stage of our present knowledge.[82] The present article, provides the first overview of the recent high-resolution state-to-state studies of proton scattering from a wide range of different types of targets.

One of the most fascinating aspects of proton scattering is the very different behaviour observed with seemingly similar molecules. This is illustrated in Fig. 2 where a series of time-of-flight spectra (closely related to energy-loss distributions) is shown for the molecules N_2, CO, NO, and O_2, Surprisingly the polar molecules CO and NO show no large difference compared to N_2, whereas the O_2 spectrum exhibits greatly enhanced vibrational excitation. As we shall see, this striking difference is now understood and can be explained in terms of a temporary charge-transfer leading to a $H\text{–}O_2^+$ transition complex. Charge transfer can also have a remarkable influence on the energy transfer of polyatomic molecules.

This chapter starts with a description of two different high-resolution differential scattering apparatus. One is designed for observing both direct and charge-transfer scattering, whereas the second has been developed to analyze direct scattering with the highest possible energy resolution. The next section deals with scattering experiments such as from N_2, CO, NO, and CF_4, all of which can be understood without taking charge transfer into account. In the third part systems involving charge transfer such as O_2 and CH_4 are discussed. Each of these two sets of data is preceded by presentation of the models used in the interpretation of the results. Virtually all of the experimental work discussed here was carried out at the Max–Planck-Institüt für Strömungsforschung in Göttingen from about 1975 to the present. Besides the authors, the principal investigators in this research were in chronological order: Manfred Faubel, Klaus Rudolph, Uwe Gierz, Martin Noll, Bretislav Friedrich, and Wolfram Maring. The theoretical calculations

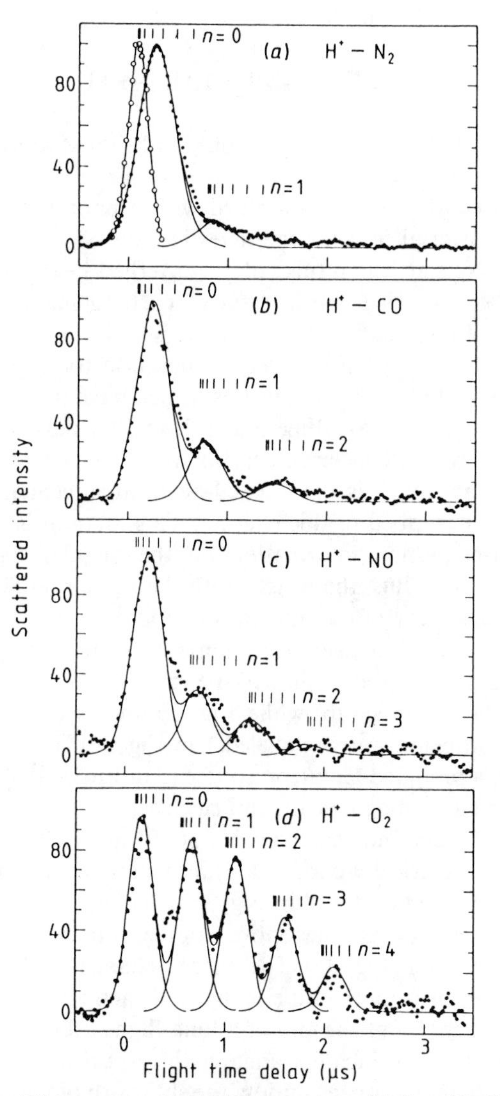

Figure 2. Time-of-flight spectra of inelastic proton scattering from N_2, CO, NO, and O_2 at $E_{\text{lab}} = 9.8$ eV and $\Theta_{\text{lab}} = 10°$ (right column) and $\Theta_{\text{lab}} = 20°$.[83] The time-of-flight channel width is 0.02 μs. The intensities are averaged over five channels plotted at the flight time of the middle channel. Vertical bars indicate the position of the peaks of inelastically scattered ions ($j_f = 0, 5, 10, \ldots,$ for each n). The spectra are fitted by Gauss curves. Also shown in the spectra of $H^+ + N_2$ is a measured spectrum for pure elastic scattering of H^+ from Ar (open circles).

were carried out in collaboration with Franco Gianturco, Ton Ellenbroek, Michael Baer, and Christoph Schlier.

II. EXPERIMENTAL METHOD

Experimental techniques for carrying out ion beam scattering studies can be classified into four categories.[58] Ion-beam–gas-cell scattering provides integral cross sections. When combined with a photoionization ion source, it becomes a powerful tool for the investigation of state-to-state scattering as surveyed in Ng's article in this volume. Merged-beam investigations have in the past been carried out to study rearrangement collisions, charge transfer, and chemiionization[32,77,223] at very low energies. As far as we are aware such apparatus are no longer in operation today. One reason might be the development of guided ion beams, which have the advantages of being technically simpler and providing higher fluxes. This experimental approach will be covered by Gerlich's article in this volume. The fourth and probably most universal approach is the crossed-beam arrangement as used in both machines to be described in this chapter. This method is the only one that provides high-resolution information on the angular distribution of the scattered reactants. Thus, the insight into the scattering event itself is more detailed than from any of the integral methods.

The first of the two scattering apparatus in which the energy-loss distributions are measured is shown in Figs. 3*a* and 3*b*. Figure 3*a* shows a side view of the entire setup, while Fig. 3*b* shows a detailed view of the rotatable-proton-beam source (labeled 1 in Fig. 3*a*). The primary H^+ beam is produced in a modified Colutron source (1 in Fig. 3b)[158] and mass as well as energy selected by magnetic (3) and electrostatic (5) deflection fields, both with radii $R = 5.0\,\text{cm}$ and sector angles of 60° and 127°, respectively. Then the ion beam is electrically chopped (7) into short pulses, which are focused (6) onto the scattering center (8 in Fig. 3*b*, 2 in Fig. 3*a*). In order to perform differential measurements, the entire source can be rotated around the target-beam axis (3 in Fig. 3*a*). The energy change of the scattered ions is determined by measuring the time of flight of each ion on arrival at an open electron multiplier (11) at the end of a long flight tube (9) while holding the ion source at a selected incident angle with respect to the flight tube.

To obtain a high-energy resolution, the slit width of the electrostatic sector field was chosen as $d = 0.2\,\text{mm}$ (0.1 mm later on), which yields a nominal resolving power of $E/\Delta E = 2R/d = 500\,(1000)$, corresponding to a resolution of 20 meV (FWHM) at a kinetic energy of 10 eV. However, the overall experimental resolution is also influenced by the finite pulse duration of typically 100 ns, which has to be compared to the total flight time of about 16 μs (at 1 m flight path and 20 eV proton energy). This effect was considerably

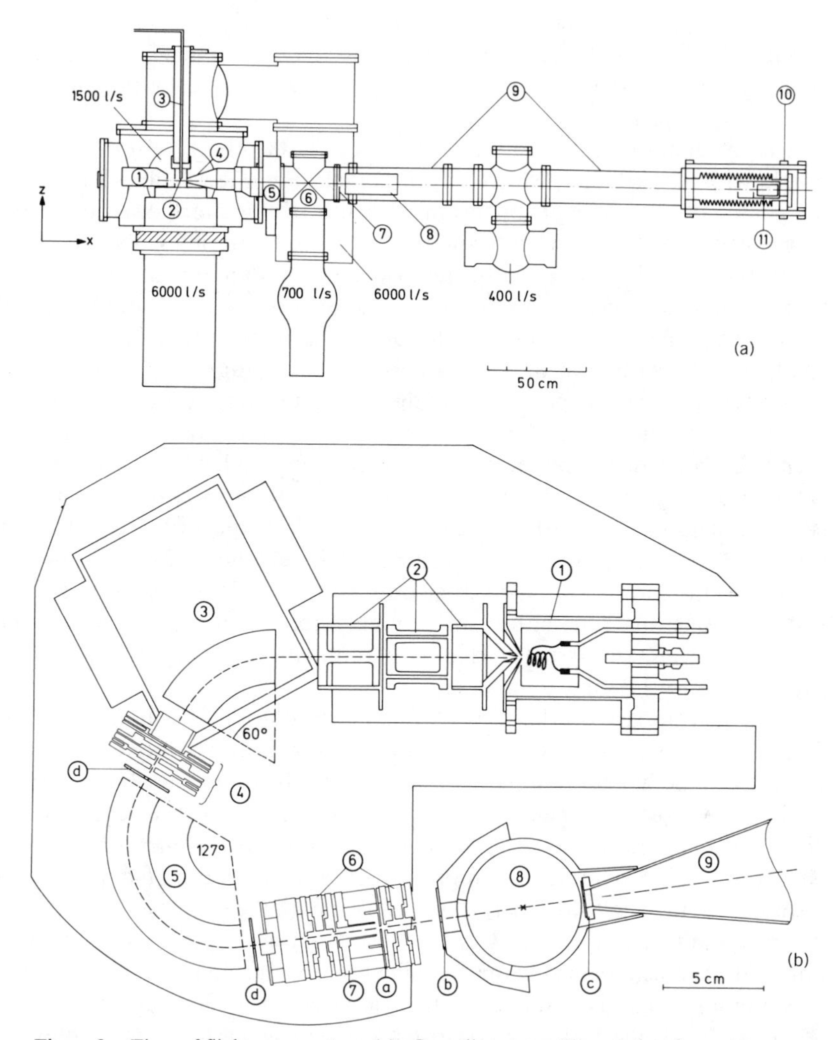

Figure 3. Time-of-flight apparatus. (*a*) Overall setup: (1) rotational proton source, (2) scattering center, (3) secondary target gas beam, (4) entrance into the flight tube, (5) gate valve, (6) differential pumping stage, (7) aperture, (8) repelling ring electrode, (9) flight tube, (10) 30 cm translator for absolute energy calibration, (11) particle-detecting multipliers. (*b*) Rotatable-proton-beam source: (1) low-pressure glow discharge, (2) extraction optics, (3) 60° magnetic mass selector, (4) deceleration optics, (5) 127° electrostatic energy analyzer, (6) beam shaping einzel lenses, (7) beam chopping pulser plates, (8) scattering center, (9) part of the flight tube, (a, b, c, d) beam defining slit apertures.[169]

reduced by consecutively lengthening the flight tube up to a maximum length of 3.74 m. The overall energy spread in the forward direction was usually about $\Delta E = 60$ meV. This is about 50% larger than calculated from the sector field geometry. The intensity of the chopped beam typically amounted to about 4×10^4 counts/s.

The secondary beam was a low-pressure (100 mbar), 50-μm-diameter nozzle without skimmer, which was mounted only a few millimeters above the center of the incident ion beam. Because of the short distance of the nozzle orifice to the primary beam and the large scattering length of about 5 mm, the overall attenuation of the primary beam amounted to 30%. The large angular divergence and the velocity spread of the target beam did not significantly affect the overall resolution owing to the favorable kinematic conditions[59] resulting from the low mass of the protons.

Scattered projectile particles can be either protons when investigating inelastic scattering or fast H atoms, which originate from charge-transfer events. Both particles are detected by the same multiplier. To measure only the H atom signal, the protons are completely suppressed by an electrical repelling field of about 100 V applied to a thick cylindrical electrode (8 in Fig. 3*a*) inside the flight tube. At an energy of, for example, 23 eV the H-atom detection efficiency is estimated to be only about 1%[186] compared to almost 100% for ions, since the latter can be highly acclerated prior to hitting the multiplier. Thus, the signal measured without the repelling field is almost entirely due to the ionic component.

With this apparatus the best obtainable resolution was limited by the length of the ion pulses, which could not be easily further reduced. Therefore, a new machine has been designed[175,230,244] as shown in Fig. 4. For energy analysis electrostatical hemispherical analyzers are used instead of time of flight. This apparatus has been designed to overcome the energy-resolution limitations by providing an improved energy resolution (< 10 meV) at medium and high collision energies (> 10 eV). With the electrostatic analyzers it is possible to select a narrow energy loss "window" for detailed investigation. The tandem hemispherical condenser arrangement for both energy selection and analysis is similar to that described in Refs. 132 and 151. The use of two hemispherical analyzers instead of a single one has the advantage of reducing the background signal by removing primary beam ions, which are reflected from the walls of the first hemispherical analyzer. In addition, it narrows the base width of the transmission function, which also leads to an increase in the available dynamical range.[123] This is important when neighboring peaks with large intensity differences are to be resolved as is the case for the very high vibrational overtones in CF_4 ($8 < n < 25$).

In the following section, we will present experimental data from both of the machines described. Throughout this review most of the angular

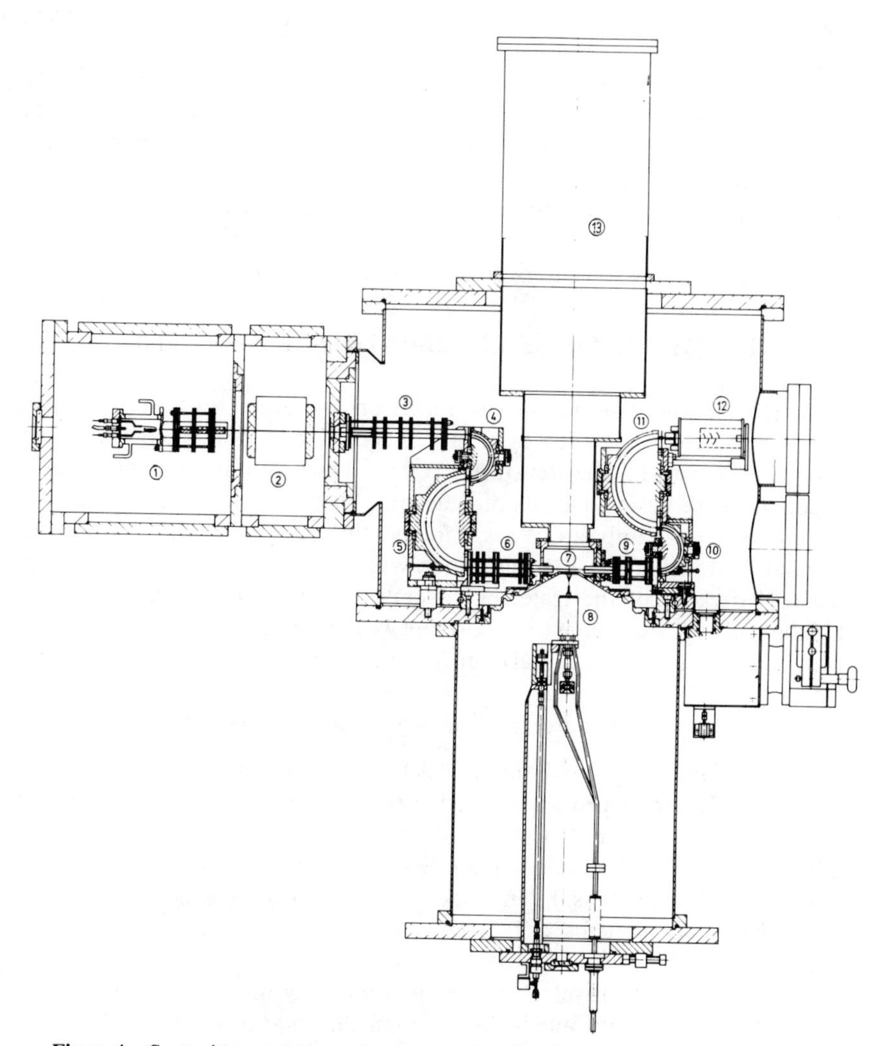

Figure 4. Scattering apparatus using two tandem hemispherical analyzer for measuring proton-energy-loss spectra.[153,230] (1) Low-pressure hydrogen discharge, (2) Wien filter, (3) beam-shaping optics, (4) preselector, (5) main energy selector, (6) beam-shaping optics, (7) scattering center, (8) secondary beam source, (9) beam-shaping optics, (10) preanalyzer, (11) main energy analyzer, (12) multiplier, (13) secondary beam sink. Note, that units (9)–(12) are rotatable around the secondary beam axis.

distributions will be presented as measured and have not been converted to the center of mass system. Because of the low mass of the proton, this transformation has only a small effect of a few percent on the scattering angles and collision energy and can therefore be safely neglected. Only in the case of the scattering from H_2 are the corrections significant. For the H^+–H_2 the collision energy is two-thirds of the laboratory proton energy and at a typical laboratory scattering angle of $\Theta_{lab} = 20°$ the center of mass angle is about 50% larger. All the time-of-flight spectra have been converted to an energy-loss scale and the intensities have been corrected with the appropriate Jacobian factors. For a detailed description of the transformations from the laboratory to center-of-mass systems see Ref. 61.

III. MECHANISMS OF VIBRATIONAL EXCITATION

Although we will also present some data on ion–atom scattering, we have not attempted to review the dynamics of elastic scattering, since it has been extensively covered in the literature.[31,178,218] Thus, we assume that the reader is familiar with basic concepts of deflection functions and special scattering features, such as rainbows. We will frequently make use of the fact that the angular position of a scattering rainbow Θ_R is proportional to the ratio of the well depth ε of the trajectory averaged interaction potential and the kinetic energy E: $\Theta_R = \text{const} \times \varepsilon/E$, where we have used the value const = 1.71, which appears to be an optimal approximation for ion-molecule systems.[15]

A. The Forced-Harmonic-Oscillator Model

The forced-harmonic-oscillator model has been found to be a very useful model for explaining proton-induced energy transfer in molecules, for which charge transfer does not occur. Gentry and Giese[93] were the first to find that this simple model could explain their observed vibrational distributions of H_2 excited by protons. Ellenbroek et al. subsequently extended the model to polyatomic systems[54,55] and found good agreement with the experiments.[91,92,167,171]

In view of the relatively high energies and low mass of the proton, the collision time is always much shorter than the rotational period. Thus, the molecule can be assumed to be stationary during the collision. If we assume a normal mode description of the harmonic molecular vibrations, then the equation of motion of each of the normal modes Q_k of a polyatomic molecule will be given by

$$Q_k + \omega_k^2 Q_k = \frac{1}{m} F_k(t), \tag{1}$$

where ω_k is the fundamental frequency of the harmonic motion, m is the reduced mass, and $F_k(t)$ is the external driving force. If it is assumed that the oscillator has no net effect on the external driving force, then the energy transferred to it is given analytically by classical mechanics as the Fourier transform of the external force,[137]

$$\Delta E_{cl}^{(k)} = \frac{1}{2m} \left| \int_{-\infty}^{+\infty} \mathbf{F}(t) e^{-i\omega_k t} dt \right|^2. \tag{2}$$

It has been shown that the classical energy transfer is identical to the expectation value of the quantum-mechanical energy transfer[16,115]:

$$\Delta E_{cl}^{(k)} \equiv \langle \Delta E^{(k)} \rangle_{qm} \tag{3}$$

and the quantum-mechanical energy transfer is given by

$$\langle \Delta E^{(k)} \rangle_{qm} = \sum_n P_k(n) n \hbar \omega_k, \tag{4}$$

where the transition probabilities $P_k(n)$ are given by a Poisson distribution

$$P_k(n) = \frac{1}{n!} \varepsilon_k{}^n e^{-\varepsilon_k} \tag{5}$$

with

$$\varepsilon_k = \frac{\langle \Delta E^{(k)} \rangle}{\hbar \omega_k} \tag{6}$$

With Eq. (3) we can obtain $\langle \Delta E^{(k)} \rangle$ from the calculated classical energy transfer $\Delta E_{cl}^{(k)}$ for a particular molecular vibrational mode (k) excited by the passing proton.

The force $F(t)$ is given by the negative gradient of the interaction potential $V(R)$ between proton and molecule at a distance R. Despite the overall complexity of this interaction, we may derive rather simple expressions for the long-range part of $V(R)$, which couples to the vibrational modes of the molecule in small-angle grazing collisions. An effective isotropic part of the interaction potential $V_0(R)$ has to be taken into account in calculating a deflection function, but by definition does not couple to the vibrations.

The forced-harmonic-oscillator model is expected to apply regardless of the coupling mechanism responsible for the force $F(t)$. Several mechanisms have been proposed to explain the observations made in proton scattering. The most widely observed mechanism with polyatomic molecules is the induced-dipole mechanism. In the next section we will illustrate the application of the forced-harmonic-oscillator model for this mechanism.

B. The Induced-Dipole Mechanism

At small scattering angles the long-range electrostatic multipole interaction contributes predominantly. The charge of the proton results in an electric field $\boldsymbol{E}(R)$, which at the center of the target molecule is given by

$$\mathbf{E}(\mathbf{R}(t)) = \frac{1}{4\pi\varepsilon_0}\frac{e}{R^3}\mathbf{R}(t), \tag{7}$$

where $\mathbf{R}(t)$ is the trajectory of the ion. For simplicity we consider only the dipole interaction, but higher-order interactions can be considered analogously.[55] Then we can write for the force $\mathbf{F}(t)$:

$$\mathbf{F}(t) = -\nabla_Q V(Q, \mathbf{R}(t)) \tag{8}$$

where

$$V(Q, \mathbf{R}(t)) = V_0(R) - \boldsymbol{\mu}(Q)\cdot\mathbf{E}(R(t)) \tag{9}$$

and expand $\boldsymbol{\mu}$ in a Taylor series with respect to Q_k. We get

$$\mathbf{F}(t) = -\nabla_Q \boldsymbol{\mu}_0 \cdot \mathbf{E}(\mathbf{R}(t)) - \nabla_Q \sum_{k=1}^{n_k} \sum_{n=1}^{\infty} Q_k{}^n \frac{\partial^n \boldsymbol{\mu}}{\partial Q_k{}^n} \cdot \mathbf{E}(R(t)). \tag{10}$$

For a molecule without a permanent dipole moment, $\boldsymbol{\mu}_0 = 0$, and neglecting higher-order expansion terms, we get finally for the external force on one mode

$$F_k(t) = -\frac{\partial \boldsymbol{\mu}}{\partial Q_k} \cdot \mathbf{E}(\mathbf{R}(t)). \tag{11}$$

The dipole moment derivative $\partial\boldsymbol{\mu}/\partial Q_k$ appearing in Eq. (11) can be obtained from the infrared line intensity A_k integrated over a vibrational band:

$$A_k = N_0 \frac{\pi d_k}{3c^2}\left|\frac{\partial \boldsymbol{\mu}}{\partial Q_k}\right|^2, \tag{12}$$

where N_0 is Avogadro's number, d_k is the degeneracy of the kth mode, and c is the velocity of light. Thus, this mechanism is restricted to the infrared-active modes of a polyatomic molecule.

The computation of the electric field $\mathbf{E}(R)$ that enters Eq. (11) requires a knowledge of the proton trajectory $\mathbf{R}(t)$. For energies much greater than the well depth ε and for angles less than the rainbow, it has been found that the trajectory R can be reasonably well approximated by a straight line

$$R(t) = \sqrt{b^2 + g^2 t^2}, \tag{13}$$

where b is the impact parameter and g is the velocity of the proton. This is a very common approximation and has been justified in many places in the literature.[94,112]

The electrical field pulse produced by the proton at the center of the molecule can be calculated as a function of time. The field components perpendicular and parallel to the straight line trajectory are shown at the

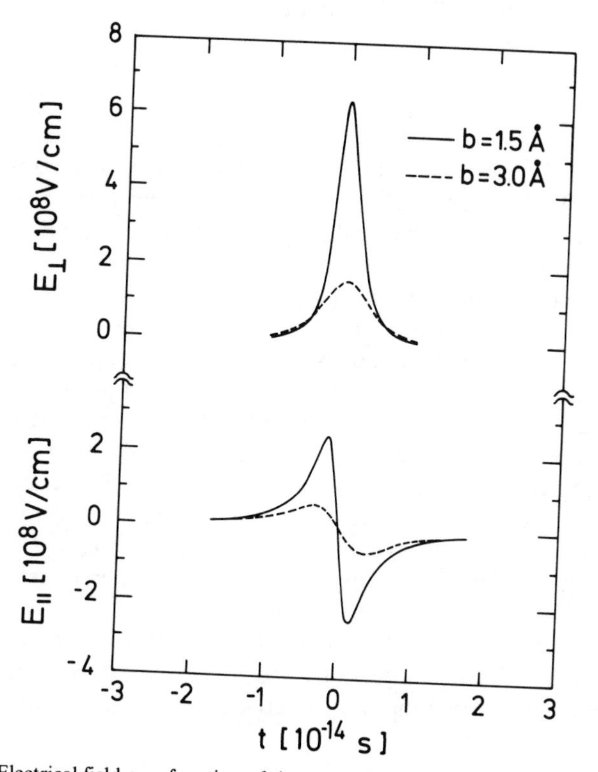

Figure 5. Electrical field as a function of time at the center of a molecule produced by the passage of a proton ($E_{cm} = 18.5\,\text{eV}$) at an impact parameter b in a grazing collision.[92] The field components perpendicular and parallel to the straight line trajectory are plotted at the top and bottom, respectively. The two impact parameters $b = 3.0$ and $1.5\,\text{Å}$ are typical for the grazing and close collisions observed in the $H^+ + CF_4$ experiments.

top and bottom of Fig. 5, respectively. The two impact parameters chosen are typical for grazing and close collisions observed in the $H^+ + CF_4$ experiments (cf. subsequent discussion). Note, that the maximum field strengths exceeds 10^8 V/cm while at the same time the pulse durations are 10^{-15}–10^{-14} s, which is of the order of or less than the vibrational periods for most vibrations. These conditions are very extreme even when compared to high-power infrared laser pulses such as those used in multiphoton excitation where the resulting field strengths are on the order of 10^5 V/cm with pulse durations around 10^{-9} s.

By the use of Eqs. (7), (11), and (13) the integral for the classical energy transfer, Eq. (2), can be solved *analytically* and leads to[90,92]

$$\Delta E_{cl}^{(k)} = 2\left(\frac{e\omega_k}{4\pi\varepsilon_0 g^2}\right)^2 \left|\frac{\partial \boldsymbol{\mu}}{\partial Q_k}\right|^2 (K_0^2(W_k) + K_1^2(W_k)), \tag{14}$$

where

$$W_k = \frac{\omega_k b}{g} \tag{15}$$

and K_0 and K_1 are modified Bessel functions (see, for example, Ref. 3, p. 374). Thus, W_k is essentially proportional to the ratio $\tau_{\text{coll}}/\tau_{\text{vib}}$ of the collision time τ_{coll} to the vibrational period τ_{vib}. The modified Bessel functions K_0 and K_1 account for the components of the time-dependent electrical field along the directions parallel and perpendicular to the collision trajectory, respectively (cf. Fig. 5). As an example we present in Fig. 6 the average energy transfers into the ν_3 and ν_4 modes of SF_6 calculated from Eq. (14) as a function of the relative velocity g. For ν_3 the contributions of the parallel and perpendicular components of the electrical field are shown individually. For small velocities g the ratio $\tau_{\text{coll}}/\tau_{\text{vib}}$ and thus W_k approaches infinity, corresponding to the adiabatic limit with negligible energy transfer. At high velocities the excitation is out of resonance leading to reduced ΔE's corresponding to the shortening in the collision time. Note, that the maximum of all the excitation curves, although they appear at different velocities, occur in fact at ratios of $2\tau_{\text{coll}}/\tau_{\text{vib}}$ close to unity.

In order to relate the b-dependent calculated average energy transfer to the observable deflection angle, one has to evaluate a deflection function $\Theta(b)$. Fortunately, the dominating attractive induction potential allows for an analytical treatment via[41]

$$\Theta = \frac{1}{4\pi\varepsilon_0}\frac{3\pi}{8}\frac{e^2\alpha_0}{Eb^4}, \tag{16}$$

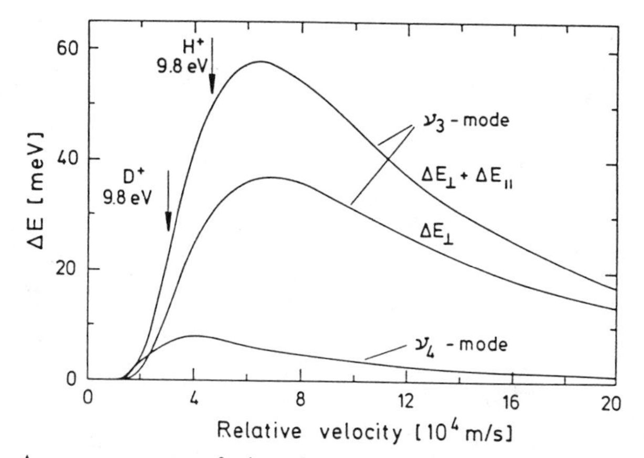

Figure 6. Average energy transfer into the ν_3 and ν_4 mode of SF_6 by proton impact as a function of the proton velocity as calculated by the straight line approximation.[92] Note that the perpendicular electric field component contributes abouy 70% for the ν_3 excitation with the remainder coming from the parallel component. The maximum of the ν_3 excitation occurs at 6×10^4 m/s ($t_{coll}/0.5\tau_{vib} = 0.96$), whereas the lower frequency ν_4 mode has its maximum at about 3.9×10^4 m/s ($t_{coll}/0.5\tau_{vib} = 0.99$).

where α_0 is the isotropic polarizability. Such a simplified treatment has its shortcomings, of course, but it is surprising to what extend it is capable of predicting reliable results when compared with experiments and with a more rigorous treatment.

Instead of approximating the required proton trajectory by a straight line, one can take the entire potential including the isotropic short-range term and perform classical trajectory calculations to get a trajectory $\mathbf{R}(t)$ that is more realistic than the straight line of Eq. (13). This new $\mathbf{R}(t)$ can then be used to evaluate the force $\mathbf{F}(t)$ as given by Eq. (11) and then the integration of Eq. (2) can be carried out. Such a treatment has been described in detail before[55,93] and shall not be repeated here. The full trajectory calculations agree remarkably well, to within better than 10%, with the simple straight line approximation as shown in Fig. 7*c*. Note, that the straight line approximation calculations can be performed with a pocket calculator and a tabulated list of the Bessel functions,[3] whereas the full trajectory calculations require a considerable amount of computational effort and time. Of course, repulsive trajectories (II and III in Fig. 7) can be accurately modeled only within the full framework. It is worth noting, however, that these two trajectories, although passing on different sides of the molecule, yield about the same amount of energy transfer, which is, moreover, nearly independent of the scattering angle.

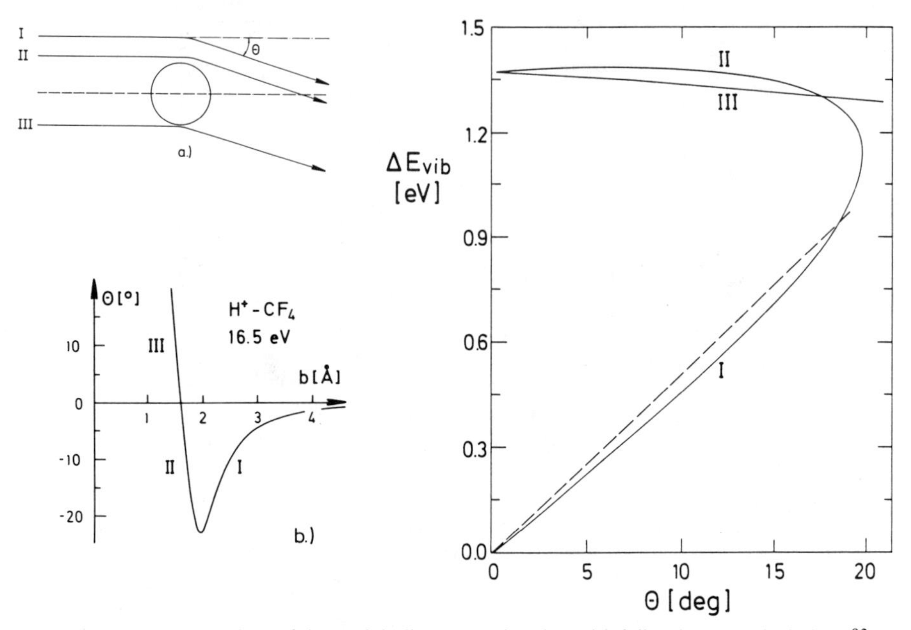

Figure 7. Comparison of the straight line approximation with full trajectory calculations.[92] In (*a*) a schematic plot shows the three different trajectories, which lead to the same deflection for scattering angles less than the rainbow. The trajectory calculations determine the classical deflection function (*b*) which is shown for $H^+ + CF_4$ at a collision energy of 16.5 eV. In (*c*) the energy transfer calculated from the full trajectories is plotted as a function of scattering angle of the contributing impact parameters (solid lines). The dashed line shows the result from the straight line approximation. The agreement is very good out to the rainbow angle at 20°.

Rotational excitation, which is only expected to be strong for polar molecules such as the hydrogen halides, has been neglected altogether. This is justified by the smallness of the ratio τ_{coll}/τ_{rot}, where τ_{rot} is the rotational period. As we shall see, the forced-harmonic-oscillator model is remarkably accurate in describing the experimental results on multimode vibrational excitation in ion scattering even from large polyatomic molecular targets such as CF_4 and SF_6.

C. Bond Dilution and the Internal Vibronic Mechanism (IVM)

An open-shell ion such as a proton is also expected to interact strongly with a closed-shell scattering partner via strong valence forces. These will tend to shift electron density from the occupied electronic orbitals of the closed-shell system into unoccupied orbitals of the open-shell system. Depending on the molecular and electronic structure of the closed-shell partner, this partial electron abstraction may lead to a weakening of some of the internuclear bonds. The temporarily formed complex will, in general, have a somewhat different equilibrium structure than the scattering partners.

As an example we note that in H_3^+ the H_2 bond distance is increased by about 10% compared to a free H_2 molecule. Thus, in this example of a homonuclear molecule where the induced-dipole mechanism is symmetry forbidden, the force acting on the molecule is entirely due to bond dilution. This model predicts that $H^+ + H_2$ scattering will show large amounts of vibrational excitation.[103,168,228] According to the forced-oscillator model, the perturbed molecule will be vibrationally excited to an extent that depends on the time scale of the interaction. In the case of H_2, an impinging proton fullfills the resonance condition $\tau_{coll} \approx \frac{1}{2}\tau_{vib}$ at $E_{cm} = 13$–$30\,eV$ assuming a typical interaction range of 2–3 Å. The observed Poisson distributions for excitation of different final vibrational states could be nicely fitted using average energy transfers calculated from classical trajectories[93] for a reactive potential hypersurface for H_3^+. More accurate close coupling calculations[197] give similar results. The most recent experimental and theoretical results on the $H^+ + H_2$ system will be presented in the chapter on charge-transfer systems.

The preceding bond dilution interaction has also been designated the internal vibronic mechanism to account for the excitation of infrared-inactive vibrational modes in proton collisions from the polyatomic target molecules like CH_4[34] and C_2H_2.[5]

D. The Quasimolecular Mechanism (QMM)

This mechanism takes into account the vibronic interaction *within* the transient collision complex (quasimolecule) invoking the pseudo-Jahn–Teller effect. The QMM was recently investigated for a polyatomic molecule, CH_4, both experimentally and theoretically.[34] A theoretical explanation in terms of a vibronic symmetry correlation theory was worked out to explain the excitation of noninfrared-active vibrational modes, which cannot be understood in terms of previously discussed mechanisms. The symmetry rules are the same as for the internal vibronic mechanism. In the QMM the resulting vibrational excitation might favor or promote charge transfer to one of the unfilled orbitals in the quasimolecule, but this is not a necessary requirement for the mechanism to work. The QMM usually makes only a small contribution in grazing collisions and for many molecules is of little importance since the potential surfaces interacting via the pseudo Jahn–Teller effect are usually several eV apart.[34,35]

E. Rovibrational Excitation via Impulsive Scattering

Finally, impulsive interaction of the proton with the steep repulsive core of the anisotropic potential can lead to additional rotational or vibrational excitation. This model has been successful in describing results for scattering of both ions[89,238] and neutral atoms[53,62] from polyatomic molecules. A type of spectator approximation is involved in which internal interactions within the target molecule are neglected and the projectile scatters from a subunit of the target that is described by an effective mass m_{eff}. Energy transfer to the entire molecule results from purely elastic scattering off that subunit and is calculated from purely kinematic considerations that account for the fact that the subunit is bound to the rest of the target molecule. In this way the kinetic energy gain ΔE of the subunit is transferred into internal degrees of freedom of the entire molecule. Thus, the average energy transfer ΔE is given by

$$\Delta E = \frac{2E_{\text{cm}}}{(1+q)^2}\left(q + \sin^2\theta_{cm} \pm (\cos\theta_{cm})\sqrt{q^2 - \sin^2\theta_{cm}}\right), \tag{17}$$

where

$$q = \frac{m_{\text{eff}}}{m_i m_R}(m_i + m_{\text{eff}} + m_R) \tag{18}$$

and m_i is the projectile ions mass, m_{eff} and m_R the mass of the subunit and of the rest of the target molecule, respectively, and E_{cm} and Θ_{cm} are the energy and the scattering angles within the center-of-mass frame, respectively.[53]

In the case of linear molecules with a large anisotropy, this impulsive mechanism leads mostly to rotational excitation. The distributions have an extremum at a small and at a large value of the rotational angular momentum and are therefore referred to as rotational rainbows.[19,216] They were first observed in ion scattering experiments on $Li^+ + N_2$ and CO,[27] but some of the prominent features were erroneously attributed to vibrational excitation.[90]

Since protons are devoid of electrons, the repulsive interactions with a target molecule are of very short range (< 1 Å). Significant contributions from impulsive interactions in proton–molecule collisions, except for collisions with hydrocarbons, are expected to be small because of the small mass of the proton. In any case, they can be expected to be most important for small impact parameters leading to large-angle scattering and less important for the small-angle scattering discussed here.

IV. EXPERIMENTAL EXAMPLES: SYSTEMS WITHOUT CHARGE TRANSFER

A. Atomic Targets

The scattering of protons by rare-gas atoms has been the subject of numerous studies, both experimental[33,37,86,104,149,160] and theoretical.[22,86,179,181,

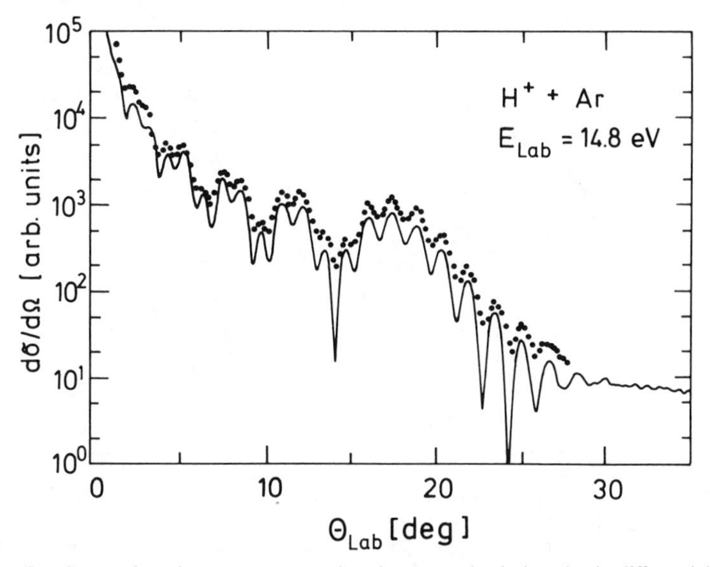

Figure 8. Comparison between computed and measured relative elastic differential cross sections for $H^+ + Ar$ collisions ($E_{lab} = 14.8$ eV).[86] The points show the experimental data, while the continuous curve has been calculated for a semiempirical potential, which also fits the available experimental scattering data in the energy range of 4.7–29.8 eV.

[189,203] With increasing collision energy these systems can also give rise to a variety of electronic processes.[2] At very low energies ($\leqslant 1$ eV) orbiting resonances have recently been observed for $H^+ + He$ and $H^+ + Ar$.[125,126]

At higher energies orbiting is no longer possible and deflection rainbows dominate the structure in the differential cross sections. As an example, Figs. 8 and 9 show the measured angular distributions for $H^+ + Ar(E_{lab} = 14.8$ eV)[86] and for $H^+ + Ne$ ($E_{lab} = 1.90$ eV),[126] respectively. The solid points in both figures show the experimental data. The fast oscillatory pattern comes from interferences between parts of the proton wave packet that have passed on opposite sides of the target atom. The oscillations with the wide angular period are primary and secondary rainbows, which result from the interference of the main parts of the wave packet on the inner and outer slopes of the potential well on one side of the target atom. The continuous curves originate from exact quantum-mechanical phase shift calculations[86] using a semiempirical best fit potential. This potential could be parametrized analytically in a modified Morse form using only six parameters, many of which are determined by the *ab initio* calculations. The cross sections calculated with this potential are in excellent agreement with the experimental points. This semiempirical potential is also in reasonable agreement with previous *ab initio* calculations. The high accuracy of the present fit justifies

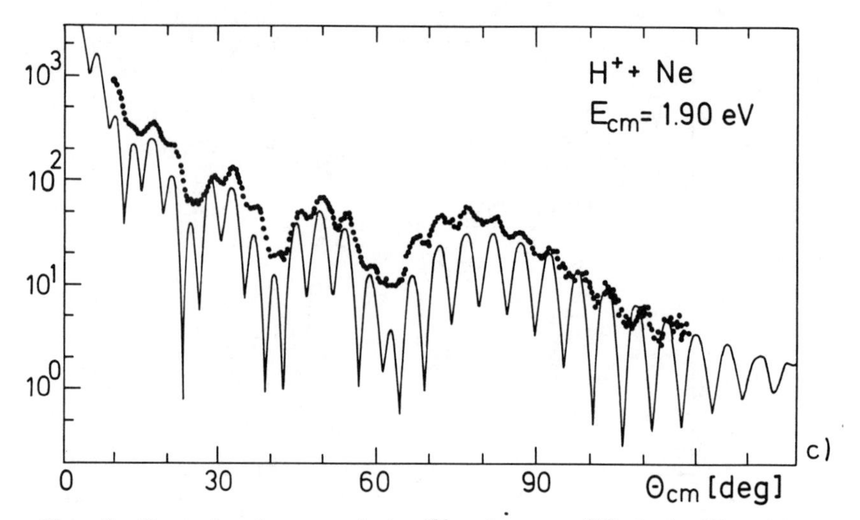

Figure 9. Comparison between calculated[86] and measured[126] elastic differential cross sections for $H^+ + Ne$ collisions ($E_{lab} = 1.90$ eV). The points show the experimental data, while the lower continuous curve has been calculated for a semiempirical potential, which also fits the available experimental scattering data in the energy range 1.9 to 14.8 eV.

the use of the potential in theoretical investigations of resonant states[88] and of ion mobilities.[87] Table I compares the potential parameters determined in this way for Ne and Ar with values for the other rare gases.

B. Diatomic Target Molecules

The first double differential cross sections for proton scattering of diatomic target molecules N_2 and HF were published in 1974.[229] The small amount of unresolved energy transfer was attributed to rotational excitation, and no vibrational excitation was observed. For an early review of inelastic ion molecule scattering, see Ref. 48. The first high-resolution studies were done shortly afterward on N_2, CO, and NO[134] followed by the investigations of our group[83] shown in Fig. 2, which also included O_2. Vibrationally resolved

TABLE I
Best Fit Parameters for the Ground-State Interactions of $H^+ + Ar$ and $H^+ + Ne$[86] Compared to $H^+ + He$ and $H^+ + Kr$, Xe[184]

Parameters	He	Ne	Ar	Kr	Xe
ε [eV]	2.0	2.300	4.147	4.45	8.5
R_m [Å]	0.824	0.996	1.282	1.536	

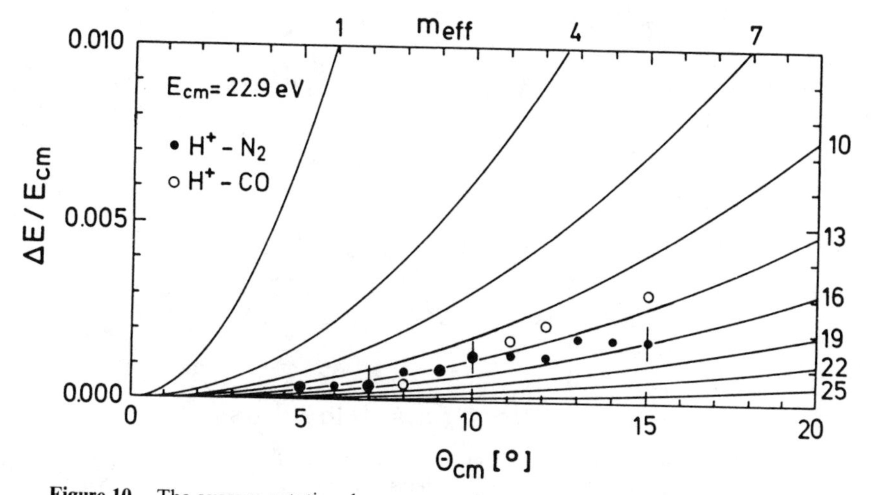

Figure 10. The average rotational energy transfer to N_2 (solid dots) and CO (open circles) in proton collisions at $E_{cm} = 22.9\,\text{eV}$ as a function of the center-of-mass scattering angle is compared to predictions of the impulsive model (solid lines) for different subunit effective masses m_{eff}.[172] The best agreement with N_2 is achieved for an effective mass $m_{eff} = 14$ and for CO for $m_{eff} = 12$, indicating an impulsive interaction with a single atom within the target molecule.

time-of-flight experiments with lithium ions were performed much earlier achieving even rotational resolution in $Li^+ + H_2$ collisions[44,60] and in $Li^+ + CO$.[52,90] More recently results have become available for $H^+ + N_2$, CO,[172] and $H^+ + HF$.[70]

The homonuclear nitrogen molecule has neither a dipole moment nor a dipole moment derivative $\partial\boldsymbol{\mu}/\partial Q$ and is therefore infrared inactive. The small amount of vibrational excitation as shown in Fig. 2 has been attributed to the bond-dilution mechanism. The coupling is weak, since the change of the bond length in N_2 upon protonation is less than 1%.[233] Impulsive rotational inelastic scattering probably accounts for the observed small energetic shifts ($\approx 10\,\text{meV}$) of both the resolved $v = 0$ and $v = 1$ peaks for $H^+ + N_2$ in Fig. 2. Figure 10 compares the measured relative rotational energy transfer (solid dots) with predictions of the impulsive model according to Eq. (17)[172] for various effective target subunit masses m_{eff} between 1 and 25 a.u. (solid lines). Since the mass $m_{eff} \approx 14$, which coincides with that of a single nitrogen atom, fits the data best, we conclude that the rotational excitation is caused by a two-particle impulsive interaction of H^+ with one of the N atoms with the second N atom serving only as a spectator. Much stronger rotational excitation has been observed in $Li^+ + N_2$ scattering, which is attributed to the greater mass of the Li^+ ion. The results in this case are also in good agreement with the impulsive mechanism.[90]

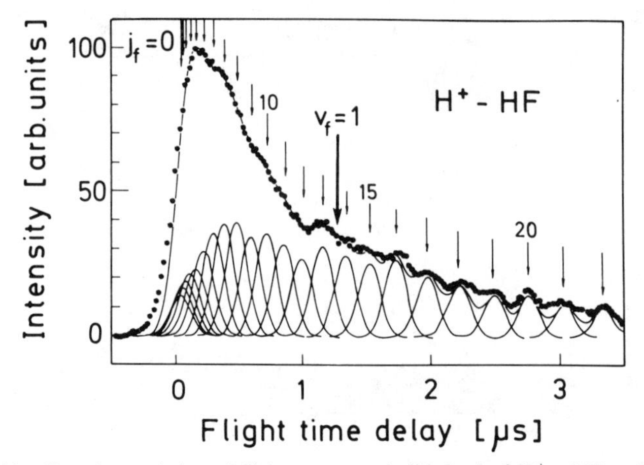

Figure 11. Experimental time-of-flight spectrum (solid dots) of $H^+ + HF$ at $\Theta = 10°$ and $E_{lab} = 9.8$ eV and simulated rotational distributions (Gaussian curves).[70] The spectrum can be explained by a progression of final rotational states up to $j_f = 22$ ($\Delta E \approx 1.3$ eV).

Owing to its permanent dipole moment HF couples directly to the electric field of the passing ion [cf. first term in Eq. (10)]. This very-long-range interaction ($\sim R^{-2}$) leads to huge amounts of rotational excitation as can be seen in Fig. 11. The arrows indicate expected positions of rotational and vibrational levels. For $j_f = 17$ and higher the observed structure although weak agrees quite well with the predicted rotational levels of the vibrational ground state. This observation was used to justify a fit of the entire spectrum including the nonresolved part with Gaussian curves of equal half widths for each possible final rotational state as shown by the solid lines in Fig. 11. The relatively low collision energy of $E_{lab} = 9.8$ eV and the wide rotational spacing bring rotational excitation closer toward the resonance condition $\tau_{coll} \approx \frac{1}{2}\tau_{rot}$ than for many other systems. As can be seen in Fig. 12 when going from $E_{lab} = 20$ to 50 eV peaks resulting from vibrational excitation rise gradually above the rotational energy-loss background and overtones become apparent. The mechanism responsible for the observed vibrational excitation is not completely understood, but the bond-dilution mechanism appears to be one possible explanation. In the protonated hydrogenfluorid complex H_2F^+, which has two indistinguishable hydrogen atoms each with half of the total positive charge, the H–F bond length is larger by about 5.4% than in an isolated HF molecule. This is attributed to a significant redistribution of electron density from the second bond.[29] The experimental scattering data in Fig. 12 indicate that it requires rather high proton velocities and deep penetration into the repulsive potential region in order for bond dilution to be effective in exciting vibrations. This strong effect of HF bond

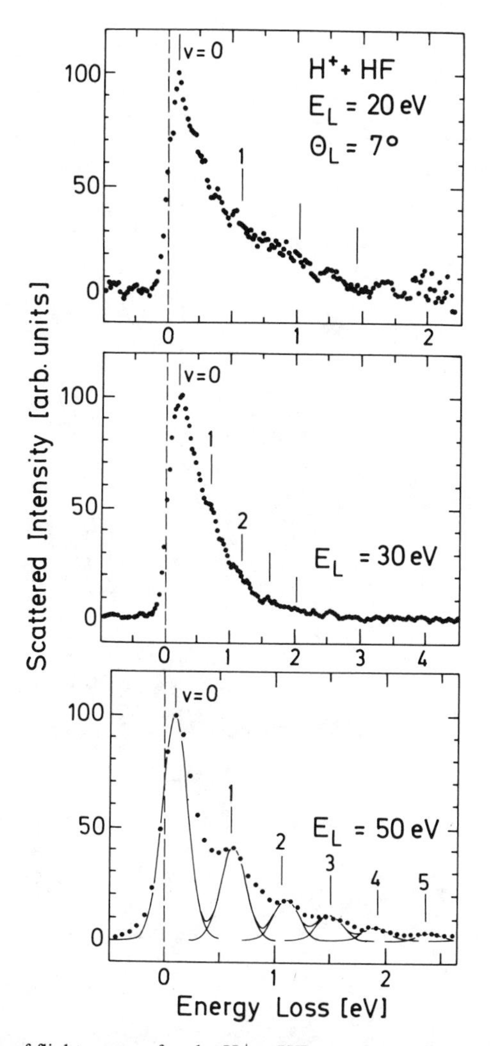

Figure 12. Time-of-flight spectra for the $H^+ + HF$ system at a fixed laboratory scattering angle ($\Theta = 7°$) and different collision energies of 20 eV (*a*), 30 eV (*b*), and 50 eV (*c*).[70] The expected positions of the vibrational levels are marked in all three diagrams. The solid curves in (*c*) show the estimated relative contributions from the various HF vibrational states assuming negligible rotational excitation at these high energies.

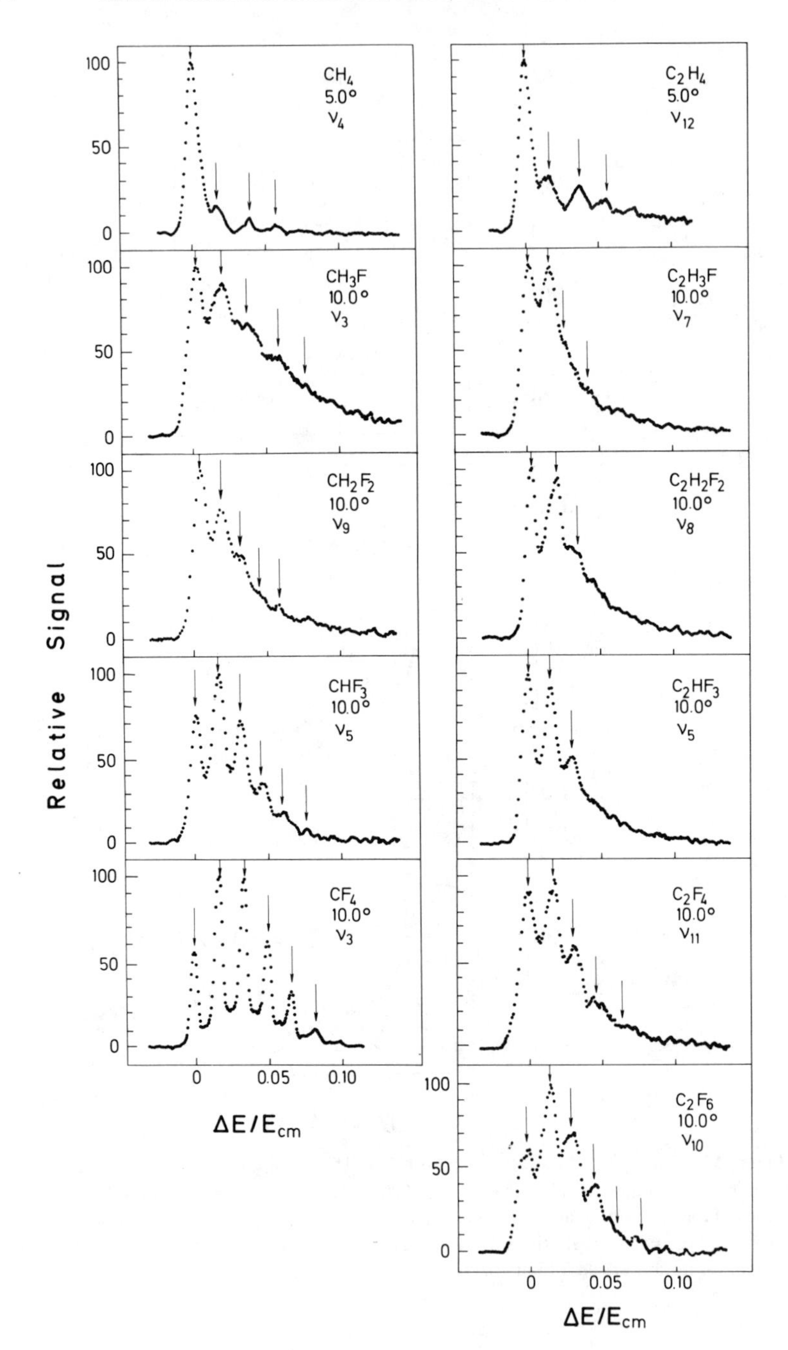

Relative Signal
CH_4 5.0° ν_4
C_2H_4 5.0° ν_{12}
CH_3F 10.0° ν_3
C_2H_3F 10.0° ν_7
CH_2F_2 10.0° ν_9
$C_2H_2F_2$ 10.0° ν_8
CHF_3 10.0° ν_5
C_2HF_3 10.0° ν_5
CF_4 10.0° ν_3
C_2F_4 10.0° ν_{11}
C_2F_6 10.0° ν_{10}
0
50
100
0.05
0.10
$\Delta E/E_{cm}$

dilution upon proton approach has not been accounted for in previous numerical scattering simulations performed so far.[70,85]

C. Polyatomic Target Molecules

In recent years the vibrationally inelastic scattering of light ions (H^+, Li^+) from simple polyatomic molecules (CO_2, N_2O, CF_4, SF_6, etc.) has been the subject of extensive experimental and theoretical studies. The data using Li^+ projectile ions have been obtained mostly at lower energies, $E_{cm} = 2.7$–$6.5\,eV$.[50] These Li^+ results were successfully interpreted in terms of a statistical model assuming a quasiequilibrium among all the vibrational degrees of freedom of the molecule and the translational energy.[51] Much of this work is surveyed in a recent review article.[128] As opposed to diatomics, polyatomic molecules have several fundamental vibrations, which, in principle, all can be excited in a scattering process. The excitation probabilities of the various possible fundamental, overtone, and combination states have been found to be quite different. Figure 13 provides a comparison of the energy-loss distribution for proton scattering from 11 different polyatomic fluorinated methane and ethane derivatives plus methane and ethane themselves.[91] At the collision energy of $E_{lab} = 9.8\,eV$ large differences are seen for the molecules within each series with the vibrational energy transfer increasing with the number of C–F bonds in the molecule. This is attributed to the larger infrared activity of the C–F bonds compared to that of the C–H bonds. At higher collision energies and at larger scattering angles this comparison would probably look quite different. Then we expect another mechanism involving charge transfer, which will be discussed in Section V, to become operative in the unsubstituted hydrocarbons or those with only one or two substituted F atoms and lead to a much larger average energy transfer. Note that in all of the molecules excited by protons discussed in this section there is no observable shift in the location of the vibrational peaks due to rotational excitation.

In the following discussion we will focus on a single target molecule, CF_4, and in some instances on SF_6. Already in 1980, Ellenbroek et al. found different propensity rules to govern the mode selectivity in $Li^+ + SF_6$ versus $H^+ + SF_6$ collision.[54] Consistent with the fact that the ionization potentials

---

Figure 13. Comparison of proton time-of-flight spectra, converted to an energy loss scale, for 11 different target molecules[91] Spectra are shown at the same angle of 10° for each target molecules except for CH_4 and C_2H_4 for which $\Theta = 5°$. The same laboratory energy of 9.8 eV has been used to permit a direct comparison. The two columns are arranged from top to bottom to show the effect of increasing F-atom substitution starting from the pure hydrocarbon molecules. A division has been made with regard to single (left) and double carbon atom (right) containing molecules. Peak locations that correspond to resolved vibrational levels are indicated by vertical arrows.

of CF_4 and of SF_6 are significantly greater than that of the H atom (cf. Section V), no charge transfer was observed up to scattering energies of 40 eV.[92] Figure. 14 gives an overview over low-energy time-of-flight spectra, converted to an energy loss scale, of $H^+(D^+) + CF_4$ and $H^+(D^+) + SF_6$ at various scattering angles. The experimental data (solid dots) show excitation of the two infrared-active modes ν_3 and ν_4 in both molecules with a dominant coupling to ν_3, which is explained by its greater infrared activity. Expected positions of vibrational levels are indicated at the top of each spectrum. The solid line fit curves are calculated from a product of two Poisson distributions in order to account for a simultaneous excitation of the ν_3 and ν_4 modes. All spectra are composed of different series of overtone and combination bands, which are only labeled in the two spectra at the top. In order to distinguish between these different contributions, the overtone progressions of the ν_3 mode are indicated by single-hatched areas, whereas the combinations with one and two quanta of the ν_4 mode are shown as cross-hatched and black areas, respectively. Note the very good agreement between predicted Poisson distributions and the measured spectra.

Since a time-of-flight apparatus achieves better resolution with heavier (slower) projectiles, measurements with He^+ were also carried out for comparison as shown in Fig. 15. In this case, even the minor ν_4 contribution is completely resolved and there is no evidence for additional structure to the left and right of the ν_4 peak. From this spectrum an upper limit for any ν_1 or ν_2 excitation is estimated to be less than about 20% of the ν_4 probability. Two effects show up in the He^+ spectrum that are not observed in proton collisions. First, impulsive rotational excitation leads to a considerable and constant shift ΔE_{rot} of all observed vibrational peak positions. This indicates that rotational and vibrational excitation are decoupled and are caused by the impulsive and the induced-dipole mechanism, respectively. Second, ν_4 is comparably stronger with respect to ν_3 than in proton collisions owing to the smaller He^+ velocity that favors the slower ν_4 vibration. More insight into

---

Figure 14. Time-of-flight spectra of $H^+(D^+) + CF_4(a)$ and $H^+(D^+) + SF_6\,(b)$ under various angles and at various low energies.[92] The dotted curves are the experimental data and the solid line fit curves are calculated from a product of two Poisson distributions in order to account for a simultaneous excitation of the two infrared-active modes, ν_3 and ν_4. All spectra are composed to different series of overtone and combination bands, which are only labeled in the two spectra at the top. In order to distinguish between these different contributions, the overtone progressions of the ν_3 mode (00*n*0) are indicated by single-hatched areas, whereas the combinations with one (two) quantum of the ν_4 mode 00*n*1 (00*n*2) are shown as cross-hatched (black) areas. Although the SF_6 spectra (*b*) are less well resolved than in the case of CF_4 (*a*), the fits with Poisson distributions allow for a distinction between contributions of the different combination and overtone transitions.

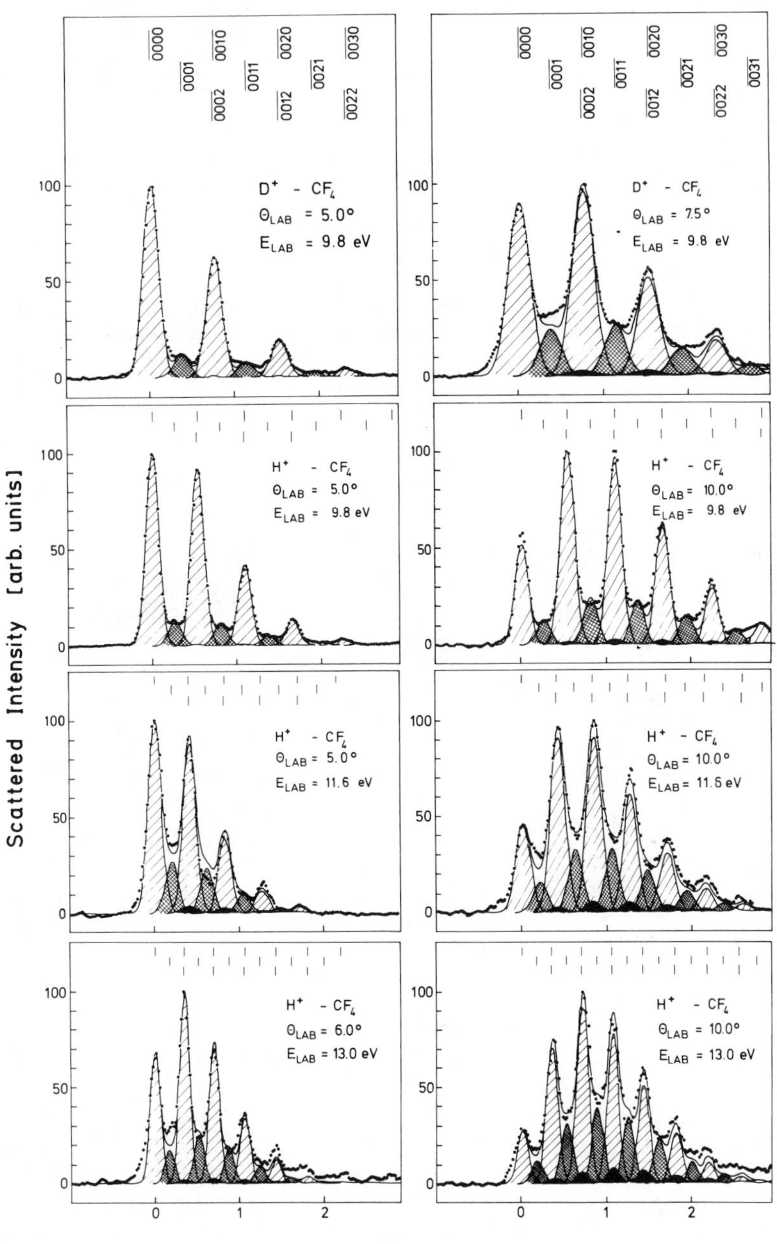

0000
0001
0002
0010
0011
0012
0020
0021
0022
0030
0031
$D^+ - CF_4$
Θ_{LAB} = 5.0°
E_{LAB} = 9.8 eV
$D^+ - CF_4$
Θ_{LAB} = 7.5°
E_{LAB} = 9.8 eV
$H^+ - CF_4$
Θ_{LAB} = 5.0°
E_{LAB} = 9.8 eV
$H^+ - CF_4$
Θ_{LAB} = 10.0°
E_{LAB} = 9.8 eV
$H^+ - CF_4$
Θ_{LAB} = 5.0°
E_{LAB} = 11.6 eV
$H^+ - CF_4$
Θ_{LAB} = 10.0°
E_{LAB} = 11.6 eV
$H^+ - CF_4$
Θ_{LAB} = 6.0°
E_{LAB} = 13.0 eV
$H^+ - CF_4$
Θ_{LAB} = 10.0°
E_{LAB} = 13.0 eV
Scattered Intensity [arb. units]
Flight time delay [μs]

(a)

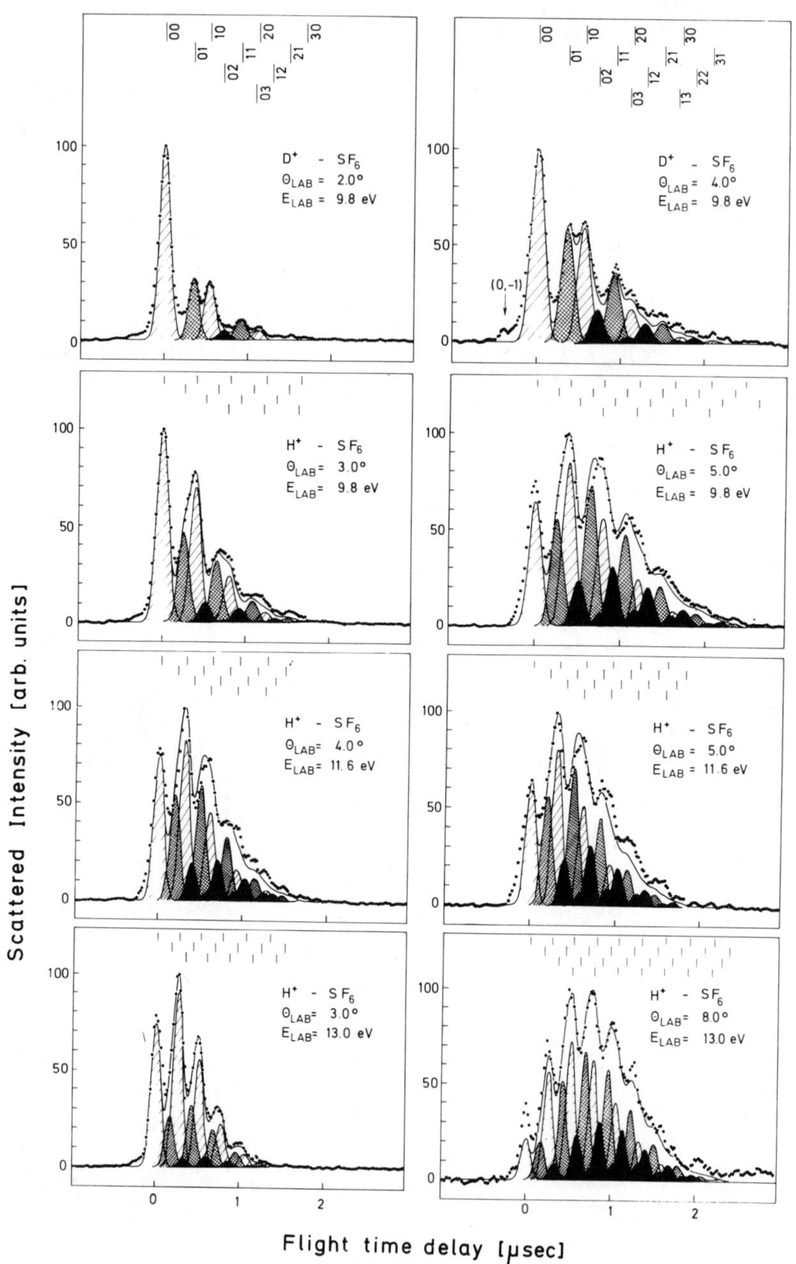

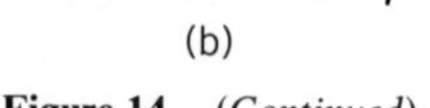

Figure 14. (*Continued*)

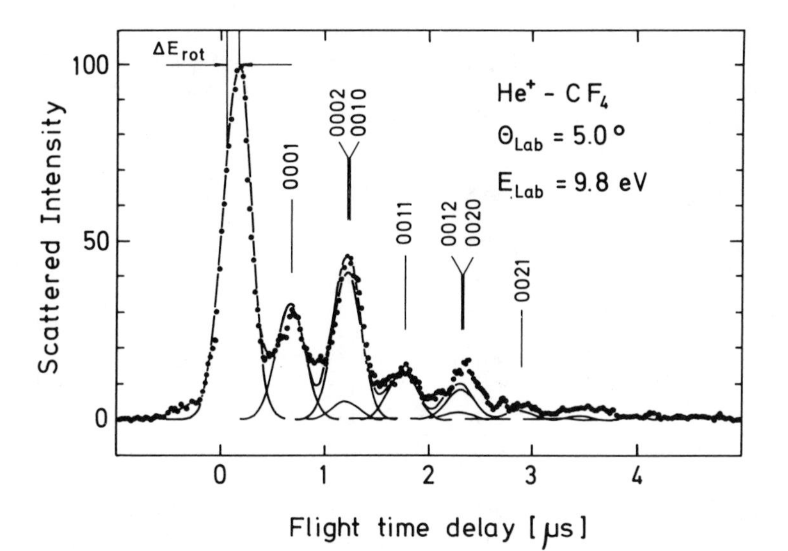

Figure 15. Time-of-flight spectrum of $He^+ + CF_4$ at $E_{lab} = 9.8$ eV and $\Theta_{lab} = 5°$.[92] Note that the whole series of calculated Poisson peaks is shifted by a constant amount of about 10 meV, which is attributed to rotational energy transfer.

contributions of weakly excited modes in proton collisions will come from better resolved spectra, which will be presented later in this chapter.

The Poisson fits of Fig. 14 allow for the accurate evaluation of the average energy transfers for both the ν_3 and ν_4 mode excitations. These experimental values have been compared with predictions from the straight line approximation [Eq. (14)]. The only unknown parameters are the dipole moment derivatives $\partial\mu/\partial Q_i$, $i = 3, 4$, which can therefore be determined from a best fit. Table II compares the values obtained from these scattering experiments with those from infrared-absorption measurements. The agreement is very good and appears to be better than the scatter in the values from different IR measurements. As discussed at some length elsewhere,[180] there are a number of difficulties associated with accurate measurements of $\partial\mu/\partial Q$ from IR data that are avoided in the proton-scattering determinations.

Investigations at higher energies and larger scattering angles reveal increasing discrepancies between predicted Poisson distributions based on straight line grazing trajectories and measured spectra. Recall that very large energy transfers are predicted by the full classical trajectory calculations from close collisions within the framework of the forced-harmonic-oscillator model (Figs. 7*a* and 7*b*, trajectories II and III). Thus, the measured time-of-flight spectra should in fact be composed of three Poisson distributions with different reduced energy transfers $\varepsilon_k = \langle \Delta E_k \rangle / \hbar\omega_k$. Since two of the

TABLE II
Compilation of Dipole Moment Derivatives Determined from Infrared Measurements and Compared to Data Derived from Proton Measurements[92]

Molecule	$\left\lvert \frac{\partial \mu}{\partial Q_3} \right\rvert$	$\left\lvert \frac{\partial \mu}{\partial Q_4} \right\rvert \left[10^{-6} \frac{As}{\sqrt{kg}} \right]$	References
CF_4	2.65	0.27	191
	2.32	0.22	194
	2.51	0.27	201
	2.22	0.25	142
	2.21	0.22	101
	2.26	0.22	97
	2.21	0.36	92
SF_6	2.38	0.58	194
	2.36	0.63	193
	2.68	0.63	116
	2.17	0.75	92

trajectories (II and III in Fig. 7) should lead to about the same energy transfer $\langle \Delta E_k \rangle$, only one additional Poisson distribution was included to account for the trajectories of types II and III. The resulting fits in the case of some high-energy spectra of $H^+ + CF_4$ are given in Fig. 16. Note, that the ν_4 mode excitation has not been accounted for, since, although present, it is not resolved. The agreement of the pure ν_3 fit with the experimental distributions is surprisingly good. This is strong evidence for the interpretation in terms of partially overlapping incoherent energy-transfer contributions from different trajectories.

The double Poisson distribution fits in Fig. 16 provide an estimate for the average energy transfer of each of the two groups of trajectories. The values for ΔE_{I} and $\Delta E_{\text{II,III}}$ are compared in Fig. 17 with those predicted by the forced-harmonic-oscillator mechanism using classical trajectory calculations at three different collision energies and for $H^+ + SF_6$ at a single energy. The dipole moment derivatives were determined from the experimental data of Fig. 14 as described above. The isotropic part $V_0(R)$ of the interaction potential, needed for the classical trajectory calculations, was assumed to have the following simple form:

$$V_0(R) = \frac{B}{R^8} - \frac{C}{R^6} - \frac{1}{4\pi\varepsilon_0}\frac{e^2\alpha_0}{2R^4}, \tag{19}$$

where α_0 is the isotropic dipole polarizability and the constants B and C are

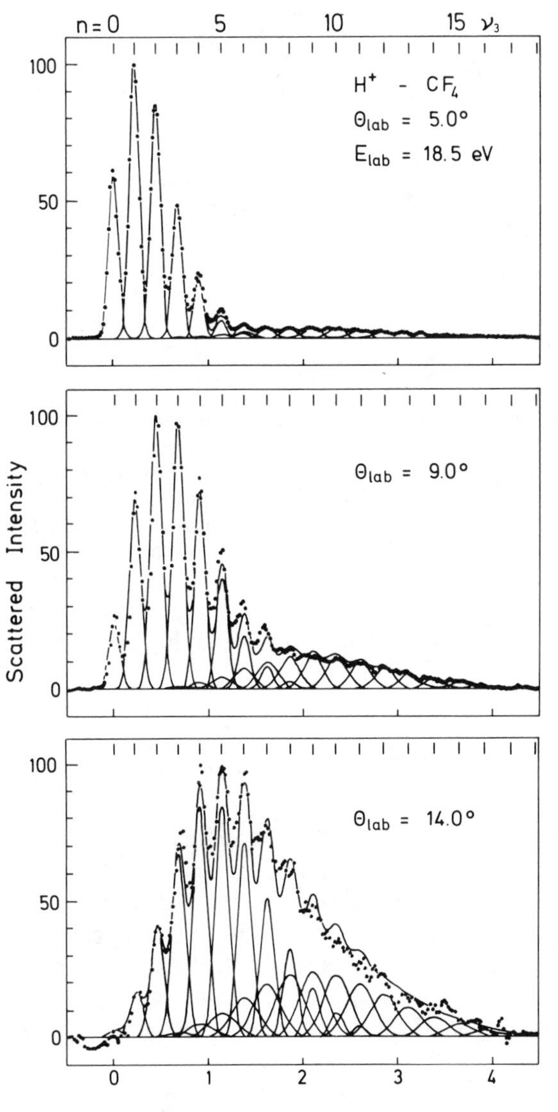

Figure 16. High-energy time-of-flight distributions for $H^+ + CF_4$ emphasizing the additional contributions of repulsive trajectories at high-energy transfers.[92] The best fit spectra (solid lines) consist of a sum of two Poisson distributions with different reduced energy transfers ε_i. The second distribution with large energy transfers, which is centered around $n = 10\nu_3$, accounts for contributions from small-impact-parameter collisions (trajectories II and III in Fig. 7). The half-widths of individual transitions were assumed to be somewhat larger than in the first distributions because of the lower resolution at higher-energy transfers.

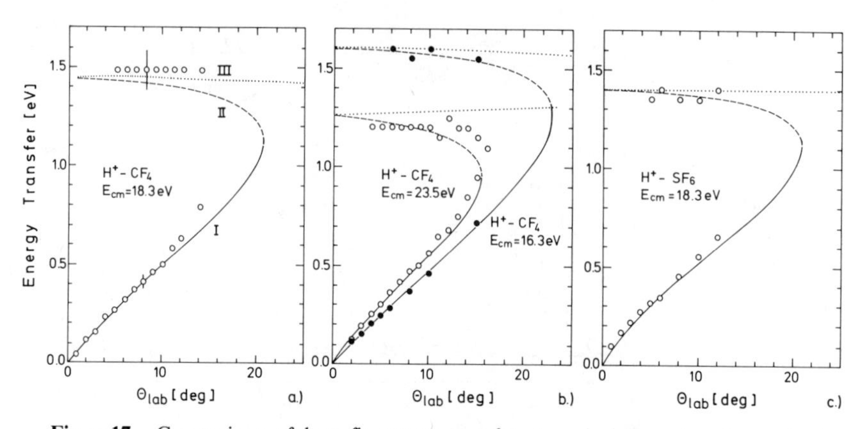

Figure 17. Comparison of best fit energy transfers to calculations based on classical trajectories for $H^+ + CF_4$ and $H^+ + SF_6$.[92] The measured best fit energy transfers $\Delta E_{\mathrm{II,III}}$ of the two Poisson distributions similar to those shown in Fig. 16 (assuming only excitation of the ν_3 mode) are compared with calculations based on classical trajectories for $H^+ + CF_4$ (SF_6) at three (one) collision energies. In the calculations the potential parameter R_m for each system was the same as obtained from energy transfer ΔE_{I} only and is listed in Table III.

calculated from the potential well depth ε and the minimum location R_m by the conditions

$$V_0(R_m) = -\varepsilon, \qquad \left.\frac{dV_0}{dR}\right|_{R_m} = 0. \tag{20}$$

Since ε [not to be confused with $\varepsilon_k = \Delta E/E$, Eq. (6)] is rather well known from rainbow scattering experiments,[161] the minimum location R_m is the only remaining adjustable parameter. It was determined by fitting the ΔE_{I} curve to the experimentally observed values of the energy transfer and values for $\Delta E_{\mathrm{II,III}}$ could be predicted. The good agreement with experiment is rather remarkable since both ΔE_{II} and ΔE_{III} are strongly dependent on R_m. The values for R_m obtained in this way are listed in Table III together with

TABLE III
Well Depth and Minimum Distance of the Interaction Potential of $H^+ + CF_4$ and $H^+ + SF_6$ as Determined from Proton Scattering Compared to Proton Affinities

Potential Parameter	CF_4	SF_6	Unit	Reference
Minimum distance R_m	1.53	1.90	Å	92
Well depth ε	3.6	3.66	eV	160, 161
Proton Affinity	5.49	5.51	eV	4

the well depths ε and values of the proton affinities of CF_4 and SF_6 from ion-flow-tube experiments.[4] Surprisingly, there is a rather large systematic discrepancy of about 35% between the well depths determined from scattering experiments and the proton affinities from flow-tube experiments. Thus, we have to conclude that a fast moving proton does not experience the same maximum potential attraction as a slowly approaching proton. Two effects are expected to contribute to this difference. For one, the attractive potential for a proton may be highly anisotropic, and this anisotropy is averaged over in the fast collisions. Moreover, in a slow approach the molecule will undergo

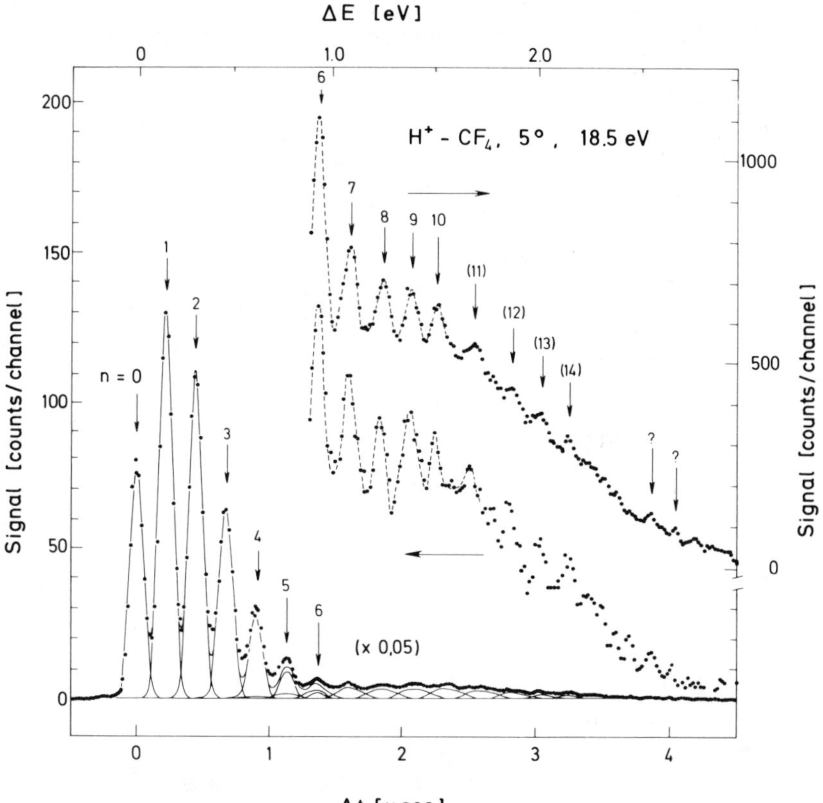

Figure 18. The high-energy transfer tails of two time-of-flight spectra for $H^+ + CF_4$ at $E_{lab} = 18.5\,eV$ and $\Theta_{lab} = 5°$.[92] The upper spectrum at the right is the integrated spectrum of eight 6 h runs, while the spectrum below is the best resolved 6 h measurement. Its low-energy transfer part is shown in the bottom left in a scale reduced by a factor 0.05. The arrows indicate reproducible peaks and overtone assignments.

a certain amount of adiabatic bond-distance readjustment. Unfortunately and as far as the authors are aware of, there are no *ab initio* studies available for the $(CF_4H)^+$ potential hypersurface.

As discussed previously the proton-scattering investigations have yielded new insight into the dynamical response of polyatomic targets and provided reliable values of interaction potentials. The large amount of energy transferred in type III trajectories (cf. Fig. 7) and the mode-selective response of the target molecule provide a new method to investigate the energy levels of very highly excited vibrational states. Note that conventional spectroscopic methods are limited to overtones up to about $v_f = 6$. Figure 18 shows an example of a $H^+ + CF_4$ time-of-flight spectrum taken at $E_{cm} = 18.5$ eV and $\Theta_{lab} = 5°$. The high-energy loss tail of the spectrum has been enlarged and both the cumulative sum of eight 6-hr. scans and a single scan are shown. The $v_f = 10$ overtone of ν_3 is clearly resolved and the observed spacings are very regular. At higher-energy transfers four additional levels are clearly apparent, although the resolution decreases and the spacings become more irregular. The Birge–Sponer plot based on those results is shown in Fig. 19

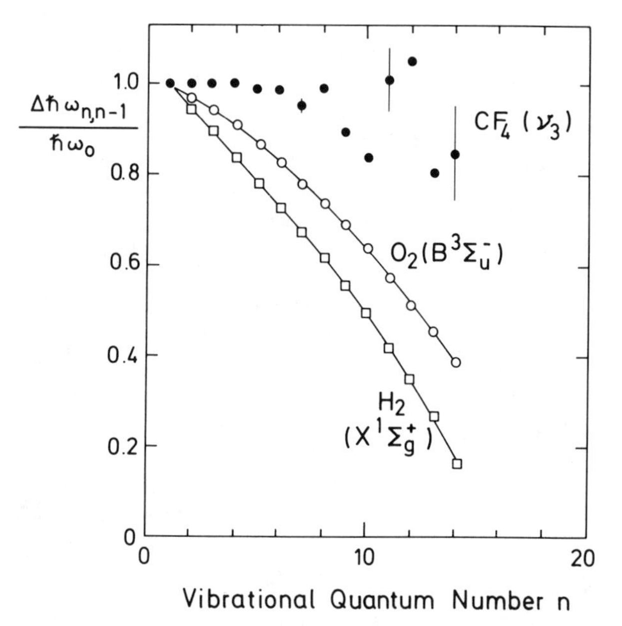

Figure 19. Birge–Sponer plot of the average level spacings determined from the data of Fig. 18, divided by the fundamental energy level, as a function of the vibrational quantum number.[92] Even after taking account of the much larger vibrational number ($\cong 34\,h\nu_3$) corresponding to dissociation of CF_4 the lack of significant anharmonicity of the CF_4 ν_3 vibrations is clearly apparent. For comparison similar spectroscopic data are shown for O_2 and H_2.

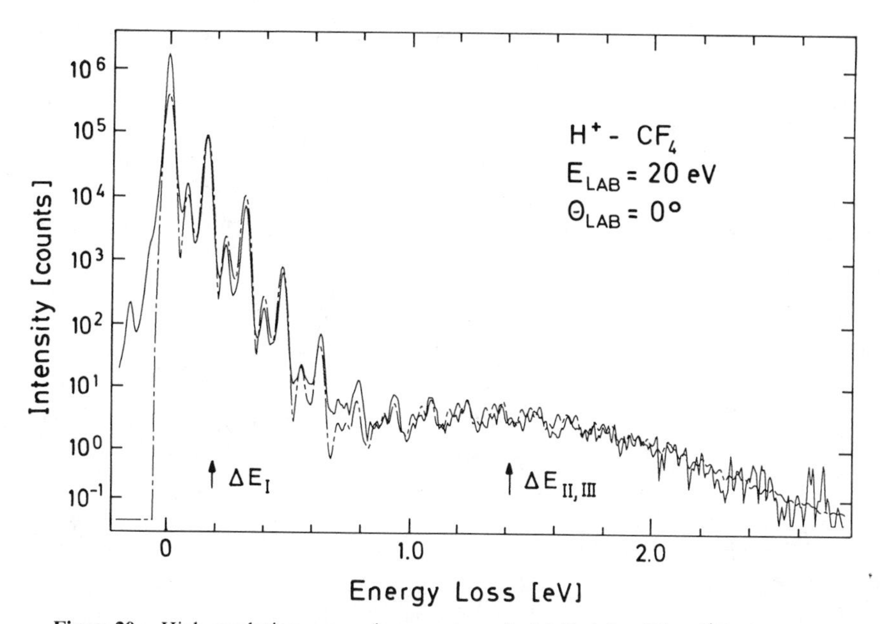

Figure 20. High-resolution energy-loss spectrum (solid line) for $H^+ + CF_4$ measured at $E_{\text{lab}} = 20\,\text{eV}$ and $\Theta_{\text{lab}} = 0°$ in the new sector field apparatus is compared to a multimode multitrajectory Poisson best fit (dashed line).[153,230,244] Note the good agreement between the experimental intensities and the fitted cuve over six orders of magnitude.

and compares the spacings of neighboring vibrational levels of the ν_3 level progression with those of the well-known diatomic molecules $O_2(B\,^3\Sigma_u^-)$[107] and ground-state H_2.[124] In comparison the CF_4 levels show surprisingly small anharmonicity. Infrared measurements also indicate a small anharmonicity for $n \leqslant 3$.[113]

Preliminary experimental results are also available for $H^+ + CF_4$ from the new sector field apparatus shown in Fig. 4.[153,174,230,244] Figure 20 shows a typical energy-loss distribution measured at $E_{\text{lab}} = 20\,\text{eV}$ and $\Theta_{\text{lab}} = 0^0$. The experimental results (solid line) are plotted on a semilogarithmic scale in order to emphasize the two separate Poisson distributions centered at low and high ΔE. Using the indicated average energy transfers ΔE_{I} and $\Delta E_{\text{II,III}}$, a predicted spectrum was obtained (dashed line) in very good agreement with experiment. Note the intensity variation by more than six orders of magnitude. Figure 21 displays on a linear scale another measurement at the same energy and $\Theta_{\text{lab}} = 4°$. A remarkable increase in resolution compared to the earlier experiments is observed (cf. Fig. 18). For example, the deexcitation of the ν_4 and ν_3 modes is clearly seen on the energy-gain side of the spectrum. The ν_4 mode is now clearly separated from the ν_3 mode

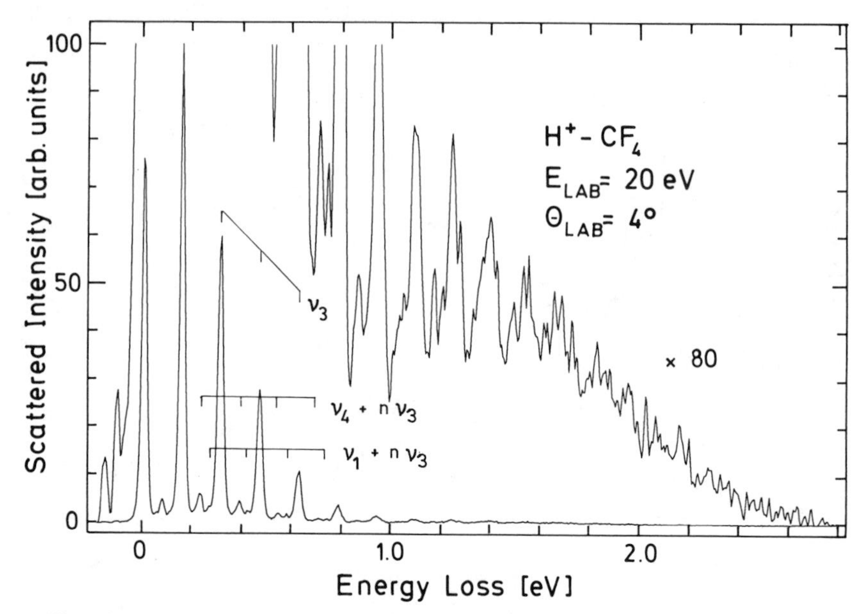

Figure 21. High-resolution energy-loss spectrum of $H^+ + CF_4$ at $E_{lab} = 20$ eV and $\Theta_{lab} = 4°$ on a linear scale measured with the new sector field apparatus.[153,230,244] Note the significantly improved resolution when compared to the older spectra (Fig. 18).

and, moreover, the ν_1 mode, which could not be observed in the earlier work, is resolved. This mode is Raman active and can be excited in proton collisions by a coupling via the polarizability gradient $\partial\alpha/\partial Q_k$.[55] Reasonable agreement is obtained with predictions based on the polarizability gradient from Raman spectroscopy measurements for CF_4.[164]

At high-energy transfer considerable additional structure is found. Although the results have not been fully analyzed, it would appear that many of the lines can be explained by a complicated set of combination bands of the ν_3, ν_4, and ν_1 progressions. Several theoretical studies have shown[24,139,141] that in addition each of the overtone levels of the ν_3 modes should be split into a sequence of closely spaced sublevels resulting from the anharmonicity. These sublevels can be classified according to their relative degree of local mode character. Modes for which most of the energy is localized in one of the bonds are energy shifted downward with respect to the harmonic overtone levels, whereas the delocalized modes are closest to the expected harmonic overtone levels. Whereas high overtone infrared spectroscopy is mostly sensitive to the local modes calculations[141] show that protons have a stronger propensity for exciting the delocalized modes. Work is in progress to further increase the resolution in order to resolve these sublevels for the first time.

V. MECHANISMS OF CHARGE TRANSFER

Charge transfer belongs to the broader class of nonadiabatic electronic transitions, which have extensively been studied in the past. Within a molecule nonadiabaticity leads to intersystem crossing, autoionization, and vibronic mixing.[237] Mutual recombination of gaseous atomic ions is one of the simplest examples of a nonadiabatic process in a collisional encounter.[150] The reverse process is ion pair formation, which has been studied, for example, in collisions of alkali and halogen atoms.[120] First insight into the involvement of molecular vibrations was gained by extending this appraoch to alkali atom–diatom systems[118,119] and by studying *E-V* transfer with optically pumped molecules.[140] Recently, sophisticated scattering laser experiments have been used to study the collision dynamics of the well-known effect of collisional quenching of alkali resonance fluorescence.[106] These crossed-beam investigations, in conjunction with *ab initio* potential calculations, revealed that the previously proposed *ionic intermediate* model[26] is inapropriate in this case and has to be replaced by a so-called *bond-stretch attraction* model.

Previous crossed ion–molecular beam scattering investigations have greatly increased the understanding of the elementary processes of charge transfer,[74] but, in general, have suffered from insufficient resolution to illucidate the role of vibrations. Thus, there have been only a few experiments with vibrational-state resolution, namely, on charge transfer in $N_2^+ + N_2$,[154,155] $Ar^+ + N_2$,[75,188], and $Ar^+ + H_2$.[109] Some recent work also achieved electronic-state resolution in crossed-beam scattering of double charged ion, namely $Ar^{2+} + He$[68] and $Hg^{2+} + Kr$.[69] A major technical breakthrough has been achieved in Orsay in developing a multicoincidence apparatus and applying it to $Cl^- + H_2$ crossed-beam scattering. These experiments will be dealt with in the chapter by Mme. Durup-Ferguson in this volume. Finally, it should be mentioned that guided-beam measurements of integral cross sections have achieved state-to-state resolution for inelastic and charge-transfer scattering as well as for rearrangement collisions. These topics will also be covered by two separate chapters within this volume.

A. Nonadiabaticity

In preparation for the discussion of the experimental results on proton-charge-transfer scattering, we review a few theoretical concepts of nonadiabatic interactions. More complete discussions of dynamical theories will be presented in the second part of this volume.

In the Born–Oppenheimer approximation,[28] the nuclear and electronic motions are dealt with separately. This is, of course, justified by the large difference in the time scales for the two motion. By choosing an appropriate basis set of electronic eigenfunctions $u_k(\zeta)$, the total wave function $\Psi(\zeta, R)$ can be expanded in the following way:

$$\Psi(\zeta, R) = \sum_{k=0}^{\infty} u_k(\zeta, R)\Psi_k(R), \tag{21}$$

where the $\Psi_k(R)$ describe the nuclear motion on well-defined potential energy surfaces $V_k(R)$, ζ denotes the vector of all electronic coordinates, R denotes the vector of all nuclear coordinates, and k is the electronic-state quantum number. In dealing with nonadiabatic processes the same expansion is adopted, but two different choices of representations have been found to be useful.

In the diabatic representation (superscript d), a set of eigenfunctions $\{u_k^d(\zeta, R_0)\}$ is chosen for a single, fixed R_0. Usually one component of R_0 is chosen at infinity and the direct product of the u_k^d's of the separated system is the basis set of the entire system. Beacause the eigenfunctions are fixed the resulting $V_k^d(R)$ can change their relative order and cross each other as R is varied.

In the adiabatic representation (superscript a), a new basis set $u_k^a(\zeta, R)$ is generated for every value of R such that the potential curves $V_k^a(R)$ are ordered consecutively according to their energies and each $V_k^a(R)$ is minimized. Such a choice of basis is, in principle, unique and none of the $V_k^a(R)$ can change position with any other $V_{k'}^a$(R). Thus, there are no curve crossings (von Neumann–Wigner noncrossing rule[165]). The adiabatic basis can be related to a diabatic representation and can also be transformed into it. For more details see Refs. 10 and 206 and the second part of this volume.

In fast collisions of $E_{\text{coll}} \gg 1$ eV repulsive potential regions where different surfaces come close to each other may be accessed. Since the nuclei move rapidly, the electrons are not able to adjust their density distributions immediately to the new situation. As a result the Born–Oppenheimer approximatation breaks down in local regions where the potential energy surfaces couple most strongly and come closest to each other. To account for this, the adiabatic Schrödinger equation has to be modified. In the diabatic representation, for example, the vector of the potential energy surfaces $\{V_k^d(R)\}$ is enlarged to a matrix $\mathbf{V}^d(R)$ where the $V_k^d(R)$'s are the diagonal elements. Nonadiabatic transitions such as those described previously are accounted for by additional offdiagonal elements $V_{ki}^d(R)$ that couple states k and i. In the adiabatic representation both wavefunctions and potentials

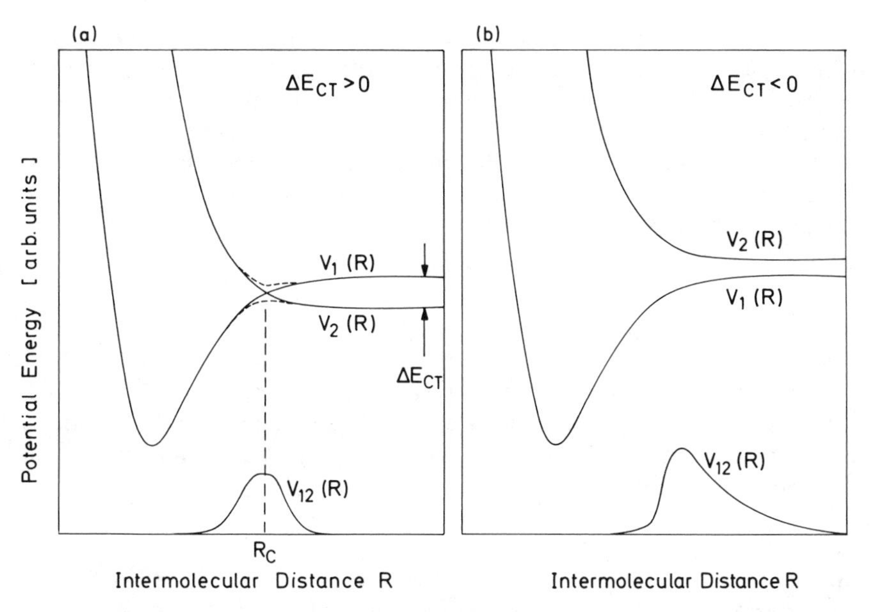

Figure 22. Potential energy schemes to illustrate different kinds of potential coupling [this work]. The potential curves $V_1(R)$ are typical for the interaction of protons, H^+, with neutral atoms or molecules, M. The figures show the dependence on the intermolecular distance R and do not take into account any internal coordinates of the target molecule. Asymptotically exoergic charge-transfer states, $H + M^+$, give rise to a direct curve crossing (*a*). Endoergic states (*b*) usually do not cross. Nonadiabatic coupling may occur in both cases. Coupling matrix elements are shown below the potential curves. Case (*a*) is approximately described by the Landau–Zener[137,241,242] model if only the potential curves cross steeply enough. Case (*b*) can be described by Demkov-type coupling.[45]

remain vectors, but additional derivative terms are introduced into the Schrödinger equation introducing nonadiabatic coupling.[10,70,225]

Two typical cases that occur in proton–molecule scattering are shown in Figs. 22*a* and 22*b*. Two diabatic potential hypersurfaces, V_1 and V_2, either cross (*a*) or do not cross (*b*). Both situations can lead, in principle, to nonadiabatic interactions. Let V_1 correspond to an initially prepared system $H^+ + M$, where M stands for any target atom or molecule and let V_2 correspond to $H + M^+$. Then, the asymptotic energy difference ΔV is given by the difference of the recombination energy E_R of the proton ($E_R = 1$ Rydberg $= 13.59\,\text{eV}$) and the ionization potential $I(M)$ of the target, $\Delta V = E_R - I(M)$. For $\Delta V < 0$ (Fig. 22*b*) the transition from V_1 to V_2, namely, charge transfer, is endoergic, for $\Delta V > 0$ (Fig. 22*a*) it is exoergic. Since V_1 is strongly attractive while V_2 is largely repulsive, the diabatic potential curves cross each other for $\Delta V > 0$ and remain parallel for $\Delta V < 0$. In the bottom of Fig. 22, the typical radial dependencies of the diabatic coupling matrix elements are displayed. The charge-transfer process becomes most likely at those distances R where the coupling matrix elements are largest.

The collision dynamics will, in general, involve both potential hypersurfaces. In a first approximatation the dynamics are determined by the electronic adiabaticity parameter[143]

$$\xi = \frac{t_c}{t_r}, \tag{22}$$

where t_c is the time required to traverse the coupling region and t_r is a measure of how long it takes the electrons to rearrange. The latter can be estimated from the adiabatic energy gap $\Delta E(R_c)$ at the crossing point R_c via the uncertainty principle:

$$t_r = \frac{h}{\Delta E(R_c)}. \tag{23}$$

It is obvious that $\xi \gg 1$ allows for an adiabatic passage, whereas $\xi \ll 1$ makes a nonadiabatic transition more likely. These two cases correspond to the high- and low-velocity limits, respectively. The probability for a nonadiabatic transition in the coupling region is given approximately by

$$P = e^{-\pi^2 \xi}. \tag{24}$$

In the case of a direct crossing (Fig. 22*a*) further assumptions lead to the semiclassical Landau–Zener model,[136,212,241,242] which can predict nonadiabatic processes reasonably well,[8,120,143] provided they are of type (*a*) in Fig. 22. This model fails completely, however, when the two diabatic curves do not cross steeply enough or even not at all. The latter is the case in Fig. 22*b*, where the Landau–Zener formula predicts vanishing nonadiabatic transition probabilties. In contrast, the Demkov–Nikitin model correctly predicts the exponential dependence on the energy gap $V_1(R)$-$V_2(R)$.[45,170]

B. Pathways

In the course of an entire collision, the coupling region is crossed twice. Thus, in the two potential schemes shown in Fig. 22, charge transfer involves an electron jump, either when proton and molecule approach or recede. Analogously, an asymptotic elastic collision requires either two electron jumps or none. In the former case the traget molecule M first is ionized and then neutralized. In this case there is an ionic intermediate and the process is referred to as intermediate charge transfer. Accordingly the following reaction scheme is expected for proton–target collisions:

$$\begin{array}{ccccc} H^+ + M & \longrightarrow & H^+ + M & \text{—}(I)\longrightarrow & H^+ + M \\ & \searrow & & \searrow (II) \nearrow & \\ & & & \nearrow (III) \searrow & \\ & & H + M^+ & \text{—}(IV)\longrightarrow & H + M^+ \\ \textit{Entrance} & & \textit{Intermediate} & & \textit{Exit} \end{array} \tag{25}$$

Altogether, there are four different pathways (I) to (IV) that connect the single entrance channel with the two exit channels. Each of these pathways occurs with different probabilities:

Pathway	**CT is Exoergic**	**Endoergic**		
I:	$(1-P)\cdot(1-P)$	$P \cdot P$	Elastic scattering	
II:	$P \cdot(1-P)$	$(1-P)\cdot P$	CT on the entrance	(26)
III:	$(1-P)\cdot P$	$P \cdot(1-P)$	CT on the exit	
IV:	$P \cdot P$	$(1-P)\cdot(1-P)$	Ionic intermediate	

Here, P is the nonadiabatic probability for a transition between two adiabatic potential curves as given in Eq. (24). Limiting cases are $P=0$ and $P=1$ where all flux exits on the entrance channel. In intermediate cases, an observed signal in either of the two exit channels consists of two contributions of varying weight. For $P=\frac{1}{2}$ all of the four possible pathways have the same weight. Interference effects are anticipated whenever multiple pathways are involved. In this case, the differential cross sections for elastic scattering as well as for charge transfer will exhibit a modulation with scattering angle, called Stückelberg oscilations.[212] We will give some examples later in the next section.

Whereas the preceding treatment has been illustrated for a two-state problem, most atomic targets, however, involve either a multiplet of spin states or several closely spaced low-lying electronic excited states, which couple directly to other states. The preceding scheme can be extended to multiple crossings in a straightforward way. Such treatments require individual well-separated curve crossings such that the coupling regions do not overlap and the scattering system stays in a well-defined state between crossings (cf., for example, Ref. 210). Certainly, this is the case for atomic targets as long as the scattering energies do not become too high.

C. Vibrational Effects

Next, we consider the effect of charge transfer on the internal vibrational motion of diatomic and polyatomic target molecules. In the nuclear wavefunction $\Psi_k(\mathbf{R})$ defined in Eq. (21) the coordinate $\mathbf{R}$ must now account for the nuclear coordinates of both the target molecule and the proton. It is convenient to introduce a second expansion in internal coordinates $\mathbf{Q}$ such that $\mathbf{R} = \{Q_1, \cdots, Q_n, R\}$ and

$$\Psi_k(\mathbf{R}) = \sum_{n=0}^{\infty} \phi_{kn}(\mathbf{Q}, R)\Psi_{kn}(R). \tag{27}$$

The expansion coefficients $\phi_{kn}(\mathbf{Q}, R)$ then describe the internal motion of the target molecule while $\Psi_{kn}(R)$ is the scattering wavefunction of a proton scattering from a target molecule in the kth electronic and nth internal (vibrational) state. Thus, the proton dynamics are governed by the potentials $V_{kn}(R)$ corresponding to the individual eigenstates $|\phi_{kn}>$ of the molecule. Applying this concept to the vibrational levels of a diatomic target molecule a picture emerges as shown in Fig. 23. The resulting manifold of potential curves for each of the eigenstates are called vibronic networks.[17] This concept

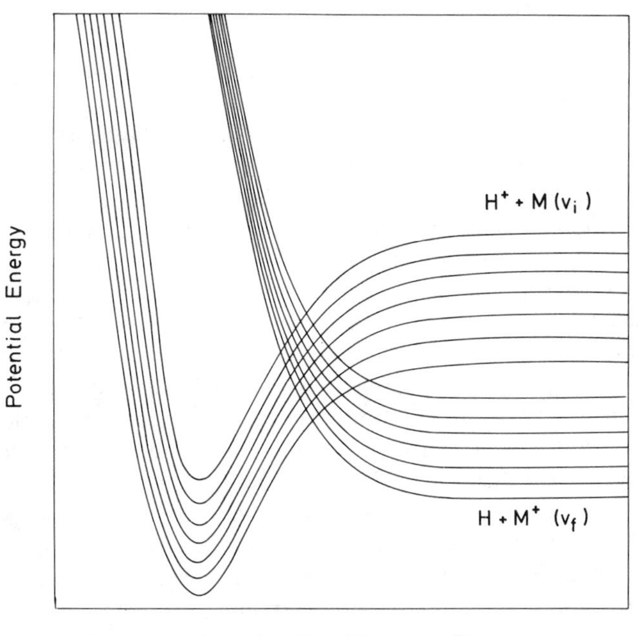

Figure 23. Schematic diagram showing the vibronic networks corresponding to different vibrational states v of the target molecule M and molecular ion M^+ produced in charge transfer with protons. In the crossing region a succession of different possible crossings can lead to a broad distribution of final states.

has been successfully applied in many treatments.[10,13] Note, that the narrowly defined coupling region is now split into multiple crossings of eigenstates. Selection rules are more complicated and depend on the detailed symmetries and properties of the proton–molecule complex. In the case of a polyatomic molecule with numerous low-lying states each with several normal vibrations, the coupling regions will become even more complicated. Therefore, it is interesting to look for mode-selective or mode-specific behavior in the experiments to help sort out the different types of coupling.

Next we briefly discuss the velocity dependence of the transition probabilities within such a vibronic network. For the limiting cases of high and low collision velocity, reasonable assumptions lead to analytical formulas for the nonadiabatic transition probabilities. Spalburg et al. investigated such limiting cases and solved the coupled Schrödinger equations analytically[121,207,210] within the framework of the Landau–Zener formalism. For large velocities in which $\xi \ll 1$, the nonadiabatic transition probability P becomes

$$P = e^{-\pi^2 \xi} |\langle \phi_{kn} | \phi_{k'n'} \rangle|^2. \tag{28}$$

where ξ stands for the electronic adiabaticity parameter as defined in Eq. (22). $|<|>|^2$ is the square of the vibrational overlap matrix element between vibronic levels $|\phi_{kn}\rangle$ and $|\phi_{k'n'}\rangle$ which are defined in Eq. (27). These matrix elements are closely related to Franck–Condon factors.[39,63] Note, however, that the matrix elements have to be evaluated in the coupling region around R_c where the target molecule geometry might be considerably distorted by the impinging proton, whereas Franck–Condon factors derived from photoelectron spectroscopy are for an undisturbed molecule that is in its equilibrium geometry. At the other extreme of very slow collision velocities, $\xi \gg 1$, the nonadiabatic transition probability becomes

$$P = e^{-\pi^2 \xi |\langle \phi_{kn} | \phi_{k'n'} \rangle|^2}. \tag{29}$$

This result is identical with that obtained earlier.[17] Unfortunately analytical solutions of the coupled equations of motion can only be obtained for these limiting cases. The most interesting case, however, both from the point of view of the theory of vibronic coupling and of the proton-scattering experiments is the intermediate region. For diatomic molecules considerable success has recently been achieved by purely quantum-mechanical approaches[10,13] and by hybrid models that solve the translational part of motion classically and the vibronic part quantum mechanically.[120,205,208,209] For triatomic and polyatomic molecules, however, it has not been possible to compare the state-to-state resolved experimental results of proton scattering with reasonably accurate and complete theoretical calculations.

VI. EXPERIMENTAL EXAMPLES: CHARGE-TRANSFER SYSTEMS

A. Atomic Targets

To illustrate the preceding discussion we present some results on electronic nonadiabatic processes in the scattering of protons from rare-gas atoms obtained with the time-of-flight apparatus (Fig. 3). We also include some data obtained with He^{2+} ions, which are also structureless and devoid of electrons but at the same impact parameter provide a factor of 2 greater electric field at the molecule. The much higher recombination energies of He^{2+} (40.77 and 24.59 eV for the first and second electron, respectively), give access to different electronic states of the target atoms. In addition, symmetry-related gerade–ungerade oscillations in the differential scattering cross section that are experimentally more difficult to investigate with protons can be investigated easily with He^{+} or He^{2+} ions. This interference arises from a splitting of the interaction potential into a gerade and an ungerade state at close distances of the scattering partners owing to electron exchange. Scattering contributions from both potential branches are shifted in phase by an amount $\Delta\alpha = \alpha_u - \alpha_g$, where α_g and α_u are the scattering phases of a collisional event taking place entirely either on the gerade or ungerade potential branch, respectively, and interfere accordingly. The interference is a slowly varying function of the scattering angle. The adiabatic probabilities for elastic scattering P_{el} and for charge transfer P_{ct} in a semiclassical approximation are given by

$$\begin{aligned} P_{el}(\theta) &= \tfrac{1}{2}[1 + \cos \Delta\eta(\theta)], \\ P_{ct}(\theta) &= \tfrac{1}{2}[1 - \cos \Delta\eta(\theta)]. \end{aligned} \tag{30}$$

The corresponding classical differential cross sections are

$$\frac{d\sigma_i}{d\theta} = \frac{b}{\sin\theta}\frac{db}{d\theta} P_i(\theta), \tag{31}$$

where i denotes *el* or *ct*, respectively. Thus opposite phase oscillations occur in both channels in the differential cross sections. A quantum-mechanical treatment gives essentially the same result.[152]

The adiabatic $(HeHe)^{2+}$ interaction potentials[38] are shown in Fig. 24 at the right. Single charge transfer, although exoergic by more than 30 eV, is not observed for scattering energies below 10 keV.[105] For $He^{+} + He$ the g–u oscillations in the differential cross sections were first observed by Aberth et al.[1] and later for $He^{2+} + He$ by Lam et al.[135] Scattering experiments

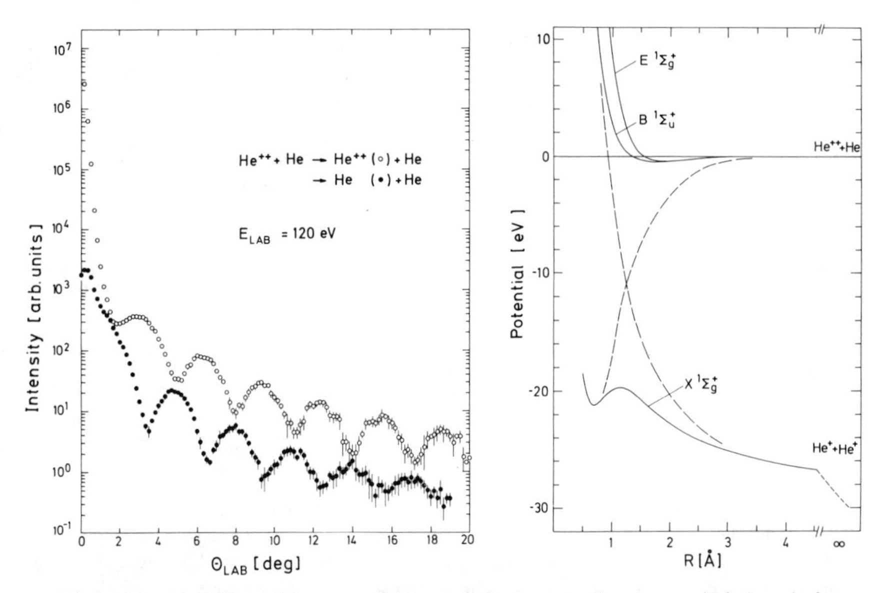

Figure 24. (*a*) Differential cross sections for elastic scattering (open circles) and charge transfer (solid dots) in $He^{2+} + He$ scattering at $E_{lab} = 120\,eV$. The charge-transfer intensities have not been corrected for the 10^{-1} detection efficiency of the He atoms for reasons for clarity. Thus, both cross sections are about equal. Note the opposite phase behavior of the gerade–underade symmetry oscillations. (*b*) Potential energy diagram for the He_2^{2+} system. Owing to the strongly avoided curve crossing, single charge transfer into $He^+ + He^+$ is impossible. The double degenerate $He^{2+} + He$ state is split at small interatomic distances. This splitting causes the oscillations shown in (*a*).[169]

performed on the apparatus of Fig. 3 provided data on both channels for the latter system,[169]

$$\begin{aligned} He_a^{2+} + He_b &\to \underline{He_a^{2+}} + He_b \quad (a), \\ &\to \underline{He_a} + He_b^{2+} \quad (b), \end{aligned} \tag{32}$$

where the detected projectile has been underlined. The measured angular distributions are shown in Fig. 24 at the right. The symmetry oscillations are clearly resolved in both channels, and they are exactly opposite in phase throughout the entire angular range covered. The signal measured with channel (*b*) has not been corrected for the lower He-atom detection probability, which is about 10^{-1} for reasons of clarity. Thus both channels have about equal probabilities as expected.

With a neon target the double charge transferred state is no longer degenerate with the elastic channel. The diabatic $He^{2+} + Ne$ potential energy

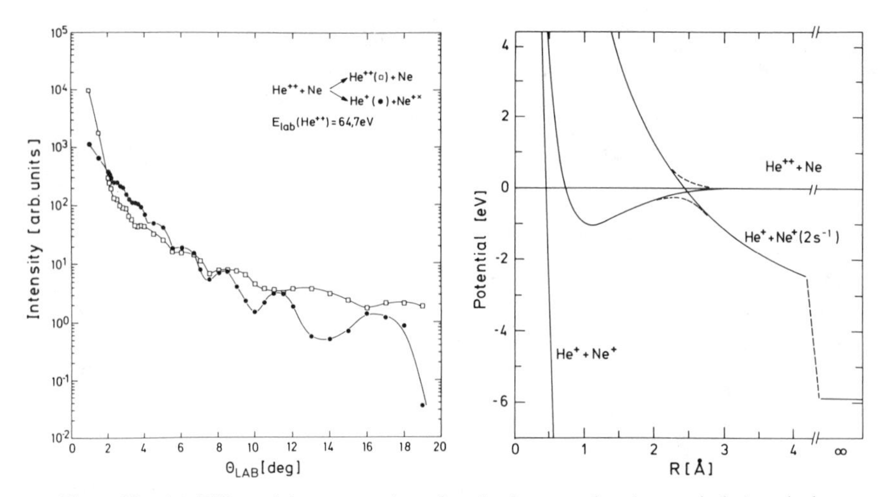

Figure 25. (*a*) Differential cross sections for elastic scattering (open circles) and charge transfer (solid dots) in He^{2+} + Ne scattering at $E_{lab} = 64.7$ eV. Note the Stückelberg oscillations in both curves. There is no fixed phase relationship between both patterns. (*b*) Potential energy diagram for the $(HeNe)^{2+}$ system. Single charge transfer into an electronically excited Ne^{+*} state occurs via a direct curve crossing with the He^{2+} + Ne state.[169]

curves from a very approximate fit of elastic experimental data at low energies[7] are shown in Fig. 25 at the right. For both the $He^+ + Ne^+(2s^2 2p^5)$ and $He^+ + Ne^{+*}(2s^1 2p^6)$ curves, pure Coulomb repulsion was assumed. The resulting avoided curve crossing is apparent at 2.5 Å and leads to nonadiabatic transitions. The measured angular intensity distributions for $He^{2+} + Ne$ scattering at $E_{lab} = 64.7$ eV are shown in Fig. 25 at the left.[169] The open squares and the solid dots denote elastic and charge-transfer scattering, respectively. Note, that no charge transfer into the Ne^+ ground state takes place. The detected products, He^{2+} and He^+, both of which are charged, could be distinguished by time-of-flight measurements. The measured increase in the energy of the He^+ peak by 5.93 eV coincides nicely with the potential asymptotes in the right part of Fig. 25. Apparently, the differential cross sections for both processes in Fig. 25 show independent oscillatory patterns. The angular spacings increase with the scattering angle in both cases, but there is no fixed-phase relation between the two oscillatory patterns. In contrast to $He^{2+} + He$, the two contributing potential branches in the $He^{2+} + Ne$ system are quite different and, therefore, exhibit different dynamics. For one, the potential curves for $R < R_c$ are not parallel but diverge. Second, the two exit channel potential functions are not parallel for $R > R_c$ but the $He^+ + Ne^{+*}$ curve is dropping continuously due to its Coulombic

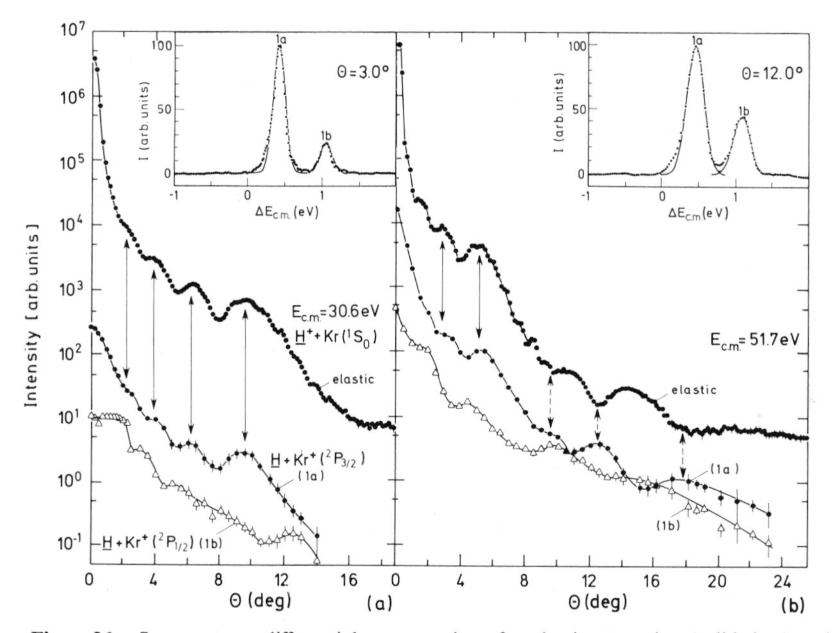

Figure 26. State-to-state differential cross scetions for elastic scattering (solid dots) and charge transfer for $H^+ + Kr$ into various electronic states (dots and triangles) at $E_{lab} = 30.6$ eV (left) and 51.7 eV (right).[11] The arrows emphasize the phase relations between elastic scattering and charge-transfer cross sections. Note the change from in-phase to an opposite-phase relation at the higher scattering energy. The insets illustrate translational energy distributions for charge transfer (detection of H atoms) at scattering angles $\Theta_{lab} = 3°$ and $12°$, respectively. The results are not corrected for the detection efficiency.

nature. Thus, the oscillations in the differential cross sections have no fixed-phase relation between the two observed channels.

The scattering of protons from the rare gases shows similar phenomena. For He and Ne charge transfer in collisions with protons is improbable at $E_{lab} < 100$ eV as their ionization potentials are considerably larger than 1 Rydberg. In contrast, for the heavier noble gases, Kr and Xe, charge transfer occurs at medium energies with quite significant cross sections. Results for both elastic scattering and charge transfer in $H^+ + Kr$ and $H^+ + Xe$ are shown in Figs. 26 and 27[11,73] for two energies, 30.6 eV (left) and 51.7 eV (right). In all four cases the uppermost experimental curve is the differential cross section for elastic scattering. The lower curves correspond to state-resolved charge-transfer scattering into different electronic states of the target ion as determined from time-of-flight measurements shown in the insets. Note, that in Figs. 26 and 27 the charge-transfer cross sections are

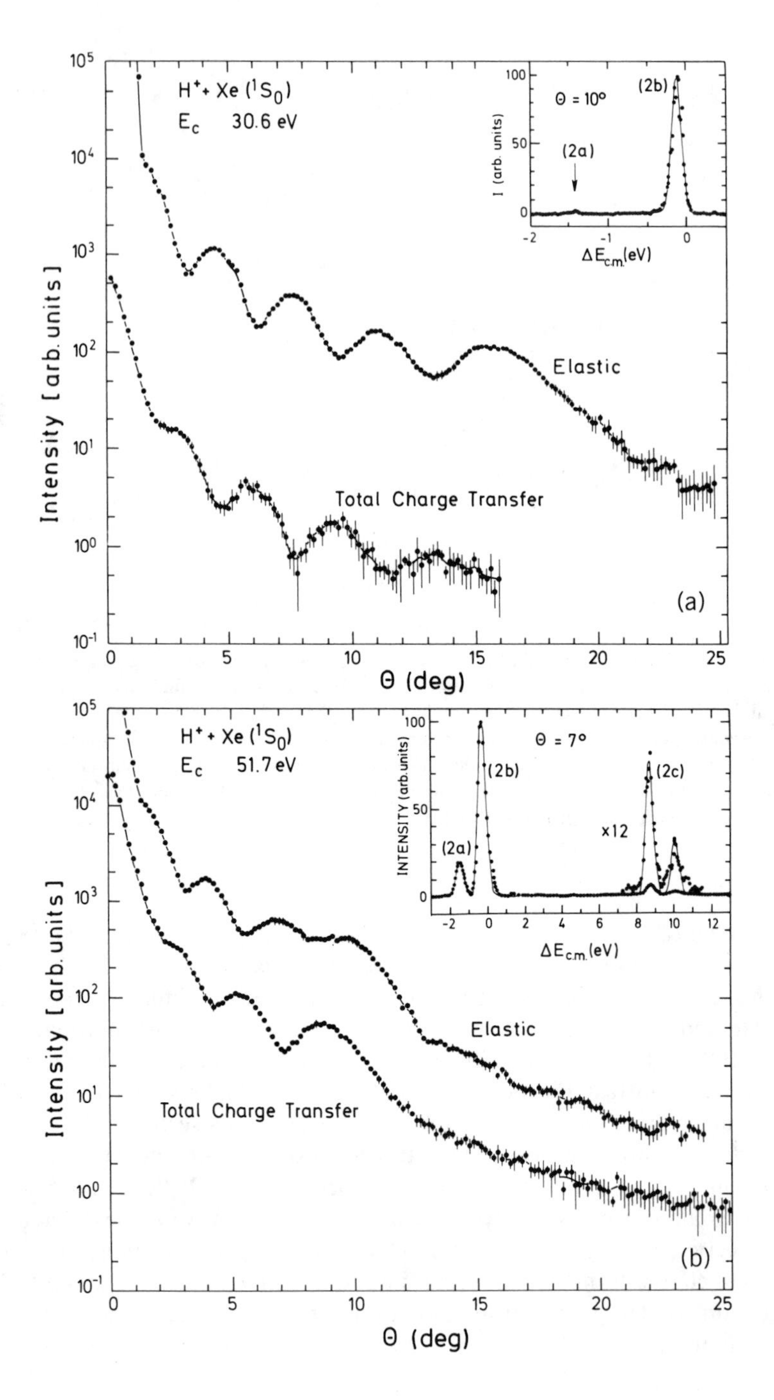

H⁺ + Xe (¹S₀)
E_c 30.6 eV
Θ = 10°
(2a)
(2b)
I (arb. units)
ΔE_c.m.(eV)
Elastic
Total Charge Transfer
Intensity [arb. units]
(a)
Θ (deg)
H⁺ + Xe (¹S₀)
E_c 51.7 eV
Θ = 7°
(2a)
(2b)
(2c)
x12
INTENSITY (arb. units)
ΔE_c.m.(eV)
Elastic
Total Charge Transfer
Intensity [arb. units]
(b)
Θ (deg)

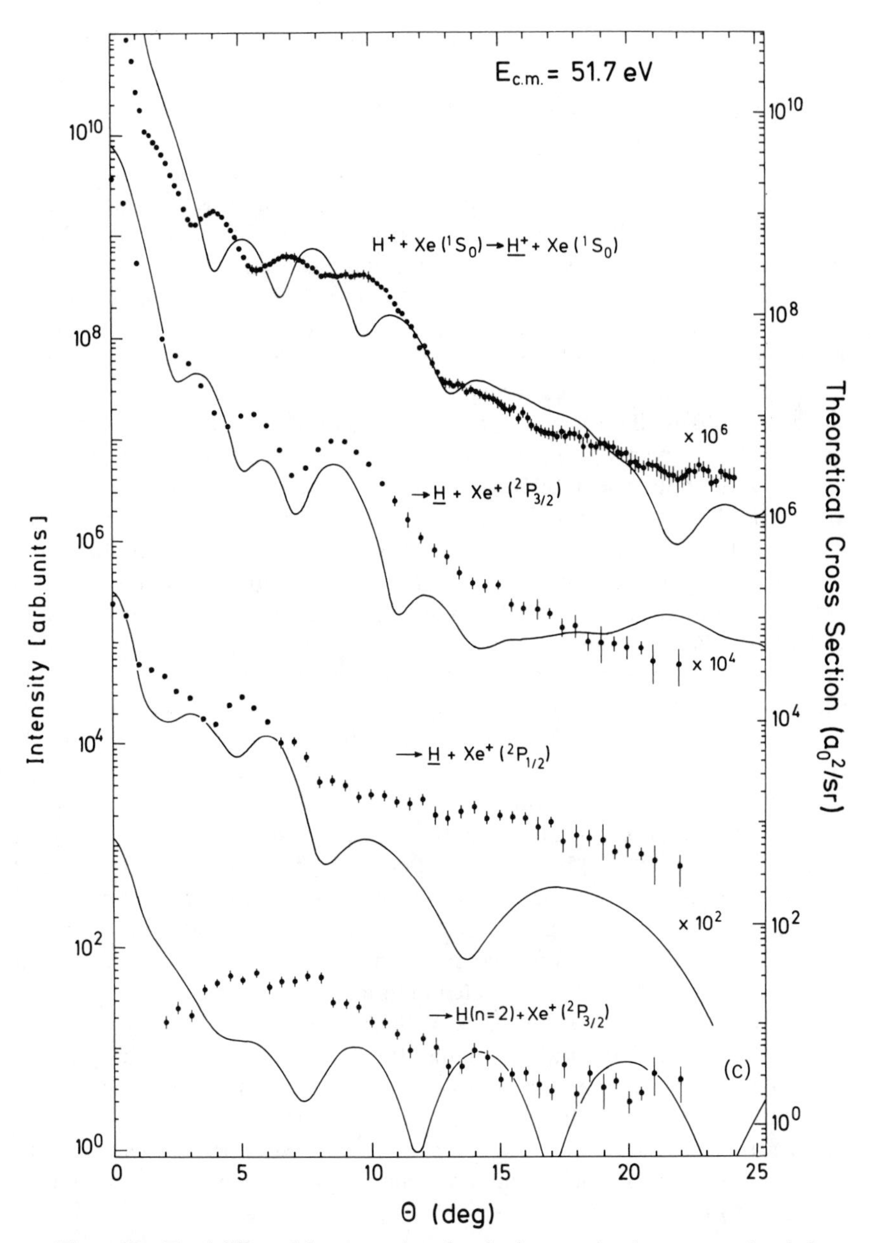

Figure 27. Total differential cross sections for elastic scattering (upper curves) and charge transfer (lower curves) in $H^+ + Xe$ at $E_{lab} = 30.6$ eV (*a*) and 51.7 eV (*b*) together with state-to-state cross sections at 51.7 eV (c).[11,73] Note the opposite phaase relations between the Stueckelberg oscillations in both cross sections of (*a*) and (*b*). The results are not corrected for detection efficiency. The insets illustrate the charge-transfer product relative energy distributions at $\Theta_{lab} = 10^\circ$ and 7°, respectively. They show various electronic channels that are evaluated explicitly in (*c*) and compared to close coupling calculations (solid lines). The left and right scales correspond to experimental and theoretical results, respectively. Differential cross sections due to different state-selective processes are shifted by a factor of 100.

not corrected for the H-atom detection probability (3% and 7% at 30.6 and 51.7 eV, respectively).

The main features observed in the differential cross sections can be understood with the aid of the potential curves for $(HKr)^+$ and $(HXe)^+$ shown in Fig. 28. All of the curves are from *ab initio* calculations with the exception of the $H(2s) + Xe^+(^2P_{3/2})$ curve that has been estimated. The entire $H^+ + Kr$ data at the lower energy and the small-scattering-angle part of the $H^+ + Kr$ data at the high scattering energy show in-phase oscillations. Connecting arrows try to emphasize the different behavior with respect to their phase relation. For scattering angles larger than $\Theta_{lab} = 8°$ the $H^+ + Kr$, $E_{lab} = 51.7$ eV oscillations are opposite in phase. This is also the case for the xenon data irrespective of energy and angle. Thus, there is a collision strength $\tau = \Theta_R \cdot E$ of about $\tau \approx 360$ eV·deg, where Θ_R denotes the rainbow angle and where the $H^+ + Kr$ system changes its behavior. For larger collision strengths opposite-phase Stückelberg-type oscillations dominate the differential cross sections in the $Kr(^1S_0)$ and $Kr^+(^2P_{3/2})$ channels. Thus, the collisions with small collision strengths τ probe mainly the deep well of the ground-state potential, whereas higher τ values manifest the strong coupling of those channels which are asymptotically close in energy (Fig. 28). Looking at Fig. 27 all the oscillations are opposite in phase, and for $\tau = > 600$ eV·deg they are totally suppressed. As no in-phase rainbowlike behavior is observed at all, we conclude that the deep well of the $H^+ + Xe(^1S_0)$ potential is in fact not probed. As seen in Fig. 28 there is a curve crossing involving the entrance channel at quite large distances. Thus, it appears that considerable flux goes into the $H + Xe^+(^2P_{1/2})$ branch at that crossing. The additional damping of the observed Stückelberg oscillations at larger scattering angles appears to be related to the opening of the $H(2s) + Xe^+(^2P_{3/2})$ scattering channel. Previously the last observed maximum in the differential cross section at about $\tau = 480$ eV·deg was interpreted in terms of a primary rainbow.[236] This can now be ruled out on the basis of the new measurements. In addition, close coupling calculations involving the four states shown in Fig. 28*b* succeeded in predicting the discussed fall-off and its energy dependence qualitatively.[11] Note, finally, that the contributions of nonresonant charge-transfer states such as $Kr^+(^2P_{1/2})$ and $Xe^+(^2P_{3/2})$ increase with increasing scattering energy and scattering angles for which the coupling is expected to increase.

B. Diatomic Target Molecules

The qualitative potential energy curves for the diatomic systems $H^+ + H_2$ and $H^+ + O_2$ to be discussed in this section are shown schematically in Fig. 29. The proton–molecule interaction potential $H^+ + M$ is strongly attractive. The charge-transfer states $H + M^+$ are purely repulsive in a first

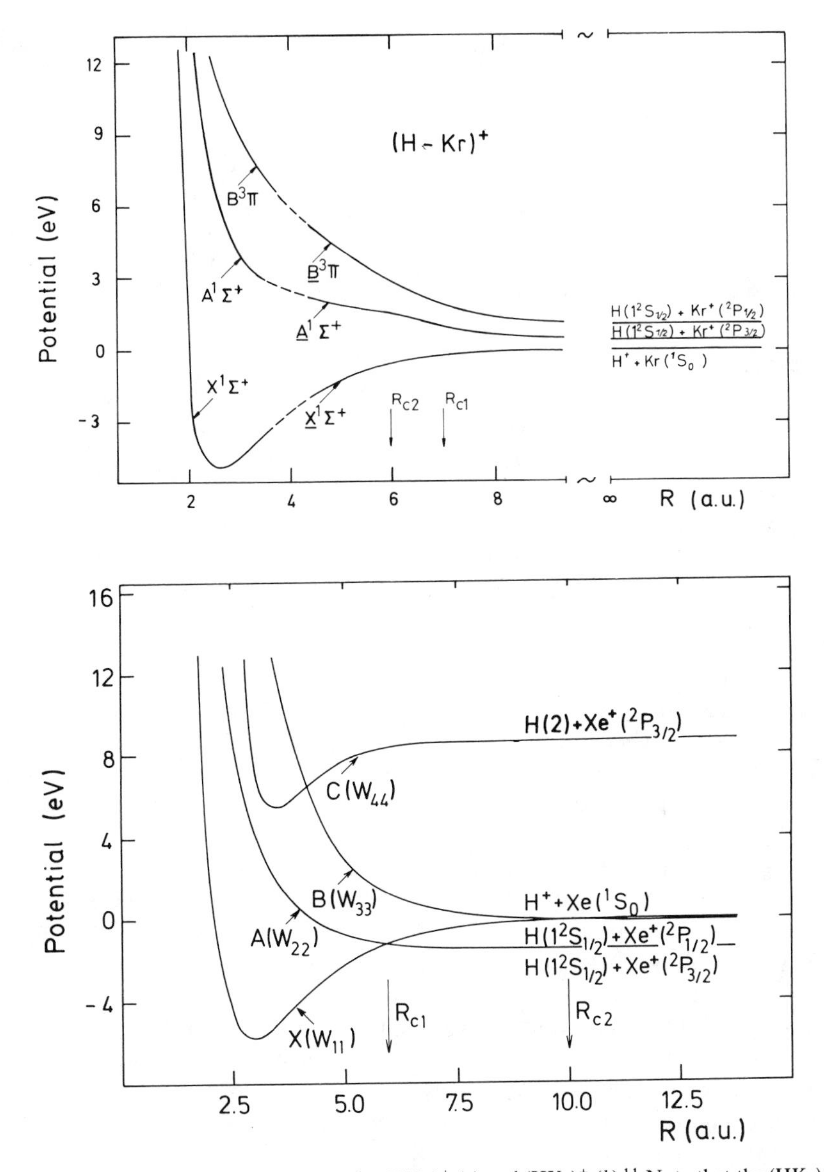

Figure 28. Potential energy curves for $(HKr)^+$ (*a*) and $(HXe)^+$ (*b*).[11] Note that the $(HKr)^+$ system nicely corresponds to the endoergic coupling case depicted in Fig. 22*b*, whereas the $(HXe)^+$ systems corresponds to the exoergic charge transfer case of Fig. 22*a*.

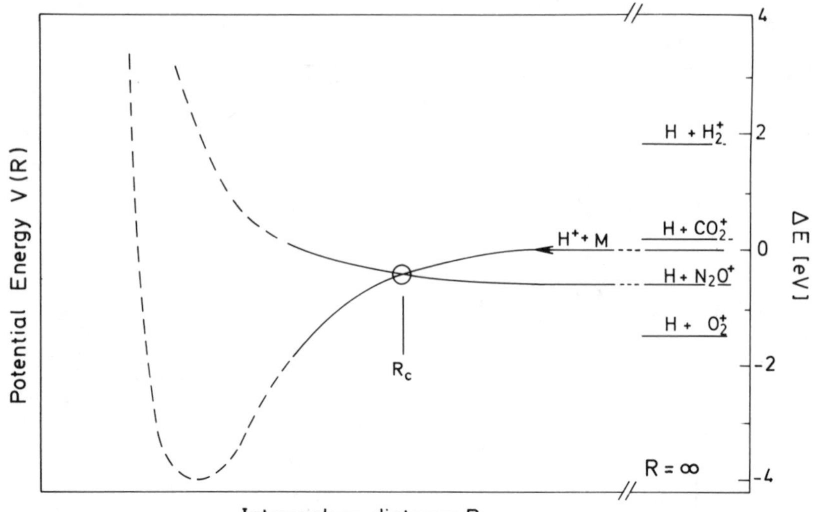

Figure 29. Schematic $(HM)^+$ potential energy curves for $M = O_2$, N_2O, CO_2, and H_2. The $H^+ + M$ curve is typical for all systems listed previously, whereas the $H + M^+$ curve is shifted in energy according to the varying ergicities and is shown explicitly only for $H + N_2O^+$.

approximation, and their asymptotes are shifted by the energy difference of the charge-transfer states. The exoergic $H^+ + O_2$ system involves a direct curve crossing, whereas for the endoergic system $H^+ + H_2$ the charge-transfer state lies higher in energy and are accessible only by Demkov coupling or via vibrational excitation of the neutral molecule.

The simplest diatomic collision system showing charge transfer is $H^+ + H_2$. This system provides a unique opportunity to test dynamical theories, since the potential surfaces have been calculated over the years with increasing precision.[57,159,183.197,215,224] At collision energies above about 5 eV the reactive scattering cross section falls off steeply (cf. Fig. 1) and can be largely neglected. The pioneering experiments[81,176,214] using guided ion beams have provided evidence for long-lived complexes at very low energies ($E = 0.1$–4 eV). Vibrational states were first resolved in 1973[228] in energy-loss measurements of the scattered protons at collision energies of $E = 4$–21 eV. Additional investigations have been reported by several groups[81,103,130] and even rotational resolution[190,198] could be achieved. Dynamical calculations based on various approximations have been used to simulate the experimental findings,[36,93,146,195,196,240] but, until recently, did not include the charge-transfer channel. The trajectory surface hopping[224] and quantum-mechanical IOS[221,222] methods were first used to predict charge transfer, but at that time no experimental data with vibrational-state resolution were available. As discussed in Section III the surprisingly large vibrational inelasticity in $H^+ + H_2$ led to the idea of the bond dilution mechanism.[228] The first detailed differential scattering experiments with vibrational state-to-state resolution for both the proton and fast H-atom channels were measured with the apparatus shown in Fig. 3.

Figure 30 shows the total differential cross section for scattering of protons (upper curve) and for charge-transfer production of H atoms (lower curve) after scattering from H_2 at $E_{lab} = 30\,eV$. Note that the lab to cm transformation (cf. Section II) yields for a stationary target $E_{cm} \cong \frac{2}{3}E_{lab} = 20\,eV$ and $\Theta_{cm} \cong \frac{3}{2}\Theta_{lab}$ for $\Theta_{lab} < 20°$. Even after correcting for the 5% H-atom detection probability, the charge-transfer cross section is much smaller than the cross section for proton scattering, which is not unexpected in view of the large adiabatic endoergicity of 1.83 eV.

Time-of-flight measurements of both the scattered protons and H atoms shown in Fig. 31 reveal clearly resolved vibrational excitation in both channels. The solid lines show Gaussian best fit distributions, which, within a given spectrum, are assumed to have the same half-width obtained from a best fit of the elastic peak. The observed H_2 vibrational transition probabilities are in remarkable agreement with the earlier experimental work of Linder et al.[195,196] The observed deviations between the measured spectra and the fit curves, which are especially noticeable in the valleys between the maxima, are attributed to significant kinematical smearing resulting from the angular distribution of the uncollimated secondary beam. The effect is more significant here since the H_2 molecules are closer in mass to the protons than for the other heavier targets. The probability distribution in the inelastic collisions shifts to higher final vibrational states with increasing scattering angles (Fig. 31*a*), whereas in the charge-transfer collisions it remains essentially the same. The charge-transfer energy-loss spectra (Fig. 31*b*) nicely show the anticipated endoergicity of $\Delta E = 1.83\,eV$ and the observed peak spacing agrees well with the H_2^+ vibrational levels.

With the vibrational transition probabilities obtained from the time-of-flight spectra (Fig. 31), it has been possible to decompose the total differential cross sections of Fig. 30 into the state-to-state relative differential cross sections shown in Figs 32*a* and 32*b* for detection of protons and in Fig. 33*c* for H atoms. The experiments yield only relative state-to-state cross sections. Since the relative inelastic cross sections show the same trends as the theoretical cross section,[13] they have been used to obtain absolute values.

The position of the primary rainbow for inelastic scattering of protons at $\Theta_{lab} = 7°$ agrees well with the position of the maximum common to the charge-transfer distributions for different vibrational states in Fig. 32*c*. Moreover, all of the state-to-state differential cross sections leading to high

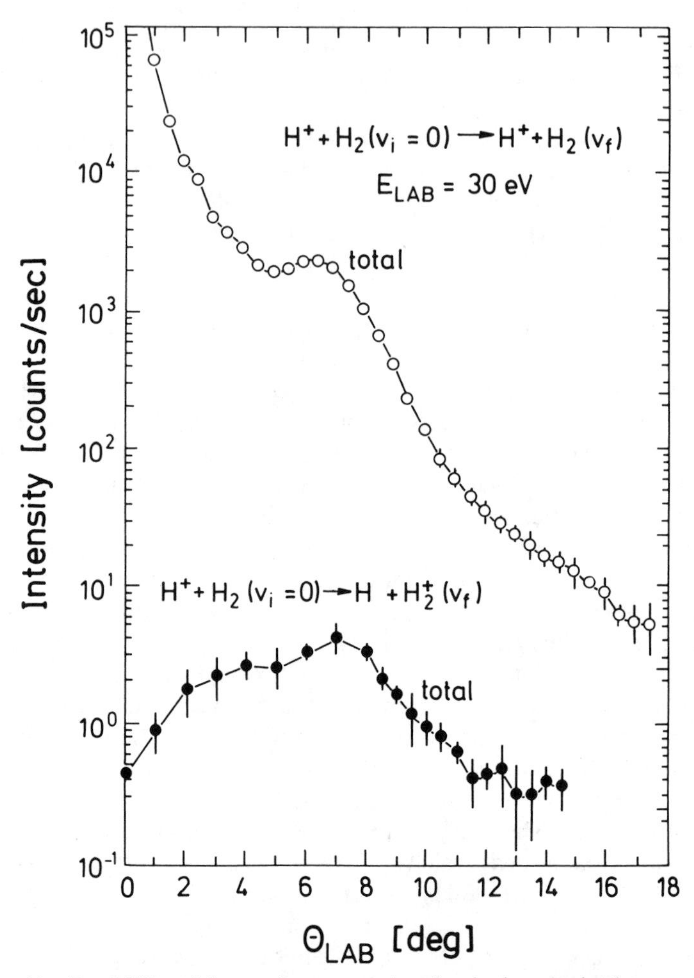

Figure 30. Total differential scattering cross sections for elastic scattering (upper curve) and charge transfer (lower curve) in $H^+ + H_2$ at $E_{cm} = 20$ eV.[168] The elastic scattering curve exhibits a clearly resolved primary rainbow at about $\Theta_{lab} = 7°$. The H-atom intensities from charge transfer are about three orders of magnitude smaller and peaked at a similar angle. They have not been corrected for the lower H-atom detection probability (5%). Note the different behavior in the small-angle region, where the charge-transfer signal decreases in contrast to the proton signal.

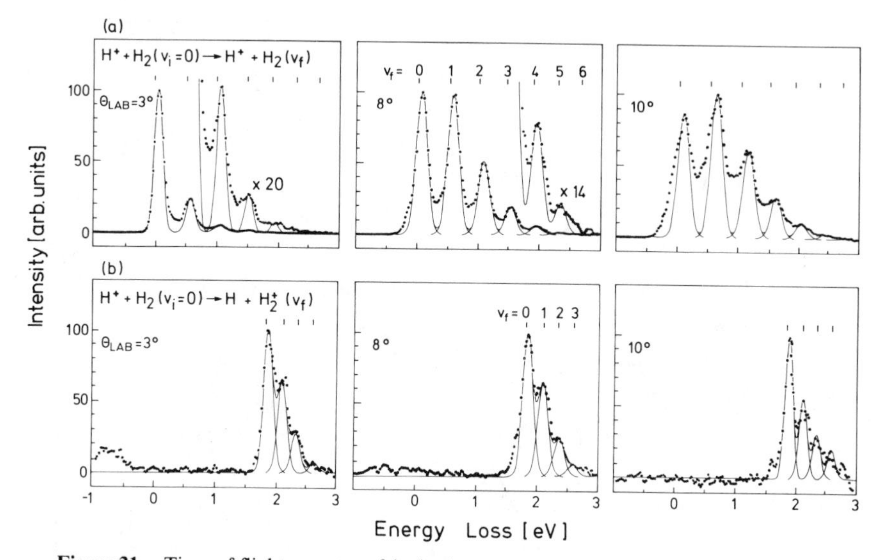

Figure 31. Time-of-flight spectra of inelastic scattering (upper row) and charge transfer (lower row) in $H^+ + H_2$ $E_{cm} = 20$ eV.[168] Three scattering angles are shown for comparison. The expected locations of the rotationless vibrational levels are indicated at the top of each spectrum. The small shifts of the observed elastic and vibrationally inelastic peaks thus indicate a small amount of rotational excitation. The endoergicity of $\Delta E = 1.83$ eV for charge transfer is clearly apparent in the spectra of the lower row, and the observed peak spacing agrees nicely with the expected $H_2{}^+$ vibrational levels.

final vibrational excitation in H_2 ($v_f \geqslant 4$) have a close similarity in shape with the corresponding charge-transfer cross sections. This is particularly apparent in comparing the H_2 ($v_f = 4$) cross section with the H_2^+ ($v_f = 0$) cross section. Since these two levels are asymptotically nearly resonant in energy, this observation suggests that, once vibrational excitation of four or more quanta in H_2 is present, the coupling to charge-transfer states becomes very strong. Thus, the small probabilities for charge transfer is related to the small probability for high vibrational excitation of H_2 by protons.

For understanding the dynamics it is sufficient to consider only a cut through the potential surface along the R coordinate and the vibronic networks (Fig. 23) of the H_2 and H_2^+ molecules as shown in Fig. 33. The close similarity in the differential cross sections for H_2 excitation to $v_f = 4$ and production of H_2^+ ($v_f = 0$) can be explained by assuming that in the first half of the collision the H_2 molecule becomes vibrational excited by the bond-dilution mechanism. Then in the second half of the collision as the proton recedes from the vibrationally excited H_2 molecule, those that have been excited to the $v_f \geqslant 4$ states at $R \geqslant 5a_0$ access the crossing seam. There

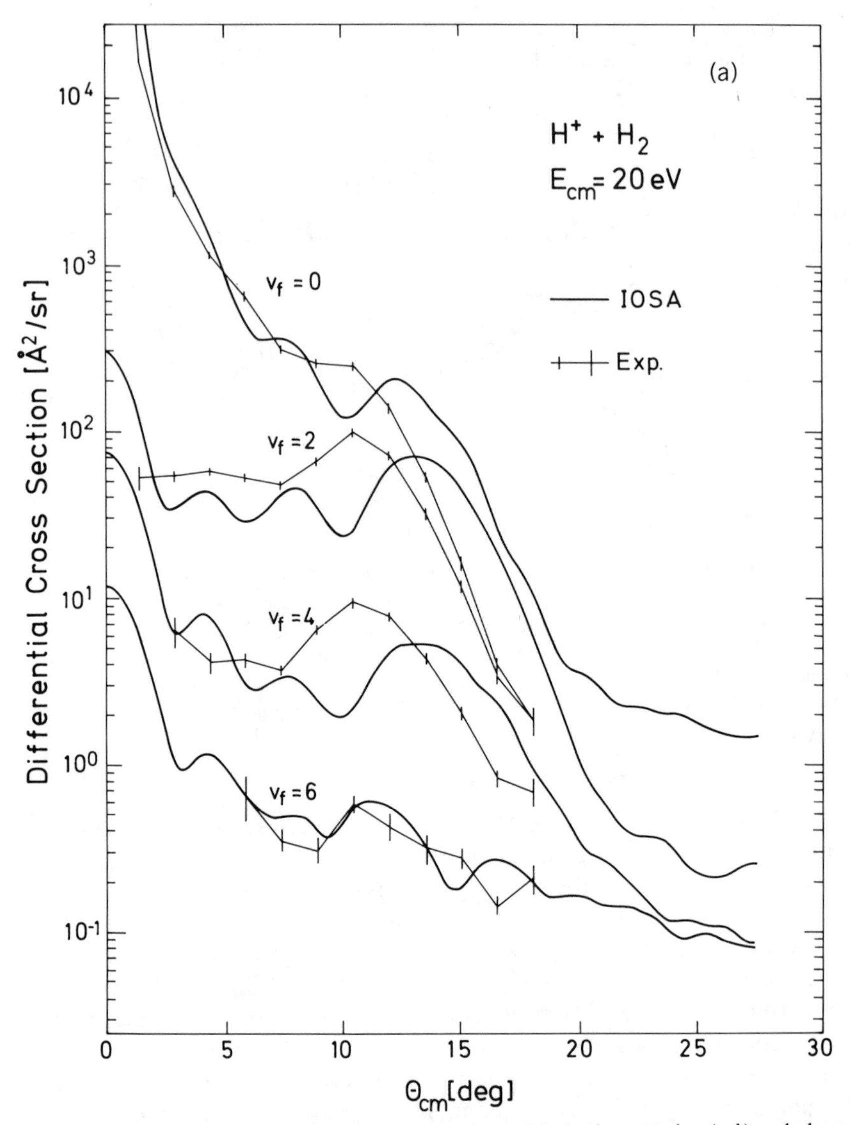

Figure 32. State-to-state differential cross sections of inelastic scattering (a, b) and charge transfer (c) in $H^+ + H_2$ at $E_{cm} = 20$ eV.[13] The experimental data are shown as straight segments connecting the individual points with error bars, whereas the quantum-mechanical calculations are shown as smooth lines. Since absolute cross sections could not be measured, the experimental intensity scale was adjusted appropriately. For reasons of clarity each set of curves in (c) has been shifted by various factors as indicated on the right side of the figure. Besides a shift in the rainbow angle the agreement between theory and experiment is noticeable.

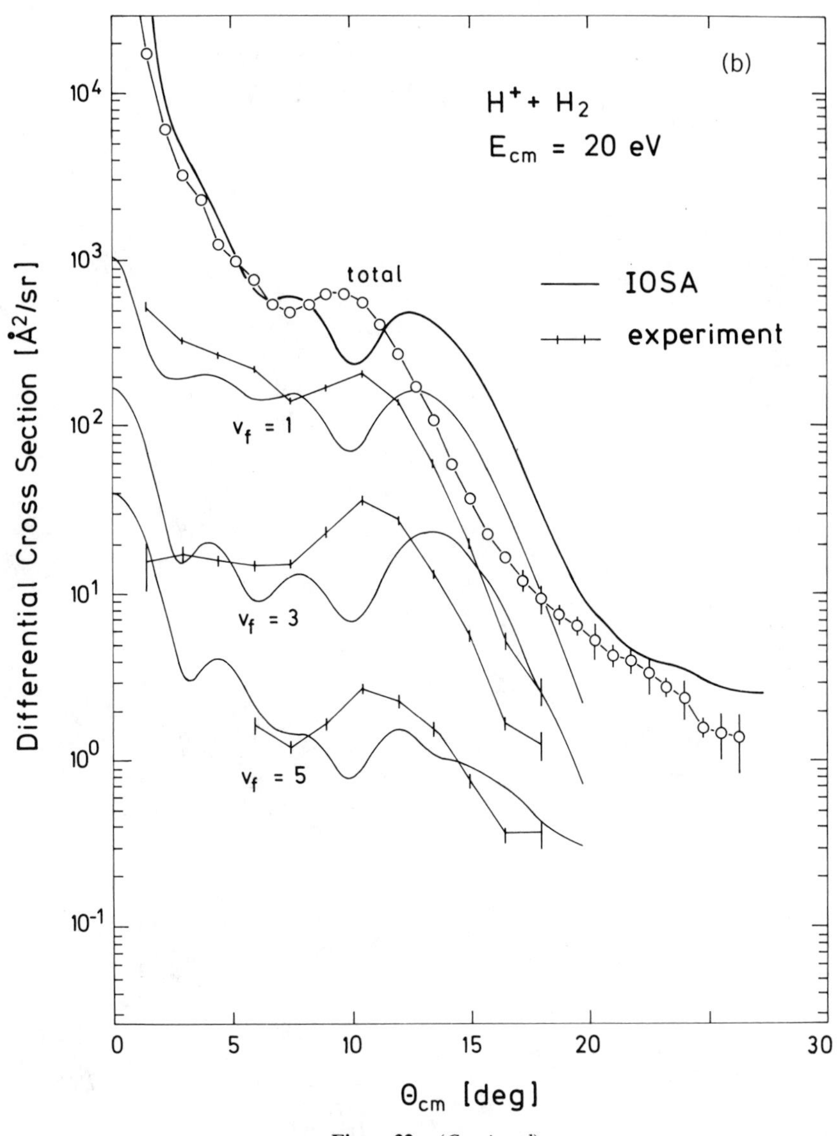

Figure 32. (*Continued*)

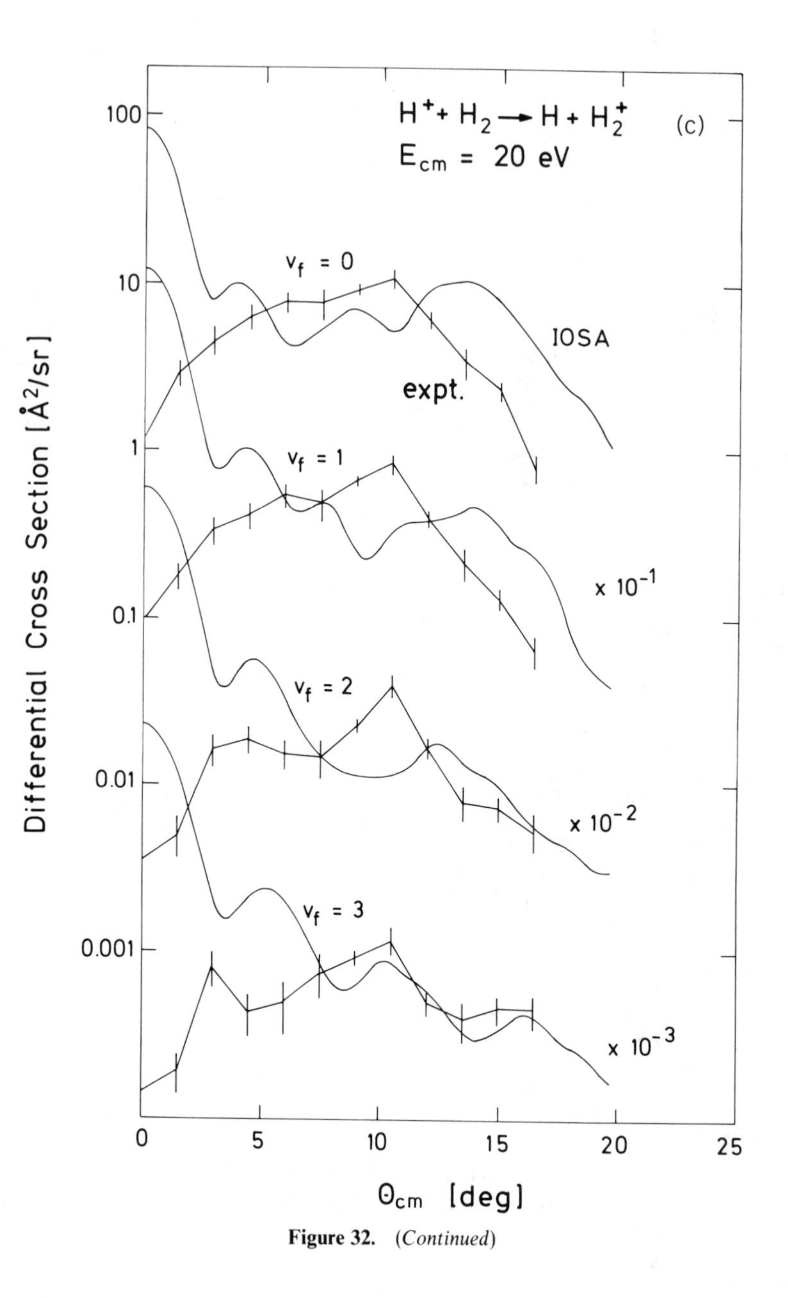

Figure 32. (*Continued*)

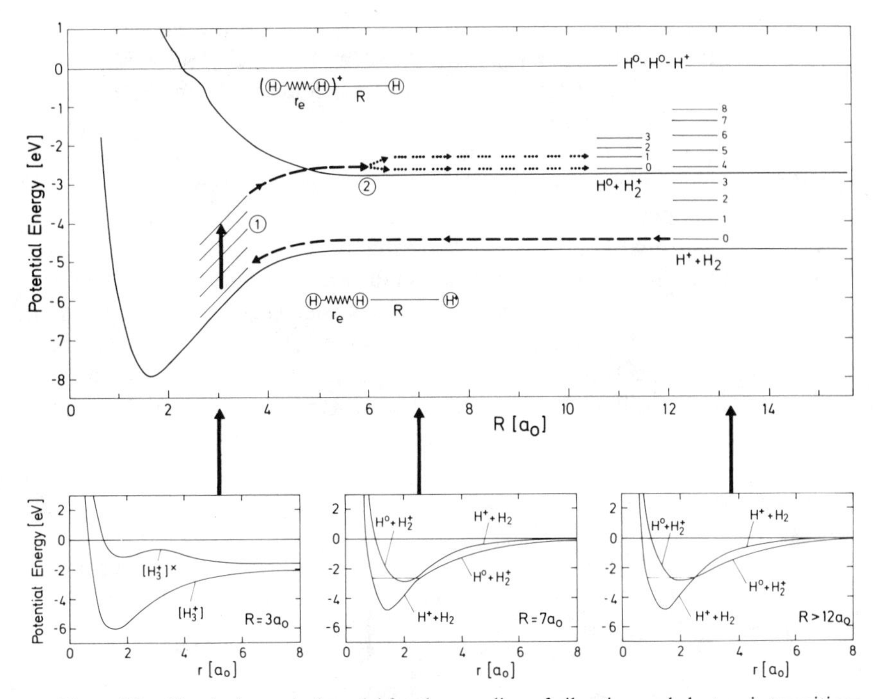

Figure 33. Simple dynamical model for the coupling of vibration and electronic transitions in $H^+ + H_2$ charge transfer.[168] The upper diagram shows a cut along the distance R between centers of masses (see inset) for fixed equilibrium bond length r_e. The asymptotic ($R = \infty$) vibrational levels of H_2 are shown at the right. The approaching unexcited reactants, $H^+ + H_2$ ($v_f = 0$), interact to produce vibrationally excited H_2 (step 1), which for $v_f \geqslant 4$ can populate $H_2{}^+$ ($v^f = 0,1$) by charge-transfer transitions (step 2). After charge transfer, no significant forces are exerted on the products, thus largely preserving the angular distribution in the charge-transfer channel. The three small figures at the bottom give cuts through the potential hypersurfaces for three different values of the scattering coordinate R as a function of the H_2 bond length r. In two of these figures the $H_2(v_f = 4)$ level is shown as well.

they undergo with high probability a curve-crossing transition into the charge-transfer state while conserving their total energy. Since charge transfer occurs at the end of the collision, the angular distributions in both channels are nearly both the same. This dynamical model in which the charge-transfer crossing between two potential hypersurfaces is accessed by vibrational motion and not by radial motion was suggested earlier.[224]

The availability of the state-to-state resolved experiments on $H^+ + H_2$ has stimulated several scattering calculations for the available DIM potential matrix, which includes the nonadiabatic coupling potentials. In the one calculation[168] based on the trajectory surface hopping model,[224] classical mechanics is used to describe the trajectories outside the crossing seam. Whenever a classical trajectory crosses the seam, it is split into two branches according to the charge-transfer probabilities calculated from the Landau–Zener formula. The weighted branches are followed out to asymptotic distances and evaluated using standard binning procedures. For comparison, quantum-mechanical IOSA (infinite order sudden approximation) calculations have been performed for the same DIM potential hypersurfaces.[13] In this approximation the rotational coupling is treated in an approximate manner, which is justified by the small anisotropy of the potential hypersurfaces. The vibrational and charge-transfer coupling are accounted for in a close-coupling calculation, which is expected to be very accurate. Figure 34 shows a comparison with the experimental results of the integral cross sections for different vibrational states for both channels calculated

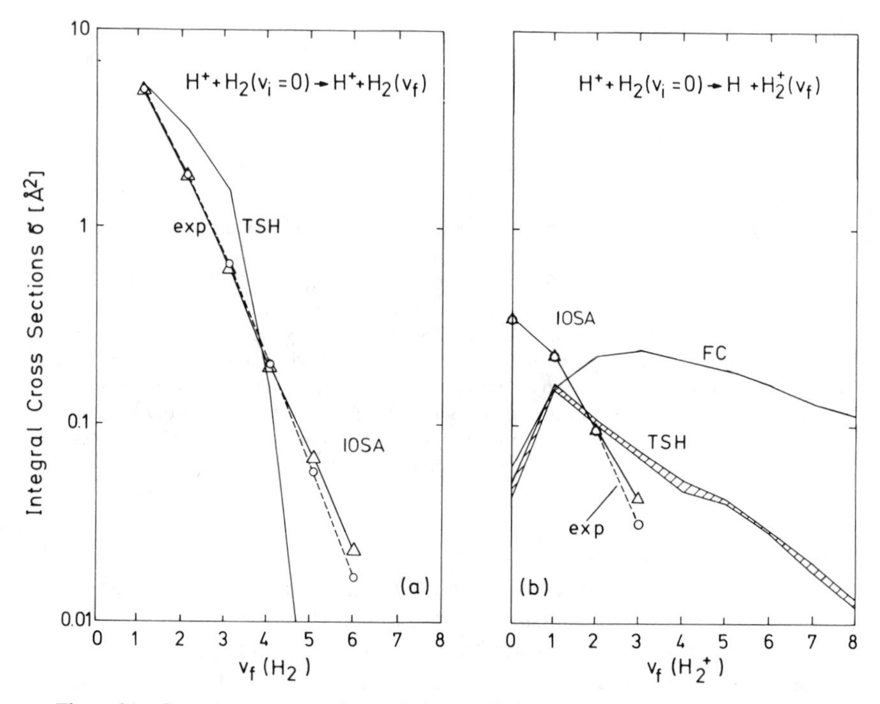

Figure 34. State-to-state experimental (open circles) and quantum-mechanical (open triangles) integral cross sections compared to a Franck–Condon distribution (FC) and predictions from a semiclassical model (TSHM).[168] Note the very good agreement between the quantum-mechanical (IOSA) calculations and the experimental data, whereas the semiclassical approach largely fails to reproduce the experiments. A Franck–Condon distribution (FC) is shown as well.

using the two theories. The charge-transfer cross sections are also compared with Franck–Condon factors from photoelectron spectra.[6] It is gratifying to observe that the IOSA calculations provide a much better fit of the data for both channels than either the classical trajectory calculations or the Franck–Condon distributions. The IOSA results reveal that charge transfer occurs over a wide range of impact parameters, whereas in the classical calculations charge transfer is largely restricted to central collisions. This observation suggests that quantum effects not included in the classical trajectories and surface hopping approximations are important. Despite the good agreement in cross sections, there are still serious discrepancies with regard to the rainbow angle, which is predicted to be larger than observed, and the shape of the charge-transfer cross sections at small angles.

As noted in Section I (Fig. 2) early proton, scattering experiments[83] indicated a remarkable increase in the vibrational inelasticity for O_2 and to a lesser extent for NO compared to N_2 and CO. This was attributed to the presence of a curve crossing with the charge-transfer state in the $H^+ + O_2$ and NO potential curves, which was not expected in $H^+ + N_2$ and CO. At the time direct measurements of charge transfer were not available. More recent, better resolved proton and H-atom time-of-flight spectra for $H^+ + CO$ confirmed the small probabilities of about 10% for $v = 1$ excitation even at large angles[172] in the proton energy range $E_{lab} = 10$–50 eV. Thus, CO can be considered as a rigid diatomic molecule that does not experience strong inelasticity in proton collisions. Since charge transfer in $H^+ + CO$ is quite endoergic (-0.42 eV), it is expected to occur with significantly less probability than elastic scattering. This is confirmed by the total differential cross sections for elastic scattering measured by proton (open circles) and H-atom detection (solid dots) shown in Fig. 35 for a scattering energy of $E_{lab} = 30$ eV. The charge-transfer data have been corrected for the small H-atom detection probability (5%). As expected, charge transfer is about one order of magnitude less likely than elastic scattering. It is remarkable, however, that the observed structures in both angular distributions have a very close similarity, similar to that found for the analogous atomic system $H^+ + Kr$ (Fig. 26*a*). With CO both differential cross sections show a primary rainbow around $\Theta_{lab} = 9^\circ$ originating from the strong proton–CO attraction. The rainbow structure is somewhat less pronounced in the charge-transfer channel. This is attributed to additional contributions from scattering events that have undergone charge transfer on the incoming trajectory, which only interacts with the purely repulsive $H + CO^+$ potential that does not give rise to any rainbow structure.

The system $H^+ + O_2$ has a highly exoergic charge-transfer channel ($\Delta E = +1.52$ eV), so that charge transfer is expected to be very probable even without vibrational excitation in the entrance channel. As with $H^+ + H_2$,

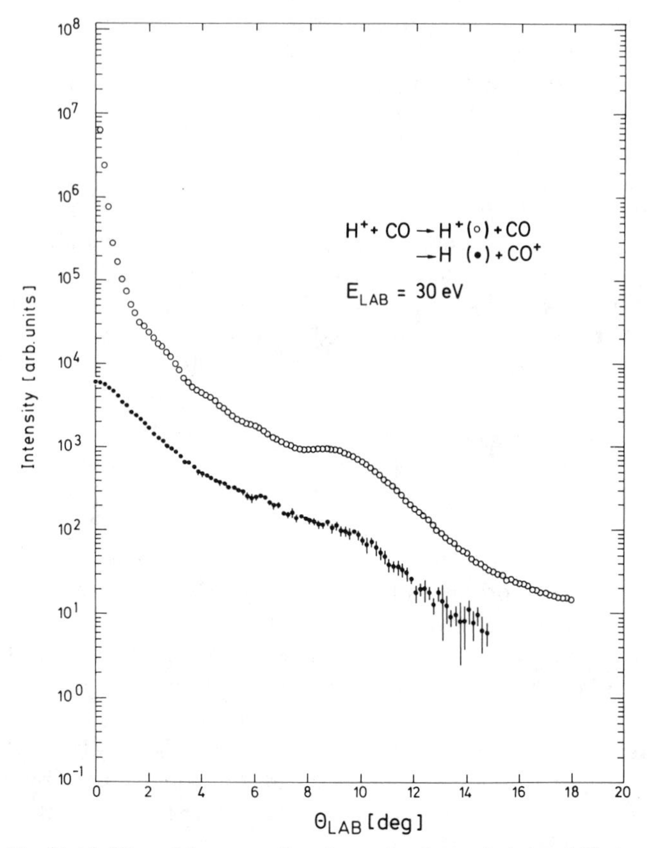

Figure 35. Total differential cross sections for proton (open circles) and H-atom detection (solid dots) in $H^+ + CO$ scattering at $E_{lab} = 30$ eV.[169] The endoergic (-0.42 eV) charge-transfer process is almost an order of magnitude less probable than elastic scattering. Note the appearance of a primary rainbow structure in both cross sections and weak secondary rainbows in the proton distribution.

the time-of-flight spectra were used to decompose the total differential angular distributions to obtain the vibrational state-to-state differential cross sections for inelastic scattering. The results shown in Fig. 36 for $E_{lab} = 23.7$ eV reveal a rainbow at about $\Theta_{lab} = 11°$, which is also observed in some of the inelastic channels.[173] The vibrational excitation probabilities show the expected fall-off with increasing final vibrational quantum number. The charge-transfer channels show much more complex angular distributions (Fig. 37). When corrected for the reduced H-atom detection efficiency, the total cross sections are nearly equivalent to those for proton scattering. This suggests that the

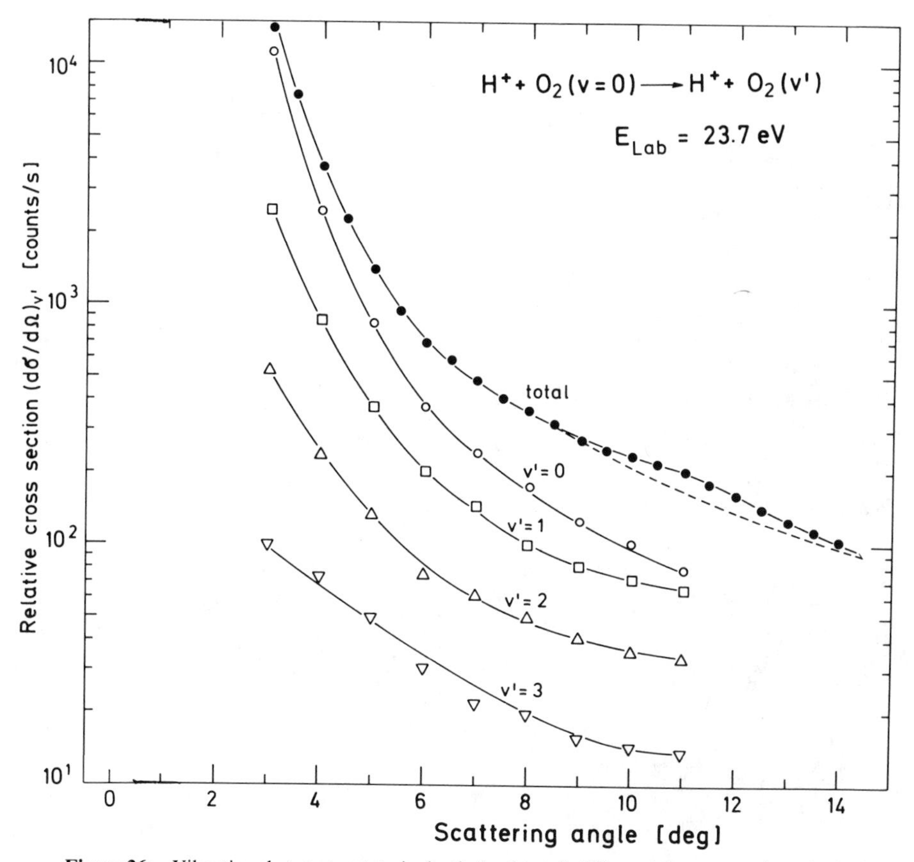

Figure 36. Vibrational state-to-state inelastic (and total) differential cross sections derived from the transition probabilities determined from time-of-flight spectra for $H^+ + O_2$ scattering at $E_{\text{lab}} = 23.7$ eV.[173] The total cross sections (upper curve) exhibit a weak rainbow maximum at about $\Theta_{\text{lab}} = 11°$. The dashed line indicates an extrapolated smooth curve to emphasize the enhancement of the total cross section in the rainbow region.

probability for charge transfer is $P_{ct} \approx \frac{1}{2}$. In charge transfer the most probable final state at small angles, $v_f = 3$, is reduced to $v_f = 1$ at intermediate angles. It is interesting to observe that transitions to higher final states are greatly enhanced near the rainbow.

According to Eqs. (28) and (29) we expect the transition probability to depend on the Franck–Condon factors. A direct comparison of the time-of-flight spectra at intermediate angles shows that, indeed, the final-state distributions are very similar to the Franck–Condon factors. Figure 38 compares the average energy transfer evaluated from the observed final-state

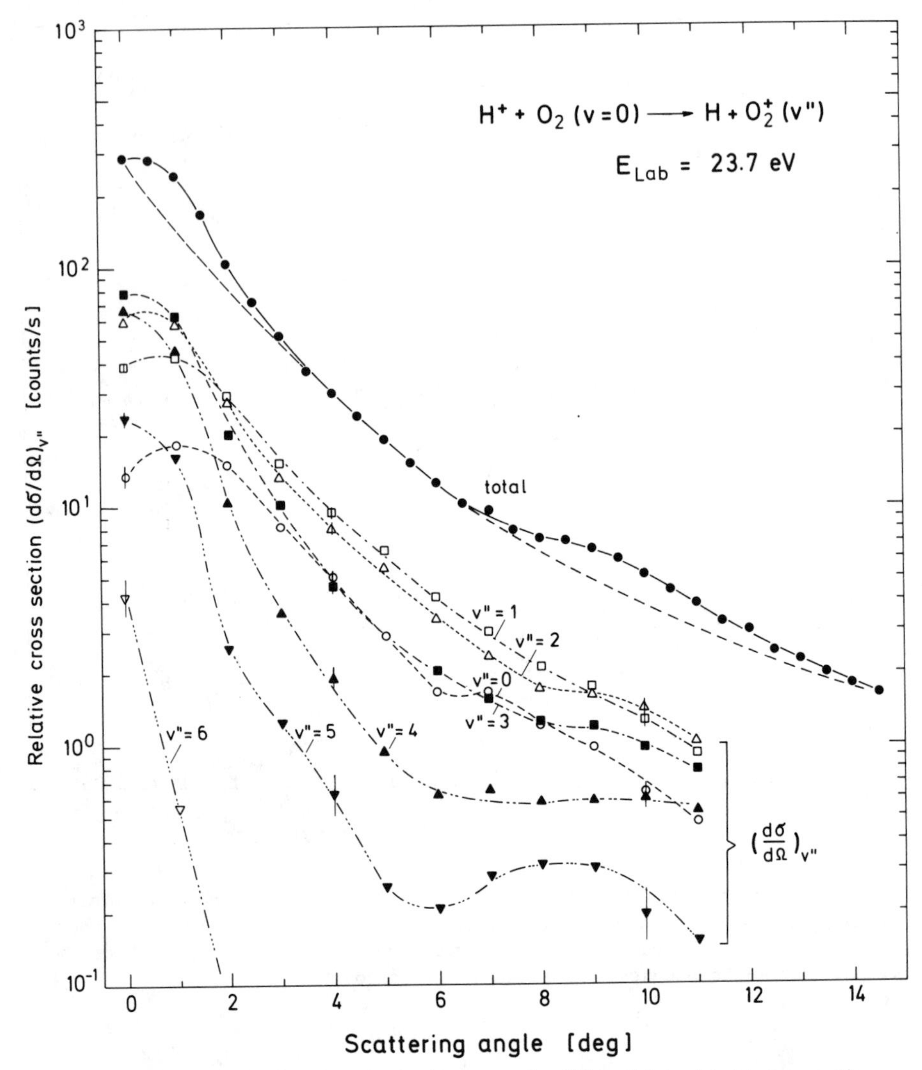

Figure 37. Vibrational state-to-state charge-transfer differential (and total cross sections derived from the transition probabilities determined from the time-of-flight spectra of H atoms in $H^+ + O_2$ scattering at $E_{lab} = 23.7$ eV.[173] As in Fig. 36 a rainbow maximum is observed in the large-angle region ($\Theta_{lab} \approx 10^\circ$) of the total-cross-section curve. At small angles ($\Theta_{lab} \approx 1^\circ$) a similar maximum is exhibited in the total as well as the lowest ($v_f = 0$–2) state-to-state cross sections. Typical error estimates of the experimental data are given by the vertical bars at $\Theta_{lab} = 0^\circ, 4^\circ$, and 10°.

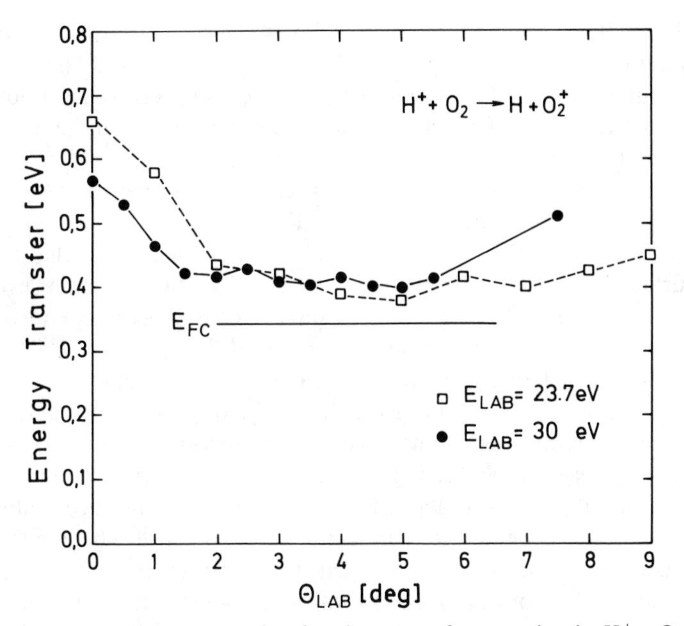

Figure 38. Average energy transfers for charge-transfer scattering in $H^+ + O_2$ at two different energies are compared to values predicted from the Franck–Condon distribution.[169,173] At intermediate scattering angles the experimental average energy transfers are quite close to the Franck–Condon value. In grazing collisions vibrationally excited ${O_2}^+$ states are enhanced by near-resonant charge transfer. The increase at large scattering angles is attributed to prior excitation of the O_2 molecule by the approaching proton.

distributions for two collision energies with the constant value predicted for a Frank–Condon distribution.[169,173] For final angles $2° \leqslant \Theta_{lab} \leqslant 6°$ the measured energy transfers are about 20% larger than predicted. Great differences are observed at smaller and larger scattering angles. The deviations to larger energy transfer at smaller angles can easily be understood by noting that for grazing collisions, corresponding to large internuclear distances, only near-resonant vibrational levels of O_2^+ are accessible via direct curve crossings in the vibronic network scheme, cf. Fig. 23. By reducing the finite angular resolution from presently $\Delta\Theta = \pm 0.5^0$ to much smaller values so that any appreciable momentum transfer can be ruled out, it should be possible to observe the resonant excitation of the $v_f = 6$ state. Thus, in the limit of $\Theta_{lab} \rightarrow 0$ very narrow final vibrational state distributions are anticipated.

The increase in energy transfer at large angles also has a simple explanation. As suggested earlier[148] at close encounters the electric field of the proton will lead to a stretching of the internuclear distance in the O_2 molecule. Trajectory calculations reveal that the bond may be stretched by as much as 0.02 Å, which is quite significant when compared to the difference in the equilibrium bond distance in O_2 ($r_e = 1.208$ Å) and O_2^+ ($r_e = 1.116$ Å), which is only 0.092 Å.[184] The stretching of the O_2 bond thus has a noticeable effect on the calculated Franck–Condon factors, which then agree very well with the experimental transition probabilities for $\Theta_{lab} \geqslant 4°$.[173] Thus, the increase in energy transfer at larger scattering angles can be attributed to temporary bond stretching during the collision.

A more quantitative theory requires a knowledge of the potential hypersurfaces. Early *ab initio* calculations of certain geometries of the interaction potential[185,211,232,234,239] have recently been extended to the most important parts of the entire potential hypersurface.[84,98,199] The coupling between charge-transfer states vanishes in C_∞ and in C_{2v} geometries and is strong for intermediate geometries, C_s. In collinear orientation C_∞ initial bond elongation and consecutive compression upon proton approach compete with each other giving rise to double-well potentials. Both the crossing seam and the well depth have much more pronounced vibrational and orientational dependence than in the $H^+ + H_2$ case. The availability of these potential hypersurfaces has led to accurate quantum-mechanical calculations.[14,204] Both approaches will be presented in the second part of this volume is detail. In contrast to the quantum-mechanical results on the $H^+ + H_2$ system,[13] the general angular behavior is simulated quite well. Although the relative magnitudes of final vibrational states v_f still show some discrepancies with the experimental data, the overall agreement is surprisingly good.

C. Triatomic Target Molecules

For triatomic target molecules several different vibrational normal modes may be excited. In the case of collisions where charge transfer can be excluded, the early state-resolved time-of-flight experiments on $Li^+ + CO_2$ revealed a propensity for the excitation of the infrared-active bending mode compared to the asymmetric and symmetric stretch modes.[49,110,122] These results were erroneously explained by a bond-dilution mechanism, but it now appears that they are more consistent with the induced-dipole mechanism discussed in connection with the proton–CF_4 experiments (cf. Sections III and IV). Proton inelastic scattering from CO_2[25,102,132,167] shows a different behavior, which has also recently been confirmed for H^-.[102] In this case the induced-dipole mechanism is not able to explain fully the high excitation probability of the infrared-active asymmetric stretch mode ν_3. Additional valence forces, which act in a concerted way on the ν_3 mode, similar to the bond-dilution mechanism have had to be invoked,[167] also, in order to explain the excitation of the ν_1 mode.[25,132]

A recent comparative study of proton and charge-transfer scattering from the equal mass targets CO_2 and N_2O provides additional insight into the dynamical details. Interesting differences are expected owing to the linear symmetric structure ($D_{\infty h}$) of CO_2 compared to the linear but asymmetric structure (C_{∞}) of N_2O. A further important aspect comes from differences in the nearby charge-transfer states for both systems. In $H^+ + CO_2$ charge transfer is slightly endoergic (−290 meV), whereas for $H^+ + N_2O$, it is exoergic (+700 meV). Thus, a Demkov–Nikitin-type coupling (Fig. 22*b*) is expected to occur in $H^+ + CO_2$, whereas in $H^+ + N_2O$ direct curve crossing occurs (Fig. 22*a*). Low-lying electronic states in CO_2^+ are of significant importance only at scattering energies in the keV region as has been shown by recent calculations.[114]

Despite these differences the total differential cross sections of inelastic scattering and charge-transfer scattering for both $H^+ + CO_2$ and $H^+ + N_2O$ exhibit essentially the same angular behavior as observed for $H^+ + CO$. The total angular distributions for both systems reveal nearly equal intensities for the two competing electronic channels throughout the entire angular range, $0° < \Theta_{lab} < 18°$. For $H^+ + CO_2$ the vibrational transition probabilities into the dominant asymmetric stretch mode ν_3 as a function of scattering angle are shown in Fig. 39. Open symbols denote inelastic scattering and solid symbols charge transfer. The small differences between the curves belonging to the same overtones in CO_2 and CO_2^+ are indicated by the hatched areas. Surprisingly, there is almost a perfect *conservation* of vibrational excitation distributions during charge transfer.

To explain this observation we note that as in the case of $H^+ + O_2$ the transition probabilities are largely governed by the Franck–Condon principle in accord with the large collision energies [cf. Eqs. (28) and (29)]. The generalization of the Franck–Condon principle to electronic transitions in *symmetric* polyatomic molecules (like CO_2) requires for a nonzero vibrational overlap integral $|\langle \phi_{kn} | \phi_{k'n'} \rangle|^2$ that the integrand $\phi_{kn}^* \cdot \phi_{k'n'}$ must be totally symmetric. Thus, for an antisymmetric stretch mode such as ν_3 the following selection rule between the *n*th overtones is expected to hold:

$$\Delta n_3 = 0, \pm 2, \pm 4, \ldots \qquad (33)$$

Furthermore, if both electronic states participating belong to the same point group, as is the case in $CO_2 \rightarrow CO_2^+$, the intramolecular potential minima along the asymmetric normal coordinate are identical irrespective of any contraction or expansion that conserves symmetry.[108] In this case, the Franck–Condon integrals will be small for any $\Delta n \neq 0$ even for large differences in the force constant, corresponding to large differences in the vibrational frequencies. As an example, even for a molecule with a difference of a factor of 2 in the respective fundamental frequencies, Herzberg calculates

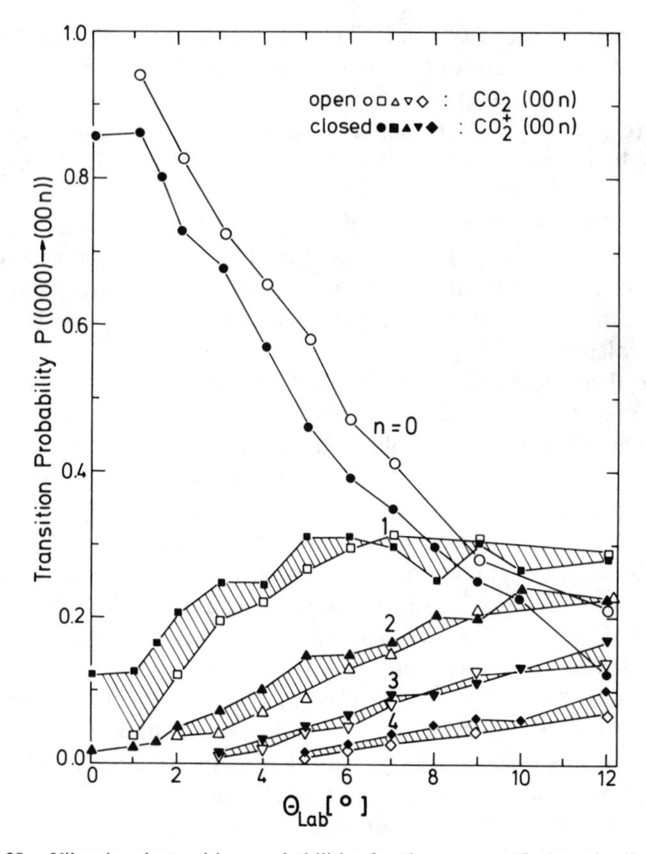

Figure 39. Vibrational transition probabilities for the asymmetric stretch vibration ν_3 of CO_2 (open symbols, inelastic scattering) and of $CO_2{}^+$ (solid symbols, charge transfer) in $H^+ + CO_2$ collisions.[167] The hatched areas denote the small deviations between the transition probabilities for the same vibrational state in the two product channels $H^+ + CO_2$ and $H + CO_2{}^+$. Thus, one can speak of mode conservation under charge transfer in the $(H + CO_2)^+$ system.

94.4% of the intensity to be localized in the $\Delta n = 0$ transition.[108] In CO_2 and CO_2^+ the frequencies of ν_3 differ only by about 35% so that the $\Delta n = 0$ selection rule should be very strong. Supporting evidence comes from photoelectron spectra showing the elastic transition dominating and revealing no noticeable $2\nu_3$ peak.[226]

Therefore, we conclude that in the $H^+ + CO_2$ system there is no change of the antisymmetric stretch mode excitation upon charge transfer. The processes of vibrational excitation and charge transfer are thus completely decoupled because of symmetry constraints. This is illustrated in Fig. 40

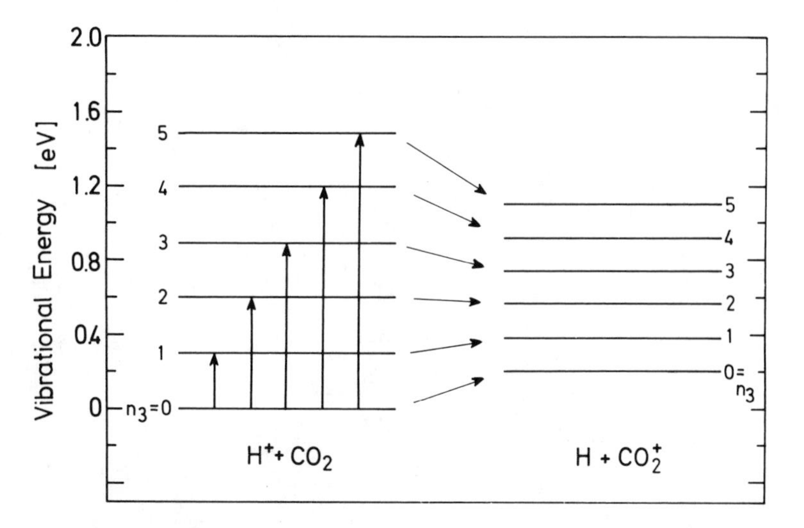

Figure 40. Simple reaction scheme for $H^+ + CO_2$ inelastic and charge-transfer collisions based on a v_3 mode energy-level diagram and emphasizing the $\Delta n_3 = 0$ selection rule for charge transfer.[167] The right part has been shifted up by 0.19 eV to take account of the charge-transfer endoergicity. The arrows indicate the assumed two-step model for explaining the close similarities between the measured H^+ and H-atom energy-loss spectra. Note the difference to the model used to explain $H^+ + H_2$ (Fig. 33).

where the overall collision is split into two steps: vibrational excitation of the dominating ν_3 mode and charge transfer conserving the vibrational quantum numbers. Note the difference with respect to $H^+ + H_2$ where vibrational excitation was required to overcome the electronic energy defect for charge transfer (Figs. 33 and 34).

The situation for scattering of protons from the asymmetric N_2O is expected to be different since charge transfer is exoergic by 700 meV and a direct curve crossing can occur at large internuclear distances. Nevertheless, the vibrational transition probabilities are again dominated by the asymmetric stretch mode ν_3 (Fig. 41). The same similarities for inelastic excitation (open symbols) and charge transfer (solid symbols) as observed for $H^+ + CO_2$ (Fig. 39) are found here as well. The very similar intramolecular bond lengths as well as vibrational frequencies of both N_2O and N_2O^+ suggest that the corresponding potentials are also very similar. As a consequence the vibrational eigenfunctions with different vibrational quantum numbers in the upper and lower state are almost orthogonal with respect to each other and the Franck–Condon overlap integrals will only be different from zero if there is no change in the quantum numbers, $|\langle \phi_{kn} | \phi_{k'n'} \rangle|^2 \approx \delta_{nn'}$. This is in agreement with a small Franck–Condon

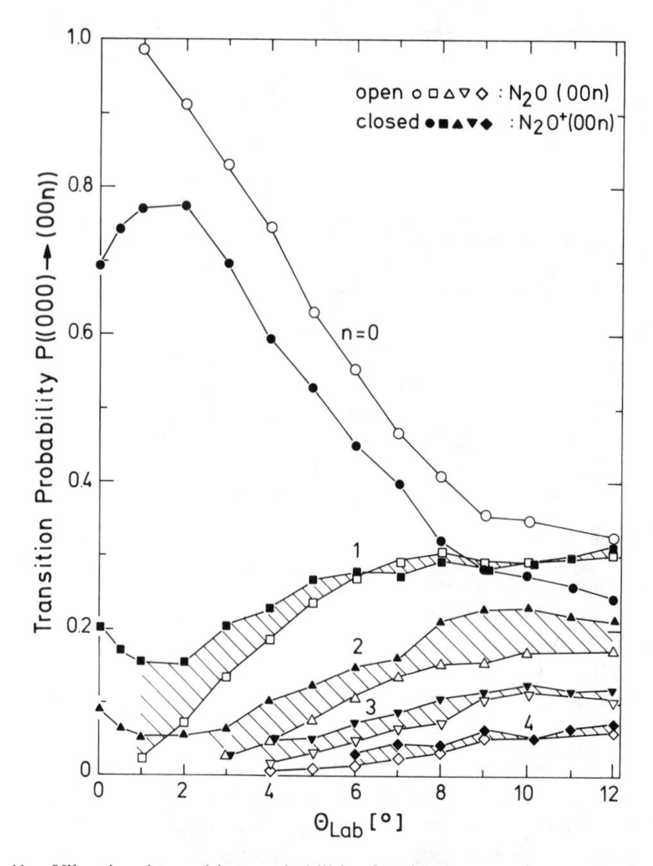

Figure 41. Vibrational transition probabilities for the asymmetric stretch vibration ν_3 of N_2O (open symbols, inelastic scattering) and of N_2O^+ (solid symbols, charge transfer) in $H^+ + N_2O$ collisions.[167] The hatched areas denote the small deviations between the transition probabilities for the same vibrational state in the two product channels $H^+ + N_2O$ and $H + N_2O^+$. Note the close agreement between the two data sets except at $\Theta_{lab} \leqslant 2°$. These deviations occur in grazing collisions where the charge-transfer exoergicity cannot be neglected anymore.

factor of only 0.03 for the photoelectron transition N_2O ($n_3 = 0$)$\rightarrow N_2O^+$ ($n'_3 = 1$) compared to 0.91 for the transition to N_2O^+ ($n'_3 = 0$).[226] Thus, the dynamics are controlled by the same constraints as in CO_2 even though the symmetry is broken. In the incoming trajectory, when the N_2O molecule is still unexcited, charge transfer is largely restricted to the ground vibrational states of the N_2O^+ molecular ion. Vibrational excitation of N_2O and N_2O^+ occurs in the subsequent close approach and is then transferred from either the molecule to the molecular ion or vice versa in the curve-crossing transition in the outgoing trajectory. With the reasonable assumption of a single passage transition probability of $P \approx \frac{1}{2}$, essentially the same vibrational distribution results in both final electronic channels as observed in Fig. 41.

Whereas the previous examples reveal the overriding importance of symmetry constraints as reflected in the Franck–Condon distributions, the next example involving H_2O indicates that other factors can also become important. In $D^+ + H_2O$ scattering the vibrational levels in H_2O^+ which exhibit a small energy defect with the vibronic entrance channel $D^+ + H_2O$ ($n = 0$) were found to be strongly enhanced. [71,72]. This is illustrated in Fig. 42 where the observed transition probabilities are compared with the photoelectron spectra. Note, that contributions come not only from the exoergic ionic ground state $\tilde{X}\,^2B_1$ but also from the first excited state $\tilde{A}\,^2B_1$, which is endoergic. In order to account for these findings, we have to realize that the observed vibrational excitation is mainly brought about by collisions with large impact parameters corresponding to the small scattering angle of $\Theta_{lab} = 1°$. Thus, the impulsive momentum transfer and the vibrational excitation are also very small. Consequently, the particular vibrational mode excited is determined predominantly by the symmetry and bonding properties of the orbital from which the electron is removed ($1b_1$ nonbonding orbital and $3a_1$ bonding orbital in the case of the $\tilde{X}$ and $\tilde{A}$ electronic states, respectively), just as in photoionization. In the charge-transfer process the Franck–Condon factors are weighted additionally by the distribution of nonadiabatic transition probabilities [cf. Eqs. (28) and (29)]. This latter distribution is centered at $\Delta E_{cm} = 0$ with an energetic width that is roughly inversely proportional to the collision time as suggested by the uncertainty principle. Thus, vibrational transitions with small energy defects are enhanced and those with large energy defects are attenuated[96] as observed.

D. Polyatomic Target Molecules

Prior to the introduction of the technique for detecting directly hydrogen atoms produced in charge-transfer collisions, there were a number of observations that were interpreted as providing evidence for the effect of charge transfer on vibrational excitation. In 1975 a large amount of unresolved energy transfer with $\Delta E = 3$–$4\,\text{eV}$ was observed in collisions of protons with CH_4 at $E_{cm} = 20\,\text{eV}$, which was attributed to an intermediate charge transfer.[78] $H^+ + CH_4$ was already known to have a rather large charge-transfer cross section of about 30 Å at $E_{lab} = 30\,\text{eV}$.[243] This explanation is also consistent with the smaller energy transfer of about $\Delta E = 1.2\,\text{eV}$ observed in $Li^+ + CH_4$ collisions at $E_{cm} = 2.7$–$6.5\,\text{eV}$.[51] Surprisingly, other experiments on $H^+ + CH_4$ at $10\,\text{eV}$[56] and at $30\,\text{eV}$ at small scattering angles[133] did not reveal the same large energy transfer as seen at $E_{cm} = 20\,\text{eV}$.

More direct access to the charge-transfer process is now available from

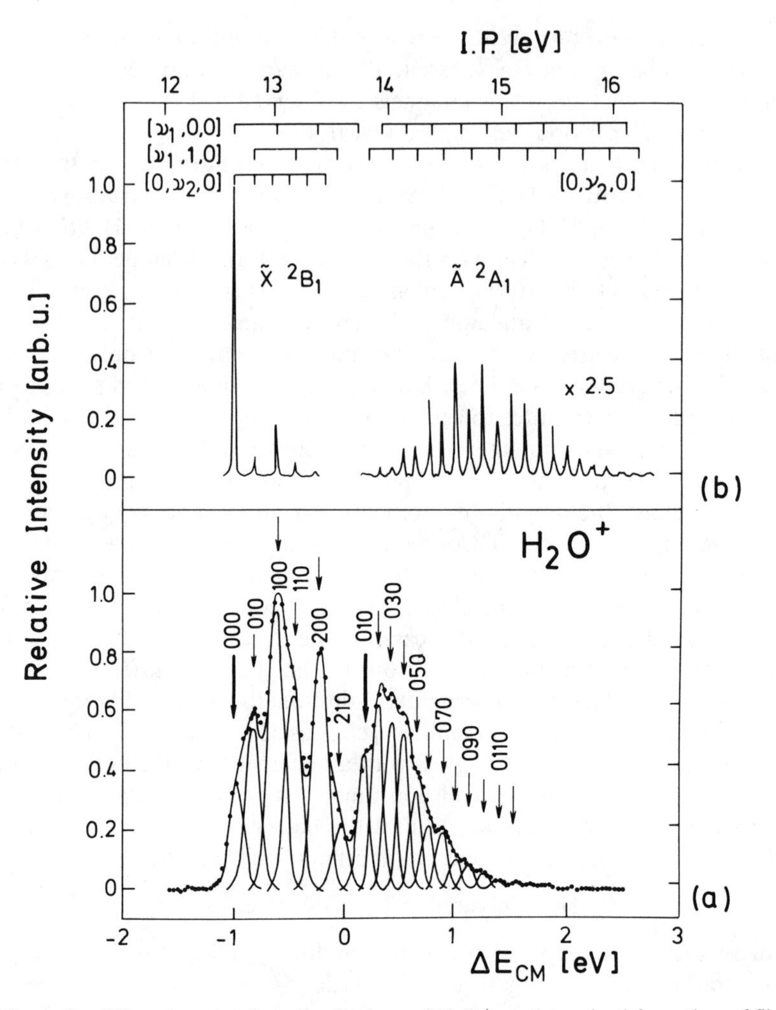

Figure 42. Vibronic-energy-loss distributions of H_2O^+ as determined from time-of-flight spectra of the D atoms produced in $D^+ + H_2O$ charge-trannsfer collisions at $E_{lab} = 30.1$ eV and $\Theta = 1^{\circ}$[72] (lower part) and from photoionization of H_2O (upper part).[187] The arrows indicate the assigned vibrational energy levels . Note the enhanced intensities of those vibrational levels, which are populated in near-resonant charge transfer (ΔE small).

the scattering cross sections measured by H-atom detection. As with the smaller targets, we expect charge transfer to depend sensitively on the extent of exoergicity for the individual system. This is clearly confirmed by the comparison of the H^+- and H-atom angular distributions scattered from CH_4, partially fluorinated methane and CF_4 shown in Fig. 43. With

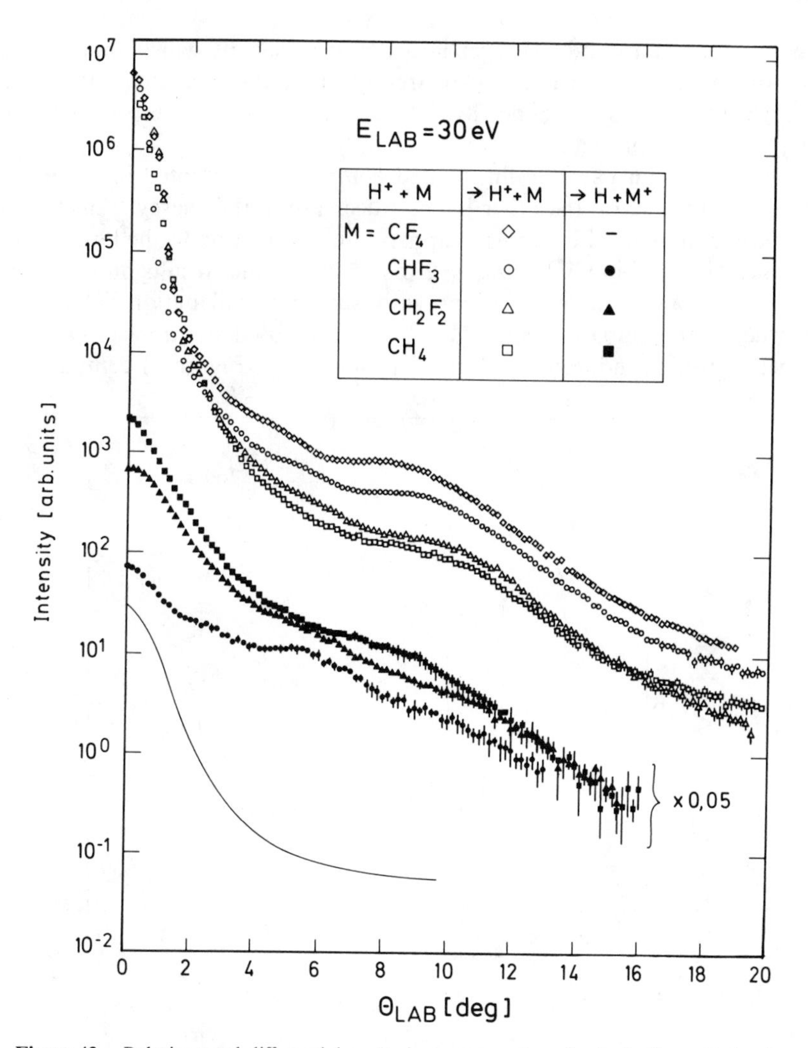

Figure 43. Relative total differential scattering cross sections for inelastic scattering (open symbols) and charge transfer (solid symbols) in $H^+ + CF_4$, CHF_3, CH_2F_2, and CH_4 at $E_{lab} = 30$ eV.[169] Decreasing fluorination increases the charge-transfer probability owing to lowering of the ionization potential. Note that the H-atom signal for charge transfer from CF_4 was less than the H-atom background and was therefore not measurable.

increasing number of fluorine atoms, the ionization potential I rises by about 2 eV from $I = 12.6$ eV for CH_4 to $I = 15.77$ eV for CF_4.[100] For CF_4 no H-atom signal could be detected above the H-atom background indicated in Fig. 43. However, for CHF_3 ($I = 13.8$ eV), a sizable H-atom signal is observed at all angles. For CH_2F_2 and CH_4 even larger H-atom signals, which approach the proton intensities, are found. Apparently in the latter systems for which charge transfer is exoergic by 0.89 eV and 1.44 eV, respectively, the charge-transfer probability approaches 50%.

The results on partially fluorinated compounds presented in Section IV (Fig. 13) indicate, on the other hand, a decrease in the energy transfer with increasing number of H atoms in apparent contradiction to the large energy transfer observed in CH_4 at $E_{cm} = 20$ eV.[78] Also, other results on CH_4[55,133] would appear to be inconsistent with such an explanation. Thus, it is instructive to compare qualitatively the measured energy transfer with that expected for the induced-dipole mechanism (IDM). Figure 44 compares the

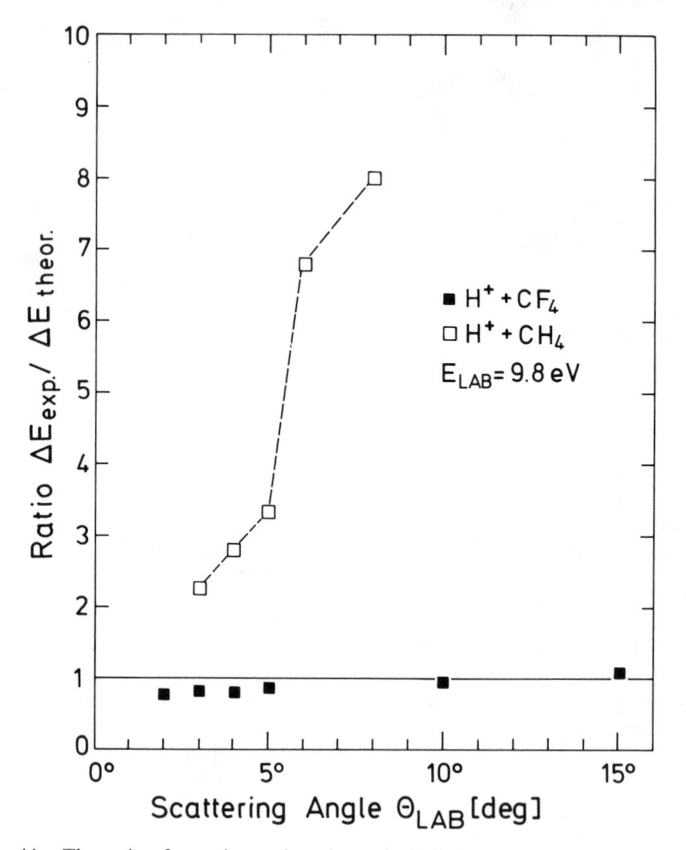

Figure 44. The ratio of experimental to theoretical vibrational energy transfer by proton impact at $E_{lab} = 9.8$ eV is compared for CF_4 and CH_4 target molecules. The calculations are based on the simple forced-harmonic-oscillator model. Whereas this model nicely explains the results for CF_4, large deviations to greater energy transfer are observed for CH_4. This has been attributed to the temporary formation of a CH_4^+ transfer state [this work].

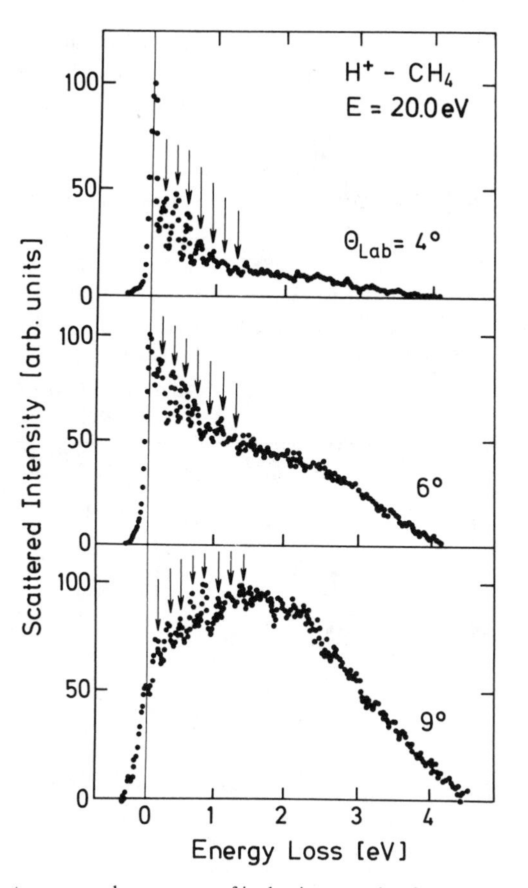

Figure 45. Proton energy-loss spectra of inelastic scattering from CH_4 at $E_{lab} = 20\,eV$. The arrows indicate the positions of the resolved energy-loss maxima.[34] Note the almost bimodal character of the distributions consisting of a resolved part at small energy transfers and a high-energy-transfer tail that increases in intensity with scattering angle. The second part is presumed to orginate from intermediate charge transfer.

measured average energy transfer for both CH_4 and CF_4 with that expected for the IDM as a function of scattering angle. There it is seen that, whereas CF_4 agrees nicely with the theory, the deviations in the case of CH_4 are very large and increase steeply with increasing scattering angles. Thus, when proper account is taken of the much smaller dipole derivative of CH_4, this system does indeed show much more energy transfer than can be explained with the IDM also at the lower energy of 10 eV. The deviations increase dramatically with scattering angles, thus also explaining the results obtained at $E_{cm} = 30\,eV$ and $\Theta_{cm} = 0°$.[133]

Despite the good resolution, the energy-loss peaks indicated by the vertical arrows in Fig. 45 cannot be uniquely assigned to specific vibrational states of the CH_4 molecule, mainly because of the near degeneracy of the ν_1 (361 meV) and ν_3 (375 meV) and the ν_2 (189 meV) and ν_4 (162 meV) manifolds. However, some preliminary data taken with the new sector field machine indicate a dominance of ν_3 mode excitation plus combination bands of the type $n_3\nu_3 + n_4\nu_4$, where $n_4 = 0, 1, 2$.[153,230] Additional, *minor* contributions of ν_2 have already been resolved previously[133] and are seen in the preliminary data as well, but there is no evidence for any major contribution of ν_2. Additional support comes from spectra taken at $E_{lab} = 30\,eV$ showing an alternating peak height distribution caused by the fact that every other peak originates from a weaker $n\nu_3 + 1\nu_4$ transition as identified in Fig. 46.[169] The

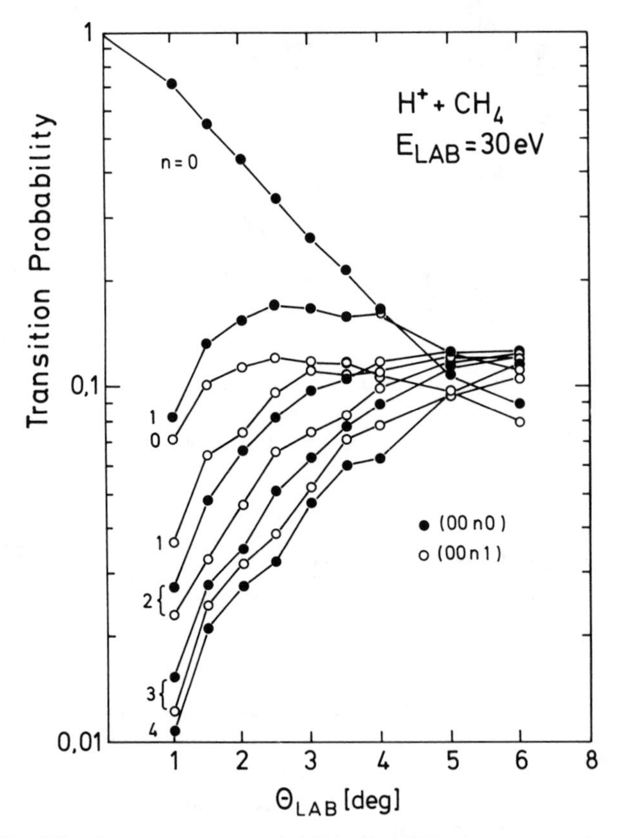

Figure 46. Vibrational transition probabilities for inelastic scattering in $H^+ + CH_4$ at $E_{lab} = 30\,eV$.[169] The dominating vibrational modes contributing are ν_3 (00*n*0) and ν_4 (00*n*1) as in the case of $H^+ + CF_4$. Note that the ν_3 excitation is slightly stronger than the ν_4 contributions throughout the entire angular range and for all overtones *n*.

two identified major contributions, ν_3 and ν_4, are those which are expected to be excited by the induced-dipole mechanism. However, whereas at small scattering angles the relative probabilities might be consistent with a Poisson distribution, considerable deviations appear at larger angles (Figs. 45 and 46).

To clarify the possible effect of charge transfer, time-of-flight spectra of H atoms produced by charge transfer in $H^+ + CH_4$ were measured at $E_{lab} = 30$ eV. The results are shown in Fig. 47 where they are compared with a photoelectron spectrum,[30] which provides an approximate measure for the Franck–Condon factors. Again, the charge-transfer distributions show a propensity for transitions with small energy difference with respect to near-resonant charge transfer. Unfortunately, no vibrational structure is resolved in the charge-transfer spectra. This is attributed to the extensive

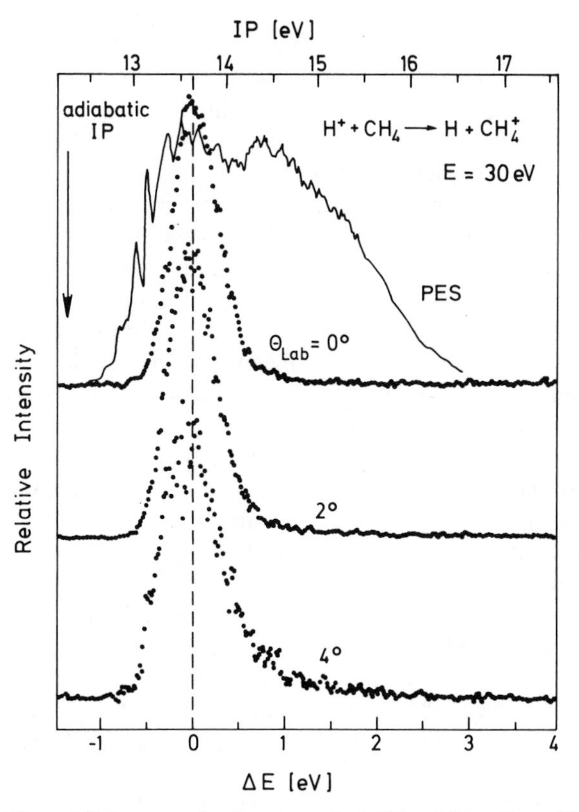

Figure 47. Time-of-flight spectra for charge transfer in $H^+ + CH_4$ at $E_{lab} = 30$ eV (solid dots) compared to a photoelectron spectrum (solid line).[34] In contrast to the photoelectron spectrum, the internal-state distributions of CH_4^+ formed in charge-transfer collisions are quite narrow and centered at the resonance energy $\Delta E = 0$ eV.

Jahn–Teller distortion[23] of the CH_4^+ (T_d) ion leading to a relaxation into lower-lying electronic states of C_{2v}, D_{2d}, or C_{3v} symmetry.[213] The vibrational patterns of all these states overlap and any structure is apparently washed out by the apparatus resolution ($\Delta E > 60$ meV).

Because of the difference of the ionization potentials ($\Delta V = 1.44$ eV), the CH_4^+ ions have a high degree of internal vibronic excitation. Part of the vibronic excitation observed in charge transfer lies beyond the dissociation limits of the CH_4^+ ion. The integral-energy-loss distribution of all scattered CH_4^+ ions at $E_{lab} = 30$ eV was derived from time-of-flight spectra for charge-transfer collisions that were weighted with the angular distributions. The results shown in Fig. 48*a* reveal clearly that the internal energy is sufficient to produce copious amounts of CH_3^+ and even CH_2^+ ions. Complementary fragmentation mass spectra of CH_4^+ ions from such charge-transfer collisions, Fig. 48*b*,[34] show a size distribution that is in remarkable agreement with the relative areas under the corresponding parts of the excitation function of Fig. 48*a*. The good agreement indicates that this combination of experimental techniques can be used for a detailed study of molecular-ion-fragmentation processes as a function of internal energy transfer for more complex molecules.[153]

Theoretical considerations in terms of vibronic symmetry correlation[34] strongly support the hypothesis of an ionic intermediate promoting vibrational excitation corresponding to an internal vibronic mechanism. Namely, $H^+ + CH_4$ couples to charge-transfer states, $H + CH_4^+(C_{2v})$, only when ν_3 or ν_4 vibrations, both of T_2 symmetry, have been excited. These infrared-active modes can be initially excited by the long-range charge-induced-dipole interaction. These vibrations distort the CH_4^+ molecular ion into C_{3v} symmetry, which favors additional excitation. In the course of the collision some of the $CH_4^+(C_{3v})$ ions charge transfer back to become CH_4 molecules, which retain the large amount of vibrational excitation of the temporary formed CH_4^+ ions. The complex mode coupling occurring in $H^+ + CH_4$ collisions has been extensively discussed in terms of the quasimolecular mechanism.[34,35]

Vibrational excitation in all the polyatomic molecules presented so far was largely dominated by the expected induced-dipole interaction. In the system to be discussed next, $H^+ + C_2H_2$, there is evidence that an additional excitation is involved[5,169] via bond dilution and IVM. Charge transfer is exoergic by the large amount of 2.2 eV and low-lying electronic states are not present.

For this system it was possible to construct a potential hypersurface for the interaction, which was based on the measured angular distributions for both channels as well as on the available quantum-chemical calculations.[43,138,245] Figure 49 shows the general features of the potential

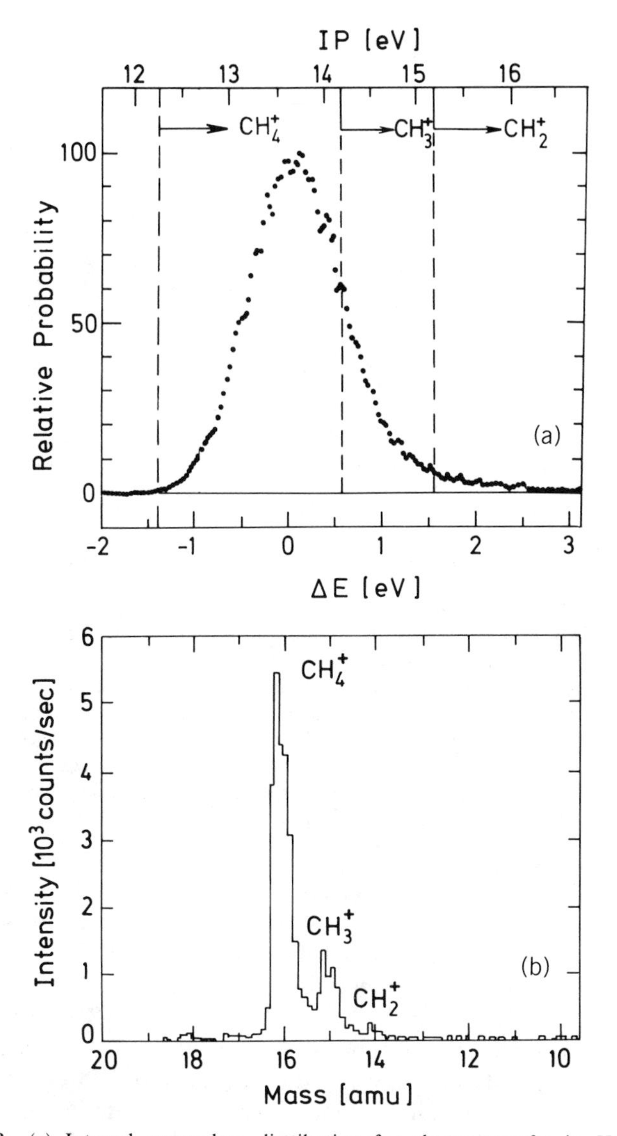

Figure 48. (*a*) Integral energy-loss distribution for charge transfer in $H^+ + CH_4$ at $E_{lab} = 30\,eV$ together with the appearance potentials of possible fragment ions emerging from CH_4^+.[34] The dashed vertical lines indicate the appearance potentials of the possible products of the CH_4^+ fragmentation processes. (*b*) Fragmentation mass spectrum of product ions out of charge-transfer collisions of protons with methane at 30 eV.[34] The observed intensities agree nicely with what one would expect from the excitation function, Fig. 48*a*.

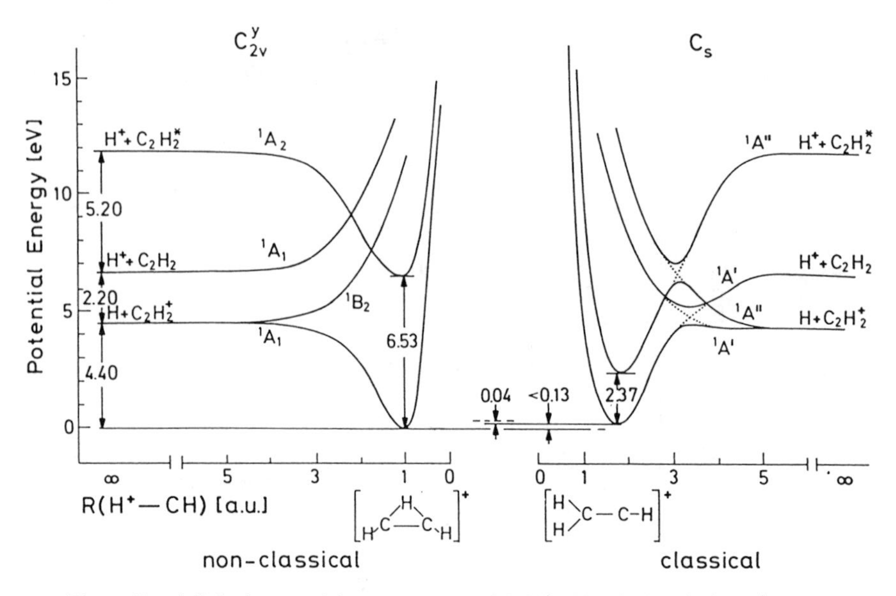

Figure 49. Adiabatic potential energy curves of $C_2H_3^+$. The abscissa is the H^+ to C atom separation in atomic units. The C_s curves are adapted from the calculations for the O_2H^+ surfaces.[98,199,239] The C_{2v}^{y} curves are taken from recent MRD-C1 calculations.[245] The full lines are meant to be qualitative representations of the curves. The proton affinity is taken from the experimental data in Ref. 34, the charge-transfer energetics from Ref. 184, and the acetylene electronic excitation energies from Ref. 117, 226. The vertical excitation energies for the $C_2H_3^+$ complexes are taken from Ref. 235. The barrier height and energy separation between the two $C_2H_3^+$ structures is still a topic of discussion. By most accounts, the bridged structure is more stable by about 0.13 eV and the barrier is about 0.04 eV.[21,245]

hypersurface for two collision geomtries, C_{2v}^{y} and C_s, with the *yz* plane taken as the collision plane, which lead to stable $C_2H_3^+$ isomers. The positions of the ground-state minima and avoided crossings for the C_s orientation are estimated from the potential energy surfaces calculated for the nearly isoelectronic O_2H^+ system.[98,199,239] These for the C_{2v}^{y} geometry were recently obtained in a very extensive, *ab initio* calculation of the full surface.[245] The abscissa in Fig. 49 has been taken as the distance between the H^+ and one of the carbon atoms. The minima in the excited-state surfaces are taken from the calculations of the geometries of $C_2H_3^+$. The equilibrium $(H\cdots C_2H_2)^+$ length increases by 3% in C_{2v}^{y} symmetry and decreases by 0.3% in C_s over those in the ground states.[138] The energetics are taken from various sources, all of which are referenced in the figure caption. For both approaches in Fig. 49 the long-range ion-induced-dipole attraction in $H^+ + C_2H_2$ leads diabatically to a deep attractive well (≈ 6.6 eV) in which the $C_2H_3^+$ complex is bound by the strong proton affinity. Whereas the C_{2v}^{y} approach leads to the nonclassical bridged $C_2H_3^+$ complex, approach through C_s leads to a collision complex resembling the other known stable structure of protonated acetylene, the classical vinyl cation of C_{2v}^z symmetry. The difference between the two $C_2H_3^+$ equilibrium conformations is between 40 and 130 meV.[21,43,138,245] The barrier for interconversion is of the same magnitude. The H atoms in the complex can participate in a sort of merry-go-round motion about the C–C bond.[35] The similarity in the potential well depths and well locations indicate that the interaction potential at least in the well region is actually quite isotropic[245] and the initial direction of approach should not be very important in determining the interaction. This is probably the reason why it has been possible to observe a rainbow in the proton angular distributions.

The other surfaces of interest are those for charge transfer, of which there are two for each collision geometry as discussed above. One leads to formation of $C_2H_2^+$ with an unpaired electron in the π_{ux} orbital. Because the polarizability of an H atom is small, the $H + C_2H_2^+$ surfaces are dominated by the repulsive parts. Only the state with the half-filled $C_2H_2^+$ π orbital in the collision plane couples with the reactant surface and leads to an avoided crossing. As a result we predict a weak potential well of about 1–2 eV in the entrance channel for the C_s geometry. This well depth correlates nicely with the observed rainbow in the proton distributions, which was attributed to a well of 1.2 eV depth.

Unlike other hydrocarbon systems there are no low-lying electronically excited states of $C_2H_2^+$ that might also couple to the reactant surface. This difference may explain the inefficient charge and energy transfer observed in this system compared to other hydrocarbons. Figure 49 indicates the estimated locations of the surfaces of the lowest excited states of C_2H_2 and $C_2H_3^+$. The experiments provide no evidence for the formation of $C_2H_2^{+*}$ in the time-of-flight data. This is agreement with earlier experiments on $H^+ + Xe$,[11] which have revealed that much higher energies are required for electronic ecitation than used here.

Despite the small amount of charge transfer (less than 10%) vibrational resolution was achieved both in charge transfer and in inelastic scattering at $E_{lab} = 30$ eV and vibrational transition probabilities to the different modes for inelastic excitation and for charge transfer was achieved. The inelastic probabilities strongly depend on the scattering angle, whereas the charge-transfer probabilities are almost invariant. The ratio of the relative probabilities for exciting the asymmetric C–H stretch mode, ν_3, and the C–C stretch mode, ν_2, are shown in Fig. 50. For inelastic scattering and at small angles the ν_3 mode has a much larger probability of excitation than the ν_2 mode. At 4^0, which is near the rainbow angle, the relative probabilities are reversed. This is consistent with the fact that the asymmetric C–H stretch mode ν_3 is infrared active and can therefore be excited by the long-range

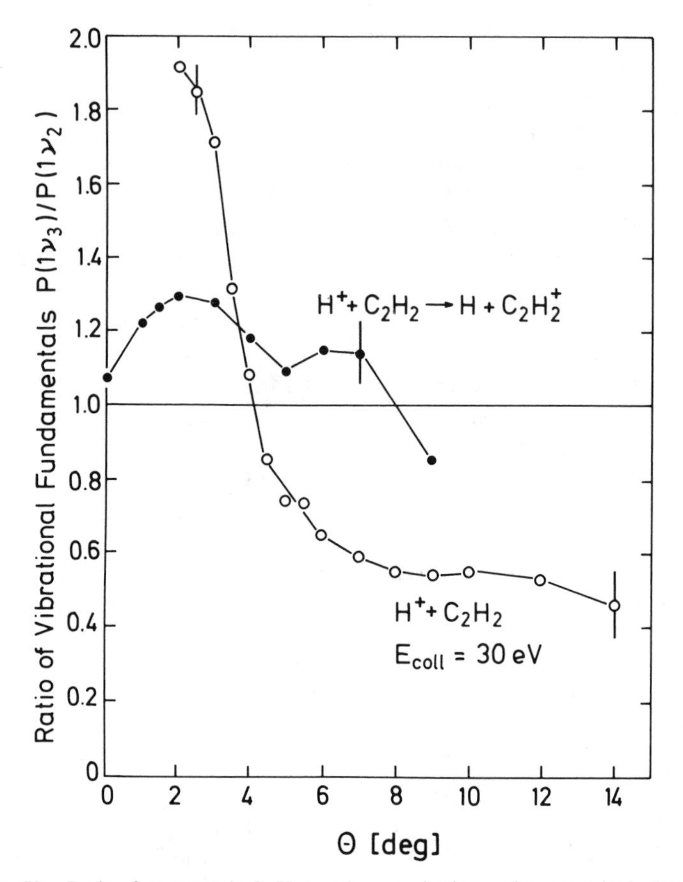

Figure 50. Ratio of asymmetric C–H stretch, ν_3 excitation and symmetric C–C stretch, ν_2 excitation in inelastic collisions (open symbols) and charge transfer (solid dots).[5,169] In grazing collisions (small scattering angles) the infrared-active ν_3 mode dominates the (in)elastic scattering by means of the induced dipole mechanism (IDM). Beyond 4° the symmetric ν_2 mode takes over owing to short-range bond-dilution interaction. In charge-transfer collisions the ratio of both modes is almost constant at an intermediate value.

charge-induced-dipole mechanism (IDM), whereas the symmetric C–C stretch mode ν_2 is inactive and cannot be excited at long ranges. Thus, it appears that another mechanism of short range is needed to explain the ν_2 excitation. This is probably the bond-dilution mechanism, since the protonated acetylene is known to have a considerably enlarged C–C bond distance[138] of 1.234 Å compared to 1.191 Å for the free C_2H_2 molecule. The expected different interaction ranges of both mechanisms can explain the different angular behavior shown in Fig. 50.

The larger hydrocarbons have additional low-lying electronic states, and these become accessible for excitation in charge-transfer collisions. Figure 51 compares a time-of-flight spectrum for charge transfer in $H^+ + C_2H_4$[169] (bottom) with the photoelectron spectrum of C_2H_4[117] (top). The photoelectron spectrum reveals excitation of a series of progressions of vibronic manifolds in different electronic states. The observed double structure in the charge-transfer spectrum coincides with excitation of the $X\,^2B_{3u}$ and $A\,B_{3g}$ electronic states of the molecular ion. As observed in the previously discussed hydrocarbons and H_2O, the energy-loss distributions are centered around $\Delta E = 0$ favoring near-resonant charge transfer. No evidence was found that any of the endoergic $C_2H_4^{+*}$ states are populated in charge-transfer collisions with protons.

As the contributions into the two lowest electronic states of $C_2H_4^+$ are well separated, it has been possible to determine from the spectra the relative total charge-transfer cross sections into the two electronic states. As shown

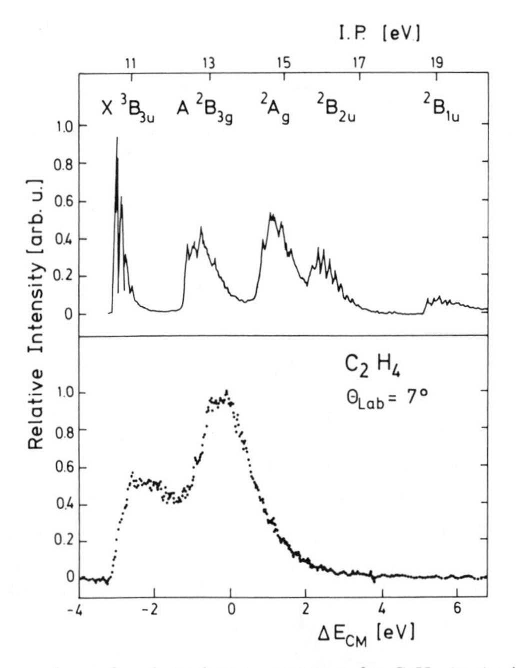

Figure 51. Comparison of a photoelectron spectrum for C_2H_4 (top) with a time-of-flight spectrum for charge transfer in $H^+ + C_2H_2$ collisions at $E_{lab} = 30\,eV$ and $\Theta_{lab} = 7°$.[169] Out of the at least five electronic states of the photoelectron spectrum, there are only two that contribute in proton-impact-induced charge transfer, namely, those that are accessible via near-resonant charge transfer with little energy transfer (ΔE small).

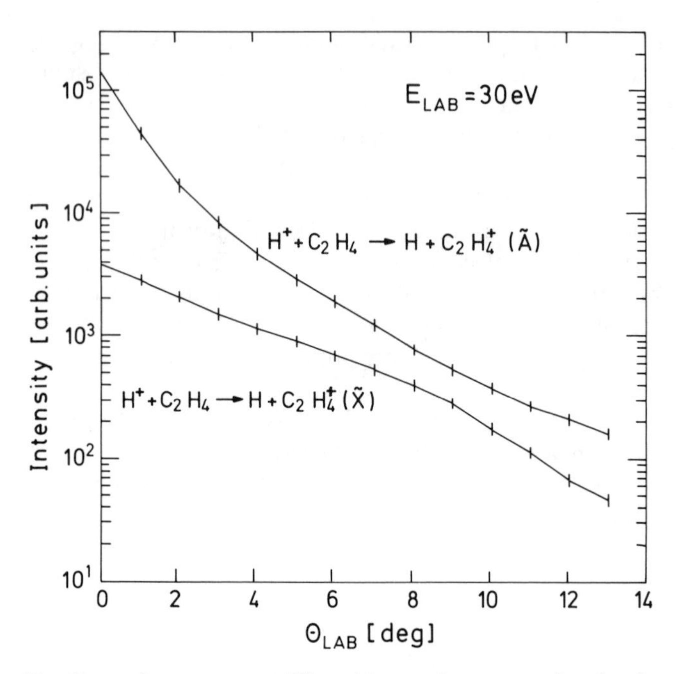

Figure 52. Electronic state-to-state differential scattering cross sections for charge transfer in $H^+ + C_2H_4$ collisions at $E_{lab} = 30\,eV$.[169] The $C_2H_4^+$ A state is enhanced in forward direction, since it can be populated by large-impact-parameter collisions owing to its small exoergicity of 1.2 eV, whereas the $C_2H_4^+$ X state's is 3.08 eV exoergic and requires small impact parameters to sense the curve crossing with the entrance channel. Thus, it shows a much weaker angular variation than the A state.

in Fig. 52 the first excited electronic state $\tilde{A}$ has a larger cross section throughout the entire angular range. It is especially enhanced in the forward direction by the near-resonance condition for charge transfer. Excitation of the electronic ground state $\tilde{X}$ of $C_2H_4^+$ is less dependent on the scattering angle, which is interpreted as an indication for a much shorter interaction range. Indeed, one expects a possible curve crossing to occur at much smaller distances than for the $\tilde{A}$ state owing to the much larger exoergicity.

Several other interesting effects occur in charge transfer in $H^+ + C_6H_6$.[169] Since the lowest ionization potential of benzene is only about 9.2 eV corresponding to an exoergicity of 4.40 eV, the ground electronic state is not excited in charge transfer with protons. Figure 53 shows once more a comparison of the energy-loss distributions with the photoelectron spectrum. The comparison is quite similar to that already observed for H_2O (Fig. 42). However, as seen for C_2H_4, transitions with a negative energy transfer are

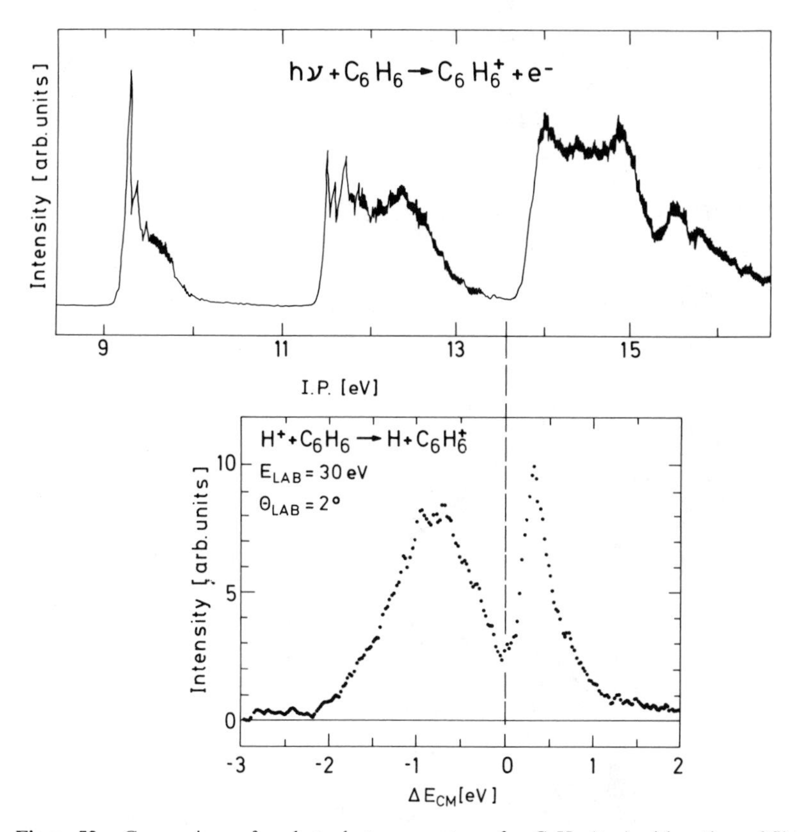

Figure 53. Comparison of a photoelectron spectrum for C_6H_6 (top) with a time-of-flight spectrum for charge transfer in $H^+ + C_6H_6$ collisions at $E_{lab} = 30\,eV$ and $\Theta_{lab} = 7°$.[169] The low-lying electronic states of $C_6H_6^+$ as shown in the photoelectron spectrum do not contribute in charge-transfer scattering with protons presumably due to the lack of a direct curve crossing with the $H^+ + C_6H_6$ state. Thus, only those excited $C_6H_6^+$ states close to resonant with $H^+ + C_6H_6$ are populated.

enhanced. Figure 54 compares the differential cross sections for inelastic and charge-transfer scattering The charge-transfer cross section has been corrected for the H-atom detection probability. Surprisingly, for scattering angles larger than 5^0 charge transfer is more probable than elastic scattering by more than an order of magnitude. This enhancement cannot be explained by the simple two-state model discussed in Section VI [cf. Eq. 26)], which predicts for the probability of charge transfer an upper limit of 50%. Thus, it appears that several (vibronic?) states are involved and that these contribute to a "trapping" of flux into the charge-transfer channels.

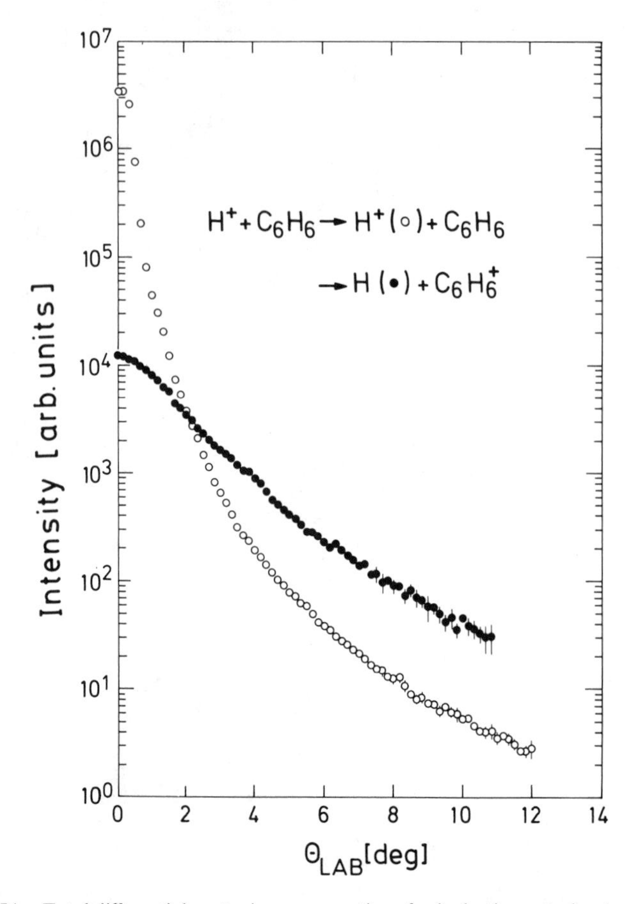

Figure 54. Total differential scattering cross sections for inelastic scattering (open symbols) and charge transfer (solid dots) in $H^+ + C_6H_6$ collisions at $E_{lab} = 30\,eV$.[169] Charge transfer becomes more probable than elastic scattering at small scattering angles, already. Both cross sections drop very steeply when compared to any other data taken with smaller target molecules. Thus, other scattering channels have to adsorb the proton flux missing here.

VII. CONCLUDING REMARKS

In this chapter we have surveyed 15 years of research on the dynamical interaction of protons with atoms, diatomic, triatomic, and polyatomic molecules. All of the results were based on high-resolution scattering experiments in which energy loss and angular distributions are measured. At collision energies of 20 and 30 eV it has also been possible to measure energy loss and angular distributions for H atoms produced in charge-transfer collisions. The small amount of rotational excitation observed can be accounted for in systems without a permanent dipole moment by an impulsive spectator mechanism. In many cases it can be neglected altogether, and this makes it possible to interpret the energy-loss distributions in a straightforward way in terms of pure vibrational excitation. Thus, the results obtained have provided a wealth of information on interaction potentials and on vibrational dynamical coupling mechanisms.

For systems that do not undergo extensive charge transfer all the observations could be explained either by an induced-dipole mechanism (IDM), where strong transient electrical fields of up to 10^8 V/cm act on the molecule for about 5×10^{-15} s, or by an internal-vibronic mechanism (IVM), where the valence forces are influenced directly by the temporary partial removal of electrons from the bonding region.

For those targets for which charge transfer with protons becomes energetically favorable, a number of additional mechanisms become operative. In a number of cases a direct transfer of vibrational excitation of the molecules to the molecular ions with an approximate preservation of the distributions is observed, for example, in CO_2 and N_2O. In the exoergic charge-transfer system $H^+ + O_2$, the O_2^+ final vibrational exciation resembles a Franck–Condon distribution at intermediate angles with large deviations at small and large scattering angles. For more complex molecules M such as H_2O, CH_4, and C_2H_2, the vibrational distributions of the molecular ions are strongly influenced not only by Franck–Condon factors but also by energy resonance, which leads to a peaking of the vibrational distributions at $\Delta E_{\text{vib}} \approx I(\text{H}) - (M)$, where I is the ionization potential. This implies an efficient transfer of electronic to vibrational energy even at long ranges. In inelastic scattering from C_2H_2 the asymmetric C–H stretching mode is excited in grazing collisions by the induced-dipole mechanism (IDM), but at angles larger than the rainbow angle the C–C bond-stretching mode is predominantly excited. This latter excitation and similar effect in CH_4 are attributed to an internal-vibronic mechanism (IVM). Finally, in the case of C_2H_4 and C_6H_6 the results reveal the resonant excitation of at least one electronically excited state in addition to the ground state of the molecular ion.

One of the most surprising observations coming out of this work is the large amount of energy transfer amounting to several eV and sometimes extending up to the dissociation limit of the molecular target. This has lead to the development of a new apparatus with improved energy resolution of about 10 meV, which is designed to investigate the high overtone levels of the molecular targets with a precision approaching that of optical spectroscopy. In principle the time-of-flight method can also be applied to the spectroscopic study of molecular ions. To improve the time-of-flight resolution it should be possible to *bunch* the proton beam prior to scattering and in this way achieve a high virtual energy transfer resolution.

The authors envisage many areas for future research in proton scattering. Linder and colleagues[147] have already demonstrated that scattering experiments are possible at sub-eV energies, a range that is also accessible to guided-beam techniques. Experiments at larger scattering angles, $\Theta > 20^0$, than studied so far, should reveal new phenomena as well. More details on the dynamics in charge-transfer collisions will be exposed by monitoring the fragment ions and their distributions as already illustrated in the case of CH_4^+ (see Section VI). Coincidence experiments involving both target ions and H atoms should also be possible. Finally, the authors call attention to related experiments on chemiluminiscence in $H^+ + NO$[18] and $H^+ + HCl$[95] at the energies of the experiments discussed previously. Angular resolved coincident detection of products and photons can reveal direct information on the impact-parameter dependence of these and similar luminiscent collisions. Such new experimental developments combined with calculations of potential surfaces and dynamics should lead to further advances in understanding the fine details of proton–molecule collisions.

Acknowledgment

One of the authors, GNS gratefully acknowledges the *Max-Planck-Gesellschaft zur Förderung der Wissenschaften* (Max-Planck society) for granting him the *Reimar–Lüst Stipendium*. He also expresses his warmest thanks to his former colleague, Dr. Martin Noll, for introducing him to the field of proton scattering. Both of the authors thank Wolfram Maring for help in preparing the figures for this article and Prof. Christoph Ottinger and Thomas Glenewinkel–Meyer for valuable comments. The authors also like to thank Prof. Arthur Phelps for making part of Fig. 1 available prior to publication.

References

1. W. Aberth, D. C. Lorents, R. P. Marchi, and F. T. Smith, *Phys. Rev. Lett.* **14**, 776 (1965).
2. M. Abignoli, J. Bandau, M. Barat, J. Fayeton, and J. Houver, *J. Phys. B* **5**, 1533 (1972).
3. M. Abramowitz and I. A. Stegun, eds., *Handbook of Mathematical Functions*, 3rd ed., NBS, Washington, 1965.
4. N. G. Adams, D. Smith, M. Tichy, G. Javahery, N. D. Twiddy, and E. E. Ferguson, *J. Chem. Phys.* **91**, 4037 (1989).
5. N. Aristov, G. Niedner, J. P. Toennies, and Y. N. Chi, *J. Chem. Phys.* **95**, in print.
6. L. Åsbrinck, *Chem. Phys. Lett.* **7**, 549 (1970).
7. T. M. Austin, J. M. Mullen, and T. L. Bailey, *Chem. Phys.* **10**, 117 (1975).
8. A. P. M. Baede, *Adv. Chem. Phys.* **30**, 463 (1975).
9. M. Baer, in *Molecular Coliision Dynamics*, Topics in Current Physics Vol. 33, J. M. Bowmann, (ed.), Springer, Berlin, 1983.
10. M. Baer, (ed.), *Theory of Chemical Reaction Dynamics*, CRC Press, Boca Raton, 1985.
11. M. Baer, R. Düren, B. Friedrich, G. Niedner, M. Noll, and J. P. Toennies, *Phys. Rev. A* **36**, 1063 (1987).

12. M. Baer, G. Niedner-Schatteburg, and J. P. Toennies, *J. Chem. Phys.* **88**, 1461 (1988).
13. M. Baer, G. Niedner-Schatteburg, and J. P. Toennies, *J. Chem. Phys.* **91**, 4169 (1989).
14. F. A. Gianturco, A. Palma, E. Semprini, F. Stefani, and M. Baer, *Phys. Rev.* **A42**, 3926 (1990).
15. G. D. Barg, G. M. Kendall, and J. P. Toennies, *Chem. Phys.* **16**, 243 (1976).
16. M. S. Bartlett and J. E. Moyal, *Proc. Cambridge Phil. Soc.* **45**, 545 (1949).
17. E. Bauer, E. R. Fisher, and F. R. Gilmore, *J. Chem. Phys.* **51**, 4173 (1969).
18. G. H. Bearman, J. D. Earl, H. H. Harris, P. B. James, and J. J. Leventhal, *Chem. Phys. Lett.* **44**, 471 (1976).
19. D. Beck, in *Physcis of Electronic and Atomic Collisions*, Invited Papers, S. Datz, (ed.). North Holland, Amsterdam 1982, p. 331.
20. J. Berkovitz and K. O. Groeneveld, (eds.), *Molecular Ions: Geometric and Electronic Structures*, NATO ASI series **B90**, Plenum, New York, 1983.
21. J. Berkowitz, C. A. Mayhew, and B. Ruscic, *J. Chem. Phys.* **88**, 7396 (1988).
22. M. Berman, U. Kaldor, J. Shmulovich, and S. Yatsiv, *Chem. Phys.* **63**, 165 (1981).
23. I. B. Bersuker, *The Jahn–Teller Effect and Vibronic Interactions in Modern Chemistry*, Plenum, New York, 1984.
24. G. D. Billing, *J. Chem. Soc. Faraday Trans.* **86**, 1663 (1990).
25. G. Bischof, V. Hermann, J. Krutein, and F. Linder, *J. Phys.* **B 15**, 249 (1982).
26. A. Bjerre and E. E. Nikitin, *Chem. Phys. Lett.* **6**, 438 (1967).
27. R. Böttner, U. Ross, and J. P. Toennies, *J. Chem. Phys.* **65**, 733 (1976).
28. M. Born and J. R. Oppenheimer, *Ann. Phys.* (*Leipzig*) **84**, 457 (1927).
29. Reference 20, p. 411.
30. C. R. Brundle, M. B. Robin, and H. Basch, *J. Chem. Phys.* **53**, 2196 (1970).
31. U. Buck, *Adv. Chem. Phys.* **30**, 313 (1975).
32. M. Burniaux, F. Brouillard, A. Jognaux, T. R. Govers, and S. Szucs, *J. Phys. B* **10**, 2421 (1977).
33. R. L. Champion, L. D. Doverspike, W. G. Rich, and S. M. Bobbio, *Phys. Rev. A* **2**, 2327 (1970).
34. Y.-N. Chiu, B. Friedrich, W. Maring, G. Niedner, M. Noll, and J. P. Toennies, *J. Chem. Phys.* **88**, 6814 (1988).
35. Y. N. Chiu, *J. Phys. Chem.* **92**, 4352 (1988).
36. B. H. Choi and K. T. Tang, *J. Chem. Phys.* **65**, 5528 (1976).
37. W. A. Chupka and M. E. Russell, *J. Chem. Phys.* **49**, 5426 (1968).
38. J. S. Cohen and J. N. Bardsley, *Phys. Rev. A* **18**, 1004 (1978).
39. E. U. Condon, *Phys. Rev.* **32**, 858 (1928).
40. R. J. Cross, *J. Chem. Phys.* **46**, 609 (1967).
41. R. J. Cross, Jr., *Accounts of Chem. Research* **8**, 225 (1975).
42. P. J. Crutzen, I. S. A. Isaksen, and G. C. Reid, *Science* **189**, 457 (1975).
43. L. A. Curtiss, and J. A. Pople, *J. Chem. Phys.* **88**, 7405 (1988).
44. R. David, M. Faubel, and J. P. Toennies, *Chem. Phys. Lett.* **18**, 37 (1973).
45. Y. N. Demkov, *Sov. Phys. JETP* **18**, 138 (1964).
46. P. F. Dittner and S. Datz, *J. Chem. Phys.* **49**, 1969 (1968).
47. D. A. Dixon and S. G. Lias, in *Molecular Structure and Energetics*, J. F. Liebman and A. Greenberg, eds., VCH, New York, 1987, Vol. 2.

48. J. P. Doering, *Ber. Bunsenges. Physik. Chemie* **77**, 593 (1973).
49. W. Eastes, U. Ross, and J. P. Toennies, *J. Chem. Phys.* **66**, 1919 (1977).
50. W. Eastes, U. Ross, and J. P. Toennies, *J. Chem. Phys.* **70**, 1652 (1979).
51. W. Eastes, and J. P. Toennies, *J. Chem. Phys.* **70**, 1644 (1979).
52. W. Eastes, U. Ross, and J. P. Toennies, *Chem. Phys.* **31**, 407 (1979).
53. J. Eccles, G. Pfeffer, E. Piper, G. Ringer, and J. P. Toennies, *Chem. Phys.* **89**, 1 (1984).
54. T. Ellenbroek, U. Gierz, and J. P. Toennies, *Chem. Pyss. Lett.* **70**, 459 (1980).
55. T. Ellenbroek and J. P. Toennies, *Chem. Phys.* **71**, 309 (1982).
56. T. Ellenbroek, U. Gierz, M. Noll, and J. P. Toennies, *J. Phys. Chem.* **86**, 1153 (1982).
57. J. O Ellison, *J. Am. Chem. Soc.* **85**, 3540, 3544 (1963).
58. J. M. Farrar and W. H. Saunders, *Techniques for the Study of Ion-Molecule Reactions*, Wiley, New York, 1988.
59. M. Faubel and J. P. Toennies, *Adv. At. Mol. Phys.* **13**, 229 (1977).
60. M. Faubel and J. P. Toennies, *J. Chem. Phys.* **71**, 3770 (1979).
61. M. Faubel, *Adv. At. Mol. Phys.* **19**, 345 (1983).
62. M. Faubel J. Frick, G. Kraft, and J. P. Toennies, *Chem. Phys Lett.* **116**, 12 (1985).
63. J. Franck, *Trans. Faraday Soc.* **21**, 536 (1925).
64. J. L. Franklin, ed., *Ion-Molecule-Reactions*, Dowden, Hutchinson & Ross, Stroudsburg, 1979, Parts I and II.
65. J. L. Franklin, ed., *Ion-Molecule-Reactions*, Plenum, New York, 1972, Vol. 182.
66. K. F. Freed, *Adv. Chem. Phys.* **47**, Part 2, 291 (1981).
67. L. Friedman and B. G. Reuben, *Adv. Chem. Phys.* **19**, 33 (1971).
68. B. Friedrich, S. Pick, L. Hladek, Z. Herman, E. E. Nikitin, A. I. Reznikov, and S. Ya. Umanski, *J. Chem. Phys.* **84**, 807 (1986).
69. B. Friedrich, J. Vancura, M. Sadilek, and Z. Herman, *Chem. Phys. Lett.* **120**, 243 (1985).
70. B. Friedrich, F. A. Gianturco, G. Niedner, M. Noll. and J. P. Toennies, *J. Phys. B* **20**, 3725 (1987).
71. B. Friedrich, G. Niedner, M. Noll, and J. P. Toennies, *J. Chem. Phys.* **87**, 1447 (1987).
72. B. Friedrich, G. Niedner, M. Noll, and J. P. Toennies, *J. Chem. Phys.* **87**, 5256 (1987).
73. B. Friedrich, G. Niedner, M. Noll, and J. P. Toennies, *Z. Phys. D* **6**, 49 (1987).
74. J. H. Futrell, ed., *Gaseous Ion Chemistry and Mass Spectrometry*, Wiley, New York, 1986.
75. J. H. Futrell, in *Structure, Reactivity and Thermochemistry of Ions*, P. Ausloos, S. G. Lias, and D. Dixon, eds., Reidel, Dordrecht, Holland, 1987, p. 57.
76. B. C. Garrett and D. G. Truhlar, in *Theoretical Chemistry: Theory of Scattering*, D. G. Henderson, ed., Academic, New York, 1981.
77. W. R. Gentry, D. J. McClure, and C. H. Douglass, *Rev. Sci. Instrum.* **46**, 367 (1975).
78. W. R. Gentry, H. Udseth, and C. F. Giese, *Chem. Phys. Lett.* **36**, 671 (1975).
79. W. R. Gentry, in *Gas Phase Ion Chemistry*, M. T. Bowers, ed., Academic Press, New York, 1979, Vol. 2.
80. W. R. Gentry, in *Kinetics of Ion-Molecule Reactions*, P. Ausloos, ed., Plenum, New York, 1979.
81. D. Gerlich, Dissertation, Universität Freiburg, Fakultät für Physik, 1977.
82. F. A. Gianturco and V. Staemmler, in *Intermolecular Interactions from Diatomics to Biopolymers*, B. Pullman, ed., Wiley, New York, 1978.

83. F. A. Gianturco, U. Gierz, and J. P. Toennies, *J. Phys. B* **14**, 667 (1981).
84. F. A. Gianturco and F. Schneider, *Chem. Phys. Lett.* **129**, 481 (1986).
85. F. A. Gianturco, A. Palma, E. Semprini, F. Stefani, H. P. Diehl, and V. Staemmler, *Chem. Phys.* **107**, 293 (1986).
86. F. A. Gianturco, G. Niedner, M. Noll, E. Semprini, F. Stefani, and J. P. Toennies, *Z. Phys. D* **7**, 281 (1987).
87. F. A. Gianturco and F. Gallese, *Chem. Phys. Lett.* **148**, 365 (1988).
88. F. A. Gianturco, M. Patriarca, and O. Roncero, Mol. Phys. **66**, 281 (1989).
89. U. Gierz, J. P. Toennies, and M. Wilde, *Chem. Phys. Lett.* **95**, 517 (1983).
90. U. Gierz, J. P. Toennies, and M. Wilde, *Chem. Phys. Lett.* **110**, 115 (1984).
91. U. Gierz, M. Noll, and J. P. Toennies, *J. Chem. Phys.* **82**, 217 (1985).
92. U. Gierz, M. Noll, and J. P. Toennies, *J. Chem. Phys.* **83**, 2259 (1985).
93. C. F. Giese and W. R. Gentry, *Phys. Rev. A* **10**, 2156 (1974).
94. R. J. Glauber, *Lect.* **1**, 315 (1959).
95. T. Glenewinkel-Meyer, B. Müller, H. Tischer, and C. Ottinger, *J. Chem. Phys.* **88**, 3475 (1988).
96. J. Glosik, B. Friedrich, and Z. Herman, *Chem. Phys.* **60**, 369 (1981).
97. W. G. Golden, C. Marcott, and J. Overend, *J. Chem. Phys.* **68**, 2081 (1978).
98. D. Grimbert, B. Lassier Govers, and V. Sidis, *Chem. Phys.* **124**, 187 (1988).
99. C. S. Gudemann and R. J. Saykally, *Ann. Rev. Phys. Chem.* **35**, 387 (1984).
100. G. Hagenow, W. Denzer, B. Brutschy, and H. Baumgärtel, *J. Phys. Chem.* **92**, 6487 (1988).
101. R. W. Hannah, Dissertation, Purdue University, 1957.
102. U. Hege and F. Linder, *Z. Phys. A* **320**, 95 (1985).
103. V. Hermann, H. Schmidt, and F. Linder, *J. Phys. B* **11**, 493 (1978).
104. F. A. Herrero, E. M. Nemeth, and T. L. Baily, *J. Chem. Phys.* **50**, 4591 (1969).
105. G. R. Hertel and W. S. Koski, *J. Chem. Phys.* **40**, 3452 (1964).
106. I. V. Hertel, *Adv. Chem. Phys.* **50**, 475 (1982).
107. G. Herzberg, *Spectra of Diatomic Molecules*, 2nd ed., Van Nostrand, Toronto, 1950.
108. G. Herzberg, *Molecular Spectra and Molecular Structure*, Van Nostrand, Princeton, 1967, Vol. III.
109. P. M. Hierl, V. Pacák, and Z. Herman, *J. Chem. Phys.* **67**, 2678 (1977).
110. Y. Itoh, N. Kobayashi, and Y. Kaneko, *J. Phys. Soc. Jpn.* **51**, 2977 (1982).
111. D. H. Jaecks, in *Electronic and Atomic Collisions*, J. Eichler, I. V. Hertel, and N. Stolterfot, eds., North-Holland, Amsterdam, 1984.
112. J. D. Jackson, *Classical Electrodynamics*, Wiley, New York, 1962.
113. A. C. Jeanotte II, C. Curtis, and J. Overend, *J. Chem. Phys.* **68**, 2076 (1978).
114. C. A. F. Johnson and J. E. Parker, *Chem. Phys.* **111**, 307 (1987).
115. E. H. Kerner, *Can. J. Phys.* **36**, 371 (1958).
116. K. Kim, R. S. McDowell, and W. T. King, *J. Chem. Phys.* **73**, 36 (1980).
117. K. Kimura, S. Katsumata, Y. Achiba, T. Yamazaki, and S. Iwata, *Handbook of He I Photoelectron Spectra of Fundamental Organic Molecues*, Japan Scientific Societies Press, Tokyo, 1981.
118. A. W. Kleyn, V. N. Khromov, and J. Los, *J. Chem. Phys.* **72**, 5282 (1980).
119. A. W. Kleyn, V. N. Khromov, and J. Los, *Chem. Phys.* **52**, 65 (1980).

120. A. W. Kleyn, J. Los, and E. A. Gislason, *Phys. Rep.* **90**, 1 (1982).
121. U. C. Klomp, M. R. Spalburg, and J. Los, *Chem. Phys.* **83**, 33 (1984).
122. N. Kobayashi, Y. Itoh, and Y. Kaneko, *J. Phys. Soc. Jpn.* **45**, 617 (1978).
123. N. Kobayashi, in *Physics of Electronic and Atomic Collisions*, S. Datz, ed., North Holland,Amsterdam, 1982.
124. W. Kolos and L. Wolniewicz, *J. Chem. Phys.* **49**, 404 (1968).
125. M. Konrad and F. Linder, *J. Phys. B* **15**, L405 (1982).
126. M. Konrad, Dissertation, Kaiserslautern, 1983.
127. W. S. Koski, *Adv. Chem. Phy.* **30**, 185 (1975).
128. D. J. Krajnovich, C. S. Parmenter, and D. L. Catlett Jr., *Chem. Rev.* **87**, 237 (1987).
129. J. Krenos and R. Wolfgang, *J. Chem. Phys.* **52**, 5961 (1968).
130. J. R. Krenos, R. K. Preston, R. Wolfgang, and J. C. Tully, *J. Chem. Phys.* **60**, 1634 (1974).
131. H. Krüger and R. Schinke, *J. Chem. Phys.* **66**, 5087 (1977).
132. J. Krutein and F. Linder, *J. Phys. B* **10**, 1363 (1977).
133. J. Krutein and F. Linder, ICPEAC, Book of Abstracts, Commissariat a l'Energie Atomique, Paris, 1977.
134. J. Krutein and F. Linder, *J. Chem. Phys.* **71**, 599 (1979).
135. S. K. Lam, L. D. Doverspike, and R. L. Champion, *Phys. Rev. A* **7**, 1595 (1973).
136. L. Landau, *Phys. Z. Sowjetunion* **1**, 88 (1932); **2**, 46 (1932).
137. L. D. Landau and E. M. Lifshitz, *Mechanics*, 2nd ed., Pergamon Press, New York, 1969.
138. T. J. Lee and H. F. Schaefer, *J. Chem. Phys.* **85**, 3437 (1986).
139. H. B. Levene and D. S. Perry, *J. Chem. Phys.* **80**, 1772 (1984).
140. S. Lemont and G. W. Flynn, *Ann. Rev. Phys. Chem.* **28**, 261 (1977).
141. H. B. Levene and D. S. Perry, *J. Chem. Phys.* **84**, 4385 (1986).
142. I. W. Levin and T. P. Lewis, *J. Chem. Phys.* **52**, 1608 (1970).
143. R. D. Levine and R. B. Bernstein, *Molecular Reaction Dynamics and Chemical Reactivity*, Oxford University Press, Oxford, 1987.
144. S. G. Lias, J. F. Liebman, and R. D. Levin, *J. Phys. Chem. Ref. Data* **13**, 695 (1984).
145. S. G. Lias, J. E. Bartmess, J. F. Liebman, J. L. Holmes, R. D. Levin, and W. G. Mallard, *J. Chem. Ref. Data* **17**, Suppl. 1, 1988.
146. Y. W. Lin, T. F. George, and K. Morokuma, *J. Chem. Phys.* **60**, 4311 (1974); *J. Phys. B* **8**, 265 (1975).
147. F. Linder, in ICPEAC XI, Kyoto 1979, N. Oda and K. Takayanagi, eds., North-Holland, Amsterdam 1980, p. 535.
148. M. Lipeles, *J. Chem. Phys.* **51**, 1252 (1969).
149. J. Lorenzen, H. Hotop, M. W. Ruf, and H. Morgner, *Z. Phys. A* **297**, 19 (1980).
150. B. H. Mahan, *Adv. Chem. Phys.* **23**, 1 (1973).
151. A. Mann and F. Linder, *J. Phys. E* **21**, 805 (1988).
152. R. P. Marchi and F. T. Smith, *Phys. Rev. A* **139**, 1025 (1965).
153. W. Maring, J. P. Toennies, R. G. Wang and H. B. Levene, *Chem. Phys. Lett.*, in print.
154. K. B. McAfee, C. R. Szmanda, and R. S. Hosack, *J. Phys. B* **14**, L243 (1981).
155. K. B. McAfee and R. S. Hosack, *J. Chem. Phys.* **83**, 5690 (1985).
156. E. W. McDaniel, V. Cermak, A. Dalgarno, E. E. Ferguson, and L. Friedman, eds., *Ion Molecule Reactions*, Wiley, New York, 1970.

157. R. D. McPeters, C. H. Jackman, and E. G. Stassinopoulos, *J. Geophys. Res.* **86**, 12071 (1981).
158. M. Menzinger and I. Wåhlin, *Rev. Sci. Instrum.* **40**, 102 (1969).
159. W. Meyer, P. Botschwina, and P. Burton, *J. Chem. Phys.* **84**, 891 (1986).
160. H.-U. Mittmann, H.-P. Weise, A. Ding, and A. Henglein, *Z. Naturforsch. A* **26**, 1112 (1971).
161. H.-U. Mittmann, H.-P. Weise, A. Ding, and A. Henglein, *Z. naturforsch. A* **26**, 1282 (1971).
162. J. H. Moore and J. P. Doering, *Phys. Rev. Lett.* **23**, 564 (1969).
163. T. F. Moran and P. C. Cosby, *J. Chem. Phys.* **51**, 5724 (1969).
164. W. F. Murphy, W. Holzer, and H. J. Bernstein, *Appl. Spectr.* **23**, 211 (1969).
165. J. von Neumann and E. P. Wigner, *Phys. Z.* **30**, 467 (1929).
166. A. Niehaus, *Adv. Chem. Phys.* **45**, 399 (1981).
167. G. Niedner, M. Noll, and J. P. Toennies, *J. Chem. Phys.* **87**, 2067 (1987).
168. G. Niedner, M. Noll, J. P. Toennies, and Ch. Schlier, *J. Chem. Phys.* **87**, 2685 (1987).
169. G. Niedner-Schatteburg, Dissertation, Georg-August-Universität Göttingen, Fachbereich Physik, 1988; also published as Bericht **13/1988**, Max-Planck-Institut für Strömungsforschung, Göttingen, 1988.
170. E. E. Nikitin and S. Y. Umanskii, *Theory of Slow Atomic Collisions*, Springer, Berlin, 1984.
171. M. Noll and J. P. Toennies, *Chem. Phys. Let.* **108**, 297 (1984).
172. M. Noll, Dissertation, Georg-August Universität Göttingen, Fachbereich Physik, 1986; also published as: Bericht **8/1986**, Max-Planck-Institut für Strömungsforschung, Göttingen, 1986.
173. M. Noll and J. P. Toennies, *J. Chem. Phys.* **85**, 3313 (1986).
174. M. Noll, *Physica Scripta* **T 23**, 151 (1988).
175. M. Noll, W. Maring, B. Ungerer, and J. P. Toennies (unpublished).
176. G. Ochs and E. Teloy, *J. Chem. Phys.* **61**, 4930 (1974).
177. I. Ozkan and L. Goodman, *Chem. Rev.* **79**, 275 (1979).
178. H. Pauly and J. P. Toennies, *Adv. At. Mol. Phys.* **1**, 195 (1965); and in *Methods of Experimental Physics*, B. Bederson and W. L. Fite, Academic Press, New York, 1968, Vol. 7A, p. 227.
179. S. Peyerimhoff, *J. Chem. Phys.* **43**, 998 (1965).
180. W. B. Person and K. C. Kim, *J. Chem. Phys.* **69**, 1764 (1978).
181. B. M. Pettitt, K. Jacobson, and R. L. Matcha, *J. Chem. Phys.* **72**, 2892 (1980).
182. A. Phelps, *J. Phys. Chem. Ref. Data* **19**, 653 (1990).
183. R. K. Preston and J. C. Tully, *J. Chem. Phys.* **54**, 4297 (1971).
184. A. A. Radzig and B. M. Smirnov, *Reference Data on Atoms, Molecules, and Ions*, Springer, Berlin, 1985.
185. G. P. Raine, H. F. Schaefer III, and N. C. Handy. *J. Chem. Phys.* **80**, 319 (1984).
186. J. A. Ray, C. F. Barnett, and B. van Zyl, *J. Appl. Phys.* **50**, 6516 (1979).
187. J. E. Reutt, L. S. Wang, Y. T. Lee, and D. A. Shirley, *J. Chem. Phys.* **85**, 6928 (1986).
188. A. L. Rockwood, S. L. Howard, W.-H. Du, W. Lindinger, and J. H. Futrell, *Chem. Phys. Lett.* **144**, 486 (1985).
189. P. Rosmus and E.-A. Reinsch, *Z. Naturforsch.* **35a**, 1066 (1980).
190. K. Rudolph and J. P. Toennies, *J. Chem. Phys.* **65**, 4483 (1976).
191. S. Säeki, M. Mizuno, and S. Kondo, *Spectrochim. Acta* **32**, 403 (1976).
192. H. P. Saha, K. S. Lam, and T. F. George, in *Gas Phase Chemoluminescence and Chemiionization,* A. Fontijn, ed., North-Holland, Amsterdam, 1985.

193. J. H. Schachtschneider, Dissertation, University of Minnesota, 1960.
194. P. N. Schatz and D. F. Hornig, *J. Chem. Phys.* **21**, 1516 (1953).
195. R. Schinke, *Chem. Phys.* **24**, 379 (1977).
196. R. Schinke, H. Krüger, V. Hermann, H. Schmidt, and F. Linder, *J. Chem. Phys.* **67**, 1187 (1977).
197. R. Schinke, M. Dupuis, and W. A. Lester, *J. Chem. Phys.* **72**, 3909 (1980).
198. H. Schmidt, V. Herrmann, and F. Linder, *Chem. Phys. Lett.* **41**, 365 (1976).
199. F. Schneider, L. Zülicke, F. DiGiacomo, F. A. Gianturco, I. Paidarova, and R. Polak, *Chem. Phys.* **128**, 311 (1988).
200. J. Schöttler and J. P. Toennies, *Z. Physik* **214**, 472 (1968).
201. G. Schurin, *J. Chem. Phys.* **30**, 1 (1959).
202. G. Scoles, ed., *Atomic and Molecular Beam Methods*, Oxford University Press, Oxford, 1988.
203. V. Sidis, *J. Phys. B* **5**, 1517 (1972).
204. V. Sidis, D. Grimbert, M. Sizun, and M. Baer, *Chem. Phys. Lett.* **163**, 19 (1989).
205. M. Sizun, D. Grimbert, and V. Sidis, *J. Phys. Chem.* **94**, 5674 (1990).
206. F. T. Smith, *Phys. Rev.* **179**, 111 (1969).
207. M. R. Spalburg and U. C. Klomp, *Comput. Phys. Commun.* **28**, 207 (1982).
208. M. R. Spalburg and E. A. Gislason, *Chem. Phys.* **94**, 327 (1985).
209. M. R. Spalburg and E. A. Gislason, *Chem. Phys.* **94**, 339 (1985).
210. M. R. Spalburg, J. Los, and A. Z. Devdariani, *Chem. Phys.* **103**, 253 (1986).
211. V. Staemmler and F. A. Gianturco, *Intern. J. Quantum Chem.* **28**, 553 (1985).
212. E. C. G. Stückelberg, *Helv. Phys. Acta* **5**, 369 (1932).
213. K. Takeshita, *J. Chem. Phys.* **86**, 329 (1987).
214. E. Teloy, *10th ICPEAC*, G. Watel, ed., North Holland, Amsterdem, 1978.
215. J. Tennyson and B. Sutcliffe, *Mol. Phys.* **51**, 887 (1984).
216. L. D. Thomas, *J. Chem. Phys.* **67**, 5224 (1977).
217. T. O. Tiernan and C. Lifshitz, *Adv. Chem. Phys.* **45**, 81 (1981).
218. J. P. Toennies, *Faraday Discuss. Chem. Soc.* **55**, 129 (1973).
219. J. P. Toennies in *Physical Chemistry—An Advanced Treatise*, H. Eyring, W. Jost, and D. Henderson, eds., Academic, New York, 1974, Vol. 5.
220. J. P. Toennies, *Annu. Rev. Phys. Chem.* **27**, 225 (1976).
221. Z. H. Top and M. Baer, *J. Chem. Phys.* **64**, 3078 (1976).
222. Z. H. Top and M. Baer, *Chem. Phys.* **25**, 1 (1977).
223. S. M. Trujillo, R. H. Neynaber, and E. W. Rothe, *Rev. Sci. Instrum.* **37**, 1655 (1966).
224. J. C. Tully and R. K. Preston, *J. Chem. Phys.* **55**, 562 (1971).
225. J. C. Tully, in *Dynamics of Molecular Collisions*, W. H. Miller (ed.), Plenum, New York, 1976.
226. D. W. Turner, C. Baker, A. D. Baker, and C. R. Brundle, *Molecular Photoelectron Spectroscopy*, Wiley-Interscience, London, 1970.
227. H. Udseth, C. F. Giese, and W. R. Gentry, *J. Chem. Phys.* **54**, 3642 (1971).
228. H. Udseth, C. F. Giese, and W. R. Gentry, *Phys. Rev.* **A8**, 2483 (1973),
229. H. Udseth, C. F. Giese, and W. R. Gentry, *J. Chem. Phys.* **60**, 3051 (1974).
230. B. Ungerer, Diplom thesis, Georg-August Universität Göttingen, Fachbereich Physik, Göttingen 1988.

231. H. E. van den Bergh, M. Faubel, and J. P. Toennies, *Faraday Discuss. Chem. Soc.* **55**, 203 (1973).
232. J. H. van Lenthe and P. J. A. Ruttnik, *Chem. Phys. Lett.* **56**, 20 1978).
233. K. Vasudevan, S. D. Peyerimhoff, and R. J. Buenker, *Chem. Phys.* **5**, 149 (1974).
234. G.J. Vazquez, R. J. Buenker, and S. D. Peyerimhoff, *Mol. Phys.* **59**, 291 (1986).
235. J. Weber, U. Yoshimine, and A. D. McLean, *J. Chem. Phys.* **64**, 4159 (1976).
236. H.-P. Weise, H.-P. Mittmann, A. Ding, and A. Henglein, *Z. Naturforsch.* **26**, 1112 (1971).
237. R. L. Whetten, G. S. Ezra, and E. R. Grant, *Annu. Rev. Phys. Chem.* **36**, 277 (1985).
238. M. Wilde, Dissertation, Georg-August Universät Göttingen, Fachbereich Physik, 1983; also published as: Bericht **6/1986**, Max-Planck-Institut für Strömungsforschung, Göttingen, 1986.
239. G. Winkelhofer, R. Janoschek, F. Fratev, and P. von Rague Schleyer, *Croatica Chemica Acta* **56**, 509 (1983).
240. O. Yenen, D. H. Jaecks, and J. Macek, *Phys. Rev.* **A30**, 597 (1984).
241. C. Zener, *Proc. Roy. Soc.* **137a**, 696 (1932).
242. C. Zener, *Proc. Roy. Soc.* **140a**, 660 (1933).
243. H. V. Koch, *Archiv Fysik* **28**, 529 (1956).
244. W. Maring, Dissertation Georg August Universität Göttingen, Fachbereich Physik, 1991.
245. F. A. Gianturco, A. Palma, and F. Schneider, *Int. J. Quantum Chem.* **37**, 729 (1990).

AUTHOR INDEX

Numbers in parentheses are reference numbers and indicate that the author's work is referred to although his name is not mentioned in the text. Numbers in *italic* show the pages on which the complete references are listed.

SUBJECT INDEX